Other Titles in This Series

148 **Vladimir I. Piterbarg,** Asymptotic methods in the theory of Gaussian processes and fields, 1996
147 **S. G. Gindikin and L. R. Volevich,** Mixed problem for partial differential equations with quasihomogeneous principal part, 1996
146 **L. Ya. Adrianova,** Introduction to linear systems of differential equations, 1995
145 **A. N. Andrianov and V. G. Zhuravlev,** Modular forms and Hecke operators, 1995
144 **O. V. Troshkin,** Nontraditional methods in mathematical hydrodynamics, 1995
143 **V. A. Malyshev and R. A. Minlos,** Linear infinite-particle operators, 1995
142 **N. V. Krylov,** Introduction to the theory of diffusion processes, 1995
141 **A. A. Davydov,** Qualitative theory of control systems, 1994
140 **Aizik I. Volpert, Vitaly A. Volpert, and Vladimir A. Volpert,** Traveling wave solutions of parabolic systems, 1994
139 **I. V. Skrypnik,** Methods for analysis of nonlinear elliptic boundary value problems, 1994
138 **Yu. P. Razmyslov,** Identities of algebras and their representations, 1994
137 **F. I. Karpelevich and A. Ya. Kreinin,** Heavy traffic limits for multiphase queues, 1994
136 **Masayoshi Miyanishi,** Algebraic geometry, 1994
135 **Masaru Takeuchi,** Modern spherical functions, 1994
134 **V. V. Prasolov,** Problems and theorems in linear algebra, 1994
133 **P. I. Naumkin and I. A. Shishmarev,** Nonlinear nonlocal equations in the theory of waves, 1994
132 **Hajime Urakawa,** Calculus of variations and harmonic maps, 1993
131 **V. V. Sharko,** Functions on manifolds: Algebraic and topological aspects, 1993
130 **V. V. Vershinin,** Cobordisms and spectral sequences, 1993
129 **Mitsuo Morimoto,** An introduction to Sato's hyperfunctions, 1993
128 **V. P. Orevkov,** Complexity of proofs and their transformations in axiomatic theories, 1993
127 **F. L. Zak,** Tangents and secants of algebraic varieties, 1993
126 **M. L. Agranovskiĭ,** Invariant function spaces on homogeneous manifolds of Lie groups and applications, 1993
125 **Masayoshi Nagata,** Theory of commutative fields, 1993
124 **Masahisa Adachi,** Embeddings and immersions, 1993
123 **M. A. Akivis and B. A. Rosenfeld,** Élie Cartan (1869–1951), 1993
122 **Zhang Guan-Hou,** Theory of entire and meromorphic functions: Deficient and asymptotic values and singular directions, 1993
121 **I. B. Fesenko and S. V. Vostokov,** Local fields and their extensions: A constructive approach, 1993
120 **Takeyuki Hida and Masuyuki Hitsuda,** Gaussian processes, 1993
119 **M. V. Karasev and V. P. Maslov,** Nonlinear Poisson brackets. Geometry and quantization, 1993
118 **Kenkichi Iwasawa,** Algebraic functions, 1993
117 **Boris Zilber,** Uncountably categorical theories, 1993
116 **G. M. Fel'dman,** Arithmetic of probability distributions, and characterization problems on abelian groups, 1993
115 **Nikolai V. Ivanov,** Subgroups of Teichmüller modular groups, 1992
114 **Seizô Itô,** Diffusion equations, 1992
113 **Michail Zhitomirskiĭ,** Typical singularities of differential 1-forms and Pfaffian equations, 1992
112 **S. A. Lomov,** Introduction to the general theory of singular perturbations, 1992
111 **Simon Gindikin,** Tube domains and the Cauchy problem, 1992
110 **B. V. Shabat,** Introduction to complex analysis Part II. Functions of several variables, 1992
109 **Isao Miyadera,** Nonlinear semigroups, 1992
108 **Takeo Yokonuma,** Tensor spaces and exterior algebra, 1992
107 **B. M. Makarov, M. G. Goluzina, A. A. Lodkin, and A. N. Podkorytov,** Selected problems in real analysis, 1992

(Continued in the back of this publication)

TRANSLATIONS OF MATHEMATICAL MONOGRAPHS
VOLUME 23

O. A. Ladyženskaja
V. A. Solonnikov
N. N. Ural′ceva

Linear and Quasi-linear Equations of Parabolic Type

American Mathematical Society

О. А. Ладыженская
В. А. Солонников
Н. Н. Уральцева

ЛИНЕЙНЫЕ
И КВАЗИЛИНЕЙНЫЕ УРАВНЕНИЯ
ПАРАБОЛИЧЕСКОГО ТИПА

Издательство «Наука»
Главная Редакция
Физико-Математической Литературы
Москва 1967

Translated from the Russian by S. Smith

2000 *Mathematics Subject Classification.* Primary 35–XX.

Library of Congress Card Number 68-19440
International Standard Book Number 978-0-8218-1573-1
International Standard Serial Number 0065-9280

∞ The paper used in this book is acid-free and falls within the guidelines
established to ensure permanence and durability.
Visit the AMS home page at http://www.ams.org/

10 9 8 7 6 16 15 14 13 12 11

PREFACE

Equations of parabolic type are encountered in many branches of mathematics and mathematical physics, and the forms in which they are investigated vary widely. The equations encountered most frequently (and in adjoining fields of study almost exclusively) are those of second order. Such equations (and certain classes of systems of second order), both linear and quasi-linear, make up the subject of investigation of the present book. Our study of these equations is concerned mainly with the solvability of their boundary value problems and with an analysis of the connections between the smoothness of the solutions and the smoothness of the known functions entering into the problem.

A basic condition that is assumed to be fulfilled for all equations considered is the condition of uniform parabolicity. For such equations we have managed to give sufficiently complete answers to central questions on the solvability of the above-indicated problems and to establish a series of exact dependences of the properties of the solutions on the properties of the known functions in terms of their mutual membership in the most commonly occurring function spaces.

For linear equations the solvability of the basic boundary value problems and of the Cauchy problem depends only on the smoothness of the functions defining the problem (i. e. the functions considered to be known in the problem, namely the coefficients and the free terms of the equations, the functions assigning the initial and boundary conditions and the boundary of the domain in which the solution exists). The smoother these known functions, the better behaved will be the solution. Conversely, if one worsens the properties of the known functions in the problem, then the differential properties of the solutions also become worse, where the deterioration (as would equally be true with an improvement) has a local character (for example, the smoothness of the solutions inside their domain of definition is determined only by the smoothness of the coefficients and free terms of the equation and does not depend on the smoothness of the boundary or of the initial and boundary functions). But one cannot arbitrarily worsen the properties of the functions defining the problem (for example, admit in the coefficients singularities of high order). There exists a limit to admissible deteriorations, beyond which such properties of the problems as uniqueness are lost. As in the analysis carried out by us for elliptic equations in the book [65 q] we begin by determining this limit, for which we construct appropriate examples. With these examples (and examples from [65 l, n, o]) we have managed to outline with sufficient accuracy the limits of a possible theory of boundary value problems for equations with discontinuous and, in general, unbounded coefficients and free terms, which is later presented in Chapter III.

As a characteristic of such "bad" known functions we have selected their membership in the spaces $L_{q,r}(Q_T)$.[1)] The solutions here fall into a certain function space, the elements of which have derivatives of first order with respect to x and of order ½ with respect to t. We then observe that the properties of these solutions improve as the differential properties of the functions defining the equation or problem improve.

A qualitatively different situation holds for nonlinear equations. For them the smoothness of the solutions and the solvability "in the large" of the boundary value problems and of the Cauchy problem is determined not only by the smoothness of the known functions $a_{ij}(x, t, u, p)$, $a(x, t, u, p)$ making up the equation but also by their behavior as u and p increase without limit. In § 3 of Chapter I we cite a number of examples elucidating certain restrictions on this behavior, the nonfulfilment of which implies a nonsolvability of these problems "in the large." And in subsequent chapters (Chapters V, VI, VII) it is proved that these restrictions, together with a certain not large smoothness, are on the whole also sufficient for the unique solvability of the basic boundary value problems and of the Cauchy problem for quasi-linear equations.

The general plan of the book is as follows. In Chapter I we present the basic notation and terminology used in the book, a description of the main results proved in it, and a number of examples indicating the exactness of these results; finally, we give a brief historical survey. In Chapter II we have assembled propositions that are used throughout the book and describe the properties, not of the solutions of any differential equations, but of arbitrary functions belonging to various function spaces or classes. It is perhaps better to treat this chapter as a reference on

---

1)For functions $u(x, t)$ of a space $L_{q,r}(Q_T)$ the norm

$$\|u\|_{q,r,Q_T} = \left(\int_0^T \left(\int_\Omega |u|^q \, dx \right)^{\frac{r}{q}} dt \right)^{\frac{1}{r}}.$$

is finite.

its different assertions. The main text begins with Chapter III. It and Chapter IV are devoted to linear equations. In Chapters V and VI we investigate quasi-linear equations. Finally, in Chapter VII we examine linear and quasi-linear systems of second order with common principal parts and give a survey of the results on general boundary-value problems for linear parabolic systems, the most general of those considered up to the present time. The main contents of each chapter can be understood independently of the others.

The contents of all the chapters, except Chapter IV and parts of Chapters II and VII, are based on the work of O. A. Ladyženskaja and N. N. Ural'ceva. These chapters were written by them. Chapter IV and §§8–10 of Chapter VII were written by V. A. Solonnikov, who is responsible for many of the results in this part of the book.

The authors are extremely grateful to Academician V. I. Smirnov for having looked over the manuscript of the entire book and having made a number of important critical remarks and suggestions. They were taken into account during the final revision.

The authors express their heartfelt thanks to their colleagues and students A. Treskunov, A. Oskolkov, M. Faddeev, I. Krol', V. Matveev and technician L. M. Dikušina for their help in the preparation of the book. A particularly large amount of quite expert assistance was rendered by A. Treskunov, a graduate student at Leningrad University, who worked with us throughout the writing of the book and obtained during this time some interesting results on linear equations (see Bibliography).

Prefatory Note to the Translation

The active cooperation of the Russian authors has made it possible to bring the present translation up-to-date and to improve it in several respects. Slight additions and corrections have been made throughout, and some of the material has been entirely rewritten, most notably Chapter II §2 on embedding theorems, Chapter IV §4 on certain supplementary theorems, and Chapter V §6 on solvability of the first boundary problem. The translator and the editorial staff wish to thank the Russian authors for their long-continued and cordial assistance.

TABLE OF CONTENTS

EDITOR'S NOTE

For reasons of economics, most displayed formulas in this translation have been inserted from the original Russian. This means that certain letters unfortunately have a different appearance in formulas from their counterparts in the text. The principal instances are summarized in the following table:

Displayed formulas	*Text*
φ	ϕ
$\mathscr{A}$	$\mathfrak{A}$
$\mathscr{B}$	$\mathfrak{B}$
$\mathscr{C}$	$\mathfrak{C}$
$\mathscr{L}$	$\mathfrak{L}$
$\mathscr{M}$	$\mathfrak{M}$
$\varkappa$	κ

CHAPTER I

INTRODUCTORY MATERIAL

This book is devoted to the basic linear and quasi-linear second-order partial differential equations of parabolic type. For them the solvability of the basic boundary value problems and of the Cauchy problem in various function spaces is studied and investigations are carried out concerning the dependences of the smoothness properties of the solutions of these equations on the known functions making up the equations and on the properties of the other known functions in the problems. We begin with a description of certain examples that permit one to outline with sufficient accuracy the contours of a possible theory for these questions, and with an enumeration of the basic results of the present book. These sections (§§ 3 and 4) may be usefully reread after making an acquaintance with Chapters III and IV. Preceding them (§2) is a description of the statements of the basic problems for parabolic equations and an account of one of the basic properties inherent in the solutions of parabolic equations of second order namely, in their classical form, the maximum principle. Later on, both the statements of the problems and the maximum principle acquire a different form, appropriate to the function space containing the solutions being investigated. These modifications in the classical form and the methods created for working with them led to success in studying quasi-linear equations "in the large" and linear equations with bad coefficients.

Although it can be read independently, the present book has much in common, in regard to the methods of investigation, with the book by O. A. Ladyženskaja and N. N. Ural'ceva, "Linear and quasi-linear equations of elliptic type" [65q].

All functions, arguments and parameters considered in this book are real. An exception occurs with §§.4 and 18 of Chapter III and §8 and 9 of Chapter VII, where Fourier and Laplace transforms are used.

§1. BASIC NOTATION AND TERMINOLOGY

1. Abridged notation.

E_n is the n-dimensional euclidean space; $x = (x_1, \cdots, x_n)$ is an arbitrary point in it.

E_{n+1} is the $(n+1)$-dimensional euclidean space; its points are denoted by (x, t), where x is in E_n and t is in $(-\infty, \infty)$.

Ω is a domain in E_n, i.e. an arbitrary open connected set of points of E_n. In all chapters except IV, unless otherwise stated, Ω is considered to be a bounded domain. In Chapter IV the letter Ω denotes an arbitrary domain.

S is the boundary of Ω.

$\bar{\Omega}$ is the closure of Ω, so that $\bar{\Omega} = \Omega \cup S$.

K_ρ is an arbitrary (open) ball in E_n of radius ρ, $\kappa_n = \operatorname{mes} K_1$, and ω_n is the surface area of K_1.

$\Omega_\rho = K_\rho \cap \Omega$.

Q_T is the cylinder $\Omega \times (0, T)$, i.e. the set of points (x, t) of E_{n+1} with $x \in \Omega$, $t \in (0, T)$.

Q' is an arbitrary open subset of Q_T.

S_T is the lateral surface of Q_T, or more precisely the set of points (x, t) of E_{n+1} with $x \in S$, $t \in [0, T]$.

$\Gamma_T = S_T \cup \{(x, t) \colon x \in \Omega,\ t = 0\}$.

$S_0 = \{(x, t) \colon x \in S,\ t = 0\}$; $\Gamma_0 = \{(x, t) \colon x \in \bar{\Omega},\ t = 0\}$.

$Q_{t_1, t_2} = \Omega \times (t_1 < t < t_2)$.

$Q(\rho, \tau)$ is an arbitrary cylinder of the form $\{(x, t) \colon |x - x^0| < \rho,\ t_0 < t < t_0 + \tau\}$.

$Q(k, \rho, \tau)$ is the set of all points $(x, t) \in Q(\rho, \tau)$ at which the investigated function $u(x, t) > k$.

$Q_{t_1}(k)$ is the set of all points $Q_{t_1} = \Omega \times (0 < t < t_1)$, at which $u(x, t) > k$.

ν, μ, ϵ, δ, δ_k, θ, θ_k, γ, α, β are positive constants, with α being assumed to belong to the interval $(0, 1)$.

$\nu(t)$ is a positive nonincreasing continuous function defined for $t \geq 0$.

$\mu(t)$ is a positive nondecreasing continuous function defined for $t \geq 0$.

δ_i^j is the Kronecker delta symbol: $\delta_i^i = 1$, $\delta_i^j = 0$ for $i \neq j$.

$$|x|=\left(\sum_{i=1}^{n} x_i^2\right)^{\frac{1}{2}}, \quad x^2=|x|^2,$$

$$p=(p_1, \ldots, p_n), \quad u_x=(u_{x_1}, \ldots, u_{x_n}), \quad |p|=\left(\sum_{i=1}^{n} p_i^2\right)^{\frac{1}{2}},$$

$$p^2=|p|^2, \quad |u_x|=\left(\sum_{i=1}^{n} u_{x_i}^2\right)^{\frac{1}{2}},$$

$$u_x^2=|u_x|^2, \quad u_{x_i}^2=(u_{x_i})^2,$$

$$|u_{xx}|=\left(\sum_{i,j=1}^{n} u_{x_ix_j}^2\right)^{\frac{1}{2}},$$

$$a(x, t, u, p)=a(x_1, \ldots, x_n, t, u, p_1, \ldots, p_n),$$

$$a(x, t, u, u_x)=a(x_1, \ldots, x_n, t, u, u_{x_1}, \ldots, u_{x_n}).$$

ocs $\{u(x); \Omega\}$ is the oscillation of $u(x)$ on Ω, i.e. the difference between vrai $\max_\Omega u(x)$ and vrai $\min_\Omega u(x)$. osc $\{u(x, t); Q_T\}$ is defined analogously.

In the equations below we will encounter such expressions as

$$\frac{d}{dx_i}[a(x, t, u(x, t), u_x(x, t))],$$

which mean that in calculating the derivative d/dx_i it is necessary to take into account the presence of x_i not only in the first group of arguments but also in the other two, i.e. in the functions $u(x, t)$ and $u_x(x, t)$, so that

$$\frac{d}{dx_i}[a(x, t, u(x, t), u_x(x, t))]=\frac{\partial a}{\partial x_i}+\frac{\partial a}{\partial u}u_{x_i}+\frac{\partial a}{\partial u_{x_k}}u_{x_kx_i}.$$

Here and elsewhere pairs of equal indices imply a summation from 1 to n; in particular,

$$\frac{\partial a}{\partial u_{x_k}}u_{x_kx_i}=\sum_{k=1}^{n}\frac{\partial a}{\partial u_{x_k}}u_{x_kx_i}.$$

Sometimes, when it does not cause confusion, the symbol for total differentiation d/dx_i will be replaced by the more widely used symbol $\partial/\partial x_i$. For example, in a linear equation we usually write a term such as $(d/dx_i)(a_{ij}(x, t)u_{x_j}(x, t))$ in the form $(\partial/\partial x_i)(a_{ij}(x, t)u_{x_j}(x, t))$, even though in differentiating here one must take account of x_i in both $a_{ij}(x, t)$ and $u_{x_j}(x, t)$.

$\mathbf{n}$ is the outward (from Ω) unit normal to S at each of its points; $\partial/\partial\mathbf{n}$ denotes differentiation along $\mathbf{n}$.

In Chapter VII the notation $\nu = -\mathbf{n}$ is also used; ν is the inward unit normal to S.

A function $u(x)$ (or $u(x, t)$) is said to be *finite in* Ω (in Q_T) if it is different from zero only on some compact set that is separated from the boundary of Ω (of Q_T) by a positive distance.

A function $\zeta(x)$ (or $\zeta(x, t)$) is said to be *a cutting function for the domain* Ω (for Q_T) if it is continuous in $\bar{\Omega}$ (in $\bar{Q}_T$), has first-order piecewise-continuous bounded derivatives, vanishes on the boundary of this domain (on Γ_T), and has its values contained between zero and one.

2. Definitions of the basic function spaces. $L_q(\Omega)$ is the Banach space consisting of all measurable functions on Ω that are qth-power ($q \geq 1$) summable on Ω. The norm in it is defined by the equalities

$$\|u\|_{q,\Omega} = \left(\int_\Omega |u(x)|^q\, dx\right)^{\frac{1}{q}} \quad \text{and} \quad \|u\|_{\infty,\Omega} = \operatorname*{vrai\,max}_{\Omega} |u|.$$

Measurability and summability are to be understood everywhere in the sense of Lebesgue. The elements of $L_q(\Omega)$ are equivalence classes of the functions on Ω.

$L_{q,r}(Q_T)$ is the Banach space consisting of all measurable functions on Q_T with a finite norm

$$\|u\|_{q,r,Q_T} = \left(\int_0^T \left(\int_\Omega |u(x, t)|^q\, dx\right)^{\frac{r}{q}} dt\right)^{\frac{1}{r}},$$

where $q \geq 1$ and $r \geq 1$.

$L_{q,q}(Q_T)$ will be denoted by $L_q(Q_T)$, and the norm $\|\cdot\|_{q,q,Q_T}$ by $\|\cdot\|_{q,Q_T}$.

Generalized derivatives are to be understood in the way that is now customary in the majority of papers on differential equations. Different but equivalent definitions and their basic properties can be found, for example, in [112, Vol. V] and [113a].

$W_q^l(\Omega)$ for l integral is the Banach space consisting of all elements of $L_q(\Omega)$ having generalized derivatives of all forms up to order l inclusively, that are qth-power summable on Ω. The norm in $W_q^l(\Omega)$ is defined by the equality

$$\|u\|_{q,\Omega}^{(l)} = \sum_{j=0}^{l} \langle\langle u\rangle\rangle_{q,\Omega}^{(j)}, \tag{1.1}$$

where

$$\langle\langle u\rangle\rangle_{q,\Omega}^{(j)} = \sum_{(j)} \| D_x^j u \|_{q,\Omega}. \tag{1.2}$$

The symbol D_x^j denotes any derivative of $u(x)$ with respect to x of order j, while $\Sigma_{(j)}$ denotes summation over all possible derivatives of u of order j. For domains with "not too bad" boundaries $W_q^l(\Omega)$ coincides with the closure in the norm (1.1) of the set of all functions that are infinitely differentiable in $\bar{\Omega}$. This will be true, for example, for domains with piecewise-smooth boundaries (a definition of which is given below). Sometimes W_q^l is written in place of $W_q^l(\Omega)$, particularly if the domain Ω is subject to a further refinement.

$\dot{W}_q^l(\Omega)$ is the set of elements of $W_q^l(\Omega)$, that are finite in Ω.

$\overset{\circ}{W}{}_q^l(\Omega)$ is the subspace of $W_q^l(\Omega)$ in which the set of all functions that are infinitely differentiable and finite in Ω is dense. It is known that $\dot{W}_q^l(\Omega) \subset \overset{\circ}{W}{}_q^l(\Omega)$.

$W_q^{2l,l}(Q_T)$ for l integral $(q \geq 1)$ is the Banach space consisting of the elements of $L_q(Q_T)$ having generalized derivatives of the form $D_t^r D_x^s$ with any r and s satisfying the inequality $2r + s \leq 2l$. The norm in it is defined by the equality

$$\|u\|_{q,Q_T}^{(2l)} = \sum_{j=0}^{2l} \langle\langle u\rangle\rangle_{q,Q_T}^{(j)}, \tag{1.3}$$

where

$$\langle\langle u\rangle\rangle_{q,Q}^{(j)} = \sum_{(2r+s=j)} \| D_t^r D_x^s u \|_{q,Q_T}. \tag{1.4}$$

The summation $\Sigma_{(2r+s=j)}$ is taken over all nonnegative integers r and s satisfying the condition $2r + s = j$.

Spaces $W_q^l(\Omega)$ and $W_q^{l,l/2}(S_T)$ with nonintegral l will be used in Chapters IV and VII. The former space is defined in §2 of Chapter II, and the latter in §3 of Chapter II.

In addition to $W_q^{2l,l}(Q_T)$, we will encounter two spaces with different ratios of the upper indices:

$W_2^{1,0}(Q_T)$ is the Hilbert space with scalar product

$$(u, v)_{W_2^{1,0}(Q_T)} = \int_{Q_T} (uv + u_{x_k} v_{x_k})\, dx\, dt$$

and $W_2^{1,1}(Q_T)$ is the Hilbert space with scalar product

$$(u, v)_{W_2^{1,1}(Q_T)} = \int_{Q_T} (uv + u_{x_k} v_{x_k} + u_t v_t)\, dx\, dt.$$

$V_2(Q_T)$ is the Banach space consisting of all elements of $W_2^{1,0}(Q_T)$ having a finite norm

$$|u|_{Q_T} = \operatorname{vrai} \max_{0<t<T} \|u(x, t)\|_{2, \Omega} + \|u_x\|_{2, Q_T}, \tag{1.5}$$

where here and below

$$\|u_x\|_{2, Q_T} = \sqrt{\int_{Q_T} u_x^2\, dx\, dt}.$$

$V_2^{1,0}(Q_T)$ is the Banach space consisting of all elements of $V_2(Q_T)$ that are continuous in t in the norm of $L_2(\Omega)$, with norm

$$|u|_{Q_T} = \max_{0<t<T} \|u(x, t)\|_{2, \Omega} + \|u_x\|_{2, Q_T}. \tag{1.6}$$

The continuity in t of a function $u(x, t)$ in the norm of $L_2(\Omega)$ means that $\|u(x, t + \Delta t) - u(x, t)\|_{2,\Omega} \to 0$ for $\Delta t \to 0$. The space $V_2^{1,0}(Q_T)$ is obtained by completing the set $W_2^{1,1}(Q_T)$ in the norm of $V_2(Q_T)$.

$V_2^{1,1/2}(Q_T)$ is the subset of those elements $u(x, t)$ of $V_2^{1,0}(Q_T)$ for which

$$\int_0^{T-h} \int_\Omega h^{-1} [u(x, t + h) - u(x, t)]^2\, dx\, dt \xrightarrow[h \to 0]{} 0.$$

A zero over $W_2^{1,0}(Q_T)$, $W_2^{1,1}(Q_T)$, $V_2(Q_T)$, $V_2^{1,0}(Q_T)$, $V_2^{1,1/2}(Q_T)$ means that only those elements of these spaces are taken which vanish on S_T.

We now define spaces consisting of functions that are continuous in the sense of Hölder.

We will say that a function $u(x)$ defined in $\bar{\Omega}$ satisfies a Hölder condition in x with exponent α, $\alpha \in (0, 1)$, and Hölder constant $\langle u\rangle_{\Omega}^{(\alpha)}$ in the domain $\bar{\Omega}$ if

$$\sup \rho^{-\alpha} \operatorname{osc}\{u; \Omega_\rho^i\} \equiv \langle u\rangle_\Omega^{(\alpha)} < \infty, \tag{1.7}$$

where the supremum is taken over all connected components Ω_ρ^i of all Ω_ρ with $\rho \le \rho_0$. If the boundary of the domain Ω is "not too bad" (for example, piecewise-smooth without double points), then $\langle u\rangle_\Omega^{(\alpha)}$ can also be defined in another way, as

$$\sup_{\substack{x,\, x' \in \Omega \\ |x-x'| < \rho_0}} \frac{|u(x)-u(x')|}{|x-x'|^\alpha} = \langle u\rangle_\Omega^{(\alpha)}. \tag{1.8}$$

For domains with a two-piece boundary, for example for the domain $\{(x_1, x_2)\colon |x_1| < 1,\ |x_2| < 1 \text{ and } x_2 \ne 0 \text{ for } |x_1| \le \tfrac{1}{2}\}$, definitions (1.7) and (1.8) are not equivalent. In such cases we will adhere to the first definition.

We proceed to define the Hölder spaces $H^l(\bar{\Omega})$ and $H^{l,l/2}(\bar{Q}_T)$. In them l is always a nonintegral positive number.

$H^l(\bar{\Omega})$ is the Banach space whose elements are continuous functions $u(x)$ in $\bar{\Omega}$ having in $\bar{\Omega}$ continuous derivatives up to order $[l]$ inclusively and a finite value for the quantity

$$|u|_\Omega^{(l)} = \langle u\rangle_\Omega^{(l)} + \sum_{j=0}^{[l]} \langle u\rangle_\Omega^{(j)}, \tag{1.9}$$

where

$$\langle u\rangle_\Omega^{(0)} = |u|_\Omega^{(0)} = \max_\Omega |u|,$$

$$\langle u\rangle_\Omega^{(j)} = \sum_{(j)} |D_x^j u|_\Omega^{(0)}, \quad \langle u\rangle_\Omega^{(l)} = \sum_{([l])} \langle D_x^{[l]} u\rangle_\Omega^{(l-[l])}.$$

Equality (1.9) defines the norm $|u|_\Omega^{(l)}$ in $H^l(\bar{\Omega})$.

$H^{l,l/2}(\bar{Q}_T)$ is the Banach space of functions $u(x, t)$ that are continuous in $\bar{Q}_T$, together with all derivatives of the form $D_t^r D_x^s$ for $2r + s < l$, and have a finite norm

$$|u|_{Q_T}^{(l)} = \langle u\rangle_{Q_T}^{(l)} + \sum_{j=0}^{[l]} \langle u\rangle_{Q_T}^{(j)}, \tag{1.10}$$

where

$$\langle u\rangle_{Q_T}^{(0)} \equiv |u|_{Q_T}^{(0)} = \max_{Q_T} |u|,$$

$$\langle u\rangle_{Q_T}^{(j)} = \sum_{(2r+s=j)} |D_t^r D_x^s u|_{Q_T}^{(0)},$$

$$\langle u\rangle_{Q_T}^{(l)} = \langle u\rangle_{x,\, Q_T}^{(l)} + \langle u\rangle_{t,\, Q_T}^{(l/2)},$$

$$\langle u\rangle_{x,\, Q_T}^{(l)} = \sum_{(2r+s=[l])} \langle D_t^r D_x^s u\rangle_{x,\, Q_T}^{(l-[l])}.$$

$$\langle u\rangle^{(l/2)}_{t,\,Q_T}=\sum_{0<l-2r-s<2}\langle D_t^rD_x^su\rangle^{\left(\frac{l-2r-s}{2}\right)}_{t,\,Q_T};$$

$$\langle u\rangle^{(\alpha)}_{x,\,Q_T}=\sup_{\substack{(x,\,t),\,(x',\,t)\in\bar Q_T\\ |x-x'|\leqslant\rho_0}}\frac{|u(x,\,t)-u(x',\,t)|}{|x-x'|^\alpha},\quad 0<\alpha<1,\tag{1.11}$$

$$\langle u\rangle^{(\alpha)}_{t,\,Q_T}=\sup_{\substack{(x,\,t),\,(x,\,t')\in\bar Q_T\\ |t-t'|\leqslant\rho_0}}\frac{|u(x,\,t)-u(x,\,t')|}{|t-t'|^\alpha},\quad 0<\alpha<1.\tag{1.12}$$

For boundaries S with double points, instead of (1.11) we take another definition for $\langle u\rangle^{(\alpha)}_{x,Q_T}$, that is analogous to (1.2), namely

$$\langle u\rangle^{(\alpha)}_{x,\,Q_T}=\sup_{t\in[0,\,T]}\sup\rho^{-\alpha}\operatorname{osc}_x\{u(x,\,t);\ \Omega^i_\rho\}.$$

Here the second supremum is taken over all connected components Ω^i_ρ of all Ω_ρ with $\rho\geq\rho_0$, while

$$\operatorname{osc}_x\{u(x,\,t);\ \Omega^i_\rho\}=\operatorname*{vrai\,max}_{x\in\Omega^i_\rho}u(x,\,t)-\operatorname*{vrai\,min}_{x\in\Omega^i_\rho}u(x,\,t).$$

For $l<1$ the spaces $H^l(\bar\Omega)$ and $H^{l,l/2}(\bar Q_T)$ will most frequently be denoted by $H^\alpha(\bar\Omega)$ and $H^{\alpha,\alpha/2}(\bar Q_T)$. The norms defined by us in $H^l(\bar\Omega)$ and in $H^{l,l/2}(\bar Q_T)$ depend on ρ_0, but for different $\rho_0>0$ they are equivalent, and therefore their dependence on ρ_0 will not be noted in the sequel.

All of the function spaces just described are complete.

$H^l(\Omega)$ is the set of functions belonging to $H^l(\bar\Omega')$ for any closed subdomain $\bar\Omega'\subset\Omega$.

$H^{l,l/2}(Q_T)$ is the set of functions belonging to $H^{l,l/2}(\bar Q')$ for any closed subdomain $\bar Q'\subset Q_T$.

$C(\bar\Omega)$ $(C(\Omega))$ is the set of all functions that are continuous in $\bar\Omega$ (in Ω).

$C(\bar Q_T)$ and $C(Q_T)$ are defined analogously. $C^l(\bar\Omega)$ $(C^l(\Omega))$ is the set of all continuous functions in $\bar\Omega$ (Ω) having continuous derivatives in $\bar\Omega$ (Ω) up to order l inclusively.

$C^{2,1}(\bar Q_T)$ $(C^{2,1}(Q_T))$ is the set of all continuous functions in $\bar Q_T$ (in Q_T) having continuous derivatives u_x, u_{xx}, u_t in $\bar Q_T$ (in Q_T).

$C^{1,1/2}(\bar Q_T)$ is the set of all continuous functions in $\bar Q_T$ satisfying a Lipschitz condition in x and a Hölder condition in t with exponent $\frac{1}{2}$.

$O^l(\bar{\Omega})$ $(l = 1, 2)$ is the set of all continuous functions in $\bar{\Omega}$ having continuous derivatives in $\bar{\Omega}$ up to order $l-1$, with the derivatives of order $l-1$ having a first differential at each point of $\bar{\Omega}$ and the derivatives of order l being bounded in $\bar{\Omega}$.

$O^{2,1}(\bar{Q}_T)$ $(O^{2,1}(Q_T))$ is the set of all continuous functions in $\bar{Q}_T$ (in Q_T) having at each point of $\bar{Q}_T$ (of Q_T) derivatives u_x and u_t with the u_x being continuous in x and having a first differential with respect to x at each point of $\bar{Q}_T$ (of Q_T) and the functions u_x, u_t, u_{xx} being bounded in $\bar{Q}_T$ (in Q_T).

3. On domains, their boundaries, and functions prescribed on these boundaries. Throughout the book we limit our considerations to domains having "piecewise-smooth boundaries with nonzero interior angles" or, more briefly, "piecewise-smooth boundaries." By this restriction we will understand a domain whose closure can be represented in the form $\bar{\Omega} = \bar{\Omega}_1 \cup \cdots \cup \bar{\Omega}_N$, $\Omega_j \cap \Omega_i = 0$, where each of the $\bar{\Omega}_k$ can be homeomorphically mapped onto the unit ball or cube by means of functions satisfying a Lipschitz condition in $\bar{\Omega}_k$ and such that the Jacobians $|\partial(z^k)/\partial(x)|$ of these transformations are bounded from below by a positive constant.

We will say that the boundary S of a domain Ω (or a part S_1 of it) satisfies condition (A) if there exist two positive numbers a_0 and θ_0 such that the inequality

$$\operatorname{mes} \Omega_\rho^i \leqslant (1 - \theta_0) \operatorname{mes} K_\rho$$

holds for any ball K_ρ with center on S (on S_1 respectively) of radius $\rho \leq a_0$ and any of the connected components Ω_ρ^i of the intersection Ω_ρ of the ball K_ρ with Ω.

Let $x^0 = (x_1^0, \cdots, x_n^0)$ be any point of the boundary S of a domain Ω. We will call $(y_1, \cdots, y_n)$ a local Cartesian coordinate system with origin at x^0 if y and x are connected by the equations $y_i = a_{ik}(x_k - x_k^0)$, $i = 1, \cdots, n$, where the a_{ik} form an orthogonal numerical matrix and the y_n axis has the direction of the outward (with respect to Ω) normal to S at x^0.

We will say that a surface S belongs to class H^l, $l > 1$ (or C^l, or O^l, $l \geq 1$) if there exists a number $\rho > 0$ such that the intersection of S with a ball K_ρ of radius ρ with center at an arbitrary point $x^0 \in S$ is a connected surface, the equation of which in the local coordinate system $(y_1, \cdots, y_n)$ with origin at x^0 has the form $y_n = \omega(y_1, \cdots, y_{n-1})$, where $\omega(y_1, \cdots, y_{n-1})$ is a function of class H^l (C^l or O^l respectively) in the domain $\bar{D}$ that is the projection of D on the surface $y_n = 0$.

Suppose a function $\phi(s)$ is given on a surface S of class H^{l_1}, $l_1 > 1$ (C^{l_1} or O^{l_1}). We will say that $\phi(s)$ is a function of class $H^l(S)$, $l \leq l_1$, if as a function of $y_1, \cdots, y_{n-1}$ it is an element of $H^l(\bar{D})$. As the norm $|\phi|_S^{(l)}$ we take the largest of the norms $|\phi(y)|_D^{(l)}$ calculated for all points x^0 of S. Functions of classes $C^l(S)$ and $O^l(S)$ are defined analogously.

If ϕ is given on all of $\bar{\Omega}$ and $\phi(x) \in H^l(\bar{\Omega})$ ($C^l(\bar{\Omega})$ or $O^l(\bar{\Omega})$), then on the boundary S (of Ω) belonging to class H^{l_1} (C^{l_1} or O^{l_1}) with $l_1 \geq \max\{1, l\}$ it defines a function $\phi(s) = \phi(x)|_{x=s \in S}$ of class $H^l(S)$ ($C^l(S)$ or $O^l(S)$ respectively). The converse is also true: if $\phi(s) \in H^l(S)$ and $S \in H^l$, $l > 1$, then $\phi(s)$ can be continued onto all of the domain Ω in such a way that the continued function $\phi(x)$ belongs to $H^l(\bar{\Omega})$. And what is more, this continuation can be performed for all functions $\phi(s)$ of $H^l(S)$ by means of one and the same construction, so that the norms $|\phi(s)|_S^l$ and $|\phi(x)|_\Omega^{(l)}$ will be equivalent. It is precisely this continuation of $\phi(s)$ onto Ω that we will have in mind in formulating boundary conditions by means of a function $\phi(x)$. Analogous facts hold for the spaces C^l and O^l with $l \geq 1$. But for $l < 1$ all of what has been said concerning functions on S that belong to $H^l(S)$ ($C^l(S)$ or $O^l(S)$) is not true. Of two possible but different definitions it is more convenient for us to take the one that preserves the equivalence of the norms $|\phi(s)|_S^l$ and $|\phi(x)|_\Omega^l$. For this reason we define $|\phi(s)|_S^{(l)}$ in the following manner:

$$|\varphi|_S^{(l)} \equiv \max_{s \in S} |\varphi(s)| + \sup(\rho^{-\alpha} \operatorname{osc}\{\varphi(s); S_\rho^i\}), \tag{1.13}$$

where the supremum is taken over all connected components S_ρ^i of all $K_\rho \cap S$ with $\rho \leq \rho_0$.

All that we have said concerning surfaces S in E_n and functions $\phi(s)$ on them carries over in a natural way to surfaces S_T in E_{n+1} and functions $\phi(x, t)$ prescribed on them.

§2. CLASSICAL STATEMENTS OF THE PROBLEMS. THE MAXIMUM PRINCIPLE

It is expedient to divide linear and quasi-linear second-order equations of parabolic type into four groups: linear equations of the form

$$\mathscr{L}u \equiv u_t - \frac{\partial}{\partial x_i}(a_{ij}(x, t)\, u_{x_j} + a_i(x, t)\, u) + b_i(x, t)\, u_{x_i} + a(x, t)\, u = \frac{\partial f_i(x, t)}{\partial x_i} - f(x, t), \tag{2.1}$$

linear equations of the form

$$\mathscr{L}u \equiv u_t - a_{ij}(x, t)\, u_{x_i x_j} + a_i(x, t)\, u_{x_i} + a(x, t)\, u = f(x, t), \tag{2.2}$$

quasi-linear equations with principal part in divergence form

$$\mathscr{L}u \equiv u_t - \frac{d}{dx_i}(a_i(x, t, u, u_x)) + a(x, t, u, u_x) = 0 \tag{2.3}$$

and quasi-linear equations of general form

$$\mathscr{L}u \equiv u_t - a_{ij}(x, t, u, u_x)\, u_{x_i x_j} + a(x, t, u, u_x) = 0. \tag{2.4}$$

In regard to these equations we assume, unless otherwise stated, that they are *uniformly parabolic*. For equations (2.1) and (2.2) this means fulfilment of the condition

$$\nu\xi^2 \leqslant a_{ij}(x, t)\, \xi_i \xi_j \leqslant \mu\xi^2, \quad (x, t) \in \bar{Q}_T, \tag{2.5}$$

where here and everywhere below ν and μ are fixed positive numbers, and $\xi = (\xi_1, \cdots, \xi_n)$ is an arbitrary real vector; for equations (2.3) and (2.4) it means fulfilment of the conditions

$$\nu(|u|)\, \xi^2 \leqslant \frac{\partial a_i(x, t, u, p)}{\partial p_j}\, \xi_i \xi_j \leqslant \mu(|u|)\, \xi^2 \tag{2.6}$$

and

$$\nu(|u|)\, \xi^2 \leqslant a_{ij}(x, t, u, p)\, \xi_i \xi_j \leqslant \mu(|u|)\, \xi^2 \tag{2.7}$$

respectively for arbitrary u and p and $(x, t) \in \bar{Q}_T$. The functions $\nu(\tau)$ and $\mu(\tau)$, as indicated in §1, are certain continuous functions of $\tau \geq 0$ that assume only positive values.

If the functions a_{ij}, a_i and f_j are differentiable, equation (2.1) can be written in the form of (2.2) and, conversely, equation (2.2) can be put in the form of (2.1).

The most common problems involving equations (2.1)–(2.4) are the Cauchy problem, which consists in determining a function $u(x, t)$ in the half-space $\{|x| < \infty, t \geq 0\}$ satisfying the corresponding equation for $t > 0$ and the initial condition

$$u|_{t=0} = \psi_0(x) \tag{2.8}$$

for $t = 0$, and the boundary value problems (or, what is the same thing, mixed boundary value problems) in cylinders of form $Q_T = \Omega \times (0, T)$, which consist in determining functions $u(x, t)$ in $\bar{Q}_T$ satisfying the corresponding equation in Q_T, the initial condition (2.8) on the lower base of Q_T, and one of the boundary conditions on the lateral surface S_T, for example, the first

$$u|_{S_T} = \psi(s, t), \tag{2.9}$$

or the second

$$\frac{\partial u}{\partial N} + \sigma u|_{S_T} \equiv a_{ij}u_{x_j}\cos(n, x_i) + \sigma u|_{S_T} = \psi(s, t). \tag{2.10}$$

We will confine ourselves on the whole to these problems, although much of what will be said will also be applicable to boundary value problems in which Q_T is not a cylinder.

In the classical formulation of the Cauchy problem it is required that the solution be continuous at all points $(x, t) \in \{|x| < \infty, t \geq 0\}$, that it have continuous derivatives u_t, u_x and u_{xx} and satisfy the equation at all interior points (x, t) (i.e., for $|x| < \infty$, $t > 0$), and that it remain bounded for $|x| \to \infty$ (or, more generally, that it not exceed some given function). In the classical formulation of the first boundary value problem it is required that the solution be continuous in $\bar{Q}_T$, have continuous derivatives u_t, u_x, u_{xx} in Q_T, satisfy the equation at all points of Q_T, and satisfy conditions (2.8) and (2.9) for $t = 0$ and $(x, t) \in S_T$. If Ω is an unbounded domain, then to these requirements must be added a condition on the behavior of the solution for $|x| \to \infty$. For the second boundary value problem it is required in addition that the derivatives u_{x_i} exist and be continuous in $Q_T \cup S_T$ (sometimes this requirement is weakened to the existence of the derivatives $\partial u/\partial N|_{S_T}$ and their continuity under certain approaches to S_T). For the nonlinear equations (2.3) and (2.4), in addition to the conditions just enumerated, one sometimes requires the continuity in $\bar{Q}_T$ of the derivatives u_x or even of all the derivatives u_t, u_x and u_{xx} (these requirements will be specially stipulated).

For all of the problems just enumerated, the uniqueness theorems in their classical statement are valid as long as the functions making up the equations have a certain smoothness.

They are proved with the help of the well-known maximum principle. In its simplest form it consists in the fact that the solutions u of equation (2.2) with $a \equiv f \equiv 0$ assume their least and greatest values in $\bar{Q}_T$ on the boundary Γ_T. In a more general form it finds its expression in the estimates of $a \equiv f \equiv 0$ given below by Theorems (2.1–2.3). It is based on the following fact: at interior extremum points of $u(x, t)$ the expressions $u_t - a_{ij}u_{x_ix_j}$ and $u_t - (\partial a_i/\partial u_{x_j})u_{x_ix_j}$, which form the principal part of the parabolic operators $\mathfrak{L}u$ of (2.1)–(2.4), are either both nonnegative or both nonpositive.

Theorem 2.1. *Suppose $u(x, t)$ is the classical solution of problem* (2.2), (2.8), (2.9), *in the cylinder Q_T, where the coefficients and free term in* (2.2) *are bounded functions and*

$$a_{ij}(x, t)\xi_i\xi_j \geqslant 0. \tag{2.11}$$

Then for any $t_1 \in [0, T]$ the estimate

$$\sup_{\lambda > a_0} \min\left\{0;\ \min_{\Gamma_{t_1}} (\psi e^{\lambda(t_1 - t)});\ \frac{1}{\lambda - a_0} \min_{Q_{t_1}} (f e^{\lambda(t_1 - t)})\right\}$$
$$\leqslant u(x, t_1) \leqslant \inf_{\lambda > a_0} \max\left\{0;\ \max_{\Gamma_{t_1}} (\psi e^{\lambda(t_1 - t)});\ \frac{1}{\lambda - a_0} \max_{Q_{t_1}} (f e^{\lambda(t_1 - t)})\right\} \tag{2.12}$$

for $u(x, t)$ is valid, where the number a_0 is equal to $\max_{Q_{t_1}}(-a(x, t)) = -\max_{Q_{t_1}} a(x, t)$, *while ψ coincides with $\psi_0(x)$ of* (2.8) *on Γ_0, and with $\psi(s, t)$ of* (2.9) *on S_T.*

We recall that we agreed to denote the cylinder $\Omega \times (0, t)$, by Q_t, and the sum of its lateral surface S_t and lower base Γ_0 by Γ_t.

For the proof of (2.12) we make a transformation which will frequently be encountered in the sequel; namely, we pass from $u(x, t)$ to a new function $v(x, t)$ connected with it by the equality

$$u(x, t) = v(x, t)e^{\lambda t}, \tag{2.13}$$

where λ is for the time being an arbitrary number. The function v, as is easily seen, satisfies the equation

$$v_t - a_{ij} v_{x_i x_j} + a_i v_{x_i} + (a+\lambda) v = f e^{-\lambda t} \tag{2.14}$$

resulting from (2.2). Take an arbitrary t_1 from $(0, T)$. There are three possible cases: either $v(x, t)$ is nonpositive in $\bar{Q}_{t_1}$, or the least positive value of $v(x, t)$ in $\bar{Q}_{t_1}$ is assumed on Γ_{t_1}, or its greatest value is assumed at some point $(x^0, t^0) \in \Omega \times (0, t_1]$. In the first case $\max_{Q_{t_1}} v(x, t) \leq 0$; in the second case $0 < \max_{Q_{t_1}} v(x, t) \leq \max_{\Gamma_{t_1}} v$; and in the third case $0 < \max_{Q_{t_1}} v(x, t) \leq v(x^0, t^0)$, with the relations

$$v_t \geqslant 0, \quad v_{x_l} = 0 \text{ and } \quad -a_{ij} v_{x_i x_j} \geqslant 0 \tag{2.15}$$

and equation (2.14) being satisfied at the point (x^0, t^0). The validity of the last of the relations (2.15) is due to the fact that at a maximum point the pure second derivatives $\partial^2 u/\partial y_k \partial y_k$ are nonpositive in any direction $y_k = \alpha_{kl}(x_l - x_l^0)$ ($\det|\alpha_{kl}| \neq 0$) while the expression $-a_{ij} v_{x_i x_j}$ is equal to $-\lambda_k v_{y_k y_k}$ for an orthogonal matrix α_{kl}, where $\lambda_1, \cdots, \lambda_n$ are the characteristic numbers of the quadratic form $a_{ij}\xi_i\xi_j$ (and consequently are nonnegative by (2.11)). By virtue of (2.15) and equation (2.14) the inequality

$$[a(x^0, t^0) + \lambda] v(x^0, t^0) \leqslant f(x^0, t^0) e^{-\lambda t^0}$$

holds at the point (x^0, t^0). We will assume that $\lambda > \max_{Q_{t_1}} [-a(x, t)] \equiv a_0$. Then in all cases the estimate

$$\max_{Q_{t_1}} v(x, t) \leqslant \max \left\{ 0;\ \max_{\Gamma_{t_1}} v;\ \max_{Q_{t_1}} \frac{f(x, t) e^{-\lambda t}}{a(x, t) + \lambda} \right\} \tag{2.16}$$

is valid, and hence

$$u(x, t_1) \leqslant e^{\lambda t_1} \max \left\{ 0;\ \max_{\Gamma_{t_1}} (u e^{-\lambda t});\ \frac{1}{\lambda - a_0} \max_{Q_{t_1}} (f e^{-\lambda t}) \right\}. \tag{2.17}$$

Analogously, by considering a point of least nonpositive value of the function $v(x, t)$, we arrive at the estimate of $u(x, t)$ from below indicated in inequality (2.12).

An estimate of u in the form

$$|u(x, t_1)| \leqslant \max_{\Gamma_{t_1}} |\psi e^{a_0(t_1 - t)}| + t_1 \max_{Q_{t_1}} |f e^{a_0(t_1 - t)}| \tag{2.18}$$

is also valid. This follows from the fact that the functions

$$\mu_1+\mu_2 t \pm u(x,t)e^{-a_0 t},$$

where

$$\mu_1=\max_{\Gamma_{t_1}}|\psi e^{-a_0 t}|,\quad \mu_2=\max_{Q_{t_1}}|fe^{-a_0 t}|,$$

are nonnegative in Q_{t_1} by virtue of the equations that they satisfy.

Corollary 2.1. *Suppose the conditions of Theorem 2.1 are satisfied. Then the solution $u(x,t)$ is nonpositive in $\bar{Q}_T$ for $u|_{\Gamma_T}\le 0$ and $f\le 0$, and is nonnegative in $\bar{Q}_T$ for $u|_{\Gamma_T}\ge 0$ and $f\ge 0$. Also, if $a(x,t)\equiv f(x,t)\equiv 0$, then*

$$\min_{\Gamma_{t_1}} u \leqslant u(x,t_1)\leqslant \max_{\Gamma_{t_1}} u \tag{2.19}$$

for any $t_1\in[0,T]$.

The first two assertions follow directly from (2.12). For a proof of (2.19) we take the function $w(x,t)=u(x,t)-u(x_0,0)$, where x_0 is a point of S, and, since it satisfies the same equation $\mathfrak{L}w=0$ as $u(x,t)$, apply estimate (2.12) to it. This gives

$$\sup_{\lambda>0}\min\left\{0;\ \min_{\Gamma_{t_1}}[(u-u(x_0,0))e^{\lambda(t_1-t)}]\right\}\leqslant u(x,t_1)-u(x_0,0)$$

$$\leqslant \inf_{\lambda>0}\max\left\{0;\ \max_{\Gamma_{t_1}}[(u-u(x_0,0))e^{\lambda(t_1-t)}]\right\}.$$

But the left side of this inequality is equal to $\min_{\Gamma_{t_1}} u-u(x_0,0)$, while the right side is equal to $\max_{\Gamma_{t_1}} u-u(x_0,0)$, and thus the inequalities (2.19) are valid for $u(x,t_1)$.

We now consider for equation (2.2) a boundary value problem in Q_T with boundary condition

$$b_i(x,t)u_{x_i}+b(x,t)u|_{S_T}=\psi(s,t), \tag{2.20}$$

in which the vector $\mathbf{b}=(b_1,\cdots,b_n)$ forms with the direction of the outer normal $\mathbf{n}$ to S_T at the same point $(x,t)\in S_T$ an angle not exceeding $\pi/2$. The following two theorems are valid for this problem.

Theorem 2.2. *Suppose $u(x,t)$ is a classical solution of problem* (2.2), (2.8), (2.20) *in the cylinder Q_T and suppose the coefficients of equation* (2.2) *and the functions f, b_i, b and ψ are bounded, with the a_{ij} being subject to con-*

dition (2.11), *while* $b(x, t)|_{S_T} > 0$. *Then for any* t_1 *from* $[0, T]$ *one has the estimates*

$$\sup_{\lambda > a_0} \min\left\{0; \min_{S_{t_1}} \frac{\psi e^{\lambda(t_1-t)}}{b}; e^{\lambda t_1} \min_{\Omega} u(x, 0); \frac{1}{\lambda - a_0} \min_{Q_{t_1}} (f e^{\lambda(t_1-t)})\right\} \leqslant$$
$$\leqslant u(x, t_1) \leqslant \inf_{\lambda > a_0} \max\left\{0; \max_{S_{t_1}} \frac{\psi e^{\lambda(t_1-t)}}{b};\right.$$
$$\left. e^{\lambda t_1} \max_{\Omega} u(x, 0); \frac{1}{\lambda - a_0} \max_{Q_{t_1}} f e^{\lambda(t_1-t)}\right\}, \tag{2.21}$$

in which a_0 *is equal to* $\max_{Q_{t_1}} (-a(x, t))$.

This theorem is proved in the same way as Theorem 2.1, i.e., the function $v(x, t) = u(x, t)e^{-\lambda t}$ is taken and all possible locations of a positive maximum and a negative minimum of it are analyzed. All of the possible situations coincide with those considered in Theorem 2.1. A difference in the estimates obtained for them exists only in the case when these extremums fall on S_{t_1}. Thus, if v takes its greatest value in $\bar{Q}_{t_1}$ at some point $(x^0, t^0) \in S_{t_1}$, then at this point $b_i v_{x_i} \geq 0$, (since the vector $\mathbf{b} = (b_1, \cdots, b_n)$ forms with the outer normal to S_T at the same point an angle not exceeding $\pi/2$); hence $v\big|_{(x^0, t^0)} \leq \psi e^{-\lambda t}/b\big|_{(x^0, t^0)}$ by virtue of the boundary condition (2.20). Analogously, if a minimum point (x^0, t^0) for v falls on S_{t_1}, then at it $b_i v_{x_i} \leq 0$, and hence $v\big|_{(x^0, t^0)} \geq \psi e^{-\lambda t}/b\big|_{(x^0, t^0)}$ by virtue of (2.20).

Theorem 2.3. *Suppose all of the conditions of Theorem 2.2 are fulfilled except* $b|_{S_T} > 0$, *which is replaced by the condition* $b|_{S_T} \geq -b_0$, $b_0 = \text{const} \geq 0$, *and suppose*

$$\max|a_{ij}, a_i| \leqslant \mu_1, \quad a_0 = \max_{Q_T}(-a(x, t))$$

and

$$b_i \cos(\mathbf{n}, x_i) \equiv \sqrt{\sum_{k=1}^{n} b_k^2} \cos(\mathbf{n}, \mathbf{b}) \geqslant \delta, \quad \delta = \text{const} > 0.$$

Then

$$\max_{Q_T} |u| \leqslant c_1 e^{cT} \max \left\{ \max_{S_T} |\psi|;\ \max_{\Omega} |u(x, 0)|;\ \max_{Q_T} |f| \right\}, \tag{2.22}$$

where c and c_1 are constants depending only on the numbers b_0, μ_1, a_0, δ and the boundary of the domain Ω, which is assumed to belong to the class O^2.

For the proof of (2.22) we introduce in place of $u(x, t)$ the function $w(x, t) = u(x, t)\phi(x)$, where $\phi(x)$ is a function of class $O^2(\bar{\Omega})$, that satisfies the conditions

$$\min_{\Omega} \varphi(x) \geqslant \frac{1}{2}, \quad \varphi|_S = 1, \quad -\frac{\partial \varphi}{\partial n}\Big|_S = m,$$

where $m = \text{const} > b_0/\delta$. Then $-1/\phi b_i \phi_{x_i}|_{S_T} = mb_i \cos(n, x_i) \geq \delta m > b_0$. The function w, as is easily seen, satisfies in Q_T an equation of the same type as that satisfied by $u(x, t)$, and on S_T it satisfies the condition

$$b_i w_{x_i} + \left(b - b_i \frac{\varphi_{x_i}}{\varphi}\right) w \Big|_{S_T} = \psi\varphi|_{S_T},$$

so that the conditions of Theorem 2.2 are fulfilled. In view of this, estimates of type (2.21) are valid for w, from which follows the estimate (2.22) for u.

Remark 2.1. As can be seen from the proof of Theorems 2.1–2.3 and Corollary 2.1, they and their assertions (2.12), (2.21), (2.22) remain in force when the coefficients $a_i(x, t)$ of equation (2.2) have isolated poles in x of less than first order. The latter means that at a singular point $x = x^0$ the product $a_i(x, t)(x - x^0)$ must tend to zero for $x \to x^0$.

From Theorems 2.1–2.3 follow uniqueness theorems for the classical solutions of the boundary value problems. Namely:

Theorem 2.4. *Problems* (2.2), (2.8), (2.9) *and* (2.2), (2.8), (2.20) *under conditions corresponding to Theorems* 2.1–2.3 *cannot have more than one classical solution.*

Indeed, the difference $u(x, t)$ of two possible solutions of any of these problems satisfies the corresponding homogeneous equation and homogeneous initial and boundary conditions, so that the corresponding inequality (2.12), (2.21) or (2.22) with $f = \phi_0 = \psi \equiv 0$ is valid for it, from which it follows that $u \equiv 0$.

We will now show that the Cauchy problem also cannot have two distinct classical solutions. For this purpose we will prove

Theorem 2.5. *Suppose the function $u(x, t)$ is continuous in the strip $\Pi_T = \{|x| < \infty,\ 0 \le t \le T\}$ and its modulus does not exceed some number M. Suppose, further, that it satisfies equation (2.2) for $0 < t \le T$, that the moduli of the coefficients a_{ij}, a_i of (2.2) do not exceed c, and that $a(x, t) \ge -a_0$, where c and a_0 are nonnegative constants. Then the estimate*

$$\max_{\Pi_T} |u(x, t)| \leqslant (\max_{E_n} |u(x, 0)| + t \max_{\Pi_T} |f|)\, e^{a_0 t} \tag{2.23}$$

is valid if condition (2.11) is fulfilled.

For the proof of (2.23) we consider the function

$$w(x, t) = u(x, t)\, e^{-(a_0+\varepsilon)t} - c_1 - c_2 t - \frac{M}{R^2}(x^2 + c_3 t),$$

where $c_1 = \max_{E_n} |u(x, 0)|$, $c_2 = \max_{\Pi_T} |f(x, t)|$, and R, ϵ and c_3 are arbitrary positive numbers. We calculate $(\mathcal{L} + a_0 + \epsilon)w$:

$$\begin{aligned}(\mathcal{L} + a_0 + \varepsilon)\, w = f e^{-(a_0+\varepsilon)t} - c_2\,[1 + (a + a_0 + \varepsilon)\, t] \\ - \frac{M}{R^2}\left[c_3 - 2\sum_i a_{ii} + 2a_i x_i + (a + a_0 + \varepsilon)\cdot(x^2 + c_3 t)\right] \\ - (a + a_0 + \varepsilon)\, c_1.\end{aligned}$$

Let us select the number c_3 for fixed $\epsilon > 0$ so large that the expression multiplied by M/R^2 is nonnegative. Then $(\mathcal{L} + a_0 + \epsilon)w \le f e^{-(a_0+\epsilon)t} - c_2 \le 0$. In addition, on the lateral surface and lower base of the cylinder $Q_T(R) = \{|x| < R,\ 0 < t < T\}$ the function w is nonpositive. Therefore, by virtue of Corollary 2.1, w is nonpositive through $Q_T(R)$. Take an arbitrary point (x^0, t^0) of the strip $0 \le t \le T$. Then according to what has been proved $w(x^0, t^0) \le 0$ for $|x^0| \le R$. Letting R tend to infinity in this inequality, and then letting ϵ tend to zero, we obtain the desired estimate for $u(x, t)$ from above:

$$u(x^0, t^0) \leqslant (c_1 + c_2 t^0)\, e^{a_0 t^0}.$$

For the estimate of $u(x, t)$ from below it is necessary to take the function

$$\hat{w}(x, t) = u(x, t)\, e^{-(a_0+\varepsilon)t} + c_1 + c_2 t + \frac{M}{R^2}(x^2 + c_3 t)$$

with the same values for the parameters as above. For it $(\mathcal{L} + a_0 + \epsilon)\hat{w} \ge 0$, and on the lower base and lateral surface of $Q_T(R)$ it is nonnegative. Therefore $\hat{w}(x^0, t^0) \ge 0$ for $|x_0| \le R$, from which, letting R tend to ∞, and to zero, we obtain the estimate (2.23) for $u(x, t)$ from below.

Remark 2.2. Theorem 2.5 may be strengthened both in regard to the conditions on the coefficients (for example, one can admit a specified rate of growth of them for $|x| \to \infty$ or the existence of singularities of the type mentioned in Remark 2.1) and in regard to a replacement of the condition $\max_{\Pi_T} |u| < \infty$ by the condition $|u(x, t)| \leq ce^{c_1 x^2}$ with c and c_1 arbitrary. We will not cite here the different generalizations and strengthenings of Theorems 2.1–2.5 but refer the reader to the relevant literature [10, 25, 31b, 34, 46, 47b, 63, 90, 93, 107, 109, 110, 123c, d, 130, etc.].

From Theorem 2.5 follows the uniqueness theorem for the Cauchy problem in its classical statement.

Theorem 2.6. *Under the conditions of Theorem 2.5 the Cauchy problem has no more than one classical solution in the class of bounded functions.*

Below, a result is established that is analogous to Theorem 2.3 in an unbounded domain Ω. It concerns an estimate for the maximum of the modulus of a bounded solution of the problem with oblique derivative (2.2), (2.20), (2.8). This result is also true for the first boundary value problem in an unbounded domain (see [62]).

Theorem 2.7. *Suppose that all of the conditions of Theorem 2.3 are fulfilled, with the domain Ω being such that there exists a function $\phi(x)$ in it having the following properties:*

$$\varphi(x) \in O^2(\bar{\Omega}), \quad \min_{\Omega} \varphi(x) \geqslant \frac{1}{2}, \quad \varphi|_S = 1, \quad -\frac{\partial \varphi}{\partial n}\Big|_S = m,$$

where $m = \text{const} > b_0/\delta$. Then the estimate

$$|u(x, t)| \leqslant c\Big(\sup_{\Omega} |\psi_0(x)| + \sup_{S_T} |\psi(x, t)| + \sup_{Q_T} |f(x, t)|\, t\Big) e^{At} \qquad (2.24)$$

is valid, where c and A are constants depending on the b_i, b, the coefficients of the operator $\mathfrak{L}$, and $\phi(x)$.

Proof. The function $v(x, t) = u(x, t)\phi(x)$ is a solution of the boundary value problem

$$\tilde{\mathcal{L}}v \equiv \mathcal{L}v - 2\varphi a_{ij}(\varphi^{-1})_{x_i} v_{x_j} + \varphi\left[a_i(\varphi^{-1})_{x_i} - a_{ij}(\varphi^{-1})_{x_i x_j}\right] v = f\varphi,$$

$$v|_{t=0} = \psi_0(x)\varphi(x),$$

$$b_i v_{x_i} + \beta v|_{S_T} = \psi\varphi|_{S_T},$$

where

$$\beta = b + \varphi b_i(\varphi^{-1})_{x_i} = b - \frac{b_i \varphi_{x_i}}{\varphi} > 0.$$

We also introduce the function

$$w(x, t) = v(x, t)\, e^{-(\tilde{a}_0+\epsilon)t} - c_1 - c_2 t - \frac{M}{R^2}(x^2 + c_3 t), \tag{2.25}$$

in which the constants c_1, c_2, M, $\tilde{a}_0$ are defined as follows:

$$c_1 = \delta + \sup_{\Omega} |\psi_0(x)|\,\varphi(x) + \sup_{S_T} \frac{|\psi(x, t)|\,\varphi(x)}{\beta(x, t)}, \tag{2.26}$$

where δ is an arbitrary positive constant,

$$c_2 = \sup_{Q_T} |f(x, t)|\,\varphi(x),$$

$$M = \sup_{Q_T} |v(x, t)|,$$

$$\tilde{a}_0 = \max\left\{0; -\inf_{Q_T}\left\{a(x, t) + \varphi\left[a_i(\varphi^{-1})_{x_i} - a_{ij}(\varphi^{-1})_{x_i x_j}\right]\right\}\right\}.$$

The constant c_3 for fixed $\epsilon > 0$ is chosen in such a way that $(\tilde{\mathcal{L}} + \tilde{a}_0 + \epsilon)w(x, t) \le 0$ as was established in investigating the function (2.25) in the proof of Theorem 2.5, such a choice of c_3 is possible.

Let Ω_R be that part of the domain Ω lying in the ball $K_R: |x| \le R$. Its boundary $S_{(R)}$ is composed of two parts: $S_{(R)} = S^1_{(R)} + S^2_{(R)}$, where $S^1_{(R)} = S \cap K_l$ and $S^2_{(R)}$ is that part of the boundary of the ball K_R lying in Ω. We will show that if R is sufficiently great, then $w(x, t) \le 0$ for $x \in \Omega_R$. Indeed, from the definition of w and the constants c_1 and M, and also from the inequality $(\tilde{\mathcal{L}} + \tilde{a}_0 + \epsilon)w \le 0$, it follows that the function $w(x, t)$ cannot achieve a positive maximum in $\Omega_R \times [0, T]$ and for $x \in S^2_{(R)}$. On $S^1_{(R)}$ the function $w(x, t)$ satisfies the boundary condition

$$b_l w_{x_l} + \beta w = \psi\varphi e^{-(\tilde{a}_0+\varepsilon)t} - b_l F_{x_l} - \beta F,$$

where

$$F(x, t) = c_1 + c_2 t + \frac{M}{R^2}(x^2 + c_3 t).$$

Therefore at a maximum point on $S^1_{(R)}$

$$\beta w \leqslant \psi\varphi e^{-(\tilde{a}_0+\varepsilon)t} - b_l F_{x_l} - \beta F. \tag{2.27}$$

But by virtue of (2.26)

$$\psi\varphi e^{-(\tilde{a}_0+\varepsilon)t} - \beta F \leqslant \psi\varphi e^{-(\tilde{a}_0+\varepsilon)t} - \beta c_1 - \frac{\beta M}{R^2} x^2 \leqslant - \frac{\beta M}{R^2} x^2 - \beta\delta,$$

$$|b_l F_{x_l}| \leqslant \max |b_l| \frac{2|x_l| M}{R^2}.$$

Thus for large R, at a maximum point on $S^1_{(R)}$

$$\beta w \leqslant 0.$$

This proves that the function $w(x, t)$ does not in general have a positive maximum point in $\bar{\Omega}_R \times [0, T]$. Thus

$$w(x, t) \leqslant 0.$$

Letting $R \to \infty$ and $\epsilon \to 0$, we see that everywhere in Q_T

$$v(x, t) \leqslant (c_1 + c_2 t) e^{\tilde{a}_0 t}.$$

Since δ is arbitrary, this gives the desired estimate (2.24) from above for $u(x, t)$. The estimate from below is obtained in an analogous manner. Theorem 2.7 is proved.

We will apply this theorem in §6 of Chapter IV to a problem with directional derivative in a half-space, where the function $\phi(x)$, whose existence is in the conditions of the theorem, is constructed very simply.

Theorems 2.4 and 2.6 permit one to draw certain conclusions concerning the uniqueness theorems for the boundary value problems and for the Cauchy problem both in their classical statement and for nonlinear parabolic equations. Thus, for example, suppose $u'(x, t)$ and $u''(x, t)$ are two solutions of equation (2.4) in which the functions $a_{ij}(x, t, u, p)$, $a(x, t, u, p)$ have partial derivatives with respect to u and p. We subtract the equation for u'' from the equation for u' and write the result first as follows:

$$(u'-u'')_t - a_{ij}(x,t,u',u'_x)(u'-u'')_{x_ix_j}$$
$$- u''_{x_ix_j}\int_0^1 \frac{d}{d\tau} a_{ij}[x,t,\tau u'+(1-\tau)u'',\tau u'_x+(1-\tau)u''_x]\,d\tau$$
$$+\int_0^1 \frac{d}{d\tau} a[x,t,\tau u'+(1-\tau)u'',\tau u'_x+(1-\tau)u''_x]\,d\tau = 0,$$

and then as a linear equation for $v = u' - u''$:

$$v_t - \hat{a}_{ij}(x,t)v_{x_ix_j} + b_i(x,t)v_{x_i} + c(x,t)v = 0, \tag{2.28}$$

in which

$$\hat{a}_{ij}(x,t) = a_{ij}(x,t,u'(x,t),u'_x(x,t)),$$
$$b_k(x,t) = -u''_{x_ix_j}(x,t)\int_0^1 \frac{\partial a_{ij}[\ldots]}{\partial u_{x_k}}\,d\tau + \int_0^1 \frac{\partial a[\ldots]}{\partial u_{x_k}}\,d\tau,$$
$$c(x,t) = -u''_{x_ix_j}(x,t)\int_0^1 \frac{\partial a_{ij}[\ldots]}{\partial u}\,d\tau + \int_0^1 \frac{\partial a[\ldots]}{\partial u}\,d\tau.$$

If the coefficients in (2.28) are bounded in $\bar{Q}_T$ (or at least satisfy the conditions of Remark 2.1), then from the homogeneity of (2.28) and the vanishing of v on Γ_T it follows by virtue of Theorem 2.4 that $v(x,t)$ is identically equal to zero in Q_T, i.e. that u' and u'' coincide. One of the sufficient conditions for this is given by the following theorem.

Theorem 2.8. *If the coefficients a_{ij} satisfy the condition*

$$a_{ij}(x,t,u,p)\xi_i\xi_j \geqslant 0$$

and $a_{ij}(x,t,u,p)$ and $a(x,t,u,p)$ and their partial derivatives with respect to u and p are bounded on any compact set of the values of their arguments, then the first boundary value problem for (2.4) has no more than one solution in the class of functions belonging to $C(\bar{Q}_T)\cap C^{2,1}(Q_T)$ and having derivatives with respect to x of first and second order that are bounded in $\bar{Q}_T$.

An analogous theorem is also true for the Cauchy problem.

Theorems 2.1–2.3 and 2.5 are easily extended to quasi-linear equations. We cite here a theorem that permits one to estimate the solutions of the first boundary value problem for equations (2.4).

Theorem 2.9. *Let* $u(x, t)$ *be a classical solution of equation* (2.4) *in* Q_T. *Suppose that the functions* $a_{ij}(x, t, u, p)$ *and* $a(x, t, u, p)$ *take finite values for any finite* u, p, *and* $(x, t) \in \bar{Q}_T$, *and that for* $(x, t) \in Q_T$ *and arbitrary* u

$$a_{ij}(x, t, u, 0)\xi_i\xi_j \geqslant 0 \tag{2.29}$$

and

$$ua(x, t, u, 0) \geqslant -b_1u^2 - b_2, \tag{2.30}$$

where b_1 *and* b_2 *are nonnegative constants. Then*

$$\max_{Q_T}|u(x, t)| \leqslant \inf_{\lambda > b_1} e^{\lambda T}\left[\max_{\Gamma_T}|u|, \sqrt{\frac{b_2}{\lambda - b_1}}\right]. \tag{2.31}$$

If in place of (2.30) *the condition*

$$ua(x, t, u, 0) \geqslant -\Phi(|u|)|u| - b_2 \tag{2.32}$$

is fulfilled, where $b_2 \geq 0$ *and* $\Phi(r)$ *is a nondecreasing positive function of* $r \geq 0$ *satisfying the condition*

$$\int_0^\infty \frac{d\tau}{\Phi(\tau)} = \infty, \tag{2.33}$$

then the estimate

$$\max_{Q_T}|u(x, t)| \leqslant \inf_{\lambda > 1}\varphi(\xi),$$

$$\xi = e^{\lambda T}\max\left\{1;\ \varphi^{-1}\left(\frac{b_2}{(\lambda - 1)\Phi(0)}\right);\ \varphi^{-1}\left(\max_{\Gamma_T}|u|\right)\right\} \tag{2.34}$$

is valid, in which $\phi^{-1}(\xi)$ *is the inverse of the function* $\phi(\xi)$ *defined by the equation*

$$\int_0^{\varphi(\xi)} \frac{d\tau}{\Phi(\tau)} = \ln\xi. \tag{2.35}$$

The proof of the first assertion of the theorem is analogous to the proof of Theorem 2.1. Essentially, the second assertion is proved in the same way.

Namely, we introduce the function v connected with u by the equation $u = \phi(v)$. From (2.4) we obtain the relation

$$v_t - a_{ij}(x, t, u, u_x)\left(v_{x_ix_j} + \frac{\varphi''}{\varphi'} v_{x_i} v_{x_j}\right) + \frac{1}{\varphi'} a(x, t, u, u_x) = 0 \tag{2.36}$$

for v. For the function $\hat{v} = ve^{-\lambda t}$, $\lambda > 0$ we have the equation

$$\hat{v}_t - a_{ij}(x, t, u, u_x)\left(\hat{v}_{x_ix_j} + \frac{\varphi''}{\varphi'} \hat{v}_{x_i} \hat{v}_{x_j} e^{\lambda t}\right) + \frac{1}{\varphi'} a(x, t, u, u_x) e^{-\lambda t} + \lambda \hat{v} = 0, \tag{2.37}$$

from (2.36). If $\hat{v}$ takes its greatest value in $\bar{Q}_T$ at some interior point (x^0, t^0) of the domain Q_T and $\hat{v}(x_0, t_0) \geq e^{-\lambda t_0}$, then at this point $\hat{v}_t = 0$, $-a_{ij}\hat{v}_{x_ix_j} \geq 0$, $\hat{v}_{x_i} = 0$, so that $u_{x_i} = \phi'(v)v_{x_i} = \phi'(v)\hat{v}_{x_i}e^{\lambda t} = 0$, and hence on the basis of (2.37) we have at this point

$$[a(x, t, u, 0) + \lambda\varphi'(v)v]|_{(x_0, t_0)} \leqslant 0. \tag{2.38}$$

Since by assumption $\hat{v}(x_0, t_0) \geq e^{-\lambda t_0}$, we have $v(x_0, t_0) \geq 1$ and hence $u(x_0, t_0) \geq 0$. Multiplying (2.38) by $u(x_0, t_0)$ and using (2.32), we obtain

$$-b_2 + [-\Phi(u) + \lambda\varphi'(v)v]u \leqslant 0. \tag{2.39}$$

As is easily seen, the function $\phi(v)$ satisfies the differential equation $\phi'(v)v = \Phi(\phi)$ and the condition $\phi(1) = 0$.

Under a variation of v from 0 to ∞ the function $\phi(v)$ increases monotonically (by virtue of condition (2.33)) from $-\infty$ to $+\infty$. For such a ϕ inequality (2.39) takes the form

$$(\lambda - 1)u\Phi(u) \leqslant b_2.$$

For $\lambda \geq 1$ this inequality gives an estimate for $u(x_0, t_0)$ from above, namely

$$u(x_0, t_0) \leqslant \frac{b_2}{(\lambda - 1)\Phi(0)}. \tag{2.40}$$

By virtue of the relations between u, v and $\tilde{v}$ and the fact that $\max_{Q_T} \hat{v}(x, t) = \hat{v}(x_0, t_0)$, we will have at any point $(x, t) \in \overline{Q}_T$ the inequality

$$\begin{aligned} v(x, t) &= \hat{v}(x, t) e^{\lambda t} \leqslant \hat{v}(x_0, t_0) e^{\lambda t} \\ &= v(x_0, t_0) e^{\lambda(t-t_0)} = \varphi^{-1}(u(x_0, t_0)) e^{\lambda(t-t_0)} \\ &\leqslant \varphi^{-1}\left(\frac{b_2}{(\lambda-1)\Phi(0)}\right) e^{\lambda(t-t_0)} \leqslant \varphi^{-1}\left(\frac{b_2}{(\lambda-1)\Phi(0)}\right) e^{\lambda T}. \end{aligned}$$

This inequality in combination with the two other possibilities for $\max_{Q_T} v$ gives the desired estimate for $u(x, t)$ from above. Analogous arguments with the function $-u(x, t)$ give the estimate for $u(x, t)$ from below. Theorem 2.9 is proved.

In §7 of Chapter V an additional theorem will be cited, which permits one to estimate the maximum of the modulus of classical solutions of equations (2.3) that satisfy other boundary conditions.

§3. ON ADMISSIBLE EXTENSIONS OF THE CONCEPT OF A SOLUTION

1. On uniqueness theorems. If the coefficients and free terms of equations (2.1) and (2.2) are not smooth functions, then these equations do not in general have classical solutions. But the possibility is not excluded that they possess solutions belonging to a class of functions $\mathfrak{M}$ that is wider than is permitted by the classical statement and yet sufficiently narrow to preserve one of the basic properties of boundary value problems for equations of parabolic type, namely, their determinateness (i.e. the uniqueness theorems). It is accordingly natural here to extend the classical statements of the problems and to replace them with investigations of solutions belonging to a class $\mathfrak{M}$. We will consider an extension (and the class $\mathfrak{M}$ of generalized solutions corresponding to it) to be *admissible* if a uniqueness theorem is retained for it.

The choice of the class $\mathfrak{M}$ is dictated by the smoothness properties of the coefficients of the equation (i.e. by the qth-power summability of them and their derivatives of a particular order for various values of q) and of the other functions that are known in the problem; but above all it must be made so that the conditions of the problem can be put in a form that has a meaning for any element of $\mathfrak{M}$.[1)] These requirements do not in general determine the class $\mathfrak{M}$ uniquely, but once a choice of $\mathfrak{M}$ is made, a transformation of the conditions of the problem is carried out according to a particular criterion. Let us illustrate this with an

1) Here we are interested only in those cases in which $\mathfrak{M}$ does not depend on the actual form of these known functions, more precisely, on the location of their singular points.

example of the problem

$$u_t - \frac{\partial}{\partial x_i}\left(a_{ij}(x,\ t)\,u_{x_j}\right) = f(x,\ t), \tag{3.1$_1$}$$

$$u\,|_{t=0} = \psi_0(x), \quad u\,|_{S_T} = 0, \tag{3.1$_2$}$$

in which the a_{ij} are nondifferentiable functions satisfying condition (2.5), $f(x, t) \in L_2(Q_T)$, and $\psi_0(x) \in L_2(\Omega)$. Clearly, for such a_{ij} equation (3.1$_1$) must be put in a form that does not contain derivatives of the a_{ij}. This is done as follows: equation (3.1$_1$) is multiplied by an arbitrary smooth function $\eta(x, t)$, then both sides are integrated over Q_T, and in the terms containing the a_{ij} a single integration by parts is carried out. As a result we arrive at the identity

$$\int\limits_{Q_T} \left(u_t\eta + a_{ij}u_{x_j}\eta_{x_i}\right) dx\,dt - \int\limits_{S_T} \frac{\partial u}{\partial N}\eta\, ds\,dt = \int\limits_{Q_T} f\eta\, dx\,dt. \tag{3.2$_1$}$$

As is well known, it is equivalent to equation (3.1$_1$) if all of the functions entering into it are sufficiently smooth. Yet in the case of nondifferentiable a_{ij} (3.2$_1$) has a meaning (for well-behaved functions u and η) while (3.1$_1$) does not. We first take as $\mathfrak{M}$ all functions from $W_2^{1,1}(Q_T)$ having derivatives u_x that are summable on any smooth manifold of dimension n. Then problem (3.1$_1$), (3.1$_2$) is reformulated as the problem of determining a function u from $\mathfrak{M}$ satisfying conditions (3.1$_2$) and the identity (3.2$_1$) for any smooth η. Such a function u is called a generalized solution of problem (3.1$_1$), (3.1$_2$) from class $\mathfrak{M}$.

If, however, there does not exist a solution in $\mathfrak{M}$ (or we are unable to prove that there exists one), then it is natural to extend $\mathfrak{M}$. For example, one takes $\mathfrak{M} = W_2^{1,1}(Q_T)$. But the integral $\int_{S_T}(\partial u/\partial N)\eta\, ds\, dt$ in (3.2$_1$) does not have a meaning for arbitrary u from $W_2^{1,1}(Q_T)$ and so, in order to get rid of it, one must require that η vanish on S_T. For such η the identity (3.2$_1$) takes the form

$$\int\limits_{Q_T} \left(u_t\eta + a_{ij}u_{x_j}\eta_{x_i}\right) dx\,dt = \int\limits_{Q_T} f\eta\, dx\,dt, \tag{3.2$_2$}$$

which is meaningful for any element u from $W_2^{1,1}(Q_T)$. This permits problem (3.1$_1$), (3.1$_2$) to be understood now as the problem of finding a function u from $W_2^{1,1}(Q_T)$ that satisfies conditions (3.1$_2$) and the identity (3.2$_2$) for any smooth

η equal to zero on S_T. Such solutions will be called *generalized solutions of problem* (3.1_1), (3.1_2) *from class* $W_2^{1,1}(Q_T)$.

If the class $W_2^{1,1}(Q_T)$ also turns out to be insufficiently wide, then one can waive the existence of the derivative u_t. But this implies the necessity of changing the form of the identity (3.2_2), since u_t enters into it. And what is more, if one takes as $\mathfrak{M}$ the space $W_2^{1,0}(Q_T)$, for example, then it becomes necessary to find another form for the initial condition (3.1_2), since the elements of $W_2^{1,0}(Q_T)$ do not have any kind of continuity (even integral) in t. Both of these problems are dealt with upon integrating the first term in (3.2_2) by parts and then substituting $\psi_0(x)$ for $u(x, 0)$. But the resulting identity contains the term $\int_\Omega u(x, T)\eta(x, T)dx$, which does not have a meaning for an arbitrary element u from $W_2^{1,0}(Q_T)$. In order to remove it one must require that $\eta(x, t)$ be equal to zero for $t = T$. For such $\eta(x, t)$ the identity (3.2_2) takes the form

$$\int_{Q_T} (-u\eta_t + a_{ij}u_{x_j}\eta_{x_i})\,dx\,dt - \int_\Omega \psi_0\eta(x, 0)\,dx = \int_{Q_T} f\eta\,dx\,dt, \qquad (3.2_3)$$

which has a meaning for any function $u \in W_2^{1,0}(Q_T)$. *A generalized solution of problem* (3.1_1), (3.1_2) *from class* $W_2^{1,0}(Q_T)$ can now be defined as an element of $\overset{\circ}{W}{}_2^{1,0}(Q_T)$ satisfying the identity (3.2_3) for all smooth $\eta(x, t)$ that vanish on S_T and for $t = T$. At this point the identity (3.2_3) has absorbed both the equation and the initial condition.

If we wanted to extend the class $W_2^{1,0}(Q_T)$ further, by waiving the existence of the derivatives u_x of a generalized solution u, we would have to transfer the derivatives $\partial/\partial x_j$ from u to $a_{ij}\eta_{x_i}$, which is impossible for nondifferentiable a_{ij}. Consequently the coefficients impose a limit on the possible extensions of a class of generalized solutions which was also noted in the Preface. One might extend the class $W_2^{1,0}(Q_T)$ by lowering the summability power of the derivatives u_x, i.e. by replacing it with $W_q^{1,0}(Q_T)$, $q < 2$, but we will not do this.

Thus, as the widest class of generalized solutions for equations (3.1_1) with nondifferentiable bounded a_{ij} we take the class $W_2^{1,0}(Q_T)$, the elements of which do not have any smoothness in t. On the other hand, for (3.1_1) it is impossible to further reduce the class $W_{2,\infty}^{1,1}(Q_T)$ consisting of those elements of $W_2^{1,1}(Q_T)$ having a bounded generalized derivative with respect to t, since for such equations there exist solutions $u(x, t)$ of class $W_{2,\infty}^{1,1}(Q_T)$ with a discontinuous derivative u_t (for example, the equation $u_t - a(t)\Delta u = 0$ has solutions $\phi(t)\phi_k(x)$, where $\phi_k(x)$ is an eigenfunction of the Laplace operator under a zero boundary condition with eigenvalue λ_k, and $\phi(t)$ is a solution of the equation $\phi' - a\lambda_k\phi = 0$).

The classes of generalized solutions of problem (3.1_1), (3.1_2) mentioned above are embedded in each other (the preceding ones in the ones following), which produces a seeming lack of uniqueness in their definition. For example, a generalized solution of problem (3.1_1), (3.1_2) from $W_2^{1,1}(Q_T)$ could be defined as a function from $W_2^{1,1}(Q_T)$ satisfying all of the conditions of the problem in the form mentioned for generalized solutions from $W_2^{1,0}(Q_T)$ of the same problem. But, as is easily seen, two such definitions are equivalent, with the first proving to be more convenient them the second. Such a nonessential ambiguity also exists in designating the class of functions $\eta(x, t)$ entering into the identities (3.2_1) – (3.2_3). For example, in defining a generalized solution from $W_2^{1,1}(Q_T)$ we required that the identity (3.2_2) be valid for any smooth function $\eta(x, t)$ that vanishes on S_T. Instead of this we could consider $\eta(x, t)$ to be an arbitrary function from $\overset{\circ}{W}{}_2^{1,0}(Q_T)$. The information which (3.2_2) gives turns out to be the same in both cases since, in the space $\overset{\circ}{W}{}_2^{1,0}(Q_T)$, the first set of η's is dense in the second.

Thus we have shown how the definition of a generalized solution of a problem is dictated by the class $\mathfrak{M}$ to which it belongs, and therefore in the sequel we will sometimes limit ourselves to merely indicating the class $\mathfrak{M}$ without citing the complete definition of a generalized solution from this class.

We will now explain what restrictions must be imposed on the coefficients of the nonprincipal terms of equation (2.1) to make some of the classes between $W_2^{1,0}(Q_T)$ and $W_{2,\infty}^{1,1}(Q_T)$ admissible for these equations. Providing a first orientation here are examples constructed by us for the equations

$$\begin{aligned}\mathcal{M}u \equiv -\frac{\partial}{\partial x_i}\left(a_{ij}(x)u_{x_j} + a_i(x)u\right) + b_i(x)u_{x_i} \\ + a(x)u = \frac{\partial f_i(x)}{\partial x_i} - f(x)\end{aligned} \tag{3.3}$$

with the $a_{ij}(x)$ satisfying only the condition of uniform ellipticity

$$\nu\xi^2 \leqslant a_{ij}(x)\xi_i\xi_j \leqslant \mu\xi^2, \quad \nu > 0.$$

The solutions $u(x)$ of these equations can be regarded as the solutions of the corresponding parabolic equations $u_t + \mathfrak{M}u = \partial f_i/\partial x_i - f$ of form (2.1) that do not depend on t. By means of these examples (described in detail in the book [65q]) it has been established that a natural class $\mathfrak{M}$ of admissible solutions for (3.3) when $a_i \equiv b_i \equiv a \equiv 0$ is the class $W_2^1(\Omega)$, and that in order for it to be admissible in the general case the nonprincipal coefficients of $\mathfrak{M}$ must satisfy the restrictions

$$\left\| \sum_{i=1}^{n} a_i^2(x),\ \sum_{i=1}^{n} b_i^2(x),\ a(x) \right\|_{q,\Omega} < \infty,\quad q > \frac{n}{2},$$

with the case $q = n/2$ being already singular. Analogous to these conditions are the conditions

$$\left\| \sum_{i=1}^{n} a_i^2(x,t),\ \sum_{i=1}^{n} b_i^2(x,t),\ a(x,t) \right\|_{q,\infty,Q_T} < \infty,\quad q > \frac{n}{2} \tag{3.4}$$

for equations (2.1). The necessity of such restrictions is further indicated by the following examples (given for the Cauchy problem, which facilitates their construction).

Example 1. It is easily seen that the bounded function $u(x,t)$, which equals 0 for $t \leq 1$ and $e^{-x^2/4(t-1)}$ for $t > 1$, satisfies at all points except $(x = 0,\ t = 1)$ the equation

$$u_t - \Delta u + n \sum_{i=1}^{n} \frac{x_i}{|x|^2} u_{x_i} = 0. \tag{3.5}$$

The function $u_{(1)}(x,t) = \int_0^t u(x,\tau)\,d\tau$ has continuous derivatives with respect to x at all points, a derivative with respect to t everywhere except the point $(x = 0,\ t = 1)$, and derivatives of second order with respect to x everywhere except on the ray $(x = 0,\ t \geq 1)$, with the derivatives $u_{(1)t}$ and $u_{(1)x_ix_j}$, $i \neq j$, being bounded in the half-space $t \geq 0$, and the derivatives $u_{(1)x_ix_i}$ having the estimate $|u_{(1)x_ix_i}(x,t)| \geq c\,|\ln|x||$ near the ray $(x = 0,\ t \geq 1)$. In addition, the function $u_{(1)}$ as an integral with respect to t of the solution u of equation (3.5), the coefficients of which do not depend on t, itself satisfies equation (3.5) everywhere except for the points $(x = 0,\ t \geq 1)$. Yet such properties are also possessed by the functions $u_{(k)}(x,t) = \int_0^t u_{(k-1)}(x,\tau)\,d\tau$, $k = 2, 3, \cdots$. All of them satisfy the homogeneous equation (3.5), are equal to zero for $t = 0$, and rapidly decrease for $|x| \to \infty$. Thus we encounter a violation of the uniqueness theorem for the Cauchy problem for equations (3.5) in the class of "almost classical solutions." From this it follows that equations (3.5), in which the coefficients $b_i(x, t) = nx_i/|x|^2$ of the u_{x_i} have first-order singularities in x, must be excluded. This also gives condition (3.4) (the case $q = n$ will not be analyzed here). If one rewrites equation (3.5) in the form

$$u_t - \Delta u + \frac{\partial}{\partial x_i}\left(\frac{nx_i}{|x|^2}u\right) - \frac{n(n-2)}{|x|^2}u = 0 \tag{3.6}$$

and identifies the coefficients $-nx_i/|x|^2$ and $-n(n-2)/|x|^2$ with the coefficients $a_i(x, t)$ and $a(x, t)$ respectively of equation (2.1), then the same functions $u_{(k)}(x, t)$ also indicate the necessity for restrictions of type (3.4) on the coefficients a_i and a.

Example 2. The same function $u(x, t)$ as in Example 1 also satisfies the homogoneous equation

$$u_t - \Delta u - \frac{n}{2(t-1)}u = 0. \tag{3.7}$$

One can easily show that the averagings of it in x

$$u_\rho(x, t) = \int_{|x-y| \leq \rho} \omega_\rho(|x-y|)\, u(y, t)\, dy$$

with infinitely differentiable kernels $\omega_\rho(\xi)$, together with their derivatives with respect to x of any order, are continuous in (x, t) in the half-space $t \geq 0$, while the derivatives $\partial u_\rho(x, t)/\partial t$ are continuous in (x, t) for $n > 2$, and for $n = 2$ have a singularity in t only at the point $(x = 0,\ t = 1)$, that is weaker than any power $(t-1)^{-\alpha}$, $\alpha > 0$. Thus the functions $u_\rho(x, t)$, $\rho > 0$ give another example of a violation of the uniqueness theorem for the Cauchy problem for equations (2.1), and this time even in the class of classical solutions (if $n > 2$). By the same token equation (3.7) indicates another case of inadmissible singularities in the coefficients. These singularities can be avoided by the requirement

$$\|a\|_{\infty, 1, Q_T} < \infty. \tag{3.8}$$

Conditions (3.4) and (3.8) represent extreme cases that serve to indicate the reasonableness of the following restrictions on the coefficients $a_i(x, t)$, $b_i(x, t)$

and $a(x, t)$ of equations (2.1):

$$\left.\begin{aligned}&\left\|\sum_{i=1}^{n} a_i^2,\ \sum_{i=1}^{n} b_i^2,\ a\right\|_{q,r,Q_T} < \infty \\ \text{with} \\ &\frac{1}{r}+\frac{n}{2q}\leqslant 1,\ r\in[1,\infty)^{1)}.\end{aligned}\right\} \qquad (3.9)$$

2. On the possible smoothness of generalized solutions of linear equations. In Chaper III it is demonstrated that the restrictions (2.5) and (3.9) on the coefficients of equations (2.1) permit one to construct for these equations a reasonable theory for generalized solutions of the boundary value problems and of the Cauchy problem. As an admissible class $\mathfrak{M}$ of these solutions we take the space $V_2^{1,1/2}(Q_T)$, which consists, roughly speaking, of functions having derivatives with respect to x of the first order and a derivative with respect to t of order $1/2$. In $V_2^{1,1/2}(Q_T)$ there exist uniqueness theorems. The restrictions on the f_i, f and ψ_0, under which these generalized solutions are obtained, are also essential and cannot be weakened in terms of the spaces $L_{q,r}(Q_T)$ selected by us. Thus the function ψ_0 must be an element of $L_2(\Omega)$, while the f_i and f must satisfy the conditions

$$\left\|\sum_{i=1}^{n} f_i^2\right\|_{1,1,Q_T},\quad \|f\|_{q_1,r_1,Q_T} < \infty \qquad (3.10)$$

with $1/r_1 + n/2q_1 \leq 1 + n/4$ and $r_1 \in [1, \infty]$. We also assume that the boundary values are reduced to being homogeneous. Later on in Chapter III it is shown that the differential properties of generalized solutions from $V_2^{1,1/2}(Q_T)$ are improved by increasing the smoothness of the coefficients and free terms of the equations, with this improvement, as in the case of classical solutions, having a local character (i.e. the smoothness of the solutions in any subdomain Q' of $\bar{Q}_T$ is determined only by the smoothness of the known functions of the problem in an ϵ-neighborhood of this subdomain). The dependences established in it are exact, as indicated by the following examples.

Example 3. Consider for $t \geq 1$ the functions

---

1) It is more convenient to regard the case $r = \infty$, $q = n/2 + \epsilon$, $\epsilon > 0$ as a special case of (3.9) with $r = 1 + n/2\epsilon$ and $q = n/2 + \epsilon$, or even with $r = 1 + n/\epsilon$, $q = n/2 + \epsilon$, for which $1/r + n/2q < 1$.

$$v(x, t) = |x|^{2\lambda} P_m(y), \tag{3.11}$$

where the $P_m(y) = y^m e^{-y}$ and $y = x^2/(t-1)$.[1] For $y \geq 0$ the functions $P_m(y)$, when $m > 0$ do not exceed $m^m e^{-m}$ respectively, while $P_0(y) \leq 1$. It is easy to show that

$$v_t = -|x|^{2\lambda} P_{m-1} y (m - y) \frac{1}{t-1},$$
$$v_{x_i} = 2P_{m-1} |x|^{2\lambda} (m + \lambda - y) \frac{x_i}{t-1},$$
$$v_{x_i x_j} = 2|x|^{2\lambda} P_{m-2} \left\{ [(m + \lambda - y)^2 - (m+\lambda)] \frac{2x_i x_j}{(t-1)^2} + (m + \lambda - y) \frac{\delta_i^j |x|^2}{(t-1)^2} \right\},$$
$$\Delta v = 2|x|^{2\lambda} P_{m-1} [2(m + \lambda - y)^2 + n(m + \lambda - y) - 2(m+\lambda)] \frac{1}{t-1}$$

and that $v(x, t)$ satisfies the equation

$$v_t - \Delta v = f(x, t) \tag{3.12}$$

with $f(x, t)$ having the form $(|x|^{2\lambda}/(t-1)) P_{m-1}(y)(c_1 + c_2 y + c_3 y^2)$. In view of the boundedness of the $P_m(y)$ for $m \geq 0$ the function $f(x, t)$ has in general a singularity at the point $(0, 1)$ that is not higher than

$$\frac{c(\mu)}{|x|^{2\mu - 2\lambda}(t-1)^{1-\mu}} \equiv \bar{f}_\mu(x, t)$$

with $\mu \geq 1 - m$ for arbitrary λ and m, and with $\mu \geq -m$ for $\lambda = -m$. The function $\bar{f}_\mu(x, t)$ is an element of $L_{q,r}(\{|x| \leq 1, 1 \leq t \leq 2\})$ with any q and r satisfying the inequalities

$$2(\mu - \lambda) q < n \text{ and } (1 - \mu) r < 1, \tag{3.13}$$

from which it follows that

$$\frac{1}{r} + \frac{n}{2q} > 1 - \lambda. \tag{3.14}$$

If $\lambda < 0$, then equation (3.12), as we see, has an unbounded solution $v = |x|^{2\lambda} P_m(y)$; but if $\lambda > 0$ then this solution is bounded and is even Hölder

1) By means of these functions A. V. Ivanov analyzed the accuracy of a number of the results in [65n] on linear equations. Concerning his examples and examples constructed earlier by the authors of [65n], see [65l,n,q].

continuous. It turns out that the latter is nonrandom and applies to all (admissible) generalized solutions of equations (2.1). In fact, we prove in Chapter III that if in equation (2.1) $f(x, t) \in L_{q,r}(Q_T)$ (we take $f_i \equiv 0$ for the present) with q and r satisfying the inequality

$$\frac{1}{r} + \frac{n}{2q} < 1, \tag{3.15}$$

and if in (3.9) $1/r + n/2q < 1$, then any generalized solution of (2.1) is Hölder continuous, and the Hölder exponent grows with an increase in the difference $1 - (1/r + n/2q)$. Moreover, it is proved in the same place that if f is worse, i.e. belongs to $L_{q,r}(Q_T)$ with q and r connected by the relation

$$\frac{1}{r} + \frac{n}{2q} = 1 - \lambda, \tag{3.16}$$

in which $\lambda < 0$, then any generalized solution of (2.1) belongs to the space $L_{q_2,r_2}(Q')$, $Q' \subset Q_T$, with q_2 and r_2 connected by the relation $1/r_2 + n/2q_2 = -\lambda$. Example (3.12) shows that the connections between q_2, r_2 and q, r established in Chapter III are exact. In fact, from the membership of f in $L_{q,r}(Q_T)$ with any r and q satisfying condition (3.14) it follows that (2.1) has a solution v not belonging to $L_{q_2,r_2}(Q')$ with $r_2 = \infty$, $q_2 = n/-2\lambda$. Analogously, by means of the same functions $v(x, t)$, it is verified that the connections proved in Chapter III between the membership in $L_{q,r}(Q_T)$ of the functions $f_i(x, t)$ in the free term of equation (2.1) and the membership in either $L_{q_2,r_2}(Q')$ or $H^l(Q')$ of the solutions of these equations are exact.

In Chapters III and IV an investigation is also made of how the differential properties of generalized solutions subsequently improve as the differential properties of the known functions in the equation improve; in particular, an explanation is given of when they have bounded or Hölder-continuous derivatives with respect to x and t. The dependences proved are exact in the same sense as the dependences just adduced by example (3.12) between the summability powers q, r for $f(x, t)$ and the summability powers q_2, r_2 for any generalized solution.

This is outlined by examples that are analogous to those just given.

In Chapter IV we prove the unique solvability of the boundary value problems and of the Cauchy problem in the spaces $H^{l,l/2}(Q_T)$, $l > 2$ and $W_q^{2m,m}(Q_T)$, $m \geq 1$. The necessity of the restrictions imposed on the coefficients and free terms of the equations is seen immediately for the case of the spaces $H^{l,l/2}$. They consist only in the requirement that all terms of the equation $Lu = f$ prove to be elements of the space $H^{l-2,l/2-1}(Q_T)$ when any element of $H^{l,l/2}(Q_T)$ is substituted for $u(x, t)$ in them. The analogous conditions for the spaces $W_q^{2m,m}(Q_T)$ are that the coefficients a_{ij} in (2.2) be bounded functions, that f be an element of $W_q^{2m-2,m-1}(Q_T)$, and that the a_i and a belong to $L_{q_i,r_i}(Q_T)$ with certain q_i and r_i. However, as we will see by Example 4, the requirement that the a_{ij} be bounded is not sufficient for preserving the uniqueness theorems in the classes $W_q^{2,1}(Q_T)$, $q < n$, and it must accordingly be supplemented, for example, by the requirement that the a_{ij} be continuous.

Example 4. It is easily verified that the equation

$$v_t - \left(\alpha\delta_j^l + \beta\frac{x_lx_j}{x^2}\right)v_{x_lx_j} = 0 \tag{3.17}$$

has for $t > 1$ a solution in the form of (3.11) as long as the numbers α, β, λ, m and n are connected either by the relations

$$\lambda = -m,\quad \alpha = \frac{2m-1}{4(n-1)},\quad \beta = \frac{n-2m}{4(n-1)}, \tag{3.18}$$

or by the relations

$$\lambda = -1,\quad \alpha = \frac{3-2m}{4(n-1)},\quad \beta = \frac{n+2m-4}{4(n-1)}, \tag{3.19}$$

in which m is arbitrary. Let us choose the parameter m so that α is positive. Then equation (3.17) will be uniformly parabolic since $\alpha + \beta = 1/4$ and the solution (3.11) will have a singularity of the order of $|x|^{2\lambda}$ with $\lambda = -m$ or $\lambda = -1$. Take the first of these: $v^{(m)}(x, t) = |x|^{-2m} P_m(y)$. It is not difficult to show that for $m = \frac{1}{2} + \epsilon$, $0 < \epsilon \ll 1$, the functions $v_1^{(m)}(x, t) = \int_0^t v^{(m)}(x, \tau)\,d\tau$, are elements of $W_q^{2,1}(\Pi)$, where Π is the strip $\{|x| < \infty,\ 0 \le t \le 2\}$, and $q = n - \epsilon_1$, $0 < \epsilon_1 \ll 1$, which satisfy at all points of Π except the points $\{x = 0,\ t \ge 1\}$, the uniformly parabolic equation (3.17) with bounded coefficients a_{ij}. The functions $v_{(1)}^{(m)}$ give an example of a violation of the uniqueness theorem for the Cauchy problem for equations of nondivergence form (2.2) in the classes $W_q^{2,1}(\Pi)$, $q < n$. On the other hand, as will be proved in Chapter IV, for continuous a_{ij} one has unique solvability of the Cauchy problem in the classes $W_q^{2,1}(\Pi)$ with any $q > 1$ for any right side from $L_q(\Pi)$ and ψ_0, the a_i and a satisfying certain (necessary and sufficient) conditions. Thus Example 4 indicates that in order to construct a reasonable theory for the solvability of the Cauchy problem for equations of form (2.2) in the classes $W_q^{2,1}(\Pi)$, $q > 1$, it is necessary to impose on the a_{ij}, besides the conditions (2.5), some supplementary restrictions. This we do by assuming the a_{ij} to be continuous.

3. On necessary conditions for boundedness in the large of Hölder norms of solutions of quasi-linear equations. We proceed now to an analysis of the quasi-linear equations (2.3), (2.4). For them we shall for the most part occupy ourselves with searches for necessary and sufficient conditions for the unique solvability in the large of boundary value problems in the classes $H^{2+\alpha,(1+\alpha)/2}(\bar{Q}_T)$. It is already known, through an example for ordinary differential equations, that the solvability in the large (i.e. solvability in domains of arbitrary size and shape, without assumptions on the smallness of the various functions or their derivatives forming the equations) of the Cauchy problem and of the boundary value problems is not merely a consequence of sufficient smoothness of the known functions forming these equations but also depends on the character of the nonlinearities of these functions. This is also true for equations in partial derivatives. But whereas in the case of ordinary differential equations in the form $du/dt = f(t, u)$ with a smooth function $f(t, u)$ the basic criterion for solvability in the large of the Cauchy problem is that the solution $u = u(t)$ does not go to infinity in a finite interval of time, for equations with partial derivatives this requirement is necessary but, generally speaking, not sufficient, since the solution $u(x, t)$ can "disappear" in a finite interval of time not only by going to infinity but also as the result of some partial derivative of it with respect to x having infinite growth.

Owing to the results on linear problems for equations (2.1), (2.2) and to the Leray-Schauder theorem on fixed points of completely continuous transformations, the question of solvability in the large of the boundary value problems and of the Cauchy problem for the quasi-linear equations (2.3), (2.4) reduces to the question of a priori boundedness of Hölder norms in $H^{1+\alpha,\frac{1}{2}+\alpha/2}(\bar{Q}_T)$ with some $\alpha > 0$ for all possible solutions of these problems. If the boundedness exists then so does solvability in the large. One of the principal results for equations (2.3) and (2.4) is that the norms in the space $H^{1+\alpha,\frac{1}{2}+\alpha/2}(\bar{Q}_T)$ for all possible solutions of these equations can be estimated in terms of $\max_{Q_T}|u|$, $\max_{Q_T}|u_x|$ and the known functions in the problems, as long as the latter have a certain not large smoothness. The only condition that is imposed on the equation (besides the conditions on the smoothness of $a_i(x, t, u, p)$, $a(x, t, u, p)$, $a_{ij}(x, t, u, p)$ as functions of their arguments (x, t, u, p)) is that it be parabolic with respect to the solution $u(x, t)$ being investigated, i.e. the fulfilment of inequalities (2.6) or (2.7) for $(x, t) \in \bar{Q}_T$ when $u(x, t)$ and its gradient are substituted for u and p

respectively. Thus this result guarantees for the entire class of quasi-linear parabolic equations the "indestructibility" of a solution $u(x, t)$ if $\max_{Q_T}|u(x, t)|$ and $\max_{Q_T}|u_x(x, t)|$ do not have unlimited growth, and by the same token the question of solvability in the large of the initial-boundary value problems reduces to the question of a proof of the boundedness of $\max_{Q_T}|u|$ and $\max_{Q_T}|u_x|$. As the examples given below show, for the entire class of equations (2.3) and (2.4) there is no a priori boundedness of $\max_{Q_T}|u_x|$ or even of $\max_{Q_T}|u|$ for the solutions $u(x, t)$ of the boundary value problems. In order for there to be one it is necessary to impose certain restrictions on the character of the nonlinear occurrences of u and p in the functions $a_i(x, t, u, p)$, $a(x, t, u, p)$ and $a_{ij}(x, t, u, p)$. Let us first explain them for $\max_{Q_T}|u|$. If u does not depend on x, then equation (2.4) acquires the form

$$u_t + a(x_0, t, u, 0) = 0. \tag{3.20}$$

It is known (see the criterion of Osgood in [117] and in [50]) that for the boundedness of all solutions $u = u(t)$ of equations (3.20) in some interval $[0, T]$ it is necessary and sufficient that

$$a(x_0, t, u, 0)\,u \geqslant -\psi(|u|)\,|u| - b, \tag{3.21}$$

where $b = \text{const} \geq 0$ and $\psi(\tau)$ is a nondecreasing positive function of $\tau \geq 0$ satisfying the condition

$$\int_0^\infty \frac{d\tau}{\psi(\tau)} = \infty. \tag{3.22}$$

A violation of (3.21), (3.22), as is easily seen, leads to the appearance of solutions of the ordinary differential equation (3.20) that are not bounded in $[0, T]$. It was proved above in §2 that the assumption (3.21), (3.22) for all $x \in \overline{\Omega}$ (together with a weakened condition of parabolicity) is sufficient to ensure the boundedness of any classical solution $u(x, t)$ of equation (2.4) (this was done in §2 for the first boundary value problem; it is done analogously for the other boundary value problems). The necessity of the condition (3.21), (3.22) is verified by the just-mentioned reference to equations of form (3.20). It is true that against this appeal to ordinary differential equations one can raise the objection: if a solution $u(x, t) = u(t)$ becomes unbounded for $t \to t_1$, then it is also unbounded on the lateral surface S_{t_1}; is it therefore possible for $u = u(x, t)$ to have unbounded growth when it is known that the boundary (and initial) values of $u(x, t)$ (prescribed in the first boundary value problem) are bounded functions? One might suspect that "the matter is straightened out" by the boundary condition and the elliptic part. The following example constructed by A. Friedman shows that this is not the case.

Example 5. Assume that the problem

$$\left.\begin{array}{c} u_t - u_{xx} - u^2 = 0, \quad u|_{t=0} = \psi_0(x), \quad u|_{x=0} = \psi_1(t), \\ u|_{x=1} = \psi_2(t) \end{array}\right\} \tag{3.23}$$

has a classical solution $u(x, t)$ in $\bar{Q}_T = \{0 \le x \le 1,\ 0 \le t \le T\}$ for smooth functions ψ_0, ψ_1 and ψ_2 greater than a certain constant

$$c = \frac{c_1}{c_2}, \quad c_1 > 0, \quad c_2 > 0.$$

It is easily verified that the function $z = c_1/c_2 - tx(1-x)$ satisfies the inequality

$$z_t - z_{xx} - z^2 \leqslant 0,$$

for $t < 4c_2$ if $c_1 \ge 1/4 + 8c_2$, and the condition $z|_{\Gamma_T} = c_1/c_2$. The function $v = (z - u)e^{-\lambda t}$ is nonpositive on Γ_T and for $t \in [0, 4c_2]$ satisfies the inequality

$$v_t - v_{xx} + (\lambda - z - u)v \leqslant 0.$$

From this it follows that for sufficiently large $\lambda > 0$ the function v cannot have a positive maximum in $Q_{4c_2} \setminus \Gamma_{4c_2}$. Therefore in Q_{4c_2} the function v is nonpositive, i.e. $u \ge z$. But $z \to \infty$ for $x = ½$ and $t \to 4c_2$, and hence u cannot be bounded in Q_{4c_2}. Thus problem (3.23) does not have a classical solution in Q_T for $T \ge 4c_2$. Condition (3.21), (3.22) is not fulfilled.

We will now occupy ourselves with determining the conditions that are necessary in order that $|u_x|$ remain bounded for bounded solutions of equations (2.4). More precisely, we wish to find conditions on the functions $a_i(x, t, u, p)$, $a(x, t, u, p)$, $a_{ij}(x, t, u, p)$ (below we will see that they concern only the behavior of these functions and their partial derivatives for $|p| \to \infty$), under which for classical solutions of equations (2.3) (or (2.4)) it is possible to estimate $\max_{Q'} |u_x|$ (i) in any domain $Q' \in Q_T$, separated from Γ_T by a positive distance d, in terms of $\max_{Q_T} |u|$ and d, and (ii) for domains $\max_{Q'} |u_x|$, adjacent to some part Γ' of the boundary Γ_T, in terms of $\max_{Q_T} |\bar{u}|$, some stronger norms of u on Γ', and the distance from Q' to $\Gamma_T\setminus\Gamma'$. For linear equations, as is shown in Chapters III and IV, they are merely conditions for a certain smoothness of the functions a_i, a, a_{ij} on Γ'. But this is not true for nonlinear equations. We will show that one of the necessary conditions is the condition

$$|a(x, t, u, p)| \leqslant \mu(|u|)(|p|+1)^2, \tag{3.24}$$

where $\mu(r)$ is a monotonic increasing continuous function of $r \geq 0$. This is proved by means of the following example, which was constructed and analyzed in §2 of Chapter 1 of [65q] in connection with resolving the same questions for elliptic equations.

Example 6. The functions $u_\delta(x) = (x + \delta)^\lambda$, $0 < \delta \leq ½$, satisfy on the closed interval $\bar{\Omega} = \{0 \leq x \leq ½\}$ the equations

$$u_{xx} + c\,|u_x|^{2(1+\epsilon)} = 0 \tag{3.25}$$

with $\epsilon = \lambda/2(1-\lambda)$ and $c = \lambda^{-1-2\epsilon}(1-\lambda)$, they are infinitely differentiable, and for $0 < \lambda < 1$ their moduli do not exceed 1. The coefficient c and the boundary values of u_δ (and even $\max_{(0,1)}|u_\delta|$) are uniformly bounded for all $\delta \in (0, ½)$ and fixed λ from (0, 1). Nevertheless the derivatives $du_\delta(x)/dx$ have unbounded growth at the point $x = 0$ for $\delta \to 0$. The reason is that condition (3.24) is not fulfilled: the exponent of $|u_x|$ in (3.25) is greater than 2.

Let us consider the functions $u_\delta(x)$ as solutions of the boundary value problems

$$\begin{gathered} u_t - u_{xx} - c\,|u_x|^{2(1+\epsilon)} = 0, \\ u|_{t=0} = (x+\delta)^\lambda, \quad u|_{x=0} = \delta^\lambda, \quad u|_{x=1} = (1+\delta)^\lambda. \end{gathered} \tag{3.26}$$

The functions $\psi_1(t) = \delta^\lambda$ and $\psi_2(t) = (1+\delta)^\lambda$ in the boundary conditions are infinitely differentiable, while the function $(x+\delta)^\lambda$ in the initial condition is bounded. The function u_δ, the coefficient c, the functions $\psi_i(t)$, and the derivatives of the $\psi_i(t)$ of all orders are uniformly bounded for all δ from (0, ½), yet the derivatives $u_{\delta x}$ are not uniformly bounded in the domains $\{0 \leq x \leq 1,\ t_1 \leq t \leq T\}$, $t_1 > 0$, and hence the desired property does not exist for equations (3.26) with $\epsilon > 0$. Condition (3.24) excludes similar equations.

In connection with the example just given one might raise the question: is the reason for the unbounded growth of $\max|u_{\delta x}|$ for $\delta \to 0$ then that it has unbounded growth at $t = 0$? The example which we give below (constructed by A. F. Filippov) shows that this is not the case.

Example 7. Consider in the domain $Q_1 = \{0 < x < 1,\ 0 < t < 1\}$ the initial-boundary value problem

$$\left.\begin{aligned} &u_t - u_{xx} - u\,|u_x|^{\frac{2}{1-\alpha}} = 0, \\ &u|_{t=0} = \left(1 + \frac{1}{\alpha}\right)x, \quad u|_{x=0} = 0, \\ &u|_{x=1} = 1 + \frac{1}{\alpha} + t\left(1 + \frac{1}{\alpha}\right)^{1+\frac{2}{1-\alpha}} \end{aligned}\right\} \tag{3.27}$$

for an arbitrary fixed value of α from $(0, 1)$. Suppose that there exists in $\bar{Q}_1$ a continuous solution $u(x, t)$ having a bounded derivative u_x in $\bar{Q}_1$ and continuous derivatives u_t, u_x, u_{xx} in Q_1. Consider in Q_1 the function

$$z(x, t) = \begin{cases} \dfrac{x}{1-t} & \text{for } 0 \leqslant x < 1-t, \\ \dfrac{1}{\alpha}(x+t-1)^{\alpha} + 1 & \text{for } 1-t \leqslant x \leqslant 1. \end{cases}$$

It is easy to see that $u \geq z$ on Γ_1, and at points of Q_1 not lying on the straight line $x = 1 - t$

$$z_t - z_{xx} - z\,|z_x|^{\frac{2}{1-\alpha}} \leqslant 0.$$

Therefore

$$v_t - v_{xx} + bv_x - |u_x|^{\frac{2}{1-\alpha}} v \leqslant 0, \tag{3.28}$$

where $v = z - u$, and

$$b(x, t) = z\frac{|u_x|^{\frac{2}{1-\alpha}} - |z_x|^{\frac{2}{1-\alpha}}}{z_x - u_x}$$

is bounded in $\bar{Q}_1$. Since $\max_{Q_1} |u_x|$ by supposition does not exceed a certain constant it follows from (3-28) that for sufficiently large λ the function $\hat{v} = ve^{-\lambda t}$ cannot have a positive maximum at an interior point of Q_1 not lying on the diagonal $x + t = 1$ (see §2). Also, at points of this straight line $\hat{v}(x, t)$ cannot have a maximum by virtue of the fact that

$$\hat{v}_x = e^{-\lambda t}(z_x - u_x)|_{x=1-t+0} = +\infty.$$

On the boundary Γ_1, as we know, $\hat{v} \leq 0$, and hence everywhere in Q_1 the function $\hat{v} \leq 0$, i.e. $u \geq z$. But this inequality contradicts the fact that $u(0, 1) = 0$, $\lim_{x \to +0} z(x, 1) = 1$ and $u(x, t)$ is continuous at the point $(0, 1)$.

Thus Examples 6 and 7 show that one cannot replace the exponent 2 in condition (3.24) by a larger one.[1] If equation (2.3) is written in the form (2.4), i.e.

$$u_t - \frac{\partial a_i(x, t, u, u_x)}{\partial u_{x_j}} u_{x_i x_j} + A(x, t, u, u_x) = 0, \tag{3.29}$$

where

$$A(x, t, u, u_x) = -\frac{\partial a_i(x, t, u, u_x)}{\partial u} u_{x_i} - \frac{\partial a_i(x, t, u, u_x)}{\partial x_i} + a(x, t, u, u_x),$$

then, since each of the terms making up A may be missing, condition (3.24), when applied to equation (3.29), assumes the form

$$\sum_{i=1}^{n}\left[\left|\frac{\partial a_i(x, t, u, p)}{\partial u}\right|(|p|+1)+\left|\frac{\partial a_i}{\partial x_i}\right|\right]+|a(x, t, u, p)| \leqslant \mu(|u|)(|p|+1)^2. \tag{3.30}$$

In Chapter V we show that the assumptions (3.30) together with the condition (2.6) on uniform parabolicity are sufficient for the existence of the desired estimates of $\max|u_x|$ in terms of $\max|u|$ for solutions of equations (2.3). For the quasi-linear equations of general form (2.4) we require, besides the fulfilment of conditions (2.7) and (3.24), the fulfilment of corresponding conditions for the partial derivatives of the functions $a_{ij}(x, t, u, p)$ and $a(x, t, u, p)$ with respect to x, u and p, namely

$$\sum_{i,j,k=1}^{n}\left[\left|\frac{\partial a_{ij}}{\partial p_k}\right|(|p|+1)^3+\left|\frac{\partial a_{ij}}{\partial u}\right|(|p|+1)^2 +\left|\frac{\partial a_{ij}}{\partial x_k}\right|(|p|+1)^2+\left|\frac{\partial a}{\partial p_k}\right|(|p|+1)+\left|\frac{\partial a}{\partial u}\right|+\left|\frac{\partial a}{\partial x_k}\right|\right] \leqslant \mu(|u|)(|p|+1)^2, \tag{3.31}$$

which are consequences of conditions (2.7) and (3.24) in the case of polynomial dependence of the $a_{ij}(x, t, u, p)$ and $a(x, t, u, p)$ on x, u and p for large $|p|$. More precisely, for an estimate of $\max|u_x|$ in terms of $|u|^{(\alpha)}_{Q_T}$ with some $\alpha > 0$ (or, more generally, in terms of $\max|u|$ and the modulus of continuity of $u(x, t)$) we impose the restrictions (3.31) on $\partial a_{ij}/\partial p_k$, $\partial a_{ij}/\partial u$, $\partial a/\partial p_k$ and $\partial a/\partial u$, but actually weaken the assumptions (3.31) concerning $\partial a_{ij}/\partial x_k$ and $\partial a/\partial x_k$

1) As is seen from the estimates for $\max|u_x|$ in Chapters V and VI, it is possible to insert in the right side of condition (3.24) a slowly increasing factor $\psi(|p|)$, namely, a factor such that $\int_0^\infty d\tau/\tau\psi(|\tau|) = \infty$.

replacing them by the inequality

$$\left|\frac{\partial a_{ij}}{\partial x_k}\right|(|p|+1)^{-1}+\left|\frac{\partial a}{\partial x_k}\right|(|p|+1)^{-3}\leqslant\mu(|u|). \qquad (3.32)$$

In estimating $\max|u_x|$ in terms of $\max|u|$ we impose in comparison to the just-enumerated conditions the additional conditions

$$\left.\begin{aligned}\left|\frac{\partial a_{ij}(x,t,u,p)}{\partial u}\right|&\leqslant\varepsilon(|u|)+P(|p|,|u|),\\ \frac{\partial a(x,t,u,p)}{\partial u}&\geqslant-[\varepsilon(|u|)+P(|p|,|u|)](|p|+1)^2,\end{aligned}\right\} \qquad (3.33)$$

in which $\epsilon(t)$ is a continuous, monotonically increasing nonnegative function of $t\geq 0$ whose value $\epsilon(\max_{Q_T}|u|)$ is sufficiently small, and $P(r,t)$ is a function that is continuous on the set $\{r\geq 0,\ t\geq 0\}$, is monotonically increasing in t, and tends to zero for $r\to\infty$ uniformly with respect to t in $[0,t_1]$, where t_1 is an arbitrary positive number.

Thus we have determined that, in the general case, for classical solvability in the large of boundary value problems for uniformly parabolic equations (2.3), (2.4) with smooth functions a_i, a and a_{ij} it is necessary to impose the restrictions (3.21), (3.22), (3.24), (3.30) and (3.31). The basic aim of Chapters V and VI is a proof of the fact that these restrictions are on the whole also sufficient for the solvability of boundary value problems in the class of smooth functions.

In Chapter V an investigation is made of generalized solutions for the quasilinear equations (2.3). In connection with this we note that assumption (3.24) is necessary in order to speak meaningfully about bounded generalized solutions $u(x,t)$ from $W_2^{1,0}(Q_T)$ for equations (2.3) with smooth functions $a_i(x,t,u,p)$, $a(x,t,u,p)$. Such solutions are defined as the elements of $W_2^{1,0}(Q_T)$ with $\operatorname{vrai\,max}_{Q_T}|u|<\infty$ that satisfy the identity

$$\int_{Q_T}[-u\eta_t+a_i(x,t,u,u_x)\eta_{x_i}+a(x,t,u,u_x)\eta]\,dx\,dt=0 \qquad (3.34)$$

for any smooth function $\eta(x,t)$ vanishing on the boundary of Q_T. In order that the integral on the left side of (3.34) be finite with respect to such solutions one must require the fulfilment of conditions (2.6) and (3.24). For a correct theory of generalized solutions that are worse than bounded generalized solutions from $W_2^{1,0}$ one must impose on the functions $a_i(x,t,u,p)$ and $a(x,t,u,p)$ even more stringent restrictions, which would guarantee the finiteness of all integrals appearing in (3.34) with respect to any such solution. We formulate these restrictions in the form of inequalities

$$
\begin{aligned}
&|a_i(x, t, u, p)| \leqslant c\left(|p|^{\beta_1} + |u|^{\beta_2}\right) + \varphi_1(x, t),\\
&|a(x, t, u, p)| \leqslant c_1\left(|p|^{\beta_3} + |u|^{\beta_4}\right) + \varphi_2(x, t),
\end{aligned} \tag{3.35}
$$

where the ϕ_i are elements of certain $L_{q_i, r_i}(Q_T)$, thereby permitting singularities of the functions a_i and a in x and t.

In §2 of Chapter V it is proved (Theorem 2.1) that if the indices β_i, q_i^{-1}, r_i^{-1} are not greater than certain numbers, then any generalized solution of equation (2.3) from $W_2^{1,0}(Q_T) \cap L_{q,r}(Q_T)$ will be bounded. Examples show that the conditions of Theorem 2.1 are sharp (one cannot increase the indices β_i, q_i^{-1}, r_i^{-1}). One of these examples is example (2.31) at the end of Chapter I of the book [65q] which was constructed by us for the case of elliptic equations.

Finally, we note that for equations (2.4) the generalized solutions $u(x, t)$, concerning which it is known only that they are elements of $W_2^{2,1}(Q_T)$, are in general inadmissible [see in this connection example (2.23) in Chapter I of [65q], and also Example 4 of the present section, in which the function v can also be considered as a solution of the quasi-linear equation

$$
v_t - \left(\alpha\delta_i^j + \beta\,\frac{v_{x_i} v_{x_j}}{|v_x|^2}\right) v_{x_i x_j} = 0
$$

with the same values for the parameters α and β as in (3.17)].

§4. BASIC RESULTS AND THEIR POSSIBLE DEVELOPMENT

We will briefly describe the basic results stated in the present book.

The various propositions of classical and functional analysis used in the book are gathered together in Chapter II. Some of these are well known and are presented without proof, while for others an explanation or proof is given. In the main (§§6–9) Chapter II is devoted to a study of functions satisfying certain integro-differential inequalities. A particularly important role in the investigation of solutions of parabolic equations is played by the function classes $\mathfrak{A}$, $\mathfrak{B}_2$ and $\overset{N}{\mathfrak{B}}{}_2^{1}$, to which, as it has turned out, belong the solutions of these equations and their derivatives. It is proved that the elements of $\mathfrak{A}$ are bounded, and that the elements of $\mathfrak{B}_2$ are bounded, and that the elements of $\mathfrak{B}_2$ and $\overset{N}{\mathfrak{B}}{}_2^{1}$ are Hölder continuous. The definitions of these classes differ somewhat from those given by us earlier in the papers [65n,o], but their investigation is carried out in almost the same way as the investigation of those originally introduced (see in connection with this §§5–8 of Chapter II of [65q] and the paper [51f]).

1. Linear equations. Chapters III and IV are devoted to linear equations of second order.

In Chapter IV linear equations with smooth coefficients in the spaces $H^{l,l/2}(\bar{Q}_T)$, $l>2$, and $W_q^{2m,m}(Q_T)$ are studied. Exact dependences of the differential properties of the solutions on the differential properties of the free terms and of the functions determining the initial and boundary conditions in the various boundary value problems and in the Cauchy problem are established for these equations, and the exact estimates corresponding to them are deduced. The epithet "exact" refers to the differential orders and means the following: if, for example, the free term $f(x, t)$ of an equation is an element of $H^{l-2,l/2-1}(\bar{Q}_T)$, $l>2$, then it is established that the solution is an element of $H^{l,l/2}(\bar{Q}_T)$ and by the same token that each term of the equation belongs to the same class as their sum $f(x, t)$. Thus, from the membership in $H^{l-2,l/2-1}(\bar{Q}_T)$ of the sum is deduced the membership in the same space of each of the summands. A result concerning a dependence of the smoothness of the solutions on the smoothness of the initial and boundary conditions is also exact. In the example just mentioned the membership of the solution $u(x, t)$ in $H^{l,l/2}(\bar{Q}_T)$ is proved in the case of the first boundary value problem under the condition that the values of $u(x, t)$ on Γ_T coincide with the values of some function $\psi(x, t)$ from $H^{l,l/2}(\bar{Q}_T)$. Analogous dependences established in the spaces $W_q^{2m,m}(Q_T)$, $m\geq 1$, as well as a dependence of the smoothness of the solutions on the smoothness of the coefficients of the differential operators, are also exact (see in this connection Example 4 in §3 of Chapter I).

The first exact estimates for parabolic operators, under various homogeneous boundary conditions, were established by O. A. Ladyženskaja in 1954 [65d,e,f] (see also [33c]) in the spaces $W_2^{2m,m}(Q_T)$. They were obtained by means of a simple device from the exact estimates given by her earlier for operators of elliptic type. The analogous estimates in $W_q^{2m,m}$ for $q\neq 2$ required a different analytic technique. These were proved by V. A. Solonnikov in [116a] (see also [116b; 111b]) for homogeneous and nonhomogeneous boundary conditions, where the dropping of the homogeneity of the boundary conditions required the introduction of new function spaces with derivatives of fractional order and different differential properties with respect to the variables x and t (in connection with the latter see [9;111d;116b]).

Estimates in $H^{2+\alpha, 1+\alpha/2}(\bar{Q}_T)$ for the first boundary value problem for (2.2) were proved by C. Ciliberto [13] in the case of one space variable and by Barrar [7] and A. Friedman [31] in the case of any number of space variables. For the second boundary value problem and for the problem with directional derivative this was done by V. A. Solonnikov [116c,d] (in which this was proved at once for general boundary problems for arbitrary systems of parabolic type) and by L. I. Kamynin and V. N. Maslennikova [56b,c]. Friedman [31b] has also given estimates of the Hölder norm of u_x for the solutions u of linear equations, the leading coefficients of which are continuous, while the coefficients ir the minor terms and free term are measurable and bounded.

In Chapter IV two distinct methods for the proof of the unique solvability of the boundary value problems and of the Cauchy problem are stated. One of them, involving exact estimates, is based on the construction of a regularizor, namely an operator that inverts the principal part with frozen coefficients and permits a reduction of the problem for $t \subset [0, \tau_0]$ with τ_0 small to an equation of the form $(I + K)u = f$, where I is the identity operator and K is an operator with small norm. By making use of estimates catching the exact character of the dependence of the smoothness of the solution on the smoothness of all of the known functions, one is then able to prove the solvability of the original problem when $0 \leq t \leq T$ for any finite T. This method permitted V. A. Solonnikov [116c] to establish the unique solvability of very general initial-boundary value problems for systems that are parabolic according to Petrovskiĭ and Shirota, and also for a wider class of systems introduced by him. Here this method is stated (with various admissible simplifications and concrete definitions) in reference to equations of second orde

It should be noted that in this method the restrictions on the coefficients of the equation and on the smoothness of the boundary and of the known functions must be exactly coordinated in dependence on the particular class of functions containing the solution. In Chapter IV the boundary value problems for equation (2.2) are considered in the classes $H^{l, l/2}(\bar{Q}_T)$ and $W_q^{2m,m}(\bar{Q}_T)$, while the Cauchy problem is considered in the classes $H^{l, l/2}(D_{n+1}^{(T)})$ and $W_q^{2m,m}(D_{n+1}^{(T)})$ $(D_{n+1}^{(T)} = E_n \times (0, T))$.

The second method for investigating boundary value problems is the classical method of the theory of heat (more precisely, parabolic) potentials, investigated in detail in the case of one space variable in papers by E. Levi and Gevrey [69;35]. Many papers [21;25;74;79;94;111a,c;etc] are devoted to it, and it has found expression in the books [25a;31;112;116c;129]. It has been used in proving the unique solvability of the Cauchy problem and of the basic boundary value problems for equations (2.2) in their classical statement.

The potentials in terms of which the solutions of these equations are expressed are formed by means of the fundamental solution of equation (2.2) with variable coefficients. For equations (2.2) of general form the fundamental solution was first constructed in the papers [21], in which it was required that the coefficients a_{ij} in (2.2) belong to $H^{2+\alpha,1+\alpha/2}$. In [94a,c; 25a,b] the restrictions on the smoothness of the coefficients are lowered to the condition that they be Hölder continuous, and in a recently published note [79] it is required that they satisfy the more general (than the Hölder condition) Dini condition. By the same token the existence of the fundamental solution and of the classical solution of the Cauchy problem for equation (2.2) is obtained under restrictions that are somewhat weaker than those in the first method. But the results obtained by means of potential theory on the solvability of boundary value problems are not exact, due to significant analytic difficulties arising in the estimate of even the first derivatives of single-and double-layer potentials near the boundary.

In the case of a half-space it is possible with the use of heat potentials to obtain exact estimates, both in the norms of $H^{l,l/2}$ and in the norms of $W_q^{2m,m}$, for solutions of boundary value problems for equations with constant coefficients. Such estimates (for the heat equation), which underlie the first method of investigation for boundary value problems, are stated in detail in §§ 3 and 2 of Chapter IV.. These estimates could be derived from general theorems on estimates of integral operators proved in [38c] and [116e] (in [116e] they are applied to parabolic systems). But these theorems rest on certain subtle and so far not yet very widely known (at least among specialists in differential equations) propositions from the theory of functions. Thus for purely practical reasons we have preferred the more customary but somewhat lengthy method of estimating integral operators arrising in the solution of boundary value problems in a half-space, that is stated in §§ 2 and 3 of Chapter IV.

Chapter III is devoted to linear equations with discontinuous and, in general, unbounded coefficients. In the main, equations of form (2.1) are investigated in it under assumption (2.5) on the uniform parabolicity of (2.1). For them it has been necessary not only to create new methods for investigating and obtaining solutions but also to develop a new attitude on the concept of a solution itself, having rejected the rigid classical framework (see in this connection § 2 of [65c]). This rejection, which contributed much to progress in the study of boundary value problems, required justification, primarily from the point of view of preserving the uniqueness theorems. For equations of elliptic type there has been available a class of generalized solutions, namely the class of generalized solutions with order of derivatives half the order of the equation, for which such a justification was clear; but for nonstationary equations the question has required studying.

Before the fifties there were two methods for proving uniqueness theorems. The first, proposed by Holmgren, consists in proving the absence of an orthogonal complement to the range of values of the adjoint of the operator for the problem being investigated; the second consists in the deduction of an a priori estimate, namely the energy inequality. Neither is suitable for the stated purpose. The first, because in order to carry it out it is necessary to know the solvability of the adjoint problem (and it is a problem of the same type as the one being investigated) in the class of sufficiently smooth functions, i.e. we must know that which we wish to prove as one of the last steps in the whole investigation; the second, because it assumes that solutions of the problem have the derivatives appearing in the equations, which is not true with generalized solutions in the most interesting cases.

Proofs of uniqueness theorems for generalized solutions of nonstationary problems, that do not make use of any information on the solvability of the problem (itself or its adjoint) in classes of functions having the derivatives appearing in the equations, were given in the fifties. The first of them [65c] applied to the so-called "generalized solutions with finite energy integral" (for parabolic equations these are the generalized solutions from $W_2^{1,0}(Q_T)$), while subsequent ones [65d–f; 73] dealt with this and certain other classes of generalized solutions (concerning them see §§ 3 and 14 of Chapter III).

The existence of generalized solutions was first proved by means of the method of finite differences [65c],[1] and then by means of functional methods and the method of Galerkin [65d–f; 124b,c,f; 73]. The passage from the classical statements to generalized ones and, connected with this, the replacement of spaces of continuously differentiable functions by the Hilbert spaces $W_2^{l_1,l_2}(Q_T)$, and also the methods developed for working with such solutions, which depend very little on the explicit form of the elliptic part of a nonstationary operator, have permitted us to proceed beyond the proper limits of boundary value problems for differential equations and to study a more general object, namely, the problems

1) The book [65c] is devoted on the whole to a more difficult subject, that of hyperbolic equations, but the methods and ideas stated in it are also directly applicable to equations of parabolic type. Strictly parabolic equations are discussed in [65d–h] and [124f].

$$\frac{du}{dt} + S_1(t)\,u = f(t), \qquad u(0) = u_0, \tag{4.1}$$

and

$$\left.\begin{aligned} &\frac{d^2u}{dt^2} + S_1(t)\frac{du}{dt} + S_2(t)\,u = f(t), \\ &u(0) = u_0, \quad \left.\frac{du}{dt}\right|_{t=0} = u_1, \end{aligned}\right\} \tag{4.2}$$

in which $u(t)$ and $f(t)$ are functions of t with values in a Hilbert space H, u_0 and u_1 are given elements of H, and the $S_i(t)$ are given unbounded operators in H depending on t as on a parameter. This was done in the papers [65d–f; 124b,c] of M. I. Višik and O. A. Ladyženskaja (see also their survey [124f]). The theorems on the unique solvability of problems (4.1), (4.2) established in them presuppose the fulfilment of only such conditions as are satisfied by the first and second boundary value problems for parabolic and hyperbolic equations and for strongly parabolic and strongly hyperbolic systems, by Schrödinger's equations and by a number of other nonstationary equations and systems of mathematical physics. This (abstract) approach later received its development in papers [60;61;114;11; 85;etc.] of several mathematicians, and the monograph [73] is devoted to it. Among the papers on the Cauchy problem for abstract equations (problem (4.1)) a paper by T. Kato [57a], having a different origin and different applications, occupies a special place. Its author started from a desire to generalize the results of the theory of semigroups (see [15b;120;etc]) to the case of infinitesimal operators S_1 depending on t, and he had before him, apparently, as litmus paper the Cauchy problem for the one-dimensional parabolic equation in its classical statement. Extensions of [57a] have been given in [57b,c;115a,b;etc.].

But let us return to the parabolic equations (2.1) with discontinuous coefficients. Essentially two results have been established for them by different methods: 1) unique solvability of the basic boundary value problems in the function classes $W_2^{1,0}(Q_T)$ and $V_2(Q_T)$ under the condition that the coefficients of equations (2.1) be bounded, that $\psi_0(x)$ belong to $L_2(\Omega)$, and that $f(x, t)$ belong to $L_2(Q_T)$ [65c–f; 73; 124a–f], and 2) unique solvability in the class $W_2^{2,1}(Q_T)$ under the additional condition that the a_{ijx} and a_{ijt} be bounded and that ψ_0 belong to $W_2^1(\Omega)$ [65d–f] (under some other assumptions concerning the a_{ij} an analogous result was proved in [33c] for equation (2.2)). In addition, methods have been given [65c;124f] permitting one to prove, roughly speaking, that an increase in the smoothness of all functions entering into the problem by unity in x and by ½ in t increases the smoothness of the solution of the problem by the same amount.

Yet everything that had been done on equations with discontinuous coefficients left a large gap in regard to revealing the true dependences of the differential properties of the solutions on the known functions in the problem. Thus it was clear that solutions from $W_2^{1,0}(Q_T)$ can be obtained under somewhat weaker assumptions than those concerning the boundedness of the minor coefficients. But will they be necessary? On the other hand, how will the smoothness of solutions from $W_2^{1,0}(Q_T)$ increase under an increase in the summability power of the free term? When do they become bounded or Hölder continuous? And so on. The first, and unexpected, result of this kind was a theorem of Nash [88] on the possibility of estimating the Hölder norm (everywhere except in a neighborhood of a singular point) for a basic singular solution of the equation

$$u_t - \frac{\partial}{\partial x_i}(a_{ij}(x, t)\, u_{x_j}) = 0 \tag{4.3}$$

in terms of only ν and μ from condition (2.5). This theorem and an analogous theorem of De Giorgi [17] on the elliptic equations $(\partial/\partial x_i)(a_{ij}(x)u_{x_j}) = 0$ laid the foundation for detecting new connections between the properties of the coefficients and the free terms of the equations and the properties of their solutions. From Nash's result followed the possibility of estimating the Hölder norm $|u|^{(\alpha)}$ for smooth solutions $u(s, t)$ of equations (2.1) under the assumption that $a_i \equiv f_i \equiv 0$, and the b_i, a and f are bounded [88;62a,e;91c]. But the complexity and opacity of his proof did not permit one to essentially sharpen his results and find exact conditions on the coefficients and free terms of (2.1) under which there would exist some smoothness of all solutions of (2.1). This was done, using another approach, by O. A. Ladyženskaja and N. N. Ural'ceva [65l,n]. They consider the complete equation (2.1) and establish when all of its solutions (including the nonclassical but generalized ones) belong to some function space, in particular, the Hölder space $H^{\alpha,\alpha/2}$. By means of specially constructed examples [65l,n,q] it was shown that the conditions imposed in this connection on the known functions and the free terms of the equation are not only sufficient but also necessary. These conditions are expressed in terms of the membership of the functions in question to the spaces $L_{q,\infty}(Q_T)$ with different q. [65c,d,n] form the basis of

Chapter III. In comparison with them the following generalization is introduced: instead of the spaces $L_{q,\infty}(Q_T)$ the full scale of spaces $L_{q,r}(Q_T)$ is considered and in terms of it the properties of the known, and in one case also the desired, functions are formulated (see in connection with this the papers [51d,e,f; 62c; 41]).

The basic contents of Chapter III are as follows. First the unique solvability of the first and second boundary value problems for (2.1) in the class $V_2^{1,\frac{1}{2}}(Q_T)$ is established under the widest possible conditions on all of the data of the problem. Then we outline step by step how the differential properties of all the solutions of equation (2.1) improve as the properties of the known functions improve, in particular, when they become classical. The proof of existence is carried out by Galerkin's method. This method is suitable both for proving the solvability of the boundary value problems and for actually determining their solutions. As a calculating method it was applied long ago to stationary and nonstationary problems. A proof of its convergence for nonstationary problems commenced with the papers [26; 39; 46b] (see also [124f]). Other possible methods for proving existence theorems are described, with references to their origins in the literature, in §§14–18. We note that in carrying out all of these methods restrictions were imposed on the coefficients of the equation that are stronger than the restrictions made in §§1–5 of Chapter III.

In Chapter III there is one substantial omission, namely the absence of a proof of Harnack's inequality (a formulation of it is given in §10 of Chapter III). This is explained in part by the fact that a proof of this beautiful inequality is for the present long and complicated, and in part by the fact that it is not used anywhere in the book. For solutions of equations (4.3) it is established by J. Moser [87b]. Its generalization to the full equation (2.1) is given by A. V. Ivanov [51e], Aronson and Serrin [132d], Kurihara [133] and Trudinger [136]. In papers [131a,b] of Aronson the existence of the Green's function is proved for the equations (2.1) with discontinuous coefficients and its behavior is investigated in a neighborhood of the singular point.

2. Quasi-linear equations and systems. Chapter V is devoted to quasi-linear parabolic equations with principal part in divergence form, i.e. to equations of form (2.3). Its main results are proofs of the unique solvability in the large, in the class of smooth functions, of the first boundary value problem, the Cauchy problem and a number of problems with nonlinear boundary conditions, and only under such restrictions as are essential. Preceding them we obtain a priori estimates of various norms of the solutions in terms of known quantities, including both interior (local) estimates, valid for all solutions of the equations considered, and total estimates in the whole domain $\bar{Q}_T$, that take into account the initial and boundary conditions. Part of the estimates are carried out for bad

(generalized) solutions, and the others under the assumption that the solutions are classical. All estimates could be carried out under not large (necessary) assumptions only on the smoothness of the solutions being investigated (as was done in [65q] for elliptic equations). But we did not do this, partly because we did not want to make more tedious the calculations that are difficult as it is, partly because we would have had to impose additional restrictions on the functions making up the equation, but mainly because our path of investigation of quasi-linear equations proceeds from good solutions to bad solutions and by the same token is not required in estimates of such a type. We first obtain classical solutions of the problems, using a priori estimates of the classical solutions and the Leray-Schauder principle, and then, by reducing the assumptions on the smoothness of the data of the problem, we obtain worse solutions as limits of the good solutions. One could also go in the opposite direction: first, under not large assumptions on the data of the problem, one establishes the existence of bad solutions, and then one proves an improvement in their properties under an improvement in the properties of the data of the problem.

In Chapter VI quasi-linear parabolic equations of general form (2.4) are investigated. The following cardinal fact is proved for the entire class of these equations: if a solution $u(x, t)$ has u_t and the u_{xx} from $L_2(Q_T)$, while the u_x are bounded and continuously depend on t in the norm of $L_2(\Omega)$ (the class of such solutions is denoted by $\mathfrak{M}$), and if the equation (2.4) is parabolic with respect to this solution, i.e. inequality (2.7) holds under substitution of $u(x, t)$ and $u_x(x, t)$ for u and p respectively, then the $u_x(x, t)$ are Hölder continuous (belong to $H^{\alpha, \alpha/2}(Q_T)$) and their norms $|u_x|_{Q'}^{(\alpha)}$ for any $Q' \subset Q_T$ are estimated only in terms of $\max_{Q_T} |u_x|$, the known parameters characterizing the functions $a_{ij}(x, t, u, p)$, $\partial a_i(x, t, u, p)/\partial p_k$, $\partial a_i(x, t, u, p)/\partial u$, $\partial a_i(x, t, u, p)/\partial x_j$, $a(x, t, u, p)$ for $u = u(x, t)$, $p = u_x(x, t)$, and the distance from Q' to Γ_T. A corresponding result is also established in a closed domain. This proposition together with the results of Chapter IV on linear equations provides the possibility of obtaining estimates of all norms $|u|_{Q_T}^{(l)}$, $l > 1$, in terms of $\max_{Q_T} |u|$ and $\max_{Q_T} |u_x|$ and known quantities.

Later in §3 of Chapter VI a priori estimates of $\max_{A'} |u_x|$ and $\max_{Q_T} |u_x|$ are established in terms of quantities characterizing known functions and in terms of weaker norms of u, namely, $|u|_{Q_T}^{(\alpha)}$ with some $\alpha > 0$ or $\max_{Q_T} |u|$. As the examples in §3 of Chapter I show, this is not possible for the whole class of equations (2.4) but in general only under definite restrictions on the behavior of the equations (2.4) but in general only under definite restrictions on the behavior of the functions $a_{ij}(x, t, u, p)$ and $a(x, t, u, p)$ for $p \longrightarrow \infty$. An estimate of $\max |u_x|$ in terms of $|u|_{Q_T}^{(\alpha)}$ is carried out under only such assumptions whose necessity is confirmed by the examples of §3. As to the estimate of $\max |u_{Q_T}^{(\alpha)}|$ in terms of $\max |u|$, there exists one condition (the smallness of ϵ in inequalities (3.33)) whose necessity is not established;[1] it is not needed for equations (2.3) but it is necessary for very close objects, such as the system of Heinz in §2 of Chapter VII. We note that for individual classes of equations (2.4), in which the functions $a_{ij}(x, t, u, p)$, $a(x, t, u, p)$ have a special form, it turns out to be possible to estimate $\max |u_x|$ without any restrictions on the orders of growth of the a_{ij} and a with respect to p or else by using only a portion of such restrictions.

The estimates of $|u_x|_{Q_T}^{(\alpha)}$ obtained in §§2 and 3 together with the well-known estimate of $\max_{Q_T} |u|$ and the estimates in §5 of Chapter IV with respect to linear equations with smooth coefficients provide all that is necessary for a proof of the solvability in the large of the boundary value problems for equations (2.4). In Chapter VI, unlike Chapter V, we restrict ourselves to only the first boundary value problem. Its solvability is proved by means of the Leray-Schauder principle. One could use other theorems on the solvability of abstract nonlinear equations, for example the theorem of L. V. Kantorovič on the convergence of Newton's method for a solution of such equations, if one observes that the uniqueness theorem holds for the problems being studied.

The results on quasi-linear equations stated in Chapters V–VII, except for the results in §9 of Chapter V and part of §5 of Chapter VI, have been obtained by the authors of [65q] (see [65h,n,o,q] and also [65p; 51f]). They provide answers to questions concerning the classical solvability in the large of boundary value problems and of the Cauchy problem for quasi-linear uniformly parabolic equations.

An investigation of quasi-linear parabolic equations was begun with papers of Gevrey [35]. In them and also in subsequent papers [25a; 31; 80; 114c] solvability in the small of the boundary value problems and of the Cauchy problem was proved. Such solvability, as in the case of nonlinear equations of other types, holds for the whole class of quasi-linear parabolic equations (2.4) as long as the

---

1) When we speak of the necessity of some such condition for the possibility of some such estimate, we mean that there exists among all equations of the form (2.4) an equation, satisfying all conditions except the one in question, for which the estimate is not correct.

functions entering into the problem are sufficiently smooth and are compatible with each other (in connection with conditions of compatibility see §5 of Chapter IV and the following text). Its proof is based essentially on properties of linear parabolic operators only and does not make use of any a priori estimates for solutions of strictly nonlinear equations (the only exception is the well-known estimate for the maximum of the modulus of a solution, based on the maximum principle). The same remarks also hold in the papers of Gagliardo [33d], Friedman [31] and P. E. Sobolevskiĭ [114b,c] devoted to different classes of equations of the form

$$u_t - a_{ij}(x, t) u_{x_i x_j} = a(x, t, u, u_x) \tag{4.4}$$

with linear principal part to which the right side is subordinate.

Strictly nonlinear estimates first appeared in the papers of Ciliberto [13], Prodi [96] and Ventcel' and Oleĭnik [123e,f] devoted to quasi-linear equations of the form

$$u_t = a_{11}(x_1, t, u) u_{x_1 x_1} + b(x_1, t, u) u_{x_1} + c(x_1, t, u) \tag{4.5}$$

and certain generalizations of them (see [40; 16; etc.]). They permitted one to prove solvability in the large of the basic boundary value problems for these equations. But in obtaining these estimates essential use was made of both the special form of equation (4.5) (namely, the independence of $a_{11}(x_1, t, u)$ from u_{x_1} and the less than quadratic growth of $a(x_1, t, u, u_x) = b(x_1, t, u) u_{x_1} + c(x_1, t, u)$ with respect to u_{x_1}) and the presence of only one space variable.

The first paper to present nonlinear estimates for many-dimensional quasi-linear equations was [65h]. The unique solvability in the large of the first boundary value problem is proved in it for many-dimensional quasi-linear equations of the form (4.5) under the condition that $\partial a_{ij}(x, t, u)/\partial u$ be sufficiently small. Methods of estimating $\max |u_x|$, that are suitable for the general quasi-linear equations (2.4), are given in it for arbitrarily small subdomains (local estimates) and for the entire domain of definition of the solution $u(x, t)$ (total estimates). These estimates have been used in subsequent investigations on the solvability of nonlinear boundary value problems both for elliptic [122a; 62e; 65q] and for parabolic [123a; 62a,e; 65n,o; 91c] equations. They and estimates of the type of De Giorgi and Nash for linear equations have permitted us to establish solvability in the large for elliptic equations of the form

$$a_{ij}(x, u) u_{x_i x_j} + a(x, u, u_x) = 0 \tag{4.6}$$

under certain natural restrictions on $a(x, u, p)$ [122a] and analogously [62e] for parabolic equations of the form

$$u_t - a_{ij}(x, t, u)\, u_{x_i x_j} + a(x, t, u, u_x) = 0 \tag{4.7}$$

under certain supplementary (nonessential) conditions on $a(x, t, u, p)$. In §3 of Chapter VI we give a derivation of these estimates, using an approach somewhat different from our original one, with which it has been possible, as in the elliptic case [65q], to effect a maximum reduction in the number of derivatives required of the solution.

The general case of quasi-linear equations required the obtainment of still other nonlinear estimates for solutions $u(x, t)$ of equations (2.3) and (2.4), namely, estimates of the Hölder constant for $u(x, t)$ and the $u_x(x, t)$. These estimates were given in [65n,o], and under conditions that were in a definite sense necessary. They are stated in Chapters II, V–VII. From the subsequent papers we mention [29], in which an estimate of $\max_{Q_T} |u_{x_1}|$ is given for solutions of general quasi-linear equations in the case of one space variable by making use of only necessary assumptions on the behavior of the known functions $a_{11}(x_1, t, u, p_1)$, $a(x, t, u, p)$ for $p \to \infty$. We present it in §5 of Chapter VI.

In Chapter VII a certain class of linear and quasi-linear parabolic systems of second order is studied. This class was chosen by us according to the following principle: if one rejects from a system all minor terms and "freezes" the leading coefficients (i.e. takes arbitrary values for their arguments), then the maximum principle must be fulfilled for the resultant linear system, i.e. any norm of a vector solution $u(x, t)$ must take its least and greatest values on Γ_T. We have retained this basic property, which is inherent in all parabolic equations of second order, for the class of systems investigated in Chapter VII. It has turned out that the same properties that are valid for one equation of second order, in particular the classical solvability in the large of the basic boundary value problems, are also valid for this class of systems.

The last sections of Chapter VII are a survey of what has been done on linear equations of high orders and systems of parabolic type. Special attention is given to papers in which exact dependences and exact estimates corresponding to them are established for their solutions in various function spaces.

3. On some unsolved problems. Let us now indicate several unsolved problems connected with a further investigation of nonlinear problems of parabolic type. For quasi-linear uniformly parabolic equations of general form (form (2.4)) it is

desirable to understand whether the assumption concerning the sufficient smallness of $\epsilon(|u|)$ in inequalities (3.33) is necessary. It is superfluous for equations of form (2.3) when $m=2$. If it turns out to be necessary for equations (2.4), then it would be interesting to have an explanation of whether or not it will be needed for equations of form (2.3) with the $a_i(x, t, u, p)$ satisfying the inequalities

$$\begin{aligned} \nu(|u|)(|p|+1)^{m-2}\xi^2 \leqslant \frac{\partial a_i(x,t,u,p)}{\partial p_j}\xi_i\xi_j \\ \leqslant \mu(|u|)(|p|+1)^{m-2}\xi^2 \end{aligned} \tag{4.8}$$

with arbitrary $m>1$. As shown in [65q], for elliptic equations

$$\frac{d}{dx_i}(a_i(x, u, u_x))+a(x, u, u_x)=0,$$

that satisfy condition (4.8), $\epsilon(|u|)$ can be arbitrary.

A second series of questions is connected with the replacement of the conditions (2.6) and (2.7) of uniform parabolicity for (2.3) and (2.4) by the conditions

$$\begin{aligned} \nu(|u|)(|p|+1)^{m_1}\xi^2 \leqslant \frac{\partial a_i(x,t,u,p)}{\partial p_j}\xi_i\xi_j \\ \leqslant \mu(|u|)(|p|+1)^{m_2}\xi^2 \end{aligned} \tag{4.9}$$

and

$$\begin{aligned} \nu(|u|)(|p|+1)^{m_1}\xi^2 \leqslant a_{ij}(x, t, u, p)\xi_i\xi_j \\ \leqslant \mu(|u|(|p|+1)^{m_2}\xi^2 \end{aligned} \tag{4.10}$$

with different m_1 and m_2. What are the conditions for solvability in the large of the boundary value problems for such equations?

A third set of questions concerns equations of forms (2.3) and (2.4) for which the $a_i(x, t, u, p)$, $a(x, t, u, p)$, $a_{ij}(x, t, u, p)$ have various singularities in the arguments u and p. Such equations are encountered, for example, in Stefan problems, which can normally be formulated as ordinary boundary value problems for equations of form (2.3) with elementary nonlinearities in the known functions but with discontinuities of the first kind in u (see §9 of Chapter V). It is comparatively simple to establish the existence of generalized solutions for Stefan problems. Yet the classicalness of these solutions, including the smoothness of the free (unknown) boundary, has been investigated only in the one-dimensional case. A number of other linear problems with unknown boundaries can be reformulated in the form of ordinary boundary value problems for quasi-linear equations of form (2.3) with the functions $a_i(x, t, u, p)$ and $a(x, t, u, p)$ having discontinuities in u. In problems of underground hydrodynamics equations of form

(2.3) are encountered with the functions $a_i(x, t, u, p)$ having unrestricted isolated singularities in p, and also with $a_i(x, t, u, p)$ for which the function $\nu(|u|)$ in (4.8) vanishes when $|u| = 0$ (for equations admitting a degeneracy, for example, filtration equations, see [6; 22; 54; 103; 130]). It is necessary to construct examples which would outline the contours of a possible theory of boundary value problems for such equations.

Further, it is desirable to investigate the general nonlinear parabolic equations

$$u_t = F(x, t, u, u_{x_i}, u_{x_k x_j}) \tag{4.11}$$

and above all to prove the following proposition concerning the entire class of equations (4.11): if $u(x, t)$ is a sufficiently smooth solution of equation (4.11) and equation (4.11) is parabolic with respect to this solution, then the Hölder norms $|u|_Q^{(l)}$ for u can be estimated in terms of $\max_{Q_T}|u, u_x, u_{xx}, u_t|$ and certain characteristics of the known function $F(x, t, u, p_i, p_{kj})$, which is assumed to be smooth. One must then find those conditions on the function $F(x, t, u, p_i, p_{kj})$ under which $\max_{Q'}|u_t, u_{xx}|$ can be estimated in terms of $|u, u_x|_{Q_T}^{(\alpha)}$, and $|u_x|_{Q'}^{(\alpha)}$ can be estimated in terms of $|u|_{Q_T}^{(\alpha)}$ and $\max_{Q_T}|u_x|$, and so on.

All of these problems are formulated by us for equations of second order. We believe that the results and methods stated in the present book will prove to be useful in solving them.

The following set of problems concerns systems and equations of high orders of parabolic type. It is desirable to carry out investigations for them with the same degree of thoroughness as was done for equations of second order. At the present time linear problems are well investigated in the case of smooth coefficients. For discontinuuous coefficients the only known results are analogous to the results in §§1–5 of Chapter III. They concern strongly parabolic systems and are proved in the same way as for one equation of second order. As to nonlinear problems, particularly their classical solvability, the results are comparatively few. Let us point out a number of papers (in addition to those mentioned above) that are devoted to an investigation of quasi-linear equations and systems.

They can be conventionally divided into three groups. To one of them we refer the papers [11b–d; 22; 68a; 84a,b; 124d,e; 134] devoted to quasi-linear equations and systems of any order. Existence theorems are established in them for generalized solutions of boundary value problems. For this purpose one makes use of Galerkin's method. Beginning with papers of Minty [84] and Browder [11] there emerged a new feature in the implementation of this method, consisting in the carrying out of weak passages to the limit under the sign of the so-called monotone operators. Questions concerning the uniqueness of such solutions under "natural" restrictions and those concerning an increase in their smoothness under an increase in the smoothness of the known functions in the problem remain unresolved.

To the second group can be referred the papers [60b; 57c; 114b,c; etc.] in which there is investigated the Cauchy problem for abstract equations of the form

$$\frac{du}{dt} + S(t)\,u = f(t,\ u), \quad u(0) = u_0,$$

where $u(t)$ is the desired function of a parameter $t \in [0,\ T]$ with values in a Banach (or Hilbert) space H, $S(t)$ is a linear, in general, unbounded operator in H depending on t as on a parameter, and $f(t,\ u)$ is a known function defined on $[0,\ T] \times H$ and subordinate in some sense to $S(t)$. The results obtained in regard to its solvability made it possible to prove the solvability of the first and second boundary value problems for equations of form (4.4) under the condition that $a(x,\ t,\ u,\ p)$ have a weak nonlinearity in u and p.

Finally, in the third group can be combined the papers [37; 65g; 123b–d; 126] in which are analyzed certain classes of quasi-linear systems encountered in the mechanics of continuous media and systems simulating them. Of these three groups of papers we give an account dealing only with the first (see the end of §6 of Chapter V).

We have also had to leave to one side many other investigations concerning the stabilization of solutions $u(x,\ t)$ for $t \to \infty$, the analyticity of solutions in x under analyticity of the coefficients of the equations, the behavior of solutions for $|x| \to \infty$ (theorems of Liouville type) and for the leading coefficients tending to zero (problems with a small parameter), the analysis of nonclassical problems for equations of parabolic type (for example, such as in [66]), and others. Also untouched are questions arising in the theory of probability in the study of continuous Markov processes (see [24; 27]), including investigations on degenerate linear parabolic equations.

The extensiveness and diversity of the material, and in certain directions a definite lack of completeness to it, has not permitted the inclusion of it in this (even without it, alas!, thick) book.

CHAPTER II

AUXILIARY PROPOSITIONS

In this chapter are collected propositions that concern arbitrary elements of various function spaces and that for us play an important but auxiliary role.

In §1 are presented the Cauchy and Hölder inequalities and various consequences of them. The second section is devoted to embedding theorems and the inequalities corresponding to them, including certain multiplicative inequalities for functions depending in the same way on all of their arguments $x_1, \cdots x_n$. Because of the need to know the character of the dependence of the constants in these multiplicative inequalities on all of the numerical parameters entering into them we have had to give a careful derivation of them. In the same section are formulated a special case of the second basic inequality for elliptic operators and a theorem on the extension of functions of class $W_q^l(S)$ from S onto all of Ω. In §3 are presented embedding theorems and extension theorems, for functions depending on x and on t, with respect to which the functions have other properties.

In §4 are proved several simple assertions on averagings of elements of the spaces $L_q(\Omega)$, $L_{q,r}(Q_T)$ and $V_2^{1,0}(Q_T)$ and their "cuts" $u^{(k)} = \max\{u - k, 0\}$. In §5 are collected the different propositions used in the proofs of Hölder continuity for functions of classes $\mathfrak{B}_2$ and in the investigation of other properties of smoothness and summability of generalized solutions with various summability powers.

The remaining sections (§§6–9) are devoted to an investigation of the function classes $\mathfrak{A}$ and $\mathfrak{B}_2$ introduced by us. The elements of these classes are functions satisfying certain systems of integro-differential inequalities. It is proved that the elements of $\mathfrak{A}$ are bounded functions (§6), while the elements of the classes $\mathfrak{B}_2$ are Hölder continuous in x and t (§§7–9). At the same time estimates are given of $\max|u|$ for $u \in \mathfrak{A}$ and of $\langle u\rangle^{(\alpha)}$ for $u \in \mathfrak{B}_2$ that depend only on the numerical parameters entering into the definitions of $\mathfrak{A}$ and $\mathfrak{B}_2$.

The propositions proved in §§6–9 are used essentially in §§7–12 of Chapter III and in the Chapters V–VII devoted to nonlinear problems, since in the majority of cases the solutions u of parabolic equations and their derivatives u_x belong to $\mathfrak{A}$ or $\mathfrak{B}_2$.

§1. SOME ELEMENTARY INEQUALITIES

We will frequently make use of certain well-known algebraic and functional inequalities. Among the algebraic inequalities we require the following.

1) Cauchy's inequality

$$a_{ij}\xi_i\eta_j \leqslant \sqrt{a_{ij}\xi_i\xi_j}\,\sqrt{a_{ij}\eta_i\eta_j}, \tag{1.1}$$

valid for any nonnegative quadratic form $a_{ij}\xi_i\xi_j$ with $a_{ij} = a_{ji}$ and any numbers $\xi_1, \cdots, \xi_n, \eta_1, \cdots, \eta_n$;

2) "Cauchy's inequality with ϵ"

$$ab \leqslant \frac{\varepsilon}{2}a^2 + \frac{1}{2\varepsilon}b^2, \tag{1.2}$$

valid for any $\epsilon > 0$ and arbitrary a and b;

3) Young's (more general than (1.2)) inequality

$$ab \leqslant \frac{1}{m}\varepsilon^m a^m + \frac{m-1}{m}\varepsilon^{-\frac{m}{m-1}} b^{\frac{m}{m-1}}, \tag{1.3}$$

valid for any positive a, b, ϵ and $m > 1$.

From the functional inequalities we require:

4) the inequalities

$$\|u+v\| \leqslant \|u\| + \|v\|,$$
$$|(u, v)| \leqslant \|u\| \cdot \|v\|,$$

the first of which is valid for the norm of any Banach space, and the second for the scalar product and norm of any Hilbert space;

5) the Cauchy-Bunjakovskiĭ-Schwarz inequality

$$\left|\int_\Omega \sum_i u_i v_i\,dx\right| \leqslant \left(\int_\Omega \sum_i u_i^2\,dx\right)^{\frac{1}{2}} \cdot \left(\int_\Omega \sum_i v_i^2\,dx\right)^{\frac{1}{2}}, \tag{1.4}$$

which is a special case of the last inequality;

6) Hölder's inequality

$$\left|\int_\Omega \prod_{i=1}^s u_i\,dx\right| \leqslant \prod_{i=1}^s \left(\int_\Omega |u_i|^{\lambda_i}\,dx\right)^{\frac{1}{\lambda_i}}, \tag{1.5}$$

in which $\lambda_i \geq 1$, $\Sigma_{i=1}^s \lambda_i^{-1} = 1$;

7) the consequences of Hölder's inequality

$$\left|\int_{Q_T} u(x, t)\,v(x, t)\,dx\,dt\right| \leqslant \int_0^T \|u\|_{q, \Omega}\cdot\|v\|_{\frac{q}{q-1}, \Omega}\,dt$$
$$\leqslant \|u\|_{q, r, Q_T}\cdot\|v\|_{\frac{q}{q-1}, \frac{r}{r-1}, Q_T}, \quad q \geqslant 1, \quad r \geqslant 1. \tag{1.6}$$

and

$$\|uv\|_{q, r, Q_T} = \left[\int_0^T\left(\int_\Omega |u|^q |v|^q\,dx\right)^{\frac{r}{q}} dt\right]^{\frac{1}{r}}$$
$$\leqslant \|u\|_{\lambda q, \mu r, Q_T}\cdot\|v\|_{\lambda' q, \mu' r, Q_T}, \tag{1.7}$$
$$\frac{1}{\lambda}+\frac{1}{\lambda'}=1, \quad \frac{1}{\mu}+\frac{1}{\mu'}=1; \quad \lambda, \lambda', \mu, \mu' \geqslant 1.$$

We write the latter in a form that is more convenient for us:

$$\|uv\|_{q, r, Q_T} \leqslant \|u\|_{q_1, r_1, Q_T}\cdot\|v\|_{\frac{qq_1}{q_1-q}, \frac{rr_1}{r_1-r}, Q_T}, \quad q_1 \geqslant q, \quad r_1 \geqslant r; \tag{1.7'}$$

8) the inequality of form (1.6) for several factors

$$\left|\int_{Q_T} \prod_{i=1}^s u_i(x, t)\,dx\,dt\right| \leqslant \prod_{i=1}^s \|u_i\|_{q_i, r_i, Q_T}, \tag{1.8}$$

where

$$q_i \geqslant 1, \quad \sum_{i=1}^s q_i^{-1} = 1, \quad r_i \geqslant 1, \quad \sum_{i=1}^s r_i^{-1} = 1.$$

Inequalities (1.4)–(1.8) are valid for arbitrary $u_i(x)$, $v_i(x)$ given in Ω, and $u(x, t)$, $v(x, t)$, $u_i(x, t)$ in Q_T, for which the norms in the right side of the corresponding inequalities are finite.

We note the following useful proposition.

Lemma 1.1. *If a summable function $u(x, t)$ in Q_T satisfies for any $\eta(x, t)$ from $L_{q,r}(Q_T)$ the inequality*

$$\left| \int_{Q_T} u\eta \, dx\, dt \right| \leqslant c \, \|\eta\|_{q,r,Q_T} \tag{1.9}$$

with the constant c *not depending on* η, *then* $u(x, t)$ *belongs to the conjugate space* $L_{q',r'}(Q_T)$, $q' = q/(q-1)$, $r' = r/(r-1)$ *and* $\|u\|_{q',r',Q_T} \le c$. *Here, as above,* $r \ge 1$, $q \ge 1$.

§2. THE SPACES $W_q^l(\Omega)$ AND $H^l(\Omega)$. EMBEDDING THEOREMS

In §1 of Chapter I were defined spaces $L_q(\Omega)$ and $W_q^l(\Omega)$ with $q \ge 1$ and integral l. These spaces were the subject of special investigations (cf. [111a; 49; 35c,d; 89; 90b; 5;9;38a,b; and others]). We present here some of the basic results of these investigations that are needed for the purposes of the present book. They concern the behavior of arbitrary elements of spaces $W_q^l(\Omega)$ on manifolds of dimensions $r \le n$ and the convergence relations in these spaces with different l, q and r.

We recall that the norm in $L_q(\Omega)$ $(q \ge 1)$ is defined by the equality

$$\| u \|_{q,\Omega} = \left(\int_\Omega | u(x) |^q \, dx \right)^{\frac{1}{q}}, \tag{2.1}$$

and the norm in $W_q^l(\Omega)$ is defined by

$$\| u \|_{q,\Omega}^{(l)} = \sum_{j=0}^{l} \langle\langle u \rangle\rangle_{q,\Omega}^{(j)} \equiv \sum_{j=0}^{l} \sum_{(j)} \| D_x^j u \|_{q,\Omega}. \tag{2.2}$$

One of the central formulas, which we will repeatedly make use of, is the formula of integration by parts

$$\int_\Omega u_{x_i} v \, dx = -\int_\Omega u v_{x_i} \, dx + \int_S uv \cos(\mathrm{n}, x_i) \, ds \tag{2.3}$$

for bounded domains Ω. In it n is the outward unit normal to the boundary S. It is valid for any functions u from $W_q^1(\Omega)$ and v from $W_p^1(\Omega)$ with $1/q + 1/p \le 1 + 1/n$ (more precisely, for specially chosen representatives of them from the function classes that are equivalent to them on Ω) as long as S has a certain regularity (for example, if S is piecewise smooth). If the function $u(x)$ belongs to $\overset{\circ}{W}{}_q^1(\Omega)$ (or $v \in \overset{\circ}{W}{}_p^1(\Omega)$), then formula (2.3) acquires the form

$$\int_\Omega u_{x_i} v \, dx = -\int_\Omega u v_{x_i} \, dx. \tag{2.4}$$

We recall that $\overset{\circ}{W}{}_q^1(\Omega)$ was defined as the closure in the norm of $W_q^1(\Omega)$ of the set of functions that are infinitely differentiable and have a compact support in Ω. Under such a definition of $\overset{\circ}{W}{}_q^1(\Omega)$ formula (2.4) is clearly valid for any boundary S of Ω. For piecewise-smooth boundaries $\overset{\circ}{W}{}_q^1(\Omega)$ can also be defined as the set of all elements from $W_q^1(\Omega)$ that vanish on S, with the norm of $W_q^1(\Omega)$.

By an embedding operator from the space $W_p^l(\Omega)$ into a space $W_q^{l_1}(S_r)$ (or $H^\alpha(\bar{S}_r)$), where S_r is a surface of dimension $r < n$, or a subdomain of Ω when $r = n$, we will mean the operator that associates with each function from $W_p^l(\Omega)$ its trace on S_r. Various sufficient conditions are known under which such operators are bounded or completely continuous. One of them is the following:

Theorem 2.1. [1)] *The embedding operator from $W_p^l(\Omega)$ (l integral) into $L_q(S_r)$, where Ω is a bounded domain in euclidean space of dimension n, and S_r is some plane r-dimensional piece belonging to Ω, is bounded if $n > lp$, $r > n - lp$ and $q \leq pr/(n - lp)$, and is completely continuous for $q < pr/(n - lp)$. For $n = lp$ it is completely continuous for any finite q. For $n < lp$ the embedding operator from $W_p^l(\Omega)$ into $H^\alpha(\bar{\Omega})$ is bounded for $\alpha \leq (lp - n)p$; for $\alpha < (lp - n)/p$ this operator is completely continuous. It is assumed here that $l \geq 1$, while $\alpha < 1$.*

The basic part of Theorem 2.1 was proved by S. L. Sobolev and V. I. Kondrašov (see [113a] and [112]). Additions to the propositions proved in [113a] on the embeddability of $W_p^l(\Omega)$ in $H^\alpha(\Omega)$ are given in the papers of several authors (see [33;90b;49a;et al.]). In these papers are described the restrictions to which the domain Ω must be subjected. For domains Ω with piecewise smooth boundaries (their definition is given in §1 of Chapter I) Theorem 2.1 is correct.

This theorem guarantees for any function $u(x)$ from $W_p^l(\Omega)$ the validity of the inequalities

$$\|u\|_{q,S_r} \leq c\,\|u\|_{p,\Omega}^{(l)} \tag{2.5}$$

and

$$|u|_{\Omega}^{(\alpha)} \leq c\,\|u\|_{p,\Omega}^{(l)}. \tag{2.6}$$

The first is valid for $q \leq pr/(n - pl)$ when $n > pl$ and for any finite q when

1) This and other analogous theorems concerning an arbitrary element $u(x)$ from $W_p^l(\Omega)$ must be understood as the existence of at least one function, equivalent to $u(x)$ on Ω, for which the assertions of the theorems are true.

$n = pl$, where $r > n - pl$. The second holds for $n < pl$ with $\alpha \leq (pl - n)/p$, $\alpha < 1$. The constants c depend in general on n, p, l, r, q, Ω, S_r, but not on $u(x)$.

From these inequalities one easily derives the two following inequalities:

$$\|u\|^2_{q,\Omega} \leqslant c(\|u_x\|^2_{p,\Omega} + \|u\|^2_{2,\Omega}) \tag{2.7}$$

with $q \leq np/(n-p)$ for $p < n$ and with any q for $p = n$, and

$$\max_{\Omega} |u|^2 \leqslant c(\|u_x\|^2_{p,\Omega} + \|u\|^2_{2,\Omega}) \tag{2.8}$$

for $p > n$.

We note that for bounded Ω and any function $u(x)$ from $L_{\tilde{p}}(\Omega)$ the inequality

$$\|u\|_{p,\Omega} \leqslant c\|u\|_{\tilde{p},\Omega}$$

with $c = (\text{mes}\,\Omega)^{1/p - 1/\tilde{p}}$, which follows from Hölder's inequality, is valid for $\tilde{p} > p$.

The following proposition is useful: for a fixed bounded Ω the norm $\|u\|^{(1)}_{p,\Omega}$ is equivalent to $\|u_x\|_{p,\Omega} + \|u\|_{1,\Omega}$, and if the function $u(x)$ vanishes on some part of the $(n-1)$-dimensional boundary of Ω that has a positive measure, then $\|u\|^{(1)}_{p,\Omega}$ is equivalent to $\|u_x\|_{p,\Omega}$.

To the assertions of Theorem 2.1 should be added the following: the norms of the embedding operators are uniformly bounded for all possible situations of S_r in Ω, and the variation of the operators is small under a small displacement of S_r and even under a sufficiently smooth small deformation of S_r. Corresponding assertions follow from Theorem 2.1 for nonplanar sections of S_r as long as they are sufficiently smooth (for example, if $S_r \in O^l$). Indeed, by straightening them with the help of l-times boundedly differentiable transformations of the independent variables, we arrive at the case described in Theorem 2.1.

Besides Theorem 2.1 we will need the following more refined characteristics of embedding operators, expressed in the form of the so-called multiplicative inequalities.

Theorem 2.2. *For any function* $u(x) \in \overset{\circ}{W}{}^{(1)}_m(\Omega)$, $m \geq 1$, *and number* $r \geq 1$ *the inequality*

$$\|u\|_{q,\Omega} \leqslant \beta \|u_x\|^{\alpha}_{m,\Omega} \|u\|^{1-\alpha}_{r,\Omega} \tag{2.9}$$

is valid, in which

$$\alpha = \left(\frac{1}{r} - \frac{1}{q}\right)\left(\frac{1}{n} - \frac{1}{m} + \frac{1}{r}\right)^{-1} \tag{2.10}$$

and: 1) *for* $m \geq n = 1$ *the number* q *can be any number from* $[r, \infty]$, *while* $\beta = (1 + (m-1)/m\,r)^{\alpha}$; *under a variation of* q *from* r *to* ∞ *the number* α *varies from zero to* $m/(m + r(m-1))$, *including both ends*;

2) *for* $n > 1$ *and* $m < n$ *the number* q *can be any number from* $[r, nm/(n-m)]$, *if* $r \leq nm/(n-m)$, *and any number from* $[nm/(n-m), r]$ *if* $r \geq nm/(n-m)$, *while* $\beta = ((n-1)m/(n-m))^{\alpha}$; *under a variation of* q *between* r *and* $nm/(n-m)$ *the number* α *varies between* 0 *and* 1, *including both ends*;

3) for $m \geqslant n > 1$ the number q can be any number from $[r, \infty)$ and $\beta = \max\{q(n-1)/n;\ 1 + r(m-1)/m\}^{\alpha}$. Under a variation of q from r to ∞ the number α varies from zero to $nm/[nm + r(m-n)]$ excluding the right end point. If $m > n > 1$, then (2.9) is also valid for $q = \infty$ with $\alpha = nm/[nm + r(m-n)]$ for some $\beta < \infty$.

Remark 2.1. The condition in Theorem 2.2 that $u(x)$ vanish on S can be replaced, for example, by the condition

$$\int_\Omega u(x)\,dx = 0. \tag{2.11}$$

In this case the constant β in (2.9) will depend on n, m, r, α and on the domain Ω, although it does not vary under a similarity transformation of Ω. The boundary S must be piecewise smooth.

The first inequalities of the type (2.9) were found independently by E. Gagliardo [33d] and O. Ladyženskaya (see her book *Mathematical problems in the dynamics of a viscous incompressible fluid*, Gos. Izdat. Fiz.-Mat. Lit., Moscow, 1961 [English transl., Gordon and Breach, New York-London, 1963]; see also [90b; 38a, b; 49a, b]). We shall prove items 1), 2) with the help of a method suggested by Ladyženskaya and extended by L. Nirenberg and K. K. Golovkin. We used the same approach in the Russian edition of this book for the case $m > n > 1$. Here, for $m \geqslant n > 1$, we demonstrate the approach of V. P. Il'in. The expression for β obtained by this method is different from the expression for β given by the first one (for $m > n > 1$).

The inequalities (2.9) for $m < n$ are a special case of more general multiplicative inequalities proved with different methods in [33;38a;49a]. We will derive inequalities (2.9) here for any $m \geq 1$ and we will observe on what and how the constant β depends.

For the case $m \geq n = 1$ we make use of the inequality

$$|u(x)|^q = |u(x)|^r \left(|u(x)|^s\right)^{\frac{q-r}{s}} \leqslant |u(x)|^r \left(\int_\Omega s\,|u(\xi)|^{s-1}\,|u_\xi(\xi)|\,d\xi\right)^{\frac{q-r}{s}},$$

which is valid for any $s \geq 1$ and $q \geq r$. We estimate the integral on the right by Hölder's inequality and then integrate both sides of the resultant inequality over Ω. This gives

$$\int_\Omega |u(x)|^q dx \leqslant \int_\Omega |u(x)|^r dx \left(s \|u_x\|_{m,\Omega} \|u\|^{s-1}_{\frac{m(s-1)}{m-1},\Omega}\right)^{\frac{q-r}{s}}. \tag{2.12}$$

Let us take $s = 1 + (m-1)r/m$ and designate $(q-r)/qs = \alpha$. Then the parameters q, r, m and α turn out to be connected by the relation $\alpha(1/r + (m-1)/m) = 1/r - 1/q$, which coincides with (2.10), and from (2.12) we obtain

$$\|u\|_{q,\Omega} \leqslant \left(1 + \frac{m-1}{m} r\right)^\alpha \|u_x\|^\alpha_{m,\Omega} \|u\|^{1-\alpha}_{r,\Omega},$$

i.e., inequality (2.9) with $\beta = (1 + (m-1)r/m)^\alpha$. Since under a variation of q from r to ∞ the constant β remains bounded, it is clear that (2.9) also holds for $q = \infty$.

Consider now the case $n > 1$ and $m < n$. We first show that for any function $u(x)$ from $\overset{\circ}{W}{}^1_p(\Omega)$ the inequality

$$\|u\|_{\bar{p},\Omega} \leqslant c_p \|u_x\|_{p,\Omega} \tag{2.13}$$

is valid, in which p is any number from $[1, n)$, $\bar{p} = np/(n-p)$ and $c_p = (n-1)p/(n-p)$.

We first establish it for $p = 1$, i.e. we verify that

$$\|u\|_{\frac{n}{n-1},\Omega} \leqslant \prod_{i=1}^n \|u_{x_i}\|^{\frac{1}{n}}_{1,\Omega}. \tag{2.14}$$

Inequality (2.14) is valid for $n = 2$ since

$$\begin{aligned}\int\int & u^2(x_1, x_2)\,dx_1\,dx_2 \\ &\leqslant \int \max_{x_2}|u(x_1, x_2)|\,dx_1 \int \max_{x_1}|u(x_1, x_2)|\,dx_2 \\ &\leqslant \int\int |u_{x_2}|\,dx_1\,dx_2 \int\int |u_{x_1}|\,dx_1\,dx_2.\end{aligned}$$

Suppose it holds for dimension $n - 1 \geq 1$. Then it also holds for dimension n, since by Hölder's inequality and our induction assumption

$$\begin{aligned}\int & |u(x_1, \ldots, x_n)|^{\frac{n}{n-1}}\,dx_1 \ldots dx_n \\ &\leqslant \int dx_n \left[\left(\int |u|\,dx_1 \ldots dx_{n-1}\right)^{\frac{1}{n-1}} \right. \\ &\quad \left. \times \left(\int |u|^{\frac{n-1}{n-2}}\,dx_1 \ldots dx_{n-1}\right)^{\frac{n-2}{n-1}}\right] \leqslant\end{aligned}$$

$$\leqslant \max_{x_n}\left(\int |u|\,dx_1 \ldots dx_{n-1}\right)^{\frac{1}{n-1}} \int \prod_{i=1}^{n-1} \|u_{x_i}\|_{1, E_{n-1}}^{\frac{1}{n-1}}\, dx_n$$

$$\leqslant \left(\int |u_{x_n}|\,dx_1 \ldots dx_n\right)^{\frac{1}{n-1}} \prod_{i=1}^{n-1}\left(\int |u_{x_i}|\,dx_1 \ldots dx_n\right)^{\frac{1}{n-1}}.$$

Inequality (2.14) is proved. To prove (2.13) for $p > 1$ we introduce a new function $v = u^{1/\kappa}$, where $1/\kappa = \bar{p}(n-1)/n > 1$. It is clear that

$$\|u\|_{\bar{p}, \Omega} = \|v\|^{\varkappa}_{\frac{n}{n-1}, \Omega}.$$

Employing inequality (2.14) and Hölder's inequality for an estimate of $\|v\|_{n/(n-1),\Omega}$, we get

$$\|u\|_{\bar{p}, \Omega} = \|v\|^{\varkappa}_{\frac{n}{n-1}, \Omega} \leqslant \prod_{i=1}^{n} \|v_{x_i}\|_{1, \Omega}^{\frac{\varkappa}{n}}$$

$$= \prod_{i=1}^{n} \left\| \frac{1}{\varkappa} u^{\frac{1}{\varkappa}-1} u_{x_i} \right\|_{1, \Omega}^{\frac{\varkappa}{n}}$$

$$\leqslant \left(\frac{1}{\varkappa}\right)^{\varkappa} \prod_{i=1}^{n} \left(\left\| u^{\frac{1}{\varkappa}-1} \right\|_{\frac{p}{p-1}, \Omega} \|u_{x_i}\|_{p, \Omega} \right)^{\frac{\varkappa}{n}}$$

$$\leqslant \varkappa^{-\varkappa} \|u\|_{\bar{p}, \Omega}^{1-\varkappa} \|u_x\|_{p, \Omega}^{\varkappa},$$

from which follows estimate (2.13).

Let us show how to prove (2.9) with it. If $m < n$, then we estimate $\|u\|_{q,\Omega}$ for $\min\{r; \bar{m}\} \le q \le \max\{r; \bar{m}\}$ by Hölder's inequality as follows:

$$\|u\|_{q, \Omega} \leqslant \|u\|^{\alpha}_{\bar{m}, \Omega} \|u\|^{1-\alpha}_{r, \Omega}, \tag{2.15}$$

where α is a number defined by the equality $\alpha/\bar{m} + (1-\alpha)/r = 1/q$. It is obvious that α belongs to the closed interval $[0, 1]$, and when q varies between r and $\bar{m}$ the number α varies between 0 and 1, and conversely. From (2.15) and (2.13) with $p = m$ we obtain inequality (2.9)

$$\|u\|_{q, \Omega} \leqslant c_m^{\alpha} \|u_x\|^{\alpha}_{m, \Omega} \|u\|^{1-\alpha}_{r, \Omega}$$

with

$$\beta = c_m^{\alpha} = \left(\frac{(n-1)\,m}{n-m}\right)^{\alpha}.$$

The case $m \geq n > 1$ is considered by following the methods of V. P. Il'in. The number q, by assumption, is taken to be an arbitrary number from $[r, \infty]$ for $m > n$, and from $[r, \infty)$ for $m = n$. We first assume that q is subject to the additional condition $q \geq m$. It is sufficient to prove inequality (2.9) for functions $u(x)$ that are infinitely differentiable in E_n and equal to zero outside of Ω (since they are dense in $\overset{\circ}{W}{}^1_m(\Omega)$). For such functions, Poisson's formula for a Newtonian potential

$$u(x) = -\frac{1}{(n-2)\omega_n} \int_{E_n} \frac{\Delta u(y)}{|x-y|^{n-2}} dy$$

implies the representation

$$u(x) = -\frac{1}{\omega_n} \int_{E_n} \frac{\partial u(x+z)}{\partial x_i} \frac{z_i}{|z|^n} dz.$$

Breaking the domain of integration E_n into two parts: $\{|z| < \sigma\}$ and $\{|z| \geq \sigma\}$, $\sigma > 0$, and carrying out an integration by parts in the second integral, we obtain

$$u(x) = -\frac{1}{\omega_n} \int_{|z|<\sigma} \frac{\partial u(x+z)}{\partial x_i} \frac{z_i}{|z|^n} dz + \frac{1}{\omega_n \sigma^{n-1}} \int_{|z|=\sigma} u(x+z) d S_z.$$

We multiply both sides of this equality by $n\sigma^{n-1}$ and integrate with respect to σ from 0 to an arbitrary $h > 0$. This gives us

$$h^n u(x) = -\frac{n}{\omega_n}\left[\int_0^h \sigma^{n-1} d\sigma \int_0^\sigma d\rho \int_{|\theta|=1} \frac{\partial u(x+\rho\theta)}{\partial x_i} \theta_i d\theta - \int_{|z|\leq h} u(x+z) dz \right].$$

Upon making some simple estimates we obtain

$$|u(x)| \leq \frac{1}{\omega_n} \int_{|z|\leq h} \left| \frac{\partial u(x+z)}{\partial x_i} \right| \frac{|z_i|}{|z|^n} dz + \frac{n}{\omega_n h^n} \int_{|z|\leq h} |u(x+z)| dz.$$

We let

$$v_1(x, h) = \int\limits_{|z|\le h} \left|\frac{\partial u(x+z)}{\partial x_i}\right| \frac{|z_i|}{|z|^n} dz,$$

$$v_2(x, h) = \frac{1}{h^n} \int\limits_{|z|\le h} |u(x+z)| dz$$

and estimate v_1 and v_2 in the following manner:

$$v_1(x, h) \le \left\{ \int\limits_{|z|\le h} |u_x(x+z)|^m dz \right\}^{1/m-1/q} \left\{ \int\limits_{|z|\le h} |u_x(x+z)|^m \frac{dz}{|z|^a} \right\}^{1/q} \left\{ \int\limits_{|z|\le h} \frac{dz}{|z|^b} \right\}^{1-1/m},$$

where the numbers a and b are connected by the relation

$$\frac{a}{q} + b\left(1 - \frac{1}{m}\right) = n - 1,$$

with $a < n$, $b < n$. Since

$$\frac{n}{q} + n\left(1 - \frac{1}{m}\right) > n\left(1 - \frac{1}{n}\right) = n - 1,$$

it is possible to select such numbers. Thus

$$v_1(x, h) \le \|u_x\|_m^{1/m-1/q} \left\{ \int\limits_{|z|\le h} |u_x(x+z)|^m \frac{dz}{|z|^a} \right\}^{1/q} \left[\frac{\omega_n}{n-b}\right]^{1-1/m} h^{(n-b)(1-1/m)},$$

$$\|v_1(x, h)\|_q \le \frac{\omega_n^{1-1/m+1/q}}{(n-b)^{1-1/m}(n-a)^{1/q}} h^{n(1/q-1/m+1/n)} \|u_x\|_m. \tag{2.16}$$

In an analogous way

$$v_2(x, h) \le \frac{1}{h^n} \left\{ \int\limits_{|z|\le h} |u(x+z)|^r dz \right\}^{1/r} \left\{ \int\limits_{|z|\le h} dz \right\}^{1/r'}$$

$$\le \frac{\omega_n^{1-1/r}}{n^{1-1/r} h^{n/r}} \|u\|_r^{1-r/q} \left\{ \int\limits_{|z|\le h} |u(x+z)|^r dz \right\}^{1/q},$$

$$\|v_2(x, h)\|_q \leq \frac{\omega_n^{1-1/r+1/q}}{n^{1-1/r+1/q} h^{n(1/r-1/q)}} \|u\|_r. \tag{2.17}$$

Combining (2.16) and (2.17) we obtain

$$\|u\|_q \leq \frac{\omega_n^{1-1/m+1/q}}{(n-b)^{1-1/m}(n-a)^{1/q}} h^{n(1/q-1/m+1/n)} \|u_x\|_m + \frac{\omega_n^{1-1/r+1/q}}{n^{1-1/r+1/q} h^{n(1/r-1/q)}} \|u\|_r$$

or more briefly

$$\|u\|_q \leq Ah^{\delta} + Bh^{-\tilde{\delta}}.$$

This inequality is valid for any $h > 0$. Minimizing the right side with respect to h, as is easily done, we obtain

$$\|u\|_q \leq \left[\left(\frac{\tilde{\delta}}{\delta}\right)^{\delta/(\delta+\tilde{\delta})} + \left(\frac{\delta}{\tilde{\delta}}\right)^{\tilde{\delta}/(\delta+\tilde{\delta})}\right] A^{\tilde{\delta}/(\delta+\tilde{\delta})} B^{\delta/(\delta+\tilde{\delta})},$$

i.e., inequality (2.9). The constant β in it tends to some finite number for $q \to \infty$ if $m > n$, and to ∞ if $m = n$.

If in case 3) our addition assumption $q \geq m$ is not fulfilled, then we take some $\tilde{q} \geq m$ and make use of the inequality

$$\|u\|_q \leq \|u\|_{\tilde{q}}^{\gamma} \|u\|_r^{1-\gamma},$$

where $1/q = \gamma/\tilde{q} + (1-\gamma)/r$, and consequently $\gamma = (1/r - 1/q)/(1/r - 1/\tilde{q})$, $\gamma \in [0, 1)$. By estimating $\|u\|_{\tilde{q}}$ here from above by inequality (2.9) (since $\tilde{q} \geq m$):

$$\|u\|_{\tilde{q}} \leq \beta \|u_x\|_m^{\tilde{\alpha}} \|u\|_r^{1-\tilde{\alpha}}, \quad \tilde{\alpha} = \frac{1/r - 1/\tilde{q}}{1/r + 1/n - 1/m},$$

we obtain (2.9) for the case $q < m$.

Theorem 2.2 is proved.

We will make use of a number of consequences of inequality (2.9). One of them is the inequality

$$\|u\|_{\frac{2p}{p-2}, \Omega} \leqslant c \|u_x\|_{2, \Omega}. \tag{2.18}$$

valid for any function $u(x) \in \overset{\circ}{W}{}_2^1(\Omega)$, in which $c = \beta_1 \operatorname{mes}^{1/n - 1/p}(\Omega)$,

$$\begin{aligned} &p \geq n \text{ for } n \geq 3, \\ &p > 2 \text{ for } n = 2, \\ &p \geq 2 \text{ for } n = 1, \end{aligned} \tag{2.19}$$

and $\beta_1 = 2(n-1)/(n-2)$ for $n \geq 3$, $\beta_1 = 2(p-1)/(p-2)$ for $n = 2$ and $\beta_1 = 2$ for $n = 1$. For $u(x)$ that do not vanish on S but satisfy condition (2.11), the constant β_1 depends on n, p and on the boundary S. Inequality (2.18) follows from (2.9) when in the latter one puts $m = 2$, $q = 2p/(p-2)$, $r < q$ and estimates $\|u\|_{r,\Omega}$ by Hölder's inequality

$$\|u\|_{r,\Omega} \leqslant \|u\|_{\frac{2p}{p-2},\Omega} (\operatorname{mes} \Omega)^{\frac{1}{r} - \frac{p-2}{2p}}.$$

Another inequality that will be made use of is

$$\|u\|^2_{\frac{2p}{p-2},\Omega} \leqslant \varepsilon \|u_x\|^2_{2,\Omega} + c_\varepsilon \|u\|^2_{2,\Omega}, \tag{2.20}$$

where $p > n$ and ≥ 2, ϵ is an arbitrary positive number, $c_\epsilon = (p-n)/p(\beta_1\sqrt{n/p\epsilon})^{2n/(p-q)}$, and the constant β_1 is the same as in (2.18). Inequality (2.20) is obtained from inequality (2.9) for $m = 2$, $r = 2$, $q = 2p/(p-2)$ and an estimate of its right side by Young's inequality (1.3).

For estimates of integrals over the $(n-1)$-dimensional boundary S of an n-dimensional domain Ω when $n \geq 2$, we need in addition to (2.5) the following inequality of type (2.9):

$$\|u\|_{q,S} \leqslant c \|u_x\|^{\alpha}_{2,\Omega} \|u\|^{1-\alpha}_{2,\Omega}, \quad \alpha = \frac{n}{2} - \frac{n-1}{q}, \tag{2.21}$$

which is valid for any function $u(x)$ from $W_2^1(\Omega)$ with $\int_\Omega u(x)dx = 0$. In it q is any number from $[2(n-1)/n, 2(n-1)/(n-2)]$ for $n \geq 3$ and any number from $[1, \infty)$ for $n = 2$, while the constant c depends only on n, q and the surface S, which is assumed to be piecewise smooth.

An important role in the investigation of operators of elliptic type is played by the inequalities $\longrightarrow$ ([65b, d])

$$\|u\|^{(2)}_{2,\Omega} \leqslant c \|\mathscr{L}u\|_{2,\Omega} + c_1 \|u\|_{2,\Omega}, \tag{2.22}$$

which are valid for an arbitrary function $u(x)$ from $W_2^2(\Omega)$ satisfying one of the homogeneous boundary conditions, $u|_S = 0$ or $\partial u/\partial l + \sigma u|_S = 0$ (l is a given vector that forms an acute angle with the direction of the normal to S at the same point), and an arbitrary uniformly elliptic operator $\mathfrak{L}$ of the form

$$\mathscr{L}u = a_{ij}(x)\,u_{x_i x_j} + a_i(x)\,u_{x_i} + a(x)\,u$$

with continuous a_{ij}, with a_i from $L_q(\Omega)$, $q > n$, and with $a \in L_{\hat{q}/2}(\Omega)$, where $\hat{q} = \max\{q;\ 4\}$. The constants c and c_1 in them are determined only by the coefficients of $\mathfrak{L}$, the boundary S (which must be from class O^2) and the known functions entering into the boundary conditions.

Inequalities (2.22) are also useful in the investigation of parabolic operators. Besides them, we need a generalization of (2.22)

$$\nu\left(\| u \|_{2,\Omega}^{(2)}\right)^2 \leqslant (\mathscr{L}u,\ \tilde{\mathscr{L}}u) + c_1 \| u \|_{2,\Omega}^2, \quad \nu = \text{const} > 0, \tag{2.23}$$

or, more precisely, a special case of it in the form

$$\frac{3}{4}\nu \| u_{xx} \|_{2,\Omega}^2 \leqslant (\mathscr{L}u,\ \Delta u) + c_1 \| u_x \|_{2,\Omega}^2, \tag{2.24}$$

where $\mathfrak{L}u = (\partial/\partial x_i)(a_{ij}(x)u_{x_j})$, while the $a_{ij}(x)$ are continuous and $\operatorname{vrai\,max}_\Omega |\partial a_{ij}/\partial x_k| < \infty$, $k, i, j = 1, \cdots, n$. Inequality (2.23) is valid for any function u from $\overset{\circ}{W}{}_2^{\prime 2}(\Omega)$ and any two elliptic operators $\mathfrak{L}$ and $\tilde{\mathfrak{L}}$ of the same type as in (2.22). Detailed proofs and an analysis of inequalities (2.22), (2.24) are given in Chapter III of the book [65q].

The space $W_q^l(\Omega)$ with nonintegral l will be used in Chapters IV and VII. It is a Banach space consisting of the elements of $W_q^{[l]}$ with finite norm

$$\| u \|_{q,\Omega}^{(l)} = \langle\langle u \rangle\rangle_{q,\Omega}^{(l)} + \| u \|_{q,\Omega}^{([l])}, \tag{2.25}$$

where

$$\| u \|_{q,\Omega}^{([l])} = \sum_{j=0}^{[l]} \sum_{(j)} \| D_x^j u \|_{q,\Omega},$$

$$\langle\langle u \rangle\rangle_{q,\Omega}^{(l)} = \sum_{(j=[l])} \left(\int_\Omega dx \int_\Omega | D_x^j u(x) - D_y^j u(y) |^q \frac{dy}{| x - y |^{n+q(l-[l])}} \right)^{\frac{1}{q}}.$$

Such spaces were introduced (first for $q = 2$, and later for any $q > 1$ [111b]) in the study of properties of functions from $W_q^l(\Omega)$ with integral l on manifolds S_r of dimensions $r < n$ and in the proof of exact converse propositions concerning the extension of a function from S_r onto all of Ω. A theorem of this type, in which functions from $W_q^l(\Omega)$ are studied on the boundary S of Ω, is cited below.

It plays an important role in the study of boundary value problems with nonhomogeneous boundary conditions, especially in the proof of exact estimates for their solutions. For a formulation of this theorem we must introduce classes W_q^l with nonintegral l on a curvilinear surface S. This is done by means of a parametrization of the surface. Let $S_1, \cdots, S_k, \cdots$ be a finite or infinite (in the case of an infinite S) family of subsets of S such that $\bigcup_k S_k = S$ and for each point $x \in S$ it is possible to find an S_k such that $x \in S_k$ and the distance from x to $S \setminus S_k$ exceeds a certain fixed number δ. Further, it is assumed that each S_k intersects with only a finite number of other subsets, not exceeding some number m, and is mapped onto some canonical domain σ of $(n-1)$-dimensional euclidean space, for example a cube, by means of a transformation of class $H^{l'}$, $l' > l$. Suppose that under a mapping of S_k onto σ a function $u(x)$ $(x \in S_k)$ goes over into $u^{(k)}(z)$ $(z \in \sigma)$. We put

$$\|u\|_{q,S}^{(l)} = \Big\{\sum_k \big(\|u^{(k)}(z)\|_{q,\sigma}^{(l)}\big)^q\Big\}^{\frac{1}{q}}.$$

The norms corresponding to the various coverings of a surface by sets S_k satisfying the above-mentioned conditions are equivalent.

Theorem 2.3 [33c; 111c]. *Suppose $u(x) \in W_q^{(l)}(\Omega)$, where l is a positive integer and $q > 1$. If $S \in O^{(l)}$, then $u(x)|_S \in W_q^{l-1/q}(S)$ and the inequality $\|u\|_{q,S}^{(l-1/q)} \le c\|u\|_{q,\Omega}^{(l)}$ is valid. Conversely, any function $u(x) \in W_q^{l-1/q}(S)$, defined on S, can be extended into the domain Ω so that $u(x) \in W_q^l(\Omega)$ and $\|u\|_{q,\Omega}^{(l)} \le c\|u\|_{q,S}^{(l-1/q)}$.*

The constants c in both inequalities are independent of u.

It was subsequently determined that the so-defined spaces $W_q^l(\Omega)$ for $q \neq 2$ do not form a "continuous" scale in l. It "tears" at integral l. In order to create such a scale the spaces $W_q^l(\Omega)$ for integral l must be replaced by the spaces $B_q^l(\Omega)$. For a convex domain Ω the norm in $B_q^l(\Omega)$ can be defined by for formula (2.25) if in the right side one takes instead of $\langle\langle u\rangle\rangle_{q,\Omega}^{(l)}$

$$\sum_{(l-1)}\Big(\int_\Omega dx \int_\Omega \Big|D^{l-1}u(x) - 2D^{l-1}u\Big(\frac{x+y}{2}\Big) + D^{l-1}u(y)\Big|^q \frac{dy}{|x-y|^{n+q}}\Big)^{\frac{1}{q}}. \quad (2.26)$$

Let us formulate three more propositions that are known from the theory of functions of a real variable.

Lemma 2.1. *If a sequence of functions $u_k(x)$, $k = 1, 2, \cdots$, converges strongly in $L_q(\Omega)$, $q \geq 1$, to a function $u(x)$, then it is possible to extract a subsequence which converges to u almost everywhere on Ω. And the convergence of the u_k to u almost everywhere on Ω implies an almost uniform convergence on Ω, i.e. one such that for any $\epsilon > 0$ there exists a set $\Omega^\epsilon \subset \Omega$ with* mes $\Omega^\epsilon \leq \epsilon$ *such that the u_k converge to u uniformly on $\Omega - \Omega^\epsilon$.*

Lemma 2.2. *If $\|u_k\|_{q,\Omega} \leq c$, $q > 1$, for a given sequence of functions $u_k(x)$, $k = 1, 2, \cdots$, then it is possible to extract a subsequence from $\{u_k\}$ that is weakly convergent in $L_q(\Omega)$. If $\|u_k\|_{q,\Omega} \leqslant c$, $q > 1$, and $\|f_k\|_{q', \Omega'} \leqslant \mu(\text{mes } \Omega')$, $q' = q/(q-1)$, where $\mu(\tau)$, $\tau \in E_1^+ \equiv [0, \infty)$, is continuous and $\mu(0) = 0$, for all $k = 1, 2, \ldots$ and arbitrary $\Omega' \subset \Omega$, and if the $u_k(x)$ converge to $u(x)$ almost everywhere on Ω and the f_k converge to f weakly in $L_{q'}(\Omega)$, then $\int_\Omega u_k f_k\, dx \longrightarrow \int_\Omega uf\, dx$, $k \longrightarrow \infty$.*

Lemma 2.3. *Let $f(x, u)$ be a measurable function on the set $\{x \in \Omega, u \in (-\infty, \infty)\}$ that is continuous in u for almost all x from Ω. If a sequence of functions $u_k(x)$, $k = 1, 2, \cdots$, from $L_1(\Omega)$ converges almost everywhere to $u(x) \in L_1(\Omega)$ and $\|f(x, u_k(x))\|_{q,\Omega} \leq c$, $q > 1$, then the functions $f(x, u_k(x))$ converge to $f(x, u(x))$ in the norms of $L_{q^*}(\Omega)$, $q^* < q$, and weakly in $L_q(\Omega)$. If, in addition, it is known that $\|f(x, u_k(x))\|_{q,\Omega'} \leq \mu(\text{mes}\, \Omega')$, where $\mu(\tau)$ is a continuous function of $\tau \geq 0$ that is equal to zero for $\tau = 0$, while Ω' is an arbitrary measurable subset of Ω, then the $f(x, u_k(x))$ converge to $f(x, u(x))$ strongly in $L_q(\Omega)$.*

Let us cite one more assertion, which is a generalization of one of the propositions of K. Friedrichs on functions from $\overset{\circ}{W}{}_2^1(\Omega)$, to functions from $\overset{\circ}{W}{}_m^1(\Omega)$ (see [15b], Chapter 7).

Lemma 2.4. *Suppose the functions $\psi_k(x)$, $k = 1, 2, \cdots$, form an orthogonal basis in $L_2(\Omega)$. Then for any $\epsilon > 0$ there exists a number N_ϵ such that the inequality*

$$\|u\|_{2,\Omega} \leq \left[\sum_{k=1}^{N_\epsilon} (u, \psi_k)^2 \right]^{1/2} + \epsilon \|u\|_{m,\Omega}^{(1)} \tag{2.27}$$

is valid for any function $u(x)$ from $\overset{\circ}{W}{}_m^1(\Omega)$ with $m > 2n/(n+2)$ for $n \geq 2$ and with $m \geq 1$ for $n = 1$. The number N_ϵ does not depend on u.

It is sufficient to prove the validity of the inequality

$$\|u\|_{2,\Omega} \le (1+\delta)\left[\sum_{k=1}^{N_{\epsilon,\delta}} (u,\psi_k)^2\right]^{1/2} + \epsilon\|u\|_{m,\Omega}^{(1)} \tag{2.27'}$$

with arbitrary $\delta > 0$, and $\epsilon > 0$, since from it, by virtue of the orthonormality of $\{\psi_k\}$ in $L_2(\Omega)$ and inequality (2.5) for $S_r = \Omega$, $q = 2$, $l = 1$, $p = m$, there follows

$$\|u\|_{2,\Omega} \le \left[\sum_{k=1}^{N_{\epsilon,\delta}} (u,\psi_k)^2\right]^{1/2} + \delta\|u\|_{2,\Omega} + \epsilon\|u\|_{m,\Omega}^{(1)}$$

$$\le \left[\sum_{k=1}^{N_{\epsilon,\delta}} (u,\psi_k)^2\right]^{1/2} + (c\delta+\epsilon)\|u\|_{m,\Omega}^{(1)},$$

i.e. the assertion of Lemma 2.4.

Let us assume that inequality (2.27') does not hold. Then there exists an $\epsilon_0 > 0$ and a sequence of functions u_r, $r = 1, 2, \cdots$, from $W_m^1(\Omega)$ such that

$$\|u_r\|_{2,\Omega} > (1+\delta)\left[\sum_{k=1}^{r} (u_r,\psi_k)^2\right]^{1/2} + \epsilon_0\|u_r\|_{m,\Omega}^{(1)}$$

for any r. Hence for the functions $v_r = u_r/\|u_r\|_{2,\Omega}$ we obtain

$$1 = \|v_r\|_{2,\Omega} > (1+\delta)\left[\sum_{k=1}^{r} (v,\psi_k)^2\right]^{1/2} + \epsilon_0\|v_r\|_{m,\Omega}^{(1)},$$

so that the norms $\|v_r\|_{m,\Omega}^{(1)}$, $r = 1, 2, \cdots$, are uniformly bounded. By virtue of the compactness of the embedding of $W_m^1(\Omega)$ into $L_2(\Omega)$ (Theorem 2.1) there exists a subsequence v_{r_p}, $p = 1, 2, \cdots$, that is strongly convergent in $L_2(\Omega)$ to some function v, with $\|v\|_{2,\Omega} = \|v_r\|_{2,\Omega} = 1$. The functions $P_{r_p} v_{r_p} \equiv \sum_{k=1}^{r_p} (v_{r_p}, \psi_k)\psi_k$ also converges strongly to v, since

$$\|v - P_{r_p} v_{r_p}\|_{2,\Omega} = \|P_{r_p}(v - v_{r_p}) + (E - P_{r_p})v\|_{2,\Omega}$$

$$\le \|v - v_{r_p}\|_{2,\Omega} + \|(E - P_{r_p})v\|_{2,\Omega} \to 0 \text{ for } p \to \infty.$$

In view of this, if in the preceding inequality we pass to the limit with respect to the subsequence r_p, $p \to \infty$, we obtain $1 \ge 1 + \delta$, which is impossible. Lemma

2.4 is proved.

§3. DIFFERENT SPACES OF FUNCTIONS DEPENDING ON x AND t. EMBEDDING THEOREMS

We will first prove some properties of the functions $u(x, t)$ from the space $V_2(Q_T)$, and consequently also from the space $\overset{\circ}{V}{}_2^{1,0}(Q_T) \subset V_2(Q_T)$. Suppose $u(x, t)$ is an arbitrary element of $\overset{\circ}{V}_2(Q_T)$. For almost all t from $[0, T]$ inequality (2.9), which when $m = r = 2$ can be rewritten in the form

$$\| u(x, t) \|_{q, \Omega} \leqslant \beta \| u_x(x, t) \|_{2, \Omega}^{\alpha} \| u(x, t) \|_{2, \Omega}^{1-\alpha},$$
$$\alpha = \frac{n}{2} - \frac{n}{q}, \tag{3.1}$$

holds for it. Here $q \in [2, 2n/(n-2)]$ for $n \geq 3$, $q \in [2, \infty)$ for $n = 2$ and $q \in [2, \infty]$ for $n = 1$, while β depends only on n and q as indicated in Theorem 2.2.

From (3.1) we have

$$\| u \|_{q, r, Q_T} = \left(\int_0^T \| u \|_{q, \Omega}^r dt \right)^{\frac{1}{r}}$$
$$\leqslant \beta \left(\int_0^T \| u_x \|_{2, \Omega}^{\alpha r} dt \right)^{\frac{1}{r}} \operatorname{vrai} \max_{0 \leqslant t \leqslant T} \| u(x, t) \|_{2, \Omega}^{1-\alpha}.$$

For $\alpha = 2/r$ we obtain the series of inequalities

$$\| u \|_{q, r, Q_T} \leqslant \beta \operatorname{vrai} \max_{0 \leqslant t \leqslant T} \| u(x, t) \|_{2, \Omega}^{1-\frac{2}{r}} \cdot \| u_x \|_{2, Q_T}^{\frac{2}{r}}. \tag{3.2}$$

In them $1/r + n/2q = n/4$, with

$$\left. \begin{array}{lll} r \in [2, \infty], & q \in \left[2, \frac{2n}{n-2}\right] & \text{for } n > 2, \\ r \in (2, \infty], & q \in [2, \infty) & \text{for } n = 2, \\ r \in [4, \infty], & q \in [2, \infty] & \text{for } n = 1. \end{array} \right\} \tag{3.3}$$

Estimating the right side of (3.2) by Young's inequality (1.3) with $m = r/2$ and $\epsilon = 1$, we obtain a second series of inequalities:

$$\|u\|_{q,r,Q_T} \leqslant \beta \frac{2}{r} \|u_x\|_{2,Q_T} + \beta\left(1-\frac{2}{r}\right) \operatorname{vrai}\max_{0\leqslant t\leqslant T} \|u\|_{2,\Omega} \leqslant \beta |u|_{Q_T}. \tag{3.4}$$

In them q and r are subject to the same relations (3.3) as in (3.2). For $q = r = 2(n+2)/n$ inequality (3.4) is given in the paper [87b].

Let us denote by $\widetilde{A}_0(t)$ the set of points x from Ω at which $|u(x, t)| > 0$, and by $\widetilde{Q}_0$ the set of points (x, t) from Q_T at which $|u(x, t)| > 0$. Take arbitrary $q \geq 1$, $r \geq 1$ and $q_1 \in [1, q]$, $r_1 \in [1, r]$ connected only by the condition $r_1/r = q_1/q$.

From Hölder's inequality (1.7′) applied to $u(x, t)$ and the characteristic function $\chi(x, t)$ for the set $\widetilde{Q}_0$ we have

$$\|u\|_{q_1,r_1,Q_T} \leqslant \|u\|_{q,r,Q_T} \|\chi\|_{\frac{qq_1}{q-q_1}, \frac{rr_1}{r-r_1}, Q_T} = \left(\int_0^T \operatorname{mes}^{\frac{r}{q}} \widetilde{A}_0(t)\, dt\right)^{\frac{r-r_1}{rr_1}} \|u\|_{q,r,Q_T}. \tag{3.5}$$

If one of the indices r_1, q_1 is infinite, then, by analogous arguments, we obtain

$$\|u\|_{q_1,\infty,Q_T} = \operatorname{vrai}\max_{0\leqslant t\leqslant T} \|u\|_{q_1,\Omega} \leqslant \operatorname{vrai}\max_{0\leqslant t\leqslant T} \operatorname{mes}^{\frac{1}{q_1}-\frac{1}{q}} \widetilde{A}_0(t) \|u\|_{q,\infty,Q_T}, \quad q \geqslant q_1,$$

and

$$\|u\|_{\infty,r_1,Q_T} = \left(\int_0^T \operatorname{vrai}\max_{\Omega} |u(x, t)|^{r_1} dt\right)^{\frac{1}{r_1}} \leqslant \widetilde{\mu}(0)^{\frac{1}{r_1}-\frac{1}{r}} \|u\|_{\infty,r,Q_T}, \quad r \geqslant r_1.$$

Here

$$\widetilde{\mu}(0) = \operatorname{mes}\{t \in [0, T] : \operatorname{mes} \widetilde{A}_0(t) > 0\}.$$

From inequality (3.4) with r and q satisfying conditions (3.3) and from inequalities (3.5) it is not difficult to deduce that for any function $u(x, t)$ from

$\overset{\circ}{V}_2(Q_T)$ with arbitrary $q_1 \geq 1$, $r_1 \geq 1$ subject to the conditions

$$\frac{1}{r_1}+\frac{n}{2q_1}=\frac{n}{4}+\varkappa_1\equiv\frac{n\lambda}{4}, \quad \lambda>1,$$

and

$$\begin{aligned} &r_1\in\left[\frac{2}{\lambda},\infty\right], && q_1\in\left[\frac{2}{\lambda},\frac{2n}{\lambda(n-2)}\right] && \text{for}\quad n\geqslant 3,\\ &r_1\in\left(\frac{2}{\lambda},\infty\right], && q_1\in\left[\frac{2}{\lambda},\infty\right) && \text{for}\quad n=2,\\ &r_1\in\left[\frac{4}{\lambda},\infty\right], && q_1\in\left[\frac{2}{\lambda},\infty\right] && \text{for}\quad n=1,\end{aligned}$$

the inequality

$$\|u\|_{q_1,r_1,Q_T}\leqslant\beta|u|_{Q_T}\tilde{\mu}(0)^{\frac{1}{r_1}\frac{\varkappa_1}{\varkappa_1+\frac{n}{4}}} \tag{3.6}$$

is valid, in which

$$\tilde{\mu}(0)=\int_0^T\operatorname{mes}^{\frac{r_1}{q_1}}\tilde{A}_0(t)\,dt.$$

The notation $\tilde{\mu}^{1/r_1}(0)$ means $\operatorname{vrai\,max}_{0\leq t\leq T}\operatorname{mes}^{1/q_1}\tilde{A}_0(t)$ for $r_1=\infty$, while $\tilde{\mu}(0)=\operatorname{mes}\{t\in[0,T]:\ \operatorname{mes}\tilde{A}_0(t)>0\}$ for $q_1=\infty$.

In the special case $q_1=r_1=2$, $\kappa_1=1/2$ it follows from (3.6) that

$$\|u\|_{2,Q_T}\leqslant\beta(\operatorname{mes}\tilde{Q}_0)^{\frac{1}{n+2}}|u|_{Q_T}. \tag{3.7}$$

Inequality (3.4), derived above from the multiplicative inequality (2.9), also holds for functions $u(x,t)$ from $V_2(Q_T)$ that do not necessarily vanish on S_T but satisfy for all $t\in[0,T]$ the condition $\int_\Omega u(x,t)dx=0$. The constant β in (3.4) depends in this case on the surface S, which is assumed to be piecewise smooth, although β does not vary under a similarity transformation of the domain Ω.

Let us now assume that $u(x,t)$ is an arbitrary element of $V_2(Q_T)$. We denote by $u_0(t)$ the average value of $u(x,t)$ in Ω. The difference

$u(x, t) - u_0(t)$ satisfies the condition $\int_\Omega (u - u_0)dx = 0$ for all $t \in [0, T]$, and hence inequality (3.4) is valid for it. On the other hand, the following estimates are obvious:

$$\|u\|_{q,r,Q_T} \leqslant \|u - u_0\|_{q,r,Q_T} + \|u_0\|_{q,r,Q_T}$$
$$\leqslant \|u - u_0\|_{q,r,Q_T} + \operatorname*{vrai\,max}_{0 \leqslant t \leqslant T} |u_0(t)| T^{\frac{1}{r}} \operatorname{mes}^{\frac{1}{q}} \Omega,$$
$$|u - u_0|_{Q_T} \leqslant |u|_{Q_T} + |u_0|_{Q_T}$$
$$= |u|_{Q_T} + \operatorname*{vrai\,max}_{0 \leqslant t \leqslant T} |u_0(t)| \operatorname{mes}^{\frac{1}{2}} \Omega,$$
$$\operatorname*{vrai\,max}_{0 \leqslant t \leqslant T} |u_0(t)| \leqslant \operatorname*{vrai\,max}_{0 \leqslant t \leqslant T} \|u\|_{2,\Omega} \operatorname{mes}^{-\frac{1}{2}} \Omega \leqslant |u|_{Q_T} \operatorname{mes}^{-\frac{1}{2}} \Omega.$$

From what has been said it is clear that for any $u(x, t)$ from $V_2(Q_T)$ with arbitrary parameters q and r satisfying relations (3.3) one has the estimate

$$\|u\|_{q,r,Q_T} \leqslant c |u|_{Q_T},$$
$$c = 2\beta + T^{\frac{1}{r}} \operatorname{mes}^{-\frac{1}{2}+\frac{1}{q}} \Omega = 2\beta + \left(T^{\frac{2}{n}} \operatorname{mes}^{-1} \Omega\right)^{\frac{1}{2}-\frac{1}{q}}. \tag{3.8}$$

The constant β in (3.8) is determined only by r, n, q and S and does not depend on T or $\operatorname{mes}\Omega$.

Inequality (3.5) holds for any $u(x, t)$ from $V_2(Q_T)$. Letting $r_1 = q_1 = 2$, $r = q = 2(n + 2)/n$ in it and using (3.8), we conclude that for an arbitrary function $u(x, t)$ from $V_2(Q_T)$ an estimate of form (3.7) is valid:

$$\|u\|_{2,Q_T} \leqslant c (\operatorname{mes} \tilde{Q}_0)^{\frac{1}{n+2}} |u|_{Q_T},$$
$$c = 2\beta + \left(T^{\frac{n}{2}} \operatorname{mes}^{-1} \Omega\right)^{\frac{1}{n+2}}, \tag{3.9}$$

where the constant β is the same as in (3.8).

An analogous estimate can be derived for the norm $\|u\|_{q,r,S_T}$, if instead of inequality (2.9) one uses inequality (2.21) with $\alpha = 2/r$. Namely, for arbitrary $u(x, t)$ from $V_2(Q_T)$ and q and r satisfying the conditions

$$\left.\begin{aligned} &\frac{1}{r}+\frac{n-1}{2q}=\frac{n}{4}, \\ &r\in[2,\infty],\ q\in\left[\frac{2(n-1)}{n},\ \frac{2(n-1)}{n-2}\right] \quad \text{for } n\geqslant 3, \\ &r\in(2,\infty],\ q\in[1,\infty) \quad \text{for } n=2, \end{aligned}\right\} \tag{3.10}$$

inequality (2.21) implies the estimate

$$\left.\begin{aligned} &\|u\|_{q,r,S_T}\leqslant c\,|u|_{Q_T}, \\ &c=\beta\left[1+\left(T^{\frac{n}{2}}\operatorname{mes}^{-1}\Omega\right)^{\frac{1}{2}-\frac{n-1}{nq}}\right], \end{aligned}\right. \tag{3.11}$$

with the constant β depending only on n, r, q and on local properties of the surface S, which is assumed to be piecewise smooth.

In the case $n=1$ we can derive an estimate of form (3.11) if instead of (2.21) we make use of the inequality

$$u^4(s,t)\leqslant u^2(s,t)\,2\int_\Omega |u_x(x,t)\,u(x,t)|\,dx,\quad s\in S,$$

which holds if $u(x,t)$ vanishes at least one point of Ω. Integrating it with respect to t, we obtain, after some elementary estimates,

$$\|u\|_{4,S_T}=\left(\int_0^T u^4(s,t)\,dt\right)^{\frac{1}{4}}\leqslant |u|_{Q_T}.$$

From this follows the inequality

$$\|u\|_{4,S_T}\leqslant c\,|u|_{Q_T},\quad c=2+T^{\frac{1}{4}}\operatorname{mes}^{-\frac{1}{2}}\Omega. \tag{3.12}$$

for an arbitrary function $u(x,t)$ from $V_2(Q_T)$.

Let us prove the following proposition, which is a special case of more general facts established in the papers [89;38b].

Lemma 3.1. *Suppose $u(x,t)$ in $\overline{Q}_T$ satisfies a Hölder condition in t with exponent α and Hölder constant μ_1 and has derivatives u_x, which for any t from $[0,T]$ are Hölder continuous in the variables x, more precisely, are such that*

$$\max_{\substack{K_\rho \\ 0\leqslant t\leqslant T}} \operatorname{osc}_x\{u_x(x,t),\ K_\rho\cap\Omega\}\leqslant \mu_2\rho^\beta.$$

Suppose, further, that the domain Ω satisfies a cone condition, i.e. there exists a fixed spherical cone $\mathcal{K}$ of some altitude d such that, no matter at what point of $\overline{\Omega}$ its vertex is placed, the cone itself can be swung so that all of it is located in $\overline{\Omega}$. Then the derivatives u_x satisfy in Q_T a Hölder condition in t with exponent $\delta = \alpha\beta/(1+\beta)$ and constant μ determined only by α, β, μ_1, μ_2, d and the solid angle at the vertex of the cone $\mathcal{K}$.

For arbitrary points x' and x'', which can be connected by a rectilinear segment $[x', x'']$ belonging to the domain $\overline{\Omega}$, one has the relation

$$\int_{x'}^{x''}\left[\frac{\partial u(x,t)}{\partial l}-\frac{\partial u(x',t')}{\partial l}\right]dl\Big|_{t'}^{t''} = u(x'',t'')-u(x',t'')-u(x'',t')+u(x',t'). \tag{3.13}$$

In it $\partial/\partial l$ denotes differentiation along the segment $[x', x'']$. By virtue of the conditions of the lemma the right side of (3.13) does not exceed $2\mu_1|t''-t'|^{\alpha}$, and

$$\left|\int_{x'}^{x''}\left[\frac{\partial u(x,t')}{\partial l}-\frac{\partial u(x',t')}{\partial l}\right]dl\right| \leqslant \mu_2|x''-x'|^{1+\beta}.$$

In addition,

$$\left|\int_{x'}^{x''}\left[\frac{\partial u(x,t'')}{\partial l}-\frac{\partial u(x',t')}{\partial l}\right]dl\right| = |x''-x'|\left|\frac{\partial u(\hat{x},t'')}{\partial l} - \frac{\partial u(x',t')}{\partial l}\right| \geqslant |x''-x'|\left[\left|\frac{\partial u(x',t'')}{\partial l}-\frac{\partial u(x',t')}{\partial l}\right| - \mu_2|x''-x'|^{\beta}\right],$$

where $\hat{x}$ is some point of the segment $[x', x'']$ lying between x' and x''. On the basis of these estimates there follows from (3.13) the inequality

$$\left|\frac{\partial u(x',t'')}{\partial l}-\frac{\partial u(x',t')}{\partial l}\right| \leqslant 2\mu_1\frac{|t''-t'|^{\alpha}}{|x''-x'|}+2\mu_2|x''-x'|^{\beta}. \tag{3.14}$$

Letting x' be the vertex of the cone $\mathcal{K}$, we minimize the right side of (3.14) with respect to x'' over a segment of some ray l of length d belonging to $\mathcal{K} \in \overline{\Omega}$. As is easily calculated, for $|t''-t'|^{\alpha} \leq (\beta\mu_2/\mu_1)d^{1+\beta}$ this minimum is equal to

$$2\left(\beta^{\frac{1}{1+\beta}}+\beta^{-\frac{\beta}{1+\beta}}\right)\mu_1^{\frac{\beta}{1+\beta}}\mu_2^{\frac{1}{1+\beta}}|t''-t'|^{\frac{\alpha\beta}{1+\beta}} \equiv \mu|t''-t'|^{\frac{\alpha\beta}{1+\beta}},$$

i.e.,

$$\left|\frac{\partial u(x',t'')}{\partial l}-\frac{\partial u(x',t')}{\partial l}\right|\leqslant\mu\,|t''-t'|^{\frac{\alpha\beta}{1+\beta}}.$$

Since the ray l can be arbitrarily directed in K, this implies the validity of the lemma.

Let us now formulate some propositions concerning the spaces $W_q^{2l,l}(Q_T)$ and $H^{2l,l}(Q_T)$ used by us in Chapter IV. Their proofs can be found in [38b;49c;116b]. A definition of the spaces $W_q^{2l,l}(Q_T)$ for integral l and of the spaces $H^{l,l/2}(Q_T)$ for nonintegral l is given in §1 of Chapter I.

Lemma 3.2. *Suppose the domain Ω satisfies a cone condition (see Lemma 3.1). Then for any function $u\in H^{l,l/2}(Q_T)$ the inequality*

$$\langle u\rangle_{Q_T}^{(r)}\leqslant c_1\delta^{l-r}\langle u\rangle_{Q_T}^{(l)}+c_2\delta^{-r}|u|_{Q_T}^{(0)}.$$

is valid; here r and δ are arbitrary numbers from the intervals $(0,l]$ and $(0,\min\{d,\sqrt{T}\})$ respectively, where d is the altitude of the cone, and c_1 and c_2 are constants depending on r, l, n and on the angle at the vertex of the cone K.

Lemma 3.3. *Let Ω be the same as in Lemma 3.2. Then for any function $u(x,t)$ from $W_q^{2l,l}(Q_T)$ with integral l the inequality*

$$\|D_t^rD_x^su\|_{p,Q_T}\leqslant c_3\delta^{2l-2r-s-\left(\frac{1}{q}-\frac{1}{p}\right)(n+2)}\langle\langle u\rangle\rangle_{q,Q_T}^{(2l)}+c_4\delta^{-\left[2r-s+\left(\frac{1}{q}-\frac{1}{p}\right)(n+2)\right]}\|u\|_{q,Q_T}\tag{3.15}$$

is valid under the conditions $p\geq q$, $2l-2r-s-(1/q-1/p)(n+2)\geq 0$. In addition, if $2l-2r-s-(n+2)/q>0$, then for $0\leq\lambda<2l-2r-s-(n+2)/q$

$$\langle D_t^rD_x^su\rangle_{Q_T}^{(\lambda)}\leqslant c_5\delta^{2l-2r-s-\frac{n+2}{q}-\lambda}\langle\langle u\rangle\rangle_{q,Q_T}^{(2l)}+c_6\delta^{-\left(2r+s+\frac{n+2}{q}+\lambda\right)}\|u\|_{q,Q_T}$$

(for nonintegral $2l-2r-s-(n+2)/q$ this inequality is also valid for $\lambda=2l-2r-s-(n+2)/q$). The constant δ here must satisfy the condition

$$0<\delta\leqslant\min\{d;\sqrt{T}\},$$

and the constants c_3–c_6 depend on l, r, s, n, q and on the angle at the vertex of the cone K (cf. Lemma 3.1).

Spaces $W_q^{l,l/2}(\sigma_T)$ with nonintegral l will be used in Chapters IV and VII. Let us first define them for functions given in a cylindrical domain $\sigma_{T_1,T_2} = \sigma \times (T_1, T_2)$, where σ is some domain in $(n-1)$-dimensional euclidean space E_{n-1} and $-\infty \leq T_1 < T_2 \leq \infty$. The space $W_q^{l,l/2}(\sigma_T)$ consists of functions with finite norm

$$\|u\|_{q,\sigma_{T_1,T_2}}^{(l)} = \langle\langle u\rangle\rangle_{q,\sigma_{T_1,T_2}}^{(l)} + \sum_{(0\leq 2r+s<l)} \|D_t^r D_x^s u\|_{q,\sigma_{T_1,T_2}},$$

where

$$\langle\langle u\rangle\rangle_{q,\sigma_{T_1,T_2}}^{(l)} = \langle\langle u\rangle\rangle_{q,x,\sigma_{T_1T_2}}^{(l)} + \langle\langle u\rangle\rangle_{q,t,\sigma_{T_1T_2}}^{(l/2)},$$

$$\langle\langle u\rangle\rangle_{q,x,\sigma_{T_1,T_2}}^{(l)} = \sum_{(2r+s=[l])} \langle\langle D_t^r D_x^s u\rangle\rangle_{q,x,\sigma_{T_1,T_2}}^{(l-[l])},$$

$$\langle\langle u\rangle\rangle_{q,t,\sigma_{T_1,T_2}}^{(l/2)} = \sum_{(0<l-2r-s<2)} \langle\langle D_t^r D_x^s u\rangle\rangle_{q,t,\sigma_{T_1,T_2}}^{((l-2r-s)/2)} \tag{3.17}$$

and for $0<\alpha<1$

$$\langle\langle v\rangle\rangle_{q,x,\sigma_T}^{(\alpha)} = \left(\int_{T_1}^{T_2} dt \int_\sigma dx \int_\sigma |v(x,t)-v(y,t)|^q \frac{dy}{|x-y|^{n-1+q\alpha}}\right)^{\frac{1}{q}},$$

$$\langle\langle v\rangle\rangle_{q,t,\sigma_T}^{(\alpha)} = \left(\int_\sigma dx \int_0^T dt \int_0^T |v(x,t)-v(x,t')|^q \frac{dt'}{|t-t'|^{1+q\alpha}}\right)^{\frac{1}{q}}.$$

Finally, let us define spaces $W_q^{l,l/2}(S_T)$, and also $H^{l,l/2}(S_T)$ with nonintegral $l>0$, in which the elements are functions given on a curvilinear surface $S_T = S \times (0, T)$. As in the case of $W_q^l(S)$, these spaces are defined with the help of a parametrization of the surface S. Let S be covered by sets $S_1, \cdots, S_k$ in the manner described in §2 in the definition of the spaces $W_q^l(S)$. We introduce the following notation:

$$S_T^{(k)} = S_k \times (0, T),$$
$$\sigma_T = \sigma \times (0, T)$$

(σ is the canonical domain into which each set S_k is mapped). Let $u^{(k)}(z,t)$ ($z \in \sigma$) be the function into which the function $u(x,t)$ ($x \in S_k$) goes under the mapping of S_k into σ. We define the space $W_q^{l,l/2}(S_T)$ as the set of functions with finite norm

$$\|u\|_{q,S_T}^{(l)}=\left\{\sum_k\left(\|u^{(k)}(z,t)\|_{q,\sigma_T}^{(l)}\right)^q\right\}^{\frac{1}{q}}, \tag{3.18}$$

and the space $H^{l,l/2}(S_T)$ as the set of functions with finite norm

$$|u|_{S_T}^{(l)}=\sup_k|u^{(k)}|_{\sigma_T}^{(l)}. \tag{3.19}$$

The norms in the right sides of (3.18) and (3.19) are defined with the help of (3.16) of the present chapter and (1.10) of Chapter I. Under different coverings of S by the S_k the corresponding norms (3.18) and (3.19) turn out to be equivalent. Finally, we note that in the definition of $H^{l,l/2}(S_T)$ it is sufficient to require that the S_k be mapped onto σ by means of a transformation of class H^l.

The spaces $W_q^{l,l/2}(S_T)$ are needed to obtain exact results on the solvability of boundary value problems for linear parabolic equations in classes $W_q^{2m,m}(Q_T)$. This is connected with the fact that the differential properties of the boundary values of functions from classes $W_q^{2m,m}(Q_T)$ and of certain of their derivatives can be exactly described in terms of the spaces $W_q^{l,l/2}(S_T)$ with nonintegral l (namely, with $l=s-1/q$, where s is an integer). We have

Lemma 3.4. *If $u\in W_q^{2m,m}(Q_T)$, then for $2r+s<2m-2/q$*

$$D_t^rD_x^su\big|_{t=0}\in W_q^{2m-2r-s-\frac{2}{q}}(\Omega)\quad\text{and}\quad\|u\|_{q,\Omega}^{\left(2m-2r-s-\frac{2}{q}\right)}\leqslant c\,\|u\|_{q,Q_T}^{(2m)}.$$

In addition, for $2r+s<2m-1/q$

$$D_t^rD_x^su\big|_{S_T}\in W_q^{2m-2r-s-\frac{1}{q},\;m-r-\frac{s}{2}-\frac{1}{2q}}(S_T)$$

and

$$\|u\|_{q,S_T}^{\left(2m-2r-s-\frac{1}{q}\right)}\leqslant c\,\|u\|_{q,Q_T}^{(2m)}. \tag{3.20}$$

§4. ON AVERAGINGS AND CUTS OF ELEMENTS OF $L_q(\Omega)$, $L_{q,r}(Q_T)$ AND $V_2^{1,0}(Q_T)$

Let $u(x)$ be a function from $L_q(\Omega)$, $q\geq 1$. We will denote by $u^{(k)}(x)$ the function $u^{(k)}(x)=\max\{u(x)-k,\,0\}$, which obviously also belongs to $L_q(\Omega)$ for any k. It is easily seen that if u and v belong to $L_q(\Omega)$, then for almost all x from Ω we have the inequality

$$|u^{(k)}(x)-v^{(k)}(x)|\leqslant|u(x)-v(x)|, \tag{4.1}$$

one of the consequences of which is the well known

Lemma 4.1. *If a sequence of functions $u_p(x)$ from $L_q(\Omega)$, $p=1, 2, \cdots$, converges to a function $u(x)$ in the norm of $L_q(\Omega)$, then the sequence $u_p^{(k)}(x)$ converges to $u^{(k)}(x)$ in the norm of $L_q(\Omega)$.*

The following assertions are also valid (see, for example, Chapter II of [65q]).

Lemma 4.2. *If $u(x)\in L_q(\Omega)$ and has a derivative $u_{x_i}\in L_q(\Omega)$, then $u^{(k)}$ has the derivative $u^{(k)}_{x_i}\in L_q(\Omega)$; in particular, if $u(x)\in W_q^1(\Omega)$, then $u^{(k)}(x)\in W_q^1(\Omega)$. If, in addition,* $\operatorname{vrai\,max}_S u\le k_0$, *then for $k\ge k_0$ the function $u^{(k)}(x)$ belongs to $\overset{\circ}{W}{}_q^1(\Omega)$.*

Lemma 4.3. *Suppose that the functions $u(x)$, $u_1(x)$, $u_2(x)$, $\cdots$ belong to $L_q(\Omega)$ and have generalized derivatives $u_{x_i} u_{px_i}$, and $p=1, 2, \cdots$, from $L_q(\Omega)$ for some i. If $\|u_p-u\|_{q,\Omega}\to 0$ and $\|u_{px_i}-u_{x_i}\|_{q,\Omega}\to 0$ for $p\to\infty$, then the functions $u_p^{(k)}$ and $\partial u_p^{(k)}/\partial x_i$ converge in $L_q(\Omega)$ to $u^{(k)}$ and $\partial u^{(k)}/\partial x_i$ respectively.*

Let us now consider a function $u(x, t)$ belonging to $V_2^{1,0}(Q_T)$. The function $u^{(k)}(x, t)=\max\{u(x, t)-k;\ 0\}$ for any t from $[0, T]$ will belong to $L_2(\Omega)$. By virtue of inequality (4.1) we have

$$\|u^{(k)}(x,\ t+\Delta t)-u^{(k)}(x,\ t)\|_{2,\,\Omega}\leqslant\|u(x,t+\Delta t)-u(x,t)\|_{2,\Omega},$$

and therefore the functions $u^{(k)}(x, t)$ are continuous in t as elements of $L_2(\Omega)$. In addition, for almost all $t\in[0, T]$ they belong to $W_2^1(\Omega)$, with $\|u_x^{(k)}(x, t)\|_{2,\Omega}\le\|u_x(x, t)\|_{2,\Omega}$. In view of this we have

Lemma 4.4. *If $u(x, t)$ belongs to $V_2^{1,0}(Q_T)$, then $u^{(k)}(x, t)$ also belongs to $V_2^{1,0}(Q_T)$.*

From Lemma 4.3 and inequality (4.1) it is easy to deduce

Lemma 4.5. *Suppose the sequence $u_p(x, t)\in V_2^{1,0}(Q_T)$, $p=1, 2, \cdots$ converges to $u(x, t)\in V_2^{1,0}(Q_T)$ in the norm of $V_2^{1,0}(Q_T)$. Then $|u_p^{(k)}-u^{(k)}|_{Q_T}\to 0$ for $p\to\infty$.*

We note also the following simple fact.

Lemma 4.6. *If on some part Γ' of the surface S_T a function $u(x, t)$ from $V_2^{1,0}(Q_T)$ does not exceed k_0 (i.e.* $\operatorname{vrai\,max}_{\Gamma'} u\le k_0$*), then for $k\ge k_0$ the functions $u^{(k)}(x, t)\zeta(x, t)$ belong to $\overset{\circ}{V}{}_2^{1,0}(Q_T)$, where $\zeta(x, t)$ is an* *arbitrary smooth function that vanishes on* $S_T\setminus\Gamma'$.

On the other hand, if Γ' *is an arbitrary part of a surface* Γ_T *and* $\zeta(x, t)$ *vanishes on* $\Gamma_T\setminus\Gamma'$, *then* $u^{(k)}\zeta$ *for* $k \geq k_0$ *belongs to* $\overset{\circ}{V}{}_2^{1,0}(Q_T)$ *and vanishes for* $t = 0$.

We will consider the averagings

$$u_\rho(x) = \int_{|x-y|\leqslant\rho} \omega_\rho(|x-y|)\,u(y)\,dy = \int_{|x-y|\leqslant\rho} \rho^{-n}\omega\left(\frac{|x-y|}{\rho}\right) u(y)\,dy \tag{4.2}$$

with a sufficiently smooth nonnegative averaging kernel $\omega(|\xi|)$ that is equal to zero for $|\xi| \geq 1$ and is such that $\int_{|\xi|\leq 1}\omega(\xi)d\xi = 1$ (concerning them, see [113a] and [112]). If $u(x)$ is defined in a domain Ω, then an averaging $u_\rho(x)$ is clearly defined in any subdomain Ω' that is separated from the boundary S of Ω by a distance not less than ρ. Suppose $u(x)$ is an element of some Banach function space $B(\Omega)$. It is known that if the elements of $B(\Omega)$ are continuous with respect to a translation in the norm of $B(\Omega)$, then the averagings $u_\rho(x)$ for $\rho \to 0$ converge to $u(x)$ in the norm of $B(\Omega')$, where Ω' is any interior subdomain of Ω, since

$$\begin{aligned}\|u_\rho - u\|_{B(\Omega')} &= \left\| \int_{|z|\leqslant\rho} \omega_\rho(z)\,[u(x-z) - u(x)]\,dz \right\|_{B(\Omega')} \\ &\leqslant \int_{|z|\leqslant\rho} \omega_\rho(z)\,\|u(x-z) - u(x)\|_{B(\Omega')}\,dz \\ &\leqslant \max_{|z|\leqslant\rho} \|u(x-z) - u(x)\|_{B(\Omega')} \to 0\end{aligned}$$

for $\rho \to 0$.

In the sequel we need averagings not in all arguments but only in part of them. If, for example, an averaging is carried out only in the variables $x_1, x_2, \cdots x_m$, then the smoothed functions corresponding to it are defined in the domains Ω' belonging to Ω together with their translation by ρ in any direction lying in the plane $\{x_{m+1} = 0, \ldots, x_n = 0\}$, and if in the norm of $B(\Omega)$ there is continuity with respect to such a translation, then the averagings will converge in the norm of a space $B(\Omega')$ to u for $\rho \to 0$.

Thus we will deal with functions $u(x, t)$, $x = (x_1, \cdots, x_n)$, defined in a cylinder $\overline{Q}_T = \overline{\Omega} \times [0, T]$, and we will consider for them averagings in the variables x of the form

$$u_\rho(x, t) = \int\limits_{|x-y|\leq\rho} \omega_\rho(|x-y|)\, u(y, t)\, dy, \tag{4.3}$$

where $\omega_\rho(|\xi|)$ is the same averaging kernel as in (4.2), and averagings in t (Steklov averagings)

$$u_h(x, t) = \frac{1}{h}\int\limits_t^{t+h} u(x, \tau)\, d\tau. \tag{4.4}$$

It is clear that the first of them is defined for all sufficiently small $\rho > 0$ in cylinders of the form $\overline{Q}'_T = \overline{\Omega}' \times [0, T]$, where Ω' is some interior subdomain of Ω, while the second is defined in cylinders $\overline{Q}_{T-\delta} = \overline{\Omega} \times [0, T-\delta]$ for $h \leq \delta$. From what has been said above, it follows that if $u(x, t)$ belongs to some $L_{q,r}(Q_T)$ with q and $r \geq 1$, then the functions (4.3) and (4.4) converge to u in the norms of $L_{q,r}(Q'_T)$ and $L_{q,r}(Q_{T-\delta})$, respectively, for ρ and $h \to 0$. Then we may conclude that for certain sequences $\rho_k \to 0$ or $h_k \to 0$ they converge to $u(x, t)$ almost everywhere in Q'_T and $Q_{T-\delta}$, respectively.

Let us assume that $u(x, t)$ belongs to the space $V_2^{1,0}(Q_T)$. The elements of this space are continuous in the norm of $V_2^{1,0}(Q_T)$ with respect to a translation in the direction of the t-axis. This follows from the definition of $V_2^{1,0}(Q_T)$ and the properties of $L_2(Q_T)$. It is therefore possible to assert that for $h \to 0$ the functions (4.4) converge to $u(x, t)$ in the norm of $V_2^{1,0}(Q_{T-\delta})$, $\delta > 0$ (we note that $u_h \in W_2^{1,1}(Q_{T-\delta})$, $h \leq \delta$).

Let us formulate this in the form of a lemma.

Lemma 4.7. *If $u(x, t)$ belongs to class $V_2^{1,0}(Q_T)$, then the averagings $u_h(x, t)$ for $h \leq \delta$ belong to class $W_2^{1,1}(Q_{T-\delta})$, with*

$$|u_h - u|_{Q_{T-\delta}} \to 0 \quad \textit{for} \quad h \to 0.$$

If $u(x, t) \in L_{q,r}(Q_T)$, then the averagings u_h converge to u for $h \to 0$ in the norms of $L_{q,r}(Q_{T-\delta})$.

Let us now consider the averagings u_ρ of form (4.3) with infinitely differentiable kernel. We have

Lemma 4.8. *Let $u(x, t) \in V_2^{1,0}(Q_T)$. Then for an arbitrary cylinder $\overline{Q}'_T = \overline{\Omega}' \times [0, T]$ and $\rho \leq d'$, where $d' > 0$ is the distance from Ω' to A, the averagings $u_\rho(x, t)$ are continuous in $\overline{Q}'_T$, have continuous derivatives $D_x^l u_\rho$ of any order, and*

$$|u_\rho - u|_{Q'_T} \to 0 \quad \textit{for} \quad \rho \to 0.$$

On the other hand, if $u \in L_{q,r}(Q_T)$, *then the* u_ρ *converges to* u *for* $\rho \to 0$ *in the norms of* $L_{q,r}(Q_T)$.

The validity of the last assertion was noted above. Assuming that $u \in V_2^{1,0}(Q_T)$, let us estimate the expression

$$|u_\rho(x, t) - u_\rho(x', t')| = \rho^{-n} \left| \int_{|x-y|\leq\rho} \omega\left(\frac{|x-y|}{\rho}\right) u(y, t)\, dy - \int_{|x'-y|\leq\rho} \omega\left(\frac{|x'-y|}{\rho}\right) u(y, t')\, dy \right|$$

for fixed ρ, under the condition that the point $(x', t') \in Q'_T$ tends to $(x, t) \in Q'_T$. It does not exceed the sum

$$\rho^{-n} \int_\Omega \left| \omega\left(\frac{|x-y|}{\rho}\right) - \omega\left(\frac{|x'-y|}{\rho}\right) \right| |u(y, t)|\, dy + \rho^{-n} \int_\Omega \omega\left(\frac{|x'-y|}{\rho}\right) |u(y, t) - u(y, t')|\, dy,$$

in which the first term tends to zero for $x' \to x$ because of the continuity of the kernel $\omega(\xi)$, while the second tends to zero for $t' \to t$ by virtue of the continuity of the function $u(y, t)$ in t as an element of $L_2(\Omega)$, since

$$\rho^{-n} \int_\Omega \omega\left(\frac{|x'-y|}{\rho}\right) |u(y, t) - u(y, t')|\, dy \leq c(\rho) \| u(y, t) - u(y, t') \|_{2, \Omega}.$$

We have proved that the u_ρ are continuous in $\overline{Q}'_T$. The continuity in $\overline{Q}'_T$ of the derivatives $D_x^l u_\rho$ is verified in exactly the same way, in view of the infinite differentiability of the kernel $\omega(\xi)$.

Let us show that as $\rho \to 0$ the $u_\rho(x, t)$ converge to $u(x, t)$ in the norm of $V_2^{1,0}(Q'_T)$ for $u \in V_2^{1,0}(Q_T)$. By virtue of what has been stated above it is sufficient to establish that the norm $|u(x, t) - u(x + \xi, t)|_{Q'_T}$ tends to zero for $|\xi| \to 0$. This is true for $\|u_x(x, t) - u_x(x + \xi, t)\|_{2,Q'_T}$ by virtue of the membership of u_x in $L_2(Q_T)$. We will verify that it is also true for $\max_{0\leq t\leq T} \|u(x, t) - u(x + \xi, t)\|_{2,\Omega'}$.

For any finite set $t_1, \cdots, t_N \in [0, T]$ it is possible to satisfy the inequality

$$\max_{k=1,\ldots,N} \| u(x, t_k) - u(x+\xi, t_k)\|_{2,\Omega'} \leqslant \varepsilon, \tag{4.5}$$

where ϵ is an arbitrary positive number, as long as $|\xi|$ is taken sufficiently small. On the other hand, by virtue of the uniform continuity of $u(x, t)$ in t from $[0, T]$ as an element of $L_2(\Omega)$ it is possible to assert that

$$\max_{k=1,\ldots,N} \max_{t\in[t_{k-1}, t_k]} \| u(x, t) - u(x, t_k)\|_{2,\Omega'} \leqslant \varepsilon, \tag{4.6}$$

if the distances between adjacent points t_{k-1} and t_k, $k = 1, \cdots, N$, are sufficiently small.

From (4.5) and (4.6) it follows that

$$\max_{0\leqslant t\leqslant T} \| u(x, t) - u(x+\xi, t)\|_{2,\Omega'} \leqslant 3\varepsilon.$$

Thus it is proved that $|u(x, t) - u(x+\xi, t)|_{Q'_T} \to 0$ for $|\xi| \to 0$, from which follows the validity of Lemma 4.8 on the convergence of the $u_\rho(x, t)$ to $u(x, t)$ in the norm of $V_2(Q'_T)$ for $\rho \to 0$. Lemma 4.8 is proved.

The following simple proposition is useful.

Lemma 4.9. *If* $u(x, t) \in V_2^{1,0}(Q_T)$ *and* $\operatorname{vrai}\max_{S_T} u \le k_0$, *then for* $k \ge k_0$ *the functions* $u_h^{(k)}(x, t) = \max\{u_h(x, t) - k; 0\}$ *belong to* $\overset{\circ}{W}{}_2^{1,1}(Q_{T-\delta})$, $h \le \delta$.

Its validity follows from the inequality

$$\operatorname{vrai}\max_{S_{T-\delta}} u_h(x, t) \le \operatorname{vrai}\max_{S_T} u(x, t)$$

and Lemmas 4.6 and 4.7.

In several sections, in the course of investigating differential properties of solutions we will deal with difference ratios. In this connection it is necessary to bear in mind that for any functions $u(x)$ and $v(x)$

$$\frac{1}{\Delta x_k}\Delta(v(x)\,u(x)) = \frac{\Delta v(x)}{\Delta x_k}u(x) + v(x+\Delta x_k)\frac{\Delta u(x)}{\Delta x_k},$$
$$x+\Delta x_k = (x_1, \ldots, x_{k-1}, x_k+\Delta x_k, \ldots x_n), \tag{4.7}$$

and for a composite function $f(x, u(x))$

$$\frac{\Delta f(x, u(x))}{\Delta x_k} = \widehat{\frac{\partial f}{\partial u}}\frac{\Delta u(x)}{\Delta} + \widehat{\frac{\partial f}{\partial x_k}}, \tag{4.8}$$

where

$$\frac{\widehat{\partial f}}{\partial x_k} = \int_0^1 \frac{\partial f[x(1-\tau) + (x + \Delta_{x_k})\tau, u(x)(1-\tau) + u(x + \Delta x_k)\tau]}{\partial x_k} d\tau,$$

$$\frac{\widehat{\partial f}}{\partial u} = \int_0^1 \frac{\partial f[\dots]}{\partial u} d\tau,$$

and

$$\frac{\Delta u(x)}{\Delta x_k} = \frac{1}{\Delta x_k}[u(x + \Delta x_k) - u(x)].$$

In the sequel we will use the formula for summation by parts

$$\int_\Omega u(x)\frac{\Delta v(x)}{\Delta x_k} dx = -\int_\Omega \frac{\Delta u(x)}{\Delta x_k} v(x + \Delta x_k)\, dx, \tag{4.9}$$

which is valid for any functions $u \in L_m(\Omega)$, $v \in L_{m'}(\Omega)$, $1/m + 1/m' = 1$ one of which has a compact support in Ω and for all sufficiently small $|\Delta x_k|$.

Lemma 4.10. *If a function $u(x)$ from $L_m(\Omega)$ has a generalized derivative $u_{x_k} \in L_m(\Omega)$, then for any strictly interior subdomain Ω of Ω*

$$\left\| \frac{\Delta u}{\Delta x_k} - u_{x_k} \right\|_{m, \Omega'} \to 0 \quad \textit{for} \quad |\Delta x_k| \to 0.$$

Indeed, for almost all x for which $x, x + \Delta x_k \in \Omega$, the ratio $\Delta u/\Delta x_k$ is equal to $(1/\Delta x_k)\int_{x_k}^{x_k+\Delta x_k} u_\xi(x_1, \dots, x_{k-1}, \xi, x_{k+1}, \dots, x_n)\,d\xi$, i.e. coincides with a Steklov averaging (cf. (4.4)) of the derivative u_{x_k}, and therefore $\Delta u/\Delta x_k$ for $|\Delta x_k| \to 0$ converges to u_{x_k} in the norm of $L_m(\Omega')$.

Sometimes, instead of proving that a function $u(x, t)$ has a generalized derivative u_t from $L_2(Q_T)$, we will verify that the integrals $\int_{Q_{T-\Delta t}} (\Delta u/\Delta t)^2 dx\, dt$ are bounded uniformly in Δt. In this connection we will have in mind Lemma 4.11 (see for example [65q]):

Lemma 4.11. *Suppose that $u(x, t) \in L_2(Q_T)$ and that for all positive $\Delta t \leq h_0$ the integrals*

$$\int_{Q_{T-\Delta t}} \left| \frac{u(x, t + \Delta t) - u(x, t)}{\Delta t} \right|^2 dx\, dt \tag{4.10}$$

do not exceed some constant c. Then in Q_T there exists a generalized derivative u_t and

$$\| u_t \|^2_{2, Q_T} \leqslant c.$$

From the uniform boundedness of the integrals (4.10) it follows that for some sequence $\Delta_k t \to 0$, $k = 1, 2, \cdots$, the difference ratios $\Delta u / \Delta_k t$ converge weakly in $L_2(Q_{T-\delta})$, $\delta > 0$ to some function $w(x, t) \in L_2(Q_{T-\delta})$. We apply the formula for summation by parts (cf. (4.9))

$$\int_{Q_T} \frac{\Delta u(x, t)}{\Delta t} v(x, t)\, dx\, dt = - \int_{Q_T} u(x, t + \Delta t) \frac{\Delta v(x, t)}{\Delta t}\, dx\, dt$$

to $u(x, t)$ and an arbitrary smooth function $v(x, t)$ that has a compact support in $Q_{T-\delta}$, and then in this equality pass to the limit with respect to the chosen sequence $\Delta_k t \to 0$. As a result we obtain an equality from which it will follow that $w(x, t) = u_t(x, t)$ in $Q_{T-\delta}$ and, in view of the arbitrariness in the choice of $\delta > 0$, in all of Q_T.

It is not difficult to prove the following proposition.

Lemma 4.12. *In the subspace of the space* $\overset{\circ}{W}{}_2^{1,1}(Q_T)$ *consisting of elements of* $\overset{\circ}{W}{}_2^{1,1}(Q_T)$ *that are equal to zero for* $t = T$, *functions of the form* $\sum_{k=1}^N d_k(t)\psi_k(x)$ *are dense, where the* $d_k(t)$ *are arbitrarily smooth functions that are equal to zero for* $t = T$, *and the* $\{\psi_k(x)\}$, $k = 1, 2, \ldots$, *is a fundamental system of functions in the space* $\overset{\circ}{W}{}_2^1(\Omega)$.

§5. SOME OTHER AUXILIARY PROPOSITIONS

In this section we will cite some important lemmas of various kinds; part of them we will prove, but for the rest, the proof of which is given in the book [65q], we will confine ourselves to an exact reference.

Lemma 5.1. *Suppose there is defined in a ball* $K_\rho = \{x: |x| < \rho\}$ *a nonnegative function* $u(x) \in W_1^1(K_\rho)$, *with* $u(x) = 0$ *on some set* $\mathfrak{S}_0$ *of positive measure. Then for any measurable set* $\mathfrak{S}$ *from* K_ρ *the inequality*

$$\int_{\mathscr{E}} u(x)\mathfrak{N}(x)\, dx \leqslant \beta_1 \frac{\rho^n}{\operatorname{mes} \mathscr{E}_0} \operatorname{mes}^{\frac{1}{n}} \mathscr{E} \int_{K_\rho} |u_x(x)| \mathfrak{N}(x)\, dx \tag{5.1}$$

is valid, where $\beta_1 = (1 + n\kappa_n) 2^n n^{-1}$, *and* $\mathfrak{N}(x) = \mathfrak{N}(|x|)$ *is an arbitrary nonincreasing function of* $|x|$ *with values from* $[0, 1]$ *that is equal to unity on* $\mathfrak{S}_0$.

Consider the case $n \geq 2$. For almost all $x \in \mathfrak{S}$ and $x' \in \mathfrak{S}_0$ we have

$$u(x) = u(x) - u(x') = -\int_0^{|x'-x|} \frac{du(x+r\omega)}{dr}\,dr \leqslant \int_0^{|x'-x|} |u_y(y)|\,dr,$$

where $y = x + r\omega$ and $\omega = (x' - x)/|x' - x|$ is a unit vector issuing from the point x. We multiply both sides of this inequality by $\mathfrak{N}(x)$ and on the right we insert $\mathfrak{N}(x)$ under the integral sign, and replace it by $\mathfrak{N}(y)$. The inequality is not violated by doing this. Then we integrate both sides of it over $x' \in \mathfrak{E}_0$ and $x \in \mathfrak{E}$. This gives us

$$\operatorname{mes} \mathscr{E}_0 \int_{\mathscr{E}} u(x)\,\mathfrak{N}(x)\,dx \leqslant \int_{\mathscr{E}} dx \int_{\mathscr{E}_0} dx' \int_0^{|x'-x|} |u_y(y)|\,\mathfrak{N}(y)\,dr. \tag{5.2}$$

Setting $\mathfrak{N}(y)$ equal to zero outside of K_ρ, we estimate the integral over $\mathfrak{E}_0$ in the following manner:

$$\int_{\mathscr{E}_0} dx' \int_0^{|x'-x|} |u_y(y)|\,\mathfrak{N}(y)\,dr \leqslant \int_{\substack{|x'-x|\leqslant 2\rho \\ x'\in K_\rho}} dx' \int_0^{|x'-x|} \frac{|u_y(y)|\,\mathfrak{N}(y)}{|y-x|^{n-1}} |y-x|^{n-1}\,d|y-x|$$

$$\leqslant \int_0^{2\rho} |x'-x|^{n-1}\,d|x'-x| \int d\omega \int_0^{|x'-x|} \frac{|u_y(y)|\,\mathfrak{N}(y)}{|y-x|^{n-1}} \times |y-x|^{n-1}\,d|y-x| \leqslant \frac{(2\rho)^n}{n} \int_{K_\rho} \frac{|u_y(y)|\,\mathfrak{N}(y)}{|y-x|^{n-1}}\,dy.$$

From this and (5.2) we deduce

$$\operatorname{mes} \mathscr{E}_0 \int_{\mathscr{E}} u\mathfrak{N}\,dx \leqslant \frac{(2\rho)^n}{n} \int_{K_\rho} |u_y(y)|\,\mathfrak{N}(y)\,dy \max_{y\in K_\rho} \int_{\mathscr{E}} \frac{dx}{|x-y|^{n-1}} \tag{5.3}$$

The integral $j = \int_{\mathfrak{E}} dx/|x-y|^{n-1}$ for any point $y \in K_\rho$ does not exceed $(1 + n\kappa_n)\operatorname{mes}^{1/n}\mathfrak{E}$. In fact, the part of the integral corresponding to the domain of integration $\mathfrak{E} \cap \{x: |x - y| \leq \delta\}$ does not exceed $\int_{|x-y|\leq\delta} dx/|x-y|^{n-1} = n\kappa_n\delta$. The remaining part obviously does not exceed $\delta^{1-n}\operatorname{mes}\mathfrak{E}$. Let us take

$\delta = \operatorname{mes}^{1/n} \mathfrak{S}$. Then it is clear that $j \leq (1 + n\kappa_n)\operatorname{mes}^{1/n} \mathfrak{S}$, and hence from (5.3) we get the desired inequality (5.1) for $n \geq 2$.

In the case $n = 1$ inequality (5.1) and the even stronger inequality

$$\int_{\mathscr{E}} u(x)\,\mathfrak{N}(x)\,dx \leqslant \operatorname{mes} \mathscr{E} \int_{K_\rho} |u_x(x)|\,\mathfrak{N}(x)\,dx \tag{5.1'}$$

are deduced directly from the Newton-Leibnitz formula

$$u(x) = u(x) - u(x') = \int_{x'}^{x} u_y(y)\,dy.$$

Concerning $\mathfrak{S}_0$ it is sufficient here to assume only that it is not empty.

We will make use of two consequences of Lemma 5.1. The first of them is the inequality

$$\int_{K_\rho} u^2(x)\,\mathfrak{N}^2(x)\,dx \leqslant \beta_2 \left(\frac{\rho^n}{\operatorname{mes} \mathscr{E}_0}\right)^2 \rho^2 \int_{K_\rho} u_x^2(x)\,\mathfrak{N}^2(x)\,dx, \qquad \beta_2 = 4\beta_1^2 \varkappa_n^{\frac{2}{n}}, \tag{5.4}$$

which is valid for an arbitrary function $u(x)$ from $W_2^1(K_\rho)$ that is equal to zero on a set $\mathfrak{S}_0 \subset K_\rho$ for the same $\mathfrak{N}(x)$ as in (5.1).

The other consequence of (5.1) is the inequality

$$(l - k) \operatorname{mes} A_{l,\rho} \leqslant \beta \frac{\rho^{n+1}}{\operatorname{mes}(K_\rho \setminus A_{k,\rho})} \int_{A_{k,\rho} \setminus A_{l,\rho}} |u_x|\,dx, \qquad \beta = \beta_1 \varkappa_n^{\frac{1}{n}}. \tag{5.5}$$

It holds for any $u(x) \in W_1^1(K_\rho)$ and arbitrary values of l and k for $l > k$. Here and below $A_{k,\rho}$ is the set of points x from K_ρ for which $u(x) > k$.

Inequality (5.4) is easily deduced from inequality (5.1) if in (5.1) instead of $u(x)$ and $\mathfrak{N}(x)$ one takes the functions $u^2(x)$ and $\mathfrak{N}^2(x)$, and for the set $\mathfrak{S}$ one takes the whole ball K_ρ. For a proof of inequality (5.5) we take in (5.1) as $u(x)$ the function

$$\hat{u}(x)=\begin{cases} l-k, & \text{for} \quad x\in A_{l,\rho}, \\ u(x)-k, & \text{for} \quad x\in A_{k,\rho}\setminus A_{l,\rho}, \\ 0, & \text{for} \quad x\in K_\rho\setminus A_{k,\rho}, \end{cases}$$

and as the sets $\mathfrak{E}_0$ and $\mathfrak{E}$ the sets $K_\rho \setminus A_{k,\rho}$ and K_ρ respectively. The function $\mathfrak{N}(x)$ is set identically equal to 1. This gives

$$\int_{K_\rho} \hat{u}(x)\,dx \leqslant \beta_1 \frac{\rho^n}{\operatorname{mes}(K_\rho\setminus A_{k,\rho})} (\varkappa_n\rho^n)^{\frac{1}{n}} \int_{K_\rho} |\hat{u}_y|\,dy.$$

From it follows inequality (5.5) if one notes that $\int_{K_\rho} \hat{u}(x)\,dx = \int_{A_{k,\rho}} \hat{u}(x)\,dx \geq (l-k)\operatorname{mes} A_{l,\rho}$. But if one takes as $\mathfrak{E}$ the set $A_{l,\rho}$, then instead of (5.5) we obtain the inequality

$$(l-k)\operatorname{mes} A_{l,\rho} \leqslant \beta_1 \frac{\rho^n}{\operatorname{mes}(K_\rho\setminus A_{k,\rho})} \operatorname{mes}^{\frac{1}{n}} A_{l,\rho} \int_{A_{k,\rho}\setminus A_{l,\rho}} |u_x|\,dx \tag{5.6}$$

established by De Giorgi [17].

For $n=1$ the stronger inequality

$$\left(\frac{\operatorname{mes} A_{l,\rho}}{2\rho}\right)^{\varepsilon_1} \left(\frac{\operatorname{mes}(K_\rho\setminus A_{k,\rho})}{2\rho}\right)^{\varepsilon_2} (l-k) \leqslant \int_{A_{k,\rho}\setminus A_{l,\rho}} |u_x|\,dx \tag{5.6$'$}$$

is valid for arbitrary positive ϵ_1 and ϵ_2. Inequality (5.6′) follows directly from the Newton-Leibnitz formula.

Remark 5.1. Inequalities (5.1), (5.4)–(5.6) are valid not only for the balls K_ρ but also for any domains that are star-shaped with respect to the set $\mathfrak{E}_0$.

The following two lemmas apply to the case $n \geq 2$.

Lemma 5.2. *Suppose that in a domain Ω_R of diameter $2R$ there is defined a nonnegative function $v(x)$ such that for any ball K_ρ with center in Ω_R the estimate*

$$\int_{K_\rho\cap\Omega_R} v(x)\,dx \leqslant c\rho^{n-2+\alpha}, \quad \alpha>0$$

holds. Then for any function $\zeta(x)$ from $\overset{\circ}{W}{}_2^1(\Omega_R)$ the inequality

$$\int_{\Omega_R} v(x)\zeta^2(x)\,dx \leqslant c_1 R^\alpha \int_{\Omega_R} \zeta_x^2(x)\,dx \tag{5.7}$$

is valid, with the constant c_1 depending only on c, α and n.

A proof of this proposition is contained in Lemmas 4.3–4.4 of Chapter II of [65q]. Inequality (5.7) also holds for functions $\zeta(x)$ that vanish on only part of the boundary of Ω_R. In particular, we will make use of the following variant of Lemma 5.2.

Lemma 5.3. *Suppose that $\Omega_R = K_R \cap \Omega$ in the conditions of Lemma 5.2, where the boundary S of the domain Ω contains a plane part S_1 and the ball K_R does not intersect with $S \setminus S_1$. Then inequality (5.7) is valid for any function $\zeta(x)$ from $W_2^1(Q_R)$ that is equal to zero on the spherical part of the boundary of Ω_R. Here, as above, the constant c_1 in (5.7) depends only on c, α and n.*

Indeed, consider the domain $\widetilde{\Omega}_R$ that is symmetric to Ω_R with respect to S_1 and extend the functions $\zeta(x)$ and $v(x)$ onto $\widetilde{\Omega}_R$ in an even manner with respect to S_1. Then all the conditions of Lemma 5.2 will be fulfilled in the domain $\Omega_R \cup \widetilde{\Omega}_R$. Consequently, inequality (5.7) will be true for the domain $\Omega_R \cup \widetilde{\Omega}_R$ and hence also for the domain Ω_R.

For $n = 1$, instead of Lemmas 5.2 and 5.3 we will make use of the following proposition.

Lemma 5.3′. *If Ω_R is a segment of length $2R$ and $u(x)$ is a nonnegative function from $L_1(\Omega_R)$, then the inequality*

$$\int_{\Omega_R} u(x)\zeta^2(x)\,dx \leqslant c_1R\int_{\Omega_R}\zeta_x^2(x)\,dx, \quad c_1 = 2\|u\|_{1,\Omega_R} \tag{5.7′}$$

is true for any function $\zeta(x)$ from $W_2^1(\Omega_R)$ that is equal to zero at at least one point $x_0 \in \overline{\Omega}_R$.

Inequality (5.7′) follows from the fact that for any x from Ω_R

$$\zeta^2(x) = \left|\int_{x_0}^{x}\zeta_y(y)\,dy\right|^2 \leqslant 2R\int_{\Omega_R}\zeta_y^2\,dy.$$

The following simple proposition is useful.

Lemma 5.4. *Suppose $u(x)$ is a bounded function from $W_{2s+2}^1(\Omega)$, $s \geq 0$, and $\zeta(x)$ is a smooth function such that the product $u(x)\zeta(x)$ vanishes on the boundary S of the domain Ω. Then*

$$\int_\Omega |u_x|^{2s+2}\zeta^2\,dx \leqslant 16\operatorname{osc}^2\{u,\Omega\}\int_\Omega\left(c|u_x|^{2s-2}u_{xx}^2\zeta^2+|u_x|^{2s}\zeta_x^2\right)dx, \qquad c=n^2+s^2. \tag{5.8}$$

In order to prove this lemma (cf. Lemma 4.5 of Chapter II of [65q]) we transform the integral $\int_\Omega |u_x|^{2s+2}\zeta^2\,dx$ by means of an integration by parts and carry out elementary estimates with the help of inequality (1.2) of Chapter II:

$$\begin{aligned}\int_\Omega |u_x|^{2s+2}\zeta^2\,dx &= \int_\Omega |u_x|^{2s}u_{x_i}u_{x_i}\zeta^2\,dx\\ &= -\int_\Omega [u(x)-u(x_0)]\big[\Delta u|u_x|^{2s}\zeta^2+u_{x_i}2s|u_x|^{2s-2}u_{x_kx_i}u_{x_k}\zeta^2\\ &\quad+2u_{x_i}|u_x|^{2s}\zeta\zeta_{x_i}\big]dx\leqslant\int_\Omega\Big[\varepsilon|u_x|^{2s+2}\zeta^2\\ &\quad+\frac{(u-u(x_0))^2}{\varepsilon}|u_x|^{2s-2}n^2u_{xx}^2\zeta^2+\varepsilon|u_x|^{2s+2}\zeta^2\\ &\quad+\frac{(u-u(x_0))^2}{\varepsilon}|u_x|^{2s}\zeta_x^2+\varepsilon|u_x|^{2s+2}\zeta^2\\ &\quad+\frac{(u-u(x_0))^2s^2}{\varepsilon}|u_x|^{2s-2}u_{xx}^2\zeta^2\Big]dx,\end{aligned}$$

where x_0 is chosen so that the product $[u(x)-u(x_0)]\zeta(x)$ vanishes on S. Letting $\epsilon=1/4$, we arrive at estimate (5.8).

The following lemma is well known.

Lemma 5.5. *Suppose that a nonnegative function $y(t)$ is absolutely continuous on $[0, T]$, equals zero for $t=0$, and satisfies for almost all t the inequality*

$$\frac{dy(t)}{dt}\le c(t)y(t)+\mathcal{F}(t) \tag{5.9}$$

with nonnegative functions $c(t)$ and $\mathcal{F}(t)$ that are summable on $[0, \tau]$. Then

$$y(t)\leqslant\exp\left\{\int_0^t c(\tau)\,d\tau\right\}\int_0^t\mathcal{F}(\tau)\,d\tau, \tag{5.10}$$

$$\frac{dy(t)}{dt}\leqslant c(t)\exp\left\{\int_0^t c(\tau)\,d\tau\right\}\int_0^t\mathcal{F}(\tau)\,d\tau+\mathcal{F}(t). \tag{5.11}$$

In particular, for $c(t)=c$ and nondecreasing $\mathcal{F}(t)$

$$y(t) \leq \frac{\mathcal{F}(t)}{c}(e^{ct}-1), \tag{5.12}$$

$$\frac{dy(t)}{dt} \leq \mathcal{F}(t)\, e^{ct}. \tag{5.13}$$

For a proof of this lemma we multiply both sides of (5.9) by $\exp\{-\int_0^t c(\tau)d\tau\}$, transfer the first term on the right side to the left side, and write the result in the form of an inequality

$$\frac{d}{dt}\left[y(t)\exp\left\{-\int_0^t c(\tau)\,d\tau\right\}\right] \leqslant \mathcal{F}(t)\exp\left\{-\int_0^t c(\tau)\,d\tau\right\}.$$

Integrating it with respect to t from 0 to t_1, we obtain the inequality

$$y(t_1) \leqslant \int_0^{t_1} \mathcal{F}(t)\,dt \exp\left\{\int_0^{t_1} c(\tau)\,d\tau\right\}.$$

From it and (5.9) follow (5.10)–(5.13).

We distinguish in the form of separate lemmas the following two propositions on numerical sequences connected by recursion inequalities. The first of them is taken from [65q] (Lemma 4.7 of Chapter II).

Lemma 5.6. *Suppose a sequence* y_h, $h = 0, 1, 2, \cdots$ *of nonnegative numbers satisfies the recursion relation*

$$y_{h+1} \leqslant cb^h y_h^{1+\varepsilon}, \qquad h = 0, 1, \ldots, \tag{5.14}$$

with some positive constants c, ϵ *and* $b \geq 1$. *Then*

$$y_h \leqslant c^{\frac{(1+\varepsilon)^h-1}{\varepsilon}} b^{\frac{(1+\varepsilon)^h-1}{\varepsilon^2}-\frac{h}{\varepsilon}} y_0^{(1+\varepsilon)^h}. \tag{5.15}$$

In particular, if

$$y_0 \leqslant \theta = c^{-\frac{1}{\varepsilon}} b^{-\frac{1}{\varepsilon^2}} \text{ and } b > 1, \tag{5.16}$$

then

$$y_h \leqslant \theta b^{-\frac{h}{\varepsilon}} \tag{5.17}$$

and consequently $y_h \to 0$ *for* $h \to \infty$.

This is proved directly by induction.

Lemma 5.7. *Suppose nonnegative numbers* y_h *and* z_h, $h = 0, 1, 2, \cdots$, *are connected by the system of recursion inequalities*

$$\begin{aligned} y_{h+1} &\leqslant cb^h (y_h^{1+\delta} + z_h^{1+\varepsilon} y_h^{\delta}), \\ z_{h+1} &\leqslant cb^h (y_h + z_h^{1+\varepsilon}), \end{aligned} \tag{5.18}$$

where c, b, ϵ *and* δ *are certain fixed positive numbers, with* $b \geq 1$. *Then*

$$y_h \leqslant \lambda b^{-\frac{h}{d}}, \quad z_h \leqslant \left(\lambda b^{-\frac{h}{d}}\right)^{\frac{1}{1+\varepsilon}}, \tag{5.19}$$

where

$$d = \min\left\{\delta, \frac{\varepsilon}{1+\varepsilon}\right\},$$

$$\lambda = \min\left\{(2c)^{-\frac{1}{\delta}} b^{-\frac{1}{\delta d}};\ (2c)^{-\frac{1+\varepsilon}{\varepsilon}} b^{-\frac{1}{\varepsilon d}}\right\}, \tag{5.20}$$

as long as

$$y_0 \leqslant \lambda \ \text{and}\ z_0 \leqslant \lambda^{\frac{1}{1+\varepsilon}}. \tag{5.21}$$

Indeed, inequalities (5.19) are by condition valid for $h = 0$. Suppose they hold for y_h and z_h. Then on the basis of (5.18)

$$y_{h+1} \leqslant cb^h 2 \left(\lambda b^{-\frac{h}{d}}\right)^{1+\delta} = 2c\lambda^{1+\delta} b^{h\left(1-\frac{1+\delta}{d}\right)},$$

$$z_{h+1} \leqslant 2c\lambda b^{h\left(1-\frac{1}{d}\right)}.$$

But, as is easily calculated, the right sides of these inequalities do not exceed $\lambda b^{-(h+1)/d}$ and $(\lambda b^{-(h+1)/d})^{1/(1+\epsilon)}$ respectively, and hence inequalities (5.19) also hold for y_{h+1} and z_{h+1}. The lemma is proved.

The following lemma is very useful in proving the Hölder continuity of functions; it is analogous to Lemma 4.8 in Chapter II of [65q].

Lemma 5.8. *Suppose a function* $u(x, t)$ *is measurable and bounded in some cylinder* $Q_{\rho_0} = Q(\rho_0, \theta_0\rho_0^2)$. *Consider cylinders* $Q_\rho = Q(\rho, \theta_0\rho^2)$ *and* $Q_{b\rho}$ *that are coaxial with* Q_{ρ_0} *and have a common vertex (center of the top) with* Q_{ρ_0}, *where* $b > 1$ *is a fixed constant, and suppose that for any* $\rho \leq b^{-1}\rho_0$ *there is fulfilled for* $u(x, t)$ *at least one of the relations,*

$$\operatorname{osc}\{u;\ Q_\rho\} \leqslant c_1\rho^{\delta} \tag{5.22}$$

or

$$\operatorname{osc}\{u;\, Q_\rho\} \leqslant \eta \operatorname{osc}\{u;\, Q_{b\rho}\} \tag{5.23}$$

with certain positive constants c_1, $\delta \le 1$ *and* $\eta < 1$. *Then for* $\rho \le \rho_0$ *the estimate*

$$\operatorname{osc}\{u;\, Q_\rho\} \leqslant c\rho_0^{-\alpha}\rho^\alpha, \tag{5.24}$$

is valid, where

$$\alpha = \min\{-\ln_b \eta;\, \delta\}; \quad c = b^\alpha \max\{\omega_0;\, c_1\rho_0^\delta\};$$
$$\omega_0 = \operatorname{osc}\{u;\, Q_{\rho_0}\}.$$

Proof. Let us take a sequence of cylinders Q_{ρ_k}, $\rho_k = b^{-k}\rho_0$, $k = 0, 1, 2, \cdots$, of the type described above and denote by ω_k the oscillation of $u(x, t)$ in Q_{ρ_k}. From the hypotheses of the lemma it follows that

$$\omega_k \leqslant \max\{c_1\rho_k^\delta;\, \eta\omega_{k-1}\}, \qquad k = 1, 2, \ldots, \tag{5.25}$$

and

$$\omega_0 \leqslant cb^{-\alpha}.$$

From this and (5.25) we have for $y_k = b^{k\alpha}\omega_k$, $k = 1, 2, \cdots$, the estimates

$$\begin{aligned} y_n &\leqslant \max\{b^{k\alpha}c_1\rho_k^\delta;\, b^{k\alpha}\eta\omega_{k-1}\} \\ &= \max\{c_1 b^{k(\alpha-\delta)}\rho_0^\delta;\, b^\alpha\eta y_{k-1}\} \leqslant \max\{cb^{-\alpha};\, y_{k-1}\} \end{aligned} \tag{5.26}$$

and

$$y_0 = \omega_0 \leqslant cb^{-\alpha}.$$

From these estimates we see that for all $k = 0, 1, 2, \cdots$

$$y_k \leqslant cb^{-\alpha},$$

i.e.

$$\omega_k \leqslant cb^{-\alpha}b^{-k\alpha} = cb^{-\alpha}\left(\frac{\rho_k}{\rho_0}\right)^\alpha.$$

Let us now consider a cylinder Q_ρ with arbitrary $\rho \le \rho_0$. For some $k \ge 1$ we have

$$\rho_k \leqslant \rho \leqslant \rho_{k-1},$$

and therefore

$$\operatorname{osc}\{u;\, Q_\rho\} \leqslant \operatorname{osc}\{u;\, Q_{\rho_{k-1}}\} \leqslant cb^{-\alpha}\rho_0^{-\alpha}\rho_{k-1}^\alpha \leqslant c\rho_0^{-\alpha}\rho^\alpha,$$

i.e. the lemma is proved.

In obtaining estimates of Hölder norms for vector functions we will make use of a more general proposition:

Lemma 5.9. *Suppose that in a cylinder* $Q_{\rho_0} = Q(\rho_0, \theta_0\rho_0^2)$ *there are given measurable bounded functions* $u^1(x, t), \cdots, u^N(x, t)$ *and* $w^1(x, t), \cdots, w^{N_1}(x, t)$ *possessing the following properties: for any pair of cylinders* $Q_\rho = Q(\rho, \theta_0\rho^2)$ *and* $Q_{b\rho}$ *having a common vertex and axis with* Q_{ρ_0}, *where* $b > 1$ *is a fixed constant and* $b\rho \le \rho_0$, *there exists a function* $w^r(x, t)$ *such that*

$$\operatorname{osc}\{w^r; Q_{b\rho}\} \geqslant \delta_1 \max_{l=1,\dots,N} \operatorname{osc}\{u^l; Q_{b\rho}\}, \tag{5.27}$$

and at least one of the inequalities

$$\operatorname{osc}\{w^r; Q_\rho\} \leqslant c_1\rho^\delta \tag{5.28}$$

and

$$\operatorname{osc}\{w^r; Q_\rho\} \leqslant \eta \operatorname{osc}\{w^r; Q_{b\rho}\} \tag{5.29}$$

is valid. Here b, δ_1, c_1, δ *and* η *are fixed contants, and* $\eta < 1$. *Then the estimates*

$$\operatorname{osc}\{u^l; Q_\rho\} \leqslant c\rho_0^{-\alpha}\rho^\alpha, \quad l = 1, \dots, N, \tag{5.30}$$

hold for $\rho \le \rho_0$, *where*

$$\left.\begin{aligned} \alpha &= \frac{1}{N_1}\min\{-\ln_b\eta;\ \delta\}, \\ c &= b^{\alpha(N_1+1)}\delta_1^{-1}\max\{c_1\rho_0^\delta;\ \omega_0 b^{\alpha N_1}\}, \\ \omega_0 &= \max_{l=1,\dots,N_1} \operatorname{osc}\{w^l; Q_{\rho_0}\}. \end{aligned}\right\} \tag{5.31}$$

The proof of this lemma is analogous to the proof of the preceding lemma. We will not reproduce it here; instead we refer the reader to [65q, Chapter II, § 4].

Remark 5.2. In the conditions and assertions of Lemmas 5.8 and 5.9 all of the cylinders can be replaced by their intersections with some fixed domain Q.

Finally, let us prove a proposition that permits us to estimate $\int_{Q_T} \exp[bu^{(0)}(x, t)]\,dx\,dt$ for generalized solutions of linear equations.

Lemma 5.10. *Suppose a function* $u(x, t)$ *belongs to* $V_2(Q_T)$, *is bounded from above on* S_T, *and for all* $k \ge \hat{k} = \operatorname{vrai\,max}_{S_T} u$ *satisfies the inequalities*

$$|u^{(k)}|_{Q_T} \leqslant \gamma\left[k\mu^{\frac{1+\varkappa}{r}}(k) + \mu^{\frac{1}{r}}(k)\right], \tag{5.32}$$

where $\mu(k) = \int_0^T \operatorname{mes}^{r/q} A_k(t)\,dt$, $A_k(t)$ *is the set of points* x *from* Ω *at which*

$u(x, t) > k$, γ and κ are positive constants, and the values of q and r are subject to conditions (3.3). Then $u^{(0)} \in L_{q_1}(Q_T)$ with any $q_1 > 1$. Moreover, there exist positive numbers b and B determined by T, $\operatorname{mes}\Omega$, γ, $\hat{k}$, q, r and κ such that

$$\int_{Q_T} e^{bu^{(0)}(x,t)}\,dx\,dt \leqslant B. \tag{5.33}$$

Remark 5.3. In conditions (3.3) r can be infinite for any n, while q can be infinite for $n = 1$. In these cases we assume that $\mu^{1/r}(k) = (\int_0^T \operatorname{mes}^{r/q} A_k(t)dt)^{1/r} = \operatorname{vrai\,max}_{t\in[0,T]} \operatorname{mes}^{1/q} A_k(t)$ for $r = \infty$ and

$$\mu^{\frac{1}{r}}(k) = \left(\int_0^T \operatorname{mes}^{\frac{r}{q}} A_k(t)\,dt\right)^{\frac{1}{r}} = \operatorname{mes}^{\frac{1}{r}}\{t \in [0, T] : \operatorname{mes} A_k(t) > 0\}$$

for $q = \infty$.

For a proof of the lemma let us consider a sequence of levels

$$k_h = k_{h-1} + k_0 \geqslant \hat{k}, \quad h = 1, 2, \ldots,$$

where k_0 is a positive number that will be chosen below, and the sequence of quantities corresponding to them is

$$z_h = \mu^{\frac{1}{r}}(k_h), \quad h = 0, 1, \ldots$$

From (5.32) on the basis of inequality (3.4) we obtain the recursion inequalities for z_h

$$\begin{aligned}(k_{h+1} - k_h)\, z_{h+1} &\leqslant \| u^{(k_h)} \|_{q,r,Q_T} \\ &\leqslant \beta \,| u^{(k_h)} |_{Q_T} \leqslant \beta\gamma(k_h z_h^{1+\varkappa} + z_h),\end{aligned}$$

from which it follows that

$$z_{h+1} \leqslant \beta\gamma\left[(h+1)\, z_h^{1+\varkappa} + k_0^{-1} z_h\right]. \tag{5.34}$$

Let us show that for sufficiently large k_0

$$z_h \leqslant \lambda e^{-h}, \quad h = 0, 1, \ldots, \tag{5.35}$$

where λ is some constant. Indeed, if (5.35) holds for z_h, then for z_{h+1} we will have by virtue of (5.34) the estimate

$$z_{h+1}\leqslant\beta\gamma e\left[(h+1)\lambda^{\varkappa}e^{-\varkappa h}+k_0^{-1}\right]\lambda e^{-(h+1)},$$

from which will follow the desired estimate (5.35) for z_{h+1} as long as

$$\beta\gamma e\left[(h+1)\lambda^{\varkappa}e^{-\varkappa h}+k_0^{-1}\right]\leqslant 1. \tag{5.36}$$

The function $(h+1)e^{-\kappa h}$ for all $h\geq 0$ does not exceed $1/\kappa e^{\kappa-1}$, and hence, in order to satisfy (5.36), we choose k_0 and λ from the condition

$$\beta\gamma\left(\frac{1}{\varkappa}\lambda^{\varkappa}e^{\varkappa}+k_0^{-1}e\right)\leqslant 1;$$

for example, we take

$$k_0\geqslant 2\beta\gamma e,\quad \lambda=\left(\frac{\varkappa}{2\beta\gamma e^{\varkappa}}\right)^{\frac{1}{\varkappa}}. \tag{5.37}$$

For such k_0 and λ inequality (5.35) will hold for all $h=0,1,2,\cdots$ as long as it is valid for $h=0$, i.e., if $z_0\leq\lambda$. But for z_0 we have the estimates

$$\begin{aligned}z_0=\left(\int_0^T \operatorname{mes}^{\frac{r}{q}} A_{k_0}(t)\,dt\right)^{\frac{1}{r}} &\leqslant (k_0-\hat{k})^{-1}\|u^{(\hat{k})}\|_{q,r,Q_T}\\ &\leqslant (k_0-\hat{k})^{-1}\beta\,|u^{(\hat{k})}|_{Q_T}\\ &\leqslant (k_0-\hat{k})^{-1}\beta\gamma\left[\hat{k}\operatorname{mes}^{\frac{1+\varkappa}{q}}\Omega T^{\frac{1+\varkappa}{r}}+\operatorname{mes}^{\frac{1}{q}}\Omega T^{\frac{1}{r}}\right],\end{aligned}$$

so that the inequality $z_0\leq\lambda$ is satisfied if we set

$$k_0=\max\left\{2\beta\gamma e;\ \hat{k}+\beta\gamma\lambda^{-1}\left[\hat{k}\operatorname{mes}^{\frac{1+\varkappa}{q}}\Omega T^{\frac{1+\varkappa}{r}}+\operatorname{mes}^{\frac{1}{q}}\Omega T^{\frac{1}{r}}\right]\right\}. \tag{5.38}$$

Thus (5.35) is proved. Let us now consider the quantities

$$\operatorname{mes}Q_{h+1}=\operatorname{mes}\{(x,t)\in Q_T: u(x,t)>k_{h+1}\}=\int_0^T \operatorname{mes}A_{k_{h+1}}(t)\,dt.$$

By virtue of

$$(k_{h+1}-k_h)\max_{0\leqslant t\leqslant T}\operatorname{mes}^{\frac{1}{2}}A_{k_{h+1}}(t)\leqslant\max_{0\leqslant t\leqslant T}\|u^{(k_h)}(x,t)\|_{2,\Omega}\leqslant|u^{(k_h)}|_{Q_T}$$

it follows from (5.32), (5.35) and (5.37) that

$$\operatorname{mes} Q_{h+1} \leqslant k_0^{-2} T \, |u^{(k_h)}|_{Q_T}^2$$
$$\leqslant k_0^{-2}\gamma^2 \left[(h+1)\,k_0(\lambda e^{-h})^{1+\varkappa} + \lambda e^{-h}\right]^2 T \leqslant \lambda^2 \beta^{-2} T e^{-2(h+1)}.$$

Because of this inequality

$$\int_{Q_T} e^{bu^{(0)}(x,t)}\,dx\,dt \leqslant \sum_{h=0}^{\infty} \int_{Q_h \setminus Q_{h+1}} e^{bu(x,t)}\,dx\,dt + e^{bk_0} \operatorname{mes} Q_T$$
$$\leqslant \sum_{h=0}^{\infty} e^{bk_{h+1}} \operatorname{mes} Q_h + e^{bk_0} \operatorname{mes} Q_T$$
$$\leqslant \lambda^2 \beta^{-2} T \sum_{h=0}^{\infty} e^{b(h+2)k_0 - 2h} + e^{bk_0} \operatorname{mes} Q_T.$$

Thus for $b < 2k_0^{-1}$ the integral $\int_{Q_T} e^{bu^{(0)}(x,t)}\,dx\,dt$ will not exceed the number $B = \lambda^2 T e^{2bk_0}/\beta^2[1 - e^{-(2-k_0 b)}] + e^{bk_0} \operatorname{mes} Q_T$. The lemma is proved.

Remark 5.4. The assertion of Lemma 5.10 remains in force if instead of the fulfilment of inequalities (5.32) one requires that the inequalities

$$|u^{(k)}|_{Q_T} \leqslant \gamma \left[\sum_{i=1}^{N} \mu_i^{\frac{1+\varkappa_i}{r_i}}(k)\, k^{\delta_i} + \sum_{i=1}^{N} \mu_i^{\frac{1}{r_i}}(k) \right],$$

be fulfilled for $k \geq \hat{k}$, where $\mu_i(k) = \int_0^T \operatorname{mes}^{r_i/q_i} A_k(t)\,dt$ (see Remark 5.3), $\kappa_i > 0$, $\delta_i < 1 + \kappa_i$, and the q_i and r_i are subject to conditions (3.3).

The proof of this assertion is carried out in exactly the same way as the proof stated above, except that a system of recursion inequalities for $z_h^{(i)} = \mu_i^{1/r_i}(k_h)$ is obtained instead of inequality (5.34).

Remark 5.5. If the requirement of boundedness for $\operatorname{vrai\,max}_{S_T} u$ is dropped from the conditions of Lemma 5.10, inequality (5.33) is retained, but the constants b and B in it will also depend on the form of the domain. Namely, this dependence will occur through the constant c from inequality (3.8), whereas above under the conditions of Lemma 5.10 we could use inequality (3.4), in which the constant β does not depend on the domain.

§6. ON ESTIMATES OF $\max|u|$. THE CLASS $\mathfrak{A}(Q_T, \gamma, r, \hat{k}, \kappa)$

In this section the boundedness of functions $u(x, t)$ satisfying certain inequalities will be proved. To such inequalities are subject the solutions $u(x, t)$ of linear and quasi-linear equations with principal part in divergence form and their derivatives u_x if it is understood that the functions forming the equations satisfy certain conditions. These inequalities in general estimate the growth of the quantities $|u^{(k)}|_{Q(\rho-\sigma_1\rho, \tau-\sigma_2\tau)}$ in terms of k, the measure of the set where $u^{(k)} > 0$, the weaker norm $\|u^{(k)}\|_{2,Q(\rho,\tau)}$, $\sigma_1\rho$ and $\sigma_2\tau$. We begin with a more simple proposition, which permits one to give a total estimate for the maximum of the modulus of solutions in the whole domain of definition.

Let $u(x, t)$ belong to $V_2(Q_T)$. We denote by $A_k(t)$ the set of points $x \in \Omega$ at which $u(x, t) > k$. The following theorem is valid.

Theorem 6.1. *Suppose that* $\operatorname{vrai\,max}_{S_T} u \leq \hat{k}$, $\hat{k} \geq 0$, *and that the inequalities*

$$|u^{(k)}|_{Q_T} \leqslant \gamma k \mu^{\frac{1+\kappa}{r}}(k) \tag{6.1}$$

hold for $k \geq \hat{k}$ *with certain positive constants* γ *and* κ. *Here* $\mu(k) = \int_0^T \operatorname{mes}^{r/q} A_k(t)dt$, *while* q *and* r *are arbitrary numbers satisfying conditions* (3.3).[1] *Then*

$$\operatorname*{vrai\,max}_{Q_T} u(x, t) \leqslant 2\hat{k}\left[1 + 2^{\frac{2}{\kappa}+\frac{1}{\kappa^2}}(\beta\gamma)^{1+\frac{1}{\kappa}} T^{\frac{1+\kappa}{r}} \operatorname{mes}^{\frac{1+\kappa}{q}} \Omega\right], \tag{6.2}$$

where β *is the constant in inequality* (3.4).

Proof. Let us take the sequence of levels $k_h = M(2 - 2^{-h})$, $h = 0, 1, \cdots$, assuming $M \geq \hat{k} > 0$).[2] It is clear that

$$(k_{h+1} - k_h)\mu^{\frac{1}{r}}(k_{h+1}) \leqslant \|u^{(k_h)}\|_{q,r,Q_T}. \tag{6.3}$$

On the other hand, by virtue of inequalities (3.4) and (6.1)

$$\|u^{(k_h)}\|_{q,r,Q_T} \leqslant \beta|u^{(k_h)}|_{Q_T} \leqslant \beta\gamma k_h \mu^{\frac{1+\kappa}{r}}(k_h) \tag{6.4}$$

and therefore

---

1) For $z = \infty$ and $r = \infty$ the quantity $\mu^{1/r}(l)$ is defined in the same way as in Remark 5.3.

2) For $\hat{k} = 0$ the theorem is obvious.

$$\mu^{\frac{1}{r}}(k_{h+1}) \leqslant \frac{\beta\gamma k_h}{k_{h+1}-k_h}\mu^{\frac{1+\varkappa}{r}}(k_h) \leqslant 4\beta\gamma 2^h \mu^{\frac{1+\varkappa}{r}}(k_h). \tag{6.5}$$

From here and from Lemma 5.6 it follows that $\mu^{1/r}(k_h)$ will tend to zero for $h \to \infty$ if $\mu^{1/r}(k_0)$ is sufficiently small, namely if

$$\mu^{\frac{1}{r}}(k_0) = \mu^{\frac{1}{r}}(M) \leqslant (4\beta\gamma)^{-\frac{1}{\varkappa}} 2^{-\frac{1}{\varkappa^2}}. \tag{6.6}$$

In order to satisfy inequality (6.6), we take $M = m\hat{k}$, $m > 1$, and substitute $\hat{k}$ for k_h and M for k_{h+1} in (6.3). This and (6.4) give

$$\mu^{\frac{1}{r}}(M) \leqslant \frac{\beta\gamma}{m-1}\mu^{\frac{1+\varkappa}{r}}(\hat{k}) \leqslant \frac{\beta\gamma}{m-1} T^{\frac{1+\varkappa}{r}} \operatorname{mes}^{\frac{1+\varkappa}{q}} \Omega.$$

Hence we see that for

$$m = 1 + \beta\gamma T^{\frac{1+\varkappa}{r}} \operatorname{mes}^{\frac{1+\varkappa}{q}} \Omega (4\beta\gamma)^{\frac{1}{\varkappa}} 2^{\frac{1}{\varkappa^2}}$$

condition (6.6) with $M = m\hat{k}$ will be fulfilled, and consequently $\mu(2M)$ will be equal to zero, i.e.

$$\operatorname*{vrai\,max}_{Q_T} u \leqslant 2M = 2m\hat{k}.$$

The latter coincides with estimate (6.2).

Remark 6.1. The assertion of the theorem remains in force if instead of inequalities (6.1) the function $u(x, t)$ satisfies the inequalities

$$|u^{(k)}|_{Q_T} \leqslant \gamma \sum_{i=1}^{N} \mu_i^{\frac{1+\varkappa_i}{r_i}}(k)\, k^{\delta_i}, \tag{6.7}$$

where

$$\varkappa_i > 0, \quad \delta_i < 1 + \varkappa_i, \quad \mu_i = \int_0^T \operatorname{mes}^{\frac{r_i}{q_i}} A_k(t)\, dt,$$

while the parameters q_i and r_i are subject to conditions (3.3). This is proved in the same way as Theorem 6.1.

Remark 6.2. In the conditions of Theorem 6.1 it is possible to drop the requirement of boundedness for $\operatorname{vrai\,max}_{S_T} u$ and to assume that inequalities (6.1) are fulfilled for k larger than a certain $\hat{k} \geq 0$. Estimate (6.2) will then be valid with the constant c from (3.8) in place of β from (3.4). Indeed, the functions $u^{(k_h)}$ in this case are in general not equal to zero on S_T, and hence in deducing (6.4) it is necessary to use inequality (3.8) in place of (3.4).

Let us turn to local estimates for $\operatorname{vrai\,max} u$. Suppose we have a cylinder $Q(\rho_0, \tau_0) = \{|x - x_0| < \rho_0, t_0 - \tau_0 < t < t_0\}$ and a family of cylinders of the form

$$Q(\rho-\sigma_1\rho,\ \tau-\sigma_2\tau) = \{|x-x_0|<\rho-\sigma_1\rho,\ t_0-(1-\sigma_2)\tau<t<t_0\},$$

that are coaxial with and have a common vertex with $Q(\rho_0, \tau_0)$ and are such that

$$\frac{\rho_0}{2}\leqslant\rho-\sigma_1\rho\leqslant\rho\leqslant\rho_0,\quad \frac{\tau_0}{2}\leqslant\tau-\sigma_2\tau\leqslant\tau\leqslant\tau_0. \tag{6.8}$$

We denote by $A_{k,\rho}(t)$ the set of points from the ball $\{|x-x_0|<\rho\}$ at which $u(x,t)>k$, and by $\mu(k,\rho,\tau)$ the integral $\int_{t_0-\tau}^{t_0}\operatorname{mes}^{r/q}A_{k,\rho}(t)dt$. The parameters q and r, as above, are assumed to be subject to conditions (3.3), and for $r=\infty$ it is assumed that

$$\mu^{\frac{1}{r}}(k,\rho,\tau)=\operatorname*{vrai\,max}_{\tau_0-\tau\leqslant t\leqslant t_0}\operatorname{mes}^{\frac{1}{q}}A_{k,\rho}(t),$$

while for $q=\infty$ it is assumed that $\mu^{1/r}(k,\rho,\tau)$ is equal to

$$\operatorname{mes}^{\frac{1}{r}}\{t\in[t_0-\tau,\ t_0]:\ \operatorname{mes}A_{k,\rho}(t)>0\}.$$

We will say that *a function* $u(x,t)$ *from* $V_2(Q(\rho_0,\tau_0))$ *belongs to class* $\mathfrak{A}(Q(\rho_0,\tau_0),\gamma,r,\hat{k},\kappa)$ if for any $k\geq\hat{k}$ and all possible pairs of cylinders $Q(\rho,\tau)$ and $Q(\rho-\sigma_1\rho,\tau-\sigma_2\tau)$ of the above-mentioned form the inequalities

$$|u^{(k)}|^2_{Q(\rho-\sigma_1\rho,\ \tau-\sigma_2\tau)}\leqslant\gamma\left\{[(\sigma_1\rho)^{-2}+(\sigma_2\tau)^{-1}]\,\|u^{(k)}\|^2_{2,\,Q(\rho,\,\tau)}+k^2\rho_0^{-\varkappa n}\mu^{\frac{2(1+\varkappa)}{r}}(k,\rho,\tau)\right\} \tag{6.9}$$

are satisfied for certain positive constants γ and κ.

Let us now consider the case when the cylinder $Q(\rho_0,\tau_0)$ intersects with a part Γ_T of the boundary of the cylinder Q_T. *We denote by* $\mathfrak{A}(Q(\rho_0,\tau_0)\cap Q_T,\gamma,r,\hat{k},\kappa)$ *the class of functions from* $V_2(Q(\rho_0,\tau_0)\cap Q_T)$, that satisfy, for $k\geq\hat{k}\geq 0$ and arbitrary cylinders $Q(\rho,\tau)$, $Q(\rho-\sigma_1\rho,\tau-\sigma_2\tau)$ of the above-mentioned form, the inequalities

$$\begin{aligned}|u^{(k)}|^2_{Q(\rho-\sigma_1\rho,\ \tau-\sigma_2\tau)\cap Q_T}&\leqslant\gamma\Big\{[(\sigma_1\rho)^{-2}+(\sigma_2\tau)^{-1}]\,\|u^{(k)}\|^2_{2,\,Q(\rho,\,\tau)\cap Q_T}\\&\quad+k^2\rho_0^{-\varkappa n}\mu^{\frac{2(1+\varkappa)}{r}}(k,\rho,\tau)\Big\}.\end{aligned} \tag{6.10}$$

Here $\mu(k,\rho,\tau)=\int_{\max\{0,t_0-\tau\}}^{t_0}\operatorname{mes}^{r/q}A_{k,\rho}(t)dt$ is defined in the same way as above; it is only necessary to understand by $A_{k,\rho}(t)$ the set of points from the intersection of the ball $\{|x-x_0|\leq\rho\}$ with Ω at which $u(x,t)>k$. The parameters γ, r, q and κ are the same as in (6.9).

The following theorem holds.

Theorem 6.2. *For any functions $u(x, t)$ from $\mathfrak{A}(Q(\rho_0, \tau_0), \gamma, r, \hat{k}, \kappa)$ one has the estimate*

$$\operatorname*{vrai\,max}_{Q\left(\frac{\rho_0}{2}, \frac{\tau_0}{2}\right)} u(x, t) \leqslant 2c\left\{\rho_0^{-\frac{n+2}{2}} \|u\|_{2, Q(\rho_0, \tau_0)}\left[1 + \tau_0^{-\frac{1}{2}}\rho_0\right] + \hat{k}\left[1 + (\tau_0\rho_0^{-2})^{\frac{1+\kappa}{r}}\right]\right\}, \tag{6.11}$$

in which the constant c is determined only by the quantities γ, q, r and κ. An analogous estimate holds for functions of class $\mathfrak{A}(Q(\rho_0, \tau_0) \cap Q_T, \gamma, r, \hat{k}, \kappa)$, it being only necessary to replace in (6.11) the cylinders $Q(\rho_0, \tau_0)$ and $Q(\rho_0/2, \tau_0/2)$ by their intersections with Q_T. In this connection the constant c depends only on γ, q, r and κ if $\operatorname{vrai\,max}_{Q(\rho_0, \tau_0) \cap S_T} u(x, t) \leq \hat{k}$, while in the general case it also depends on S.

Let us verify the validity of the first part of Theorem 6.2. The second part is proved analogously.

We first of all make a change of variables: $x - x_0 = \rho_0 \tilde{x}$, $t - t_0 = \rho_0^2 \tilde{t}$. In the new variables $\tilde{x}$ and $\tilde{t}$ the inequalities (6.9), after a cancellation by ρ_0^n, take the form

$$|u^{(k)}|^2_{Q(\tilde{\rho} - \sigma_1\tilde{\rho}, \tilde{\tau} - \sigma_2\tilde{\tau})} \leqslant \gamma\left\{[(\sigma_1\tilde{\rho})^{-2} + (\sigma_2\tilde{\tau})^{-1}]\|u^{(k)}\|^2_{2, Q(\tilde{\rho}, \tilde{\tau})} + k^2\mu^{\frac{2(1+\kappa)}{r}}(k, \tilde{\rho}, \tilde{\tau})\right\}, \tag{6.12}$$

where $\tilde{\rho} = \rho\rho_0^{-1}$ and $\tilde{\tau} = \tau\rho_0^{-2}$, while conditions (6.8) take the form

$$\frac{1}{2} \leqslant \tilde{\rho} - \sigma_1\tilde{\rho} \leqslant \tilde{\rho} \leqslant 1, \quad \frac{\theta}{2} \leqslant \tilde{\tau} - \sigma_2\tilde{\tau} \leqslant \tilde{\tau} \leqslant \theta \equiv \tau_0\rho_0^{-2}, \tag{6.13}$$

so that in the new variables the cylinder $Q(1, \theta) = \{|\tilde{x}| < 1, -\theta < \tilde{t} < 0\}$ corresponds to the cylinder $Q(\rho_0, \tau_0)$, while the cylinders $Q(\tilde{\rho}, \tilde{\tau}) = \{|\tilde{x}| < \tilde{\rho}, -\tilde{\tau} < \tilde{t} < 0\}$, correspond to the cylinders $Q(\rho, \tau)$.

In the sequel we will omit the wave over x, ρ, t, etc. in writing the formulas.

Let us take a sequence of decreasing cylinders

$$Q_h = Q(\rho_h, \tau_h) \text{ with } \rho_h = \frac{1}{2} + \frac{1}{2^{h+2}}, \; \tau_h = \frac{\theta}{2} + \frac{1}{2^{h+2}},$$

$h = 0, 1, 2, \cdots$, and a sequence of increasing levels $k_h = M + N(1 - 1/2^h)$, $h = 0, 1, 2, \cdots$, where M and N are certain positive numbers.

As a preliminary we will establish the following lemma.

Lemma 6.1. *If a function $u(x, t)$ satisfies inequalities (6.12) for all k lying in the range $[M, M + N]$, then for the quantities*

$$y_h = N^{-2} \int_{-\tau_h}^{0} \int_{A_{k_h, \rho_h}(t)} (u - k_h)^2 \, dx \, dt$$

and

$$z_h = \mu^{\frac{2}{r}} (k_h, \rho_h, \tau_h) \equiv \mu_h^{\frac{2}{r}}, \quad h = 0, 1, \ldots,$$

with the k_h and ρ_h indicated above there is valid the system of recursion inequalities

$$y_{h+1} \leqslant \gamma_1 2^{4h} \left[y_h^{1+\delta} + z_h^{1+\varkappa} y_h^{\delta} \left(\frac{M}{N} + 1 \right)^2 \right], \tag{6.14}$$

$$z_{h+1} \leqslant \gamma_1 2^{2h} \left[2^{2h} y_h + z_h^{1+\varkappa} \left(\frac{M}{N} + 1 \right)^2 \right], \tag{6.15}$$

where $\gamma_1 = 2^6 \beta^2 [\gamma (2^5 + \theta^{-1}) + 2^5]$, $\delta = 2/n + 2$, and β is the constant in (3.4).

After we prove this lemma it will be possible, by choosing sufficiently large and equal numbers M and N, to conclude with the help of relations (6.14)–(6.15) that the y_h and z_h tend to zero for $h \to \infty$. This gives the desired estimate: $\operatorname{vrai\,max}_{Q(\rho_0/2, \tau_0/2)} u \leq 2M$.

Thus, let us turn to the proof of Lemma 6.1. Take the sequence of continuous cut functions $\zeta_h(|x|)$ that are equal to 1 for $|x| \leq \rho_{h+1}$, equal to zero for $|x| \geq \bar{\rho}_h = \frac{1}{2}(\rho_h + \rho_{h+1})$, and linear on the segment $|x| \in [\rho_{h+1}, \bar{\rho}_h]$, so that $|\zeta_{hx}| \leq 2^{h+4}$. Let

$$\lambda_h = \int_{-\tau_{h+1}}^{0} \operatorname{mes} A_{k_{h+1}, \bar{\rho}_h}(t) \, dt.$$

We estimate the y_{h+1}, using inequality (3.7), in the following manner:

$$y_{h+1} \leqslant N^{-2} \int_{-\tau_{h+1}}^{0} \int_{A_{k_{h+1}, \bar{\rho}_h}(t)} (u - k_{h+1})^2 \zeta_h^2 \, dx \, dt$$

$$\leqslant \beta^2 N^{-2} \lambda_h^{\frac{2}{n+2}} \left| u^{(k_{h+1})} \zeta_h \right|^2_{Q(\bar{\rho}_h, \tau_{h+1})} \leqslant$$

$$\leqslant \beta^2 N^{-2}\lambda_h^{\delta}\left\{ \operatorname*{vrai\,max}_{-\tau_{h+1}\leqslant t\leqslant 0} \int_{A_{k_{h+1},\bar\rho_h}(t)} (u-k_{h+1})^2\zeta_h^2\,dx + 2\int_{-\tau_{h+1}}^{0}\int_{A_{k_{h+1},\bar\rho_h}(t)} \left[u_x^2\zeta_h^2+(u-k_{h+1})^2\zeta_{hx}^2\right]dx\,dt\right\}$$
$$\leqslant 2\beta^2\lambda_h^{\delta}\left[N^{-2}\left|u^{(k_{h+1})}\right|^2_{Q(\bar\rho_h,\tau_{h+1})}+4^{h+4}y_h\right]. \tag{6.16}$$

The right side of (6.16) is estimated by means of the obvious inequality

$$\lambda_h\leqslant (k_{h+1}-k_h)^{-2}N^2y_h \tag{6.17}$$

and inequality (6.12) is applied to the functions $u^{(k_{h+1})}$ and the two cylinders $Q(\bar\rho_h,\tau_{h+1})$ and $Q(\rho_h,\tau_h)$. This gives

$$y_{h+1}\leqslant 2\beta^2\left[(k_{h+1}-k_h)^{-1}N\right]^{2\delta}y_h^{\delta}\left\{\gamma\left[(4^{h+4}+\theta^{-1}2^{h+3})\times N^{-2}\left\|u^{(k_{h+1})}\right\|^2_{2,Q(\rho_h,\tau_h)}+k_{h+1}^2N^{-2}\mu_h^{\frac{2}{r}(1+\varkappa)}\right]+4^{h+4}y_h\right\}$$
$$\leqslant 2^{-2}\gamma_1\left[(k_{h+1}-k_h)^{-1}N\right]^{2\delta}\left[2^{2h}y_h^{1+\delta}+k_{h+1}^2N^{-2}z_h^{1+\varkappa}y_h^{\delta}\right]. \tag{6.18}$$

Taking into account the choice of the k_h, it is easy to see that inequalities (6.14) follow from (6.18). In order to prove inequalities (6.15) we note that

$$(k_{h+1}-k_h)^2z_{h+1}=(k_{h+1}-k_h)^2\mu_{h+1}^{\frac{2}{r}}$$
$$\leqslant\left\|u^{(k_h)}\zeta_h\right\|^2_{q,r,Q(\bar\rho_h,\tau_{h+1})}\leqslant\beta^2\left|u^{(k_h)}\zeta_h\right|^2_{Q(\bar\rho_h,\tau_{h+1})}. \tag{6.19}$$

Estimating the right side by means of (6.12), in the same way as was done in (6.16)–(6.18), we obtain

$$(k_{h+1}-k_h)^2z_{h+1}\leqslant 2\beta^2 4^{h+4}N^2y_h + 2\beta^2\gamma\left\{\left[(\bar\rho_h-\rho_h)^{-2}+(\tau_{h+1}-\tau_h)^{-1}\right]\left\|u^{(k_h)}\right\|^2_{2,Q(\rho_h,\tau_h)} + k_h^2\mu_h^{\frac{2(1+\varkappa)}{r}}\right\}\leqslant 2\beta^2\gamma\left[(4^{h+4}(1+\gamma^{-1})+\theta^{-1}2^{h+3})N^2y_h+k_h^2z_h^{1+\varkappa}\right]. \tag{6.20}$$

From this, by virtue of the choice of the k_h, follow inequalities (6.15). Lemma 6.1 is proved.

Let us continue with the proof of Theorem 6.2. Put $N=M$. From the system of inequalities (6.14)–(6.15) it is possible to conclude on the basis of Lemma 5.7

that the quantities y_h and z_h tend to zero for $h \to \infty$ as long as the values of y_0 and z_0 are sufficiently small, namely, if

$$y_0 \leqslant \min\left\{(2c)^{-\frac{1}{\delta}} b^{-\frac{1}{d\delta}};\ (2c)^{-\frac{1+\varkappa}{\varkappa}} b^{-\frac{1}{\varkappa d}}\right\} \equiv \gamma_3,$$
$$z_0 \leqslant \gamma_3^{\frac{1}{1+\varkappa}}, \tag{6.21}$$

where $d = \min\{\delta;\ \kappa/(1+\kappa)\}$, $b = 16$ and $c = 4y_1$. We achieve the fulfilment of inequalities (6.21) by choosing a large number M. In fact, for y_0 we have the estimate

$$y_0 = M^{-2} \int_{-\theta}^{0} \int_{A_{M,1}} (u - M)^2\, dx\, dt \leqslant M^{-2} \|u\|^2_{2, Q(1,\theta)}. \tag{6.22}$$

In order to estimate $\hat{z}_0$ we note that inequality (6.20) holds for $h = -1$ if one assumes that $k_{-1} = \hat{k}$, $\rho_{-1} = 1$, $\tau_{-1} = \theta$. This gives

$$z_0 \leqslant \frac{\beta^2\gamma}{(M-\hat{k})^2}\left\{(4^3(1+\gamma^{-1}) + \theta^{-1}2^2)\|u^{(\hat{k})}\|^2_{2, Q(1,\theta)}\right.$$
$$\left. + \hat{k}^2\left[\mu(M,\ 1,\ \theta)\right]^{\frac{2(1+\varkappa)}{r}}\right\} \leqslant \frac{\beta^2\gamma}{(M-\hat{k})^2}$$
$$\times\left[(4^3(1+\gamma^{-1}) + \theta^{-1}2^2)\|u\|^2_{2, Q(1,\theta)} + \hat{k}^2\theta^{2\frac{(1+\varkappa)}{r}} \varkappa_n^{\frac{2(1+\varkappa)}{q}}\right]. \tag{6.23}$$

From (6.22) and (6.23) it follows that for

$$M \geqslant \max\left\{\gamma_3^{-\frac{1}{2}}\|u\|_{2, Q(1,\theta)};\ \hat{k} + \gamma_3^{-\frac{1}{2(1+\varkappa)}}\beta\gamma^{\frac{1}{2}}\right.$$
$$\left.\times\left[(4^3(1+\gamma^{-1}) + \theta^{-1}2^2)\|u\|^2_{2, Q(1,\theta)} + \hat{k}^2\theta^{\frac{2(1+\varkappa)}{r}} \varkappa_n^{\frac{2(1+\varkappa)}{q}}\right]^{\frac{1}{2}}\right\} \tag{6.24}$$

requirements (6.21) will be satisfied. Thus, for example, if one takes

$$M = c\left[\|u\|_{2, Q(1,\theta)}\left(1 + \theta^{-\frac{1}{2}}\right) + \hat{k}\left(1 + \theta^{\frac{1+\varkappa}{r}}\right)\right],$$
$$c = \max\left\{\gamma_3^{-\frac{1}{2}} + \gamma_3^{-\frac{1}{2(1+\varkappa)}}\beta\gamma^{\frac{1}{2}}2^3;\ 1 + \gamma_3^{-\frac{1}{2(1+\varkappa)}}\beta\gamma^{\frac{1}{2}}\varkappa_n^{\frac{1+\varkappa}{q}}\right\}, \tag{6.25}$$

then the y_h and z_h tend to zero for $h \to \infty$ and, consequently,

$$\operatorname*{vrai\,max}_{Q\left(\frac{\rho_0}{2},\, \frac{\tau_0}{2}\right)} u(x,\ t) \leqslant 2M.$$

Theorem 6.2 is proved.

Remark 6.3. The last term in inequality (6.9) can be replaced by the term of more general form

$$\sum_{i=1}^{N} k^{2(1+\delta_i)}\rho_0^{-\varkappa_i n}\mu_i^{\frac{2(1+\varkappa_i)}{r_i}}(k, \rho, \tau),$$

where $\mu_i = \int_{t_0-\tau}^{t_0} \operatorname{mes}^{r_i/q_i} A_{k,\rho}(t)\,dt$, $\delta_i \le \min\{\delta(1+\kappa_i); \kappa_i\}$, $\delta = 2/(n+2)$, and γ, κ_i, q_i and r_i are fixed positive parameters, with the q_i and r_i being subject to conditions (3.3). Also, the restrictions (6.8) can be replaced by the requirement that

$$\rho_0 - \sigma_1^0\rho_0 \leqslant \rho - \sigma_1\rho \leqslant \rho \leqslant \rho_0,$$
$$\tau_0 - \sigma_2^0\tau \leqslant \tau - \sigma_2\tau \leqslant \tau \leqslant \tau_0$$

with some fixed positive numbers σ_1^0 and σ_2^0. In this connection an estimate analogous to (6.11) is valid for $\operatorname{vrai}\max u(x, t)$.
$$Q(\rho_0-\sigma_1^0\rho_0,\ \tau_0-\sigma_2^0\tau_0)$$

The same remark holds for cylinders intersecting with Γ_T. All of these assertions are proved by the same method that was used in the proof of Theorem 6.2.

Remark 6.4. It is not difficult to verify that a function $u(x, t)$ from $V_2(Q_T)$ belongs to class $\mathfrak{B}(Q(\rho_0, \tau_0) \cap Q_T, \gamma, r, \hat{k}, \kappa)$ for any cylinder $Q(\rho_0, \tau_0)$ having its vertex in $\overline{Q}_T$ if $u(x, t)$ for all $k \geq \hat{k}$ and $0 \leq t_0 - \tau \leq t_0 \leq T$ satisfies the inequalities

$$\begin{aligned}\operatorname*{vrai\,max}_{t_0-\tau\leqslant t\leqslant t_0} \| u^{(k)}(x, t)\zeta(x, t)\|_{2,\Omega}^2 + \int_{t_0-\tau}^{t_0}\int_\Omega |u_x^{(k)}|^2\zeta^2\,dx\,dt \\ \leqslant \gamma_1\Bigg\{\int_{t_0-\tau}^{t_0}\int_\Omega (\zeta_x^2+\zeta|\zeta_t|)|u^{(k)}|^2\,dx\,dt \\ + k^2\Bigg[\int_{t_0-\tau}^{t_0}\Bigg(\int_{A_k(t)}\zeta\,dx\Bigg)^{\frac{r}{q}}dt\Bigg]^{\frac{2(1+\varkappa)}{r}}\Bigg\},\end{aligned} \tag{6.26}$$

in which γ_1 and κ are fixed positive numbers, r and q are subject to conditions (3.3), and $\zeta(x, t)$ is an arbitrary piecewise-smooth continuous nonnegative function such that $u^{(k)}(x, t)\zeta(x, t)$ is equal to zero on the lateral surface and lower base of the cylinder $Q(\rho_0, \tau_0)$ and $\zeta(x, t) \leq 1$.

For $q = \infty$ it is assumed that

$$\left(\int_{A_k(t)} \zeta\, dx \right)^{\frac{1}{q}} = \begin{cases} 1 & \text{for} \quad \operatorname{mes} A_k(t) > 0, \\ 0 & \text{for} \quad \operatorname{mes} A_k(t) = 0, \end{cases}$$

while for $r = \infty$ the last term in (6.26) is understood to mean

$$\operatorname*{vrai\,max}_{t_0 - \tau \leqslant t \leqslant t_0} \left(\int_{A_k(t)} \zeta\, dx \right)^{\frac{2(1+\varkappa)}{q}}.$$

§7. THE CLASS $\mathfrak{B}_2(Q_T, M, \gamma, r, \delta, \kappa)$

In this and the next sections we consider classes of bounded-in-modulus functions satisfying inequalities similar to the inequalities of the preceding section. Concerning the elements $u(x, t)$ of these classes, it is proved that they are Hölder continuous and an estimate is given of the Hölder norm $|u|^{(\alpha)}$. To such classes belong the solutions and the derivatives of solutions of linear and quasi-linear equations with principal part in divergence form.

We will say that *a function* $u(x, t)$ *belongs to class* $\mathfrak{B}_2(Q_T, M, \gamma, r, \delta, \kappa)$ if $u(x, t) \in V_2^{1,0}(Q_T)$, [1] $\operatorname{vrai\,max}_{Q_T} |u| \leq M$ and the functions $u(x, t) = \pm\, u(x, t)$ satisfy the inequalities

$$\max_{t_0 \leqslant t \leqslant t_0+\tau} \|w^{(k)}(x, t)\|^2_{2, K_{\rho - \sigma_1 \rho}} \leq \|w^{(k)}(x, t)\|^2_{2, K_\rho} + \gamma \left[(\sigma_1 \rho)^{-2} \|w^{(k)}\|^2_{2, Q(\rho, \tau)} + \mu^{\frac{2}{r}(1+\varkappa)}(k, \rho, \tau) \right] \tag{7.1}$$

and

$$\|w^{(k)}\|^2_{Q(\rho - \sigma_1 \rho, \tau - \sigma_2 \tau)} \leqslant \gamma \left\{ \left[(\sigma_1 \rho)^{-2} + (\sigma_2 \tau)^{-1} \right] \|w^{(k)}\|^2_{2, Q(\rho, \tau)} + \mu^{\frac{2}{r}(1+\varkappa)}(k, \rho, \tau) \right\}, \tag{7.2}$$

in which $w^{(k)}(x, t) = \max\{u(x, t) - k;\, 0\}$; $Q(\rho, \tau) = K_\rho \times (t_0, t_0 + \tau) = \{|x - x_0| < \rho,\ t_0 < t < t_0 + \tau\}$ is an arbitrary cylinder belonging to Q_T; ρ and τ are arbitrary positive numbers; σ_1 and σ_2 are arbitrary numbers from the interval $(0, 1)$;

---

[1] Instead of membership in the class $V_2^{1,0}(Q_T)$, in essence it is required only that $u \in V_2(Q_T)$ and that the norm $\|u(x, t)\|_{2,\Omega}$ be finite for each t from $[0, T]$.

$\mu(k, \rho, \tau) = \int_{t_0}^{t_0+\tau} \operatorname{mes}^{r/q} A_{k,\rho}(t)\,dt$, with $A_{k,\rho}(t)$ being the set of points x from K_ρ at which $u(x, t) > k$; M, γ, q, r, δ. and κ are fixed positive numbers, with q and r satisfying the conditions

$$\left.\begin{array}{llll} \frac{1}{r}+\frac{n}{2q}=\frac{n}{4}, & & & \\ q\in\left(2, \frac{2n}{n-2}\right], & r\in[2, \infty) & \text{for} & n\geqslant 3, \\ q\in(2, \infty), & r\in(2, \infty) & \text{for} & n=2, \\ q\in(2, \infty], & r\in[4, \infty) & \text{for} & n=1; \end{array}\right\} \tag{7.3}$$

and k is an arbitrary number subject only to the condition

$$\operatorname*{vrai\,max}_{Q(\rho,\tau)} w(x, t) - k \leqslant \delta. \tag{7.4}$$

In the case $q = \infty$ (for $n = 1$) by $\mu(k, \rho, \tau)$ we understand $\operatorname{mes}\{t \in (t_0, t_0 + \tau);\ \operatorname{mes} A_{k,\rho}(t) > 0\}$.

Remark 7.1. In the definition of class $\mathfrak{B}_2$ one could in (7.1) and (7.2) consider the number γ to be different in front of all five terms of the right sides, with the factor in the first summand of (7.1) being necessarily close to unity, and in this connection observe a greater differentiation in the dependences of the constants calculated below on the γ_i, $i = 1, \cdots, 5$. But this does not interest us, and therefore we have put one and the same γ in (7.1) and (7.2). Furthermore, all of the results do not change essentially if one imposes on ρ and τ in (7.1) and (7.2) the additional restrictions $\rho \le \rho_0$, $\tau \le \tau_0$, where ρ_0 and τ_0 are fixed numbers.

The basic aim of the present section is a proof of the embeddability of $\mathfrak{B}_2(Q_T, M, \gamma, r, \delta, \kappa)$ in $H^{\alpha,\alpha/2}(Q_T)$ with some positive α determined only by the parameters M, γ, r, δ and κ.

Remark 7.2. It is easily verified that inequalities (7.1) and (7.2) follow from the inequalities

$$\begin{aligned} &\|w^{(k)}(x, t_0+\tau)\zeta(x, t_0+\tau)\|_{2,K_\rho}^2 + \nu\|w_x^{(k)}\zeta\|_{2,Q(\rho,\tau)}^2 \\ &\quad\leqslant \|w^{(k)}(x, t_0)\zeta(x, t_0)\|_{2,K_\rho}^2 \\ &\quad + \gamma_1\Bigg\{\int_{Q(\rho,\tau)} (\zeta_x^2+\zeta|\zeta_t|)(w^{(k)})^2\,dx\,dt \\ &\qquad + \left[\int_{t_0}^{t_0+\tau}\left(\int_{A_{k,\rho}(t)} \zeta\,dx\right)^{\frac{r}{q}} dt\right]^{\frac{2}{r}(1+\kappa)}\Bigg\} \end{aligned} \tag{7.5}$$

in which the same notation as above is used and $\zeta(x, t)$ is an arbitrary piecewise-smooth continuous function, with values between 0 and 1, that is equal to zero on the lateral surface of the cylinder $Q(\rho, \tau)$. For $q=\infty$ the last integral is understood in the same way as indicated in Remark 6.4.

As will be shown in the following chapters, the solutions of linear and quasilinear equations, and also their derivatives with respect to x, belong to classes $\mathfrak{B}_2$ or certain generalizations of them that will be defined below. This circumstance permits one to conclude that all of these functions belong to Hölder spaces and to obtain the corresponding estimates for them

We will first prove four auxiliary propositions. They will consist of assertions on the existence of certain constants. It is essential that they all be independent of the element $u(x, t)$ in question in class $\mathfrak{B}_2$ and that there be defined only n and the numerical parameters entering into the definition of class $\mathfrak{B}_2$.

Lemma 7.1. *Suppose a function $u(x, t)$ satisfies inequalities (7.1) and $\operatorname{mes} A_{k,\rho}(t_0) \le (1/p)\operatorname{mes} K_\rho = (\kappa_n/p)\rho^n$, $p>1$. Then for any ξ from $(\sqrt{1/p}, 1)$ it is possible to find positive numbers $\theta(\xi)$ and $b(\xi)$ such that if*

$$\delta \geqslant H = \operatorname*{vrai\,max}_{\substack{x\in K_\rho \\ t_0<t<t_0+\theta(\xi)\rho^2}} u(x, t) - k > \rho^{\frac{n\varkappa}{2}},$$

then

$$\operatorname{mes}(K_\rho \setminus A_{k+\xi H,\rho}(t)) \geqslant b(\xi)\varkappa_n\rho^n \tag{7.6}$$

for

$$t \in [t_0, t_0+\theta(\xi)\rho^2].$$

By the conditions of the lemma we have

$$\int_{A_{k,\rho-\sigma_1\rho}(t)} (u(x, t)-k)^2\,dx \leqslant \frac{1}{p}H^2\varkappa_n\rho^n$$
$$+\gamma\left\{H^2\sigma_1^{-2}(t-t_0)\varkappa_n\rho^{n-2} + \left[(t-t_0)(\varkappa_n\rho^n)^{\frac{r}{q}}\right]^{\frac{2}{r}(1+\varkappa)}\right\}.$$

On the other hand

$$(\xi H)^2 \operatorname{mes} A_{k+\xi H,\rho-\sigma_1\rho}(t) \leqslant \int_{A_{k,\rho-\sigma_1\rho}(t)} (u-k)^2\,dx. \tag{7.7}$$

Therefore

$$\operatorname{mes} A_{k+\xi H,\, \rho-\sigma_1\rho}(t) \leqslant \frac{1}{p\xi^2}\varkappa_n\rho^n + \gamma\Big\{(\sigma_1\xi)^{-2}(t-t_0)\rho^{-2} + (\xi H)^{-2}\varkappa_n^{\frac{2}{q}(1+\varkappa)-1}(t-t_0)^{\frac{2}{r}(1+\varkappa)}\rho^{n\frac{2}{q}(1+\varkappa)-n}\Big\}\varkappa_n\rho^n. \tag{7.8}$$

For $t \leq t_0 + \theta(\xi)\rho^2$, taking into account the assumption $H > \rho^{n\kappa/2}$, we obtain from (7.8) the inequality

$$\operatorname{mes} A_{k+\xi H,\, \rho-\sigma_1\rho}(t) \leqslant \xi^{-2}\Big\{p^{-1} + \gamma\Big[\sigma_1^{-2}\theta(\xi) + \varkappa_n^{\frac{2}{q}(1+\varkappa)-1}\theta(\xi)^{\frac{2}{r}(1+\varkappa)}\Big]\Big\}\varkappa_n\rho^n;$$

here we have made use of the fact that q and r are connected by equality (7.3). For any $\xi > \sqrt{1/p}$ it is obviously possible to select positive numbers σ_1, $\theta(\xi)$ and $b(\xi)$ so that

$$n\sigma_1 + \xi^{-2}\Big\{p^{-1} + \gamma\Big[\sigma_1^{-2}\theta(\xi) + \varkappa_n^{\frac{2}{q}(1+\varkappa)-1}\theta(\xi)^{\frac{2}{r}(1+\varkappa)}\Big]\Big\} = 1 - b(\xi) < 1$$

and hence

$$\operatorname{mes} A_{k+\xi H,\, \rho}(t) \leqslant \operatorname{mes} A_{k+\xi H,\, \rho-\sigma_1\rho}(t) + \operatorname{mes} K_\rho \setminus K_{\rho-\sigma_1\rho} \leqslant (1-b(\xi))\operatorname{mes} K_\rho.$$

The lemma is proved.

For the subsequent arguments let us fix p, ξ, $\theta(\xi)$ and $b(\xi)$, for example, as follows: $p = 2$, $\xi = 3/4$, $b(\xi) = 1/18$ and $\theta(\xi) = \theta$ is a positive solution of the equation

$$\gamma\Big\{(36n)^2\theta + \varkappa_n^{\frac{2}{q}(1+\varkappa)-1}\theta^{\frac{2}{r}(1+\varkappa)}\Big\} = \frac{1}{64}.$$

In addition, we introduce "standard cylinders," namely, cylinders of the form

$$Q_\rho = \{|x - x_0| \leq \rho,\ t_0 \leq t \leq t_0 + \theta\rho^2\}$$

with arbitrarily situated "vertex" $(x_0, t_0 + \theta\rho^2)$.

Remark 7.3. Lemma 7.1 is also valid for the intersections of the cylinders considered in it with some fixed cylinder Q_T. It is only necessary in all of the conditions and assertions to take in place of the balls K_ρ their intersections with Ω.

We introduce one more notation: $Q_\rho(k)$ is the set of points (x, t) from Q_ρ at which $u(x, t) > k$. It is easy to prove the following relations between the measure of $Q_\rho(k)$ and the quantity $\mu(k, \rho, \theta\rho^2)$, relative to one and the same cylinder Q_ρ:

$$\left.\begin{aligned}(\varkappa_n\rho^n)^{\frac{2}{q}-\frac{2}{r}}\operatorname{mes}^{\frac{2}{r}} Q_\rho(k) &\leqslant \mu^{\frac{2}{r}}(k,\ \rho,\ \theta\rho^2)\\ &\leqslant (\theta\rho^2)^{\frac{2}{r}-\frac{2}{q}}\operatorname{mes}^{\frac{2}{q}} Q_\rho(k),\quad \text{for}\ \ r\leqslant q,\\ (\varkappa_n\rho^n)^{\frac{2}{q}-\frac{2}{r}}\operatorname{mes}^{\frac{2}{r}} Q_\rho(k) &\geqslant \mu^{\frac{2}{r}}(k,\ \rho,\ \theta\rho^2)\\ &\geqslant (\theta\rho^2)^{\frac{2}{r}-\frac{2}{q}}\operatorname{mes}^{\frac{2}{q}} Q_\rho(k),\quad \text{for}\ \ r\geqslant q.\end{aligned}\right\} \tag{7.9}$$

Indeed, the right inequalities of (7.9) follow from Hölder's inequality, while the left follow from the fact that $\operatorname{mes} Q_\rho(k) = \int_{t_0}^{t_0+\theta\rho^2} \operatorname{mes}^{r/q} A_{k,\rho}(t) \operatorname{mes}^{1-r/q} A_{k,\rho}(t)\,dt$ is estimated in terms of $\mu(k,\ \rho,\ \theta\rho^2)\operatorname{mes}^{1-r/q} K_\rho$ from above for $r \leq q$ and from below for $r \geq q$.

In the following two lemmas we will assume that the parameter q entering into the definition of $\mu(k,\ \rho,\ \tau)$ is finite. As is seen from conditions (7.3), q can be infinite only for $n = 1$. In this case it is possible to modify Lemmas 7.2 and 7.3 so as to include the entire range $q \in (2,\ \infty]$. We will do this at the end of the section.

Lemma 7.2. *Suppose $u(x,\ t)$ is an arbitrary function subject to inequalities (7.2). There exists a $\theta_1 > 0$ such that, whatever the cylinder $Q_{\rho_0} \subset Q_T$ and the number $k_0 \geq \operatorname{vrai\,max}_{Q_{\rho_0}} u(x,\ t) - \delta$, from the inequality*

$$\operatorname{mes} Q_{\rho_0}(k_0) \leq \theta_1 \rho_0^{n+2} \tag{7.10}$$

follows

$$\operatorname{mes} Q_{\frac{\rho_0}{2}}\left(k_0 + \frac{H}{2}\right) = 0. \tag{7.11}$$

as long as

$$H = \max_{Q_{\rho_0}} u(x,\ t) - k_0 \geqslant \rho_0^{\frac{n\varkappa}{2}} \tag{7.12}$$

The cylinder $Q_{\rho_0/2}$ in (7.11) has the same vertex as Q_{ρ_0}.

Thus, suppose the conditions of the lemma are fulfilled. We pass from $x,\ t$ to new independent variables $\tilde{x},\ \tilde{t}$ by means of the formulas

$$x - x_0 = \rho_0 \tilde{x},\quad t - t_0 = \rho_0^2 \tilde{t}.$$

Inequalities (7.2) in this connection take the form

$$|u^{(k)}|^2_{Q(\tilde\rho-\sigma_1\tilde\rho,\ \tilde\tau-\sigma_2\tilde\tau)}$$
$$\leqslant \gamma\left\{[(\sigma_1\tilde\rho)^{-2}+(\sigma_2\tilde\tau)^{-1}]\,\|u^{(k)}\|^2_{2,\,Q(\tilde\rho,\ \tilde\tau)}+\rho_0^{\varkappa n}\mu^{\frac{2}{r}(1+\varkappa)}(k,\tilde\rho,\tilde\tau)\right\},$$

where $\tilde\rho = \rho/\rho_0$, $\tilde\tau = \tau/\rho_0^2$, while $Q_{\tilde\rho_0} = Q(1, \theta)$. Let us pass from u to the function $v = u/H$. For it, dividing both sides of the latter inequality by H^2 and taking into account that $H^{-2}\rho_0^{\kappa n} \le 1$, by virtue of (7.12), we obtain

$$|v^{(\tilde k)}|^2_{Q(\tilde\rho-\sigma_1\tilde\rho,\ \tilde\tau-\sigma_2\tilde\tau)}\leqslant\gamma\left\{[(\sigma_1\tilde\rho)^2+(\sigma_2\tilde\tau)^{-1}]\,\|v^{(\tilde k)}\|^2_{2,\,Q(\tilde\rho,\ \tilde\tau)}\right.$$
$$\left.+\mu^{\frac{2}{r}(1+\varkappa)}(\tilde k,\tilde\rho,\tilde\tau)\right\} \tag{7.13}$$

where $\tilde k = k/H$.

The function $v(\tilde x, \tilde t)$ satisfies the conditions of Lemma 6.1. Let us take as the range $[M, M+N]$ of variation of the parameter $\tilde k$ the interval

$$\left[\tilde k_0=\frac{k_0}{H},\ \ \tilde k_0+\frac{1}{2}\right].$$

Then, according to this lemma, for the quantities

$$y_h = 4\int_{-\tilde\tau_h}^{0}\int_{A_{\tilde k_h,\tilde\rho_h}(\tilde t)}(v-\tilde k_h)^2\,dx\,dt$$

and

$$z_h=\left(\int_{-\tilde\tau_h}^{0}\operatorname{mes}^{\frac{r}{q}}A_{\tilde k_h,\,\tilde\rho_h}(\tilde t)\,d\tilde t\right)^{\frac{2}{r}}$$

the system of recursion inequalities (6.14), (6.15) is valid. From this system of inequalities it follows by Lemma 5.7 that the y_h and z_h will tend to zero for $h\to\infty$ if y_0 and z_0 are sufficiently small:

$$y_0\leqslant\lambda,\quad z_0\leqslant\lambda^{\frac{1}{1+\varkappa}} \tag{7.14}$$

(an expression for λ is written out in (5.20)).

But by virtue of condition (7.10)

$$y_0=4\int_{-\theta}^{0}\int_{A_{\tilde k_0,1}}(v-\tilde k_0)^2\,d\tilde x\,d\tilde t\leqslant\int_{-\theta}^{0}\operatorname{mes}A_{\tilde k_0,\,1}(\tilde t)\,d\tilde t$$
$$=\rho_0^{-n-2}\operatorname{mes}Q_{\rho_0}(k_0)\leqslant\theta_1,$$

while $z_0 = (\int_{-\theta}^{0} \mathrm{mes}^{r/q} A_{k_0,1}(\tilde{t})\,d\tilde{t})^{2/r}$ according to estimates (7.9) and (7.10) does not exceed

$$\varkappa_n^{\frac{2}{q}-\frac{2}{r}}\theta_1^{\frac{2}{r}} \quad \text{for } r \geqslant q \text{ and } \theta^{\frac{2}{r}-\frac{2}{q}}\theta_1^{\frac{2}{q}} \quad \text{for } r \leqslant q.$$

In view of this we will satisfy conditions (7.14) if we take

$$\theta_1 \leqslant \min\left(\lambda;\ \varkappa_n^{1-\frac{r}{q}}\lambda^{\frac{r}{2(1+\varkappa)}}\right) \quad \text{for } r \geqslant q,$$

$$\theta_1 \leqslant \min\left(\lambda;\ \theta^{1-\frac{q}{r}}\lambda^{\frac{q}{2(1+\varkappa)}}\right) \quad \text{for } r \leqslant q. \tag{7.15}$$

For such a choice of θ_1 the numbers $y_h \to 0$ for $h \to \infty$. But this means that $\operatorname{vrai\,max}_{Q_{\rho_0/2}} u \leq k_0 + H/2$, i.e. (7.11) is valid. Lemma 7.2 is proved.

Let us take an arbitrary (standard) cylinder $Q_{2\rho} = K_{2\rho} \times [t_0, t_0 + 4\theta\rho^2] \in Q_T$ and the coaxial cylinder Q_ρ having a common vertex with $Q_{2\rho}$. Let

$$\mu_1 = \max_{Q_{2\rho}} u, \quad \mu_2 = \min_{Q_{2\rho}} u, \quad \omega = \mu_1 - \mu_2.$$

We will prove the following lemma.

Lemma 7.3. *Suppose $u(x, t)$ is an arbitrary element of $\mathfrak{B}_2(Q_T, M, \gamma, r, \delta, \kappa)$. For any $\theta_1 > 0$ it is possible to find an $s = s(\theta_1) > 0$ such that either*

$$\omega = \operatorname{osc}\{u,\ Q_{2\rho}\} \leqslant 2^s \rho^{\frac{n\varkappa}{2}} \tag{7.16}$$

or

$$\mathrm{mes}\, Q_\rho\left(\mu_1 - \frac{\omega}{2^s}\right) \leqslant \theta_1 \rho^{n+2} \tag{7.17}$$

or

$$\mathrm{mes}\left\{(x, t) \in Q_\rho : u(x, t) < \mu_2 + \frac{\omega}{2^s}\right\} \leqslant \theta_1 \rho^{n+2}. \tag{7.18}$$

Let $\omega \geq 2^s \rho^{n\kappa/2}$, where s will be chosen below. Obviously we have at least one of the inequalities

$$\mathrm{mes}\, A_{\mu_1 - \frac{\omega}{2},\,\rho}(t_0 + 3\theta\rho^2) \leqslant \frac{1}{2}\varkappa_n \rho^n \tag{7.19}$$

or

$$\mathrm{mes}\left\{x \in K_\rho : u(x,\ t_0 + 3\theta\rho^2) < \mu_1 - \frac{\omega}{2}\right\} \leqslant \frac{1}{2}\varkappa_n \rho^n. \tag{7.19'}$$

Assume that (7.19) holds, and therefore that

$$\mathrm{mes}\, A_{\mu_1 - \frac{\omega}{2^r},\,\rho}(t_0 + 3\theta\rho^2) \leqslant \frac{1}{2}\varkappa_n \rho^n, \qquad r = 1,\ 2,\ \ldots,\ s-1.$$

Then all of the subsequent arguments are carried out with the function $u(x, t)$. Otherwise we would consider the function $-u(x, t)$. We denote $\max_{Q_\rho} u$ by μ_1'. If $\mu_1' \leq \mu_1 - \omega/2^s$, then $\operatorname{mes} Q_\rho(\mu_1 - \omega/2^s) = 0$ and (7.17) is valid for any θ_1. On the other hand, if $\mu_1' > \mu_1 - \omega/2^s$, then $H = \mu_1' - (\mu_1 - \omega/2^r) \geq \omega/2^r - \omega/2^s \geq \omega/2^s \geq \rho^{n\kappa/2}$ for any $r \leq s-1$, and therefore Lemma 7.1 is applicable to $u(x, t)$ in the cylinder Q_ρ for the level $k = \mu_1 - \omega/2^r$, as long as $\omega/2^r \leq \delta$. It guarantees that

$$\operatorname{mes}\left[K_\rho \setminus A_{\mu_1 - \frac{\omega}{2^r} + \frac{3}{4}H,\ \rho}(t)\right] \geqslant \frac{1}{18}\varkappa_n \rho^n$$

for

$$t \in [t_0 + 3\theta\rho^2,\ t_0 + 4\theta\rho^2]$$

and all the more so that

$$\operatorname{mes}\left[K_\rho \setminus A_{\mu_1 - \frac{\omega}{2^{r+2}},\ \rho}(t)\right] \geqslant \frac{1}{18}\varkappa_n \rho^n,$$
$$t \in [t_0 + 3\theta\rho^2,\ t_0 + 4\theta\rho^2]. \tag{7.20}$$

since $\mu_1 - \omega/2^r + 3H/4 \leq \mu_1 - \omega/2^r + (3\omega/4)/2^r = \mu_1 - \omega/2^{r+2}$. Inequalities (7.20) hold for $r \in [\lg_2 \omega/\delta,\ s-1]$. We will take r from the interval $[[2M/\delta] + 1,\ s-1] \subset [\lg_2 \omega/\delta,\ s-1]$. (The symbol $[a]$ means, by convention, the greatest integer not exceeding the number a.) Let us apply inequality (5.5) to the function $u(x, t)$ and the levels $l = \mu_1 - \omega/2^{p+1}$, $k = \mu_1 - \omega/2^p$ for $t \in [t_0 + 3\theta\rho^2,\ t_0 + 4\theta\rho^2]$ and $p \in [[2M/\delta] + 3,\ s+1]$ and take into account (7.20). This gives

$$\frac{\omega}{2^{p+1}} \operatorname{mes} A_{\mu_1 - \frac{\omega}{2^{p+1}},\ \rho}(t) \leqslant 18\beta\varkappa_n^{-1}\rho \int_{\mathscr{D}_p(t)} |u_x(x, t)|\, dx,$$

where

$$\mathscr{D}_p(t) = A_{\mu_1 - \frac{\omega}{2^p},\ \rho}(t) \setminus A_{\mu_1 - \frac{\omega}{2^{p+1}},\ \rho}(t).$$

We integrate both sides of this inequality with respect to t, using the limits indicated in (7.20); then we square both sides, after which we estimate the right side by Cauchy's inequality

$$\left(\frac{\omega}{2^{p+1}}\right)^2 \operatorname{mes}^2 Q_\rho\left(\mu_1 - \frac{\omega}{2^{p+1}}\right)$$
$$\leqslant \left(18\varkappa_n^{-1}\beta\right)^2 \rho^2 \int_{t_0+3\theta\rho^2}^{t_0+4\theta\rho^2} \int_{\mathscr{D}_p(t)} u_x^2\,dx\,dt \int_{t_0+3\theta\rho^2}^{t_0+4\theta\rho^2} \operatorname{mes} \mathscr{D}_p(t)\,dt. \tag{7.21}$$

For an estimate of the first integral on the right side we use inequality (7.2), choosing for $Q(\rho, \tau)$ the cylinder $Q_{2\rho}$, and for $Q(\rho - \sigma_1\rho, \tau - \sigma_2\tau)$ the cylinder Q_ρ. After obvious simplifications it gives

$$\int_{t_0+3\theta\rho^2}^{t_0+4\theta\rho^2} \int_{\mathscr{D}_p(t)} u_x^2\,dx\,dt \leqslant \left| u^{\left(\mu_1 - \frac{\omega}{2^p}\right)} \right|_{Q_\rho}^2$$
$$\leqslant \gamma\left\{\left[\rho^{-2} + (3\theta\rho^2)^{-1}\right]\left(\frac{\omega}{2^p}\right)^2 \varkappa_n (2\rho)^n\, 4\theta\rho^2 + \left[[\varkappa_n (2\rho)^n]^{\frac{r}{q}}\, 4\theta\rho^2\right]^{\frac{2}{r}(1+\varkappa)}\right\}. \tag{7.22}$$

By assumption $\omega \geq 2^s\rho^{n\kappa/2}$, so that for $p \in [[2M/\delta] + 3, s + 1]$ it follows from (7.21) and (7.22) that

$$\operatorname{mes}^2 Q_\rho\left(\mu_1 - \frac{\omega}{2^{p+1}}\right) \leqslant c\rho^{n+2} \int_{t_0+3\theta\rho^2}^{t_0+4\theta\rho^2} \operatorname{mes} \mathscr{D}_p(t)\,dt, \tag{7.23}$$

where

$$c = \left(18\varkappa_n^{-1}\beta\right)^2 \gamma\left\{\left(\theta + \frac{1}{3}\right) 2^{n+2}\varkappa_n + \left(2^n \varkappa_n^{\frac{2}{q}} \theta^{\frac{2}{r}}\right)^{1+\varkappa}\right\}.$$

Let us sum inequalities (7.23) with respect to p from $[2M/\delta] + 3$ to $s - 1$, first replacing the left sides by $\operatorname{mes}^2 Q_\rho(\mu_1 - \omega/2^s)$. This gives the inequality

$$\left(s - \left[\frac{2M}{\delta}\right] - 3\right) \operatorname{mes}^2 Q_\rho\left(\mu_1 - \frac{\omega}{2^s}\right)$$
$$\leqslant c\rho^{n+2} \int_{t_0+3\theta\rho^2}^{t_0+4\theta\rho^2} \operatorname{mes} K_\rho\,dt = c\theta\varkappa_n\rho^{2n+4},$$

from which it is seen that for

$$s = \left[\frac{2M}{\delta}\right] + 4 + \left[\frac{c\theta\varkappa_n}{\theta_1^2}\right] \tag{7.24}$$

we will have (7.17). The lemma is proved.

Remark to Lemma 7.3. In the case when in $\mathfrak{B}_2(Q_T, M, \gamma, r, \delta, \kappa)$ the parameter $\delta = \infty$, i.e. the functions $u(x, t)$ satisfy inequalities (7.1) and (7.2) for all k, the parameter $s = 4 + [c\theta\kappa_n/\theta_1^2]$ by the same token does not depend on M.

Lemmas 7.2 and 7.3 imply the following proposition.

Lemma 7.4. *For any element $u(x, t)$ from $\mathfrak{B}_2(Q_T, M, \gamma, r, \delta, \kappa)$ and any standard coaxial cylinders $Q_{\rho/2}$ and $Q_{2\rho} \in Q_T$ having a common vertex, there is valid at least one of two inequalities: either*

$$\operatorname{osc}\left\{u;\ Q_{\frac{\rho}{2}}\right\} \leqslant 2^s \rho^{\frac{n\kappa}{2}} \tag{7.25}$$

or

$$\operatorname{osc}\left\{u;\ Q_{\frac{\rho}{2}}\right\} \leqslant \left(1 - \frac{1}{2^s}\right) \operatorname{osc}\{u;\ Q_{2\rho}\}, \tag{7.26}$$

where s is the number $s(\theta_1)$ in Lemma 7.3 while θ_1 is the number θ_1 in Lemma 7.2.

We recall that the numbers θ_1 and $s = s(\theta_1)$ are determined only by the parameters of the class $\mathfrak{B}_2$.

Suppose θ_1 is the number defined in Lemma 7.2, while $s = s(\theta_1)$ is the number determined by θ_1 in Lemma 7.3. We assume that $\omega_1 = \operatorname{osc}\{u;\ Q_{\rho/2}\} > 2^s \rho^{n\kappa/2}$. Then all the more so $\omega = \operatorname{osc}\{u;\ Q_{2\rho}\} > 2^s \rho^{n\kappa/2}$.

According to Lemma 7.3 this guarantees the validity of inequality (7.17) or (7.18). Let us consider the case when relation (7.17) is fulfilled (the second case is considered analogously with the function $u(x, t)$ replaced by $-u(x, t)$). By virtue of Lemma 7.2 applied to the function $u(x, t)$ in Q_ρ and to the level $k_0 = \mu_1 - \omega/2^{s-1}$, where $\mu_1 = \max_{Q_{2\rho}} u$, either $H = \max_{Q_\rho} u - (\mu_1 - \omega/2^{s-1}) < \rho^{n\kappa/2}$, or $\operatorname{mes} Q_{\rho/2}(k_0 + H/2) = 0$. (Note that the condition $k_0 \geq \operatorname{vrai\,max}_{Q_\rho} u(x, t) - \delta$ of Lemma 7.2 is fulfilled.) The first assertion of the alternative guarantees the inequality $\max_{Q_{\rho/2}} u \leq \mu_1 - \omega/2^{s-1} + \rho^{\kappa n/2} \leq \mu_1 - \omega/2^s$, and by the same token also inequality (7.26). The second assertion guarantees that

$$\begin{aligned}\max_{Q_{\frac{\rho}{2}}} u &\leqslant \mu_1 - \frac{\omega}{2^{s-1}} + \frac{H}{2} \\ &= \mu_1 - \frac{\omega}{2^{s-1}} + \frac{1}{2}\left[\max_{Q_\rho} u - \left(\mu_1 - \frac{\omega}{2^{s-1}}\right)\right] \leqslant \mu_1 - \frac{\omega}{2^s}\end{aligned}$$

and all the more so (7.26). Lemma 7.4 is proved.

From this lemma and Lemma 5.8 follows one of the principal assertions on the embeddability of $\mathfrak{B}_2(Q_T, M, \gamma, r, \delta, \kappa)$ in $H^{\alpha, \alpha/2}(Q_T)$; namely,

Theorem 7.1. *Suppose $u(x, t)$ is an arbitrary function from $\mathfrak{B}_2(Q_T, M, \gamma, r, \delta, \kappa)$ and $Q_{\rho_0} = Q(\rho_0, \theta\rho_0^2)$ is a cylinder of standard form belonging to Q_T. Then for any $Q_\rho = Q(\rho, \theta\rho^2)$, coaxial and having a common vertex with Q_{ρ_0}, the oscillation of $u(x, t)$ in Q_ρ is estimated as follows:*

$$\operatorname{osc}\{u, Q_\rho\} \leqslant c\rho^\alpha\rho_0^{-\alpha}, \tag{7.27}$$

where

$$\alpha = \min\left\{-\ln_4\left(1 - \frac{1}{2^s}\right); \frac{n\kappa}{2}\right\}, \quad c = 4^\alpha \max\left\{2M, 2^s\rho_0^{\frac{n\kappa}{2}}\right\},$$

and the numbers θ, $s = s(\theta_1)$ and θ_1 are defined in Lemmas 7.1–7.3. The number α for $\delta = \infty$ does not depend on M, and for $\delta < \infty$ is a monotonically decreasing function of M/δ. The number θ in all cases does not depend on M. Thus the class $\mathfrak{B}_2(Q_T, M, \gamma, r, \delta, \kappa)$ is embedded in $H^{\alpha, \alpha/2}(Q_T)$, with α independent of M for $\delta = \infty$.

Remark to Theorem 7.1. If inequalities (7.1) and (7.2) are valid only for $r \leq \theta_2\rho^2 \leq \theta_2\rho_0^2$, then inequalities (7.27) will hold for Q with $\rho \leq \rho_0$, where instead of θ in the definition of Q_ρ and in other places it is necessary to take $\theta' = \min\{\theta, \theta_2\}$.

Theorem 7.1 is completely proved for dimensions $n \geq 2$. In the case $n = 1$ we have not considered all of the values of the parameter q that are admissible by conditions (7.3). Namely, in the proof of Lemma 7.2 the assumption that $q < \infty$ was essentially used, although in conditions (7.3) for $n = 1$ the parameter $q \in (2, \infty]$. We will show how Lemmas 7.2 and 7.3 must be altered to make the proof suitable for all $n \geq 1$ and all values of q and r from (7.3).

Instead of Lemma 7.2 one must use the following proposition.

Lemma 7.2′. *If $u(x, t)$ satisfies inequalities (7.2), then there exists a $\theta_1 > 0$ such that from the inequality*

$$\mu^{\frac{2}{r}}(k_0, \rho_0, \theta\rho_0^2) \leqslant \theta_1\rho_0^n \tag{7.28}$$

for the quantity $\mu(k_0, \rho_0, \theta\rho_0^2)$ corresponding to the cylinder Q_{ρ_0} of Lemma 7.2 there follows (7.11) as long as (7.12) is fulfilled and $k_0 \geq \operatorname{vrai\,max}_{Q_{\rho_0}} u(x, t) - \delta$.

The proof of this lemma is based on the use of the same recursion inequalities (6.14), (6.15) that were used in the proof of Lemma 7.2. The difference consists in the fact that now the smallness of the quantity z_0 is guaranteed directly by condition (7.28):

$$z_0 \leqslant \theta_1,$$

while the smallness of y_0 is deduced from (7.28) with the help of estimates (7.9):

$$y_0 \leqslant 4 \begin{cases} \varkappa_n^{1-\frac{r}{q}} \theta_1^{\frac{r}{2}} & \text{for} \quad r \leqslant q, \\ \theta^{1-\frac{q}{r}} \theta_1^{\frac{q}{2}} & \text{for} \quad r \geqslant q. \end{cases}$$

It is seen that, in contrast to Lemma 7.2, unrestricted growth of q does not represent a hazard for estimates of z_0 and y_0.

Lemma 7.3 must be replaced by Lemma 7.3'.

Lemma 7.3'. *Suppose $u(x, t) \in \mathfrak{B}_2(Q_T, M, \gamma, r, \delta, \kappa)$. For any $\theta_1 > 0$ it is possible to find a number $s = s(\theta_1) > 0$ such that if the inequality opposite to (7.16) is fulfilled, then either*

$$\mu^{\frac{2}{r}}\left(\mu_1 - \frac{\omega}{2^s}, \rho, \theta\rho^2\right) = \left[\int_{t_0+3\theta\rho^2}^{t_0+4\theta\rho^2} \operatorname{mes}^{\frac{r}{q}} A_{\mu_1 - \frac{\omega}{2^s}, \rho}(t)\, dt\right]^{\frac{2}{r}} \leqslant \theta_1 \rho^n, \tag{7.29}$$

or

$$\left[\int_{t_0+3\theta\rho^2}^{t_0+4\theta\rho^2} \operatorname{mes}^{\frac{r}{q}} \left\{x \in K_\rho\colon u(x, t) < \mu_2 + \frac{\omega}{2^s}\right\} dt\right]^{\frac{2}{r}} \leqslant \theta_1 \rho^n. \tag{7.30}$$

Here μ_1, μ_2, ω denote the same quantities as in Lemma 7.3.

Relations (7.9) show that for $q < \infty$ the assertions of Lemma 7.3' follow from Lemma 7.3. But q can be unbounded only for $n = 1$. For this case, i.e. for $n = 1$, we prove Lemma 7.3' somewhat differently, omitting estimates (7.17) and (7.18). Namely, as in Lemma 7.3, we will assume without loss of generality that inequality (7.19), and hence also (7.20), is fulfilled.

Instead of applying inequality (5.5) to the function $u(x, t)$ and the levels $l = \mu_1 - \omega/2^{p+1}$, $k = \mu_1 - \omega/2^p$ we apply the stronger inequality (5.6) with $\epsilon_1 = r/q$, $\epsilon_2 = 1$. For $q < \infty$ this is possible and by virtue of assumption (7.20) gives

$$\frac{\omega}{2^{p+1}} \left(\frac{\operatorname{mes} A_{\mu_1 - \frac{\omega}{2^{p+1}}, \rho}(t)}{2\rho}\right)^{\frac{r}{q}} \leqslant 18 \int_{\mathscr{D}_\rho(t)} |u_x(x, t)|\, dx. \tag{7.31}$$

But this is also true for $q=\infty$ if it is agreed that in the case $q=\infty$ the quantity $\mathrm{mes}^{r/q} A_{k,\rho}(t)$ will be regarded as being equal to zero for $\mathrm{mes}\, A_{k,\rho}(t)=0$ and equal to one for $\mathrm{mes}\, A_{k,\rho}(t)>0$.

Integrating inequality (7.31) with respect to t from $t_0+3\theta\rho^2$ to $t_0+4\theta\rho^2$, we get

$$\frac{\omega}{2^{p+1}}\,\mu\left(\mu_1-\frac{\omega}{2^{p+1}},\,\rho,\,\theta\rho^2\right) \leqslant 18\,(2\rho)^{-\frac{r}{q}}\int_{t_0+3\theta\rho^2}^{t_0+4\theta\rho^2}\int_{\mathcal{D}_\rho(t)}|u_x(x,t)|\,dx\,dt, \tag{7.32}$$

where for $q=\infty$ the quantity $\mu(k,\rho,\theta\rho^2)$ on the left side is equal to $\mathrm{mes}\{t\in[t_0+3\theta\rho^2,\,t_0+4\theta\rho^2]\colon\ \mathrm{mes}\,A_{k,\rho}(t)>0\}$, which conforms to the notation adopted from the very beginning.

Squaring both sides of inequality (7.32) and extending the same arguments used in Lemma 7.3, we come directly to estimate (7.29), by-passing the estimate for $\mathrm{mes}\,Q_\rho(\mu_1-\omega/2^{s+1})$; in this connection it is necessary to take

$$s=\left[\frac{2M}{\delta}+\theta_1^{-r}\gamma\theta 2^r\left(\theta+\frac{4}{3}+2^{\varkappa}\theta^{\frac{2}{r}(1+\varkappa)}\right)\right]+4.$$

Lemma 7.3′ can be considered proved. This completes the proof of Theorem 7.1.

Corollary 7.1. *If a function $u(x,t)$ belongs to class $\mathfrak{B}_2(Q_T,M,\gamma,r,\delta,\kappa)$, then for any cylinder $\overline{Q}(\rho,\rho^2)$ belonging to Q_T the estimate*

$$\int_{Q(\rho,\rho^2)} u_x^2(x,t)\,dx\,dt\leqslant c_1\rho^{n+2\alpha} \tag{7.33}$$

is valid where the constant c_1 is determined only by the parameters of class $\mathfrak{B}_2$, n and the distance from $Q(\rho,\rho^2)$ to Γ_T, while α is the same as in (7.27).

The validity of this assertion follows from inequalities (7.2) and (7.27).

§8. THE FUNCTION CLASSES $\mathfrak{B}_2(Q_T\cup\Gamma',\cdots)$ AND $\hat{\mathfrak{B}}_2(Q_T\cup\Gamma',\cdots)$

Functions of class $\mathfrak{B}_2(Q_T,M,\gamma,r,\delta,\kappa)$, generally speaking, cannot be smooth near the boundary Γ_T of the cylinder Q_T. Let us distinguish from it two subclasses $\mathfrak{B}_2(Q_T\cup\Gamma',\cdots)$ and $\hat{\mathfrak{B}}_2(Q_T\cup\Gamma',\cdots)$, where Γ' is a subdomain on the surface $\Gamma_T\equiv S_T\cup\Gamma_0$, consisting of S_T' (part of S_T) and Γ_0') (part of Γ_0). The elements $u(x,t)$ of these subclasses, as will be proved in the present section, are continuous (in the sense of Hölder) functions in $Q_T\cup\Gamma'$ as long as S_T' and $u(x,t)$ on Γ' have a certain regularity. It has turned out that $\mathfrak{B}_2(Q_T\cup\Gamma',\cdots)$ and $\hat{\mathfrak{B}}_2(Q_T\cup\Gamma',\cdots)$ contain the solutions of the basic boundary value problems for parabolic equations.

The class $\hat{\mathfrak{B}}_2(Q_T \cup \Gamma', M, \gamma, r, \delta, \kappa)$ *consists of all elements* $u(x, t)$ *of class* $\mathfrak{B}_2(Q_T, M, \gamma, r, \delta, \kappa)$ *for which the functions* $w(x, t) = \pm u(x, t)$ *satisfy the inequalities*

$$\|w^{(k)}\|^2_{Q(\rho-\sigma_1\rho,\, \tau-\sigma_2\tau) \cap Q_T} \leqslant \gamma \left\{ \left[(\sigma_1\rho)^{-2} + (\sigma_2\tau)^{-1} \right] \| w^{(k)} \|^2_{2,\, Q(\rho,\, \tau) \cap Q_T} + \hat{\mu}^{\frac{2}{r}(1+\varkappa)}(k, \rho, \tau) \right\} \tag{8.1}$$

for any cylinders $Q(\rho, \tau)$ *not intersecting* $\Gamma_T \setminus \Gamma'$. *In them all quantities have the same meaning as in* (7.2), *except that the numbers* k *are subject to the additional restriction*

$$k \geqslant \max_{Q(\rho,\, \tau) \cap \Gamma'} w(x, t), \tag{8.2}$$

and $\hat{\mu}(k, \rho, \tau) = \int_{\max\{0;\, t_0\}}^{t_0+\tau} \operatorname{mes}^{r/q} A_{k,\rho}(t)\, dt$, *where* $A_{k,\rho}(t)$ *is the set of points* x *from* $\Omega_\rho \equiv K_\rho \cap \Omega$ *at which* $u(x, t) > k$.

Inequalities (8.1) coincide with (7.2) if the function $u(x, t)$ is considered to be extended onto all of the cylinder $Q(\rho, \tau)$ in such a way that its values in $Q(\rho, \tau) \setminus Q(\rho, \tau) \cap Q_T$ do not exceed the maximum of $u(x, t)$ on $Q(\rho, \tau) \cap \Gamma'$. For functions of class $\hat{\mathfrak{B}}_2(Q_T \cup \Gamma', \cdots)$ the following theorem is valid.

Theorem 8.1. *Suppose the projection of* S'_T *onto* S_0 [1] *satisfies condition* (A). *Then any element* $u(x, t)$ *of class* $\hat{\mathfrak{B}}_2(Q_T \cup \Gamma', M, \gamma, r, \delta, \kappa)$ *satisfying a Hölder condition on* Γ':

$$\operatorname{osc}\{u;\, Q(\rho, \rho^2) \cap \Gamma'\} \leqslant c\rho^{\varepsilon}, \quad \varepsilon > 0, \tag{8.3}$$

belongs to $H^{\alpha, \alpha/2}(Q_T \cup \Gamma')$, *where* $\alpha > 0$ *is determined by* n, *the parameters of class* $\hat{\mathfrak{B}}_2$, *the numbers* c *and* ϵ *from* (8.3) *and* a_0, θ_0 *from condition* (A). *For* $\delta = \infty$ *the quantity* α *does not depend on* M. *A norm* $|u|^{(\alpha)}_{Q'}$ *in any domain* $Q' \subset Q_T$ *separated from* $\Gamma_T \setminus \Gamma'$ *by a positive distance* d *is estimated from above by a constant depending only on* d *and the quantities affecting* α.

This theorem is proved in the same way as Theorem 7.1. Some small modifications are necessary only in Lemma 7.3. In this lemma we were choosing the number $s = s(\theta_1)$ by making use of inequalities (7.2) and the presence, for at least one of the functions $\pm u(x, t)$, of a range $[\mu_1 - \omega/2^{r+2}, \mu_1]$, $r \geq r_0$, in the vicinity of its maximum μ_1 for which the inequalities

1) That is, the projection of S'_T onto $S_0 = \{(x, t):\ x \in S,\ t = 0\}$ by means of straight lines parallel to the t axis.

$$\operatorname{mes}\left[K_\rho \setminus A_{\mu_1-\frac{\omega}{2^r+2},\,\rho}(t)\right] \geqslant b\varkappa_n\rho^n, \quad t\in[t_0+3\theta\rho^2,\ t_0+4\theta\rho^2], \tag{8.4}$$

are valid with positive constants r_0, b and θ determined by known parameters (in Lemma 7.3 one has $b = 1/18$, $r_0 = [2M/\delta] + 1$, while θ is chosen in Lemma 7.1).

For boundary cylinders we achieve the same result in the following manner. We separate the boundary cylinders $Q_{2\rho} \equiv K_{2\rho}\times[t_0, t_0 + 4\theta\rho^2]$ into two types: the first do not intersect with S_T but intersect with Γ_0', and the second have an intersection with S_T'. It is not difficult to see that for a proof of the assertions of Theorem 8.1 it is sufficient to consider from the boundary cylinders $Q_{2\rho}$ of the first type only those for which $t_0 + 3\theta\rho^2 \le 0$, and of the second type only those whose axes lie on S_T'. Let us take a boundary cylinder $Q_{2\rho}$ with $K_{2\rho}\subset\Omega$ and $t_0 + 3\theta\rho^2 \le 0$. We assume that

$$\omega \equiv \operatorname{osc}\{u;\ Q_{2\rho}\} \geqslant 2^s\rho^{\epsilon_1}, \tag{8.5}$$

where $\epsilon_1 = \min\{\epsilon, n\kappa/2\}$, while the number s remains to be defined. The oscillation of the functions $\pm u(x, 0)$ on $K_{2\rho}$ will by virtue of (8.3) be less than $\omega/4$ if we assume $\rho \le 1$ and subject the number s to the condition

$$2^s \geqslant 2^{2+\epsilon}c. \tag{8.6}$$

In view of this, for at least one of the functions $\pm u(x, t)$ the range $[\mu_1 - \omega/4, \mu_1]$ in the vicinity of its maximum μ_1 in $Q_{2\rho}$ will be free from its values on $K_{2\rho}$ for $t = 0$. Let this be true, for example, for the function $u(x, t)$, i.e. let

$$u(x, 0) \leqslant \mu_1 - \frac{\omega}{4}, \quad x\in K_{2\rho}. \tag{8.7}$$

Then the subsequent arguments will be carried out with the function $u(x, t)$ and on the level $k \ge \mu_1 - \omega/4$, assuming that $u(x, t)$ is extended onto the domain of negative values of t in such a way that for them $u(x, t) \le \mu_1 - \omega/4$. All of these arguments are based on inequalities (8.1), which, as noted above, for $u(x, t)$ and $k \ge \mu_1 - \omega/4$ coincide with (7.2), and on Lemma 7.1 derived from inequality (7.1). But for $u(x, t)$ and $k \ge \mu_1 - \omega/4$ inequalities (7.1) in $Q_\rho = K_\rho \times [t_0 + 3\theta\rho^2, t_0 + 4\theta\rho^2]$ are consequences of inequalities (8.2) (in fact, inequalities (8.2) applied, for example, to the cylinders Q_ρ and $K_\rho\times[t_0 + 3\theta\rho^2 - \rho^2, t_0 + 4\theta\rho^2]$ give (7.1) for Q_ρ, although with γ twice its former value), and hence for $u(x, t)$ and $k \ge \mu_1 - \omega/4$ in Q_ρ the assertion of Lemma 7.1 guaranteeing inequalities (8.4) is valid. One can take for θ the value fixed on after Lemma 7.1. All subsequent arguments and results are the same as in Lemma 7.3. The only difference is quantitive: in the value of $s = s(\theta_1)$ itself, which this time depends on c and ϵ from (8.3) (see condition (8.6) on s), and in the alternate premise (7.16), which is replaced by

$$\omega = \operatorname{osc}\{u;\ Q_{2\rho}\} \leqslant 2^s \rho^{\varepsilon_1}, \quad \varepsilon_1 = \min\left\{\frac{n\varkappa}{2},\ \varepsilon\right\}. \tag{8.7'}$$

In addition, we assume $\rho \leq 1$.

Lemma 7.2 is applied to $u(x, t)$ and $k \geq \mu_1 - \omega/4$ in Q_ρ without change. From it and Lemma 7.3 follows Lemma 7.4 with the replacement in condition (7.25) of $2^s \rho^{n\kappa/2}$ by $2^s \rho^{\epsilon_1}$. Also Lemma 7.4 together with Lemma 5.8 of Chapter II guarantees the desired estimate for the oscillations of $u(x, t)$ in boundary cylinders of the first type. For boundary cylinders of the second type the arguments are almost the same as for cylinders of the first type. It is true that for them we do not in general enlist Lemma 7.1, but we guarantee inequalities (8.4) because of the membership of the center of K_ρ to the boundary S and because of condition (A), which gives

$$\operatorname{mes}\left[K_\rho \setminus A_{\mu_1 - \frac{\omega}{4},\, \rho}(t)\right] \geqslant \operatorname{mes}\left[K_\rho \setminus \Omega_\rho\right] \geqslant \theta_0 \varkappa_n \rho^n.$$

Thus Theorem 8.1 is proved. Let us introduce the following class of functions, which are used in the investigation of solutions of the second boundary value problem.

The class $\hat{\mathfrak{B}}_2(Q_T \cup \Gamma', M, \gamma, r, \delta, \kappa)$ *consists of elements* $u(x, t)$ *of class* $\mathfrak{B}_2(Q_T \cup \Gamma', M, \gamma, r, \delta, \kappa)$ *for which inequalities* (8.1) *are fulfilled for all* k *satisfying the conditions*

$$k \geqslant \max_{Q(\rho,\tau) \cap Q_T} w(x, t) - \delta \quad \text{и} \quad k \geqslant \max_{Q(\rho,\tau) \cap \Gamma_0'} w(x, t), \tag{8.8}$$

and the inequalities of type (7.1)

$$\max_{t_0 \leqslant t \leqslant t_0 + \tau} \|w^{(k)}(x, t)\|^2_{2, K_{\rho - \sigma_1\rho} \cap \Omega} \leqslant \|w^{(k)}(x, t_0)\|^2_{2, K_\rho \cap \Omega}$$
$$+ \gamma\left[(\sigma_1\rho)^{-2} \|w^{(k)}\|^2_{2, Q(\rho,\tau) \cap Q_T} + \hat{\mu}^{\frac{2}{r}(1+\varkappa)}(k, \rho, \tau)\right] \tag{8.9}$$

for all k *satisfying the conditions*

$$k \geqslant \max_{Q(\rho,\tau) \cap Q_T} w(x, t) - \delta. \tag{8.10}$$

In regard to $\Gamma' = S'_T \cup \Gamma'_0$ we assume the following. Suppose S' is the projection of S'_T onto S_0, $\Omega_\rho = K_\rho \cap \Omega$, and K_ρ is a ball with center at an arbitrary point $x_0 \in S'$ and with radius ρ not exceeding some fixed number ρ_0 and the distance from x_0 to $S \setminus S'$. We require that there exist positive constants β_1 and θ_2 such that

1) $\operatorname{mes}\Omega_\rho \geq \theta_2 \operatorname{mes} K_\rho$

and

2) for any function $u(x)$ from $W_2^1(\Omega_\rho)$ the inequalities of form (5.5)

$$(l-k)\operatorname{mes} A_{l,\rho} \leqslant \frac{\beta_1 \rho^{n+1}}{\operatorname{mes}(\Omega_\rho \setminus A_{k,\rho})} \int_{A_{k,\rho} \setminus A_{l,\rho}} |u_x|\, dx \tag{8.11}$$

are valid for arbitrary numbers l and k, $l > k$. As was indicated in §5, inequalities (8.11) hold, for example, for the cases when Ω_ρ is a convex domain. It is not difficult to show that properties 1) and 2) are fulfilled for domains satisfying the conditions that are usually imposed in the proof of embedding theorems (see [113a]). For functions of class $\hat{\mathfrak{B}}_2(Q_T \cup \Gamma', \cdots)$ we have

THEOREM 8.2. *Suppose a function $u(x, t)$ belongs to $\hat{\mathfrak{B}}_2(Q_T \cup \Gamma', M, \gamma, r, \delta, \kappa)$ and satisfies on Γ'_0 a Hölder condition with exponent $\epsilon > 0$. Suppose, further, that conditions* 1) *and* 2) *formulated above are fulfilled for S'_T. Then $u(x, t)$ is Hölder continuous in $Q_T \cup \Gamma'$ and for any domain $Q' \subset Q_T$ separated from $\Gamma_T \setminus \Gamma'$ by a positive distance d the norm $|u|_{Q'}^{\alpha}$ is estimated from above by a constant c determined only by the parameters $M, \gamma, r, \delta, \kappa$ of class $\hat{\mathfrak{B}}_2$, d, ϵ, the norm $|u(x, 0)|_{\Gamma'_0}^{\epsilon}$ and the constants ρ_0, β_1 and θ_2 from conditions* 1) *and* 2). *The quantity $\alpha > 0$ depends on the same quantities as c, except for d.*

The proof of this theorem is carried out in the same way as the proof of Theorems 7.1 and 8.1. For boundary cylinders of the first type the arguments coincide with those just described in the proof of Theorem 8.1. For boundary cylinders of the second type the arguments are analogous to those given for interior cylinders in the proof of Theorem 7.1, it being only necessary to take instead of balls and cylinders their intersections with Ω and Q_T, respectively, and in this connection to replace $\operatorname{mes} K_\rho = \kappa_n \rho^n$ in Lemmas 7.1 and 7.3 by $\operatorname{mes}\Omega_\rho$. The analytic inequalities on which this substitution is predicated are inequalities (3.7) and (5.5). The first of them was applied above in the proof of Lemma 6.1, and in the proof of Lemma 7.2 following from it, to the functions $u^{(k)}\zeta$ from $\overset{\circ}{V}_2(Q(\rho, \tau))$ (see (6.16)). In the present case for boundary cylinders the functions $u^{(k)}\zeta$ do not vanish on all of the lateral surface of the domain $Q(\rho, \tau) \cap Q_T$. In view of this, instead of inequality (3.7) one should use the analogous inequality (3.9), which is valid for any function from $V_2(Q_T)$. Since by virtue of condition 1) we have $(\theta\rho^2)^{n/2} \operatorname{mes}^{-1} \Omega_\rho \le \theta^{n/2} \theta_2^{-1} \kappa_n^{-1}$ for the domains $\Omega_\rho \times (t_0, t_0 + \theta\rho^2)$ considered in Lemma 7.2 it follows that for $\Omega_\rho \times (t_0, t_0 + \theta\rho^2)$ one can consider the constant c in (3.9) to be independent of ρ.

As to inequality (5.5), it was applied to those levels k for which $\operatorname{mes}(K_\rho \setminus A_{k,\rho}) \ge b\rho^n$, $b > 0$ (see (7.20)), and was somewhat simplified to the inequality

$$(l - k) \operatorname{mes} A_{l,\rho} \leqslant \frac{\beta}{b} \rho \int\limits_{A_{k,\rho} \setminus A_{l,\rho}} |u_x| \, dx. \tag{8.12}$$

In the present case one should use (8.11) instead of (5.5), and one should also use (8.11) for those levels k for which $\operatorname{mes}(\Omega_\rho \setminus A_{k,\rho}) \ge b \operatorname{mes} \Omega_\rho$. This, by virtue of condition 1), gives an inequality of form (8.12). In other respects the proof of Theorem 8.2 is carried out in the same way as the proof of Theorems 7.1 and 8.1.

Let us make one more remark that will be useful in the sequel.

Remark 8.1. The membership of a function $u(x, t)$ in this or that class $\mathfrak{B}_2$ follows from the function $u(x, t)$ satisfying inequalities (7.5) with the functions $\zeta(x, t)$ indicated there for such distributions of cylinders $Q(\rho, \tau)$ and such levels k as are required in the definition of the class $\mathfrak{B}_2$. In this connection, if $Q(\rho, \tau)$ intersects with Γ_T, then all of the integrals in (7.5) are considered to be extended onto the intersections of the domains indicated in (7.5) with Q_T, whereas the function $\zeta(x, t)$ as before must vanish only on the lateral surface and lower base of $Q(\rho, \tau)$ (but not on $Q(\rho, \tau) \cap Q_T$!). It is easy to see that from inequalities (7.5) follow the inequalities entering into the definition of the different classes $\mathfrak{B}_2$.

Lemma 8.1. *Suppose*

$$u(x, t) \in V_2^{1,0}(Q_T), \quad \operatorname*{vrai\,max}_{Q_T} |u| \leqslant M,$$

and the functions $w(x, t) = \pm u(x, t)$ *satisfy inequalities* (7.5) *for any* $Q(\rho, \tau)$ *not intersecting with* $\Gamma_T \setminus \Gamma'$ *under arbitrary values of* k *satisfying the inequality* $k \ge \operatorname{vrai\,max}_{Q(\rho,\tau) \cap Q_T} w - \delta$. *In this connection it is assumed that all of the integrals in* (7.5) *are extended only onto the intersections of the domains indicated in* (7.5) *with* Q_T, *while the numbers* $\kappa > 0$, *and* r *and* q *are subject to restrictions* (7.3). *Suppose, further, that* Γ' *satisfies the conditions of Theorem* 8.2 *and for any* K_ρ

$$\int_{K_\rho \cap \Gamma_0'} u_x^2(x, 0)\, dx \leqslant c\rho^{n-2+2\varepsilon}, \qquad \varepsilon > 0. \tag{8.13}$$

Then for an arbitrary cylinder $Q_\rho = K_\rho \times (t_0, t_0 + \rho^2)$ *separated from* $\Gamma_T \setminus \Gamma'$ *by a positive distance d the estimate*

$$\int_{Q_\rho \cap Q_T} u_x^2\, dx\, dt \leqslant c_1 \rho^{n+2\alpha} \tag{8.14}$$

is valid, where the constants c_1 *and* $\alpha > 0$ *are determined by* n, M, δ, *the numbers* γ_1, ν, q, r *and* κ *from* (7.5), c *and* ϵ *from* (8.13) *and the properties of* Γ', *with* c_1 *depending also on* d.

Inequality (8.14) is proved in the same way as the analogous inequality (7.33), it being only necessary, for cylinders intersecting the plane $\{t = 0\}$, to use instead of (7.2) the inequalities (7.5) directly with the function $\zeta = \zeta(x)$ equal to zero outside of a ball K_{ρ_1} concentric with the ball K_ρ.

§9. THE FUNCTION CLASSES $\mathfrak{B}_2^{N_1}$

In the investigation of quasi-linear equations of general nondivergence form, and also in the investigation of systems of equations, we encounter the need for further extensions of the classes $\mathfrak{B}_2$.

We will say that a vector function $\mathbf{u}(x, t) = (u^1(x, t), \cdots, u^N(x, t))$, defined and measurable on Q_T, belongs to class $\mathfrak{B}_2^{N_1}(Q_T, M_1, \delta_1, \delta_2, \delta_3, \gamma, r, \delta, \kappa)$ if one can construct for it N_1 functions $\phi^1(u^1, \cdots, u^N), \cdots, \phi^{N_1}(u^1, \cdots, u^N)$ that are continuous and continuously differentiable in the domain

$$\left\{ u \in E_N :\ |\mathbf{u}| \equiv \sqrt{\sum_{k=1}^{N} (u^k)^2} \leqslant \operatorname{vrai}\max_{Q_T} |\mathbf{u}(x, t)| \right\}$$

and such that the functions $w^l(x, t) = \phi^l(u^1(x, t), \cdots, u^N(x, t))$, $l = 1, \cdots, N_1$, possess the following properties.

1) $\operatorname{vrai}\max_{Q_T} |w^l(x, t)| \le M_1$ and $w^l \in V_2^{1,0}(Q_T)$.

2) For any cylinder $Q = K_{2\rho} \times [t_0, t_0 + \tau] \subset Q_T$ and any $t_1 \in (t_0, t_0 + \tau)$ there exists an index p for which

$$\operatorname{osc}\{w^p(x, t);\, Q\} \geqslant \delta_1 \max_{i=1,\ldots,N} \operatorname{osc}\{u^i(x, t);\, Q\} \tag{9.1}$$

and

$$\operatorname{mes}\{x\in K_\rho:\ w^p(x,\ t_1)\leqslant \operatorname*{vrai\,max}_Q w^p(x,\ t) \\ -\delta_2 \operatorname{osc}\{w^p(x,\ t);\ Q\}\}\geqslant (1-\delta_3)\varkappa_n\rho^n, \qquad (9.2)$$

where K_ρ is a ball concentric with the ball $K_{2\rho}$, and the δ_i, $i = 1, 2, 3$, are fixed positive numbers of which δ_2 and δ_3 are less than unity.

3) For each of the functions $w = w^l(x, t)$, $l = 1, \cdots, N_1$, inequalities (7.1) and (7.2) are fulfilled for arbitrary $Q(\rho, \tau) \subset Q_T$ and $\sigma_1, \sigma_2 \in (0, 1)$ and k satisfying inequality (7.4). In the same way as in §7, $\mu(k, \rho, \tau) = \int_{t_0}^{t_0+\tau} \operatorname{mes}^{r/q} A_{k,\rho}(t)dt$, with $A_{k,\rho}(t)$ being the set of points x from K_ρ at which $u(x, t) > k$; $M, \gamma, q, r, \delta, \kappa$ are fixed positive numbers, with q and r satisfying conditions (7.3). For $q = \infty$ we have

$$\mu(k,\ \rho,\ \tau)=\operatorname{mes}\{t\in[t_0,\ t_0+\tau]:\ \operatorname{mes} A_{k,\rho}(t)>0\}.$$

In the same way as in §7, one can assume that inequalities (7.1), (7.2) are fulfilled only for $\rho \le \rho_0$, $\tau \le \tau_0$, where ρ_0 and τ_0 are fixed numbers. It is not difficult to see that the function class $\mathfrak{B}_2(Q_T, M, \gamma, r, \delta, \kappa)$ is the class $\mathfrak{B}_2^{N1}(Q_T, M_1, \delta_1, \delta_2, \delta_3, \gamma, r, \delta, \kappa)$ for which $\phi^1(u) = u$, $\phi^2(u) = -u$, $w^1(x, t) = u(x, t)$, $w^2(x, t) = -u(x, t)$, $N_1 = 2$, $M_1 = M$, $\delta_1 = 1$, $\delta_2 = \delta_3 = 1/2$, while γ, δ, r and κ are the same as for $\mathfrak{B}_2(Q_T, M, \gamma, r, \delta, \kappa)$.

We will prove that the functions $u(x, t)$ of classes $\mathfrak{B}_2^{N1}$ belong to $H^{\alpha,\alpha/2}(Q_T)$.

Theorem 9.1. *Suppose* $u(x, t) = (u^1, \cdots, u^N)$ *is an arbitrary function of class* $\mathfrak{B}_2^{N1}(Q_T, M_1, \delta_1, \delta_2, \delta_3, \gamma, r, \delta, \kappa)$ *and* $Q_{\rho_0} = Q(\rho_0, \theta\rho_0^2)$ *is a cylinder belonging to* Q_T. *Then for any cylinder* Q_ρ, *coaxial with* Q_{ρ_0} *and having a common vertex with it, the oscillation of* $u^i(x, t)$ *in* Q_ρ *is estimated as follows:*

$$\operatorname{osc}\{u^i;\ Q_\rho\}\leqslant c\rho_0^{-\alpha}\rho^\alpha,\quad i=1,\ \ldots,\ N, \qquad (9.3)$$

where

$$\left.\begin{aligned} \alpha&=\frac{1}{N_1}\min\left\{-\lg_4\left(1-\frac{1}{2^s}\right);\ \frac{n\varkappa}{2}\right\},\\ c&=\frac{4^{\alpha(N_1+1)}}{\delta_1}\max\left\{2^{2\alpha N_1+1}M_1;\ 2^s\rho_0^{\frac{n\varkappa}{2}}\right\},\end{aligned}\right\} \qquad (9.4)$$

the number $\theta = \theta((\sqrt{\delta_3} + 1)/2)$ *is taken from Lemma* 7.1, *while the number* s *is defined below in Lemma* 9.2. *The number* α *for* $\delta = \infty$ *does not depend on* M_1, *and for* $\delta < \infty$ *is a decreasing function of* $M_1\delta^{-1}$.

The proof of this theorem, like that of Theorem 7.1, is based on Lemmas

7.1–7.4. Lemma 7.1 is valid for any function $w^l(x, t)$, $l = 1, \cdots, N_1$, since the w^l satisfy inequalities (7.1). Let us fix in Lemma (7.1) the number $\xi = (\sqrt{\delta_3} + 1)/2$ and the values of $\theta = \theta(\xi)$ and $b = b(\xi)$ corresponding to this ξ; we will consider the standard cylinders $Q_\rho = \{|x - x_0| < \rho, t_0 < t < t_0 + \theta\rho^2\}$. For each of the $w^l(x, t)$, $l = 1, \cdots, N_1$, Lemma 7.2 will also hold, since the w^l satisfy inequalities (7.2). The situation is different with Lemma 7.3. Its assertions can be incorrect for any w^l, since we have not assumed in this case that together with w^l the function $-w^l$ also satisfies inequalities (7.2), and therefore we cannot assume that inequality (7.19) (or (7.19′)) is fulfilled, which was the starting point for the derivation of estimate (7.17). Instead of (7.19) inequality (9.2) is known to us, but not for all w^l; only for w^l with the special index $l = p$. Namely, for w^l with this index $l = p$ we have

Lemma 9.1. *Consider an arbitrary function* $u(x, t)$ *from* $\mathfrak{B}_2^{N_1}(Q_T, M_1, \delta_1, \delta_2, \delta_3, \gamma, r, \delta, \kappa)$ *and the functions* $w^l(x, t)$, $l = 1, \cdots, N_1$, *corresponding to it. With respect to any* $\theta_1 > 0$ *it is possible to find an* $s = s(\theta_1) > 0$ *such that for any cylinder* $Q_{2\rho} = K_{2\rho} \times (t_0, t_0 + 4\theta\rho^2) \subset \Omega_T$ *and* $t_1 = t_0 + 3\theta\rho^2$ *and function* $w^p(x, t)$ *corresponding to it by condition* 2) *at least one of two inequalities holds: either*

$$\omega = \operatorname{osc}\{w^p;\ Q_{2\rho}\} \leqslant 2^s \rho^{\frac{\varkappa n}{2}}, \tag{9.5}$$

or

$$\operatorname{mes} Q_\rho\left(\mu_1 - \frac{\omega}{2^s}\right) \leqslant \theta_1 \rho^{n+2}, \tag{9.6}$$

where $\mu_1 = \operatorname{vrai\,max}_{Q_{2\rho}} w^p$, *while* $Q_\rho = K_\rho \times [t_0 + 3\theta\rho^2, t_0 + 4\theta\rho^2]$ *is a cylinder that is coaxial with* $Q_{2\rho}$. *The number* s *is determined by the parameters of classes* $\mathfrak{B}_2^{N_1}$, *and for* $\delta = \infty$ *does not depend on* M_1.

The proof of this lemma is the same as for Lemma 7.3, it being only necessary to use instead of inequality (7.19) the inequality following from (9.2)

$$\operatorname{mes} A_{\mu_1 - \delta_2\omega}(t_0 + 3\theta\rho^2) \leqslant \delta_3 \varkappa_n \rho^n. \tag{9.7}$$

Let us turn to the last of the lemmas, namely Lemma 7.4, which is a direct consequence of Lemmas 7.1–7.3. It takes the following form.

Lemma 9.2. *Suppose* $u(x, t)$ *is an arbitrary element of* $\mathfrak{B}_2^{N_1}(Q_T, M_1, \delta_1, \delta_2, \delta_3, \gamma, r, \delta, \kappa)$ *and the* $w^l(x, t)$, $l = 1, \cdots, N_1$, *are the functions corresponding to* $u(x, t)$. *For any standard coaxial cylinders* $Q_{\rho/2}$ *and* $Q_{2\rho} \subset Q_T$ *having a common vertex, and function* $w^p(x, t)$ *distinguished by condition* 2) *for the cylinder* $Q_{2\rho}$ *and* $t_1 = t_0 + 3\theta\rho^2$, *there is valid at least one of two inequalities: either*

$$\operatorname{osc}\{w^p;\, Q_{\rho/2}\} \le 2^s \rho^{n\kappa/2},$$

or

$$\operatorname{osc}\{w^p;\, Q_{\rho/2}\} \le \left(1 - \frac{1}{2^s}\right) \operatorname{osc}\{w^p,\, Q_{2\rho}\},$$

where s is the number $s(\theta_1)$ from Lemma 9.1, while the number θ_1 is taken from Lemma 7.2. The numbers θ_1 and $s(\theta_1)$ are determined only by the parameters of class $\mathfrak{B}_2^{N1}$.

From this lemma and Lemma 5.9 follows Theorem 9.1. It will be used in proving the smoothness of the derivatives u_{x_i} of solutions $u(x, t)$ of general quasi-linear equations and the smoothness of solutions (and their derivatives) of the systems of linear and quasi-linear equations considered in the last chapter, at interior points of the domain of definition. For an investigation of the smoothness of all of these functions in the vicinity of the boundary of Q_T we need subclasses of class $\mathfrak{B}_2^{N1}$, analogous to the subclasses $\mathfrak{B}_2(Q_T \cup \Gamma', \cdots)$ and $\hat{\mathfrak{B}}_2(Q_T \cup \Gamma', \cdots)$ of class $\mathfrak{B}_2(Q_T, \cdots)$. introduced in the preceding section.

We will say that $u(x, t)$ belongs to class $\mathfrak{B}_2^{N1}(Q_T \cup \Gamma', M_1, \delta_1, \cdots, \delta_6, \gamma, r, \delta, \kappa)$, if $u(x, t) \in \mathfrak{B}_2^{N1}(Q_T, M_1, \delta_1, \delta_2, \delta_3, \gamma, r, \delta, \kappa)$ and if the functions $w^l(x, t)$, $l = 1, \cdots, N_1$, constructed with respect to it possess in addition the following properties.

4) *They satisfy inequalities* (8.1) *for all $Q(\rho, \tau)$ and k indicated under* (8.1).

5) *For any boundary cylinder $Q = K_\rho \times (t_0, t_0 + \tau)$ with axis on S'_T ($S'_T = \Gamma' \cap S_T$) there exists an index p such that*

$$\operatorname{osc}\{w^p;\, Q \cap Q_T\} \geqslant \delta_4 \max_{l=1,\ldots,N} \operatorname{osc}\{u^l;\, Q \cap Q_T\}, \tag{9.8}$$

and if

$$\operatorname{osc}\{w^p;\, Q \cap Q_T\} \geqslant \delta_5 \max_{l=1,\ldots,N_1} \operatorname{osc}\{w^l;\, Q \cap \Gamma'\}, \tag{9.9}$$

then

$$\operatorname*{vrai\,max}_{Q \cap \Gamma'} w^p \leqslant \operatorname*{vrai\,max}_{Q \cap Q_T} w^p - \delta_6 \operatorname{osc}\{w^p;\, Q \cap Q_T\}. \tag{9.10}$$

where $\delta_4 > 0$, $\delta_5 > 1$, $\delta_6 \in (0, 1)$.

For functions of this class we have

Theorem 9.2. *Suppose the projection of S'_T anto S_0 satisfies condition* (A). *If $u(x, t) \in \mathfrak{B}_2^{N1}(Q_T \cup \Gamma', M_1, \delta_1, \cdots, \delta_6, \gamma, r, \delta, \kappa)$ and the functions $w^l(x, t)$, $l = 1, \cdots, N_1$, corresponding to it belong to $H^{\epsilon, \epsilon/2}(\Gamma')$, then $u(x, t) \in$*

$H^{\alpha, \alpha/2}(Q_T \cup \Gamma')$, *for any domain* $Q' \subset Q_T$ *separated from* $\Gamma \setminus \Gamma'$ *by a positive distance* d *the norm* $|u|_{Q'}^{\alpha}$ *is estimated from above by a constant depending only on the parameters of class* $\mathfrak{B}_2^{N_1}$, ϵ *the norm* $|u|_{\Gamma'}^{(\epsilon)}$, *the constants* a_0 *and* θ_0 *from condition* (A) *and the distance* d. *The exponent* α *is determined by the same quantities, with the exception of* d; *for* $\delta = \infty$ *it does not depend on* M_1.

This theorem is proved according to the same plan as Theorem 9.1. The necessary changes in the arguments concerning boundary cylinders are analogous to the changes introduced in the proof of Theorem 8.1 in comparison with Theorem 7.1. Theorems 9.1 and 9.2 are generalizations of Theorems 7.1 and 8.1, in which were considered elementary variants of classes $\mathfrak{B}_2^{N_1}$ with $N_1 = 2$, $\phi_1 = u$ and $\phi_2 = -u$. An analogous generalization can be made for Theorem 8.2 by introducing classes of type $\hat{\mathfrak{B}}_2^{N_1}(Q_T \cup \Gamma', \delta_1, \delta_2, \delta_3, r, \delta, \kappa)$.

Let us formulate one more lemma, to be used in Chapters VI and VII.

Lemma 9.3. *Suppose* $\mathrm{u}(x, t) = (u^1(x, t), \cdots, u^N(x, t))$ *is an arbitrary vector function from* $V_2^{1,0}(Q_T)$ *with* $\min_{i=1,\cdots,N} \operatorname{vrai\,min}_{Q_T} u^i$, $\max_{i=1,\cdots,N} \operatorname{vrai\,max}_{Q_T} u^i \le M$. *Then the functions*

$$\varphi_+^l(\mathrm{u}) \equiv \varphi_+^l(u^1, \ldots, u^N) = \lambda \frac{u^l - m}{M - m} + \sum_{i=1}^N \left(\frac{u^i - m}{M - m}\right)^2,$$

$$\varphi_-^l(\mathrm{u}) \equiv \varphi_-^l(u^1, \ldots, u^N) = \lambda \frac{M - u^l}{M - m} + \sum_{i=1}^N \left(\frac{u^i + M}{M - m}\right)^2,$$

$$l = 1, \ldots, N,$$

for $\lambda > 4N$ *possess properties* 1), 2) *and* 5) *from the definition of classes* $\mathfrak{B}_2^{N_1}(Q_T \cup \Gamma', M_1, \delta_1, \cdots, \delta_6, \gamma, r, \delta, \kappa)$ *with* $M_1 = \lambda + N$, $N_1 = 2N$ *and certain* $\delta_1, \cdots, \delta_6$. *For* $\lambda = 10N$ *these parameters have the following values:* $\delta_1 = 8N/(M - m)$, $\delta_2 = 1/4$, $\delta_3 = 1/2$, $\delta_4 = 8N/(M - m)$, $\delta_5 = 3$, $\delta_6 = 1/24$, $M_1 = 11N$, $N_1 = 2N$.

A proof of this proposition is available in the book [65q] (see Lemmas 8.1 and 8.4 of Chapter II). The only difference consists in the fact that in [65q] balls $K_{2\rho}$ are considered instead of cylinders Q; but it is nonessential.

CHAPTER III

LINEAR EQUATIONS WITH DISCONTINUOUS COEFFICIENTS

The present chapter and Chapter IV are devoted to linear equations and above all to the question of solvability of their basic boundary value problems. Equations with discontinuous coefficients are studied in Chapter III. This is explained by the fact that the existence theorems are most easily established in classes of generalized solutions having only the derivatives u_x of first order, where there is no need to assume that the coefficients are smooth. Thus the case of discontinuous coefficients is studied first, with the coefficients being assumed as bad as permitted by the uniqueness theorem (see §3 of Chapter I). The properties of the coefficients and free terms are characterized by their membership in the spaces $L_{q,r}(Q_T)$ with this or that q and r.

The solvability of boundary value problems in $V_2(Q_T)$ is established by means of Galerkin's method, and it is then shown that any generalized solution from $V_2(Q_T)$ belongs to $V_2^{1,\frac{1}{2}}(Q_T)$.

Under a somewhat different form of the conditions on part of the coefficients and free terms of the equations, generalized solutions from $V_2(Q_T)$ also belong to $W_2^{1,\frac{1}{2}}(Q_T)$.

The results of §§1–5 are easily carried over to the so-called strongly parabolic systems (defined in [65h]), a special case of which is one parabolic equation of high order.

After proving the solvability of the boundary value problems in $V_2^{1,\frac{1}{2}}(Q_T)$ (§§1–5) we investigate the progressive increase in the smoothness of the solutions as the properties of the known functions in the problems are improved, giving interior estimates for generalized solutions of the equations and estimates right up to the boundary for generalized solutions of the boundary value problems (§§6–12). The dependences proved in these sections are exact (see, for example, §3 of Chapter I). In §12 the results of Chapter III are linked with the results of Chapter IV, which are derived without any reference to Chapter III. Sections 14–18 are devoted to a description of the other known methods for proving solvability of the basic boundary value problems. In §13 it is shown that diffraction problems (in which there are in the general case several heterogeneous media) are a special case of the problems on determining generalized solutions from $V_2^{1,1/2}(Q_T)$ considered in the present chapter.

§1. STATEMENT OF THE PROBLEM. GENERALIZED SOLUTIONS

The present chapter is devoted to a study of the linear equations

$$\mathscr{L}u \equiv u_t - \mathscr{M}u = \frac{\partial f_i}{\partial x_i} - f, \tag{1.1}$$

where

$$\mathscr{M}u = \frac{\partial}{\partial x_i}(a_{ij}(x, t)\, u_{x_j} + a_i(x, t)\, u) - b_i(x, t)\, u_{x_i} - a(x, t)\, u,$$

with discontinuous and, generally speaking, nondifferentiable and unbounded coefficients satisfying the condition of uniform parabolicity, namely,

$$\nu \sum_{i=1}^{n} \xi_i^2 \leqslant a_{ij}(x, t)\, \xi_i \xi_j \leqslant \mu \sum_{i=1}^{n} \xi_i^2, \quad \nu, \mu = \text{const} > 0. \tag{1.2}$$

In §3 of Chapter I we discussed the question of admissible extensions of the concept of a solution for these equations and their boundary value problems, and we determined what restrictions must be imposed on the coefficients and free terms of the equations in order that one may begin to construct a theory of these generalized solutions that preserves the basic features inherent in parabolic equations with smooth coefficients and smooth solutions. We will show here that it is not necessary to impose any further restrictions for this purpose.

The most extensive of these restrictions are the following:[1] the coefficients a_i, b_i, a are measurable functions with finite norms

$$\left\| \sum_{i=1}^{n} a_i^2;\ \sum_{i=1}^{n} b_i^2;\ a \right\|_{q, r, Q_T} \leqslant \mu_1. \tag{1.3}$$

---

1) The assertions of certain theorems remain valid under the condition that the summability powers of the $b_i(x, t)$ are equal to the values of the ends of admissible intervals. We indicate these cases in the Remarks. Their proof does not require any new methods, but it violates the uniformity of the calculations and dependences. This violation is connected with an essential point: for these extreme cases the dependences of the properties of a solution on the b_i have a different form.

in which q and r are arbitrary numbers satisfying the conditions

$$\left.\begin{aligned} &\frac{1}{r}+\frac{n}{2q}=1, \\ &q\in\left(\frac{n}{2},\ \infty\right],\quad r\in[1,\ \infty)\ \text{ for }\ n\geqq 2, \\ &q\in[1,\ \infty],\ r\in[1,\ 2]\ \text{ for }\ n=1 \end{aligned}\right\} \tag{1.4}$$

(the case $r=\infty$, $q=n/2+2\epsilon$, $\epsilon>0$, in (1.3) is contained in (1.4) since $L_{q,\infty}\subset L_{n/2+\epsilon,\, n/2\epsilon+1}$ and the indices $q=n/2+\epsilon$, $r=n/2\epsilon+1$ satisfy (1.4)), while the free terms have finite norms

$$\|\mathbf{f}\|_{2,\,Q_T}\equiv\left(\int\limits_{Q_T}\sum_{i=1}^{n}f_i^2\,dx\,dt\right)^{1/2}\leqslant\mu_2,\quad \|f\|_{q_1,\,r_1,\,Q_T}\leqq\mu_2 \tag{1.5}$$

in which

$$\left.\begin{aligned} &\frac{1}{r_1}+\frac{n}{2q_1}=1+\frac{n}{4}, \\ &q_1\in\left[\frac{2n}{n+2},\ 2\right],\ r_1\in[1,\ 2]\quad \text{for}\ \ n\geqq 3, \\ &q_1\in(1,\ 2],\ r_1\in[1,\ 2)\qquad\ \ \text{for}\ \ n=2, \\ &q_1\in[1,\ 2],\ r_1\in[1,\ 4/3]\qquad \text{for}\ \ n=1. \end{aligned}\right\} \tag{1.6}$$

Remark 1.1. The cases considered in [65n] are contained here since $L_{q,\infty}$ with $q>n/2$ is embedded for $n\geq 2$ in $L_{\hat q,\hat r}$, $\hat q=n/2+\epsilon$, $\hat r=n/2\epsilon+1$, $\epsilon=q-n/2>0$, and $\hat q$, $\hat r$ satisfy conditions (1.4). It can be assumed in conditions (1.3) that q and r are different for different coefficients. But we will consider them to be equal. The case of different q and r is treated analogously and the final results are the same. In addition, since the functions from $L_{q,r}(Q_T)$ belong to $L_{q',r'}(Q_T)$ with $q'\leq q$ and $r'\leq r$, in conditions (1.4) and (1.6) one could write instead of the equalities

$$\frac{1}{r}+\frac{n}{2q}=1,\quad \frac{2}{r_1}+\frac{n}{q_1}=\frac{n+4}{2}$$

the inequalities

$$\frac{1}{r}+\frac{n}{2q}\leqslant 1,\quad \frac{2}{r_1}+\frac{n}{q_1}\leqslant\frac{n+4}{2}$$

and not place any upper bounds on the choice of q_1 and r_1. But for the sake of convenience and uniformity in the writing of all of the relations it is more convenient for us to assume that one of the parameters (q and r) is arbitrarily chosen from the indicated interval, while the other is determined from the equalities of

(1.4) and (1.6).

Under fulfilment of conditions (1.3)–(1.6), as will be proved in §§4 and 5, the basic boundary value problems for (1.1) are uniquely solvable in the class $V_2(Q_T)$. Furthermore, under these conditions any solution from $V_2(Q_T)$ of equation (1.1) is found to possess a certain additional (in comparison with arbitrary elements of $V_2(Q_T)$) smoothness in t; namely, it belongs to $V_2^{1,1/2}(Q_T)$.

We begin with definitions of the generalized solutions of equations (1.1) and of the first boundary value problem for them that belong to $V_2(Q_T)$, $V_2^{1,0}(Q_T)$ and $V_2^{1,1/2}(Q_T)$. In accordance with the basic purpose described in §3 of Chapter I, we will mean by *a generalized solution from $V_2(Q_T)$ of equation* (1.1) a function $u(x, t)$ belonging to $V_2(Q_T)$ and satisfying for almost all $t_1 \in [0, T]$ the identity

$$I(t_1; u, \eta) \equiv \int_\Omega u(x, t_1)\eta(x, t_1)\,dx - \int_0^{t_1}\int_\Omega u\eta_t\,dx\,dt + \int_0^{t_1} [\mathscr{L}_1(u, \eta) + \mathscr{L}_2(\mathbf{f}, \eta)]\,dt = 0 \tag{1.7}$$

for all $\eta(x, t)$ from $\overset{\circ}{W}{}_2^{1,1}(Q_T)$ that are equal to zero for $t = 0$. Here and below we make use of the abridged notation

$$\mathscr{L}_1(u, \eta) = \int_\Omega [(a_{ij}u_{x_j} + a_i u)\eta_{x_i} + (b_i u_{x_i} + au)\eta]\,dx,$$

$$\mathscr{L}_2(\mathbf{f}, \eta) = \int_\Omega (f_i\eta_{x_i} + f\eta)\,dx. \tag{1.8}$$

By *a generalized solution from $V_2(Q_T)$ of the problem*

$$\mathscr{L}u = \frac{\partial f_i}{\partial x_i} - f, \quad u|_{S_T} = 0, \quad u|_{t=0} = \psi_0(x) \tag{1.9}$$

we will mean a function $u(x, t)$ from $\overset{\circ}{V}_2(Q_T)$ satisfying for almost all t_1 from $[0, T]$ the identity

$$I(t_1; u, \eta) = \int_\Omega \psi_0(x)\eta(x, 0)\,dx \tag{1.10}$$

for all η from $\overset{\circ}{W}{}_2^{1,1}(Q_T)$.

Let us verify that if conditions (1.3)–(1.6) are fulfilled and $\psi_0 \in L_2(\Omega)$, then all of the integrals in these identities are finite for any functions u and η from the indicated classes. This is so by virtue of the following estimates (see inequalities (1.8) and (3.4) of Chapter II):

$$\int_{Q_T} |a_i u \eta_{x_i}| \, dx\, dt \leqslant \|\eta_x\|_{2,Q_T} \left\| \sum_i a_i^2 \right\|_{q,r,Q_T}^{1/2} \|u\|_{\bar q,\bar r,Q_T} \leqslant \mu_1^{1/2}\beta \|\eta_x\|_{2,Q_T} |u|_{Q_T}; \tag{1.11}$$

$$\int_{Q_T} |b_i u_{x_i} \eta| \, dx\, dt \leqslant \left\| \sum_i b_i^2 \right\|_{q,r,Q_T}^{1/2} \|u_x\|_{2,Q_T} \|\eta\|_{\bar q,\bar r,Q_T} \leqslant \mu_1^{1/2}\beta \|u_x\|_{2,Q_T} |\eta|_{Q_T}; \tag{1.12}$$

$$\int_{Q_T} |au\eta| \, dx\, dt \leqslant \|a\|_{q,r,Q_T} \|u\|_{\bar q,\bar r,Q_T} \|\eta\|_{\bar q,\bar r,Q_T} \leqslant \mu_1\beta^2 |u|_{Q_T} |\eta|_{Q_T}; \tag{1.13}$$

$$\int_{Q_T} |f\eta| \, dx\, dt \leqslant \|f\|_{q_1,r_1,Q_T} \|\eta\|_{q_1',r_1',Q_T} \leqslant \beta \|f\|_{q_1,r_1,Q_T} |\eta|_{Q_T} \tag{1.14}$$

In all of these inequalities q and r are connected with $\bar q$ and $\bar r$ by the equalities $q = \bar q/(\bar q - 2)$, $r = \bar r/(\bar r - 2)$, while q_1 and r_1 are connected with q_1' and r_1' by the equalities $q_1' = q_1/(q_1 - 1)$, $r_1' = r_1/(r_1 - 1)$.

The remaining integrals in (1.7) and (1.10) also obviously exist and are finite for any u and η from the indicated classes. Thus the definitions given above are correct. We note in addition that the values $r = \infty$, $q = n/2$ are admissible for $n \geq 3$.

It is possible to define the generalized solutions from $V_2(Q_T)$ somewhat differently. Namely, $u(x, t)$ is a generalized solution from $V_2(Q_T)$ of equation (1.1) if $u(x, t) \in V_2(Q_T)$ and satisfies the identity

$$I_0(u, \eta) \equiv -\int_{Q_T} u\eta_t \, dx\, dt + \int_0^T [\mathscr{L}_1(u, \eta) + \mathscr{L}_2(\hat f, \eta)]\, dt = 0 \tag{1.15}$$

for all $\eta(x, t)$ from $\overset{\circ}{W}{}_2^{1,1}(Q_T)$ that are equal to zero for $t = 0$ and $t = T$, and $u(x, t)$ is a generalized solution from $V_2(Q_T)$ of problem (1.9) if $u \in \overset{\circ}{V}_2(Q_T)$ and satisfies the identity

$$I_0(u, \eta) = \int_\Omega \psi_0(x)\eta(x, 0)\, dx \tag{1.16}$$

for all $\eta(x, t)$ from $\overset{\circ}{W}{}_2^{1,1}(Q_T)$ that are equal to zero for $t = T$.

Let us show that these definitions are equivalent to the ones given above. The fact that the latter two follow from the first two is obvious, since, taking η

equal to zero in (1.7), (1. 10) for $t \geq T-\epsilon$, $\epsilon > 0$, and the number $t_1 > T-\epsilon$, we obtain (1.15), (1.16) for such η, and hence for all η required in the latter definitions. Now suppose u is a generalized solution of equation (1.1) in the sense of the second definition. In (1.15) we take η in the form $\Phi(t, t_1, \epsilon)\hat{\eta}(x, t)$, where $\hat{\eta}(x, t)$ is an arbitrary function from $\overset{\circ}{W}{}_2^{1,1}(Q_T)$ that is equal to zero for $t = 0$, while $\Phi(t, t_1, \epsilon)$ is equal to 1 for $t \leq t_1 - \epsilon$, to zero for $t \geq t_1$, and to $(t_1 - t)/\epsilon$ for $t_1 - \epsilon \leq t \leq t_1$. We substitute this η into (1.15) and pass to the limit as $\epsilon \to 0$. It is easy to see that the second integral in (1.15) gives in the limit the last integral in (1.7). And the first integral in (1.15) is representable in the form

$$-\int_{Q_T} u(\Phi\hat{\eta}_t + \Phi_t\hat{\eta})\,dx\,dt$$

$$= -\int_{Q_{t_1}} u\Phi\hat{\eta}_t\,dx\,dt + \frac{1}{\varepsilon}\int_{t_1-\varepsilon}^{t_1}\int_\Omega u\hat{\eta}\,dx\,dt. \qquad (1.17)$$

The function $\psi(t) = \int_\Omega u(x, t)\hat{\eta}(x, t)\,dx$ is summable on $[0, T]$, so that (see §4 of Chapter II) there exists a sequence $\epsilon_k \to 0$ for which the integrals $(1/\epsilon_k)\int_{t_1-\epsilon_k}^{t_1}\psi(t)\,dt$ converge to $\psi(t_1)$ for almost all t_1 from $[0, T]$. The limit of the integral $-\int_{Q_{t_1}} u\Phi\hat{\eta}_t dx\,dt$ will obviously be $-\int_{Q_{t_1}} u\hat{\eta}_t dx\,dt$, and hence in the limit relation (1.15) goes over into relation (1.7).

Thus the equivalence of the two definitions of generalized solutions from $V_2(Q_T)$ of equations (1.1) is proved. The equivalence of the two definitions given above of generalized solutions from $V_2(Q_T)$ of problem (1.9) is proved in the same way.

For an investigation of the properties of generalized solutions of equations (1.1) it is convenient to introduce two additional classes of solutions: generalized solutions from the space $V_2^{1,0}(Q_T)$ and from the space $V_2^{1,1/2}(Q_T)$. They are defined in the same way as the generalized solutions from $V_2(Q_T)$ of an equation (or problem), with the only difference that they are elements of $V_2^{1,0}(Q_T)$ or $V_2^{1,1/2}(Q_T)$ ($\overset{\circ}{V}{}_2^{1,0}(Q_T)$ or $\overset{\circ}{V}{}_2^{1,1/2}(Q_T)$ respectively) and satisfy identity (1.7) (or (1.10)) for all t_1 from $[0, T]$. Both of the definitions of generalized solutions from $V_2^{1,0}(Q_T)$ and $V_2^{1,1/2}(Q_T)$ corresponding to the two definitions of generalized solutions from $V_2(Q_T)$ given above are equivalent to each other.

It is clear that a generalized solution from $V_2^{1,0}(Q_T)(V_2^{1,1/2}(Q_T))$ of equation (1.1) or problem (1.9) is a generalized solution from $V_2(Q_T)$ of equation (1.1) or problem (1.9). And, conversely, if $u(x, t)$ is a generalized solution from $V_2(Q_T)$

(of an equation or problem) and if additionally it is known that $u(x, t)$ is an element of $V_2^{1,0}(Q_T)(V_2^{1,1/2}(Q_T))$, then it will also be a generalized solution from $V_2^{1,0}(Q_T)(V_2^{1,1/2}(Q_T))$ (of the equation or problem respectively).

§2. THE ENERGY INEQUALITY

We will show that for any generalized solution $u(x, t)$ from $V_2^{1,0}(Q_T)$ of problem (1.9) one can estimate $|u|_{Q_T}$ in terms of the known quantities in this problem. For this purpose we will prove as a preliminary the following lemma.

Lemma 2.1. *Suppose $u \in \overset{\circ}{V}_2(Q_T)$ and satisfies for almost all t_1 and t_2 from $[0, T]$, including $t_1 = 0$, the inequalities*

$$\frac{1}{2}\int u^2(x, t)\,dx\Big|_{t=t_1}^{t=t_2} + \int_{t_1}^{t_2} [\mathscr{L}_1(u, u) + \mathscr{L}_2(\hat{\mathbf{f}}, u)]\,dt \leqslant 0, \tag{2.1}$$

in which the functions a_i, b_i, a, f_i and f satisfy conditions (1.2)–(1.6). *Then*

$$|u|_{Q_T} \leqslant c\,[\,\|u(x, 0)\|_{2,\Omega} + \|\mathbf{f}\|_{2,Q_T} + \|f\|_{q_1, r_1, Q_T}], \tag{2.2}$$

with the constant c depending only on n, ν, μ, μ_1 and q from (1.2) *and* (1.3).

From (2.1) by virtue of (1.2) we have

$$\frac{1}{2}\int_\Omega u^2(x, t)\,dx\Big|_{t=t_1}^{t=t_2} + \nu\int_{t_1}^{t_2}\int_\Omega u_x^2\,dx\,dt$$
$$\leqslant \int_{t_1}^{t_2}\int_\Omega \Big[\frac{\nu}{2}u_x^2 + \Big(\frac{1}{\nu}\sum_i a_i^2 + \frac{1}{\nu}\sum_i b_i^2 + |a|\Big)u^2 + |f_i u_{x_i}| + |fu|\Big]\,dx\,dt,$$

from which, using conditions (1.3)–(1.6) and inequalities (1.6) and (3.4) of Chapter II, we obtain (in a manner analogous to the derivation of (1.11)–(1.14)) the inequalities

$$\int_\Omega u^2(x, t)\,dx\Big|_{t=t_1}^{t=t_2} + \nu\int_{t_1}^{t_2}\int_\Omega u_x^2\,dx\,dt \leqslant 2\int_{t_1}^{t_2}\int_\Omega (\mathscr{D}u^2 + |f_i u_{x_i}|$$
$$+ |fu|)\,dx\,dt \leqslant 2\beta^2\|\mathscr{D}\|_{q, r, Q_{t_1, t_2}}|u|^2_{Q_{t_1, t_2}}$$
$$+ 2\,(\|\mathbf{f}\|_{2, Q_{t_1, t_2}} + \beta\|f\|_{q_1, r_1, Q_{t_1, t_2}})\,|u|_{Q_{t_1, t_2}},$$

in which

$$\mathscr{D}=\frac{1}{\nu}\sum_i a_i^2+\frac{1}{\nu}\sum_i b_i^2+|a|.$$

From these inequalities it follows that

$$\min(1;\ \nu)|u|^2_{Q_{t_1,t_2}}\leqslant\|u(x,\ t_1)\|^2_{2,\Omega}+\mu(t_1,\ t_2)|u|^2_{Q_{t_1,t_2}}+\mathscr{F}(t_1,\ t_2)|u|_{Q_{t_1,t_2}},\tag{2.3}$$

where

$$\mu(t_1,\ t_2)=2\beta^2\|\mathscr{D}\|_{q,r,Q_{t_1,t_2}}$$

and

$$\mathscr{F}(t_1,\ t_2)=2\left(\|\mathbf{f}\|_{2,Q_{t_1,t_2}}+\beta\|f\|_{q_1,r_1,Q_{t_1,t_2}}\right).$$

If $\mu(t_1,\ t_2)$ is less than $\min\{1,\ \nu\}$, then with the use of (2.3) it is possible to estimate $|u|_{Q_{t_1,t_2}}$ from above in terms of $\|u(x,\ t)\|_{2,\Omega}$ and known quantities. In order to do this, we divide the interval $[0,\ T]$ into a finite number of intervals $[0,\ t_1],\ [t_1,\ t_2],\cdots,\ [t_{s-1},\ t_s=T]$ of such small length that

$$\mu(t_{k-1},\ t_k)\leqslant\frac{\nu_1}{2}\equiv\frac{1}{2}\min(1;\ \nu),\tag{2.4}$$

with the division points t_k having been selected so that they can be taken as limits of integration in (2.1) (we denote the set of such admissible t_k by $\mathfrak{S}$). This is possible for $r<\infty$, with the number of divisions not exceeding some known quantity. In fact,

$$\sum_{k=1}^{s}\mu^r(t_{k-1},\ t_k)=(2\beta^2)^r\|\mathscr{D}\|^r_{q,r,Q_T},$$

and therefore, if a partitioning is selected so that all $\mu(t_{k-1},\ t_k)$, except possibly the last, are not less than $\nu_1/4$, then

$$(s-1)\left(\frac{\nu_1}{4}\right)^r\leqslant(2\beta^2)^r\|\mathscr{D}\|^r_{q,r,Q_T},$$

and consequently the number of divisions s will be bounded from above:

$$s\leqslant 1+\left(\frac{8\beta^2}{\nu_1}\|\mathscr{D}\|_{q,r,Q_T}\right)^r\tag{2.5}$$

Since it is necessary that no $\mu(t_{k-1},\ t_k)$ be greater than $\nu_1/2$, it is possible, without increasing the number of divisions, to place the ends t_k in $\mathfrak{S}$.

For each of these divisions we have by virtue of (2.3) the inequalities

$$\frac{\nu_1}{2}|u|^2_{Q_{t_{k-1},t_k}} \leqslant \|u(x,\ t_{k-1})\|^2_{2,\Omega}+\mathscr{F}(t_{k-1},\ t_k)|u|_{Q_{t_{k-1},t_k}}$$
$$\leqslant \|u(x,\ t_{k-1})\|^2_{2,\Omega}+\frac{\nu_1}{4}|u|^2_{Q_{t_{k-1},t_k}}+\frac{1}{\nu_1}\mathscr{F}^2(t_{k-1},\ t_k),$$

from which follows assertion (2.2) of Lemma 2.1.

Remark 2.1. From the given derivation it is seen that if

$$2\beta^2\|\mathscr{D}\|_{q,\,r,\,Q_T}<\min\{1;\ \nu\}, \tag{2.6}$$

then inequality (2.2) also holds for $r=\infty$, since in this case it is deduced directly from (2.3) without a partitioning of the interval $[0,\ T]$.

Now suppose $u(x,\ t)$ is a generalized solution from $\overset{\circ}{V}{}_2^{1,0}(Q_T)$ of problem (1.9). We take as $\eta(x,\ t)$ in (1.15) the function

$$\widehat{\eta}_{\bar h}(x,\ t)=\frac{1}{h}\int\limits_{t-h}^{t}\widehat{\eta}(x,\ \tau)\,d\tau, \tag{2.7}$$

where $\widehat{\eta}(x,\ t)$ is an arbitrary element of $\overset{\circ}{W}{}_2^{1,1}(Q_{-h,T})$ that is equal to zero for $t\geq T-h$ and for $t\leq 0$, and we transform the first term in it in the following manner:

$$-\int\limits_{Q_T}u\widehat{\eta}_{\bar h t}\,dx\,dt=-\int\limits_{Q_T}u_h\widehat{\eta}_t\,dx\,dt=\int\limits_{Q_T}u_{ht}\widehat{\eta}\,dx\,dt. \tag{2.8}$$

Here we have used the relation

$$\int\limits_0^T u(t)\,\widehat{\eta}_{\bar h}(t)\,dt=\int\limits_0^{T-h}u_h(t)\,\widehat{\eta}(t)\,dt, \tag{2.9}$$

valid for any square-summable functions $u(t)$ and $\widehat{\eta}(t)$ on $[-h,\ T]$, one of which is equal to zero on the intervals $[-h,\ 0]$ and $[T-h,\ T]$, and the notation

$$u_h(x,\ t)=\frac{1}{h}\int\limits_t^{t+h}u(x,\ \tau)\,d\tau. \tag{2.10}$$

Equality (2.9) is the result of interchanging the order of integration with respect to t and τ. In all other terms of (1.16) we also transfer the averaging $(\ )_{\bar h}$ from $\widehat{\eta}$ to its coefficient, taking into account the permutability of this averaging with differentiation with respect to x. This gives the identity

$$\int\limits_{Q_{T-h}}[u_{ht}\widehat{\eta}+(a_{ij}u_{x_j}+a_iu+f_i)_h\widehat{\eta}_{x_i}+(b_iu_{x_i}+au+f)_h\,\widehat{\eta}]\,dx\,dt=0. \tag{2.11}$$

This identity is actually valid for a class of functions $\hat{\eta}(x, t)$ that is more extensive than the class just considered; namely, it is valid for any function $\hat{\eta}$ that is equal to zero for $t > t_1$ $(t_1 \leq T - h)$ and is equal to some function $\eta(x, t)$ from $\overset{\circ}{V}{}_2^{1,0}(Q_{t_1})$ for $t \in [0, t_1]$. Indeed, the set $\overset{\circ}{W}{}_2^{1,1}(Q_{-h,T})$ is dense in the space $\overset{\circ}{V}{}_2^{1,0}(Q_{-h,T})$. Thus for any $\eta(x, t)$ from $\overset{\circ}{V}{}_2^{1,0}(Q_{-h,T})$ there is a sequence of functions $\eta_m(x, t)$ from $\overset{\circ}{W}{}_2^{1,1}(Q_{-h,T})$, that is strongly convergent to it for $m \to \infty$ in the norm of $\overset{\circ}{V}{}_2^{1,0}(Q_{-h,T})$. We denote by $\chi_k(t)$ the continuous piecewise-linear functions

$$\chi_k(t) = \begin{cases} 0 & \text{for} \quad t \leqslant 0, \\ kt & \text{for} \quad 0 \leqslant t \leqslant \frac{1}{k}, \\ 1 & \text{for} \quad \frac{1}{k} \leqslant t \leqslant t_1 - \frac{1}{k}, \\ k(t_1 - t) & \text{for} \quad t_1 - \frac{1}{k} \leqslant t \leqslant t_1, \\ 0 & \text{for} \quad t \geqslant t_1. \end{cases}$$

Identity (2.11) is established for $\hat{\eta}_{m,k}(x, t) = \eta_m(x, t)\,\chi_k(t)$ when $t_1 \leq T - h$. It is easy to see that one can pass to the limit in it as m and $k \to \infty$ and thereby prove the identity

$$\int\limits_{Q_{t_1}} [u_{ht}\eta + (a_{ij}u_{x_j} + a_i u + f_i)_h \eta_{x_i} + (b_i u_{x_i} + au + f)_h \eta]\, dx\, dt = 0 \tag{2.12}$$

for any function $\eta(x, t)$ from $\overset{\circ}{V}{}_2^{1,0}(Q_{t_1})$ when $t_1 \leq T - h$.

In (2.12) we take $\eta(x, t) = u_h(x, t)$ and represent the first term in the form

$$\int\limits_{Q_{t_1}} u_{ht}u_h\, dx\, dt = \frac{1}{2}\int\limits_{\Omega} u_h^2(x, t)\, dx \Big|_{t=0}^{t=t_1},$$

after which we let h tend to zero. This gives

$$\frac{1}{2}\int\limits_{\Omega} u^2(x, t)\, dx \Big|_{t=0}^{t=t_1} + \int\limits_0^{t_1} [\mathscr{L}_1(u, u) + \mathscr{L}_2(\mathfrak{f}, u)]\, dt = 0. \tag{2.13}$$

Indeed, by virtue of our assumptions (1.2)–(1.6) on the coefficients and free terms and by virtue of the membership of u in $\overset{\circ}{V}{}_2^{1,0}(Q_T)$ it follows that the functions $a_{ij}u_{x_j} + a_i u + f_i$ are elements of $L_2(Q_T)$, the functions au and $b_i u_{x_i}$ are elements of $L_{2q/(q+1),2r/(r+1)}(Q_T)$, while the function f is by condition an element of $L_{q_1,r_1}(Q_T)$. This follows from conditions (1.5), estimates (1.11)–(1.14) and

Theorem 1.1 of Chapter II. By Lemma 4.7 of Chapter II the averagings of functions from some space $L_{q,r}(Q_T)$ converge to them in the norm of this space. But for elements from $V_2^{1,0}(Q_T)$ their averagings converge to them in the norm of $V_2^{1,0}(Q_{T-\delta})$ and in any of the norms of $L_{q,r}(Q_{T-\delta})$ with q and r subject to conditions (3.3) of Chapter II, including the norm of the space $L_{2q/(q-1),2r/(r-1)}(Q_T)$, which is conjugate to the space $L_{2q/(q+1),2r/(r+1)}(Q_T)$, with q and r taken from conditions (1.3), (1.4). Therefore in the integrals

$$\int_{Q_{t_1}} (a_{ij}u_{x_j} + a_i u + f_i)_h \eta_{hx_i}\, dx\, dt, \quad \int_{Q_{t_1}} (b_i u_{x_i} + au + f)_h \eta_h \, dx\, dt \tag{2.14}$$

we can pass to the limit as $h \to 0$ for any η from $V_2^{1,0}(Q_T)$ and obtain in the limit

$$\begin{gathered}\int_{Q_{t_1}} (a_{ij}u_{x_j} + a_i u + f_i)\eta_{x_i}\, dx\, dt, \\ \int_{Q_{t_1}} (b_i u_{x_i} + au + f)\eta\, dx\, dt.\end{gathered} \tag{2.15}$$

This is true, in particular, for $\eta = u$. Finally, the integrals $\int_\Omega u_h^2(x, t)\, dx$ converge to $\int_\Omega u^2(x, t)\, dx$ for all $t \in [0, T-h]$ since $u(x, t)$ depends continuously on t in the norm of $L_2(\Omega)$. Thus (2.13) is proved.

From (2.13) and Lemma 2.1 we get

Theorem 2.1. *For any generalized solution u from $\overset{\circ}{V}{}_2^{1,0}(Q_T)$ of problem* (1.9) *the inequality*

$$|u|_{Q_T} \leqslant c\,[\,\|\psi_0\|_{2,\Omega} + \|\mathbf{f}\|_{2,Q_T} + \|f\|_{q_1,r_1,Q_T}] \tag{2.16}$$

is valid if assumptions (1.2)–(1.6) *are fulfilled with respect to equation* (1.1). *The constant c in it depends only on n, ν, μ, μ_1 and q from* (1.2) *and* (1.3).

Remark 2.2. Estimate (2.16) also holds for $r = \infty$ if it is known in addition that inequality (2.6) is valid.

Let us now assume that $u(x, t)$ is a generalized solution from $V_2^{1,0}(Q_T)$ of equation (1.1), so that in comparison with the case just considered it does not vanish on S_T. It is obvious that for it identity (2.12) is valid for all η from $\overset{\circ}{V}{}_2^{1,0}(Q_T)$. We put $\eta(x, t) = u_h(x, t)\zeta^2(x, t)$, in it, where $\zeta(x, t)$ is an arbitrary continuous nonnegative function from $\overset{\circ}{W}{}_2^{1,1}(Q)$, and, transforming the first term as above, we pass to the limit as $h \to 0$. This gives the relation

$$\frac{1}{2}\int_{\Omega} u^2(x, t)\zeta^2(x, t)\,dx\Big|_{t=0}^{t=t_1}$$
$$+\int_0^{t_1} [\mathscr{L}_1(u, u\zeta^2) + \mathscr{L}_2(\hat{\mathrm{f}}, u\zeta^2) - (u, u\zeta\zeta_t)]\,dt = 0, \tag{2.17}$$

which is similar to (2.13). From it follows the estimate

$$u\zeta|_{Q_T} \leqslant c\big[\,\| u(x, 0)\zeta(x, 0)\|_{2, \Omega} + \|\mathrm{f}\zeta\|_{2, Q_T}$$
$$+ \| f\zeta\|_{q_1, r_1, Q_T} + \| u\zeta_x\|_{2, Q_T} + \big\| u\sqrt{\zeta|\zeta_t|}\,\big\|_{2, Q_T}\big], \tag{2.18}$$

in which the constant c depends only on n, ν, μ, μ_1 and q from (1.2) and (1.3). It is derived from (2.17) in almost the same way as the energy equality (2.2) was derived from (2.1). The terms in which ζ is not differentiated differ from the corresponding terms of (2.1) only by the factor ζ^2 and are estimated in the same way as above. The remaining terms (not represented in (2.1)) have the form

$$I_1 = \int_{Q_{t_1}} \big[(a_{ij}u_{x_j} + a_i u + f_i)\, u 2\zeta\zeta_{x_i} - u^2\zeta\zeta_t\big]\,dx\,dt$$

and are estimated by means of Cauchy's inequality as follows:

$$|I_1| \leqslant 2\,(\mu\,\| u_x\zeta\|_{2, Q_{t_1}}$$
$$+ \Big\|\sqrt{\textstyle\sum_i a_i^2}\,u\zeta\Big\|_{2, Q_{t_1}} + \|\mathrm{f}\zeta\|_{2, Q_{t_1}})\,\| u\zeta_x\|_{2, Q_{t_1}} + \big\| u\sqrt{\zeta|\zeta_t|}\,\big\|^2_{2, Q_{t_1}}. \tag{2.19}$$

The quantity

$$\Big\|\sqrt{\textstyle\sum_i a_i^2}\,u\zeta\Big\|_{2, Q_{t_1}}$$

as is seen from (1.11), does not exceed $\mu_1^{1/2}\beta|u\zeta|_{Q_{t_1}}$, and thus (2.19) implies

$$|I_1| \leqslant \varepsilon\,(\mu\,\| u_x\zeta\|_{2, Q_{t_1}} + \mu_1\beta\,|u\zeta|_{Q_{t_1}} + \|\mathrm{f}\zeta\|_{2, Q_{t_1}})^2$$
$$+ \varepsilon^{-1}\| u\zeta_x\|^2_{2, Q_{t_1}} + \big\| u\sqrt{\zeta|\zeta_t|}\,\big\|^2_{2, Q_{t_1}}. \tag{2.20}$$

This estimate in combination with the estimate of the other (basic) terms carried out above gives inequality (2.18).

§3. UNIQUENESS THEOREMS

Uniqueness theorems for solutions of boundary value problems and of the Cauchy problem for the parabolic equations (1.1) in the space $C^{2,1}(Q_T)\cap C(\bar{Q}_T)$ were proved in §2 of Chapter I. This was done on the basis of the so-called "maximum principle," which requires the existence of those derivatives of the solution that enter into the equation. We will prove here that a uniqueness theorem for problem (1.9) is preserved for solutions not having the derivatives $\partial/\partial t$ and $\partial^2/\partial x_i \partial x_j$.

The first theorems of this type were established by one of the authors of the present book (see [65c–f; 124f]). The uniqueness of generalized solutions of boundary value problems was proved in them for second order equations of hyperbolic and parabolic types in the spaces $W_2^{1,1}(Q_T)$ and $W_2^{1,0}(Q_T)$ respectively. We will give here the original proof for the case of parabolic equations (Theorem 3.2). In the process of carrying it out an inequality giving an estimate of the norm of the solution is established that is weaker than the energy inequality (such inequalities, following Gårding, have been called "dual"). But it is necessary in this connection to assume that the coefficients a_{ij} and b_i have derivatives a_{ijt} and b_{ix_i}. Ladyženskaja, Lions and Prodi have since given other methods of proof, in which the only requirement imposed on the coefficients of equation (1.1) is that they be bounded. Theorem 3.3 is a generalization of these theorems to the case of unbounded coefficients. A desire to include the more general case, when the coefficients a_i, b_i, a belong to classes $L_{q_k,r_k}(Q_T)$, led to generalized solutions from $V_2^{1,0}(Q_T)$: a somewhat narrower class of solutions than the solutions from $W_2^{1,0}(Q_T)$, in which nevertheless, as we will show below, one has unique solvability of initial-boundary value problems.

Uniqueness in the class $V_2^{1,0}(Q_T)$ is a simple consequence of Theorem 2.1, which was proved in the preceding section. From Theorems 2.1 and 4.2, the latter of which will be proved below, there follows a stronger assertion:

Theorem 3.1. *If the coefficients of equation* (1.1) *satisfy conditions* (1.3) *and* (1.4), *then the first boundary value problem for* (1.1) *cannot have two distinct generalized solutions from* $V_2(Q_T)$.

Indeed, the difference $u(x, t)$ of two such solutions would be a generalized solution from $V_2(Q_T)$ of the homogeneous problem (1.9) (i.e. problem (1.9) with $\psi_0 = f = f \equiv 0$). By virtue of Theorem 4.2 the function $u(x, t)$ will be an element of $V_2^{1,1/2}(Q_T)$, and hence inequality (2.18) is valid for it, so that $u \equiv 0$.

In the classes of bounded generalized solutions uniqueness is preserved under broader assumptions concerning the coefficients $a_i(x, t)$. One of the variants of theorems of this type is Theorem 3.4. For its proof we employ the classical method proposed by Holmgren at the beginning of the century, which makes use of one's ability to solve the adjoint problem in a class of sufficiently well-behaved functions (we note that the adjoint of an initial-boundary-value problem is essentially a problem of the same type). We will also mention uniqueness theorems for generalized solutions of boundary value problems from the space $L_2(Q_T)$ [65d, f; 124f]. These theorems are distinguished by the fact that existence theorems are readily deduced consequences of them, just as in linear algebra uniqueness implies solvability of a linear algebraic system for any right sides. We will discuss them through the example of problem (14.2) in §14 of the present chapter. These theorems are valid if the coefficients of the equations possess a certain smoothness greater than that used in the proof of unique solvability in the space $V_2^{1,0}(Q_T)$. But in return solvability is also proved in a class of better-behaved functions. This conforms to a general tendency: the better the coefficients of an equation are, the better is any solution of the equation corresponding to sufficiently smooth free terms and the worse are the solutions admissible for such equations, the deterioration being effected at the expense of extending the class of free terms and functions determining the initial and boundary conditions.

Let us proceed to a proof of the theorems.

Theorem 3.2. *Suppose the coefficients of equation* (1.1) *satisfy the conditions* (1.2),

$$\max_{Q_T}\left| a_i - b_i,\ \frac{\partial a_{ij}}{\partial t}\right| \leqslant \mu_1 \tag{3.1}$$

and

$$\max_{0\leq t\leq T}\left\| a - \frac{\partial b_i}{\partial x_i}\right\|_{q,\,\Omega} \leqslant \mu_2, \tag{3.2}$$

while the free terms satisfy the conditions

$$\mathbf{f}\in L_2(Q_T),\quad f\in L_{\frac{2q}{q+2},\,2}(Q_T), \tag{3.3}$$

in which $q = n$ for $n \geq 3$, $q > n$ for $n = 2$ and $q = 2$ for $n = 1$. Then for any generalized solution $u(x, t)$ from $W_2^{1,0}(Q_T)$ of problem (1.9) *with $\psi_0(x) = \partial\psi_i(x)/\partial x_i$, $\psi(x) = (\psi_1, \cdots, \psi_n) \in L_2(\Omega)$, the estimate (the inequality dual to the energy inequality)*

$$\max_{0\leq t\leq t_1}\frac{\nu}{2}\|\zeta_x(x,t)\|^2_{2,\Omega}+2\|\zeta_t\|^2_{2,Q_{t_1}}$$
$$\leqslant 2e^{\frac{4c_1}{\nu}t_1}\left[\frac{4}{\nu}\|\boldsymbol{\psi}\|^2_{2,\Omega}+\|\mathbf{f}\|^2_{2,Q_{t_1}}+\|f\|^2_{\frac{2q}{q+2},2,Q_{t_1}}\right] \tag{3.4}$$

holds for $t_1\in[0,\nu/4c_1]$, *in which* $\zeta(x,t)=\int_0^t u(x,\xi)\,d\xi$, $c_1=10n\mu_1+4+4c^2(1+\mu_2)$ *and the constant* c *is taken from inequality* (2.18) *of Chapter* II. *In particular, for* $\psi_0(x)=u|_{t=0}=0$ *and* $\mathbf{f}=f\equiv 0$ *the solution* $u(x,t)\equiv 0$, *i.e. a uniqueness theorem holds for generalized solutions from* $W_2^{1,0}(Q_T)$ *of problem* (1.9).

By *a generalized solution from* $W_2^{1,0}(Q_T)$ *of problem* (1.9), in accordance with the general concept of a generalization of the notion of a solution described in §3 of Chapter I, it is necessary to understand a function $u(x,t)$ from $\overset{\circ}{W}{}_2^{1,0}(Q_T)$ satisfying the integral identity (1.16) or the identity

$$\int_{Q_T}\{-u\eta_t+[a_{ij}u_{x_j}+(a_i-b_i)u+f_i]\eta_{x_i}$$
$$+(a-b_{ix_i})u\eta+f\eta\}\,dx\,dt-\int_\Omega\psi_0(x)\eta(x,0)\,dx=0 \tag{3.5}$$

for any function $\eta(x,t)$ from $\overset{\circ}{W}{}_2^{1,1}(Q_T)$ that is equal to zero for $t=T$ if $\psi_0(x)\in L_2(\Omega)$. If on the other hand $\psi_0(x)$ is a generalized function of the form ψ_{ix_i} with $\psi_i(x)\in L_2(\Omega)$, then in identity (3.5) we consider the term $-\int_\Omega\psi_0(x)\eta(x,0)\,dx$ to be equal to $\int_\Omega\psi_i(x)\eta_{x_i}(x,0)\,dx$ and we assume that $\eta(x,t)$ has the derivatives η_{x_it} from $L_2(Q_T)$. Under the conditions of Theorem 3.2 we assume that u satisfies an integral identity in the form of (3.5).

For a proof of (3.4) we set $\eta(x,t)$ equal to zero for $t\geq t_1$ and to $-\int_{t_1}^t u(x,\tau)\,d\tau$ for $t\in[0,t_1]$, where t_1 is an arbitrary number from $(0,T]$, and write the resultant equality in the form

$$\int_{Q_{t_1}}\{\eta_t^2-[a_{ij}\eta_{tx_j}+(a_i-b_i)\eta_t-f_i]\eta_{x_i}$$
$$-(a-b_{ix_i})\eta_t\eta+f\eta\}\,dx\,dt+\int_\Omega\psi_i\eta_{x_i}(x,0)\,dx=0. \tag{3.6}$$

We transform the remaining terms of this equality, bearing in mind that $\eta(x,t_1)=0$, and then pass from the equality to the corresponding inequality:

$$\frac{1}{2}\int_\Omega a_{ij}(x,0)\eta_{x_i}(x,0)\eta_{x_j}(x,0)\,dx+\int_{Q_{t_1}}\eta_t^2\,dx\,dt$$
$$=\int_{Q_{t_1}}\Big[-\frac{1}{2}\frac{\partial a_{ij}}{\partial t}\eta_{x_i}\eta_{x_j}+(a_i-b_i)\eta_t\eta_{x_i}-f_i\eta_{x_i}$$
$$+(a-b_{ix_i})\eta\eta_t-f\eta\Big]dx\,dt-$$
$$-\int_\Omega \psi_i\eta_{x_i}(x,0)\,dx\leqslant\int_{Q_{t_1}}\Big[\frac{n}{2}\mu_1\eta_x^2+\frac{1}{4}\eta_t^2+n\mu_1^2\eta_x^2$$
$$+\frac{1}{4}\mathrm{f}^2+\eta_x^2+\frac{1}{4}\eta_t^2+(a_i-b_{ix_i})^2\eta^2\Big]dx\,dt$$
$$+\int_0^{t_1}\|f\|_{\frac{2q}{q+2},\Omega}\|\eta\|_{\frac{2q}{q-2},\Omega}dt+\frac{\nu}{4}\|\eta_x(x,0)\|^2_{2,\Omega}+\frac{1}{\nu}\|\psi\|^2_{2,\Omega}. \tag{3.7}$$

Further, we estimate the term containing η^2, using Hölder's inequality and the embedding theorem (2.18) of Chapter II, as follows:

$$\int_\Omega (a-b_{ix_i})^2\eta^2\,dx\leqslant\|a-b_{ix_i}\|^2_{q,\Omega}\|\eta\|^2_{\frac{2q}{q-2},\Omega}\leqslant c^2\mu_2^2\|\eta_x\|^2_{2,\Omega},$$

and analogously

$$\int_0^{t_1}\|f\|_{\frac{2q}{q+2},\Omega}\|\eta\|_{\frac{2q}{q-2},\Omega}dt\leqslant\frac{1}{4}\|f\|^2_{\frac{2q}{q+2},2,Q_{t_1}}+c^2\|\eta_x\|^2_{2,Q_{t_1}}.$$

Substituting these estimates into (3.7), using condition (1.2) and reducing similar terms, we obtain

$$\nu\|\eta_x(x,0)\|^2_{2,\Omega}+2\|\eta_t\|^2_{2,Q_{t_1}}$$
$$\leqslant c_1\|\eta_x\|^2_{2,Q_{t_1}}+\|\mathrm{f}\|^2_{2,Q_{t_1}}+\|f\|^2_{\frac{2q}{q+2},2,Q_{t_1}}+\frac{4}{\nu}\|\psi\|^2_{2,\Omega}, \tag{3.8}$$

where

$$c_1=4\Big(\frac{n}{2}\mu_1+n\mu_1^2+1+c^2\mu_2^2+c^2\Big).$$

The function $\eta(x,t)$ depends on the parameter t_1, which had not been mentioned explicitly. We now take advantage of the arbitrariness in the choice of t_1. We introduce in place of η another function $\zeta(x,t)=\int_0^t u(x,\xi)\,d\xi$. It is clear that $\eta(x,t)=\zeta(x,t_1)-\zeta(x,t)$,

and

$$\|\eta_x\|^2_{2,Q_{t_1}} \leqslant 2\|\zeta_x\|^2_{2,Q_{t_1}} + 2t_1\|\zeta_x(x,t_1)\|^2_{2,\Omega}.$$

In view of this we have from (3.8) the inequality

$$(\nu - 2c_1t_1)\|\zeta_x(x,t_1)\|^2_{2,\Omega} + 2\|\zeta_t\|^2_{2,Q_{t_1}} \leqslant 2c_1\|\zeta_x\|^2_{2,Q_{t_1}} + \mathcal{F}(t_1), \tag{3.9}$$

in which

$$\mathcal{F}(t_1) = \|\mathbf{f}\|^2_{2,Q_{t_1}} + \|f\|^2_{\frac{2q}{q+2},2,Q_{t_1}} + \frac{4}{\nu}\|\psi\|^2_{2,\Omega}.$$

For a t_1 satisfying the condition

$$0 \leqslant t_1 \leqslant \frac{\nu}{4c_1}, \tag{3.10}$$

we will have from (3.9), on the basis of Lemma 5.5 of Chapter II,

$$\frac{\nu}{2}\|\zeta_x(x,t_1)\|^2_{2,\Omega} + 2\|\zeta_t\|^2_{2,Q_{t_1}} \leqslant e^{\frac{4c_1}{\nu}t_1}\mathcal{F}(t_1),$$

and therefore also (3.4). If $\psi_i = \mathbf{f} = f = 0$, then $u(x,t) \equiv 0$ for $t_1 \in [0, \nu/4c_1]$, and therefore identity (3.5) will take the form

$$\int\limits_{\frac{\nu}{4c_1}}^{T}\int\limits_{\Omega}\{-u\eta_t + \ldots\}\,dx\,dt = 0,$$

i.e. $u(x,t)$ will be a generalized solution from $\overset{\circ}{W}{}_2^{1,0}(Q_{\nu/4c_1,T})$ of problem (1.9) in $Q_{\nu/4c_1,T}$ with zero initial condition for $t = \nu/4c_1$. Applying estimate (3.4) to it again in the cylinder $\Omega \times [\nu/4c_1, \nu/2c_1]$, we find that $u \equiv 0$ in it also, and so on, until we exhaust the whole cylinder Q_T. Theorem 3.2 is proved.

Let us now give another proof of the uniqueness theorem in the space $W_2^{1,0}(Q_T)$, not using the existence of derivatives of the coefficients of $\mathfrak{L}$.

Theorem 3.3. *For equations* (1.1), *the coefficients of which satisfy conditions* (1.2) *and the conditions*

$$\left\|\sum_{i=1}^n a_i^2,\ \sum_{i=1}^n b_i^2,\ a\right\|_{q,\infty,Q_T} \leqslant \mu_1,\quad q > \frac{n}{2}\quad \text{for}\quad n \geqslant 2, \tag{3.11$_1$}$$

$$\left\|\sum_{i=1}^n a_i^2,\ \sum_{i=1}^n b_i^2,\ a\right\|_{1,2,Q_T} \leqslant \mu_1, \qquad \text{for}\quad n = 1, \tag{3.11$_2$}$$

problem (1.9) *cannot have more than one generalized solution from* $\overset{\circ}{W}{}_2^{1,0}(Q_T)$.

This theorem is a consequence of Lemma 4.3 and Theorem 3.1 of the present section. Lemma 4.3 guarantees the membership of the difference u of two possible generalized solutions from $W_2^{1,0}(Q_T)$ of problem (1.9) in $\overset{\circ}{V}{}_2^{1,\frac{1}{2}}(Q_{T-\delta})$, where δ is an arbitrary small positive number. It is applied to the function $v(x, t) = u^*(x, t)\omega(t)$, where $u^*(x, t)$ is constructed with respect to $u(x, t)$ as described in (4.9). The function v satisfies identity (4.11) with $f_i = f \equiv 0$, i.e., an identity of type (4.8). It is not difficult to verify that the functions F_i and F corresponding to the case considered by us will satisfy all of the requirements of the first part of Lemma 4.3, and that in (3.11_1) the index q can be taken equal to even $n/2$.

Thus, $u \in \overset{\circ}{V}{}_2^{1,\frac{1}{2}}(Q_{T-\delta})$, $\delta > 0$. Further, conditions (3.11_i) imply conditions (1.3) and (1.4), since for $n \geq 2$ we have $L_{q,\infty}(Q_T) \subset L_{n/2+\epsilon,\, n/2\epsilon+1}(Q_T)$, where $\epsilon = q - n/2 > 0$ and the indices $\hat{q} = n/2 + \epsilon$, $\hat{r} = n/2\epsilon + 1$ satisfy conditions (1.4), while for $n = 1$ condition (3.11_2) simply coincides with (1.3) and (1.4). In view of this, Theorem 3.1 guarantees the equality of $u(x, t)$ to zero in $Q_{T-\delta}$, and since δ is an arbitrary positive number, $u = 0$ in Q_T.

Theorem 3.4. *Suppose the coefficients of* $\mathfrak{L}u$ *satisfy the conditions* (1.2),

$$\left\| \sum_i (a_i - b_i)^2 \right\|_{1,\, Q_T}^{1/2} \qquad \left\| a - \frac{\partial b_l}{\partial x_l} \right\|_{1,\, Q_T} \leqslant \mu_1 \tag{3.12}$$

and[1])

$$a(x, t) - \frac{\partial b_l(x, t)}{\partial x_l} \geqslant -\mu_2, \qquad \mu_2 \geqslant 0. \tag{3.13}$$

Then a uniqueness theorem holds for the generalized solutions u *of problem* (1.9) *from* $W_2^{1,0}(Q_T)$ *with finite* $\operatorname{vrai\,max}_{Q_T}|u|$.

---

1) Condition (3.13) can be replaced by the condition $a(x, t) - \partial b_i(x, t)/\partial x_i \geq -\Theta(t)$, where $\Theta(t)$ is a positive function that is summable on $[0, T]$. It will be seen from the proof of Theorem 3.4 that uniqueness is preserved if the condition of finiteness for $\operatorname{vrai\,max}_{Q_T}|u|$ is weakened to a condition of finiteness for $\|u\|_{q,r,Q_T}$, $q, r \geq 2$, and in return the restrictions on the coefficients of $\mathfrak{L}$ are strengthened by replacing (3.12) with the conditions

$$\left\| \sum_{l=1}^n (a_l - b_l)^2 \right\|_{\frac{q}{q-2},\, \frac{r}{r-2},\, Q_T}, \quad \|a - b_{lx_l}\|_{\frac{q}{q-1},\, \frac{r}{r-1},\, Q_T} \leqslant \mu_1.$$

Such generalized solutions must satisfy identity (3.5) with any $\eta(x, t)$ from $\overset{\circ}{W}{}_2^{1,1}(Q_T)$ equal to zero for $t = T$ and having $\operatorname{vrai\,max}_{Q_T}|\eta| < \infty$. Suppose u is the difference of two possible solutions. Then

$$\int_{Q_T} u\mathscr{L}_m^*\eta\, dx\, dt = l_m(u, \eta), \tag{3.14}$$

where

$$\mathscr{L}_m^*\eta = -\eta_t - \frac{\partial}{\partial x_j}\left(a_{ij}^m\eta_{x_i}\right) + \left(a_i^m - b_i^m\right)\eta_{x_i} + \left(a^m - \frac{\partial b_i^m}{\partial x_i}\right)\eta,$$

$$l_m(u, \eta) = \int_{Q_T}\Big\{\left(a_{ij}^m - a_{ij}\right)u_{x_j}\eta_{x_i} + \left[\left(a_i^m - b_i^m\right) - \left(a_i - b_i\right)\right]u\eta_{x_i}$$
$$+ \left[\left(a^m - \frac{\partial b_i^m}{\partial x_i}\right) - \left(a - \frac{\partial b_i}{\partial x_i}\right)\right]u\eta\Big\}\, dx\, dt,$$

and the a_{ij}^m, b_i^m, a_i^m and a^m are averagings with a smooth kernel of radius $1/m$ of the corresponding coefficients, which we can assume beforehand to be extended onto the whole space with preservation of the properties distinguished by conditions (1.2), (3.12) and (3.13). Let us take as $\eta(x, t)$ the solution of the adjoint boundary value problem in Q_T:

$$\mathscr{L}_m^*\eta = f(x, t), \quad \eta|_{S_T} = 0, \quad \eta|_{t=T} = 0, \tag{3.15}$$

where $f(x, t)$ is an arbitrary smooth function that is equal to zero in a neighborhood of the base $\{x \in \Omega,\ t = T\}$. This problem (it is a problem of the same type as the original one and is changed into the usual form by replacing the independent variable t with $\tau = T - t$), as will be proved in §5 of Chapter IV, has a unique solution $\eta^m(x, t)$ in $H^{2+\alpha, 1+\alpha/2}(\bar{Q}_T)$. Let $m = 1, 2, \cdots$. There are two norms that are uniformly bounded for the full set of solutions $\{\eta^m\}$:

$$\max_{Q_T}|\eta^m| \leq ce^{\mu_2 T}\max_{Q_T}|f| \equiv c_1, \tag{3.16}$$

and

$$\|\eta_x^m\|_{2, Q_T} \leq c_2. \tag{3.17}$$

The first of these inequalities is a consequence of estimate (2.12) of Chapter I and assumptions (1.2) and (3.13), which are preserved for averaged functions (we assume that the averaging kernel is nonnegative). The second is easily deduced from the relation

$$\int_t^T \int_\Omega f\eta^m \, dx \, dt = \int_t^T \int_\Omega \mathscr{L}^*_m \eta^m \eta^m \, dx \, dt$$

$$= \frac{1}{2} \| \eta^m(x, t) \|^2_{2, \Omega} + \int_t^T \int_\Omega \Big[a^m_{ij} \eta^m_{x_i} \eta^m_{x_j} + (a^m_i - b^m_i) \eta^m_{x_i} \eta^m + \left(a^m - \frac{\partial b^m_i}{\partial x_i}\right) \eta^m \eta^m \Big] dx \, dt.$$

if one takes into account assumptions (1.2), (3.12) and the just proved estimate (3.16).

Let us now pass to the limit in (3.14) as $m \to \infty$. The right side of (3.14) will tend to zero in this case, since

$$\left| \int_{Q_T} (a^m_{ij} - a_{ij}) u_{x_j} \eta^m_{x_i} \, dx \, dt \right| \leqslant \sum_{i,j} \| \eta^m_{x_i} \|_{2, Q_T} \| (a^m_{ij} - a_{ij}) u_{x_j} \|_{2, Q_T} \leqslant c_2 \sum_{i,j} \| (a^m_{ij} - a_{ij}) u_{x_j} \|_{2, Q_T}.$$

and the integral $\int_{Q'_T} (a^m_{ij} - a_{ij})^2 u^2_{x_j} dx \, dt \equiv I^m(Q'_T)$ tends to zero on a set $Q'_T \subset Q_T$ of measure differing arbitrarily little from the measure of all of Q_T, owing to the almost uniform convergence of the a^m_{ij} to a_{ij}, while with respect to any set of small measure $I^m(Q'_T)$ is uniformly small owing to the summability of u^2_x on Q_T. The integral

$$\left| \int_{Q_T} [(a^m - b^m_{ix_i}) - (a - b_{ix_i})] u \eta^m \, dx \, dt \right| \leqslant c_1 \max_{Q_T} |u| \int_{Q_T} |(a^m - b^m_{ix_i}) - (a - b_{ix_i})| \, dx \, dt$$

tends to zero owing to the convergence of the $a^m - b^m_{ix_i}$ to $a - b_{ix_i}$ in the norm of $L_1(Q_T)$. Finally,

$$\left| \int_{Q_T} [(a^m_i - b^m_i) - (a_i - b_i)] u \eta^m_{x_i} \, dx \, dt \right| \leqslant c_2 \max_{Q_T} |u| \left(\int_{Q_T} \sum_i [(a^m_i - b^m_i) - (a_i - b_i)]^2 \, dx \, dt \right)^{1/2} \to 0$$

because of the convergence of the $a_i^m - b_i^m$ to $a_i - b_i$ in the norm of $L_2(Q_T)$. Thus we obtain from (3.14) in the limit

$$\int_{Q_T} uf\,dx\,dt = 0,$$

from which, in view of the sufficient arbitrariness in the choice of f, it follows that $u \equiv 0$. The theorem is proved.

§4. SOLVABILITY AND STABILITY OF THE FIRST BOUNDARY VALUE PROBLEM IN THE CLASSES $V_2^{1,½}(Q_T)$ AND $W_2^{1,½}(Q_T)$

Let us first show that problem (1.9) is solvable in the class $V_2(Q_T)$; namely, let us prove the following proposition:

Theorem 4.1. *If conditions* (1.2)–(1.6) *are fulfilled and* $\psi_0(x) \in L_2(\Omega)$, *then problem* (1.9) *has a solution from* $V_2(Q_T)$.

For the prove we make use of Galerkin's method. We take a fundamental system of functions $\psi_k(x)$, $k = 1, 2, \cdots$, in $\overset{\circ}{W}{}_2^1(\Omega)$ and for the sake of nonessential simplifications we will assume it to be orthonormalized in $L_2(\Omega)$, so that $(\psi_k, \psi_l) = \delta_k^l$. We will seek an approximate solution in the form

$$u^N(x, t) = \sum_{k=1}^{N} c_k^N(t)\,\psi_k(x),$$

where the $c_k^N(t) = (u^N, \psi_k)$ are determined from the conditions

$$\frac{d}{dt}(u^N, \psi_k) + \mathscr{L}_1(u^N, \psi_k) + \mathscr{L}_2(\mathrm{f}, \psi_k) = 0 \tag{4.1}$$

and

$$c_k^N(0) = (\psi_0, \psi_k) \tag{4.2}$$

for $k = 1, \cdots, N$. Conditions (4.1) are a system of linear ordinary differential equations of the form

$$\frac{dc_k^N(t)}{dt} + A_{kl}(t)\,c_l^N(t) + F_k(t) = 0, \qquad k = 1, \ldots, N, \tag{4.3}$$

with summable coefficients and free terms on $[0, T]$. (The summability of the $A_{kl}(t)$ and $F_k(t)$ on $[0, T]$ is seen from estimates (1.11)–(1.14).) In view of this problems (4.3), (4.2) have a unique solution on $[0, T]$, and by the same token the $u^N(x, t)$ are determined uniquely for any N. Let us show that the norms of the u^N in $V_2(Q_T)$ are uniformly bounded, i.e.

$$|u^N|_{Q_T} \leqslant c_1 \tag{4.4}$$

for all $N = 1, 2, \cdots$. For this purpose we multiply each of the equations (4.1) by its own $c_k^N(t)$, then sum the obtained equalities over all k from 1 to N, and integrate the result with respect to t from 0 to t_1. This gives

$$\frac{1}{2}\| u^N(x, t)\|_{2,\Omega}^2 \Big|_{t=0}^{t=t_1} + \int_0^{t_1} [\mathscr{L}_1(u^N, u^N) + \mathscr{L}_2(\hat{\mathbf{f}}, u^N)]\, dt = 0,$$

from which, in consequence of Lemma 2.1, we obtain estimate (4.4) (in this connection we took into account that $\|u^N(x, 0)\|_{2,\Omega}^2 = \Sigma_{k=1}^N (c_k^N(0))^2 \leq \|\psi_0\|_{2,\Omega}^2$). It follows from (4) that the functions $l_{N,k}(t) = (u^N(x, t), \psi_k(x))$ are uniformly bounded. Let us show that for fixed k and arbitrary $N \geq k$ they are equicontinuous on $[0, T]$. Indeed, from (4.1) we have

$$|l_{N,k}(t + \Delta t) - l_{N,k}(t)| \leqslant \int_t^{t+\Delta t} [|\mathscr{L}_1(u^N, \psi_k)| + |\mathscr{L}_2(\hat{\mathbf{f}}, \psi_k)|]\, dt.$$

For an estimate of the right side we use inequalities (1.11)–(1.14), (4.4), the inequality

$$\int_t^{t+\Delta t}\int_\Omega |a_{ij} u^N_{x_j} \psi_{kx_i}|\, dx\, dt \leqslant \mu \| u^N_x \|_{2, Q_{t,t+\Delta t}} \| \psi_{kx} \|_{2, Q_{t,t+\Delta t}}$$

and the fact that all integrals of ψ_k and the known functions (the coefficients and free terms) encountered in this connection are small over a small volume $Q_{t,t+\Delta t}$ (more precisely, they tend to zero for $\Delta t \to 0$). This gives $|l_{N,k}(t + \Delta t) - l_{N,k}(t)| \leq \epsilon(\Delta t)\|\psi_{kx}\|_{2,\Omega}$ with $\epsilon(\Delta t)$ not depending on N and tending to zero for $\Delta t \to 0$, i.e. the equicontinuity of the $l_{N,k}(t)$, $N = k, k + 1, \cdots$, in t. By the usual diagonal process we select a subsequence N_m, $m = 1, 2, \cdots$, with respect to which the functions $l_{N_m,k}(t)$ converge uniformly on $[0, T]$ to some continuous function $l_k(t)$ for each $k = 1, 2, \cdots$. The functions $l_k(t)$, $k = 1, 2, \cdots$, determine a function $u(x, t) = \Sigma_{k=1}^\infty l_k(t)\psi_k(x)$. To this function the u^{N_m} converge weakly in $L_2(\Omega)$ and uniformly with respect to t from $[0, T]$. Indeed, for any function $\psi(x)$ from $L_2(\Omega)$ we have

$$(u^{N_m} - u, \psi) = \sum_{k=1}^{s} (\psi, \psi_k)(u^{N_m} - u, \psi_k) + \left(u^{N_m} - u, \sum_{k=s+1}^{\infty} (\psi, \psi_k)\psi_k\right). \tag{4.5}$$

with

$$\left|\left(u^{N_m}-u,\ \sum_{k=s+1}^{\infty}(\psi,\psi_k)\psi_k\right)\right|\leqslant c_2\left(\sum_{k=s+1}^{\infty}(\psi,\psi_k)^2\right)^{1/2}\equiv c_2R(s),$$

where c_2 does not depend on N_m.

We choose s so large that $c_2R(s)$ becomes less than a preassigned $\epsilon>0$. On the other hand, for fixed s, the first sum in (4.5) will be, as proved above, less than ϵ for all t from $[0, T]$. Thus $|(u^{N_m}-u,\psi)|$ can be made less than ϵ for all t from $[0, T]$. By the same token it is shown that the u^{N_m}, $m=1,2,\cdots$, converge to u weakly in $L_2(Q_T)$, uniformly with respect to $t\in[0,T]$. Owing to (4.4) one can select a subsequence from $\{u^{N_m}\}$ that converges to u weakly in $L_2(Q_T)$ together with its derivatives $u_x^{N_m}$. Without loss of generality we will assume that the whole sequence u^{N_m} converges to u in this way. By virtue of a well-known property of weak convergence, inequality (4.4) is preserved for the limit function u and hence u will be an element of $V_2(Q_T)$. Let us verify that it satisfies (1.16). For this purpose we multiply each equation of (4.1) by a smooth function $d_k(t)$ that is equal to zero for $t=T$, then sum over all k from 1 to $N'\leq N$, and integrate the result with respect to t from 0 to T. After an integration by parts in the first term we obtain

$$I_0(u^N,\Phi^{N'})\equiv\int_0^T[-(u^N,\Phi_t^{N'})+\mathscr{L}_1(u^N,\Phi^{N'})+\mathscr{L}_2(\mathrm{f},\Phi^{N'})]\,dt=(u^N(x,0),\Phi^{N'}(x,0)),\tag{4.6}$$

where $\Phi^{N'}(x,t)=\sum_{k=1}^{N'}d_k(t)\psi_k(t)$. In this equality one can pass to the limit with respect to the subsequence $N=N_m$, $m=1,2,\cdots$, selected above, assuming $\Phi^{N'}$ fixed, and thereby arrive at (4.6) for u (instead of u^N).

Indeed, the functions $\Phi_t^{N'}$, $a_{ij}\Phi_{x_i}^{N'}$ are elements of $L_2(Q_T)$, while the functions $a_i\Phi_{x_i}^{N'}$, $b_i\Phi^{N'}$ and $a\Phi^{N'}$ belong (as follows from estimates (1.11)–(1.14) of the present chapter and Theorem 2.1 of Chapter II) to be the spaces $L_{q_2,r_2}(Q_T)$, $L_{q_3,r_3}(Q_T)$ and $L_{q_2,r_2}(Q_T)$ with $q_2=\bar q'=\bar q/(\bar q-1)$, $r_2=\bar r'=\bar r/(\bar r-1)$ and $q_3=r_3=2$, respectively, and the indices $\bar q,\bar r$ conjugate with q_2, r_2 satisfy the conditions under which $L_{\bar q,\bar r}(Q_T)$ is embedded in $V_2(Q_T)$. In view of this and the above-mentioned weak convergence of the u^{N_m} to u one can pass to the limit in all volume integrals as $m\to\infty$. But the integral $(u^{N_m}(x,0),\Phi^{N'}(x,0))$ has for its limit $(\psi_0(x), \Phi^{N'}(x, 0))$, since the $u^{N_m}(x, 0)$ converge to $\psi_0(x)$ in $L_2(\Omega)$.

Thus we have proved that the limit function $u(x, t)$ satisfies identity (1.16), i.e.

$$I_0(u, \Phi^{N'}) = (\psi_0, \Phi^{N'}(x, 0)) \tag{4.7}$$

for all $\Phi^{N'}$ of the form $\Sigma_{k=1}^{N'} d_k(t)\psi_k(x)$ with the above-mentioned properties of $d_k(t)$ and $\psi_k(x)$. But such $\Phi^{N'}$ are dense in the space of all Φ required in the second definition of a generalized solution from $V_2(Q_T)$ of problem (1.9) (see Lemma 4.12 of Chapter II). In view of this and estimates (1.11)–(1.14) for the remaining terms, the identities (4.7) hold for all Φ from $\overset{\circ}{W}{}_2^{1,1}(Q_T)$ that are equal to zero for $t = T$; i.e. $u(x, t)$ is a generalized solution from $V_2(Q_T)$ of problem (1.9). The theorem is proved.

Let us now carry out some additional investigations concerning the smoothness of generalized solutions in t. Suppose u is a generalized solution from $V_2(Q_T)$ of problem (1.9), and suppose conditions (1.2)–(1.6) are fulfilled for (1.1). The function u satisfies identity (1.16), which we rewrite in the following manner:

$$\int_{Q_T} -u\eta_t\,dx\,dt - \int_{\Omega} \psi_0(x)\,\eta(x, 0)\,dx = \int_{Q_T} [F_i\eta_{x_i} + (F - f)\eta]\,dx\,dt. \tag{4.8}$$

Here

$$F_i = -a_{ij}u_{x_j} - a_i u - f_i, \tag{4.8$_1$}$$

$$F = -b_i u_{x_i} - au, \tag{4.8$_2$}$$

$\psi_0 \in L_2(\Omega)$, while η, as in (1.16), is an arbitrary function from $\overset{\circ}{W}{}_2^{1,1}(Q_T)$ that is equal to zero for $t = T$. From estimates (1.11)–(1.14) of the present chapter and Theorem 1.1 of Chapter II it follows that the $F_i(x, t)$ belong to $L_2(Q_T)$, while $F(x, t) \in L_{q_2, r_2}(Q_T)$ with $q_2 = 2q/(q+1)$, $r_2 = 2r/(r+1)$ (an analogous calculation of the summability indices was made in §2 of the present chapter in carrying out the passage to the limit from (2.12) to (2.13)). The indices q_2, r_2 satisfy the same relations as the indices q_1, r_1 of the space $L_{q_1, r_1}(Q_T)$ containing f.

Let us show that any function u from $\overset{\circ}{V}_2(Q_T)$, satisfying identity (4.8) with the just-mentioned properties of the functions entering into it, is strongly continuous in t in the norm of $L_2(\Omega)$ and belongs to $\overset{\circ}{V}{}_2^{1,1/2}(Q_T)$[1] (the definition of $\overset{\circ}{V}{}_2^{1,1/2}(Q_T)$ was given in §1 of Chapter I). For this purpose we will express identity (4.8) in a more convenient form. Namely, we extend the definition of the functions u, F_i, F and f onto the infinite cylinder $Q_{-\infty,\infty} = \Omega \times (-\infty, \infty)$ as follows:

$$\left.\begin{aligned} u^*(x,t) &= \begin{cases} u(x,t), & t \in [0,T], \\ u(x,-t), & t \in [-T,0], \\ 0, & |t| > T, \end{cases} \\ F_i^*(x,t) &= \begin{cases} F_i(x,t), & t \in [0,T], \\ -F_i(x,-t), & t \in [-T,0), \\ 0, & |t| > T, \end{cases} \\ F^*(x,t) &= \begin{cases} F(x,t), & t \in [0,T], \\ -F(x,-t), & t \in [-T,0), \\ 0, & |t| > T, \end{cases} \\ f^*(x,t) &= \begin{cases} f(x,t), & t \in [0,T], \\ -f(x,-t), & t \in [-T,0), \\ 0, & |t| > T, \end{cases} \end{aligned}\right\} \tag{4.9}$$

and until the end of the section, in place of $Q_{-\infty,\infty}$ we will write simply Q. By virtue of (4.8) we have for these functions the identity

$$\int_Q -u^*\eta_t\,dx\,dt = \int_Q \left[F_i^*\eta_{x_i} + (F^* - f^*)\,\eta\right] dx\,dt, \tag{4.10}$$

with any function η from $\overset{\circ}{W}{}_2^{1,1}(Q)$ that is equal to zero for $|t| \geq T$. We take a smooth function $\omega(t)$ that is equal to 1 for $t \in [-T+\delta, T-\delta]$, where δ is some positive number, and to zero for $|t| \geq T$. We put $\eta(x,t) = \omega(t)\,\Phi(x,t)$ in (4.10), where Φ is an arbitrary element of $\overset{\circ}{W}{}_2^{1,1}(Q)$, and we rewrite the resultant equality in the form

$$\int_Q -v\Phi_t\,dx\,dt = \int_Q \left[F_i^*\omega\Phi_{x_i} + \left(F^*\omega - f^*\omega + u\omega_t\right)\Phi\right] dx\,dt, \tag{4.11}$$

where $v(x,t) = u(x,t)\omega(t)$. Identity (4.11) is a particular case of the identity

---

[1] Concerning $\psi_0(x)$ here and below, it is sufficient to assume only that the integral $\int_\Omega \psi_0(x)\,\eta(x,0)\,dx$ is finite for any function $\eta(x,0)$ from $\overset{\circ}{W}{}_2^{1}(\Omega)$.

$$\int\limits_Q -v\Phi_t\,dx\,dt = \int\limits_Q \left(F_i\Phi_{x_i} + \sum_{i=1}^m F^i\Phi\right) dx\,dt, \tag{4.12}$$

in which v, F_i and F^i are given elements of spaces $\overset{\circ}{V}_2(Q)$, $L_2(Q)$ and $L_{q_i,r_i}(Q)$ with q_i and r_i satisfying conditions (1.6), respectively, that are equal to zero for $|t| \geq T$, while Φ is an arbitrary element of $\overset{\circ}{W}{}_2^{1,1}(Q)$. Let us prove the following lemma.

Lemma 4.1. *If an element v of the space $\overset{\circ}{V}_2(Q)$ satisfies identity (4.12), in which the F_i are elements of $L_2(Q)$, the F^i are elements of $L_{q_i,r_i}(Q)$ with q_i and r_i subject to conditions (1.6), and Φ is an arbitrary element of $\overset{\circ}{W}{}_2^{1,1}(Q)$, and if v and ϕ are finite in t (that is, equal to zero for large $|t|$), then $v \in \overset{\circ}{V}{}_2^{1,1/2}(Q)$. The proposition remains valid if the "$\circ$" of V_2, $W_2^{1,1}$ and $V_2^{1,1/2}$ means the vanishing of the elements on only a part $S_1 \times (-\infty, \infty)$, $S_1 \subset S$, of the lateral surface of Q.*

We take as Φ in (4.12) the function

$$\eta_{\bar h}(x, t) = h^{-1}\int\limits_{t-h}^{t} \eta(x, \tau)\,d\tau, \quad \eta \in \overset{\circ}{W}{}_2^{1,0}(Q),$$

and in all of the terms of (4.12) we transfer the averaging $(\ )_{\bar h}$ from the second factor to the first according to formula (2.9). This gives the relation

$$\int\limits_Q -v_h\eta_t\,dx\,dt = \int\limits_Q \left(F_{ih}\eta_{x_i} + \sum_{i=1}^m F_h^i\eta\right) dx\,dt,$$

which for $\eta(x, t) = \chi(t)\phi(x)$, (where $\chi(t)$ is some smooth function of t, while $\phi(x)$ is an arbitrary element of $\overset{\circ}{W}{}_2^1(\Omega)$) can be rewritten as follows:

$$\int\limits_{-\infty}^{\infty} -\chi_t(v_h, \varphi)\,dt = \int\limits_{-\infty}^{\infty} \chi\left[(F_{ih}, \varphi_{x_i}) + \sum_{i=1}^m (F_h^i, \varphi)\right] dt. \tag{4.13}$$

Hence by the definition of the generalized derivative d/dt it follows that for almost all t we have

$$\frac{d}{dt}(v_h, \varphi) = (F_{ih}, \varphi_{x_i}) + \sum_{i=1}^m (F_h^i, \varphi) \tag{4.14}$$

for any $\phi(x) \in \overset{\circ}{W}{}_2^1(\Omega)$.

The function v_h is strongly continuous in t in the norm of $L_2(\Omega)$. All the more so the scalar product (v_h, ϕ) is continuous in t. In addition, v_h has the generalized derivative v_{ht} from $L_2(Q)$, and so for almost all t

$$\frac{d}{dt}(v_h, \varphi) = (v_{ht}, \varphi)$$

and

$$(v_{ht}, \varphi) = (F_{ih}, \varphi_{x_i}) + \sum_{l=1}^{m} (F_h^l, \varphi). \tag{4.15}$$

Hence for any positive h_1 and h_2 we have

$$(v_{h_1 t} - v_{h_2 t}, \varphi) = (F_{ih_1} - F_{ih_2}, \varphi_{x_i}) + \sum_{l=1}^{m} (F_{h_1}^l - F_{h_2}^l, \varphi). \tag{4.16}$$

We put $\phi = v_{h_1} - v_{h_2}$ in it (which, clearly, is permissible), and present the left side in the form of a derivative with respect to t of the continuous function $\|v_{h_1} - v_{h_2}\|_{2,\Omega}^2$; namely,

$$\frac{1}{2}\frac{d}{dt}\|v_{h_1} - v_{h_2}\|_{2,\Omega}^2$$

$$= (F_{ih_1} - F_{ih_2}, v_{h_1 x_i} - v_{h_2 x_i}) + \sum_{l=1}^{m} (F_{h_1}^l - F_{h_2}^l, v_{h_1} - v_{h_2}).$$

Integrating this equality with respect to t, we obtain

$$\frac{1}{2}\|v_{h_1} - v_{h_2}\|_{2,\Omega}^2 \Big|_{t=t_1}^{t=t_2} = \int_{t_1}^{t_2} \Big[(F_{ih_1} - F_{ih_2}, v_{h_1 x_i} - v_{h_2 x_i})$$

$$+ \sum_{l=1}^{m} (F_{h_1}^l - F_{h_2}^l, v_{h_1} - v_{h_2}) \Big] dt, \tag{4.17}$$

which is valid for all t_1 and t_2. All of the functions here are finite in t. As h_1 and h_2 tend to zero we get

$$\|F_{ih_1} - F_{ih_2}\|_{2,Q} + \|F_{h_1}^l - F_{h_2}^l\|_{q_l, r_l, Q}$$

$$+ \|v_{h_1 x} - v_{h_2 x}\|_{2,Q} + \|v_{h_1} - v_{h_2}\|_{q'_l, r'_l, Q} \to 0$$

(see Lemma 4.7 of Chapter II). In view of this, from (4.17) for $t_1 = -\infty$, and arbitrary t_2 follows the strong convergence of the v_h in $L_2(\Omega)$, uniformly in $t \in (-\infty, \infty)$. Therefore the limit function v is equivalent in Q to a function that is strongly continuous in t in the norm of $L_2(\Omega)$.

Let us now prove that

$$\int_{-\infty}^{\infty} \| v(x, t+h) - v(x, t) \|_{2, \Omega}^2 dt = o(h).$$

For this purpose we take $\phi(x) = \Delta_h v(x, t) \equiv v(x, t+h) - v(x, t)$ in (4.15) and integrate both sides of the equality with respect to t from $-\infty$ to $+\infty$. Then in the right side of the resultant equality we transfer the averaging $(\ )_h$ from the first factors to the second, and the difference Δ_h from the second factors to the first. As a result we obtain

$$\int_{-\infty}^{\infty} h^{-1} \| \Delta_h v \|_{2, \Omega}^2 dt = \int_{-\infty}^{\infty} \left[(\Delta_h F_i, v_{hx_i}) + \sum_{l=1}^{m} (\Delta_h F^l, v_h) \right] dt. \tag{4.18}$$

Estimating the right side by Hölder's inequality and making use of the uniform boundedness of $\|v_{hx}\|_{2,Q}$ and $\|v_h\|_{q_i', r_i', Q}$ and the integral continuity of the elements F_i and F^i of spaces $L_2(Q)$ and $L_{q_i, r_i}(Q)$, we show that

$$h^{-1} \int_{-\infty}^{\infty} \| \Delta_h v \|_{2, \Omega}^2 dt \leqslant \varepsilon(h) \to 0 \quad \text{for} \quad h \to 0.$$

This together with the strong continuity of v in t in the norm of $L_2(\Omega)$ and the membership of v in $\overset{\circ}{V}_2(Q)$ means that $v \in \overset{\circ}{V}{}_2^{1,1/2}(Q)$. From here and from Theorem 4.1 follows the main theorem of the present section.

Theorem 4.2. *Under the fulfilment of conditions* (1.2)–(1.6) *any generalized solution* u *of problem* (1.9) *from* $\overset{\smile}{V}_2(Q_T)$ *belongs to* $\overset{\circ}{V}{}_2^{1,1/2}(Q)$, *and problem* (1.9) *is uniquely solvable in* $\overset{\circ}{V}{}_2^{1,1/2}(Q_T)$, *if* $\psi_0(x) \in L_2(\Omega)$.

Remark 4.2. Strictly speaking, we have proved the membership of u in $\overset{\circ}{V}{}_2^{1,1/2}(Q_{T-\delta})$, where δ is some positive number, but not in $\overset{\circ}{V}{}_2^{1,1/2}(Q_T)$. This is not important, however, since the solution u and the equation itself can be extended beforehand onto the larger cylinder $Q_{T+\delta}$ with preservation of all properties of the functions in question.

Let us explain when a generalized solution u of problem (1.9) is an element of $\overset{\circ}{W}{}_2^{1,1/2}(Q_T)$.

The space $\overset{\circ}{W}{}_2^{1,1/2}(Q)$, where $Q = \Omega \times (-\infty, \infty)$, consists of all elements u of $\overset{\circ}{W}{}_2^{1,0}(Q)$ having a finite integral

$$|||u|||_Q = \left(\int_0^{\infty} h^{-2} \| u(x, t+h) - u(x, t) \|_{2, Q}^2 dh \right)^{1/2}.$$

It will be a Hilbert space if the norm in it is defined by the equality

$$\| u \|_{W_2^{1,1/2}(Q)} = [\| u \|^2_{2,Q} + \| u_x \|^2_{2,Q} + ||| u |||^2_Q]^{1/2}.$$

It is not difficult to calculate that

$$||| u |||^2_Q = \int_{-\infty}^{\infty} \int_{\Omega} | \tilde{u}(x, \alpha)|^2 |\alpha| \, dx \, d\alpha,$$

where $\tilde{u}(x, \alpha)$ is the Fourier transform of $u(x, t)$ with respect to t, i.e.,

$$\tilde{u}(x, \alpha) = \frac{1}{\sqrt{2\pi}} \int_{-\infty}^{\infty} u(x, t) e^{i\alpha t} \, dt.$$

The membership of a function u in $\overset{\circ}{W}{}_2^{1,1/2}(Q_T)$ can be defined, for example, as follows: *u is an element of* $\overset{\circ}{W}{}_2^{1,1/2}(Q_T)$ if and only if the function $\hat{u}(x, t) = u'(x, t)\omega(t)$, where

$$u'(x, t) = u'(x, -t) = \begin{cases} u(x, t), & t \in [0, T], \\ u(x, 2T - t), & t \in [T, 2T], \\ 0, & t > 2T, \end{cases}$$

and

$$\omega(t) = \omega(-t) = \begin{cases} 1, & t \in [0, T], \\ \dfrac{(3T - 2t)}{T}, & t \in \left[T, \dfrac{3}{2}T\right], \\ 0, & t > \dfrac{3}{2}T, \end{cases}$$

is an element of $\overset{\circ}{W}{}_2^{1,1/2}(Q)$. As the norm of u in $\overset{\circ}{W}{}_2^{1,1/2}(Q_T)$ one can take the number $\|\hat{u}'\|_{W_2^{1,1/2}(Q)}$.

Let us also introduce the *Banach spaces* $L^*_{r,q}(Q_T)$, $r, q \geq 1$. They consist of measurable functions u having a finite integral

$$\| u \|^*_{r,q,Q_T} = \left(\int_{\Omega} \left(\int_0^T | u |^r \, dt \right)^{\frac{q}{r}} dx \right)^{\frac{1}{q}}. \tag{4.19}$$

The value of $\|u\|^*_{r,q,Q_T}$ is taken as the norm of u in $L^*_{r,q}(Q_T)$. It is well known that

$$\| u \|^*_{r,q,Q_T} \leqslant \| u \|_{q,r,Q_T}, \quad r \leqslant q. \tag{4.20}$$

and

$$\|u\|_{q,r,Q_T} \leqslant \|u\|^*_{r,q,Q_T}, \quad r \geqslant q \tag{4.20$_1$}$$

(see [42], p. 169, problem 245). In view of this and inequality (3.4) of Chapter II, we get

$$\|u\|^*_{r,q,Q_T} \leqslant \|u\|_{q,r,Q_T} \leqslant \beta |u|_{Q_T}, \tag{4.21}$$

if $r \leq q$ and if r and q satisfy conditions (3.3) of Chapter II guaranteeing the validity of the second of the inequalities (4.21).

Let us assume that the coefficients a_{ij} and a_i and the functions f_i satisfy conditions (1.2)–(1.5), while the coefficients b_i and a and the function f satisfy the conditions

$$\left\| \sum_{i=1}^{n} b_i^2, \ a \right\|^*_{r,q,Q_T} \leqslant \mu_1, \quad r \geqslant q, \tag{4.22}$$

and

$$\|f\|^*_{r_1,q_1,Q_T} \leqslant \mu_1, \quad r_1 \geqslant q_1, \tag{4.23}$$

in which the indices q, r, q_1 and r_1 are subject to restrictions (1.4) and (1.6). By virtue of (4.21) these requirements imply the previous assumptions (1.2)–(1.6), and therefore all of the conclusions made in this chapter are valid. In particular, the function $v(x, t) = u(x, t)\omega(t)$, where $u(x, t)$ is a generalized solution of problem (1.9) from $\overset{\circ}{V}_2(Q_T)$, satisfies an identity of type (4.12) (namely (4.11)) with the F_i and F^i having the properties described there. But, as is easily calculated in the present case, the functions standing for Φ will also be elements of $L^*_{r_i,q_i}(Q)$ with r_i and q_i subject to the same restrictions as the indices r_1 and q_1 for f. Let us show that in view of this the function v will be an element of $W_2^{1,\frac{1}{2}}(Q)$.

Lemma 4.2. *If an element v of the space $\overset{\circ}{V}_2(Q)$ satisfies identity* (4.12) (*in which $F_i \in L_2(Q)$, $F^i \in L^*_{r_i,q_i}(Q)$ with q_i and r_i satisfying conditions* (1.6) *and $q_i \leq r_i$) for any Φ from $\overset{\circ}{W}{}_2^{1,1}(Q)$, and if v and ϕ are finite in t, then v belongs to $\overset{\circ}{W}{}_2^{1,\frac{1}{2}}(Q)$. The proposition remains valid if the "o" over V_2, $W_2^{1,1}$ and $W_2^{1,\frac{1}{2}}$ means the vanishing of all of the elements on only a part $S_1 \times (-\infty, \infty)$, $S_1 \subset S$, of the lateral surface of Q, which can even be empty.*

For a proof of this theorem we make use of the Fourier transform with respect to t, denoting, as above, the Fourier transform of a function $\psi(x, t)$ with respect to t by $\widetilde{\psi}(x, \alpha)$, so that, in particular,

$$\widetilde{v}(x, \alpha)=\frac{1}{\sqrt{2\pi}}\int_{-\infty}^{\infty} v(x, t)\, e^{i\alpha t}\, dt.$$

It is necessary to establish the finiteness of the integral

$$\int_{-\infty}^{\infty}\int_{\Omega} |\alpha|\,|\widetilde{v}(x, \alpha)|^2\, dx\, d\alpha.$$

It is not difficult to see that the Fourier transform $\widetilde{v}_h(x, \alpha)$ of the function $v_h(x, t)$ is for each α an element of $\overset{\circ}{W}{}_2^1(\Omega)$, while the Fourier transform of the function $v_{ht}(x, t)$ is equal to $-i\alpha\widetilde{v}_h(x, \alpha)$. Therefore (4.15) implies the identity

$$-i\alpha(\widetilde{v}_h, \varphi)=(\widetilde{F}_{lh}, \varphi_{x_l})+\sum_{l=1}^{m}(\widetilde{F}^l_h, \varphi). \tag{4.24}$$

Let us take as ϕ the function $\chi(\alpha)\,\overline{\widetilde{v}}_h(x, \alpha)$, where $\chi(\alpha)$ is equal to i for $\alpha > 0$ and to $-i$ for $\alpha < 0$, and integrate both sides with respect to α from $-\infty$ to $+\infty$:

$$\int_{-\infty}^{\infty}\int_{\Omega} |\alpha|\,|\widetilde{v}_h(x, \alpha)|^2\, dx\, d\alpha$$

$$=\int_{-\infty}^{\infty}\int_{\Omega}\left(\chi\widetilde{F}_{lh}\overline{\widetilde{v}}_{hx_l}+\sum_{l=1}^{m}\chi\widetilde{F}^l_h\overline{\widetilde{v}}_h\right) dx\, d\alpha. \tag{4.25}$$

The functions $\chi(\alpha)\widetilde{F}_{lh}(x, \alpha), \chi(\alpha)\widetilde{F}^l_h(x, \alpha)$ are the Fourier transforms of the functions $\hat{F}_{lh}(x, t)$, $\hat{F}^l_h(x, t)$, which are obtained from the functions $F_{lh}(x, t)$, $F^l_h(x, t)$ by means of the Hilbert transform (see the book [120]). It is known [120] that the Hilbert transform is a bounded linear operator from $L_r(-\infty, \infty)$ into $L_r(-\infty, \infty)$ for any $r > 1$, so that

$$\left(\int_{-\infty}^{\infty}|\hat{F}_{lh}(x, t)|^r\, dt\right)^{\frac{1}{r}} \leqslant c\left(\int_{-\infty}^{\infty}|F_{lh}(x, t)|^r\, dt\right)^{\frac{1}{r}}, \tag{4.26}$$

and analogously for F^l_h. For $r = 2$, in place of (4.26) we have the equality sign if c is put equal to 1. In view of this and Parseval's equality, from (4.25) we get

$$\int_{-\infty}^{\infty}\int_{\Omega}|\alpha||\tilde{v}_h(x,\alpha)|^2\,dx\,d\alpha=\int_{-\infty}^{\infty}\int_{\Omega}\left(\hat{F}_{lh}v_{hx_l}+\sum_{l=1}^{m}\hat{F}^l_h v_h\right)dx\,dt$$

$$\leqslant\|\hat{F}_{lh}\|_{2,Q}\|v_{hx_l}\|_{2,Q}+\sum_{l=1}^{m}\|\hat{F}^l_h\|^*_{r_l,q_l,Q}\|v_h\|^*_{r'_l,q'_l,Q}$$

$$\leqslant\|F_{lh}\|_{2,Q}\|v_{hx_l}\|_{2,Q}+c\beta\sum_{l=1}^{m}\|F^l_h\|^*_{r_l,q_l,Q}|v_h|_Q, \tag{4.27}$$

since the indices r'_l, q'_l are such that estimate (4.22) holds for them. Passing to the limit in this inequality as $h\to 0$, we prove the finiteness of the integral on the left, and thereby also prove the membership of v in $\overset{\circ}{W}{}_2^{1,1/2}(Q)$. Lemma 4.2 is proved. From it we get

Theorem 4.3. *Suppose u is a generalized solution from $\overset{\circ}{V}_2(Q_T)$ of problem (1.9), and suppose the a_{ij}, a_i and f_i satisfy conditions (1.2)–(1.5), while the b_i, a and f satisfy conditions (4.22), (4.23) with the meaning of the indices the same as indicated there. Then u is an element of $\overset{\circ}{W}{}_2^{1,1/2}(Q_T)$.*

A proof analogous to that of Lemmas 4.1 and 4.2 exists for the following lemma.

Lemma 4.3. *If an element v of the space $\overset{\circ}{W}{}_2^{1,0}(Q)$ satisfies the identity*

$$\int_Q -v\Phi_t\,dx\,dt=\int_Q(F_i\Phi_{x_i}+F\Phi)\,dx\,dt, \tag{4.28}$$

in which Φ is an arbitrary element of $\overset{\circ}{W}{}_2^{1,1}(Q)$, while the F_i and F are given functions from $L_2(Q)$ and $L_{q,2}(Q)$, where $q=2n/(n+2)$ for $n\geq 3$, $q>1$ for $n=2$ and $q=1$ for $n=1$, and if v and ϕ are finite in t, then $v\in\overset{\circ}{V}{}_2^{1,1/2}(Q)$. But if in this connection $F\in L^*_{2,q}(Q)$, then $V\in\overset{\circ}{W}{}_2^{1,1/2}(Q)$. The proposition remains valid if the "o" of $W_2^{1,0}$, $W_2^{1,1}$, $V_2^{1,1/2}$ and $W_2^{1,1/2}$ means the vanishing of the elements on only a part $S_1\times(-\infty,\infty)$, $S_1\subset S$, of the lateral surface of Q, which can even be empty.

For its deduction the term $\int_Q F\Phi\,dx\,dt$ is estimated in the first case as follows:

$$\left|\int_Q F\Phi\,dx\,dt\right|\leqslant\|F\|_{q,2,Q}\|\Phi\|_{q',2,Q}\leqslant c\|F\|_{q,2,Q}\|\Phi\|^{(1)}_{2,Q}.$$

while in the second case

$$\left| \int_{-\infty}^{\infty} \int_{\Omega} \chi \tilde{F}_h \tilde{\tilde{v}}_h \, dx \, d\alpha \right| = \left| \int_{-\infty}^{\infty} \int_{\Omega} \hat{F}_h v_h \, dx \, dt \right|$$
$$\leqslant \| \hat{F}_h \|_{2,q,Q} \| v_h \|_{2,q',Q} \leqslant c \| F_h \|_{2,q,Q} \| v_h \|_{2,Q}^{(1)}.$$

In both cases we have made use of the embeddability of $\overset{\circ}{W}{}_2^1(\Omega)$ in $L_{q'}(\Omega)$ (see Theorem 2.1 of Chapter II).

Let us now assume that u is a generalized solution from $V_2(Q)$ of equation (1.1). Then the function $v(x, t) = u(x, t)\,\omega(t)$ satisfies (4.11) for all $\Phi \in \overset{\circ}{W}{}_2^{1,1}(Q)$ if $\omega(t)$ is a smooth function that is equal to zero for $t \le \delta$ and $t \ge T - \delta$, where δ is some positive number. From this identity, as was shown above, follow identities (4.15), (4.16) and (4.24). If in (4.16) we set $\phi(x) = [v_{h_1}(x, t) - v_{h_2}(x, t)]\zeta^2(x)$, where $\zeta(x)$ is a smooth nonnegative function of x that is equal to zero on S, we arrive at an equality from which we deduce the strong continuity in t of the function $v(x, t)\,\zeta(x)$ in the norm of $L_2(\Omega)$. From equality (4.15) with $\phi(x) = \zeta^2(x)\,\Delta_h v(x, t)$ we see that the integral

$$\int_{-\infty}^{\infty} h^{-1} \| \zeta \Delta_h v \|_{2,\Omega}^2 \, dt$$

tends to zero for $h \to 0$. Finally, equality (4.24) with $\phi(x) = \chi(\alpha)\,\tilde{\tilde{v}}_h(x, \alpha)\zeta^2(x)$ permits us to conclude that $v(x, t)\,\zeta(x)$ is an element of $\overset{\circ}{W}{}_2^{1,\frac{1}{2}}(Q)$. If $u \in V_2(Q_T)$ and satisfies identity (1.16), then the function $\omega(t)$ in v can be taken equal to unity for $|t| \le T - \delta$, $\delta > 0$. Thus we have proved

Theorem 4.4. *If the coefficients and free terms of equation* (1.1) *satisfy the conditions of Theorem* 4.2, *then any generalized solution u of it from $V_2(Q_T)$ belongs to $V_2^{1,\frac{1}{2}}(\Omega' \times [\epsilon, t_1])$, where Ω' is an arbitrary strictly interior subdomain of the domain Ω, $\epsilon > 0$, and $t_1 < T$. And any generalized solution u of equation* (1.1) *from $V_2(Q_T)$, satisfying* (1.16) *with η from $\overset{\circ}{W}{}_2^{1,1}(Q_T)$ and equal to zero for $t = T$, belongs to $V_2^{1,\frac{1}{2}}(\Omega' \times [0, t_1])$. If the coefficients and free terms of equation* (1.1) *satisfy the conditions of Theorem* 4.3, *then any generalized solution u of it from $V_2(Q_T)$ belongs to $W_2^{1,\frac{1}{2}}(\Omega' \times [\epsilon, t_1])$, $\epsilon > 0$, $t_1 < T$, and any generalized solution u from $V_2(Q_T)$, satisfying identity* (1.16) *with η from $\overset{\circ}{W}{}_2^{1,1}(Q_T)$ and equal to zero for $t = T$, belongs to $W_2^{1,\frac{1}{2}}(\Omega' \times [0, t_1])$.*

A generalized solution of problem (1.9) is stable in the norm of the space $V_2^{1,0}(Q_T)$ with respect to variations of the coefficients and free terms of the equation, and also of the function $\psi_0(x)$ determining the initial condition. More precisely, we have

Theorem 4.5. *Suppose that for all operators*

$$\mathscr{L}^m u \equiv u_t - \frac{\partial}{\partial x_i}(a_{ij}^m u_{x_j} + a_i^m u) + b_i^m u_{x_i} + a^m u,$$

$$m = 1, 2, \ldots,$$

the conditions of Theorem 4.1 are fulfilled with one and the same constants. Suppose the $a_{i,j}^m(x, t)$, remaining uniformly bounded, converge almost everywhere to a_{ij}, while the functions a_i^m, b_i^m, a^m, f_i^m, f^m and ψ_0^m converge to a_i, b_i, a, f_i, f and ψ_0 in the norms of the spaces to which they belong according to the conditions of Theorem 4.1. Then the generalized solutions u^m from $V_2^{1,0}(Q_T)$ of the problems

$$\mathscr{L}^m u = \frac{\partial f_i^m}{\partial x_i} - f^m, \quad u|_{S_T} = 0, \quad u|_{t=0} = \psi_0^m, \tag{4.29}$$

converge strongly in $V_2^{1,0}(Q_T)$ to the generalized solution u of the limit problem (1.9).

For a proof of this proposition we subtract from the integral identity (1.10) for u^m the integral identity (1.10) for u and write the result of this subtraction in the form of an integral identity for the function $v^m = u^m - u$ in the following manner:

$$\begin{aligned}\int_\Omega v^m \eta\, dx \Big|_{t=0}^{t=t_1} &+ \int_{Q_{t_1}} [-v^m\eta_t + a_{ij}^m v_{x_j}^m \eta_{x_i} + a_i^m v^m \eta_{x_i} \\ &+ b_i^m v_{x_i}^m \eta + a^m v^m \eta]\, dx\, dt = \int_{Q_{t_1}} [(a_{ij} - a_{ij}^m) u_{x_j} \eta_{x_i} \\ &+ (a_i - a_i^m) u \eta_{x_i} + (b_i - b_i^m) u_{x_i} \eta + (a - a^m) u \eta \\ &+ (f_i - f_i^m) \eta_{x_i} + (f - f^m) \eta]\, dx\, dt.\end{aligned}$$

We express the right side of this identity as $\int_{Q_{t_1}} [F_i^m \eta_{x_i} + F^m \eta + (f - f^m)\eta]\, dx\, dt$. The functions F_i^m and F^m are composed of terms of the same type as F_i and F in (4.81–2), and hence, for the reasons mentioned there, satisfy the conditions of Theorem 4.1. Consequently v^m can be considered as a generalized solution from $V_2^{1,0}(Q_T)$ of problem (4.16), in which $v^m|_{t=0} = \psi_0^m - \psi_0$, f_i^m is $-F_i^m$, and f^m

is equal to $-F^m - f + f^m$. According to Lemma 2.1 we have the energy estimate

$$|v^m|_{Q_T} \leqslant c\Big[\|\psi_0^m - \psi_0\|_{2,\Omega} + \|F^m\|_{2,Q_T} + \|F^m\|_{q_2, r_2, Q_T} + \|f^m - f\|_{q_1, r_1, Q_T}\Big],$$

from which it is seen that $|v^m|_{Q_T} \to 0$ for $m \to \infty$. And what is more, $\sup_{0\le h\le T} h^{-1}\|v^m(x, t+h) - v^m(x, t)\|_{2,Q_{T-h}} \to 0$, and under fulfilment of the conditions of Lemma 4.2 also $|||v^m|||_{Q_T} \to 0$ for $m \to \infty$.

§5. ON THE SOLVABILITY OF OTHER BOUNDARY VALUE PROBLEMS. THE CAUCHY PROBLEM

Let us consider the second boundary value problem for equations (1.1), i.e. the problem of determining $u(x, t)$ in Q_T from the conditions

$$\mathscr{L}u = -f; \quad \frac{\partial u}{\partial \mathscr{N}} + \sigma(s, t)\, u|_{S_T} = \psi(x, t)$$

$$u|_{t=0} = \psi_0(x), \tag{5.1}$$

where $\sigma(s, t)$, as well as $\psi_0(x)$, and the coefficients and free term f of equation (1.1), are given functions belonging to well-defined classes, $\partial u/\partial\mathscr{N} = a_{ij}u_{x_j}\cos\alpha_i$, with α_i being the angle formed by the outward normal to S with the x_i axis, while the $a_{ij}(x, t)$ are the same as in equation (1.1).

Concerning equation (1.1) and the function $\psi_0(x)$ we make the same assumptions as in §§1, 2 and 4. The boundary S of the domain Ω is assumed to be piecewise smooth, while the function $\sigma(s, t)$ is assumed to satisfy the condition

$$\|\sigma\|_{q_1, r_1, S_T} \leqslant \mu_1, \tag{5.2}$$

where q_1 and r_1 are subject to the restrictions

$$\left.\begin{array}{lll} \dfrac{1}{r_1} + \dfrac{n-1}{2q_1} = \dfrac{1}{2}, & & \\ q_1 \in (n-1, \infty], \; r_1 \in [2, \infty) & \text{for} & n > 2, \\ q_1 \in (1, \infty), \; r_1 \in (2, \infty) & \text{for} & n = 2. \end{array}\right\} \tag{5.2'}$$

In the case $n = 1$ we assume that $\|\sigma\|_{r_1, S_T} \le \mu_1$, $r_1 = 2.$ [1)]

[1)] For $n = 1$ it is necessary that norms of the form $\|\cdot\|_{q,r,S_T}$ be replaced everywhere by norms of the form $\|\cdot\|_{r,S_T}$.

Finally, concerning the function $\psi(s, t)$ we assume that

$$\|\psi\|_{q_2,r_2,S_T} \le \mu_1, \tag{5.3}$$

where q_2 and r_2 are subject to the restrictions

$$\left.\begin{aligned} &\frac{1}{r_2} + \frac{n-1}{2q_2} = \frac{n}{4} + \frac{1}{2},\\ &q_2 \in \left[\frac{2n-1}{n}, \frac{2n-2}{n-2}\right], \ r_2 \in [1, 2] \ \text{for} \ n > 2,\\ &q_2 \in (1, \infty], \ r_2 \in [1, 2) \qquad \text{for} \ n = 2. \end{aligned}\right\} \tag{5.3'}$$

In the case $n = 1$ we assume that $\|\psi\|_{4/3,S_T} \le \mu_1$.

With the use of inequalities (1.8) and (3.11), (3.12) of Chapter II it is not difficult to verify that under such conditions on S and σ the integrals $\int_0^T\int_S|\sigma u\eta|\,ds\,dt$ and $\int_0^T\int_S|\psi\eta|\,ds\,dt$ will be finite for any $u(x, t)$ and $\eta(x, t)$ from $V_2(Q_T)$ and admit the estimates

$$\begin{aligned} \int_0^T\int_S |\sigma u\eta|\,ds\,dt &\le \|\sigma\|_{q_1,r_1,S_T}\|u\|_{\frac{2q_1}{q_1-1},\frac{2r_1}{r_1-1},S_T}\\ &\times\|\eta\|_{\frac{2q_1}{q_1-1},\frac{2r_1}{r_1-1},S_T} \le c^2\|\sigma\|_{q_1,r_1,S_T}|u|_{Q_T}|\eta|_{Q_T},\\ \int_0^T\int_S|\psi\eta|\,ds\,dt &\le c\,\|\psi\|_{q_2,r_2,S_T}|\eta|_{Q_T}, \end{aligned} \tag{5.4}$$

where the constant c is taken from inequalities (3.11), (3.12) of Chapter II.

By a *generalized solution from* $V_2(Q_T)$ (*from* $V_2^{1,0}(Q_T)$ *or* $V_2^{1,1/2}(Q_T)$) *of problem* (5.1) we will mean a function $u(x, t)$ from $V_2(Q_T)$ (respectively from $V_2^{1,0}(Q_T)$ or $V_2^{1,1/2}(Q_T)$ satisfying the identity

$$\begin{aligned} -\int_{Q_T} u\eta_t\,dx\,dt &+ \int_0^T [\mathscr{L}_1(u, \eta) + (\hat{f}, \eta)]\,dt\\ &+\int_{S_T}(\sigma u - \psi)\eta\,ds\,dt = \int_\Omega \psi_0(x)\eta(x, 0)\,dx \end{aligned} \tag{5.5}$$

for any $\eta(x, t)$ from $W_2^{1,1}(Q_T)$ that is equal to zero for $t = T$. This identity is obtained from the identity

$$\int_0^T\int_\Omega\left(\mathscr{L}u + f\right)\eta\,dx\,dt = 0$$

as the result of a single integration by parts with respect to t and with respect to x_i in certain terms and a taking into account of the initial and boundary conditions from (5.1). In the same way as in §1 we verify that all of the integrals in (5.5) are finite for any function u from $V_2(Q_T)$ and any function η from $W_2^{1,1}(Q_T)$, and that in place of (5.5) one could postulate the fulfilment of the identities

$$\int_\Omega u(x, t)\eta(x, t)\,dx + \int_0^t \int_\Omega -u\eta_t\,dx\,dt + \int_0^t [\mathscr{L}_1(u, \eta) + (f, \eta) + \mathscr{L}_3(u, \eta)]\,dt = \int_\Omega \psi_0(x)\eta(x, 0)\,dx \tag{5.6}$$

with

$$\mathscr{L}_3(u, \eta) = \int_S (\sigma u - \psi)\eta\,ds$$

for almost all t (respectively for all t) from $[0, T]$ and an arbitrary function $\eta(x, t)$ from $W_2^{1,1}(Q_T)$.

For a proof of the solvability of problem (5.1) in $V_2(Q_T)$ we again make use of Galerkin's method. In contrast to §4 the functions $\psi_k(x)$, $k = 1, 2, \cdots$, must form a fundamental system in $W_2^1(\Omega)$ but not in $\overset{\circ}{W}{}_2^1(\Omega)$. Suppose, in addition, $(\psi_k, \psi_l) = \delta_k^l$. We seek an approximate solution u^N in the form

$$u^N(x, t) = \sum_{k=1}^N c_k^N(t)\psi_k(x),$$

where $c_k^N(t) = (u^N, \psi_k)$ are determined from the conditions

$$\frac{d}{dt}(u^N, \psi_k) + \mathscr{L}_1(u^N, \psi_k) + \mathscr{L}_3(u^N, \psi_k) + (f, \psi_k) = 0,$$
$$c_k^N(0) = (\psi_0, \psi_k).$$

The subsequent arguments and estimates are analogous to those mentioned in §4. The fact that all of the functions encountered this time do not vanish on S_T is not important, it being only necessary in estimating $\|u\|_{q,r,Q_T}$ in terms of $|u|_{Q_T}$ to use, instead of inequality (3.4) of Chapter II, inequality (3.8) of Chapter II for functions not equal to zero on S_T. For a justification of the passage to the limit in the term $\int_{S_T}\sigma u^N\Phi^{N'}\,ds\,dt$ we note that the uniform unboundedness of the norms $|u^N|_{Q_T}$ implies by virtue of inequalities (3.1)–(3.12) of

Chapter II the uniform boundedness and weak compactness of the $u^N(x, t)$ in the norm of $L_{q_3,r_3}(S_T)$, where $n - 1/2q_3 + 1/r_3 = n/4$ with $r_3 \in [2, \infty]$, $q_3 \in [2(n-1)/n, 2(n-1)/(n-2)]$ for $n \geq 3$, with $r_3 \in (2, \infty]$, $q_3 \in [1, \infty)$ for $n = 2$, and in the norm of $L_4(S_T)$ for $n = 1$. This is true, in particular, for $q_3 = \bar{q}_1 \equiv 2q_1/(q_1 - 1)$, $r_3 = \bar{r}_1 \equiv 2r_1/(r_1 - 1)$, where q_1 and r_1 are taken from (5.3). On the other hand, Theorem 2.1 of Chapter II and estimate (5.4) imply the membership of the product $\sigma\Phi^{N_1}$ in the conjugate space $L_{\bar{q}_4,\bar{r}_4}(S_T)$ $(L_{4/3}(S_T))$ of $L_{q_4,r_4}(S_T)$ (of $L_4(S_T)$). All of this makes it possible to pass to the limit in the integral $\int_{S_T} \sigma u^N \Phi^{N'}\, ds\, dt$ as $N \to \infty$ with respect to a weakly convergent subsequence u^{N_k} in $L_{\bar{q}_1,\bar{r}_1}(S_T)$ (in $L_4(S_T)$), and then to pass to the limit as $N' \to \infty$.

In this way, by means of Galerkin's method we have proved the existence of at least one generalized solution u of problem (5.1) in the class $V_2(Q_T)$. The membership of such solutions in $V_2^{1,1/2}(Q_T)$ and $W_2^{1,1/2}(Q_T)$ follows from Lemmas 4.1 and 4.2. Thus we have

Theorem 5.1. *Suppose the coefficients and free terms in the equation and boundary condition of* (5.1) *are subject to restrictions* (1.2)–(1.6) *and* (5.2)–(5.3), *while* Ω *is a domain with piecewise-smooth boundary* S. *Then for any* $\psi_0(x)$ *there exists a unique solution of problem* (5.1) *in the class* $V_2^{1,1/2}(Q_T)$. *Any solution from* $V_2(Q_T)$ *of problem* (5.1) *belongs to* $V_2^{1,1/2}(Q_T)$. *If furthermore the* b_i, a *and* f *satisfy conditions* (4.22) *and* (4.23) *with the meaning of the indices indicated there, while* $\sigma(s, t) \in L^*_{r_1,q_1}(S_T)$ *and* $\psi(s, t) \in L^*_{r_2,q_2}(S_T)$ *with* r_k *and* q_k, $k = 1, 2$, *subject to conditions* (5.2′), (5.3′) *and* $q_k \leq r_k$, $k = 1, 2$, *then* u *will be an element of* $W_2^{1,1/2}(Q_T)$.

The same result is also valid for the boundary value problem

$$\mathfrak{L}u = f,$$

$$u\big|_{S'_T} = 0, \quad \frac{\partial u}{\partial \nu} + \sigma u\big|_{S''_T} = \psi, \quad u\big|_{t=0} = \psi_0(x), \tag{5.7}$$

where $S'_T = S' \times [0, T]$, $S''_T = S'' \times [0, T]$ and $S' \cup S'' = S$. Its solution, if the assumptions of Theorem 5.1 are complied with, will be an element of $V_2^{1,1/2}(Q_T)$ $(W_2^{1,1/2}(Q_T))$, that is equal to zero on S'_T. This is proved in the same way as Theorem 5.1, except that the functions $\psi_k(x)$, $k = 1, 2, \cdots$, in Galerkin's method must form a fundamental system not in $W_2^1(\Omega)$ but in the subspace of $W_2^1(\Omega)$ consisting of all elements of it that are equal to zero on S'.

The Cauchy problem. Let us consider in the layer $R_T = E_n \times [0, T]$ the Cauchy problem for equation (1.1) with initial condition

$$u|_{t=0} = \psi_0(x), \quad x \in E_n. \tag{5.8}$$

By a *generalized solution from* $V_2^{1,\frac{1}{2}}(R_T)$ *of the Cauchy problem* (1.1), (5.8) we will mean a function $u(x, t)$ from $V_2^{1,\frac{1}{2}}(R_T)$ satisfying the integral identity

$$-\int_0^T \int_{E_n} u\eta_t \, dx \, dt + \int_0^T [\mathscr{L}_1(u, \eta) + \mathscr{L}_2(\mathbf{f}, \eta)] \, dt = \int_{E_n} \psi_0(x) \eta(x, 0) \, dx \tag{5.9}$$

with

$$\mathscr{L}_1(u, \eta) = \int_{E_n} [(a_{ij}u_{x_j} + a_i u)\eta_{x_i} + (bu_{x_i} + au)\eta] \, dx,$$

$$\mathscr{L}_2(\mathbf{f}, \eta) = \int_{E_n} (f_i \eta_{x_i} + f\eta) \, dx$$

for any $\eta(x, t)$ from $V_2^{1,1}(R_T)$ that is equal to zero for $t = T$.

It is possible to make other correct statements of the Cauchy problem in classes of functions with weaker restrictions at infinity, for example, in a class of functions admitting exponential growth $e^{c_1 x^2}$ in x for $|x| \to \infty$. We do not occupy ourselves with an investigation of these questions here (in this connection see the references given on p. 19) and we restrict ourselves to an examination of the solutions of class $V_2^{1,\frac{1}{2}}(R_T)$.

The following existence and uniqueness theorem holds.

Theorem 5.2. *Suppose the coefficients and free terms of equation* (1.1) *satisfy conditions* (1.2)–(1.6) *with* $\overline{Q}_T = R_T$. *Then for any function* $\psi_0(x)$ *from* $L_2(E_n)$ *there exists a unique generalized solution* $u(x, t)$ *from class* $V_2^{1,\frac{1}{2}}(R_T)$ *of the Cauchy problem* (1.1), (5.8).

Under the fulfilment of the conditions of Theorem 4.3 in R_T a solution of problem (1.1), (5.8) will also be an element of $W_2^{1,\frac{1}{2}}(R_T)$. The proofs given above for bounded domains Ω are carried over directly to unbounded Ω, in particular to $\Omega = E_n$, since the estimates on which they are based do not depend on the dimensions of Ω.

§6. ON ESTIMATES IN THE SPACE $W_2^{2,1}(Q_T)$ AND THEIR CONSEQUENCES

In the paper [65d] (see also [65f, 124f] and [33d]) it was shown that operators $\mathfrak{L}$ of parabolic type possess the same properties as elliptic ones: the membership of $\mathfrak{L}u$ in $L_2(Q_T)$ implies (under the fulfilment of some homogeneous boundary conditions) the membership in $L_2(Q_T)$ of each of the terms making up $\mathfrak{L}u$. Although this fact was later generalized to the spaces $L_p(Q_T)$ and $H^{l,l/2}(\bar{Q}_T)$ (the proof of these more general propositions is allotted a central place in Chapter IV) we will give a proof of it here both because of the simplicity of the proof itself and because of a number of consequences that arise in this connection and do not result from the theorems of Chapter IV. It is based on analogous assertions concerning the elliptic part of the operator $\mathfrak{L}u$ ([65b, c, k]; see also §2 of Chapter II and §8 of Chapter III of the book [65q]) that are formulated below (in a particular case) in the form of inequality (6.6).

In order not to overburden the book with calculations analogous to those carried out in previous sections we will restrict ourselves to a "shortened" equation $\mathfrak{L}_0 u = f$, retaining only the principal terms. The symmetry of the resultant equation is unimportant, as is also the fact that we take the first boundary condition.

Thus, let us take the differential operator

$$\mathcal{L}_0 u \equiv \alpha u_t - \mathcal{M} u \equiv \alpha(x,t)\, u_t - \frac{\partial}{\partial x_i}\left(a_{ij}(x,t)\, u_{x_j}\right)$$

with $\alpha > 0$ and the a_{ij} satisfying inequalities (1.2), and the boundary condition

$$u\big|_{S_T} = 0. \tag{6.1}$$

We assume from the beginning that $\alpha(x, t)$ satisfies the inequalities $0 < \nu_1 \leq \alpha(x, t) \leq \mu_1$ and has a derivative α_t, with

$$\operatorname*{vrai\,max}_{x\in\Omega} |\alpha_t(x,t)| \equiv \mu_2(t), \quad \int_0^T \mu_2(t)\, dt < \infty.$$

We consider the expression $\int_{Q_t} \mathfrak{L}_0 u \cdot u\, dx\, dt$ and transform it in the following manner:

$$\int_{Q_t} \mathcal{L}_0 u \cdot u\, dx\, dt = \int_{Q_t} \left[\frac{\partial}{\partial t}\left(\frac{\alpha}{2} u^2\right) - \frac{1}{2}\alpha_t u^2 + a_{ij} u_{x_j} u_{x_i}\right] dx\, dt.$$

Hence by virtue of our assumptions on α and the a_{ij} we have

$$\nu_1 \int_\Omega u^2(x, t)\,dx + 2\nu \int_{Q_t} u_x^2\,dx\,dt \leqslant \mu_1 \int_\Omega u^2(x, 0)\,dx$$

$$+ \int_0^t [1 + \mu_2(t)] \int_\Omega u^2(x, t)\,dx\,dt + \int_{Q_t} (\mathscr{L}_0 u)^2\,dx\,dt. \tag{6.2}$$

Let

$$1 + \mu_2(t) = \mu_3(t), \quad \nu_1 \int_0^t \mu_3(t) \int_\Omega u^2(x, t)\,dx = y(t),$$

and

$$\mu_1 \int_\Omega u^2(x, 0)\,dx + \int_{Q_t} (\mathscr{L}_0 u)^2\,dx\,dt = \mathscr{F}(t).$$

For $y(t)$ there follows from (6.2) the inequality

$$\frac{dy(t)}{dt} \leqslant \nu_1^{-1}\mu_3(t)\,y(t) + \mu_3(t)\,\mathscr{F}(t),$$

from which, in turn, by virtue of Lemma 5.5 of Chapter II, we get

$$y(t) \leqslant \exp\left\{\nu_1^{-1} \int_0^t \mu_3(\tau)\,d\tau\right\} \int_0^t \mu_3(\tau)\,\mathscr{F}(\tau)\,d\tau.$$

Substituting this estimate into (6.2), we obtain the first of the inequalities of interest to us:

$$\nu_1 \| u(x, t)\|_{2,\Omega}^2 + 2\nu \| u_x \|_{2, Q_t}^2$$

$$\leqslant \left[1 + \exp\left\{\nu_1^{-1} \int_0^t \mu_3(\tau)\,d\tau\right\} \nu_1^{-1} \int_0^t \mu_3(\tau)\,d\tau\right]$$

$$\times \left[\mu_1 \| u(x, 0)\|_{2,\Omega}^2 + \| \mathscr{L}_0 u \|_{2, Q_t}^2\right]. \tag{6.3}$$

An analogous inequality also holds for the complete operator $\mathfrak{L}$ in (1.1) as long as its coefficients satisfy conditions (1.3) and (1.4).

Suppose now that the coefficients $a_{ij}(x, t)$ have the derivative a_{ijt}, with

$$\operatorname*{vrai\,max}_{x \in \Omega} | a_{ijt}(x, t)| \leqslant \mu_4(t), \quad \int_0^T \mu_4(\tau)\,d\tau < \infty.$$

while α is a bounded function satisfying the inequalities $0 < \nu_1 \le \alpha(x, t) \le \mu_1$.

We consider $\int_{Q_t} \mathfrak{L}_0 u \cdot u_t \, dx\, dt$ and transform it in the following manner:

$$\int_{Q_t} \mathscr{L}_0 u \cdot u_t \, dx\, dt = \int_{Q_t} (\alpha u_t^2 + a_{ij} u_{x_j} u_{x_i t})\, dx\, dt$$
$$= \frac{1}{2} \int_{\Omega} a_{ij} u_{x_j} u_{x_i} \, dx \Big|_{t=0}^{t=t} + \int_{Q_t} \left(\alpha u_t^2 - \frac{1}{2} a_{ijt} u_{x_j} u_{x_i}\right) dx\, dt,$$

from which in view of our assumptions on α and the a_{ij} it follows that

$$\frac{\nu}{2} \| u_x(x, t) \|_{2, \Omega}^2 + \int_{Q_t} \alpha u_t^2 \, dx\, dt \leqslant \frac{\mu}{2} \| u_x(x, 0) \|_{2, \Omega}^2$$
$$+ \frac{\nu_1}{2} \| u_t \|_{2, Q_t}^2 + \frac{1}{2\nu_1} \| \mathscr{L}_0 u \|_{2, Q_t}^2 + \frac{n}{2} \int_0^t \mu_4(t) \| u_x(x, t) \|_{2, \Omega}^2 \, dt,$$

and hence

$$\nu \| u_x(x, t) \|_{2, \Omega}^2 + \nu_1 \| u_t \|_{2, Q_t}^2$$
$$\leqslant \mu \| u_x(x, 0) \|_{2, \Omega}^2 + \nu_1^{-1} \| \mathscr{L}_0 u \|_{2, Q_t}^2 + n \int_0^t \mu_4(t) \| u_x \|_{2, \Omega}^2 \, dt. \tag{6.4}$$

We introduce the notation

$$\nu \int_0^t \mu_4(\tau) \| u_x(x, \tau) \|_{2, \Omega}^2 \, d\tau = y(t)$$

and

$$\mu \| u_x(x, 0) \|_{2, \Omega}^2 + \nu_1^{-1} \| \mathscr{L}_0 u \|_{2, Q_t}^2 = \mathscr{F}(t).$$

Inequality (6.4) implies the inequality

$$\frac{dy(t)}{dt} \leqslant n\nu^{-1} \mu_4(t) y(t) + \mu_4(t) \mathscr{F}(t), \tag{6.5}$$

which, as proved in Lemma 5.5 of Chapter II, guarantees the estimate

$$y(t) \leqslant \exp\left\{ n\nu^{-1} \int_0^t \mu_4(\tau)\, d\tau \right\} \int_0^t \mu_4(\tau) \mathscr{F}(\tau)\, d\tau.$$

Substituting it into (6.4), we obtain the second of the inequalities of interest to us:

$$\nu \| u_x(x, t) \|_{2, \Omega}^2 + \nu_1 \| u_t \|_{2, Q_t}^2$$
$$\leqslant \left[1 + \exp\left\{ n\nu^{-1} \int_0^t \mu_4(\tau)\, d\tau \right\} n\nu^{-1} \int_0^t \mu_4(\tau)\, d\tau \right]$$
$$\times \left[\mu \| u_x(x, 0) \|_{2, \Omega}^2 + \nu_1^{-1} \| \mathcal{L}_0 u \|_{2, Q_t}^2 \right]. \tag{6.6}$$

Remark 6.1. An analogous inequality also holds for the complete equation (1.1) as long as

$$\int_0^T \operatorname{vrai}\max_{x \in \Omega} | a_{tt}(x, t) |\, dt < \infty$$

and the coefficients a_i, b_i and a, for example, are bounded by an absolute quantity. This is proved in the same way as inequality (6.6).

Let us assume now that the a_{ij} have the derivatives $\partial a_{ij}/\partial x_k$, with

$$\operatorname{vrai}\max_{Q_T} \left| \frac{\partial a_{ij}}{\partial x_k} \right| \leqslant \mu_5 < \infty,^{1)} \quad \alpha(x, t) \equiv 1$$

and $S \in O^2$. We consider the integral $\int_{Q_t} \mathcal{L}_0 u(-\Delta u)\, dx\, dt$ and transform it in the following manner:

$$\int_{Q_t} \mathcal{L}_0 u (-\Delta u)\, dx\, dt = \int_{Q_t} \left(u_{x_k t} u_{x_k} + \mathcal{M} u\, \Delta u \right) dx\, dt. \tag{6.7}$$

But by virtue of (2.24) of Chapter II, for any function u from $\overset{\circ}{W}{}_2^2(\Omega)$

$$\int_\Omega \mathcal{M} u\, \Delta u\, dx \geqslant \frac{3}{4} \nu \| u_{xx} \|_{2, \Omega}^2 - c_1 \| u_x \|_{2, \Omega}^2, \tag{6.8}$$

where c_1 is a constant depending only on ν, μ, μ_5 and S. Therefore (6.7) implies that

$$\| u_x(x, t) \|_{2, \Omega}^2 + \frac{3\nu}{2} \| u_{xx} \|_{2, Q_t}^2 \leqslant \| u_x(x, 0) \|_{2, \Omega}^2$$
$$+ 2c_1 \| u_x \|_{2, Q_t}^2 + \varepsilon \| \Delta u \|_{2, Q_t}^2 + \varepsilon^{-1} \| \mathcal{L}_0 u \|_{2, Q_t}^2,$$

from which, choosing $\epsilon = \nu/2n$ and using Lemma 5.5 of Chapter II, we obtain the inequality

1) This assumption on the a_{ijx_k} could be weakened somewhat.

$$\| u_x(x,\ t)\|^2_{2,\Omega} + \nu \| u_{xx}\|^2_{2,Q_t} \leqslant e^{2c_1 t}\Big[\| u_x(x,\ 0)\|^2_{2,\Omega} + \frac{2n}{\nu}\| \mathcal{L}_0 u\|^2_{2,Q_t}\Big], \tag{6.9}$$

which in combination with the equation $u_t = \mathfrak{M}u + \mathfrak{L}_0 u$ gives the third inequality of interest to us:

$$\| u_x(x,\ t)\|^2_{2,\Omega} + \| u_{xx}\|^2_{2,Q_t} + \| u_t\|^2_{2,Q_t} \leqslant c_2 e^{2c_1 t}(t+1)\Big[\| u_x(x,\ 0)\|^2_{2,\Omega} + \| \mathcal{L}_0 u\|^2_{2,Q_t}\Big] \tag{6.10}$$

with c_2 depending only on n, ν, μ, μ_5 and S.

The derivation of inequalities (6.6), (6.10) is based on two facts: inequality (6.8) for elliptic operators and the positiveness of the principal part of the integrals $\int_{Q_t} -u_t \Delta u\,dx\,dt$ and $\int_{Q_t} -u_t (a_{ij}u_{x_j})_{x_i} dx\,dt$ combining the terms u_t and either $-\Delta u$ or $-\partial(a_{ij}u_{x_j})/\partial x_i$ in a parabolic operator.

If the boundary of the domain is not smooth, then inequality (6.8) cannot hold In this case it is possible to obtain interior estimates for u_t and u_{xx} by considering the integrals

$$\int_{Q_t} \Big[u_t - \frac{\partial}{\partial x_i}(a_{ij}u_{x_j})\Big](u_t - \Delta u)\zeta^2(x)\,dx\,dt,$$

where $\zeta(x)$ is a smooth function that is equal to zero on S, and assuming $\operatorname{vrai\,max}_{Q_T} |\partial a_{ij}/\partial t| < \infty$. This leads to the inequality

$$\| u_x(x,\ t)\zeta(x)\|^2_{2,\Omega} + \| u_t\zeta\|^2_{2,Q_t} + \| u_{xx}\zeta\|^2_{2,Q_t} \leqslant c_3 e^{c_4 t}\Big[\| u_x(x,\ 0)\zeta(x)\|^2_{2,\Omega} + \|(\mathcal{L}_0 u)\zeta\|^2_{2,Q_t} + \| u_x\zeta_x\|^2_{2,Q_t}\Big], \tag{6.11}$$

where the integral $\|u_x\zeta_x\|_{2,Q_t}$ can be estimated from the energy inequality. If the initial values of u do not belong to $\overset{\circ}{W}{}^1_2(\Omega)$, then, by considering the integrals

$$\int_{Q_t} \mathcal{L}_0 u(u_t - \Delta u)\zeta^2(x)\chi^2(t)\,dx\,dt \quad \text{c} \quad \chi(0) = 0,$$

we obtain the estimate

$$\| u_x(x,\ t)\zeta(x)\chi(t)\|^2_{2,\Omega} + \|u_t\zeta\chi\|^2_{2,Q_t} + \| u_{xx}\zeta\chi\|^2_{2,Q_t} \leqslant c_5 e^{c_6 t}\Big[\|(\mathcal{L}_0 u)\zeta\chi\|^2_{2,Q_t} + \| u_x\zeta_x\chi\|^2_{2,Q_t} + \| u_x^2\zeta^2\chi'\chi\|_{1,Q_t}\Big]. \tag{6.12}$$

In accordance with the inequalities derived here one can prove that the generalized solutions from $V_2^{1,0}(Q_T)$ found in §4 have finite integrals in the left sides of these inequalities as long as all of the conditions under which these inequalities are derived are fulfilled. This can be done in various ways. For example, it can be done at once in the process of proving the convergence of Galerkin's method. In this regard it is necessary to show that analogous estimates hold for all of the approximations u^N with constants not depending on the order of approximations. Estimate (6.6) can be established without introducing any changes in the plan of Galerkin's method described above. Here it is necessary to multiply each of the equations (4.1) by the corresponding $dc_k^N(t)/dt$, then sum the resultant equalities from $k = 1$ to $k = N$ and integrate with respect to t from zero to t. This gives the relation

$$\int\limits_{Q_t} [(u_t^N)^2 + (a_{ij}u_{x_j}^N + a_i u^N)\, u_{tx_i}^N + (b_i u_{x_i}^N + au^N)\, u_t^N]\, dx\, dt$$
$$= - \int\limits_{Q_t} (f_i u_{tx_i}^N + f u_t^N)\, dx\, dt, \tag{6.13}$$

the left side of which coincides with the integral $\int_{Q_t} \mathfrak{L}u^N \cdot u_t^N dx\, dt$, from a consideration of which inequality (6.6) was derived for the case $\mathfrak{L} = \mathfrak{L}_0$. In an analogous manner one deduces from (6.13) an estimate that is uniform in N:

$$\max_{0 \leqslant t \leqslant T} \| u_x^N(x, t)\|_{2,\Omega}^2 + \| u_t^N \|_{2, Q_t}^2$$
$$\leqslant c_7 \left[\| u_x^N(x, 0)\|_{2,\Omega}^2 + \left\| \frac{\partial f_i}{\partial x_i} - f \right\|_{2, Q_T}^2 \right] \leqslant c_8, \tag{6.14}$$

from which it follows that the limit function u for the u^{N_m} will have the derivative u_t from $L_2(Q_T)$.

For another case (under which inequality (6.10) was derived) we take as the basis functions $\{\psi_k(x)\}$ in Galerkin's method the eigenfunctions for the Laplace operator under a zero boundary condition on the boundary S of the domain Ω, so that

$$\Delta\psi_k = -\lambda_k \psi_k, \quad \psi_k|_S = 0. \tag{6.15}$$

For a deduction of estimate (6.9) we multiply each of the equations (4.4) by the corresponding $\lambda_k c_k^N(t)$, then sum the resultant equalities over k from 1 to N and integrate with respect to t. In view of (6.15) this leads to the relation

$$-\int\limits_{Q_t} \mathcal{L}u^N \cdot \Delta u^N \, dx\, dt = \int\limits_{Q_t} \left(-\frac{\partial f_i}{\partial x_i} + f\right) \Delta u^N \, dx\, dt, \tag{6.16}$$

the left side of which coincides for $\mathfrak{L} = \mathfrak{L}_0$ with the integral from which inequality (6.9) was derived. In the same way one deduces from (6.16) an estimate that is uniform in N:

$$\max_{0 \leqslant t \leqslant T} \| u_x^N(x, t)\|^2_{2, \Omega} + \| u^N_{xx} \|^2_{2, Q_T} \leqslant c_9 \left[\| u_x^N(x, 0)\|^2_{2, \Omega} + \left\| \frac{\partial f_i}{\partial x_i} - f \right\|^2_{2, Q_T} \right] \leqslant c_{10}. \tag{6.17}$$

It implies that the limit function u for the u^{N_m} will have derivatives u_{xx} from $L_2(Q_T)$. From this and from the integral identity (1.16), which the function u satisfies, it follows that u has the derivative u_t and satisfies for almost all (x, t) from Q_T the equation $\mathfrak{L}u = \partial f_i/\partial x_i - f$; the derivative u_t in this connection turns out to be an element of $L_2(Q_T)$ (which is seen from the equation). Let us formulate the proven assertions in the form of a theorem.

Theorem 6.1. *The problem*

$$\left. \begin{aligned} &u_t - \frac{\partial}{\partial x_i}(a_{ij}(x, t) u_{x_j}) = \mathcal{F}(x, t), \\ &u|_{S_T} = 0, \quad u|_{t=0} = \psi_0(x), \end{aligned} \right\} \tag{6.18}$$

has a unique solution from $W_2^{1,1}(Q_T)$ *if* $\mathfrak{F} \in L_2(Q_T)$, $\psi_0 \in \overset{\circ}{W}{}_2^1(\Omega)$, a_{ij} *satisfies condition* (1.2) *and*

$$\int\limits_0^T \operatorname{vrai}\max_{x \in \Omega} \left| \frac{\partial a_{ij}}{\partial t} \right| dt < \infty.$$

If the last condition is replaced by the condition $\operatorname{vrai}\max_{Q_T}|\partial a_{ij}/\partial x_k| < \infty$ *and it is assumed that* $S \in O^2$, *then the solution will belong to* $W_2^{2,1}(Q_T)$ *and will satisfy the equation for almost all* (x, t) *from* Q_T.

Remark 6.2. The first part of Theorem 6.1 concerning solvability in $W_2^{1,1}(Q_T)$ also holds for the equation

$$\mathcal{L}_0 u \equiv \alpha(x, t) u_t - \frac{\partial}{\partial x_i}(a_{ij}(x, t) u_{x_j}) = \mathcal{F}(x, t) \tag{6.19}$$

with an arbitrary measurable function $\alpha(x, t)$ satisfying the condition $0 < \nu_1 \leq \alpha(x, t) \leq \mu_2 < \infty$. This can be proved by various methods based on inequality (6.6).

For example, let us consider the auxiliary problems

$$\alpha^{(m)}(x, t)u_t - \frac{\partial}{\partial x_i}(a_{ij}^{(m)}u_{x_j}) = \mathcal{F}(x, t),$$
$$u|_{S_T} = 0, \quad u|_{t=0} = \psi_0(x), \tag{6.20}$$

where $\alpha^{(m)}(x, t)$ and $a_{ij}^{(m)}(x, t)$ are sequences of infinitely differentiable functions converging almost everywhere to α and a_{ij} respectively and such that $\nu_1/2 \le \alpha^{(m)} \le 2\mu_2$ and $\int_0^T \operatorname{vrai\,max}_{x\in\Omega} |\partial a_{ij}^{(m)}/\partial t|\, dt \le 2\mu$. If $S \bar{\in} O^2$, then we approximate S by smooth contours S^m from O^2. By virtue of the second part of Theorem 6.1 the problems (6.20) are uniquely solvable in $W_2^{2,1}(Q_T^{(m)})$, and for their solutions u^m we have inequalities (6.6), which guarantee the uniform estimate

$$\| u^m \|_{W_2^{1,1}(Q_T^m)} \leqslant c.$$

Because of it one can select from $\{u^m\}$ a subsequence giving in the limit the desired solution of equation (6.19).

Let us note one more particular case for the equations

$$\mathcal{L}_1 u \equiv u_t - \alpha(x, t)\Delta u = \mathcal{F}(x, t) \tag{6.21}$$

with $\alpha(x, t)$ satisfying the conditions $0 \le \alpha(x, t) \le \mu_2 < \infty$ and with $\mathcal{F}(x, t) \in \overset{\circ}{W}{}_2^{1,0}(Q_T)$. We consider for them the problem

$$\mathcal{L}_1 u = \mathcal{F}, \quad u|_{S_T} = 0, \quad u|_{t=0} = \psi_0(x)$$

with $\psi_0 \in \overset{\circ}{W}{}_2^1(\Omega)$. From the relations

$$-\int_{Q_t} \mathcal{L}_1 u \,\Delta u \, dx\, dt = -\int_{Q_t} \mathcal{F} \Delta u\, dx\, dt = \int_{Q_t} \mathcal{F}_{x_i} u_{x_i}\, dx\, dt \tag{6.22}$$

one can deduce the inequality

$$\| u_x(x, t)\|_{2,\Omega}^2 + \int_{Q_t} \alpha(\Delta u)^2\, dx\, dt$$
$$\leqslant e^{c_1 t}\left(\| u_x(x, 0)\|_{2,\Omega}^2 + \|\mathcal{F}_x\|_{2,Q_t}^2\right). \tag{6.23}$$

in a way similar to that in which (6.9) was deduced from (6.7). From it and (6.21) follows the inequality

$$\|u_t\|_{2,Q_t}^2 \leqslant 2\|\mathcal{F}\|_{2,Q_t}^2 + 2\mu_2 e^{c_1 t}\left(\|u_x(x,0)\|_{2,\Omega}^2 + \|\mathcal{F}_x\|_{2,Q_t}^2\right). \tag{6.24}$$

On the basis of these a priori estimates (6.23) and (6.24), in the same way as above, it is proved that problem (6.22) has a solution u possessing derivatives u_x and u_t, as well as u_{xx} when $\alpha(x, t) > 0$, and satisfying the inequality

$$\int_{Q_T} [u_t^2 + \alpha(\Delta u)^2]\,dx\,dt + \operatorname{vrai}\max_{0 \leqslant t \leqslant T} \|u_x\|_{2,\Omega}^2 \leqslant c(T)\left(\int_{Q_T} \mathcal{F}_x^2\,dx\,dt + \int_\Omega \psi_{0x}^2(x,0)\,dx\right), \tag{6.25}$$

in which $c(T)$ depends only on T and μ_2.

Remark 6.3. In a manner analogous to that of Theorem 6.1 the unique solvability of the first boundary value problem (and the other basic boundary value problems) is proved for the complete equation (1.1) under $\mathfrak{F} = \partial f_i/\partial x_i - f \in L_2(Q_T)$ and the following assumptions regarding its coefficients: solvability in $W_2^{1,1}(Q_T)$ under the conditions

$$\operatorname{vrai}\max_{x \leqslant \Omega} \left| b_i(x,t),\ \frac{\partial a_{ij}(x,t)}{\partial t} \right| \leqslant \mu_1(t),$$

$$\int_0^T \mu_1(t)\,dt < \infty, \quad a \in L_{n,2}(Q_T);$$

solvability in $W_2^{2,1}(Q_T)$ under the conditions

$$\frac{\partial a_{ij}}{\partial x_k},\ b_i \in L_{q,r}(Q_T),\ \frac{1}{r} + \frac{n}{2q} = \frac{1}{2}, \quad r < \infty \tag{6.26}$$

and $a \in L_{q,r}(Q_T)$, $r < \infty$ and

$$\left.\begin{array}{lll} \frac{1}{r} + \frac{n}{2q} = 1, & & \text{for } n \geqslant 4, \\ \frac{1}{r} + \frac{n}{2q} = 1, & q > 2 & \text{for } n = 3, \\ r > 4, & q = 2 & \text{for } n = 3, \\ r > 2, & q = 2 & \text{for } n = 2. \end{array}\right\} \tag{6.27}$$

We assume that the coefficients $a_i(x, t)$ are equal to zero (which is equivalent to adjoining the terms

$$a_l u_{x_l} + \frac{\partial a_l}{\partial x_l} u = \frac{\partial}{\partial x_l}(a_l u)$$

to $b_i u_{x_i}$ and au respectively). The conditions on ψ_0 are the same as in Theorem 6.1. For $r \to \infty$ conditions (6.26) and (6.27) must be replaced by the condition of uniform smallness of the norms $\|\partial a_{ij}/\partial x_k,\ b_i\|_{n,K_\rho}$ and $\|a\|_{n/2,K_\rho}$ for small ρ and any t from $[0, T]$.

Estimate (6.10) is relative to the number of exact estimates. In §§14 and 15 we show how it may be used to prove the solvability of problem (1.9) in $W_2^{2,1}(Q_T)$ without resorting to approximate methods.

§7. AN ESTIMATE OF $\max_{Q_T}|u|$. THE MAXIMUM PRINCIPLE

Let us assume that the free terms and the coefficients of the minor terms of equation (1.1) possess somewhat better properties than in §1, namely,

$$\left\| \sum_{i=1}^n a_i^2;\ \sum_{i=1}^n b_i^2;\ \sum_{i=1}^n f_i^2;\ a;\ f \right\|_{q,\,r,\,Q_T} \leqslant \mu_1, \tag{7.1}$$

where q and r are arbitrary positive numbers satisfying the condition

$$\left.\begin{aligned} &\frac{1}{r} + \frac{n}{2q} = 1 - \varkappa_1, \\ \text{with}\quad & \\ &q \in \left[\frac{n}{2(1-\varkappa_1)},\ \infty\right],\ r \in \left[\frac{1}{1-\varkappa_1},\ \infty\right], \\ &\qquad\qquad 0 < \varkappa_1 < 1, \text{ for } n \geqslant 2, \\ &q \in [1,\ \infty],\ r \in \left[\frac{1}{1-\varkappa_1},\ \frac{2}{1-2\varkappa_1}\right], \\ &\qquad\qquad 0 < \varkappa_1 < \frac{1}{2}, \quad \text{for}\quad n = 1. \end{aligned}\right\} \tag{7.2}$$

We keep the same assumptions concerning the coefficients a_{ij}, namely (1.2). Let us show that in this case any generalized solution u from $V_2^{1,0}(Q_T)$ of equation (1.1). is a bounded function. More precisely, let us prove the following theorem.

Theorem 7.1. *If the coefficients and free terms of equation* (1.1) *satisfy conditions* (1.2) *and* (7.1), (7.2), *then, for any generalized solution* $u(x, t)$ *from* $V_2^{1,0}(Q_T)$ *of equation* (1.1) *not exceeding* $\hat{k}$ *on* Γ_T, *the quantity* $\operatorname{vrai}\max_{Q_T} u(x,t)$ *is finite and is estimated from above by a constant c determined only by* n, $\hat{k}$ *and* *the parameters* ν, μ_1, q, r *entering into conditions* (1.2), (7.1) *and* (7.2), *with the dependence on* $\hat{k}$ *being linear.*

Remark 7.1. The constant c also depends on $\operatorname{mes} Q_T$: it increases with an increase in $\operatorname{mes} Q_T$. Below (§8) we give local estimates of $\max |u|$, thanks to which $\max_Q |u|$ can be estimated for domains Q of any measure, including unbounded ones.

Remark 7.2. In conditions (7.1) and (7.2) q and r can be different for different coefficients and free terms. The necessary generalization of Theorem 6.1 of Chapter II is given by Remark 6.1 to it.

Remark 7.3. The assertion of the theorem concerning the boundedness from above of $\operatorname{vrai\,max}_{Q_T} u(x, t)$ remains in force if instead of (7.1), (7.2) the b_i are subject to the requirement

$$\left\| \sum_{i=1}^{n} b_i^2 \right\|_{q_1, r_1, Q_T} \leqslant \mu_2, \tag{7.3}$$

where q_1 and r_1 are arbitrary numbers satisfying the conditions

$$\left.\begin{aligned} &\frac{1}{r_1} + \frac{n}{2q_1} = 1, \\ &q_1 \in \left(\frac{n}{2}, \infty\right], r_1 \in [1, \infty) \qquad \text{for } n \geqslant 2, \\ &q_1 \in [1, \infty], r_1 \in [1, 2] \qquad \text{for } n = 1. \end{aligned}\right\} \tag{7.4}$$

This generalization is proved in the same way as Theorem 7.1. The necessary small modifications will be indicated at the end of the section.

In the same place we will also describe the case when $r_1 = \infty$, $q_1 = n/2$ and $n \geq 3$ in condition (7.3).

Remark 7.4. The assertions of Theorem 7.1 and Remarks 7.1–7.3 remain valid for functions $u(x, t)$ from $V_2^{1,0}(Q_T)$ satisfying instead of (1.16) the corresponding inequality

$$l_0(u, \eta) \leq \int_\Omega \psi_0(x) \eta(x, 0)\, dx,$$

in which $\eta(x, t)$ is an arbitrary nonnegative function from $\overset{\circ}{W}{}_2^{1,1}(Q_T)$ that is equal to zero for $t = T$. The proof of this is the same as the proof given below of Theorem 7.1.

Remark 7.5. Since $-u(x, t)$ satisfies an equation of the same type as $u(x, t)$, all theorems concerning an estimate of vrai max u from above give corresponding assertions concerning an estimate of vrai min u from below.

For a proof of the theorem we use identity (2.12), in which we take

$$\eta(x, t) = u_h^{(k)}(x, t) = \max\{u_h(x, t) - k;\ 0\},\quad k \geqslant \hat{k}.$$

This is possible since the function u_h and consequently the function $u_h^{(k)}$ (see Lemma 4.9 of Chapter II) belong to $\overset{\circ}{V}{}_2^{1,0}(Q_{T-h})$ and $\overset{\circ}{W}{}_2^{1,1}(Q_{T-h})$. Since for almost all t

$$\int_\Omega u_{ht}(x, t)\, u_h^{(k)}(x, t)\, dx = \frac{1}{2}\frac{d}{dt}\int_\Omega [u_h^{(k)}(x, t)]^2\, dx,$$

we can rewrite the resultant equalities in the form

$$\frac{1}{2}\int_\Omega [u_h^{(k)}(x, t)]^2\, dx\,\Big|_{t=0}^{t=t_1} + \int_0^{t_1}\int_\Omega [(a_{ij}u_{x_j} + a_i u + f_i)_h\, u_{hx_i}^{(k)} + (b_i u_{x_i} + au + f)_h\, u_h^{(k)}]\, dx\, dt = 0. \tag{7.5}$$

We let h tend to zero in (7.5). Since $u \in \overset{\circ}{V}{}_2^{1,0}(Q_T)$, it follows that u_h converges to u in the norm of $V_2^{1,0}(Q_{T-\delta})$, $\delta > 0$, and therefore the "undercut" functions $u_h^{(k)}$ converge to $u^{(k)}$ in the same norm (see Lemma 4.5 of Chapter II). As was shown in §2 in the deduction of (2.13) from (2.12), from here and from the assumed properties of the known functions the limit relation for (7.5) will be

$$\frac{1}{2}\int_\Omega [u^{(k)}(x, t)]^2 dx\,\Big|_{t=0}^{t=t_1} + \int_0^{t_1}\int_\Omega [(a_{ij}u_{x_j} + a_i u + f_i)\, u_{x_i}^{(k)} + (b_i u_{x_i} + au + f)\, u^{(k)}]\, dx\, dt = 0,\quad k \geqslant \hat{k}. \tag{7.6}$$

We denote the set of points x from Ω at which $u(x, t) > k$ by $A_k(t)$. Since, by condition, $u(x, 0) \le \hat{k}$ it follows that $\int_\Omega [u^{(k)}(x, 0)]^2 dx = 0$. We leave in the left side of (7.6) the first two trivially nonnegative terms and estimate them from below, using the condition of ellipticity, while the remaining terms are transferred to the right side and estimated from above by means of Cauchy's inequality ((1.2) of Chapter II). Namely,

$$\frac{1}{2}\int_{\Omega}[u^{(k)}(x, t_1)]^2\,dx+\nu\int_0^{t_1}\int_{A_k(t)} u_x^2\,dx\,dt$$

$$\leqslant -\int_0^{t_1}\int_{A_k(t)}[(a_i u+f_i)\,u_{x_i}+(b_i u_{x_i}+au+f)(u-k)]\,dx\,dt$$

$$\leqslant \int_0^{t_1}\int_{A_k(t)}\Big[\frac{\nu}{4}u_x^2+\frac{2}{\nu}\sum_i(a_i^2u^2+f_i^2)+\frac{\nu}{4}u_x^2$$

$$+\frac{1}{\nu}\sum_i b_i^2(u-k)^2+|au|(u-k)+|f|(u-k)\Big]\,dx\,dt, \tag{7.7}$$

from which, assuming $k\geq 1$, we get

$$\min\left\{\frac{1}{2};\ \frac{\nu}{2}\right\}|u^{(k)}|_{Q_{t_1}}^2\leqslant\int_0^{t_1}\int_{A_k(t)}\mathscr{D}(x, t)[(u-k)^2+k^2]\,dx\,dt,$$

where

$$\mathscr{D}(x, t)=2\left[\frac{4}{\nu}\sum_{i=1}^n a_i^2+\frac{4}{\nu}\sum_{i=1}^n f_i^2+\frac{2}{\nu}\sum_{i=1}^n b_i^2+2|a|+|f|\right].$$

For an estimate of the right side we use Hölder's inequalities ((1.6) of Chapter II). This gives

$$\min\left\{\frac{1}{2};\ \frac{\nu}{2}\right\}|u^{(k)}|_{Q_{t_1}}^2$$

$$\leqslant\|\mathscr{D}\|_{q, r, Q_{t_1}(k)}\|(u-k)^2+k^2\|_{\frac{q}{q-1},\ \frac{r}{r-1},\ Q_{t_1}(k)} \tag{7.8}$$

Here $Q_{t_1}(k)$ is the set of points (x, t) of the cylinder $Q_{t_1}=\Omega\times(0, t_1)$, at which $u(x, t)>k$. We estimate the norm of the function u standing in the right side of (7.8) by means of inequality (3.5) of Chapter II in the following manner:

$$\|(u-k)^2\|_{\frac{q}{q-1},\ \frac{r}{r-1},\ Q_{t_1}(k)}=\|u^{(k)}\|_{\bar{q}, \bar{r}, Q_{t_1}}^2$$

$$\leqslant\|u^{(k)}\|_{\hat{q}, \hat{r}, Q_{t_1}}^2\,\mu^{\frac{2}{\bar{r}}-\frac{2}{\hat{r}}}(k). \tag{7.9}$$

Here $\mu(k) = \int_0^{t_1} \operatorname{mes}^{\hat{r}/\hat{q}} A_k(t)\,dt$,[1] $\bar{q} = 2q/(q-1)$, $\bar{r} = 2r/(r-1)$, $\hat{q} = \bar{q}(1+\kappa)$, $\hat{r} = \bar{r}(1+\kappa)$, with $\kappa = 2\kappa_1/n$. It is easy to calculate that, by virtue of assumptions (7.2), $1/\bar{r} + n/2\bar{q} = n/4 + \kappa_1/2$, $1/\hat{r} + n/2\hat{q} = 1$, and for $n \geq 2$ we have

$$\hat{q} \in \left[2(1+\varkappa),\ \frac{2n(1+\varkappa)}{n-2+2\varkappa_1}\right],\ \hat{r} \in \left[2(1+\varkappa),\ \frac{2(1+\varkappa)}{\varkappa_1}\right],$$

while for $n = 1$ we have

$$\hat{q} \in [2(1+\varkappa),\ \infty],\ \hat{r} \in \left[4,\ \frac{4(1+\varkappa)}{\varkappa}\right].$$

Therefore $V_2^{1,0}(Q_{t_1})$ is embedded in $L_{\hat{q},\hat{r}}(Q_{t_1})$ and $\|u^{(k)}\|_{\hat{q},\hat{r},Q_{t_1}}$ can be estimated in terms of $|u^{(k)}|_{Q_{t_1}}$, using inequality (3.4) of Chapter II; namely,

$$\|(u-k)^2\|_{\frac{q}{q-1},\ \frac{r}{r-1},\ Q_{t_1}(k)} \leqslant \beta^2 |u^{(k)}|^2_{Q_{t_1}} \mu^{\frac{2\varkappa}{\hat{r}}}(k), \tag{7.10}$$

where $\mu(k)$ has the same meaning as in (7.9). We write the second summand in the right side of (7.8) as follows:

$$\begin{aligned}\|k^2\|_{\frac{q}{q-1},\ \frac{r}{r-1},\ Q_{t_1}(k)} &= k^2 \|1\|^2_{\bar{q},\ \bar{r},\ Q_{t_1}(k)} \\ &= k^2 \left(\int_0^{t_1} \operatorname{mes}^{\frac{\bar{r}}{\bar{q}}} A_k(t)\,dt\right)^{\frac{2}{\bar{r}}} = k^2 \mu^{\frac{2(1+\varkappa)}{\hat{r}}}(k),\end{aligned} \tag{7.11}$$

since $\bar{r}/\bar{q} = \hat{r}/\hat{q}$, while $\hat{r} = \bar{r}(1+\kappa)$. Taking into account estimates (7.10) and (7.11), we obtain from (7.8)

$$\begin{aligned}&\min\left(\frac{1}{2},\ \frac{\nu}{2}\right) |u^{(k)}|^2_{Q_{t_1}} \\ &\leqslant \|\mathscr{D}\|_{q,\ r,\ Q_{t_1}(k)} \left[\beta^2 \mu^{\frac{2\varkappa}{\hat{r}}}(k) |u^{(k)}|^2_{Q_{t_1}} + k^2 \mu^{\frac{2(1+\varkappa)}{\hat{r}}}(k)\right].\end{aligned} \tag{7.12}$$

Let us choose t_1 so small that

$$\beta^2 \|\mathscr{D}\|_{q,\ r,\ Q_{t_1}(k)}\ t_1^{\frac{2\varkappa}{\bar{r}}} (\operatorname{mes}\Omega)^{\frac{2\varkappa}{\bar{q}}} \leqslant \frac{1}{4} \min\{1;\ \nu\}. \tag{7.13}$$

---

[1] For $q = n = 1$ the numbers $\bar{q}$ and $\hat{q}$ will be equal to ∞ and instead of $\mu(k) = \int_0^{t_1} \operatorname{mes}^{\hat{r}/\hat{q}} A_k(t)\,dt$ we will have (here and below) the measure of the set of points t from the interval $[0, t_1]$ for which $\operatorname{mes} A_k(t) > 0$.

Then from (7.12) it follows that for $k \geq \max\{1, \hat{k}\} \equiv k_1$

$$\min\left(\frac{1}{4}, \frac{\nu}{4}\right) \| u^{(k)} \|^2_{Q_{t_1}} \leqslant \gamma k^2 \mu^{\frac{2}{r}(1+\varkappa)}(k), \tag{7.14}$$

where

$$\gamma = \| \mathscr{D} \|_{q, r, Q_{t_1}}, \quad \varkappa = \frac{2\varkappa_1}{n}.$$

Since κ in (7.14) is positive we conclude from (7.14) on the basis of Theorem 6.1 of Chapter II that $\operatorname{vrai\,max}_{Q_{t_1}} u(x, t)$ does not exceed a known constant multiplied by k_1.

Analogous arguments are valid for the cylinder $Q_s = \Omega \times [t_s, t_{s+1}]$, as long as their altitudes are subject to requirements (7.13). Thus, after a finite number of steps we obtain the estimate

$$\operatorname*{vrai\,max}_{Q_T} u(x, t) \leqslant c k_1 \tag{7.15}$$

with a constant c that could be explicitly expressed in terms of known characteristics (see inequality (6.2) from Theorem 6.1 of Chapter II). Theorem 7.1 is proved.

Corollary 7.1. *Suppose all of the conditions of Theorem* 7.1 *are fulfilled except* $u|_{\Gamma_T} \leq \hat{k}$, *which is replaced by the condition* $u|_{\Gamma_T} \geq \hat{k}$. *Then a solution* $u(x, t)$ *is bounded from below and* $\operatorname{vrai\,min}_{Q_T} u$ *is estimated from below by a constant determined, like* c *in Theorem* 7.1, *by known qualities.*

This assertion follows from Theorem 7.1 if it is applied to the function $v(x, t) = -u(x, t)$.

As was noted in Remark 7.3, the assumptions (7.1), (7.2) concerning the b_i in Theorem 7.1 can be replaced by the somewhat broader assumptions (7.3), (7.4). Let us explain what modifications must be made in this connection in the estimates just given. The term

$$j = \int_0^{t_1} \int_{A_k(t)} \frac{1}{\nu} \sum_{i=1}^n b_i^2 (u-k)^2 \, dx \, dt = \frac{1}{\nu} \int_{Q_{t_1}} \sum_{i=1}^n b_i^2 (u^{(k)})^2 \, dx \, dt$$

in (7.7) is estimated by means of inequalities (1.6), (3.4) of Chapter II as follows:

$$|j| \leqslant \frac{1}{\nu} \left\| \sum_{i=1}^{n} b_i^2 \right\|_{q_1, r_1, Q_{t_1}(k)} \| u^{(k)} \|^2_{\frac{2q_1}{q_1-1}, \frac{2r_1}{r_1-1}, Q_{t_1}}$$

$$\leqslant \frac{\beta^2}{\nu} \left\| \sum_{i=1}^{n} b_i^2 \right\|_{q_1, r_1, Q_{t_1}(k)} | u^{(k)} |^2_{Q_{t_1}}. \tag{7.16}$$

We wish to cancel the expression on the right side of (7.16) by means of the similar expression on the left side of (7.14). This is possible, if, for example,

$$\frac{\beta^2}{\nu} \left\| \sum_{i=1}^{n} b_i^2 \right\|_{q_1, r_1, Q_{t_1}(k)} \leqslant \frac{1}{8} \min\{1; \nu\}. \tag{7.17}$$

For $r_1 < \infty$ we obtain (7.17) at the expense of choosing a sufficiently small t_1. As was shown in §2 in reducing the energy inequality, in this case the whole cylinder Q_T can be partitioned into a finite number of cylinders $Q_s = \Omega \times [t_{s-1}, t_s]$, for each of which (7.17) will hold, with the number of cylinders being dependent only on μ_2 and r_1 from (7.3), (7.4).

But if $r_1 = \infty$, then such a partitioning gives nothing, since

$$\left\| \sum_{i=1}^{n} b_i^2 \right\|_{\frac{n}{2}, \infty, Q_{t_1}(k)} = \operatorname*{vrai\,max}_{0 \leqslant t \leqslant t_1} \left\| \sum_{i=1}^{n} b_i^2 (x, t) \right\|_{\frac{n}{2}, A_k(t)}$$

We assume in this case that we know the quantity $\mu_3 = \|u\|_{1,\infty,Q_T}$ and the fact that

$$\Delta(k) = \operatorname*{vrai\,max}_{0 \leqslant t \leqslant T} \left\| \sum_{i=1}^{n} b_i^2 (x, t) \right\|_{\frac{n}{2}, A_k(t)} \to 0 \tag{7.18}$$

for $\operatorname{vrai\,max}_{0 \leq t \leq T} \operatorname{mes} A_k(t) \to 0$. By virtue of the inequality $k \operatorname{mes} A_k(t) \leq \|u(x, t)\|_{1,\Omega}$ the quantity $\operatorname{vrai\,max}_{0 \leq t \leq T} \operatorname{mes} A_k(t) \leq k^{-1}\mu_3 \to 0$ for $k \to \infty$, and therefore inequality (7.17) will be fulfilled for all $k \geq k_2$, where k_2 is such that

$$\frac{\beta^2}{\nu} \Delta(k_2) \leqslant \frac{1}{8} \min\{1; \nu\}.$$

Thus, suppose (7.17) is valid. Then, substituting the resultant estimate of $|j|$ and the estimates of the remaining terms obtained earlier into (7.7), we arrive at an inequality differing from (7.14) only by the coefficient in the left side. And this inequality, by virtue of Theorem 6.1 of Chapter II, implies an estimate from above for $\operatorname{vrai\,max}_{Q_T} u$. For $r_1 = \infty$ it will also depend on k_2.

Let us show that for generalized solutions $u(x, t)$ of equation (1.1) from $V_2^{1,0}(Q_T)$ the maximum principle is valid in the following form.

Theorem 7.2. *Suppose $u(x, t)$ is a generalized solution from $V_2^{1,0}(Q_T)$ of equation (1.1) whose coefficients a_{ij}, b_i and a satisfy the conditions (1.2)–(1.4), $a(x, t) \geq 0$, and $a_i \equiv f_i \equiv f \equiv 0$. Then for almost all (x, t) from Q_T*

$$\min\left\{0;\ \operatorname{vrai\,min}_{\Gamma_T} u(x, t)\right\} \leqslant u(x, t) \leqslant \max\left\{0;\ \operatorname{vrai\,max}_{\Gamma_T} u(x, t)\right\}. \tag{7.20}$$

For a proof of the right inequality of (7.20) we take inequality (7.7) for $k = \max\{0;\ \operatorname{vrai\,max}_{\Gamma_{t_1}} u\}$. From it, by virtue of our assumptions and the choice of k, we get

$$\frac{1}{2}\left\|u^{(k)}(x, t_1)\right\|_{2,\Omega}^2 + \nu\left\|u_x^{(k)}\right\|_{2,Q_{t_1}}^2 \leqslant -\int_{Q_{t_1}} b_i u_{x_i} u^{(k)}\,dx\,dt. \tag{7.21}$$

We estimate the right side in the same way as the term $b_i u_{x_i} u$ was estimated in §2 in deducing the energy inequality; namely,

$$\begin{aligned}\left|\int_{Q_{t_1}} b_i u_{x_i} u^{(k)}\,dx\,dt\right| &\leqslant \int_{Q_{t_1}}\left(\frac{\nu}{2}\,u_x^{(k)2} + \frac{1}{2\nu}\sum_{i=1}^n b_i^2 u^{(k)2}\right)dx\,dt \\ &\leqslant \frac{\nu}{2}\left\|u_x^{(k)}\right\|_{2,Q_{t_1}}^2 + \frac{1}{2\nu}\left\|\sum_{i=1}^n b_i^2\right\|_{q,r,Q_{t_1}} \left\|u^{(k)}\right\|^2_{\frac{2q}{q-1},\frac{2r}{r-1},Q_{t_1}} \\ &\leqslant \frac{\nu}{2}\left\|u_x^{(k)}\right\|_{2,Q_{t_1}}^2 + \frac{\beta^2}{2\nu}\left\|\sum_{i=1}^n b_i^2\right\|_{q,r,Q_{t_1}} \left|u^{(k)}\right|^2_{Q_{t_1}}.\end{aligned}$$

From this and from (7.21) it follows that

$$\min\{1, \nu\}\left|u^{(k)}\right|^2_{Q_{t_1}} \leqslant \frac{2\beta^2}{\nu}\left\|\sum_{i=1}^n b_i^2\right\|_{q,r,Q_{t_1}} \left|u^{(k)}\right|^2_{Q_{t_1}}. \tag{7.22}$$

For t_1 satisfying the condition

$$\frac{2\beta^2}{\nu}\left\|\sum_{i=1}^{n} b_i^2\right\|_{q,r,Q_{t_1}} < \min\{1, \nu\}, \tag{7.23}$$

we obtain from (7.22)

$$|u^{(k)}|_{Q_{t_1}} \leqslant 0,$$

i.e. for almost all (x, t) from Q_{t_1} the function $u(x, t)$ does not exceed $k = \max\{0; \operatorname{vrai\,max}_{\Gamma_{t_1}} u\}$. Repeating this argument for the cylinders $\Omega \times [t_1, t_2], \cdots$ and so on, we find that the right inequality of (7.20) is valid. Analogous arguments with the function $-u(x, t)$ lead to a proof of the left inequality of (7.20). If $a \equiv 0$, then the zero can be removed from (7.20).

Moser proposed a different method of estimating $\operatorname{vrai\,max}|u(x, t)|$ for solutions of linear equations of elliptic and parabolic type [87a, b], which consists in obtaining recursion relations between unboundedly increasing powers of u that turn out to yield a conclusion on the boundedness of $\operatorname{vrai\,max}|u(x, t)|$. He carried out this idea for equations (1.1) with $a_i \equiv b_i \equiv a \equiv f \equiv f_i \equiv 0$. And with the use of the estimates of minor terms given above it can be applied to the complete equation in the following way.

For the sake of simplicity we will consider the solution $u(x, t)$ to be bounded and equal to zero on S_T. In identity (2.12) let us put

$$\eta(x, t) = \varphi'(u_h(x, t))\,\varphi(u_h(x, t))\,\zeta^2(x, t),$$

where u_h is the averaging of u with respect to t defined in (2.10), $\zeta(x, t)$ is an arbitrary smooth function, and $\phi(r) = |r|^s$ with $s > 1$. It is easy to see that such a function is admissible for (2.12). We let $\phi(u_h(x, t)) = v_{(h)}(x, t)$ and represent the first term of (2.12) in the form

$$\int_{Q_{t_1}} u_{ht}\varphi'(u_h)\,v_{(h)}\,\zeta^2\,dx\,dt = \frac{1}{2}\int_{\Omega} v_{(h)}^2\zeta^2\,dx\,\Big|_0^{t_1} - \int_{Q_{t_1}} v_{(h)}^2\zeta\zeta_t\,dx\,dt.$$

After this we pass to the limit as $h \to 0$ in all of the terms of the equality. This is possible, and the limit equality can be written in the form

$$\begin{aligned}\frac{1}{2}\int_{\Omega} v^2\zeta^2\,dx\,\Big|_0^{t_1} &+ \int_{Q_{t_1}} [-v^2\zeta\zeta_t + a_{ij}u_{x_i}u_{x_j}v\varphi''\zeta^2 \\ &+ a_{ij}v_{x_i}v_{x_j}\zeta^2 + 2a_{ij}v_{x_j}v\zeta\zeta_{x_i} + (a_i u + f_i)(\varphi'' u_{x_i} v\zeta^2 \\ &+ \varphi' v_{x_i}\zeta^2 + 2\varphi' v\zeta\zeta_{x_i}) + (b_i u_{x_i} + au + f)\,\varphi' v\zeta^2]\,dx\,dt = 0,\end{aligned} \tag{7.24}$$

where $v = \phi(u)$. We put $\zeta \equiv 1$ in (7.24) and transfer to the right side all of the terms except the trivially nonnegative ones: the first integral and the second and third summands in the second integral. Then we estimate the terms transferred to the right side in the same way as was done above with the use of assumptions (7.1) and (7.2). This leads to the inequality

$$|v|^2_{Q_T} \leqslant \gamma^2 s^2 \left(\|v\|_{\bar{q}, \bar{r}, Q_T} + 1 \right)^2, \tag{7.25}$$

in which the constant γ is determined by the quantities ν, μ_1, q and r from conditions (1.2), (7.1)–(7.2), T and $\operatorname{mes}\Omega$, while $\bar{r} = 2r/(r-1)$, $\bar{q} = 2q/(q-1)$.

As above, we let $\hat{q} = \bar{q}(1+\kappa)$, $\hat{r} = r(1+\kappa)$, $\kappa = 2\kappa_1/n$. The parameters $\hat{q}$ and $\hat{r}$ are subject to restrictions (3.3) of Chapter II, and therefore by virtue of inequality (3.4) of Chapter II we get

$$\|v\|_{\hat{q}, \hat{r}, Q_T} \leqslant \beta |v|_{Q_T}.$$

Hence (7.25) implies the estimate

$$\|v\|_{\hat{q}, \hat{r}, Q_T} \leqslant \beta\gamma s \left[\left\| v^{\frac{1}{1+\varkappa}} \right\|^{1+\varkappa}_{\bar{q}, \bar{r}, Q_T} + 1 \right]. \tag{7.26}$$

We will take as s the sequence of numbers $(1+\kappa)^k$ with $k = 1, 2, \cdots$. If the norm

$$\left\| u^{(1+\varkappa)^k} \right\|_{\hat{q}, \hat{r}, Q_T} = \|u\|^{(1+\varkappa)^k}_{(1+\varkappa)^k \hat{q}, (1+\varkappa)^k \hat{r}, Q_T},$$

is denoted by Φ_k, then inequalities (7.26) for $v = |u|^{(1+\kappa)^k}$ are written as

$$\Phi_k \leqslant (1+\varkappa)^k \beta\gamma \left(\Phi^{1+\varkappa}_{k-1} + 1 \right), \quad k = 1, 2, \dots. \tag{7.27}$$

Hence we conclude (see Lemma 5.6 of Chapter II) that

$$\Phi_k \leqslant [2(1+\varkappa)\max\{1, \beta\gamma\}]^{\frac{(1+\varkappa)^k - 1}{\varkappa}} \times (1+\varkappa)^{\frac{(1+\varkappa)^k - 1}{\varkappa^2} - \frac{k}{\varkappa}} (\Phi_0 + 1)^{(1+\varkappa)^k},$$

and, consequently,

$$\begin{aligned} \operatorname*{vrai\,max}_{Q_T} |u| \equiv \|u\|_{\infty, Q_T} &= \lim_{k\to\infty} \Phi_k^{(1+\varkappa)^{-k}} \\ &\leqslant 2^{\frac{1}{\varkappa}} \max\{1, \beta\gamma\}^{\frac{1}{\varkappa}} (1+\varkappa)^{\frac{1+\varkappa}{\varkappa^2}} (\Phi_0 + 1). \end{aligned} \tag{7.28}$$

The quantity $\Phi_0 = \|u\|_{\hat{q},\hat{r},Q_T}$ in the right side of the inequality does not exceed $\beta|u|_{Q_T}$, and therefore by means of the energy inequality it can be estimated in terms of the known constants ν, μ_1, q and r from conditions (1.2), (7.1) and (7.2). In this way the desired estimate of $\operatorname{vrai}\max_{Q_T}|u|$ is obtained.

§8. LOCAL ESTIMATES OF $\max|u|$

Let us now obtain local estimates for arbitrary generalized solutions $u(x, t)$ from $V_2^{1,0}(Q_T)$ of equation (1.1) without an assumption concerning their boundedness on Γ_T. Suppose conditions (1.2) and (7.1)–(7.2) are fulfilled. We take an arbitrary cylinder $Q(\rho_0, \tau_0) = \{|x - x_0| < \rho_0;\ t_0 < t < t_0 + \tau_0\} \in Q_T$ and in (2.12) we put $\eta(x, t) = \zeta^2(x, t)u_h^{(k)}(x, t) = \zeta^2(x, t)\max\{u_h(x, t) - k; 0\}$, where $\zeta(x, t)$ is a nonnegative, continuous, piecewise-smooth function that does not exceed 1 and is equal to zero on the lateral surface $S_{t_0,t_0+\tau_0}$ of the cylinder $Q(\rho_0, \tau_0)$ and outside it. After substituting this function into (2.12) and passing to the limit as $h \to 0$ we obtain an equality similar to (7.6). From it, since the upper limit of integration with respect to t is arbitrary, we will have

$$\frac{1}{2}\|u^{(k)}(x, t)\zeta(x, t)\|_{2,\Omega}^2\Big|_{t_1}^{t_2} + \int_{t_1}^{t_2}\int_{A_k(t)}\{(a_{ij}u_{x_j} + a_iu + f_i)[u_{x_i}\zeta^2 + 2(u-k)\zeta\zeta_{x_i}] + (b_iu_{x_i} + au - f)(u-k)\zeta^2 - (u-k)^2\zeta\zeta_t\}\,dx\,dt = 0,$$
$$t_0 < t_1 < t_2 < t_0 + \tau_0. \tag{8.1}$$

This equality implies the inequalities

$$\max_{t_1 \leqslant t \leqslant t_2}\|u^{(k)}(x, t)\zeta(x, t)\|_{2,\Omega}^2 + \nu\int_{t_1}^{t_2}\int_{\Omega}(u_x^{(k)})^2\zeta^2\,dx\,dt$$
$$\leqslant \|u^{(k)}(x, t_1)\zeta(x, t_1)\|_{2,\Omega}^2 + \gamma_1\left\{\int_{t_1}^{t_2}\int_{\Omega}(\zeta_x^2 + \zeta|\zeta_t|)(u^{(k)})^2\,dx\,dt + k^2\left[\int_{t_1}^{t_2}\left(\int_{A_k(t)}\zeta^{\bar{q}}\,dx\right)^{\frac{\hat{r}}{\bar{q}}}dt\right]^{\frac{2(1+\varkappa)}{\hat{r}}}\right\}^{1)} \tag{8.2}$$

1) For $\bar{q} = \infty$ (which occurs only for $n=1$), in place of the term on the last line we will have $k^2[\int_{t_1}^{t_2}(\operatorname{vrai}\max_{x\in A_k(t)}\zeta)^{\bar{r}}\,dt]^{2(1+\kappa)/\bar{r}}$, where for all t for which $\operatorname{mes}A_k(t) = 0$ it is necessary to assume that $\operatorname{vrai}\max_{x\in A_k(t)}\zeta(x, t)$ is equal to zero.

with the same values of the parameters $\hat{q}$, $\hat{r}$, κ as in (7.14). The symbol $A_k(t)$, as above, denotes the set of points x of Ω at which $u(x, t) > k$. The constant γ_1 depends on known quantities in roughly the same way as γ in (7.14). The deduction of (8.2) from (8.1) is the same as the deduction of (7.14) from (7.6). Only this time we can satisfy condition (7.13), more precisely, the condition

$$\beta^2 \| \mathscr{D}\zeta \|_{q, r, Q_{t_0, t_0+\tau}(k)} \left[\int_{t_0}^{t_0+\tau_0} \left(\int_{A_k(t)} \zeta \, dx \right)^{\frac{\hat{r}}{\hat{q}}} dt \right]^{\frac{2\kappa}{\hat{r}}} \leq \min\left\{ \frac{1}{4};\ \frac{\nu}{4} \right\},$$

at the expense, not necessarily of smallness in t_0, but, more generally, of smallness in the measure of the set where $\zeta(x, t)$ is different from zero. As was indicated in Remark 6.4 of Chapter II, from (8.2) we get inequalities (6.9) of Chapter II, and consequently $u(x, t) \in \mathfrak{A}(Q(\rho_0, \tau_0), \gamma, r, \hat{k}, \kappa)$ and ρ_0 and τ_0 satisfying the condition

$$\beta^2 \| \mathscr{D}\zeta \|_{q, r, Q(\rho_0, \tau_0)} \tau_0^{\frac{2\kappa}{\hat{r}}} (\varkappa_n \rho_0)^{\frac{2\kappa}{\hat{q}}} \leq \min\left\{ \frac{1}{4};\ \frac{\nu}{4} \right\}. \tag{8.3}$$

This is also true for the function $-u(x, t)$.

Suppose $u(x, t)$ does not exceed some number k on $\Gamma' \subset \Gamma_T$. We take an arbitrary cylinder $Q(\rho_0, \tau_0)$, the intersection of which with Γ_T is not empty and belongs to Γ', and a function $\eta(x, t)$ that is equal to $\zeta^2(x, t) u_h^{(k)}(x, t)$ in $Q(\rho_0, \tau_0) \cap Q_T$ and to zero in the remaining part of $Q(\rho_0, \tau_0)$. Such a function is admissible for (2.12) when $k \geq \hat{k}$, since it is an element of $\overset{\circ}{W}{}_2^{1,1}(Q(\rho_0, \tau_0) \cap Q_T)$. Substituting it into (2.12) and carrying out the same estimates as for interior cylinders, we see that $u(x, t)$ satisfies inequalities (6.10) of Chapter II and is by the same token an element of $\mathfrak{A}(Q(\rho_0, \tau_0) \cap Q_T, \gamma, \hat{r}, \hat{k}, \kappa)$.

The membership of $u(x, t)$ and $-u(x, t)$ in classes $\mathfrak{A}(Q(\rho_0, \tau_0), \cdots)$, $\mathfrak{A}(Q(\rho_0, \tau_0) \cap Q_T, \cdots)$ implies, on the basis of Theorem 6.2 of Chapter II, the following theorem.

Theorem 8.1. *Suppose conditions* (1.2), (7.1) *and* (7.2) *are fulfilled for equation* (1.1). *Then any generalized solution* $u(x, t)$ *from* $V_2^{1,0}(Q_T)$ *of it has a finite* $\operatorname{vrai\,max}_{Q'} |u|$ *for any domain* $Q' \subset Q$ *separated from* Γ_T *by a positive distance* d. *The quantity* $\operatorname{vrai\,max}_{Q'} |u|$ *is estimated from above by a constant depending only on* n, $\|u\|_{2,Q_T}$, *the constants* ν, μ, μ_1, r *and* q *in conditions* (1.2), (7.1), (7.2) *and the distance* d. *If furthermore* $\operatorname{vrai\,max}_{\Gamma'} u = \hat{k} < \infty$ *for some piece* Γ' *of the surface* Γ_T, *then* $\operatorname{vrai\,max}_{Q_1} u$ *is finite for any subdomain* $Q_1 \subset Q_T$

separated by a positive distance d from $\Gamma_T \setminus \Gamma'$ and is estimated from above by a constant depending only on $\|u\|_{2,Q_T}$, $\hat{k}$, ν, μ, μ_1, r, q and d. An analogous fact holds for the function $-u(x, t)$.

The character of the dependence of an estimate for $u(x, t)$ from above and from below on the quantities enumerated in the theorem can be seen from Theorem 6.2 of Chapter II.

Remark 8.1. The assertions of Theorem 8.1 on the finiteness of $\operatorname{vrai\,max}_{Q'}|u|$ and $\operatorname{vrai\,max}_{Q_1}(\pm u)$ remain valid if in regard to the coefficients b_i one assumes only (7.3), (7.4). This is proved in the same way as Remark 7.3 at the end of §7.

Requirement (7.17) [more precisely, the requirement

$$\frac{\beta^2}{\nu}\left\|\sum_{i=1}^{n} b_i^2\right\|_{q_1,\, r_1,\, Q(k,\, \rho_0,\, \tau_0)} \leqslant \frac{1}{8}\min\{1;\ \nu\}, \tag{8.4}$$

where $Q(k, \rho_0, \tau_0)$ is the set of points $(x, t) \in Q(\rho_0, \tau_0)$ at which $u(x, t) > k$] can be satisfied this time for $r_1 = \infty$ not at the expense of choosing a high level k but at the expense of sufficient smallness in the radius ρ_0 (for $r_1 < \infty$ (8.4) will hold if $\operatorname{mes} Q(\rho_0, \tau_0)$ is taken sufficiently small). The influence of the coefficients b_i on an estimate of $\operatorname{vrai\,max}_{Q'}|u|$ and $\operatorname{vrai\,max}_{Q_1}(\pm u)$ manifests itself only through restrictions (8.4) on the cylinders $Q(\rho_0, \tau_0)$ to which Theorem 6.2 is applied. The other constants are not affected by the characteristics of the b_i.

Remark 8.2. In conditions (7.1), (7.2) and (7.3), (7.4) of Theorem 8.1 the numbers q and r can be assumed to be different for different coefficients and free terms as long as they satisfy requirements (7.2), (7.4).

A different proof of Theorem 8.1 can be obtained by using the idea of Moser stated at the end of the preceding section. For this purpose it is necessary to take a chain of coaxial cylinders

$$Q_{\rho_k} = K_{\rho_k} \times (t_0 - \rho_k^2,\ t_0),\ \rho_k = \frac{\rho}{2} + \frac{\rho}{2^{k+1}}, \qquad k = 0,\ 1,\ \ldots,$$

belonging to Q_T and a corresponding sequence of cutting functions

$$\zeta_k(x,\ t) = \zeta\left(2^{k+1}\left(\frac{|x - x_0|}{\rho} - \frac{1}{2}\right),\quad 2^{k+1}\left(\frac{\sqrt{t_0 - t}}{\rho} - \frac{1}{2}\right)\right),$$

$k = 1, 2, \cdots$, where x_0 is the center of the balls $K_{\rho k}$ and $\zeta(\xi, \eta)$ is some smooth nonnegative function, not exceeding unity, that is equal to unity for $\xi \leq 1/2$, $\eta \leq 1/2$ and to zero for $\xi \geq 1$, $\eta \geq 1$.

In identity (7.24) it is necessary to assume, as above, that $v = |u|^s$, $s = (1+\kappa)^k$, and to take $\zeta(x, t)$ equal to $\hat{\zeta}_{k-1}(x, t)$ $k = 1, 2, \cdots$. Then estimating all of the terms in much the same way as above in §§7 and 8, we arrive at a system of recursion inequalities for

$$\Phi_k = \left\| |u|^{(1+\varkappa)^k} \right\|_{\hat{q}, \hat{r}, Q_{\rho_k}} = \| u \|^{(1+\varkappa)^k}_{(1+\varkappa)^k \hat{q}, (1+\varkappa)^k \hat{r}, Q_{\rho_k}},$$

namely,

$$\Phi_k \leqslant c^k \left(\Phi_{k-1}^{1+\varkappa} + 1\right), \quad k = 1, 2, \ldots, \tag{8.5}$$

with a known constant $c \geq 1$ determined, as is also κ, by the parameters from conditions (7.1) and (7.2). From it follows, as we saw at the end of §7, the desired estimate

$$\operatorname{vrai\,max}_{Q_\rho} |u| = \lim_{k \to \infty} \Phi_k^{(1+\varkappa)^{-k}} \leqslant c^{\frac{1+\varkappa}{\varkappa^2}} \left(\| u \|_{\hat{q}, \hat{r}, Q_{2\rho}} + 1 \right). \tag{8.6}$$

§9. ESTIMATES OF SOME NORMS OF ORLICZ FOR GENERALIZED SOLUTIONS

In the two preceding sections estimates were obtained for the maximums of the moduli of the generalized solutions $u(x, t)$ from $V_2^{1,0}(Q_T)$ of equation (1.1) under conditions (7.1), (7.2) on the free terms f_i. But if these conditions are weakened, then the solutions still remain better than what is required for the energy inequality (2.2). More precisely, the following theorem holds.

Theorem 9.1. *Suppose conditions* (1.2)–(1.4) *are fulfilled for the coefficients of equation* (1.1), *while the free terms satisfy the conditions*

$$\| f \|_{q_3, r_3, Q_T}, \quad \left\| \sum_{i=1}^{n} f_i^2 \right\|_{q_4, r_4, Q_T} \leqslant \mu_2, \tag{9.1}$$

in which

$$\frac{1}{r_3} + \frac{n}{2q_3} = 1 + \frac{n\theta}{4},$$

with

$$\left.\begin{array}{lr}
q_3\in\left[\frac{2n}{4+\theta(n-2)},\ \frac{2}{\theta}\right],\ r_3\in\left[1,\ \frac{2}{\theta}\right] & \text{for}\quad n\geqslant 3,\\
q_3\in\left(1,\ \frac{2}{\theta}\right],\ r_3\in\left[1,\ \frac{2}{\theta}\right) & \text{for}\quad n=2,\\
q_3\in\left[1,\ \frac{2}{\theta}\right],\ r_3\in\left[1,\ \frac{4}{2+\theta}\right] & \text{for}\quad n=1,\\
\qquad\frac{1}{r_4}+\frac{n}{2q_4}=1+\frac{n\theta}{2}, & \\
\text{with} & \\
q_4\in\left[\frac{n}{2+\theta(n-2)},\ \frac{1}{\theta}\right],\ r_4\in\left[1,\ \frac{1}{\theta}\right] & \text{for}\quad n\geqslant 3,\\
q_4\in\left(1,\ \frac{1}{\theta}\right],\ r_4\in\left[1,\ \frac{1}{\theta}\right) & \text{for}\quad n=2,\\
q_4\in\left[1,\ \frac{1}{\theta}\right],\ r_4\in\left[1,\ \frac{2}{1+\theta}\right] & \text{for}\quad n=1,
\end{array}\right\}\tag{9.2}$$

and θ being some number from the interval $(0,\ 1)$.[1] *If* $|u||_{S_T}$ *is bounded and* $|u(x,\ 0)|^{1/\theta}\in L_2(\Omega)$, *then*

$$|u(x,\ t)|^{\frac{1}{\theta}}\in V_2(Q_T).$$

But if nothing is known concerning $|u||_{\Gamma_T}$, *then the membership of* $|u|^{1/\theta}$ *in* $V_2(Q_T)$ *holds for any strictly interior subdomain* $Q'\subset Q_T$.

Corollary. *From Theorem* 9.1 *and inequality* (3.4) *of Chapter* II *it follows that under conditions* (9.1) *and* (9.2) *any generalized solution of equation* (1.1) *belongs to* $L_{q,r}(Q_T)$ *or* $L_{q,r}(Q')$ *such that*

$$\frac{1}{r}+\frac{n}{2q}=\frac{n}{4}\theta.$$

We will assume that u is equal to zero on Γ_T. (The proof is not essentially changed if this requirement is replaced by the boundedness of u on Γ_T.)

For the proof we make use of identity (2.12), established for u and any function η from $\overset{\circ}{V}{}_2^{1,0}(Q_T)$.

Let us take in it

$$\eta(x,\ t)=u_h(x,\ t)\left[\psi_M^{(h)}(x,\ t)\right]^{\frac{2}{\theta}-2}.$$

---

[1] As is easily seen, for $\theta=1$ relations (9.2) go over into relations (1.6).

where

$$\psi_M^{(h)}(x,\ t) = \begin{cases} M & \text{for} \quad u_h \geqslant M, \\ u_h(x,\ t) & \text{for} \quad 0 \leqslant u_h \leqslant M, \\ 0 & \text{for} \quad u_h < 0. \end{cases} \tag{9.3}$$

All of the summands of (2.12), except the first, have a limit for $h \to 0$. But it is immediately obvious that the first summand can not have a limit, since u_t, in general, does not exist. We therefore transform the first summand into the more suitable form

$$\int_0^{t_1}\int_\Omega u_{ht}u_h\left(\psi_M^{(h)}\right)^{\frac{2}{\theta}-2}dx\,dt = \frac{1}{2}\int_0^{t_1}\int_\Omega (u_h^2)_t\left(\psi_M^{(h)}\right)^{\frac{2}{\theta}-2}dx\,dt$$

$$= -\frac{1}{2}\int_0^{t_1}\int_\Omega u_h^2\left[\left(\psi_M^{(h)}\right)^{\frac{2}{\theta}-2}\right]_t dx\,dt$$

$$+\frac{1}{2}\int_\Omega u_h^2\left(\psi_M^{(h)}\right)^{\frac{2}{\theta}-2}dx\,\Big|_{t=0}^{t=t_1}. \tag{9.4}$$

In the first summand of (9.4) the integration with respect to t can be carried out in the following manner:

$$\int_0^{t_1}\int_\Omega u_h^2\left[\left(\psi_M^{(h)}\right)^{\frac{2}{\theta}-2}\right]_t dx\,dt = \left(\frac{2}{\theta}-2\right)\int_0^{t_1}\int_\Omega \left(\psi_M^{(h)}\right)^{\frac{2}{\theta}-1}\psi_{Mt}^{(h)}\,dx\,dt$$

$$= (1-\theta)\int_\Omega \left[\psi_M^{(h)}\right]^{\frac{2}{\theta}}dx\,\Big|_{t=0}^{t=t_1}.$$

Therefore $\int_0^{t_1}\int_\Omega u_{ht}\eta(x,\ t)\,dx\,dt$ is equal to

$$\frac{1}{2}\int_\Omega u_h^2\left(\psi_M^{(h)}\right)^{\frac{2}{\theta}-2}dx\,\Big|_{t=0}^{t=t_1} - \frac{1}{2}(1-\theta)\int_\Omega \left[\psi_M^{(h)}\right]^{\frac{2}{\theta}}dx\,\Big|_{t=0}^{t=t_1}. \tag{9.5}$$

In these integrals it is now possible to pass to the limit as $h \to 0$. Substituting them into (2.12) and passing to the limit in (2.12) as $h \to 0$, we obtain the following equality for u:

$$\frac{1}{2}\int_\Omega u^2\psi_M^{\frac{2}{\theta}-2}dx\,\Big|_{t=0}^{t=t_1} - \frac{1}{2}(1-\theta)\int_\Omega \psi_M^{\frac{2}{\theta}}dx\,\Big|_{t=0}^{t=t_1} +$$

$$+\int_0^{t_1}\int_\Omega a_{ij}u_{x_j}\left(u\psi_M^{\frac{2}{\theta}-2}\right)_{x_i}dx\,dt=-\int_0^{t_1}\int_\Omega\Big[(a_iu+f_i)\left(u\psi_M^{\frac{2}{\theta}-2}\right)_{x_i}$$
$$+(b_iu_{x_i}+au+f)u\psi_M^{\frac{2}{\theta}-2}\Big]dx\,dt. \qquad (9.6)$$

In it $\psi_M(x, t)$ is equal to M for $u(x, t) > M$, to $u(x, t)$ for $0 \le u(x, t) \le M$, and to 0 for $u(x, t) < 0$. It is clear that $0 \le \psi_M(x, t) \le |u(x, t)|$ everywhere, and that $\psi_{Mx} = 0$ at those points where $\psi_M(x, t) \ne u(x, t)$. In view of this sum of the first two terms of the left side of (9.6) is not less than $(\theta/2)\int_\Omega u^2\psi_M^{2/\theta-2}dx\,|^{t_1=t} - \frac{1}{2}\int_\Omega|u(x, 0)|^{2/\theta}dx$. The third term is estimated from below, using the fact that at the points where $\psi_M(x, t) = u(x, t)$

$$a_{ij}u_{x_j}\left(u\psi_M^{\frac{2}{\theta}-2}\right)_{x_i}=\left(\frac{2}{\theta}-1\right)a_{ij}u_{x_j}u^{\frac{2}{\theta}-2}u_{x_i}$$
$$\geqslant\left(\frac{2}{\theta}-1\right)\nu u_x^2\psi_M^{\frac{2}{\theta}-2}=(2-\theta)\,\theta\nu\left[\left(u\psi_M^{\frac{1}{\theta}-1}\right)_x\right]^2,$$

and where $\psi_M(x, t) \ne u(x, t)$

$$a_{ij}u_{x_j}\left(u\psi_M^{\frac{2}{\theta}-2}\right)_{x_i}=a_{ij}u_{x_j}u_{x_i}\psi_M^{\frac{2}{\theta}-2}\geqslant\nu\left[\left(u\psi_M^{\frac{1}{\theta}-1}\right)_x\right]^2,$$

so that everywhere

$$a_{ij}u_{x_j}\left(u\psi_M^{\frac{2}{\theta}-2}\right)_{x_i}\geqslant\nu\theta\left[\left(u\psi_M^{\frac{1}{\theta}-1}\right)_x\right]^2.$$

Thus the left side of (9.6) is not less than

$$\theta\min\left\{\nu;\ \frac{1}{2}\right\}\left\{\int_\Omega\left(u\psi_M^{\frac{1}{\theta}-1}\right)^2dx\,\Big|^{t=t_1}+\int_{Q_{t_1}}\left[\left(u\psi_M^{\frac{1}{\theta}-1}\right)_x\right]^2dx\,dt\right\}$$
$$-\frac{1}{2}\int_\Omega|u(x,0)|^{\frac{2}{\theta}}dx. \qquad (9.7)$$

All of the terms in the right side of (9.6) are estimated from above by considering all of the expressions, in a manner analogous to the arguments just carried out, at the points where $\psi_M = u$, and where $\psi_M = \text{const}$ (i.e., where $\psi_M \ne u$), and then choosing a majorant, common for both categories of points (x, t). This leads, as is easily calculated, to the inequalities

$$\left|a_iu\left(u\psi_M^{\frac{2}{\theta}-2}\right)_{x_i}\right|\leqslant(2-\theta)\left|a_i\left(u\psi_M^{\frac{1}{\theta}-1}\right)\left(u\psi_M^{\frac{1}{\theta}-1}\right)_{x_i}\right|,$$

$$\left| b_i u_{x_i} u\psi_M^{\frac{2}{\theta}-2} \right| \leqslant \theta \left| b_i \left(u\psi_M^{\frac{1}{\theta}-1}\right) \left(u\psi_M^{\frac{1}{\theta}-1}\right)_{x_i} \right|,$$

$$\left| au^2\psi_M^{\frac{2}{\theta}-2} \right| = |a| \left(u\psi_M^{\frac{1}{\theta}-1}\right)^2,$$

$$\left| f_i \left(u\psi_M^{\frac{2}{\theta}-2}\right)_{x_i} \right| \leqslant (2-\theta) \left| f_i \psi_M^{\frac{1}{\theta}-1} \left(u\psi_M^{\frac{1}{\theta}-1}\right)_{x_i} \right|$$

$$\leqslant 2|f_i| \left| u\psi_M^{\frac{1}{\theta}-1} \right|^{1-\theta} \left| \left(u\psi_M^{\frac{1}{\theta}-1}\right)_{x_i} \right|,$$

$$\left| fu\psi_M^{\frac{2}{\theta}-2} \right| \leqslant |f| \left| u\psi_M^{\frac{1}{\theta}-1} \right|^{2-\theta}.$$

Each of their right sides contains the function $u\psi_M^{1/\theta-1}$, which we will denote by v. Substituting the inequalities obtained into (9.6) and estimating the integrals in the right side by Cauchy's inequalities with arbitrary $\epsilon > 0$, we get

$$\theta \min\left\{\frac{1}{2};\, \nu\right\}\left(\|v(x,\, t_1)\|_{2,\,\Omega}^2 + \|v_x\|_{Q_{t_1}}^2\right)$$
$$\leqslant \int_{Q_{t_1}} \Big[(2|a_i| + |b_i|)|vv_{x_i}| + |a|v^2$$
$$+ 2|f_i|\left|\,|v|^{1-\theta} v_{x_i}\right| + |f|\,|v|^{2-\theta}\Big]\, dx\, dt + \frac{1}{2}\int_\Omega |u(x,\, 0)|^{\frac{2}{\theta}}\, dx$$
$$\leqslant \int_{Q_{t_1}} \Bigg\{ 2\varepsilon v_x^2 + \left[\frac{1}{4\varepsilon}\sum_{i=1}^n (2|a_i| + |b_i|)^2 + |a|\right] v^2$$
$$+ \frac{1}{\varepsilon}\sum_{i=1}^n f_i^2 |v|^{2(1-\theta)} + |f|\,|v|^{2-\theta}\Bigg\}\, dx\, dt + \frac{1}{2}\int_\Omega |u(x,\, 0)|^{\frac{2}{\theta}}\, dx, \tag{9.8}$$

Let us denote $\theta \min\{1/2;\, \nu\}$ by $2\nu_1$ and take $\epsilon = \nu_1/2$. Then, after reducing similar terms and maximizing over t, we deduce from (9.8) the inequality

$$\nu_1 |v|_{Q_{t_1}}^2 \leqslant 2\int_{Q_{t_1}} \left(\mathscr{D}v^2 + \frac{2}{\nu_1}\sum_{i=1}^n f_i^2 |v|^{2(1-\theta)} + |f|\,|v|^{2-\theta}\right) dx\, dt$$
$$+ \int_\Omega |u(x,\, 0)|^{\frac{2}{\theta}}\, dx, \tag{9.9}$$

where $\mathfrak{D} = (1/2\nu_1)\sum_{i=1}^n(2|a_i| + |b_i|)^2 + |a|$. The terms on the right side are estimated from above by Hölder's inequality, the embedding theorem (3.4) of Chapter II and the assumptions on the coefficients and free terms of equation (1.1) in the following manner:

$$\int_{Q_{t_1}} \mathscr{D}v^2 dx\,dt \leqslant \|\mathscr{D}\|_{q,r,Q_{t_1}} \|v\|^2_{\bar{q},\bar{r},Q_{t_1}} \leqslant \beta^2 \|\mathscr{D}\|_{q,r,Q_{t_1}} |v|^2_{Q_{t_1}},$$

$$\int_{Q_{t_1}} \sum_{i=1}^n f_i^2 |v|^{2(1-\theta)} dx\,dt$$

$$\leqslant \left\|\sum_{i=1}^n f_i^2\right\|_{q_4,r_4,Q_{t_1}} \|v\|^{2(1-\theta)}_{\bar{q}_4(1-\theta),\bar{r}_4(1-\theta),Q_{t_1}}$$

$$\leqslant \beta^{2(1-\theta)} \left\|\sum_{i=1}^n f_i^2\right\|_{q_4,r_4,Q_{t_1}} |v|^{2(1-\theta)}_{Q_{t_1}},$$

$$\int_{Q_{t_1}} |f|\,|v|^{2-\theta} dx\,dt \leqslant \|f\|_{q_3,r_3,Q_{t_1}} \|v\|^{2-\theta}_{\bar{q}_3\frac{(2-\theta)}{2},\bar{r}_3\frac{(2-\theta)}{2},Q_{t_1}}$$

$$\leqslant \beta^{2-\theta}\|f\|_{q_3,r_3,Q_{t_1}} |v|^{2-\theta}_{Q_{t_1}}.$$

Here, as elsewhere, the parameters with a bar and without a bar are connected by the equality $\bar{q} = 2q/(q-1)$. For these estimates we made use of the fact that by virtue of (9.2) the space $V_2(Q_{t_1})$ is embedded in the spaces

$$L_{\bar{q},\bar{r}}(Q_{t_1}),\ L_{\bar{q}_4(1-\theta),\bar{r}_4(1-\theta)}(Q_{t_1}) \text{ and } L_{\bar{q}_3\frac{2-\theta}{2},\bar{r}_3\frac{2-\theta}{2}}(Q_{t_1}),$$

and that for them inequality (3.4) of Chapter II is valid. Because of these inequalities and Young's inequality it follows from (9.9) that

$$\frac{\nu_1}{2}|v|^2_{Q_{t_1}} \leqslant \beta^2\|\mathscr{D}\|_{q,r,Q_{t_1}}|v|^2_{Q_{t_1}} + \frac{1}{2}\||u(x,0)|^{\frac{1}{\theta}}\|^2_{2,\Omega}$$

$$+\beta^{2(1-\theta)}\left\|\sum_{i=1}^n f_i^2\right\|_{q_4,r_4,Q_{t_1}}|v|^{2(1-\theta)}_{Q_{t_1}} + \beta^{2-\theta}\|f\|_{q_3,r_3,Q_{t_1}}|v|^{2-\theta}_{Q_{t_1}}$$

$$\leqslant \left[\beta^2\|\mathscr{D}\|_{q,r,Q_{t_1}} + (1-\theta)\varepsilon^{\frac{1}{1-\theta}} + \left(1-\frac{\theta}{2}\right)\varepsilon^{\frac{2}{2-\theta}}\right]|v|^2_{Q_{t_1}} + \mathscr{F}(t_1), \quad (9.10)$$

where

$$\mathcal{F}(t_1)=\theta\varepsilon^{-\frac{1}{\theta}}\beta^{2\left(\frac{1}{\theta}-1\right)}\left\|\sum_{i=1}^{n} f_i^2\right\|_{q_4, r_4, Q_{t_1}}^{\frac{1}{\theta}} + \frac{1}{2}\left\|\, |u(x,\ 0)|^{\frac{1}{\theta}}\right\|_{2,\Omega}^{2}+\frac{\theta}{2}\varepsilon^{-\frac{2}{\theta}}\beta^{2\left(\frac{2}{\theta}-1\right)}\|f\|_{q_3, r_3, Q_{t_1}}^{\frac{2}{\theta}},$$

and ϵ is an arbitrary positive number. Let us take an ϵ such that

$$(1-\theta)\,\varepsilon^{\frac{1}{1-\theta}}+\left(1-\frac{\theta}{2}\right)\varepsilon^{\frac{2}{2-\theta}}=\frac{\nu_1}{8},$$

and choose t_1 so small that

$$\beta^2\|\mathcal{D}\|_{q,\,r,\,Q_{t_1}}\leqslant\frac{\nu_1}{8}. \tag{9.11}$$

This is possible since $r<\infty$. For such t_1 and ϵ inequality (9.10) gives

$$\frac{\nu_1}{4}|v|_{Q_{t_1}}^2=\frac{\nu_1}{4}\left|u\psi_M^{\frac{1}{\theta}-1}\right|_{Q_{t_1}}^2\leqslant\mathcal{F}(t_1).$$

Letting M go to infinity, we obtain from this the desired estimate for $u^{(0)}=\max\{u(x,\ t);\ 0\}$.

$$\left|(u^{(0)})^{\frac{1}{\theta}}\right|_{Q_{t_1}}\leqslant\left(\frac{4}{\nu_1}\mathcal{F}(t_1)\right)^{\frac{1}{2}}\leqslant c. \tag{9.12}$$

Proceeding by small increments in t, we obtain an estimate of $|(u^{(0)})^{1/\theta}|_{Q_T}$ for all of Q_T in much the same way as was done above in §4. Let us explain.

Inequality (9.12), more precisely, the inequality

$$\left|(u^{(0)})^{\frac{1}{\theta}}\right|_{Q_{t_{k-1},\,t_k}}\leqslant\left(\frac{4}{\nu_1}F(t_{k-1},\ t_k)\right)^{\frac{1}{2}} \tag{9.12$'$}$$

is proved by us for essentially any cylinder $Q_{t_{k-1},t_k}=\Omega\times(t_{k-1},\ t_k)$, $0\le t_{k-1}\le t_k\le T$, as long as condition (9.11) is fulfilled for Q_{t_{k-1},t_k}, i.e., as long as

$$\mu(t_{k-1},\ t_k)\equiv\beta^2\|\mathcal{D}\|_{q,\,r,\,Q_{t_{k-1},\,t_k}}\leqslant\frac{\nu_1}{8}. \tag{9.11$'$}$$

The quantity $F(t_{k-1},\ t_k)$ in (9.12$'$) is calculated for Q_{t_{k-1},t_k} in the same way as $F(t_1)$ was calculated above for $Q_{t_1}=Q_{0,t_1}$. We partition $[0,\ T]$ into a finite number of closed intervals $[t_0=0,\ t_1]$, $[t_1,\ t_2],\cdots,[t_{s-1},\ t_s=T]$ in such a way that (9.11$'$) is fulfilled for each of them, with the equality sign in (9.11$'$)

holding for all except possibly the last. In view of this we have

$$(s-1)\left(\frac{\nu_1}{8}\right)^r \leqslant \sum_{k=1}^{s} \mu^r(t_{k-1}, t_k) = \beta^{2n} \|\mathcal{D}\|^r_{q,r,Q_T},$$

i.e., the number of divisions s in such a partitioning will not exceed the known quantity $1 + ((8\beta^2/\nu_1)\|\mathfrak{D}\|_{q,r,Q_T})^r$. Because of this we obtain from inequalities (9.12′) the desired estimate of $|u^{(0)}|_{Q_T}$ for all of Q_T if we note that the quantities $\max_{t_{k-1}\leq t\leq t_k} \int_\Omega [u^{(0)}(x,t)]^{2/\theta} dx$ in them are estimated in terms of $\int_\Omega [u^{(0)}(x, t_{k-1})]^{2/\theta} dx$ and the known norms of the free terms of equation (1.1).

Thus the membership of $[u^{(0)}(x,t)]^{1/\theta}$ in $V_2(Q_T)$ is established. Analogously, by considering the function $(\max\{-u(x,t); 0\})^{1/\theta}$, we prove its membership in $V_2(Q_T)$ and obtain an estimate of its norm in $V_2(Q_T)$. By the same token the assertion of Theorem 9.1 concerning a global estimate is proved. If in regard to the solution u it is known that $|u|\,|_{\Gamma_T}$ is bounded, then, taking $\eta(x,t)$ in (2.12) equal to $u_h(\psi_M^{(h)})^{2/\theta-2}\zeta^2$ with a smooth $\zeta(x,t)$ equal to zero on Γ_T, and carrying out the analogous estimates, we obtain the validity of the second assertion of the theorem.

Thus Theorem 9.1 is proved.

If $\theta = 0$ in conditions (9.1), then Theorem 9.1 implies the summability of the solution u with any positive power. But if the coefficients of equation (1.1) are subject to requirements (7.1), then it is possible to say more concerning the solution: it is exponentially summable. Namely, we have

Theorem 9.2. *Suppose the coefficients of equation* (1.1) *satisfy conditions* (1.2), (7.1) *and* (7.2), *while the free terms f_i and f satisfy conditions* (9.1), (9.2) *with $\theta = 0$. Then for any generalized solution from $V_2^{1,0}(Q_T)$ of equation* (1.1) *with* $\operatorname{vrai\,max}_{\Gamma_T}|u| \equiv k < \infty$ *there exist constants $c > 0$ and c_1 such that*

$$\int_{Q_T} \exp[c\,|u(x,t)|]\,dx\,dt \leqslant c_1,$$

with c and c_1 being determined by the known parameters in the enumerated conditions.

The proof of this assertion is carried out in almost the same way as the proof of Theorem 7.1: we prove a certain integral inequality (of the same type as inequality (5.32) of Chapter II), from which by means of Lemma 5.10 of Chapter II we reach the desired conclusion.

The inequality, as in §7, is deduced from equality (7.6) with the use of estimates of the various integrals by Hölder's inequality. All of the summands except those containing the factors f_i and f are estimated in exactly the same way as in §7. And the summands with f_i and f are estimated in the following manner:

$$\int_{Q_{t_1}(k)} |f|(u-k)\,dx\,dt \leqslant \mu_3^{\frac{1}{\bar{r}_3}}(k)\,\|f\|_{q_3,r_3,Q_{t_1}}\|u^{(k)}\|_{\bar{q}_3,\bar{r}_3,Q_{t_1}},$$

$$\int_{Q_{t_1}(k)} \sum_{i=1}^n f_i^2\,dx\,dt \leqslant \mu_4^{\frac{2}{\bar{r}_4}}(k)\left\|\sum_{i=1}^n f_i^2\right\|_{q_4,r_4,Q_{t_1}} \tag{9.13}$$

where

$$\bar{q}_i=\frac{2q_i}{q_i-1},\quad \bar{r}_i=\frac{2r_i}{r_i-1},\quad i=3,\ 4,$$

$$\mu_i(k)=\int_0^{t_1}\operatorname{mes}^{\frac{\bar{r}_i}{\bar{q}_i}}A_k(t)\,dt,\qquad A_k(t)=\{x\in\Omega:\ u(x,t)>k\}\,^{1)}.$$

As a result we obtain the following estimate for $k\geq\max\{\hat{k};\,1\}$:

$$\min\left\{\frac{1}{2};\,\frac{\nu}{2}\right\}\left[\int_{A_k(t_1)}(u-k)^2\,dx+\|u_x\|^2_{2,Q_{t_1}(k)}\right]$$

$$\leqslant\int_0^{t_1}\int_{A_k(t)}\left(\frac{4}{\nu}\sum_{i=1}^n a_i^2+\frac{2}{\nu}\sum_{i=1}^n b_i^2+2|a|\right)[(u-k)^2+k^2]\,dx\,dt$$

$$+\frac{\varepsilon}{2}\|u^{(k)}\|^2_{\bar{q}_3,\bar{r}_3,Q_{t_1}}+\frac{1}{2\varepsilon}\mu_3^{\frac{2}{\bar{r}_3}}(k)\|f\|^2_{q_3,r_3,Q_{t_1}}$$

$$+\frac{4}{\nu}\mu_4^{\frac{2}{\bar{r}_4}}(k)\left\|\sum_{i=1}^n f_i^2\right\|_{q_4,r_4,Q_{t_1}}. \tag{9.14}$$

The first term in the right side is estimated in the same way as the right side of (7.8) was estimated in §7, while the integral $\|u^{(k)}\|^2_{\bar{q}_3,\bar{r}_3,Q_{t_1}}$ is estimated in terms

1) For $\bar{r}_i=\infty$ one must assume $\mu_i^{1/\bar{r}_i}(k)=\operatorname{vrai\,max}_{0\leq t\leq t_1}\operatorname{mes}^{1/\bar{q}_i}A_k(t)$, while for $\bar{q}_i=\infty$ one must assume that $\mu_i(k)$ is equal to the measure of the points of the interval $[0,t_1]$ for which $\operatorname{mes}A_k(t)>0$. The case $\bar{r}_i=\infty$ occurs for $r_i=1$, $n\geq 1$, while the case $\bar{q}_i=\infty$ is encountered only for $n=1$, when $q_i=1$.

of $|u^{(k)}|^2_{Q_{t_1}}$, using inequality (3.4) of Chapter II (it is applicable since $\bar{q}_3$ and $\bar{r}_3$ satisfy conditions (3.3) of Chapter II). After this we carry out a maximization over t, taking into account that all of our inequalities are valid for any t_1 from $[0, T]$. This leads to the inequality

$$\min\left\{\frac{1}{2};\ \frac{\nu}{2}\right\}|u^{(k)}|^2_{Q_{t_1}} \leqslant \beta^2\|\mathscr{D}_1\|_{q,r,Q_{t_1}(k)}\mu^{\frac{2\varkappa}{\hat{r}}}(k)|u^{(k)}|^2_{Q_{t_1}} + k^2\|\mathscr{D}_1\|_{q,r,Q_{t_1}}\mu^{\frac{2(1+\varkappa)}{\hat{r}}}(k) + \varepsilon\beta^2|u^{(k)}|^2_{Q_{t_1}} + \frac{1}{\varepsilon}\|f\|^2_{q_3,r_3,Q_{t_1}}\mu_3^{\frac{2}{\bar{r}_3}}(k) + \frac{8}{\nu}\left\|\sum_{i=1}^n f_i^2\right\|_{q_4,r_4,Q_{t_1}}\mu_4^{\frac{2}{\bar{r}_4}}(k),$$

where

$$\mathscr{D}_1 = 2\left(\frac{4}{\nu}\sum_{i=1}^n a_i^2 + \frac{2}{\nu}\sum_{i=1}^n b_i^2 + 2|a|\right),$$

while $\hat{r}$, κ and $\mu(k)$ are the same as in inequality (7.9). We put $\epsilon = (1/\beta^2)\min\{1/8;\ \nu/8\}$ and find t_1 from the condition

$$\beta^2\|\mathscr{D}_1\|_{q,r,Q_{t_1}(k)}\, t_1^{\frac{2\varkappa}{\hat{r}}}\operatorname{mes}\Omega^{\frac{2\varkappa}{\hat{q}}} \leqslant \min\left\{\frac{1}{8};\ \frac{\nu}{8}\right\}.$$

As a result, for $k \geq \max\{\hat{k};\ 1\}$ we obtain

$$\min\left\{\frac{1}{4};\ \frac{\nu}{4}\right\}|u^{(k)}|^2_{Q_{t_1}} \leqslant c\left[k^2\mu^{\frac{2(1+\varkappa)}{\hat{r}}}(k) + \mu_3^{\frac{2}{\bar{r}_3}}(k) + \mu_4^{\frac{2}{\bar{r}_4}}(k)\right], \tag{9.15}$$

with

$$c = \max\left\{\|\mathscr{D}_1\|_{q,r,Q_{t_1}};\ \frac{8\beta^2}{\min\{1;\ \nu\}}\|f\|^2_{q_3,r_3,Q_{t_1}};\ \frac{8}{\nu}\left\|\sum_{i=1}^n f_i^2\right\|_{q_4,r_4,Q_{t_1}}\right\}.$$

The conclusion of the theorem now follows from Remark 5.4 to Lemma 5.10 of Chapter II.

One can also obtain local estimates of $\int_{Q'}\exp[c|u(x,t)|]\,dx\,dt$ for solutions $u(x,t)$ from $V_2^{1,0}(Q_T)$ concerning which the boundedness of $|u|$ on Γ_T is known. The necessary modifications for this are analogous to those carried out in §8 in estimating $\max_{Q'}|u|$ for any u from $V_2^{1,0}(Q_T)$. Local estimates of $\|u\|_{q,r,Q'}$ and $\int_{Q'}\exp[c|u(x,t)|]\,dx\,dt$ are given in [51e].

§10. AN ESTIMATE OF HÖLDER'S CONSTANT. HARNACK'S INEQUALITY

In §§7 and 8 we proved that under conditions (1.2), (7.1) and (7.2) each of the generalized solutions $u(x, t)$ of equation (1.1) from the space $V_2^{1,0}(Q_T)$ is bounded. It turns out that under these same conditions u belongs to $H^{\alpha,\alpha/2}(Q_T)$ with some $\alpha > 0$ and the norms $|u|_{Q'}^{(\alpha)}$ of it can be estimated in terms of $\|u\|_{2,Q_T}$ and the known parameters in conditions (1.2), (7.1), (7.2). Namely, the following assertion is valid.

Theorem 10.1. *Suppose $u(x, t)$ is a generalized solution from $V_2^{1,0}(Q_T)$ of equation (1.1), the coefficients and free terms of which satisfy conditions (1.2) and (7.1), and suppose $\text{vrai}\max_{Q_T}|u| = M$. Then $u \in H^{\alpha,\alpha/2}(Q_T)$, and the norm $|u|_{Q'}^{(\alpha)}$, for any $Q' \subset Q_T$ separated from Γ_T by a positive distance d, is estimated from above by a constant depending only on n, M, the parameters ν, μ, μ_1, q and r from conditions (7.1), (7.2), and the distance d. The exponent $\alpha > 0$ is determined only by the numbers n, ν, μ, q and r.*

If in addition it is known that $u|_{\Gamma'} \in H^{\beta,\beta/2}(\Gamma')$ on some piece $\Gamma' \subset \Gamma_T$ satisfying condition (A),[1] then $u(x, t) \in H^{\alpha,\alpha/2}(Q_T \cup \Gamma')$ and the norm $|u|_{Q'}^{(\alpha)}$, for any Q' belonging to Q_T and separated from $\Gamma_T \setminus \Gamma'$ by a positive distance d, is estimated from above by a constant depending only on n, M, ν, μ, μ_1, q, d, β, $|u|_{\Gamma'}^{(\beta)}$ and the constants a_0 and θ_0 of condition (A). The exponent α belongs to the half-open interval $(0, \beta]$ and is determined by n, ν, μ, r and q.

Thus, suppose u is a generalized solution from $V_2^{1,0}(Q_T)$ of equation (1.1) with $\text{vrai}\max_{Q_T}|u| = M$. Equality (8.1) is valid for it. We will assume that $\zeta(x, t)$ in (8.1) is different from zero only for $x \in K_\rho$. From (8.1), using inequalities (1.1) and (1.2) of Chapter II, we obtain for $k \in [-M, M]$

$$\frac{1}{2}\| u^{(k)}(x, t)\zeta(x, t)\|_{2, K_\rho}^2\Big|_{t_0}^{t} + \nu \int_{t_0}^{t}\int_{K_\rho} (u_x^{(k)})^2\zeta^2 dx\, dt$$

$$\leqslant \int_{t_0}^{t}\int_{A_{k,\rho}(t)} \Big[\varepsilon_1 \mu u_x^2\zeta^2 + \frac{1}{\varepsilon_1}\mu(u-k)^2\zeta_x^2 + \varepsilon_1 u_x^2\zeta^2$$

$$+ \frac{1}{2\varepsilon_1}\sum_i (a_i^2M^2 + f_i^2)\zeta^2 + (u-k)^2\zeta_x^2 + 2\sum_i (a_i^2M^2 + f_i^2\zeta^2)$$

---

[1] Condition (A) is defined on p. 9 in § 1 of Chapter I.

$$+\varepsilon_1 u_x^2\zeta^2+\frac{1}{4\varepsilon_1}\sum_i b_i^2(u-k)^2\zeta^2+(|a|M+|f|)\,2M\zeta^2$$
$$+(u-k)^2\zeta|\zeta_t|\Big]\,dx\,dt. \qquad (10.1)$$

Hence, choosing $\epsilon_1 = \nu/2(\mu+2)$, reducing similar terms and maximizing over t, we will have

$$\frac{1}{2}\max_{t_0\leqslant t\leqslant t_0+\tau}\|u^{(k)}(x,t)\zeta(x,t)\|^2_{2,K_\rho}+\frac{\nu}{2}\int\limits_{Q(\rho,\tau)}(u_x^{(k)})^2\zeta^2dx\,dt$$
$$\leqslant\frac{1}{2}\|u^{(k)}(x,t_0)\zeta(x,t_0)\|^2_{2,K_\rho}+c\int\limits_{Q(\rho,\tau)}(u^{(k)})^2(\zeta_x^2+\zeta|\zeta_t|)\,dx\,dt$$
$$+(M^2+1)\int\limits_{Q(k,\rho,\tau)}\mathscr{D}_1(x,t)\,dx\,dt, \qquad (10.2)$$

where

$$c=2\Big[1+\frac{2\mu(\mu+2)}{\nu}\Big],$$
$$\mathscr{D}_1(x,t)=2\Big(2+\frac{\mu+2}{\nu}\Big)\sum_{i=1}^n(a_i^2+b_i^2+f_i^2)\zeta^2+\frac{4(\mu+2)}{\nu}\sum_{i=1}^n b_i^2\zeta^2$$
$$+4(|a|+|f|)\zeta^2,$$
$$Q(\rho,\tau)=K_\rho\times(t_0,t_0+\tau);$$
$$Q(k,\rho,\tau)=\{(x,t)\in Q(\rho,\tau):\ u(x,t)>k\}.$$

The last integral in (10.2) is estimated, using inequalities (1.6) and (3.5) of Chapter II, in the same way as the integral $\int\mathfrak{D}(x,t)\,dx\,dt$ from (7.7) is estimated in (7.8) and (7.11). Namely,

$$\int\limits_{Q(k,\rho,\tau)}\mathscr{D}_1\,dx\,dt\leqslant\|\mathscr{D}_1\|_{q,r,Q(k,\rho,\tau)}\|1\|^2_{\bar q,\bar r,Q(k,\rho,\tau)}$$
$$\leqslant c_1\mu^{\frac{2(1+\varkappa)}{\hat r}}\Big(k,\frac{\hat r}{\hat q},\rho,\tau\Big), \qquad (10.3)$$

where

$$c_1=\|\mathscr{D}_1\|_{q,r,Q_T},\quad \mu\Big(k,\frac{\hat r}{\hat q},\rho,\tau\Big)=\int\limits_{t_0}^{t_0+\tau}\operatorname{mes}^{\frac{\hat r}{\hat q}}A_{k,\rho}(t)\,dt\,^{1)},$$

---

1) For $\bar q=\hat q=\infty$ one must understand by $\mu(k,0,\rho,\tau)$ the measure of the set of points t from $[t_0,t_0+\tau]$ for which $\operatorname{mes}A_{k,\rho}(t)>0$.

while the numbers $\bar{q}$, $\bar{r}$, $\hat{q}$ and $\hat{r}$ are the same as in (7.9)–(7.12).

Substituting this estimate into (10.2), we arrive at the inequalities

$$\max_{t_0 \leqslant t \leqslant t_0+\tau} \| u^{(k)}(x, t)\zeta(x, t)\|_{2, K_\rho}^2 + \nu \| |u_x^{(k)}|\zeta\|_{2, Q(\rho,\tau)}^2 \leqslant \| u^{(k)}(x, t_0)\zeta(x, t_0)\|_{2, K_\rho}^2 + c \int_{Q(\rho,\tau)} (u^{(k)})^2 (\zeta_x^2 + \zeta|\zeta_t|)\, dx\, dt + c_1 (M^2+1)\mu^{\frac{2(1+\varkappa)}{\hat{r}}}\left(k, \frac{\hat{r}}{\hat{q}}, \rho, \tau\right). \tag{10.4}$$

The same inequalities are also valid for the function $-u(x, t)$. It follows from them that the solution $u(x, t)$ is an element of class $\mathfrak{B}_2(Q_T, M, \gamma, \hat{r}, \delta, \kappa)$ with $\delta = \infty$ and γ determined by ν, c and c_1 from (10.4). By Theorem 7.1 of Chapter II u will be an element of $H^{\alpha, \alpha/2}(Q_T)$ with some $\alpha > 0$. But the α guaranteed by Theorem 7.1 depends on γ, while γ in our case depends on M and μ_1. In order to achieve independence for the Hölder exponent of u from M and μ_1 we consider (10.4) only for $\tau \le \rho^2 \le \rho_0^2$, where ρ_0 is determined by the equality

$$\frac{c_1}{c}(M^2+1)\varkappa_n^{\frac{\varkappa}{\hat{q}}}\rho_0^{\frac{n\varkappa}{2}} = 1. \tag{10.5}$$

For such ρ and τ inequalities (10.4) imply the inequalities

$$\max_{t_0 \leqslant t \leqslant t_0+\tau} \| u^{(k)}(x, t)\zeta(x, t)\|_{2, K_\rho}^2 + \nu \| |u_x^{(k)}|\zeta\|_{2, Q(\rho,\tau)}^2 \leqslant \| u^{(k)}(x, t_0)\zeta(x, t_0)\|_{2, K_\rho}^2 + c\left\{ \int_{Q(\rho,\tau)} (u^{(k)})^2 (\zeta_x^2 + \zeta|\zeta_t|)\, dx\, dt + \mu^{\frac{2}{\hat{r}}\left(1+\frac{\varkappa}{2}\right)}\left(k, \frac{\hat{r}}{\hat{q}}, \rho, \tau\right)\right\}. \tag{10.6}$$

By virtue of Theorem 7.1 of Chapter II and the remark to it, from these inequalities it follows that $u(x, t)$ belongs to $H^{\tilde{\alpha}, \tilde{\alpha}/2}(Q_T)$ with some positive $\tilde{\alpha}$ (in general, less than α) already not depending on M and μ_1. But the Hölder constant $\langle u\rangle_Q^{(\tilde{\alpha})}$ will depend on M and μ_1. Thus the first part of Theorem 10.1 is proved. The two other assertions of the theorem are consequences of Theorem 8.1 of Chapter II and the fact that inequalities (10.4) and (10.6) are valid for $v(x, t) = \pm u(x, t)$ and for the cylinders $Q(\rho, \tau)$ intersecting Γ_T, as long as the levels k in them are subject to the condition $k \ge \max_{Q(\rho,\tau)\cap\Gamma_T}(\pm u)$, and all of the domains of integration are replaced by their intersections with Q_T.

In §10 of Chapter V we give another proof of the Hölder continuity of generalized solutions, and we give it directly for quasi-linear equations, of which equations (1.1) are a special case. We will illustrate it here by the example of the simplest representative of equations (1.1), namely, the equation

$$u_t - \frac{\partial}{\partial x_i}\left(a_{ij}(x, t)\, u_{x_j}\right) = 0, \tag{10.7}$$

assuming only that condition (1.2) is fulfilled.

As is known (see Lemma 5.8 of Chapter II), for a proof of the Hölder continuity of $u(x, t)$ and to obtain an estimate of the Hölder constant $\langle u\rangle_{Q_T}^{(\alpha)}$ it is sufficient to show that the oscillation of u in an arbitrary cylinder of standard form is strictly larger than its oscillation in a smaller coaxial cylinder of the same form with the same vertex.

Cylinders of standard form are cylinders $\{|x - x^0| < \rho,\ 0 < t - t_0 < \theta_1\rho^2\}$ with some $\theta_1 > 0$, which in the argument is fixed, and arbitrary x^0, t_0 and ρ. Their vertices are considered to be the points $(x^0, t_0 + \theta_1\rho^2)$. Since equation (10.7) and condition (1.2) are invariant with respect to the transformation of coordinates $\widetilde{x} = x - x^0/\rho$, $\widetilde{t} = t - t_0/\rho^2$ and the transformation of functions $\widetilde{u}(x, t) = c_1(u(x, t) - c)$, instead of studying u with arbitrary $\operatorname{vrai\,max}_{Q_T}|u|$ in arbitrary cylinders of standard type it is sufficient to prove that for any solution $u(x, t)$ of equation (10.7) defined in the cylinder $Q_2 = K_2 \times (0, \theta) = \{|x| < 2;\ 0 < t < \theta\}$ and varying in the range $[0, 1]$, with the cylinder $Q_1 = K_1 \times (3/4\,\theta, \theta) = \{|x| < 1,\ 3/4\,\theta < t < \theta\}$, one has the inequality

$$\operatorname{osc}\{u, Q_1\} \leqslant (1 - \delta_0)\operatorname{osc}\{u, Q_2\} \equiv 1 - \delta_0, \tag{10.8}$$

where θ and δ_0 are some positive constants.

Consider $u(x, 0)$ in K_1. Let

$$\operatorname{mes}\left\{x \in K_1:\ u(x, 0) \leqslant \frac{1}{2}\right\} \geqslant \frac{1}{2}\operatorname{mes} K_1; \tag{10.9}$$

then all subsequent arguments will be carried out with the function $u(x, t)$. Otherwise it is necessary to take the function $1 - u(x, t)$. Since $u(x, t)$ is a solution of equation (10.7), inequality (10.4) with $\zeta = \zeta(x)$ has for it the form

$$\max_{0 \leqslant t \leqslant \theta} \| u^{(k)}(x, t)\zeta(x)\|_{2, K_2}^2 + \nu \| |u_x^{(k)}|\zeta\|_{2, Q(2, \theta)}^2$$
$$\leqslant \| u^{(k)}(x, 0)\zeta(x)\|_{2, K_2}^2 + 4\mu \int_{Q(2, \theta)} (u^{(k)})^2 |\zeta_x|^2\, dx\, dt;$$

from this and from (10.9), by virtue of Remark 7.2 of Chapter II and Lemma 7.1 of Chapter II, it follows that

$$\operatorname{mes}\left\{x \in K_1:\ u(x, t) \leqslant \frac{7}{8}\right\} \geqslant b \operatorname{mes} K_1, \quad t \in [0, \theta], \tag{10.10}$$

with some positive b.

Let us consider in Q_2 the function $\phi(u) = -\ln[8(1 - u + \epsilon)]$, where ϵ is an arbitrary positive number, which we will later let tend to zero. The function $\phi(u)$ is bounded from below and is negative for $u \leq 7/8$. If we prove that $v(x, t) = \phi(u(x, t))$ is bounded in Q_1 for all $\epsilon > 0$ by some constant M_1, then in Q_1 we will have

$$8(1 - u) \geq e^{-M_1}$$

and by the same token (10.8) with $\delta_0 = e^{-M_1}/8$.[1)]

For the sake of simplicity we will assume that $u(x, t)$ is a classical solution (in Chapter V a direct examination will be conducted for bad solutions). We multiply (10.7) by $\phi'(u)\, \zeta(x, t)$, where $\zeta \in \overset{\circ}{W}{}_2^1(K_2)$, and, integrating the result over K_2, we obtain, after an integration by parts in the second term, the equation

$$\int_{K_2} \left(v_t\zeta + a_{ij} v_{x_i} v_{x_j} \zeta + a_{ij} v_{x_j} \zeta_{x_i}\right) dx = 0. \tag{10.11}$$

From (10.11) we easily obtain for v the inequality

$$\max_{0 < t < \theta} \| v^{(k)} \zeta \|^2_{2, K_t} + \nu \| | v_x^{(k)} | \zeta \|^2_{2, Q(2, \theta)} \leqslant \max\left(1, \frac{2\mu}{\nu}\right) \int_{Q(2, \theta)} (v^{(k)})^2 (|\zeta_x|^2 + \zeta|\zeta_t|)\, dx\, dt,$$

which implies, by Theorem 6.2 of Chapter II, the boundedness of v from above in $Q(1, \theta/4) = K_1 \times (3\theta/4, \theta)$, as long as the norm $\|v\|_{2, Q(3/2, 3\theta/4)}$ is finite. In order to obtain an estimate from above for $\|v\|_{2, Q(3/2, 3\theta/4)}$ we put $\zeta(x, t) = \mathfrak{N}^2(x)\chi^2(t)$ in (10.11) where the function $\chi(t)$ is equal to zero for $t \leq 0$, to $4t/\theta$ for $0 \leq t \leq \theta/4$, and to 1 for $\theta/4 \leq t \leq \theta$, while $\mathfrak{N}(x) = \mathfrak{N}(|x|)$ is a monotonically increasing function of $|x|$ for $|x| \in [0, 2]$, that is equal to 1 for $|x| \leq 3/2$ and to zero for $|z| = 2$. Let us integrate both sides of (10.11) with respect to t and in

1) The idea of introducing similar functions $\phi(u)$ for proving inequalities (10.8) belongs to J. Moser (see [87a], §5 of Chapter IX in [65q], and also the end of §8 and the end of §10 of the present chapter).

the first summand carry out an integration by parts:

$$\int_{K_2} v\mathfrak{N}^2\chi^2\,dx\Big|^{t=\theta} + \int_0^\theta \int_{K_2} a_{ij} v_{x_i} v_{x_j} \mathfrak{N}^2\chi^2\,dx\,dt$$
$$= \int_0^\theta \int_{K_2} [v\mathfrak{N}^2 2\chi\chi' - a_{ij} v_{x_j} 2\mathfrak{N}\mathfrak{N}_{x_i}\chi^2]\,dx\,dt.$$

Hence, using inequality (1.2) of Chapter II, we obtain

$$\int_{K_2} v\mathfrak{N}^2\chi^2\,dx\Big|^{t=\theta} + \int_0^\theta \int_{K_2} a_{ij} v_{x_i} v_{x_j} \mathfrak{N}^2\chi^2\,dx\,dt$$
$$\leqslant \varepsilon_1 \int_0^\theta \int_{K_2} a_{ij} v_{x_i} v_{x_j} \mathfrak{N}^2\chi^2\,dx\,dt + \frac{1}{\varepsilon_1} \int_0^\theta \int_{K_2} a_{ij} \mathfrak{N}_{x_i}\mathfrak{N}_{x_j}\chi^2\,dx\,dt$$
$$+ \int_0^\theta \int_{K_2} \left(\varepsilon_2 v^2 \mathfrak{N}^2\chi^2 + \frac{1}{\varepsilon_2}\mathfrak{N}^2\chi'^2\right)dx\,dt. \qquad (10.12)$$

For subsequent estimates of the terms of (10.12) we use condition (10.10) and inequality (5.4) of Chapter II. The numbers ϵ_1 and ϵ_2 are chosen as follows: $\epsilon_1 = 1/4$, $\epsilon_2 = \nu b^2\kappa_n^2/4\beta_2$, where β_2 is a constant from inequality (5.4) of Chapter II. As a result of this and the reduction of similar terms in (10.12) we deduce the desired estimate:

$$\int_{\frac{\theta}{4}}^{\theta} \int_{K_{3/2}} (v^2 + v_x^2)\,dx\,dt \leqslant c.$$

This, as mentioned above, completes the proof of inequality (10.8).

We have stated here two methods for obtaining estimates of $|u|_Q^{(\alpha)}$. Another proof of this estimate for equation (10.7) follows from the Harnack's inequality established by Moser [87b]. This inequality asserts that for nonnegative solutions in $\hat{Q}_\rho$ of equation (10.7) we have

$$\operatorname{vrai}\max_{Q_\rho^-} u(x,t) \leqslant c\,\operatorname{vrai}\min_{Q_\rho^+} u(x,t), \qquad (10.13)$$

where the constant c depends only on ν and μ from condition (1.2),

$$Q_\rho^+ = K_{\frac{\rho}{2}} \times \left(t_0 + \frac{3}{4}\rho^2,\ t_0 + \rho^2\right),$$
$$Q_\rho^- = K_{\frac{\rho}{2}} \times \left(\bar{t}_0 - \frac{3}{4}\rho^2,\ \bar{t}_0 - \frac{\rho^2}{2}\right).$$

$$\bar{t}_0 < t_0 < \bar{t}_0 + \rho^2,$$

and $\hat{Q}_\rho = K_\rho \times (\bar{t}_0 - \rho^2, t_0 + \rho^2) \subset Q_T$. From (10.13) we easily obtain an inequality of the form (10.8) for arbitrary solutions of (10.7), and therefore also an estimate of the Hölder norm $|u|_{Q'}^{(\alpha)}$ for any domain $Q' \subset Q_T$.

The basic step in Moser's proof consists in the following. First, by the method described briefly at the end of §8, one deduces the estimate

$$\operatorname*{vrai\,max}_{Q_\rho^-} u \equiv \|u\|_{\infty, Q_\rho^-} \leqslant c\|u\|_{\varepsilon, Q_\rho(x_0, \bar{t}_0)}, \quad \varepsilon > 0,$$

$$Q_\rho(x_0, t_0) = K_\rho(x_0) \times (t_0 - \rho^2, t_0).$$

Then in an analogous fashion one obtains the estimate

$$\|u\|_{-\varepsilon, Q_\rho(x_0, t_0+\rho^2)} \leqslant c \operatorname*{vrai\,min}_{Q_\rho^+} u.$$

Finally, for certain $\epsilon > 0$ one establishes the inequality

$$\|u\|_{\varepsilon, Q_\rho(x_0, t_0)} \leqslant c\,\|u\|_{-\varepsilon, Q_\rho(x_0, t_0+\rho^2)}.$$

Its derivation constitutes the most complicated part of the entire proof.

Moser's idea has been developed in the papers of S. N. Kružkov [62b-d], A. V. Ivanov [51e], Aronson [132], Kurihara [133], and Trudinger [136] along lines that apply to linear equations of general form (1.1), and also to certain classes of parabolic quasi-linear equations. In [51e], for nonnegative generalized solutions of a certain class of parabolic quasi-linear equations and, in particular, linear equations of general form, the following generalization of inequality (10.13) is established:

$$\operatorname*{vrai\,max}_{Q_\rho^-} u(x, t) \leqslant c_1 \left(\operatorname*{vrai\,min}_{Q_\rho^+} u(x, t) + c_2\rho^\delta \right),$$

where $\delta \in (0, 1]$ and where c_1 and c_2 are determined only by known quantities, with $c_2 = 0$ in the case of a homogeneous equation.

§11. AN ESTIMATE OF $\max_{Q'}|u_x|$ AND $\langle u_x \rangle_{Q'}^{(\alpha)}$

We will estimate $\max_{Q'}|u_x|$ and $\langle u_x \rangle_{Q'}^{(\alpha)}$ for the solutions u of equations (1.1), restricting ourselves on the whole to interior estimates and assuming for the sake of simplicity that u belongs to $C^{2,1}(Q_T)$. Let us prove the following theorem.

Theorem 11.1. *Suppose the coefficients and free terms of equation* (1.1) *satisfy conditions* (1.2) *and*

$$\left\| \frac{\partial a_{ij}}{\partial x_k},\ a_i,\ \frac{\partial a_i}{\partial x_k},\ b_i,\ a,\ \frac{\partial f_i}{\partial x_k},\ f_i,\ f \right\|_{2q,\,2r,\,Q_T} \leqslant \mu_1, \tag{11.1}$$

where q and r are arbitrary numbers satisfying relations (7.2) *for $n \geq 2$ and the relation*

$$\frac{1}{r}+\frac{1}{2q}=1-\varkappa_1,\quad q\in[1,\infty],\quad r\in\left[\frac{1}{1-\varkappa_1},\ \frac{2}{1-2\varkappa_1}\right],$$
$$0<\varkappa_1<\frac{1}{2}\quad for\quad n=1. \tag{11.2}$$

Then for any solution u of equation (1.1) *belonging to $C^{2,1}(Q_T)$ the quantity $|u_x|_{Q'}^{(\alpha)}$ is estimated from above by a constant depending only on $\max_{Q_T}|u|$, the constants ν, μ, μ_1, q and r of conditions* (1.2), (11.1) *and the distance from Q' to Γ_T. The exponent α is determined by n, ν, μ, q and r.*

Remark 11.1. In conditions (11.1) q and r can be different for different functions. But merely for the sake of simplicity in notation we will assume that they are the same.

Proof. We differentiate equation (1.1) with respect to each of the x_k and write the result in the form

$$u_{kt}-\frac{\partial}{\partial x_i}\left(a_{ij}u_{kx_j}+\sum_{m=0}^{n} A_{im}^{k}u_m\right)=\frac{\partial}{\partial x_i}\left[\delta_i^k f+\frac{\partial f_i}{\partial x_j}\right], \tag{11.3}$$
$$k=1,\ 2,\ \ldots,\ n,$$

where

$$u_0=u,\quad u_k=u_{x_k},\quad A_{im}^{k}=\frac{\partial a_{im}}{\partial x_k}+\delta_m^k a_i-\delta_i^k b_m$$

for $m=1,\cdots,n$ and $A_{i0}^{k}=\partial a_i/\partial x_k-\delta_i^k a$, while δ_i^k is the Kronecker delta symbol. Relations (11.3) and (1.1) are regarded as a system of equations for the u_k, $k=0,1,\cdots,n$. Its distinctive feature is the fact that the principal part consisting of terms of the form $\partial/\partial t-\partial(a_{ij}\partial/\partial x_j)/\partial x_i$ is diagonal and is the same in all equations of the system. Such systems have been successfully investigated in as much detail as a single equation of second order; §§1–4 of Chapter VII are devoted to them. There we obtain, in particular, estimates of $\max_{Q'}|u_k|$ and $\langle u_k\rangle_{Q'}^{(\alpha)}$ in terms of $\sum_{k=0}^{n}\|u_k\|_{2,Q_T}$, the distance from Q' to Γ_T and known parameters (Theorems 2.1 and 3.1), which when applied to our case gives an estimate of $|u_x|_{Q'}^{(\alpha)}$ in terms of $\|u_x\|_{2,Q_T}$.

Let us give another derivation of the estimate of $|u_x|_{Q'}^{(\alpha)}$, without using the results of Chapter VII on systems. We first estimate the norms $\|u_x\|_{s,Q'}$ for any s not exceeding some number s_0. For this purpose we consider the equality

$$-\int\limits_{Q_t}\sum_{k=1}^{n}\mathscr{L}u\frac{d}{dx_k}\left(|u_x|^{2s}u_{x_k}\zeta^2\right)dx\,dt=0, \tag{11.4}$$

where $\zeta(x, t)$ is a smooth nonnegative function, not exceeding 1 and equal to zero in the vicinity of S_T, and we transform it by means of a double integration by parts to the form

$$\begin{aligned}\frac{1}{2s+2}\int\limits_{\Omega}|u_x(x,t)|^{2s+2}\zeta^2(x,t)\,dx+\int\limits_{Q_t}\Big\{&-\frac{1}{s+1}|u_x|^{2s+2}\zeta\zeta_t\\&+\left(a_{ij}u_{x_j}+a_iu+f_i\right)_{x_k}\left(|u_x|^{2s}u_{x_k}\zeta^2\right)_{x_l}\\&-\left(b_lu_{x_l}+au+f\right)\frac{d}{dx_k}\left(|u_x|^{2s}u_{x_k}\zeta^2\right)\Big\}\,dx\,dt=0.\end{aligned} \tag{11.5}$$

Relation (11.5) holds for u from $C^{2,1}(Q_T)$, although in deriving it (in the integration by parts), we made use of the derivatives u_{xt} and D_x^3u. Indeed, let us take the sequence of infinitely differentiable functions $u^m(x, t)$ converging to u in the norm of $C^{2,1}(Q')$, where Q' is the subdomain of Q_t on which ζ is different from 0. For them the left side of (11.4) is equal to the left side of (11.5), and as m tends to ∞ they converge to the same expressions for u. Consequently, if the left side of (11.4) is equal to zero for u, then the left side of (11.5) is also equal to zero for u.

We leave the trivally nonnegative terms in the left side of (11.5), while the rest are transferred to the right and estimated from above in the following manner:

$$\begin{aligned}\frac{1}{2s+2}&\int\limits_{\Omega}|u_x(x,t)|^{2s+2}\zeta^2(x,t)\,dx+\int\limits_{Q_t}\Big(a_{ij}u_{x_kx_j}u_{x_kx_l}|u_x|^{2s}\zeta^2\\&+2sa_{ij}|u_x|^{2s-2}u_{x_k}u_{x_kx_j}u_{x_l}u_{x_lx_i}\zeta^2\Big)\,dx\,dt\\=\frac{1}{2s+2}&\int\limits_{\Omega}|u_x(x,0)|^{2s+2}\zeta^2(x,0)\,dx-\int\limits_{Q_t}\Big\{-\frac{1}{s+1}|u_x|^{2s+2}\zeta\zeta_t\\&+a_{ij}u_{x_kx_j}|u_x|^{2s}u_{x_k}2\zeta\zeta_{x_l}+\Big[\Big(\frac{\partial a_{lj}}{\partial x_k}u_{x_j}+a_lu_{x_k}+\frac{\partial a_l}{\partial x_k}u+\frac{\partial f_l}{\partial x_k}\Big)\\&-\delta_k^l\left(b_ju_{x_j}+au+f\right)\Big]\left(|u_x|^{2s}u_{x_k}\zeta^2\right)_{x_l}\Big\}\,dx\,dt\\\leqslant\frac{1}{2s+2}&\int\limits_{\Omega}|u_x(x,0)|^{2s+2}\zeta^2(x,0)\,dx+\int\limits_{Q_t}\Big[\frac{1}{s+1}|u_x|^{2s+2}\zeta\zeta_t+\end{aligned}$$

$$
\begin{aligned}
&+\varepsilon u_{xx}^2|u_x|^{2s}\zeta^2+\frac{\mu^2}{\varepsilon}|u_x|^{2s+2}\zeta_x^2\\
&+\sum_{i,k}\Big(\sum_j\Big|\frac{\partial a_{ij}}{\partial x_k}\Big||u_x|+|a_i||u_x|+\Big|\frac{\partial a_i}{\partial x_k}\Big|M+\Big|\frac{\partial f_i}{\partial x_k}\Big|\\
&+|b_i||u_x|+|a|M+|f|\Big)\Big|(|u_x|^{2s}u_{x_k}\zeta^2)_{x_i}\Big|\Big]\,dx\,dt. \qquad (11.6)
\end{aligned}
$$

We choose $\epsilon=\nu/2$ and, using the condition of ellipticity (1.2), we pass from (11.6) to the inequality

$$
\begin{aligned}
&\frac{1}{2s+2}\int_\Omega |u_x(x,t)|^{2s+2}\zeta^2(x,t)\,dx+\int_{Q_t}\Big[\frac{\nu}{2}u_{xx}^2|u_x|^{2s}\zeta^2\\
&\qquad+2s\nu|u_x|^{2s-2}\sum_i\Big(\sum_k u_{x_k}u_{x_kx_i}\Big)^2\zeta^2\Big]\,dx\,dt\\
&\qquad\leqslant\frac{1}{2s+2}\int_\Omega |u_x(x,0)|^{2s+2}\zeta^2(x,0)\,dx\\
&\qquad+\int_{Q_t}\Big[|u_x|^{2s+2}\Big(\frac{1}{s+1}|\zeta_t|\zeta+\frac{2\mu^2}{\nu}\zeta_x^2\Big)\\
&\qquad+\mathscr{D}(|u_x|+1)\sum_{i,k}\Big|(|u_x|^{2s}u_{x_k}\zeta^2)_{x_i}\Big|\Big]\,dx\,dt, \qquad (11.7)
\end{aligned}
$$

where

$$
\begin{aligned}
\mathscr{D}=\sum_{i,k}\Big[\sum_j\Big|\frac{\partial a_{ij}}{\partial x_k}\Big|+|a_i|+\Big|\frac{\partial a_i}{\partial x_k}\Big|M\\
+\Big|\frac{\partial f_i}{\partial x_k}\Big|+|b_i|+|a|M+|f|\Big].
\end{aligned}
$$

We estimate the last term in the right side of (11.7) by means of Cauchy's inequality (1.2) of Chapter II, relinquishing a small ϵ to terms with second derivatives u_{xx}. Namely,

$$
\begin{aligned}
&\mathscr{D}(|u_x|+1)\sum_{i,k}\Big|(|u_x|^{2s}u_{x_k}\zeta^2)_{x_i}\Big|\\
&=\mathscr{D}(|u_x|+1)\sum_{i,k}\Big||u_x|^{2s}u_{x_kx_i}\zeta^2\\
&+2s|u_x|^{2s-2}\sum_l u_{x_l}u_{x_lx_i}u_{x_k}\zeta^2+2|u_x|^{2s}u_{x_k}\zeta\zeta_{x_i}\Big|\\
&\leqslant\varepsilon u_{xx}^2|u_x|^{2s}\zeta^2+\frac{1}{4\varepsilon}\mathscr{D}^2(|u_x|+1)^2n^2|u_x|^{2s}\zeta^2+
\end{aligned}
$$

$$+ n\varepsilon_1 s \sum_i \Big(\sum_l u_{x_l} u_{x_l x_i}\Big)^2 |u_x|^{2s-2}\zeta^2 + \mathscr{D}^2(|u_x|+1)^2 \frac{sn}{\varepsilon_1}|u_x|^{2s}\zeta^2$$
$$+ |u_x|^{2s+2}\zeta_x^2 + n^2\mathscr{D}^2(|u_x|+1)^2|u_x|^{2s}\zeta^2.$$

We substitute this estimate into (11.7) and reduce similar terms, assuming $\epsilon = \nu/4$, while $\epsilon_1 = \nu/n$. This gives

$$j(t, s) \equiv \frac{1}{2s+2}\int_\Omega |u_x(x, t)|^{2s+2}\zeta^2(x, t)\,dx$$
$$+ \int_{Q_t}\Big[\frac{\nu}{4}u_{xx}^2|u_x|^{2s}\zeta^2 + s\nu|u_x|^{2s-2}\sum_l\Big(\sum_k u_{x_k}u_{x_k x_l}\Big)^2\zeta^2\Big]dx\,dt$$
$$\leqslant \frac{1}{2s+2}\int_\Omega |u_x(x, 0)|^{2s+2}\zeta^2(x, 0)\,dx$$
$$+ \int_{Q_t}\Big[|u_x|^{2s+2}\Big(\frac{1}{s+1}|\zeta_t|\zeta + \Big(1+\frac{2\mu^2}{\nu}\Big)\zeta_x^2\Big)$$
$$+ n^2\Big(\frac{s+1}{\nu}+1\Big)\mathscr{D}^2(|u_x|+1)^2|u_x|^{2s}\zeta^2\Big]dx\,dt$$
$$\leqslant \frac{1}{2s+2}\int_\Omega |u_x(x, 0)|^{2s+2}\zeta^2(x, 0)\,dx$$
$$+ c\int_{Q_t}\big[|u_x|^{2s+2}(|\zeta_t|\zeta+\zeta_x^2) + (s+1)\mathscr{D}^2(|u_x|^{2s+2}\zeta^2+\zeta^2)\big]dx\,dt, \qquad (11.8)$$

where

$$c = \max\Big\{1+\frac{2\mu^2}{\nu};\ 4n^2\Big(1+\frac{1}{\nu}\Big)\Big\}.$$

The term containing $\mathfrak{D}^2$ is estimated by Hölder's inequality:

$$\int_{Q_t}\mathscr{D}^2(|u_x|^{2s+2}\zeta^2+\zeta^2)\,dx\,dt$$
$$\leqslant \|\mathscr{D}^2\|_{q, r, Q_t}\big\||u_x|^{2s+2}\zeta^2+\zeta^2\big\|_{\frac{q}{q-1},\,\frac{r}{r-1},\,Q_t}$$
$$\leqslant c_1\Big(\big\||u_x|^{s+1}\zeta\big\|^2_{\frac{2q}{q-1},\,\frac{2r}{r-1},\,Q_t} + \operatorname{mes}^{1-\frac{1}{q}}\Omega t^{1-\frac{1}{r}}\Big), \qquad (11.9)$$

where c_1 depends only on μ_1 from condition (11.1), n and M. The first term of the right side is estimated in the same way as the expression

$$\|(u-k)^2\|_{\frac{q}{q-1},\ \frac{r}{r-1},\ Q_t(k)}$$

in (7.9) was estimated in §7. Namely,

$$\||u_x|^{s+1}\zeta\|^2_{\frac{2q}{q-1},\ \frac{2r}{r-1},\ Q_t}$$

$$\leqslant \||u_x|^{s+1}\zeta\|^2_{\bar q,\ \bar r,\ Q_t}\left(\int_0^t \operatorname{mes}^{\frac{\hat r}{\hat q}}\Omega\, dt\right)^{\frac{2}{\bar r}-\frac{2}{\hat r}}$$

$$\leqslant \beta^2\||u_x|^{s+1}\zeta|^2_{Q_t}\operatorname{mes}^{\frac{2}{\bar q}-\frac{2}{\hat q}}\Omega t^{\frac{2}{\bar r}-\frac{2}{\hat r}}.\text{[1)]} \tag{11.10}$$

Here the numbers $\bar q$, $\bar r$, $\hat q$ and $\hat r$ are constructed from q and r in the way described in §7, and we were justified in using the embedding theorem (3.4) of Chapter II.

The expression $j(t, s)$ is estimated from below, using the relation

$$|(|u_x|^{s+1}\zeta)_x|^2=\sum_{i=1}^n\left[(s+1)|u_x|^s\frac{u_{x_k}u_{x_kx_i}}{|u_x|}\zeta+|u_x|^{s+1}\zeta_{x_i}\right]^2$$

$$\leqslant 2(s+1)^2|u_x|^{2s-2}\sum_{i=1}^n\left(\sum_{k=1}^n u_{x_k}u_{x_kx_i}\right)^2\zeta^2+2|u_x|^{2s+2}\zeta_x^2 \tag{11.11}$$

in the following manner:

$$j(t, s)\geqslant\frac{\min(8;\nu)}{16(s+1)}\left[\||u_x(x, t)|^{s+1}\zeta\|^2_{2,\Omega}+\|(|u_x|^{s+1}\zeta)_x\|^2_{2,Q_t}\right]$$

$$+\frac{\nu}{8}\int_{Q_t}u_{xx}^2|u_x|^{2s}\zeta^2\,dx\,dt-\frac{\nu}{s+1}\int_{Q_t}|u_x|^{2s+2}\zeta_x^2\,dx\,dt. \tag{11.12}$$

From inequality (11.8) by virtue of inequalities (11.2), (11.9) and (11.10) we obtain

1) We note that by virtue of the assumptions of the theorem for (1.1) the numbers $\bar r$, $\bar q$, $\hat r$ and $\hat q$ are finite for all n.

$$\frac{\min(8;\nu)}{16(s+1)}\left[\| |u_x(x,t)|^{s+1}\zeta(x,t)\|^2_{2,\Omega} + \|(|u_x|^{s+1}\zeta)_x\|^2_{2,Q_t}\right] + \frac{\nu}{8}\int_{Q_t} u^2_{xx}|u_x|^{2s}\zeta^2\,dx\,dt$$

$$\leqslant \frac{1}{2s+2}\int_\Omega |u_x(x,0)|^{2s+2}\zeta^2(x,0)\,dx$$

$$+\left[c+\frac{\nu}{s+1}\right]\int_{Q_t}|u_x|^{2s+2}(|\zeta_t|\zeta+\zeta_x^2)\,dx\,dt$$

$$+c(s+1)c_1\beta^2\| |u_x|^{s+1}\zeta\|^2_{Q_t}\operatorname{mes}^{\frac{2}{\tilde q}-\frac{2}{\hat q}}\Omega t^{\frac{2}{\tilde r}-\frac{2}{\hat r}}$$

$$+c(s+1)c_1\operatorname{mes}^{1-\frac{1}{q}}\Omega t^{1-\frac{1}{r}}. \tag{11.13}$$

For positive t_1 satisfying the condition

$$c(s+1)c_1\beta^2\operatorname{mes}^{\frac{2}{\tilde q}-\frac{2}{\hat q}}\Omega t^{\frac{2}{\tilde r}-\frac{2}{\hat r}} \leqslant \frac{\min(8,\nu)}{32(s+1)}. \tag{11.14}$$

(11.13) implies the inequality

$$\| |u_x|^{s+1}\zeta\|^2_{Q_{t_1}} + \int_{Q_{t_1}} u^2_{xx}|u_x|^{2s}\zeta^2\,dx\,dt$$

$$\leqslant \frac{16}{\min\{8;\nu\}}\| |u_x(x,0)|^{s+1}\zeta(x,0)\|^2_{2,\Omega}$$

$$+(s+1)^2c_2\left[1+\int_{Q_{t_1}}|u_x|^{2s+2}(|\zeta_t|\zeta+\zeta_x^2)\,dx\,dt\right] \tag{11.15}$$

with the constant c_2 depending only on $\operatorname{mes}\Omega$, t_1 and the numerical parameters in conditions (1.2), (11.1). The number s can assume any of the values $s = 0, 1, 2, \cdots$.

On the other hand, for an arbitrary smooth function we have the inequality

$$\int_\Omega |u_x|^{2s+4}\zeta^2\,dx \leqslant$$

$$\leqslant c_3(s)\max_\Omega|u|^2\int_\Omega\left[u^2_{xx}|u_x|^{2s}\zeta^2+|u_x|^{2s+2}(\zeta_x^2+\zeta^2)\right]dx, \tag{11.16}$$

following from inequality (5.8) of Chapter II, with the constant $c_3(s)$ depending only on s and n. Therefore, assuming $\max_{Q_T}|u| = M$ to be known, from (11.15) and (11.16) we obtain

$$\int_{Q_{t_1}} |u_x|^{2s+4}\zeta^2\,dx\,dt \leqslant c_4(s)\left[\||u_x(x,0)|^{s+1}\zeta(x,0)\|^2_{2,\Omega} + \int_{Q_{t_1}} |u_x|^{2s+2}(|\zeta_t|\zeta + \zeta_x^2 + \zeta^2)\,dx\,dt + 1\right] \tag{11.17}$$

with a certain known constant depending on s.

We fix $s_0 > 0$ (its value will be chosen below as a function only of the parameters q and r of (11.1)) and in accordance with it we partition the closed interval $[0, T]$ into a finite number of closed intervals $[t^0 = 0, t^1], [t^1, t^2], \cdots, [t^{m-1}, t^m = T]$ of length $t_1/2$ (the last interval can have a smaller length), where t_1 satisfies condition (11.14) with $s = s_0$. An inequality of type (11.17) is valid for each cylinder $Q^k = \Omega \times [t^{k-1}, t^{k+1}]$. Choosing $\zeta(x, t) = \hat{\zeta}(x)\chi(t)$, in this cylinder, where for $k \geq 2$ $\chi(t)$ is equal to zero for $t = t^{k-1}$ and to unity for $t \in [t^k, t^{k+1}]$, but for the time being is arbitrary on the first interval, and summing the inequalities (11.17) over all (overlapping) cylinders Q^k, $k = 1, 2, \cdots$, we obtain

$$\int_{Q_T} |u_x|^{2s+4}\zeta^2\,dx\,dt \leqslant c_5(s_0)\left[\||u_x(x,0)|^{s+1}\zeta(x,0)\|^2_{2,\Omega} + \int_{Q_T} |u_x|^{2s+2}(|\zeta_t|\zeta + \zeta_x^2 + \zeta^2)\,dx\,dt + 1\right] \tag{11.18}$$

for $s = 0, 1, \cdots, s_0$ and any smooth nonnegative function ζ not exceeding 1 and equal to zero in the vicinity of S_T. These inequalities permit us to estimate an integral of $|u_x|^{2s+4}$ in terms of an integral of $|u_x|^{2s+2}$ over a somewhat more extensive domain. As shown at the end of §2, the integral $\int_{Q(0)}|u_x|^2dx\,dt$, $Q(0) \subset Q_T$, is estimated from above by a constant depending only on M (even just on $\|u\|_{2,Q_T}$), the constants in conditions (1.2)–(1.6), and the distance from $Q(0)$ to Γ_T. From (11.18) for $s = 0$ it follows that the integral $\int_{Q(1)}|u_x|^4dx\,dt$, $Q(1) \subset Q(0)$, is estimated in terms of $\int_{Q(0)}|u_x|^2dx\,dt$ and the distance from $Q(1)$ to $\Gamma_T(0)$ ($\Gamma_T(0)$ is the lower base and lateral surface of $Q(0)$). Contracting the domain step by step, we estimate the integrals

$$\int_{Q(s+1)} |u_x|^{2s+4}\,dx\,dt \leqslant c_6(s_0), \qquad s = 0, 1, \ldots, s_0, \tag{11.19}$$

where $c_6(s_0)$ depends only on s_0, M, the constants of conditions (1.2) and (11.1) and the distance from $Q(s+1)$ to Γ_T (which, as always, is assumed to be positive). If $\||u(x,0)|^{s_0+1}\zeta(x,0)\|_{2,\Omega}<\infty$, then one can take the cylinders $Q(0)\supset Q(1)\supset\cdots\supset Q(s_0+1)$ with the same lower base as that of Q_T, i.e. $Q(s)=\Omega(s)\times[0,T]$, $\Omega(0)\supset\Omega(1)\supset\cdots$, and thereby obtain estimates (11.19) also for such cylinders.

Let us return to the system (11.3). Each equation of it can be regarded as an equation relative to $u_k=u_{x_k}$ of the form

$$u_{kt}-\frac{\partial}{\partial x_i}\left(a_{ij}u_{kx_j}\right)=\frac{\partial F_i^k(x,t)}{\partial x_i}, \tag{11.20}$$

while u_k is like a generalized solution of it from $V_2^{1,0}(Q')$ (it is even better: u_k is a continuous function in $\overline{Q}'$ with continuous derivatives u_{kx} in $\overline{Q}'$). It follows from assumptions (11.1) and inequalities (11.19) that

$$\|F_i^k\|_{2q_2,\,2r_2,\,Q'}\leqslant\mu_2, \tag{11.21}$$

where s_0 can be taken so large that the numbers q_2 and r_2 satisfy the conditions $1/r_2+n/2q_2<1$, with

$$q_2\in\left(\frac{n}{2},\,q\right),\quad r_2\in(1,\,r)\quad\text{for}\quad n\geqslant 2,$$
$$q_2\in(1,\,q),\quad r_2\in(1,\,r)\quad\text{for}\quad n=1.$$

Let us take such an s_0 (it depends, obviously, only on q and r). Then Theorems 8.1 and 10.1 guarantee estimates of $\max_{Q''}|u_x|$ and $\langle u_x\rangle_{Q''}^{(\alpha)}$ for $Q''\subset Q'$ in terms of $\|u_x\|_{2,Q'}$, known constants, and the distance from Q'' to the lower base and lateral surface of Q'. These same theorems and inequalities (11.19) for the cylinders $\Omega(s)\times(0,T)$ give estimates of $\max_{Q''}|u_x'|$ and $\langle u_x\rangle_{Q'}^{(\alpha)}$ for $Q''=\Omega''\times(0,T)$ in terms of $\|u_x\|_{2,Q'}$, $Q'=\Omega'\times(0,T)$, $\max_{\Omega'}|u(x,0)|$ and $\langle u_x(x,0)\rangle_{\Omega'}^{(\beta)}$ respectively and the distance from Ω'' to the boundary of Ω'. Theorem 11.1 is proved.

It is more difficult to obtain estimates of $\max_{Q_T}|u_x|$ and $\langle u_x\rangle_{Q_T}^{(\alpha)}$ for all of Q_T (i.e., in the vicinity of S_T). We will do this in §§4 and 5 of Chapter V directly for quasi-linear equations.

Remark 11.2. The estimate of $|u_x|_{Q_T}^{(\alpha)}$ (and analogously of $|u_x|_{Q_T}^{(\alpha)}$) in terms of known quantities can also be deduced differently: from Theorem 9.1 of Chapter IV on an estimate of $\|u_t,\,u_{xx}\|_{q,Q_T}$ in terms of $\|\partial f_i/\partial x_i-f\|_{q,Q_T}$ and from a theorem of Chapter II for embedding the space $W_q^{2,1}(Q_T)$ into $H^{2-(n+2)/2,\,1-(n+2)/2q}(Q_T)$ for $q > n+2$ (see Lemma 3.3 of Chapter II). However, in this connection one must assume in contrast to Theorem 11.1 that the coefficients $a_{ij}(x, t)$ are continuous in (x, t).

§12. ON THE DEPENDENCE OF THE SMOOTHNESS OF GENERALIZED SOLUTIONS ON THE SMOOTHNESS OF THE DATA OF THE PROBLEM

We will show that the smoothness of generalized solutions from $V_2^{1,0}(Q_T)$ of equations (1.1) in an open domain Q_T is determined only by the smoothness of the coefficients and free terms of these equations; it is increased with an increase in the latter. Let us suppose that $u(x, t)$ is an arbitrary generalized solution from $V_2^{1,0}(Q_T)$ of equation (1.1), the coefficients and free terms of which satisfy conditions (1.2)–(1.6), while Q' is an arbitrary subdomain of Q_T separated from Γ_T by a positive distance.

It was proved in the preceding sections that (i) if in place of (1.3)–(1.6) the stronger conditions (1.3), (1.4) and (9.1), (9.2) are fulfilled, then the norms $\| |u|^{\theta-1} \|_{Q'}$ with θ from (0, 1) are finite; (ii) if conditions (7.1), (7.2) are fulfilled for the coefficients, while conditions (9.1), (9.2) with $\theta = 0$ are fulfilled for the f_i and f, then the integrals $\int_{Q'} \exp\{c\,|u(x, t)|\}\,dx\,dt$ with some $c > 0$ are finite; and (iii) if conditions (7.1), (7.2) are valid, then the norms $|u|_{Q'}^{(\alpha)}$ with some $\alpha > 0$ are finite, with all of these norms being effectively estimated in terms of $\|u\|_{2,Q_T}$, the distance from Q' to Γ_T and the known parameters in the corresponding conditions on the given functions (i.e. the coefficients and free terms of (1.1)).

Suppose that the given functions possess better properties, for example, that they satisfy conditions (11.1), (7.2), (11.2). Let us show that the norms $|u_x|_{Q'}^{(\alpha)}$ with some $\alpha > 0$ are then finite. For this purpose we take advantage of the a priori estimate of this norm established in §11. It is obtained by us, however, under the assumption that $u \in C^{2,1}(Q_T)$ and therefore can not be applied directly to the generalized solution in question, concerning which it is known for the present only that it belongs to $H^{\alpha,\alpha/2}(Q_T)$ and has u_x from $L_2(Q_T)$.

Let us consider in some cylinder $Q'_{t_0,t_1} = \Omega' \times (t_0, t_1)$, where Ω' is an interior subdomain of Ω with very smooth boundary S', while $0 < t_0 < t_1 \le T$, the auxiliary problem

$$\left.\begin{aligned} \mathscr{L}^\rho v \equiv v_t - \frac{\partial}{\partial x_i}\left(a_{ij\rho} v_{x_j} + a_{i\rho} v + f_{i\rho}\right) \\ + b_{i\rho} v_{x_i} + a_\rho v + f_\rho = 0, \\ v\,|_{\Gamma'_{t_0,t_1}} = u_\rho|_{\Gamma'_{t_0,t_1}}, \end{aligned}\right\} \tag{12.1}$$

where Γ'_{t_0,t_1} is the lateral surface and lower base of the cylinder Q'_{t_0,t_1}, while $a_{ij\rho}, \cdots, u_\rho$ are averagings of $a_{ij}, \cdots, u$ over x and t with an infinitely differentiable kernel of radius ρ (see §4 of Chapter II). Problem (12.1), as will be proved in Theorem 5.2 of Chapter IV, has a unique classical (and even infinitely differentiable in Q'_{t_0,t_1}) solution v^ρ. It is known that the u_ρ converge to u in the norm of $V_2^{1,0}(Q'_{t_0,t_1})$ for $\rho \to 0$. We will show that the v^ρ also converge to $u(x, t)$ in the norm of $V_2^{1,0}(Q'_{t_0,t_1})$ (and by the same token that the $|v^\rho|_{Q'_{t_0,t_1}}$ are uniformly bounded). In addition, from the estimates given in the preceding sections and the uniform (in ρ) boundedness of $|u_\rho|^{(\alpha)}_{Q'_{t_0,t_1}}$ one has the uniform boundness of the norms $|v^\rho|^{(\beta)}_{Q'_{t_0,t_1}}$ and $|v^\rho_x|^{(\beta)}_{Q''_{t_0,t_1}}$ with some $\beta > 0$, in which Q''_{t_0,t_1} is an arbitrarily chosen cylinder belonging to Q'_{t_0,t_1} and separated from Γ'_{t_0,t_1} by a positive distance. By virtue of this, for $\lim_{\rho\to 0} v^\rho$, i.e. for u, the norms $|u_x|^{(\beta)}_{Q''_{t_0,t_1}}$ will be bounded and the estimates established for them in §11 are valid. Analogously, by using the a priori estimates in §10 of Chapter IV, we establish a subsequent improvement in the differential properties of solutions u within Q_T under an increase in the smoothness of the coefficients and free terms of equation (1.1).

Thus it remains for us to prove that $|v^\rho - u_\rho|_{Q'_{t_0,t_1}} \to 0$ for $\rho \to 0$.

Let us consider the integral identity (1.15), which is satisfied by a generalized solution u, and put $\eta = \zeta_\rho$, where $\zeta(x, t)$ is a function from $W_2^{1,1}(Q_T)$ that is equal to zero in the domain $Q_T \setminus Q'_{t_0,t}$, $t_0 < t \leq t_1$, and its ρ-neighborhood.

Transferring in the integrals the operation of averaging from ζ to its coefficients and carrying out an integration by parts in the term $\int - u_\rho \zeta_t \, dx\, dt$, we arrive at the equality

$$\int\limits_{Q'_{t_0,t}} \left[u_{\rho t}\zeta + (a_{ij}u_{x_j} + a_i u + f_i)_\rho \zeta_{x_i} \right. \\ \left. + (b_i u_{x_i} + au + f)_\rho \zeta\right] dx\, dt = 0.$$

Taking equation (12.1) into account, we can write the following relation for the difference $w = u_\rho - v^\rho$:

$$\int_{Q'_{t_0,t}} \{ w_t\zeta + [(a_{ij}u_{x_j})_\rho - a_{ij\rho}v^\rho_{x_j} + (a_i u)_\rho - a_{i\rho}v^\rho]\zeta_{x_i} + [(b_i u_{x_i})_\rho - b_{i\rho}v^\rho_{x_i} + (au)_\rho - a_\rho v^\rho]\zeta\}\, dx\, dt = 0,\ t_0 < t \leqslant t_1.$$

It is not difficult to see that this equality is valid for any function $\zeta(x, t)$ from $W_2^{1,1}(Q'_{t_0,t_1})$ that is equal to zero on S'_{t_0,t_1} (see in this connection the derivation of (2.12)). We put $\zeta = w$ and under the integral sign we add and subtract the terms $a_{ij\rho}u_{\rho x_j}w_{x_i}$, $a_{i\rho}u_\rho w_{x_i}$, $b_{i\rho}u_{\rho x_i}w$, $a_\rho u_\rho w$ and write the result in the form

$$\frac{1}{2}\int_{\Omega'} w^2(x, t)\, dx + \int_{Q'_{t_0,t}} a_{ij\rho}w_{x_i}w_{x_j}\, dx\, dt = -j_1 - j_2, \tag{12.2}$$

$$j_1 = \int_{Q'_{t_0,t}} (a_{i\rho}ww_{x_i} + b_{i\rho}w_{x_i}w + a_\rho w^2)\, dx\, dt,$$

$$j_2 = \int_{Q'_{t_0,t}} \{[(a_{ij}u_{x_j})_\rho - a_{ij\rho}u_{\rho x_j}]w_{x_i} + [(a_i u)_\rho - a_{i\rho}u_\rho]w_{x_i} + [(b_i u_{x_i})_\rho - b_{i\rho}u_{\rho x_i}]w + [(au)_\rho - a_\rho u_\rho]w\}\, dx\, dt.$$

The integral j_1 admits the estimate

$$|j_1| \leqslant \int_{Q'_{t_0,t}} \left[\varepsilon w_x^2 + \frac{1}{2\varepsilon}\left(\sum_{i=1}^n a_{i\rho}^2 + \sum_{i=1}^n b_{i\rho}^2 + 2|a_\rho|\right)w^2\right] dx\, dt$$

$$\leqslant \varepsilon |w|^2_{Q'_{t_0,t}} + \frac{1}{2\varepsilon}\|\mathscr{D}\|_{q,r,Q'_{t_0,t}}\|w\|^2_{\bar q,\bar r,Q'_{t_0,t}} \leqslant \varepsilon|w|^2_{Q'_{t_0,t}}$$

$$+ \frac{\beta^2}{2\varepsilon}\|\mathscr{D}\|_{q,r,Q'_{t_0,t}}|w|^2_{Q'_{t_0,t}}\left(\operatorname{mes}\Omega'^{\frac{2}{\bar q}-\frac{2}{\hat q}}(t-t_0)^{\frac{2}{\bar r}-\frac{2}{\hat r}}\right),$$

where $\epsilon \in (0, 1)$, $\mathfrak{D} = \sum_{i=1}^n a_{i\rho}^2 + \sum_{i=1}^n b_{i\rho}^2 + 2|a_\rho|$, while the numbers $\bar q$, $\bar r$, $\hat q$ and $\hat r$ have the same values as in estimate (7.9).

For j_2, by making use of the property of averagings, conditions (7.1), (7.2) and the boundedness of $|u|_{Q_T}$, it is not difficult to deduce that

$$|j_2| \leqslant \delta(\rho)|w|_{Q'_{t_0,t}} \leqslant \frac{1}{4\varepsilon}\delta^2(\rho) + \varepsilon|w|^2_{Q'_{t_0,t}},$$

where $\delta(\rho) \to 0$ for $\rho \to 0$.

In equality (12.2) we estimate the second term from below with the use of condition (1.2), and the integrals j_1 and j_2 from above, as indicated above, and then we maximize the obtained inequality over $t \in [t_0, t_1]$. As a result we get

$$\nu_0 |w|^2_{Q'_{t_0,t_1}} \leqslant 2|w|^2_{Q'_{t_0,t_1}} \Big[2\varepsilon + \frac{\beta^2}{2\varepsilon} \|\mathscr{D}\|_{q,r,Q'_{t_0,t_1}} \times (\operatorname{mes}\Omega')^{\frac{2}{\overline{q}}-\frac{2}{\hat{q}}} (t_1 - t_0)^{\frac{2}{\overline{r}}-\frac{2}{\hat{r}}}\Big] + \frac{1}{2\varepsilon}\delta^2(\rho), \nu_0 = \min\left\{\frac{1}{2}, \nu\right\}. \tag{12.3}$$

Let us take $\epsilon = \nu_0/8$, in (12.3), and fix $t_1 - t_0$ so small that

$$\frac{8\beta^2}{\nu_0} \|\mathscr{D}\|_{q,r,Q'_T} (\operatorname{mes}\Omega')^{\frac{2}{\overline{q}}-\frac{2}{\hat{q}}} (t_1 - t_0)^{\frac{2}{\overline{r}}-\frac{2}{\hat{r}}} \leqslant \frac{\nu_0}{4}. \tag{12.4}$$

Then (12.3) will yield the estimate

$$|w|^2_{Q'_{t_0,t_1}} \leqslant 16\nu_0^{-2}\delta^2(\rho).$$

In this way we have proved that $|w|_{Q'_{t_0,t_1}} = |u_\rho - v^\rho|_{Q'_{t_0,t_1}} \to 0$ for $\rho \to 0$ as long as $t_1 - t_0$ is subject to restriction (12.4).

As indicated above, this fact and the estimates established in §11 are sufficient, as long as conditions (11.1), (11.2) are fulfilled, for one to determine whether or not the generalized solutions of class $V_2^{1,0}(Q_T)$ of equation (1.1) possess Hölder-continuous derivatives with respect to x.

If the coefficients and free terms of equation (1.1) satisfy, in addition to conditions (1.2)–(1.6), the conditions of Theorem 9.1 of Chapter IV guaranteeing the estimate (10.12) of Chapter IV for $\|u_t, u_{xx}\|_{q,Q'}$ in terms of $\|f - \partial f_i/\partial x_i\|_{q,Q_T}$, then, upon applying this estimate to the v^ρ with $Q''_{t_0,t_1} \subset Q'_{t_0,t_1}$ and observing that the constant in it does not depend on ρ, we establish that for the limit function u there also exist derivatives u_t and u_{xx} belonging to $L_q(Q''_{t_0,t_1})$. If the functions a_{ij}, $\partial a_{ij}/\partial x_i$, a_i, $\partial a_i/\partial x_i$, b_i, a, f and $\partial f_i/\partial x_i$ are elements of $H^{2m+\alpha, m+\alpha/2}(Q_T)$, $m \geq 0$, then any generalized solution of equation (1.1) is an element of $H^{2m+2+\alpha, m+1+\alpha/2}(Q_T)$. This follows from the fact that by virtue of estimate (10.5) of Chapter IV the norms in the space $H^{2m+2+\alpha, m+1+\alpha/2}(Q''_{t_0,t_1})$, are uniformly bounded for the v^ρ, and consequently are also finite for u.

In this way we prove

Theorem 12.1. *Suppose u is a generalized solution from $V_2^{1,0}(Q_T)$ of equation* (1.1), *the coefficients and free terms of which satisfy conditions* (1.2)–(1.6). *If furthermore conditions* (11.1), (7.2) *and* (11.2) *are fulfilled, then u_{x_k} belongs to $H^{\alpha,\alpha/2}(Q_T)$ with some $\alpha > 0$ determined by n, ν, μ and the numerical parameters q, r from conditions* (11.1), (7.2) *and* (11.2). *If the coefficients and free terms satisfy the conditions of Theorem* 9.1 *of Chapter* IV, *then u has derivatives u_t and u_{xx} that are qth-power summable on any domain Q' separated from Γ_T by a positive distance. If, finally, all of the coefficients and free terms, and also their derivatives, in equation* (1.1) *are elements of $H^{2m+\alpha,m+\alpha/2}(Q_T)$, $m \geq 0$, then u will belong to $H^{2m+2+\alpha,m+1+\alpha/2}(Q_T)$.*

The a priori estimates of the present chapter and Chapter IV obtained for the whole domain Q_T permit us to establish the membership of generalized solutions from $V_2^{1,0}(Q_T)$ in the spaces $H^{2m+2+\alpha,m+1+\alpha/2}(\overline{Q}_T)$, $H^{2m+2+\alpha,m+1+\alpha/2}(Q_T \cup \Gamma')$, $m \geq 0$, $W_q^{2m,m}(Q_T)$, $m \geq 1$, et al. if, in addition to the smoothness of the functions in equation (1.1), a corresponding smoothness is possessed by the boundary values and the surface S_T (or part of it), and if also the compatibility conditions on $\Gamma' \cap S_0$ are fulfilled up to the necessary order.

We will not cite here the exact formulations of all of these propositions since it is not difficult to do so independently and the number of different possibilities is rather large.

We stress only the fact that the smoothness properties of generalized solutions from $V_2^{1,0}(Q_T)$ of equation (1.1) have a local character: the smoothness of u in some interior subdomain $Q' \subset Q_T$ is determined only by the smoothness of the functions in Q' making up equation (1.1), while the smoothness of them in domains Q' adjacent to $\Gamma' \subset \Gamma_T$, is determined, in addition, by the smoothness of Γ', the smoothness of the values of u on Γ' and the compatibility conditions on $\Gamma' \cap S_0$.

Let us consider the analogous questions for the equations

$$\mathscr{L}u \equiv u_t - a_{ij}(x, t)u_{x_ix_j} + b_i(x, t)u_{x_i} + a(x, t)u = F(x, t). \tag{12.5}$$

For these equations it is proved in Chapter IV (see Theorem 9.1) that if their coefficients a_{ij} are continuous, $b_i \in L_r(Q_T)$, $a \in L_s(Q_T)$ with certain r and $s > 1$, while $F \in L_q(Q_T)$, $q > 1$, then the basic initial-boundary-value problems are uniquely solvable in $W_q^{2,1}(Q_T)$ (naturally, for a specific surface S and boundary function). The differential properties of such solutions are improved as the coefficients and free term of (12.5) are improved, and this improvement has a local character. Thus, for example, on has

Theorem 12.2. *Suppose $u(x, t)$ is a generalized solution from $W_q^{2,1}(Q_T)$, $q > 1$, of equation* (12.5). *If the coefficients and free term of* (12.5) *satisfy the conditions of Theorem* 9.1 *of Chapter* IV *with index $q_1 > q$, then $u(x, t)$ will be an element of $W_{q_1}^{2,1}(Q')$ in any subdomain Q' of the domain Q_T separated from Γ_T by a positive distance. But if the coefficients and free term of* (12.5) *are elements of $H^{\alpha,\alpha/2}(Q_T)$, then $u(x, t)$ will belong to $H^{2+\alpha,1+\alpha/2}(Q_T)$.*

This theorem is proved in the same way as Theorem 12.1, it being only necessary to use, instead of the energy inequality, inequality (9.3) of Chapter IV.

§13. ON DIFFRACTION PROBLEMS

By proper diffraction problems we mean boundary value problems in domains consisting of two or more heterogeneous media. On an interface of these media certain compatibility conditions must be fulfilled. There are two such conditions for equations of second order. These are usually the continuity of the desired solution and the continuity of its derivative along the conormal to the interface (in the corresponding physical problems the first condition reflects a natural requirement, namely the absence of discontinuities in the medium, while the second reflects the requirement that the forces acting on the interface be in equilibrium). It was shown in the note [65q] that problems of this type for equations of different types can be reduced by means of a simple technique to problems on the determination of generalized solutions (with finite energy integral) of ordinary boundary value problems for one equation. The solvability of these latter problems is proved in various ways, in the course of which various approximate methods of solution are justified (the method of finite differences, Galerkin's method, et al.). The unique solvability of diffraction problems was thereby proved for equations of basic types within a generalized statement of the problem involving very weak assumptions on all of the data.

Moreover, in the papers [65', f; 124f] techniques are described for investigating the improvement in the differential properties of these generalized solutions as the differential properties of the coefficients and free terms of the equations, the interfaces et al. are improved (this is stated in more detail in [91a,b]). Later these techniques were substantially augmented by new methods of investigation that permitted one to prove the classicalness of generalized solutions under nearly necessary conditions for them [65r].

We will dwell here only on how the existence of generalized solutions of diffraction problems for parabolic equations is deduced from the theorems proved above on the solvability of ordinary boundary value problems. But we will not carry out an investigation of their smoothness, referring instead to [65r] since in essence it is similar to what was stated in §§6–12, while being quite cumbersome technically.

We begin with the following case. Suppose the domain $Q_T = \Omega \times (0, T)$ is partitioned into several domains $Q^{(k)}$, $k = 1, 2, \cdots, N$ (so that $\overline{Q}_T = \overline{Q}^{(1)} \cup \cdots \cup \overline{Q}^{(N)}$), in each of which there is given an equation of the form

$$\mathscr{L}^{(k)} u \equiv u_t - \frac{\partial}{\partial x_i}\left(a^k_{ij}(x, t) u_{x_j}\right) + b^k_i(x, t) u_{x_i} + a^k(x, t) u = f^k(x, t) \tag{13.1}$$

with smooth coefficients and free terms. We wish to find in $\overline{Q}_T$ a function $u(x, t)$ satisfying: (i) in $Q^{(k)}$, $k = 1, \cdots, N$, the corresponding equation (13.1); (ii) on the lateral surface S_T of the domain Q_T one of the basic boundary conditions, for example,

$$u|_{S_T} = 0; \tag{13.2}$$

(iii) on the lower base of Q_T the initial condition

$$u|_{t=0} = \psi_0(x); \tag{13.3}$$

and (iv) on the interfaces $S^{(k,l)}$ of the domains $Q^{(k)}$ and $Q^{(l)}$, $k, l = 1, \cdots, N$, the two compatibility conditions

$$[u]|_{S^{(k,l)}} = 0 \tag{13.4}$$

and

$$\left[\frac{\partial u}{\partial \mathscr{N}}\right]\Big|_{S^{(k,l)}} = 0. \tag{13.5}$$

Here $\partial u/\partial \mathscr{N} = a_{ij} u_{x_j} \cos(\mathbf{n}, x_i)$, $\mathbf{n}$ is the normal to $S^{(k,l)}$ directed into $Q^{(l)}$, while the symbol $[v]$ denotes the jump in the function v as it passes through $S^{(k,l)}$ from $Q^{(k)}$ into $Q^{(l)}$, and

$$\left[\frac{\partial u}{\partial \mathscr{N}}\right]\Big|_{(x^0, t^0) \in S^{(k,l)}} = \Big(a^k_{ij} u_{x_j}\Big|_{\substack{(x,t) \in Q^{(k)} \\ (x,t) \to (x^0, t^0)}} - a^l_{ij} u_{x_j}\Big|_{\substack{(x,t) \in Q^{(l)} \\ (x,t) \to (x^0, t^0)}}\Big) \cos(\mathbf{n}, x_i).$$

The diffraction problems usually considered are a special case of this problem. Let us show that this problem is itself a special case of an ordinary boundary value problem for equations (1.1) with arbitrary discontinuous coefficients, that is studied in the present chapter. Indeed, we consider equation (13.1) as one equation

$$\mathscr{L}u \equiv u_t - \frac{\partial}{\partial x_i}(a_{ij}(x, t)u_{x_j}) + b_i(x, t)u_{x_i} + a(x, t)u = f(x, t) \tag{13.6}$$

with coefficients and free term equal in each of the $Q^{(k)}$ to the corresponding coefficients and free term of the equation $L^{(k)}u = f^{(k)}$ (their definition can be arbitrarily completed on $S^{(k,l)}$, for example, as half-sums of the limiting values on $S^{(k,l)}$, if they exist, of the corresponding functions), and for this equation we pose in Q_T the first boundary value problem (13.6), (13.2), (13.3). As proved in the preceding sections, this problem has a unique generalized solution u in the class $\overset{\circ}{V}{}_2^{1,1/2}(Q_T) \cap H^{\alpha,\alpha/2}(Q_T)$ for any $\psi_0(x) \in L_2(\Omega)$, and in each of the domains $Q^{(k)}$ its smoothness is completely determined by the smoothness of the coefficients and free term of the corresponding equation (13.1) in $Q^{(k)}$. All of the requirements formulated above for a diffraction problem, except (13.5), are satisfied in an obvious way as long as the coefficients and free terms of (13.1) in $Q^{(k)}$, $k = 1, \cdots, N$, and also the $a^k_{ijx_l}$ are, for example, elements of $H^{\alpha,\alpha/2}(Q^{(k)})$. Let us show that condition (13.5) is a consequence of the integral identity.

$$\int_{Q_T} (-u\eta_t + a_{ij}u_{x_j}\eta_{x_i} + b_i u_{x_i}\eta + au\eta - f\eta)\,dx\,dt - \int_\Omega \psi_0\eta(x, 0)\,dx = 0, \tag{13.7}$$

which is satisfied by a generalized solution u of problem (13.6), (13.2), (13.3) as long as in each of the $Q^{(k)}$ this solution belongs, for example, to the space $W_2^{2,1}(Q^{(k)})$, and the surfaces $S^{(k,l)}$ are piecewise smooth without tangent planes parallel to the plane $t = 0$. For this purpose we transform (13.7) into the form

$$\sum_{k=1}^{N} \int_{Q^{(k)}} (\mathscr{L}^{(k)}u - f^{(k)})\eta\,dx\,dt + \sum_{k,\,l=1}^{N} \int_{S^{(k,l)}} \left[\frac{\partial u}{\partial \mathscr{N}}\right]\eta\,ds = 0, \tag{13.8}$$

which is possible by virtue of the assumptions made on u and the $S^{(k,l)}$. But from it, since the choice of $\eta(x, t)$ is sufficiently arbitrary, we obtain equation

(13.1) and equality (13.5). Thus, in fact, a generalized solution from $V_2^{1,1/2}(Q_T)$ of problem (13.6), (13.2), (13.3) is a solution of the problem (13.1)–(13.5) formulated above as long as it possesses the above-indicated smoothness and the $S^{(k,l)}$ do not have horizontal tangent planes. As is easily seen, we also have the converse assertion: a classical solution of problem (13.1)–(13.5) (if it exists) is a generalized solution from $V_2^{1,1/2}(Q_T)$ of problem (13.6), (13.2), (13.3). And since the latter is unique, we can replace the original (classical) statement of a diffraction problem by a generalized one: the determination of solutions from $V_2^{1,1/2}(Q_T)$ of problem (13.6), (13.2), (13.3). By the same token the unique solvability of problem (13.1)–(13.5) in this generalized statement follows from Theorem 4.1, the assumptions on all of the data of the problem being reducible to the conditions of Theorem 4.1. The theorems of §§6–12 give sufficient conditions for when this solution belongs to the classes $H^{\alpha,\alpha/2}(Q_T)$, $H^{\alpha,\alpha/2}(\overline{Q}_T)$, and also to $H^{l,l/2}(Q^{(k)})$, $l > 1$, and to $W_q^{l,l/2}(Q^{(k)\prime})$, $l \geq 2$, $k = 1, \cdots, N$, where $Q^{(k)\prime}$ is an arbitrary interior subdomain of $Q^{(k)}$. But an investigation of the behavior of the u_{x_i} near the surfaces $S^{(k,l)}$ requires special considerations, which are developed in [65r]. Since all of these investigations bear a local character, the assumption that the boundaries $S^{(k,l)}$ have the form $S_0^{(k,l)} \times (0, T)$, where the $S_0^{(k,l)}$ are smooth surfaces in the x space that divide Ω into the domains $\Omega^{(k)}$, $k = 1, \cdots, N$, does not constitute a loss in generality: in the small any smooth piece of $S^{(k,l)}$ not having normals parallel to the t axis can be converted by means of a change in the variables (x, t) of the form $y_i = t - \phi_i(x)$, $i = 1, \cdots, n$, $\tau = t$, into a piece of a surface (and even of a plane) with elements parallel to the t axis, that does not violate the character of the problem. We cite here one of the basic results on the solvability of problem (13.1)–(13.5).

Theorem 13.1. *Problem* (13.1)–(13.5) *has a unique generalized solution* $u(x, t)$ *from* $V_2^{1,1/2}(Q_T)$ *if* $\psi_0 \in L_2(\Omega)$ *and the coefficients and free terms of equations* (13.1) *satisfy the conditions of Theorem* 4.1. *This solution* u *is the same as the generalized solution from* $V_2^{1,1/2}(Q_T)$ *of problem* (13.6), (13.2), (13.3). *For it one has the assertions of Theorems* 7.1 *and* 9.1 *on when it is summable on* Q_T *with larger summability powers or when it is Hölder continuous in* Q_T *or* $\overline{Q}_T$ *et al. Moreover, outside of the interfaces* $S^{(k,l)}$ *its smoothness is guaranteed by the theorems of* §§10–12; *in particular, it belongs to the classes* $H^{2+\alpha,1+\alpha/2}(Q^{(k)})$ *in each of the domains* $Q^{(k)}$ *as long as the coefficients and free terms of equations* (13.1) *and also the* $a^k_{ijx_l}$ *are elements of* $H^{\alpha,\alpha/2}(Q^{(k)})$. *And near any twice continuously differentiable piece* $S_1^{(k,l)}$ *of the surface* $S^{(k,l)}$ *not having normals parallel to the* t *axis*[1)] *the derivatives of the solution* u *are (Hölder) continuous on both sides of* $S_1^{(k,l)}$ *right up to the surface* $S_1^{(k,l)}$ *itself as long as in some neighborhood* $\Omega_1 \times [t_1, t_2]$ *of it one has*

$$\left\| \frac{\partial a_{ij}}{\partial t};\ \frac{\partial a_{ij}}{\partial x_k},\ b_i,\ \frac{\partial b_i}{\partial t},\ a,\ \frac{\partial a}{\partial t},\ f,\ \frac{\partial f}{\partial t} \right\|_{q,\Omega_1} \leqslant \text{const}$$

for $q > n$ *when* $t \in [t_1, t_2]$ *and if, as always, the condition of ellipticity* (1.2) *is fulfilled. Thus in this case both of the requirements* (13.4), (13.5) *are fulfilled in the classical sense.*

In view of the stability of the boundary value problems with respect to variations in the coefficients and free terms of the equations, the solutions of diffraction problems can be obtained as limits of good solutions of equations with smoothed coefficients. Thus the solution u of problem (13.1)–(13.5) or, equivalently, problem (13.6), (13.2), (13.3) (under the conditions of Theorem 4.1) is the limit in the norm of $V_2^{1,0}(Q_T)$ of the solutions u^m of the boundary value problems

$$u_t - \frac{\partial}{\partial x_i}\left(a_{ijm} u_{x_j}\right) + b_{im} u_{x_i} + a_m u = f_m,$$

$$u|_{S_T} = 0, \quad u|_{t=0} = \psi_{0m}(x),$$

in which the a_{ijm}, b_{im}, a_m, f_m, ψ_{0m} are smooth functions approximating the a_{ij}, b_i, a, f, and ψ_0 respectively in the norms of those spaces to which these (latter) functions belong (an exact formulation is given in Theorem 4.5). If furthermore the conditions of Theorem 9.1 are fulfilled, then the u^m will converge to u uniformly.

We note that the case of nonhomogeneous conditions on S_T and the $S^{(k,l)}$ is reduced to the homogeneous case by introducing in place of u a new unknown function $v(x, t) = u(x, t) - \psi(x, t)$, where $\psi(x, t)$ is some smooth function satisfying on S_T and the $S^{(k,l)}$ the same nonhomogeneous conditions as u. The other boundary conditions on S_T and the conditions

$$\left[\frac{\partial u}{\partial \mathcal{N}} + \sigma u\right]\Big|_{S^{(k,l)}} = 0 \tag{13.9}$$

in place of (13.5) are considered analogously. An important fact concerning the present method is that condition (13.5) on a discontinuity has a special form connected with the principal terms of the equation; namely, it involves the derivatives $\partial u/\partial \mathfrak{N}$ along a conormal but not along arbitrary directions l in the x space. By virtue of this fact the terms on an interface that are singled out in

1) These conditions are weakened somewhat in [65r].

passing from the equation to the integral identity either completely vanish (see (13.8) and (13.7)) or the principal terms among them vanish.

Analogous considerations permit the following generalizations of problem (13.1)–(13.5). Suppose instead of equations (13.1) we have the equations

$$\begin{aligned}\mathscr{L}^{(k)}u \equiv \alpha^k(x, t)\, u_t - \frac{\partial}{\partial x_i}\left(a^k_{ij}(x, t)\, u_{x_j} + a^k_i(x, t)\, u\right)\\ + b^k_i(x, t)\, u_{x_i} + a^k(x, t)\, u = f^k(x, t),\end{aligned} \tag{13.10}$$

the coefficients and free terms in which are good functions in each of the $\overline{Q}^{(k)}$, having discontinuities of the first kind in passing through the $S^{(k,l)}$, with the $\alpha^k(x, t)$ being strictly positive, and instead of conditions (13.5) we have the conditions

$$\left[ba_{ij}u_{x_j} + ba_iu\right]\cos(\mathbf{n}, x_i)\big|_{S^{(k,l)}} = 0, \quad k, l = 1, \ldots, N, \tag{13.11}$$

or the conditions

$$[b\alpha u]\cos(\mathbf{n}, t) - \left[ba_{ij}u_{x_j} + ba_iu\right]\cos(\mathbf{n}, x_i)\big|_{S^{(k,l)}} = 0, \tag{13.12}$$

in which $b(x, t)$ is a smooth positive function in each of the $\overline{Q}^{(k)}$, having discontinuities of the first kind in passing through the $S^{(k,l)}$. In the first problem, in conditions (13.4) and (13.11), the jumps are determined in the same way as above, while in the second problem they are determined differently, namely,

$$[u]\big|_{(x^0, t^0) \in S^{(k,l)}} = u\big|_{\substack{(x, t^0) \in Q^{(k)} \\ (x, t^0) \to (x^0, t^0)}} - u\big|_{\substack{(x, t^0) \in Q^{(l)} \\ (x, t^0) \to (x^0, t^0)}}$$

in (13.4), and also for $[ba_{ij}u_{x_j} + ba_iu]|_{S(k,l)}$ in (13.12), while

$$[b\alpha u]\big|_{(x^0, t^0) \in S^{(k,l)}} = b\alpha u\big|_{\substack{(x^0, t) \in Q^{(k)} \\ (x^0, t) \to (x^0, t^0)}} - b\alpha u\big|_{\substack{(x^0, t) \in Q^{(l)} \\ (x^0, t) \to (x^0, t^0)}}.$$

Let us show how it is possible to transform these problems into problems on the determination of generalized solutions of a boundary value problem for one equation with discontinuous coefficients and to thereby replace the classical statements of these problems with generalized ones.

For problem (13.10), (13.2)–(13.4), (13.11) we have the determination of a generalized solution u from $W_2^{1,1}(Q_T)$ of the first boundary value problem (13.2), (13.3) for the equation

$$\mathcal{L}_1 u \equiv bau_t - \frac{\partial}{\partial x_i}(ba_{ij}u_{x_j} + ba_i u) + B_j u_{x_j} + Au = bf, \qquad (13.13)$$

in which $B_j = a_{ij}b_{x_i} + b_j b$, $A = a_i b_{x_i} + ab$, while the functions $\alpha(x, t)$, $a_{ij}(x, t)$ et al. coincide with $\alpha^{(k)}(x, t)$, $a_{ij}^k(x, t)$ et al. in the domains $Q^{(k)}$. Such a solution, in accordance with our conception of the definition of generalized solutions of boundary value problems, is a function from $\overset{\circ}{W}{}_2^{1,1}(Q_T)$ satisfying condition (13.3) and the integral identity

$$\int_{Q_T} \left[bau_t\eta + (ba_{ij}u_{x_j} + ba_i u)\eta_{x_i} + B_j u_{x_j}\eta + Au\eta\right] dx\, dt = \int_{Q_T} bf\eta\, dx\, dt \qquad (13.14)$$

for any $\eta(x, t)$ from $\overset{\circ}{W}{}_2^{1,0}(Q_T)$. It is not difficult to see that there are two pieces of information stored in identity (13.14): in the $Q^{(k)}$ the function u satisfies equation (13.13) or, equivalently, equation (13.10), while on the $S^{(k,l)}$ it satisfies conditions (13.11). The remaining requirements of problem (13.2)–(13.4) are contained in the fact that u must be an element of $\overset{\circ}{W}{}_2^{1,1}(Q_T)$ satisfying condition (13.2).

Problem (13.10), (13.2)–(13.4), (13.12) is reformulated as a problem on determining a generalized solution u from $W_2^{1,0}(Q_T)$ of the first boundary value problem (13.2), (13.3) for the equation

$$\mathcal{L}_2 u \equiv \frac{\partial(bau)}{\partial t} - \frac{\partial}{\partial x_i}(ba_{ij}u_{x_j} + ba_i u) + B_j u_{x_j} + \hat{A}u = bf \qquad (13.15)$$

with the same B_j as above and with $\hat{A} = -(ba)_t + a_i b_{x_i} + ab$. This means that u must be an element of $\overset{\circ}{W}{}_2^{1,0}(Q_T)$ satisfying the identity

$$\int_{Q_T} [-bau\eta_t + (ba_{ij}u_{x_j} + ba_i u)\eta_{x_i} + B_j u_{x_j}\eta + \hat{A}u\eta]\, dx\, dt - \int_{\Omega} ba\psi_0\eta dx\Big|_{t=0} = \int_{Q_T} bf\eta\, dx\, dt \qquad (13.16)$$

for any $\eta(x, t)$ from $\overset{\circ}{W}{}_2^{1,1}(Q_T)$ that is equal to zero for $t = T$. In the present case identity (13.16) contains equation (13.15) or, equivalently, (13.10), and conditions (13.3) and (13.12).

The existence of generalized solutions of both of the problems just formulated can be proved by different methods, the first being based on inequality (6.6) and the second on an energy-type inequality. We will show how this is done, using as an example the ability to solve the first boundary value problem for equations with smooth coefficients (the corresponding theorems are proved in §5 of Chapter IV). We assume for the sake of nonessential simplifications that all of the known functions in identities (13.14) and (13.16), except f and ψ_0, are bounded in modulus by some number M. We approximate the coefficients ba, ba_{ij}, ba_i, B_j, A and $\hat{A}$ by smooth functions $(ba)_m$, $(ba_{ij})_m$ and so on in $\overline{Q}_T$ in such a way that the conditions of Theorem 5.2 in Chapter IV are fulfilled. Moreover, this is done in such way that we have the inequalities

$$\begin{gathered}\nu_1\xi^2 \leqslant (ba_{ij})_m \xi_i\xi_j \leqslant \mu_1\xi^2, \quad \nu_1 > 0, \\ \int_0^T \max_{x\in\Omega}\left|\frac{\partial (ba_{ij})_m}{\partial t}\right| dt \leqslant \mu_1, \\ \int_0^T \max_{x\in\Omega}\left|\frac{\partial (ba_i)_m}{\partial t}\right| dt \leqslant \mu_1, \qquad \int_0^T \max_{x\in\Omega}\left|\frac{\partial (ba)_m}{\partial t}\right| dt \leqslant \mu_1,\end{gathered} \tag{13.17}$$

with one and the same positive constants ν_1 and μ_1, and that the approximations for all of the other coefficients are uniformly bounded, while the $(bf)_m$ and $(b\psi_0)_m$ converge to bf and $b\psi_0$ in $L_2(Q_T)$ and $L_2(\Omega)$ respectively. Then for the whole family of solutions u^m of the auxiliary problems we will have in the first case inequality (6.6), which gives the uniform boundedness of the norms

$$\| u^m \|_{W_2^{1,1}(Q_T)} \leqslant \mu_2, \tag{13.18}$$

and in the second case inequality (6.3) guaranteeing the uniform boundedness of the

$$\| u^m \|_{W_2^{1,0}(Q_T)} \leqslant \mu_2. \tag{13.19}$$

As a result we can pass from identities (13.14) and (13.16) for the u^m to the same identities for the functions u that are the limit functions for $\{u^m\}$ in the weak topology of $\overset{\circ}{W}{}_2^{1,1}(Q_T)$ and $\overset{\circ}{W}{}_2^{1,0}(Q_T)$ respectively. These limit functions u will also be the desired generalized solutions.

The uniqueness of the generalized solutions of these two problems is established on the basis of the fact that these problems are the adjoints of each other. Namely, suppose u is a solution of the first boundary value problem with f and ψ_0 equal to zero, so that identity (13.14) with $f=0$ is valid for u. As η in it we take a generalized solution $\overset{\circ}{W}{}_2^{1,0}(Q_T)$ of the problem

$$-\frac{\partial(ba\eta)}{\partial t}-\frac{\partial}{\partial x_j}(ba_{ij}\eta_{x_i}+B_j\eta)+ba_i\eta_{x_i}+A\eta=F(x,t), \tag{13.20}$$

$$\eta|_{S_T}=0,\quad \eta|_{t=T}=0,$$

where $F(x, t)$ is an arbitrary smooth function in $\overline{Q}_T$. The existence of at least one such solution is proved above. From a comparison of the integral identity (13.14) for u with $f\equiv 0$ and η equal to a solution of problem (13.20), and the identity (13.16) for η, as solutions of problem (13.20), in which the solution u is taken as an arbitrary function, it is seen that

$$\int_{Q_T} uF\,dx\,dt=0,$$

and thus, since F is sufficiently arbitrary, we have $u\equiv 0$. The uniqueness of a generalized solution of the second problem is proved analogously. For the sake of simplicity in the calculations we assume in this connection that in each of the domains $Q^{(k)}$ of all the coefficients of $\mathfrak{L}^{(k)}$ and the function $b(x, t)$ together with the derivatives of it encountered in our arguments are bounded.

The classicalness of the obtained generalized solutions is established in principle in the same way as the classicalness of the generalized solutions of problem (13.1)–(13.5).

Let us sum up what has been proved in the form of a theorem.

Theorem 13.2. *Problem* (13.10), (13.2)–(13.4), (13.11) *has a unique solution from* $W_2^{1,1}(Q_T)$ *while problem* (13.10), (13.2)–(13.4), (13.12) *has a unique solution from* $W_2^{1,0}(Q_T)$ *for any* f *from* $L_2(Q_T)$, *with* ψ_0 *from* $L_2(\Omega)$ *in the first problem and* ψ_0 *from* $\overset{\circ}{W}{}_2^1(\Omega)$ *in the second. Concerning the functions* a_{ij}, a_{ijt}, a_i, a_{it}, b_i, b_{it}, a, α, α_t, b, b_x, b_t *and* b_{xt} *it is assumed in this connection that they are bounded by an absolute quantity in each of the domains* $\overline{Q}^{(k)}$, *that* α *and* b *are strictly positive, and that the* a_{ij} *are subject to inequalities* (1.2).

§14. FUNCTIONAL METHODS FOR THE SOLUTION OF BOUNDARY VALUE PROBLEMS

In this and the following sections of Chapter III we will set forth different methods for proving the solvability of boundary value problems for linear equations. All of them are applicable to the Cauchy problem. The majority of them can be used for the actual determination of solutions (more precisely, for the determination of the approximations u^m that converge to the desired solution for $m \to \infty$). In §4 of the present chapter Galerkin's method was investigated. In Chapter IV two other methods will be given: the classical method of the theory of potentials and one of the latest methods, which is based on exact a priori estimates and the construction of the regularizor, i.e. the operator inverting the linear principal part of an equation with frozen coefficients. In the present section we describe three methods, which are called functional in view of the fact that they make use of only the most general characteristics of the operators corresponding to the problem, and not their particular functional structure. These methods were used to prove the unique solvability of the abstract Cauchy problem

$$\frac{du(t)}{dt} + S(t)\,u(t) = f(t), \quad u\big|_{t=0} = u_0, \tag{14.1}$$

in which u_0 is a given element of a Hilbert space $\mathcal{H}$, $u(t)$ and $f(t)$ are unknown and known functions of t with values in $\mathcal{H}$, while $S(t)$ is a family of semibounded linear operators on $\mathcal{H}$ depending on the parameter t. An effort was made to formulate the conditions which were imposed in this connection on $S(t)$, $f(t)$ and u_0, in such a way that they would be fulfilled in the basic boundary value problems for linear parabolic equations. Thus problem (1.9) can be interpreted, for example, as problem (14.1) with $\mathcal{H} = L_2(\Omega)$ and with $S(t)$ defined by the elliptic operator

$$\mathcal{M}(t)\,u \equiv -\frac{\partial}{\partial x_i}\left(a_{ij}(x, t)\,u_{x_j} + a_i(x, t)\,u\right) + b_i(x, t)\,u_{x_i} + a(x, t)\,u$$

on the set $\mathfrak{D}(S(t)) = \overset{\circ}{W}{}_2^2(\Omega)$. The condition $u|_{S_T} = 0$ is "absorbed" into the designation of the domain of definition of $S(t)$. The solution $u(x, t)$ and the free term $F(x, t) = \partial f_i(x, t)/\partial x_i - f(x, t)$ are regarded as functions of t with values in $L_2(\Omega)$, while $\psi_0(x)$ is regarded as a known element of $L_2(\Omega)$ defining the initial value of $u(x, t)$.

The first two functional methods for (1.9) and for (14.1) were given by M. I. Višik [124b] and O. A. Ladyženskaja [65d] (for detailed publications and subsequent generalizations of them see [65e,f; 124c] and the survey [124f]). After that several other methods were proposed ([65f; 73; 114a; 124f] et al.; the book [73] is mainly devoted to abstract methods for the solution of the Cauchy problem for equations of form (14.1)). The results obtained with these methods for parabolic equations differ from each other both in the assumptions on the known functions and in those properties of the solutions that have been established with these methods. Some of them give more exact information on the solution, while the others are somewhat simpler, with the alternatives involving less (or more) assumptions on the data of the problem. We will not present and exactly formulate everything that has been obtained with these methods in connection with the problems of interest to us in their general form. Instead we will set forth the basic ideas of these methods (in the present and following sections) with the example of an elementary problem

$$\mathscr{L}_0 u \equiv u_t - \Delta u = f(x, t), \quad u|_{\Gamma_T} = 0, \tag{14.2}$$

or, in those methods in which problem (14.2) is considered already solved and is used as a starting point, with the example of a somewhat more general problem

$$\mathscr{L}_1 u \equiv u_t - \frac{\partial}{\partial x_i}\left(a_{ij}(x, t)\, u_{x_j}\right) + \lambda u = f(x, t), \quad u|_{\Gamma_T} = 0. \tag{14.3}$$

We note only the following: we chose Galerkin's method as the basic method in the present chapter because it does not require that the space in which the solution is found (the space $V_2^{1,0}(Q_T)$) be either a Hilbert or a reflexive space. And the choice of the nonreflexive Banach space $V_2^{1,0}(Q_T)$ was determined by the desire to maximally extend the class of equations (i.e. to weaken the assumptions on their coefficients), but in such a way that the existence and uniqueness theorems are still retained for it. The result established by us in §4 does not follow directly from the other methods.

The method proposed by O. A. Ladyženskaja [65d–f] in connection with problem (14.2) goes as follows. In the Hilbert space $\mathcal{H} = L_2(Q_T)$ one considers the operator A associating with each function $u(x, t)$ from $W_2^{2,1}(Q_T)$ and equal to zero on Γ_T the function $Au = \mathfrak{L}_0 u(x, t)$. We denote its domain of definition by $\mathfrak{D}(A)$, and its range by $\mathfrak{R}(A)$.

From the relation

$$\int_{Q_T} (Au)^2\,dx\,dt = \int_{Q_T} [u_t^2 + (\Delta u)^2]\,dx\,dt + \int_\Omega u_x^2(x,\ T)\,dx$$

$$\geqslant c\,(\|u\|^{(2)}_{Q_T})^2,\quad c = \mathrm{const} > 0, \tag{14.4}$$

it is seen that the closing of the operator A (that it is closable is easily verified directly or is established as a consequence of the fact that the domain of definition of its adjoint is dense in $L_2(Q_T)$) reduces to the closing of its range, i.e.

$$\mathscr{R}(\bar{A}) = \overline{\mathscr{R}(A)}.$$

Let us show that $\overline{\mathfrak{R}(A)}$ coincides with all of $L_2(Q_T)$. Indeed, there would otherwise exist an element $v(x, t) \in L_2(Q_T)$ orthogonal to all of $\mathfrak{R}(A)$, i.e., such that

$$\int_{Q_T} v\mathscr{L}_0 u\,dx\,dt = 0 \tag{14.5}$$

for all $u \in \mathfrak{D}(A)$.

Let us construct with respect to $v(x, t)$ a function $w(x, t)$ as a solution of the problem

$$-\Delta w(x,\ t) = \int_0^t v(x,\ \tau)\,d\tau,\quad w|_{S_T} = 0. \tag{14.6}$$

It has been proved ([65b]; see also [65q], §9 of Chapter III) that this problem is uniquely solvable in $W_2^2(\Omega)$ for any right side from $L_2(\Omega)$. In view of this $w = (-\Delta)^{-1}\int_0^t v(x, \tau)\,d\tau$ belongs to $\mathfrak{D}(A)$, and hence in (14.5) one can substitute w for u. We write the result of this substitution in the form

$$\int_{Q_T} v\left[(-\Delta)^{-1}v + \int_0^t v\,d\tau\right]dx\,dt = 0,$$

from which

$$\left\|(-\Delta)^{-\frac{1}{2}}v\right\|^2_{2,\,Q_T} + \frac{1}{2}\left\|\int_0^t v\,d\tau\right\|^2_{2,\,\Omega}\Bigg|_{t=0}^{t=T} = 0$$

and consequently $v \equiv 0$. Thus it is proved that

$$\mathscr{R}(\bar{A}) = L_2(Q_T).$$

But this also means that the equation

$$\bar{A}u = f \tag{14.7}$$

has a solution u from $\mathfrak{D}(\bar{A})$ for any f from $L_2(Q_T)$.

From relation (14.4) it is seen that the elements $u(x, t)$ from $\mathfrak{D}(\bar{A})$ have generalized derivatives u_t, u_x and $u_{x_ix_j}$ from $L_2(Q_T)$. It is assumed here that the boundary S is a surface of class O^2.

The operator $\bar{A}$ acting on them is calculated as an ordinary differential operator $\mathfrak{L}_0 u$, in which all of the derivatives are regarded as generalized, while the equation $\mathfrak{L}_0 u = f$ is satisfied almost everywhere in Q_T. And the boundary and initial conditions on Γ_T are fulfilled "in the mean".

Thus it is proved that for any $f(x, t)$ from $L_2(Q_T)$ problem (14.2) has a solution u belonging to $W_2^{2,1}(Q_T)$ (if $S \in O^2$). It is clear that this result is exact and that the solution is unique in this class of functions. Moreover, we have actually proved a significantly stronger uniqueness theorem: *uniqueness for generalized solutions of problem* (14.2) (*more precisely, its adjoint*

$$-v_t - \Delta v = \psi, \quad v|_{S_T} = 0, \quad v(x, T) = 0,$$

which goes over into problem (14.2) *if* t *is replaced by* $\tau = T - t$) *from class* $L_2(Q_T)$. Such solutions are determined in the following manner:

By a *generalized solution of problem* (14.2) *from the class* $L_2(Q_T)$ we mean a function $u(x, t)$ from $L_2(Q_T)$ satisfying the identity

$$\int_{Q_T} u(-\eta_t - \Delta\eta)\, dx\, dt = \int_{Q_T} f\eta\, dx\, dt \tag{14.8}$$

for any η from $W_2^{2,1}(Q_T)$ that is equal to zero both on S_T and for $t = T$.

The method proposed by M. I. Višik [124b,f] is based on the following easily proved proposition of functional analysis.

Lemma 14.1. *Suppose A is an operator in a Hilbert space $\mathfrak{H}$ with a dense domain of definition $\mathfrak{D}(A)$ and with a bounded inverse on its range $\mathfrak{R}(A)$. Then the range $\mathfrak{R}(A^*)$ of its adjoint coincides with all of $\mathfrak{H}$.*

Indeed, let us take an arbitrary element f from $\mathfrak{H}$. The equation $A^*x = f$, by virtue of the definition of the adjoint operator A^*, is equivalent to the identity $(x, Au) = (f, u)$ for any u from $\mathfrak{D}(A)$ or, what is the same thing, to the identity $(x, v) = (f, A^{-1}v)$ for any v from $\mathfrak{R}(A)$. But the latter is solvable for any f from $\mathfrak{H}$, since $(f, A^{-1}v)$ is a linear functional over the v's (its linearity is obvious, while its boundedness follows from the boundedness of the operator A^{-1} on $\mathfrak{R}(A)$), and hence by virtue of the theorem of Riesz on the form of a linear functional there exists at least one element x from $\overline{\mathfrak{R}(A)}$ for which $(f, A^{-1}v) = (x, v)$ for all v from $\mathfrak{R}(A)$. This element also gives one of the solutions of the equation $A^*x = f$. The lemma is proved.

It can be applied in a different way to prove the solvability of problem (14.2). Višik takes as the basic space $\mathfrak{H}$ the function space $\overset{\circ}{W}{}_2^{1,0}(Q_T)$ with scalar product $(u, v)_{\mathcal{H}} = \int_{Q_T}(uv + v_x v_x)\,dx\,dt$. Equation (14.2) and the initial condition are replaced by the integral identity

$$I_0(u, \eta) \equiv \int_{Q_T} (-u\eta_t + u_x\eta_x)\,dx\,dt = \int_{Q_T} f\eta\,dx\,dt, \tag{14.9}$$

where η is an arbitrary function from the set $\mathfrak{D}$ consisting of all elements of $W_2^{1,1}(Q_T)$ that are equal to zero both on S_T and for $t = T$. The function u from $\overset{\circ}{W}{}_2^{1,0}(Q_T)$ satisfying (14.9) for all such η's is called in accordance with §3 a generalized solution of problem (14.2) from the space $W_2^{1,0}(Q_T)$.

The integral $I_0(u, \eta)$ for any fixed η from $\mathfrak{D}$ defines, as is easily seen, a linear functional over the u's in $\mathfrak{H}$, which by virtue of the theorem of Riesz can be written in the form of a scalar product of u by some element $A\eta$ from $\mathfrak{H}$, i.e.

$$I_0(u, \eta) = (u, A\eta)_{\mathcal{H}}, \quad u \in \mathcal{H}. \tag{14.10}$$

The operator A defined by this identity has as its domain of definition $\mathfrak{D}(A)$ the set $\mathfrak{D}$. It satisfies the conditions of Lemma 14.1 since, substituting η for u in (14.10), we obtain

$$(\eta, A\eta)_{\mathcal{H}} = I_0(\eta, \eta) = \int_{Q_T} (-\eta\eta_t + \eta_x^2)\,dx\,dt$$
$$= \frac{1}{2}\|\eta(x, 0)\|_{2,\Omega}^2 + \|\eta_x\|_{2,Q_T}^2 \geq c\|\eta\|_{\mathcal{H}}^2, \quad c = \text{const} > 0,$$

from which it follows that

$$\|A^{-1}\|_{\mathcal{H}} \leq \frac{1}{c}.$$

The right side of identity (14.9) can also be represented in the form of a scalar product in $\mathfrak{H}$:

$$\int_{Q^T} f\eta\,dx\,dt = (F, \eta)_{\mathcal{H}},$$

where F is uniquely determined by f, since for any f from $L_2(Q_T)$ the integral $\int_{Q_T} f\eta dx\, dt$ defines a linear functional over the η's in H. Hence (14.9) takes the form

$$(u, A\eta)_{\mathcal{H}} = (F, \eta)_{\mathcal{H}}. \tag{14.11}$$

If we now apply Lemma 14.1 to the operator A, we will find that the equation $A^* u = F$ or, what is the same thing, identity (14.11), has a solution u from $\mathcal{H}$ for any F from $\mathcal{H}$. By the same token we have proved the existence of a generalized solution from $W_2^{1,0}(Q_T)$ of problem (14.2) for any f from $\overset{\circ}{L}_2(Q_T)$.

As $\mathcal{H}$ one could take the subspace of the Hilbert space $W_2^{1,1}(Q_T)$ consisting of functions that are equal to zero on Γ_T. In this connection it is necessary to replace equation (14.2) by the identity

$$\int_{Q_T} (u_t\zeta_t + u_x\zeta_{xt})\, dx\, dt = \int_{Q_t} f\zeta_t\, dx\, dt, \tag{14.12}$$

in which $\zeta(x, t)$ runs through the whole set $\mathfrak{D}$ of elements of $\mathcal{H}$ having the derivatives $\zeta_{x_it} \in L_2(Q_T)$, and then to transform (14.12) into an equality for scalar products in $\mathcal{H}$:

$$(u, A\zeta)_{\mathcal{H}} = (F, \zeta)_{\mathcal{H}}. \tag{14.13}$$

The solvability of (14.13) is deduced on the basis of Lemma 14.1. This reasoning yields the solvability of problem (14.2) in the space $W_2^{1,1}(Q_T)$ for any f from $L_2(Q_T)$.

Lions establishes the solvability of boundary value problems on the basis of a theorem proved by him on abstract bilinear forms (Theorem 1.1 of Chapter III in his book [73]). In connection with problem (1.9) his approach essentially coincides with the method just mentioned for proving solvability in the space $W_2^{1,0}(Q_T)$ on the basis of Lemma 14.1.

We note once more that all of the functional methods presented here apply to problem (1.9) (and to several other boundary value problems) in its general form, with the assumptions on the boundedness of the coefficients of minor terms (made in the references given above) being replaceable by their membership in certain $L_{q,\infty}(Q_T)$ (the q's are different for different coefficients and in different methods).

In between the first of the stated methods and the method of continuity in a parameter presented in the next section, there is the following method, which is based on the same a priori estimates as the first two [144a]. Suppose $S \in O_2$ and suppose it is known that problem (14.2) is uniquely solvable in $\overset{\circ}{W}{}_2^{2,1}(Q_T)$ for any f from $L_2(Q_T)$ (we recall that $\overset{\circ}{W}{}_2^{2,1}(Q_T)$ is the subspace of the space $W_2^{2,1}(Q_T)$ consisting of the elements of $W_2^{2,1}(Q_T)$ that vanish on S_T). Let us show that the same is then true for problem (14.3) as well. For this purpose we make use of a generalization of (14.4), consisting in the fact that for any function v from $\overset{\circ}{W}{}_2^{2,1}(Q_T)$

$$\int_{Q_T} \mathscr{L}_1 v \mathscr{L}_0 v\, dx\, dt \geqslant c_1 \left(\|v\|_{2, Q_T}^2 \right)^2, \quad c_1 = \text{const} > 0, \tag{14.14}$$

as long as the λ in (14.3) is a sufficiently large positive number (cf. §6). It can be assumed without any loss in generality that the λ in $\mathfrak{L}_1$ is such that (14.14) is also valid.

In addition, $\mathfrak{L}_1$ satisfies a relation of type (14.4), namely,

$$\int_{Q_T} (\mathscr{L}_1 v)^2\, dx\, dt \geqslant c_2 \left(\|v\|_{2, Q_T}^{(2)} \right)^2, \quad c_2 = \text{const} > 0 \tag{14.15}$$

(see (6.10)) for any v from $\overset{\circ}{W}{}_2^{2,1}(Q_T)$. From this it follows that the operator $\mathfrak{L}_1$ is closed on $\overset{\circ}{W}{}_2^{2,1}(Q_T)$, has a bounded inverse on its range $\mathfrak{R}(\mathfrak{L}_1)$, and $\mathfrak{R}(\mathfrak{L}_1)$ is a closed subspace of $L_2(Q_T)$. Let us show that $\mathfrak{R}(\mathfrak{L}_1)$ coincides with all of $L_2(Q_T)$. Suppose $v(x, t)$ belongs to $L_2(Q_T)$ and is orthogonal to $\mathfrak{R}(\mathfrak{L}_1)$, i.e.,

$$\int_{Q_T} \mathscr{L}_1 u v\, dx\, dt = 0 \tag{14.16}$$

for all $u \in \overset{\circ}{W}{}_2^{2,1}(Q_T)$.

We construct with respect to it the element $w = \mathfrak{L}_0^{-1} v \in \overset{\circ}{W}{}_2^{2,1}(Q_T)$ and put $u = w$ in (14.16). This gives

$$\int_{Q_T} \mathscr{L}_1 w \mathscr{L}_0 w\, dx\, dt = 0.$$

But by virtue of (14.14) it follows that $w = 0$, and hence $v = \mathfrak{L}_0 w = 0$. Thus $\mathfrak{R}(\mathfrak{L}_1) = L_2(Q_T)$, i.e. problem (14.3) is uniquely solvable in $\overset{\circ}{W}{}_2^{2,1}(Q_T)$ for any f from $L_2(Q_T)$.

§15. THE METHOD OF CONTINUITY IN A PARAMETER

Let us prove the solvability of problem (14.3), assuming that problem (14.2) has already been investigated. For this purpose we consider the family of problems

$$\mathscr{L}_\tau u = u_t - \frac{\partial}{\partial x_i}\left[\tau a_{ij}(x, t)\, u_{x_j} + (1-\tau)\, u_{x_i}\right] + \lambda\tau u = f,$$
$$u|_{\Gamma_T} = 0, \quad 0 \leqslant \tau \leqslant 1. \tag{15.1}$$

Suppose $S \in O^2$, $f \in L_2(Q_T)$, and the $a_{ij}(x, t)$ satisfy, besides the condition of ellipticity

$$\nu\xi^2 \leqslant a_{ij}\xi_i\xi_j \leqslant \mu\xi^2, \tag{15.2}$$

the conditions

$$\left|\frac{\partial a_{ij}}{\partial x_k}\right| \leqslant \mu_1. \tag{15.3}$$

Then the problems (15.1) are uniquely solvable in the space $W_2^{2,1}(Q_T)$.

The proof (it is taken from [65i]) is based on the exact estimate (6.10) of the present chapter, which is uniform for the whole family of operators $\mathfrak{L}_\tau$ and which for the solutions u^τ of problems (15.1) has the form

$$\|u^\tau\|_{2,Q_T}^{(2)} \leqslant c\|f\|_{2,Q_T}, \tag{15.4}$$

and on the fact that for the operator $\mathfrak{L}_0 = \partial/\partial t - \Delta$ the solvability of problem (15.1) in the class $W_2^{2,1}(Q_T)$ is considered established for at least a dense set of functions f in $L_2(Q_T)$. The latter follows, for example, from the results in §6 of the present chapter, the results of Chapter IV, from the first functional method presented in §14, and from Foruier's method, which will be described in §17. But if problem (15.1) for $\tau = 0$ is solvable in $W_2^{2,1}(Q_T)$ for a dense set of f's from $L_2(Q_T)$, then by virtue of (15.4) for $\tau = 0$ it is uniquely solvable in $W_2^{2,1}(Q_T)$ for any f from $L_2(Q_T)$ [because the convergence of f_m to f in $L_2(Q_T)$ implies according to (15.4) the convergence of the corresponding solutions u_m in $W_2^{2,1}(Q_T)$ to some function u from $W_2^{2,1}(Q_T)$, which will obviously be a solution of the problem corresponding to f]. Thus the operator $\mathfrak{L}_0$ establishes a one-to-one correspondence between the subspace W of the space $W_2^{2,1}(Q_T)$, consisting of the elements of $W_2^{2,1}(Q_T)$ that are equal to zero on Γ_T, and all of $L_2(Q_T)$. Let us show that this is true of all the operators $\mathfrak{L}_\tau$. For this purpose we write problem (15.1) in the form of an operator equation

$$\mathscr{L}_0 u + \tau(\mathscr{L}_1 - \mathscr{L}_0)\, u = f, \quad f \in L_2(Q_T), \quad u \in W, \tag{15.5}$$

which, by virtue of the just-proved property of $\mathfrak{L}_0$, is equivalent to the equation

$$u + \tau\mathscr{L}_0^{-1}(\mathscr{L}_1 - \mathscr{L}_0)\, u = \mathscr{L}_0^{-1} f \tag{15.5'}$$

in the space W. The operator $\mathfrak{L}_0^{-1}(\mathfrak{L}_1 - \mathfrak{L}_0)$ is bounded in W by virtue of assumptions (15.2), (15.3) and inequality (15.4) for $\tau = 0$. Therefore equation (15.5) is uniquely solvable (at least by the method of successive approximations) for $\tau < 1/\|\mathfrak{L}_0^{-1}(\mathfrak{L}_1 - \mathfrak{L}_0)\|_W \equiv a$, and by the same token the operators $\mathfrak{L}_\tau$ for $\tau > 0$ establish a one-to-one correspondence between W and $L_2(Q_T)$. In view of this, equation (15.5) is equivalent to the equation

$$u + (\tau - \tau_1)\mathscr{L}_{\tau_1}^{-1}(\mathscr{L}_1 - \mathscr{L}_0)u = \mathscr{L}_{\tau_1}^{-1}f \tag{15.6}$$

in W for any $\tau_1 < a$. The norms of the operators $\mathfrak{L}_\tau^{-1}(\mathfrak{L}_1 - \mathfrak{L}_0)$ in W for all τ from $[0, 1]$ for which $\mathfrak{L}_\tau^{-1}$ exists do not exceed a certain number a_1 that is independent of τ (which follows from (15.2)–(15.4)), and hence equations (15.6) are solvable for all τ from $[\tau_1, \tau_1 + 1/a_1)$. By repeating this argument, we come after a finite number of steps to the operator $\mathfrak{L}_1$ and we establish the unique solvability in W of problem (15.1) for $\tau = 1$ (or, what is the same thing, problem (14.3)) for any f from $L_2(Q_T)$.

Continuity in a parameter can be realized in another way, not with respect to the operators $\mathfrak{L}$ but with respect to bilinear forms corresponding to them (see [65f]).

§16. ROTHE'S METHOD AND THE METHOD OF FINITE DIFFERENCES

Rothe's method and the method of finite differences may be used for solving the Cauchy problem and the initial-boundary-value problems for equations of general form (1.1), where the domain Q of variation of (x, t) can be noncylindrical. The first of these methods was proposed by the German mathematician Rothe in the year 1930 [101] and was demonstrated by him in connection with a one-dimensional parabolic equation of second order. This demonstration was given by him, naturally, within the limits of classical solvability under the condition of sufficient (but not overly excessive) smoothness of the coefficients. Subsequent results on the solvability of boundary value problems for many-dimensional elliptic equations permitted one to carry over without difficulty the results of Rothe to the case of parabolic equations (1.1) with any $n > 1$ (see [65h], in which this is done for a more complicated object, namely a quasi-linear equation with nonlinearity in u, and also [16a] and [123a]). At the basis of the proof of classical solvability by Rothe's method there lies (as in many other methods) the maximum principle, which is involved in the approximation of parabolic equations proposed by

Rothe. We will describe Rothe's method by using as an example the problem

$$u_t - \Delta u = f(x, t), \quad u|_{S_T} = 0, \quad u|_{t=0} = \psi_0(x) \tag{16.1}$$

in the cylindrical domain $Q_T = \Omega \times [0, T]$. In this section we will denote differentiation with respect to x_i and t by $\partial/\partial x_i$ and $\partial/\partial t$, and use the symbols $(\cdot)_{x_i}$ and $(\cdot)_t$ for denoting divided differences. We divide the (x, t) space by the planes $t = kh$, $k = 0, \pm 1, \cdots$, into layers and assume for the sake of brevity in notation that $T/h = m$ is an integer. Suppose Ω_k is the intersection of the plane $t = kh$ with Q_T, while S_k is its boundary. On all of the Ω_k, $k = 1, 2, \cdots, m$, we consider the problems

$$u_{\bar{t}}(x, kh) - \Delta u(x, kh) = f(x, k), \quad u|_{S_k} = 0, \tag{16.2}$$

assuming

$$u(x, 0) = \psi_0(x). \tag{16.3}$$

Here

$$u_{\bar{t}}(x, kh) = \frac{1}{h}[u(x, kh) - u(x, kh - h)],$$

while

$$f(x, k) = f(x, kh).$$

Problems (16.2) are solved consecutively from the layer Ω_k to the layer Ω_{k+1}, beginning with $k = 1$, where each time (even in the case of the general equation (1.1)) there exists a unique solution of the corresponding boundary value problem for an elliptic equation owing to the presence in the equation of the term $(1/h)u(x, kh)$ with a large positive coefficient h^{-1}. It is proved that in the limit as $h \to 0$ the functions $u(x, kh)$ [or, for example, their linear interpolations $u^h(x, t)$, $(x, t) \in Q_T$, coinciding with $u(x, kh)$ for $t = kh$ and linearly depending on t within the layers $kh \leq t \leq (k+1)h$] give a solution $u(x, t)$ of problem (16.1). For this purpose estimates not depending on h are established for various norms of $u(z, kh)$. Some of them are based on the maximum principle inherent in the difference-differential equation (16.2). In its "pure" form this principle holds, as always, for the homogeneous equation when $f = 0$. In this case the maximal and minimal values of the functions $u(x, kh)$, $k = 0, 1, \cdots, m$, occur either on the lower base Ω_0 or on the S_k. Indeed, if a greatest value were to occur at some interior point $x_0 \in \Omega_k$ for $k \geq 1$, we would have at this point $u_{x_i}(x_0, kh) = 0$, $-\Delta u(x_0, kh) \geq 0$ and $u_{\bar{t}}(x_0, kh) \geq 0$, and hence by virtue of equation (16.2) $u_{\bar{t}}(x_0, kh) = 0$, i.e. $u(x_0, kh) = u(x_0, kh - h)$. (Thus a strict maximum at (x_0, kh) is impossible.)

Now taking the point $(x_0, kh - h)$ and carrying out the same arguments as at the point (x_0, t_0), we get $u(x_0, kh - h) = u(x_0, kh - 2h)$ and so on up to $u(x_0, 0)$. It is thereby proved that on the S_k, $k = 0, 1, \cdots, m$, or on Ω_0 there exists a point at which the functions $u(x, kh)$, $k = 0, 1, \cdots, m$, take their greatest value. An analogous assertion also holds for their least value. By complicating the reasoning somewhat it is not difficult to show that if $f = 0$, then the greatest and least values can not be taken at interior points if $u \not\equiv \text{const}$ [this is done in the same way as for the usual differential equation of the form

$$\frac{\partial u}{\partial t} - a_{ij}(x, t) \frac{\partial^2 u}{\partial x_i \partial x_j} + a_i \frac{\partial u}{\partial x_i} = 0;$$

see in this connection [90a]].

The maximum principle gives an estimate of

$$\max_{\substack{x \in \Omega \\ k=0, 1, \ldots, m}} |u(x, kh)|$$

in terms of the known quantities in problem (16.1). The case $f \not\equiv 0$ is considered in almost the same way, with a preliminary replacement of the function $u(x, kh)$ by $v(x, kh)\, e^{\lambda kh}$, $\lambda > 0$. An estimate of $\max |\partial u(x, kh)/\partial x|$, is more complicated, but it too is based on the maximum principle. For equations with coefficients not depending on x this is comparatively easy to perceive, since the derivatives $\partial u/\partial x_j$ satisfy the same system of equations as u, namely,

$$\left(\frac{\partial u(x, kh)}{\partial x_j}\right)_{\bar{t}} - \Delta\left(\frac{\partial u(x, kh)}{\partial x_j}\right) = \frac{\partial f(x, kh)}{\partial x_j}, \qquad (16.4)$$
$$k = 1, 2, \ldots, m.$$

In view of this the quantities

$$\max_{\substack{x \in \Omega \\ k=0, 1, \ldots, m}} \left| \frac{\partial u(x, kh)}{\partial x_j} \right|$$

are estimated in terms of

$$\max_{\substack{x \in \Omega \\ k=1, \ldots, m}} \left| \frac{\partial f(x, kh)}{\partial x_j} \right|$$

and upper bounds on the S_k, $k = 0, 1, \cdots, m$, and Ω_0. The upper bounds on Ω_0 are known, while those on the S_k are estimated in the same way as for a differential equation, with the use of barriers (see the proof of Lemma 3.1 in Chapter IV; it simplifies in an obvious manner for the linear case). In the case of equations with coefficients depending on x one cannot compose equations from which it would be possible to estimate the magnitude of each of the derivatives $\partial u/\partial x_j$ separately. But one can do this directly for all of the derivatives $\partial u/\partial x_j$, $j = 1, \cdots, n$, taking the functions

$$v(x, kh) = \left[\mu u^2(x, kh) + \left(\frac{\partial u(x, kh)}{\partial x}\right)^2\right] e^{-2\lambda kh},$$

$$k = 0, 1, \ldots, m,$$

with sufficiently large numbers μ and λ and composing equations of type (16.2) for them. This technique, proposed by S. N. Bernšteĭn (see [8a]) for estimating $\max|u_x|$ for the solutions u of linear parabolic equations of second order, is also suitable for the difference-differential equations (16.2) (see §17 of Chapter IV). Higher-ordered derivatives of $u(x, kh)$ with respect to x are estimated analogously. Thus, for derivatives of second order one must consider the functions

$$w(x, kh) = \left[\mu \left(\frac{\partial u(x, kh)}{\partial x}\right)^2 + \left(\frac{\partial^2 u(x, kh)}{\partial^2 x}\right)^2\right] e^{-2\lambda kh}$$

with sufficiently large numbers μ and λ. On the other hand, the smoothness with respect to t is established directly from equation (16.2). We will not explicitly carry out all of these estimates and the passage to the limit $h \to 0$; instead we refer the reader to the paper [65h] see also [65c]). We note only that since in this paper an equation depends on u nonlinearly, in order to estimate $\max|\partial u/\partial x|$ one must introduce a more complicated function v than the one mentioned above. But an estimate of $\max|\partial^2 u/\partial x^2|$ is carried out with Bernšteĭn's technique in the way just described. Here the difference of equation (16.2) from a differential equation appears to be somewhat greater than in the above estimate of $\max|u(x, kh)|$.

The final results on the classical solvability of boundary value problems in an investigation by this method turn out to be somewhat cruder than those already established and presented in Chapter IV.

Rothe's method makes it possible to very simply, and under comparatively broad assumptions on the data in boundary value problems, prove the existence of their generalized solutions (this is done for quasi-linear equations in [65h]). However, the process of making these assumptions as broad as possible (as done in §4 of the present chapter) requires additional arguments. We will show how to prove the presence of generalized solutions with Rothe's method by using as an example the problem

$$\frac{\partial u}{\partial t} - \mathcal{M}(t)\, u \equiv \frac{\partial u}{\partial t} - \frac{\partial}{\partial x_i}\left(a_{ij}(x, t)\frac{\partial u}{\partial x_j}\right) = f(x, t), \tag{16.5}$$

$$u|_{S_T} = 0, \quad u|_{t=0} = \psi_0(x),$$

for $\psi_0 \in L_2(\Omega)$, $f \in L_2(Q_T)$ and the $a_{ij}(x, t)$ satisfying the condition

$$\nu\xi^2 \leqslant a_{ij}(x, t)\,\xi_i\xi_j \leqslant \mu\xi^2, \quad \nu = \mathrm{const} > 0. \tag{16.6}$$

We replace equation (16.5) by the difference-differential equation

$$u_{\bar t}(k) - \mathcal{M}(kh)\, u(k) \equiv u_{\bar t}(x, kh) - \frac{\partial}{\partial x_i}\left(a^h_{ij}(x, k)\frac{\partial u}{\partial x_j}\right) = f^h(x, k), \tag{16.7}$$

in which

$$f^h(x, k) = \frac{1}{h}\int\limits_{kh-h}^{kh} f(x, \tau)\, d\tau,$$

and

$$a^h_{ij}(x, k) = \frac{1}{h}\int\limits_{kh-h}^{kh} a_{ij}(x, \tau)\, d\tau.$$

To it we add the boundary and initial conditions

$$u(x, 0) = \psi_0(x) \text{ and } u(x, kh)|_{S_T} = 0. \tag{16.8}$$

Since $f^h(x, k) \in L_2(\Omega)$, on each layer Ω_k problem (16.7), (16.8) is uniquely solvable in $\overset{\circ}{W}{}^1_2(\Omega)$. Let us obtain for its solutions $u(x, kh)$ an energy-type estimate not depending on h. For this purpose we multiply (16.7) by $2hu(x, kh)$, sum the obtained equality over k from 1 to some $k_0 \leq m$ and integrate the resulting equality over Ω. After an integration by parts in the second term of the left side, this brings us to the relation

$$2h\sum_{k=1}^{k_0}\int_\Omega u_{\bar t}(x, kh)\, u(x, kh)\, dx + 2h\sum_{k=1}^{k_0}\int_\Omega a^h_{ij}(x, k)\, u_{x_i}(x, kh)\, u_{x_j}(x, kh)\, dx = 2h\sum_{k=1}^{k_0}\int_\Omega f^h(x, k)\, u(x, kh)\, dx. \tag{16.9}$$

For a transformation of the first term we use an elementary but highly useful identity, indicated in [65h] for investigating Rothe's method and an implicit difference scheme for equations of parabolic type,

$$2h \sum_{k=1}^{k_0} u_{\bar{t}}(k)\, u(k) = u^2(k_0) - u^2(0) + h^2 \sum_{k=1}^{k_0} (u_{\bar{t}}(k))^2, \tag{16.10}$$

which is easily verified directly. In addition, we estimate the second term from below with the use of (16.6). This brings us to the inequality

$$\begin{aligned} I_h(k_0) &\equiv \int_\Omega u^2(x, k_0 h)\, dx + 2\nu h \sum_{k=1}^{k_0} \int_{\Omega_k} \left(\frac{\partial u}{\partial x}\right)^2 dx \\ &+ h^2 \sum_{k=1}^{k_0} \int_{\Omega_k} (u_{\bar{t}})^2\, dx \leqslant \int_\Omega u^2(x, 0)\, dx + 2h \sum_{k=1}^{k_0} \int_{\Omega_k} f^h u\, dx \\ &\leqslant \int_\Omega u^2(x, 0)\, dx + h \sum_{k=1}^{k_0} \int_{\Omega_k} (f^h)^2\, dx + h \sum_{k=1}^{k_0} \int_{\Omega_k} u^2\, dx, \end{aligned} \tag{16.11}$$

from which one readily obtains the desired estimate for $I_h(k_0)$:

$$I_h(k_0) \leqslant c\left(\|\psi_0\|^2_{2,\Omega} + \|f\|^2_{2, Q_{k_0 h}} \right), \quad k_0 = 0, 1, \ldots, m, \tag{16.12}$$

with the constant c not depending on h. Because of it one can pass to the limit as $h \to 0$ and prove that the limit function $u(x, t)$ for $\{u(x, kh), k = 0, 1, \cdots, m\}$ is a generalized solution of problem (16.5) from $W_2^{1,0}(Q_T)$. For this purpose we denote by $\bar{u}^h(x, t)$ the function that is equal on each layer $x \in \bar{\Omega}$, $t \in [kh - h, kh)$, $k = 1, 2, \cdots$, to the function $u(x, kh - h)$. It is easy to see that the functions $\bar{u}^h(x, t) \in W_2^{1,0}(Q_T)$, are equal to zero on S_T and by virtue of (16.12) have uniformly bounded norms in $W_2^{1,0}(Q_T)$:

$$\|\bar{u}^h\|_{2, Q_T} + \left\| \frac{\partial \bar{u}^h}{\partial x} \right\|_{2, Q_T} \leqslant c_1. \tag{16.13}$$

Because of (16.13) we can choose a sequence h_l, $l = 1, 2, \cdots$, tending to zero, for which the functions $\{\bar{u}^{h_l}\}$ and their derivatives $\{\partial \bar{u}^{h_l}/\partial x_k\}$ converge weakly in $L_2(Q_T)$ to some function $u(x, t)$ from $W_2^{1,0}(Q_T)$, that is equal to zero on S_T, and its derivatives $\partial u/\partial x_k$ respectively. Let us show that $u(x, t)$ satisfies the integral identity

$$\int_{Q_T}\left(-u\frac{\partial\eta}{\partial t}+a_{ij}\frac{\partial u}{\partial x_j}\frac{\partial\eta}{\partial x_i}\right)dx\,dt - \int_{\Omega}\psi_0\eta(x,0)\,dx = \int_{Q_T} f\eta\,dx\,dt \qquad (16.14)$$

for any $\eta(x, t)$ from $W_2^{1,1}(Q_T)$ that is equal to zero both on S_T and for $t = T$, and is thereby a generalized solution of problem (16.5) from $W_2^{1,0}(Q_T)$. It is sufficient to establish relation (16.14) only for continuously differentiable $\eta(x, t) \in \overset{\circ}{W}{}_2^{1,1}(Q_T)$ that are equal to zero in the vicinity of the layer $t = T$. For this purpose we multiply equation (16.7) by $h\eta(x, kh)$, sum the resulting equality over k from 1 to m and integrate over Ω. We transform the first term of the resulting relation by means of the formula for "summation by parts"

$$h\sum_{k=1}^{m} u_{\bar t}(k)\eta(k) = -h\sum_{k=0}^{m-1} u(k)\eta_t(k) + u(m)\eta(m) - u(0)\eta(0), \qquad (16.15)$$

which is easily verified directly. In it is used the notation $u(x, t) = (1/h)[u(x, t+h) - u(x, t)]$. And in the second term we transfer the derivative $\partial/\partial x_i$ to η. In this way we obtain from (16.7) the identity

$$-h\sum_{k=0}^{m}\int_{\Omega_k} u\eta_t\,dx - \int_{\Omega}\psi_0\eta(x,0)\,dx + h\sum_{k=1}^{m}\int_{\Omega_k} a_{ij}^h\frac{\partial u}{\partial x_j}\frac{\partial\eta}{\partial x_i}\,dx = h\sum_{k=1}^{m}\int_{\Omega_k} f^h\eta\,dx,$$

in which we take $\eta(x, t)$ to be equal to zero for $t > mh - h$. Let us rewrite it in the form

$$-\int_{Q_T}\overline{u_h(x,t)}\,\overline{\eta_t(x,t)}\,dx\,dt - \int_{\Omega}\psi_0(x)\eta(x,0)\,dx + \int_h^T\int_{\Omega}\overline{a_{ij}^h(x,t)}\,\overline{\left(\frac{\partial u^h(x,t)}{\partial x_j}\right)}\,\overline{\left(\frac{\partial\eta(x,t)}{\partial x_i}\right)}\,dx\,dt = \int_h^T\int_{\Omega}\overline{f^h(x,t)}\,\overline{\eta(x,t)}\,dx\,dt, \qquad (16.16)$$

where the line over some of the functions means that at the point (x, t) of each layer $x \in \overline{\Omega}$, $t \in [kh - h, kh]$ the function in question is equal to its value at the point $(x, kh - h)$. For a continuously differentiable function $\eta(x, t)$ the piecewise-continuous functions $\overline{\eta(x, t)}$ converge uniformly to $\eta(x, t)$, while the functions $\overline{(\eta_t(x, t))}$ uniformly approximate its derivative $\partial u/\partial t$. And for the functions $a_{ij}^h(x, t)$ and $f^h(x, t)$ one can choose from $\{h_l\}$ a subsequence $\{h_{l_p}\}$, $p = 1, 2, \cdots$, with respect to which these functions converge to $a_{ij}(x)$ and $f(x, t)$ almost everywhere and strongly in $L_2(Q_T)$ respectively,[1] with $|a_{ij}^h| \leq \mu$. In view of this and the weak convergence of the $u^h(x, t)$ and the $\overline{(\partial u^h(x, t)/\partial x_k)}$ to u and $\partial u/\partial x_k$ in (16.16), one can pass to the limit with respect to the subsequence $\{h_{l_p}\}$ and thereby show that the limit function $u(x, t)$ for the $u^h(x, t)$ actually satisfies identity (16.14). Thus the presence of generalized solutions from $W_2^{1,0}(Q_T)$ is proved. Because of the uniqueness of a generalized solution u from $W_2^{1,0}(Q_T)$ of problem (16.5) the entire aggregate of the $\overline{u^h}$ converges to u. Using (16.13), one can also deduce a stronger assertion, on the strong convergence of the $\overline{u^h}(x, t)$ to u in $L_2(Q_T)$ (see in this connection the paper [65m]), and thereby prove that the solution u obtained by us actually belongs to $V_2(Q_T)$.

Let us now proceed to the method of finite differences. For equations of parabolic type there exist explicit and implicit difference schemes (see [15; 30; 52; 65e, f, j; 97; 99; 104a, b; 105; 121] et al.) leading to the solutions of boundary value problems under specific relations between the lattice spacings Δx and Δt. We will consider here only one of the implicit schemes, which is analogous in a certain respect to Rothe's method and converges under any relation for the spacings Δx and Δt. The proof of its convergence is very similar to the proof just given for the convergence of the approximations u^h to a generalized solution of problem (16.5). Let us show this by using as an example problem (16.5), assuming that condition (16.6) is fulfilled and $\psi_0(x) \in L_2(\Omega)$, $f(x, t) \in L_2(Q_T)$. We divide the whole (x, t) space by the planes $x_i = k_i\Delta x$, $i = 1, \cdots, n$, $t = k_0\Delta t$, $k_\alpha = 0, \pm 1, \pm 2, \cdots$, into elementary cells, namely parallelepipeds with sides of length Δx in the direction of the x_k axes and of length Δt in the direction of the t axis, where for the sake of nonessential simplicity in notation we will assume that $m = T/\Delta t$ is an integer. We denote by Q_T^Δ the union of all elementary

---

[1] The last integral on the left side and the integral on the right side can be written as integrals over all of Q_T, putting $\eta(x, t) = 0$ in them for $t \in [0, h)$. This extension of the definition of η, as is easily seen, "disappears" in the limit, not having manifested itself on the other terms of the identity.

cells belonging to $\overline{Q}_T$, by Ω_k^Δ the intersection of Q_T^Δ with the plane $t = k\Delta t$, and by S_k^Δ the boundary of Ω_k^Δ. The same notation Q_k^Δ, Ω_k^Δ and S_k^Δ is retained for aggregates of lattice points (i.e., the points with coordinates $(k_1\Delta x, \cdots, k_n\Delta x, k_0\Delta t)$ belonging to the sets Q_T^Δ, Ω_k^Δ and S_k^Δ respectively). We replace equation (16.5) at all interior lattice points of Q_T^Δ by the difference equation

$$u_{\bar{t}}(x, t) - (a_{ij}^\Delta(x, t)\, u_{x_j}(x, t))_{\bar{x}_i} = f^\Delta(x, t), \tag{16.17}$$

in which the following notation is used: $u_{\bar{t}}(x, t)$, as above, is equal to $(1/\Delta t)[u(x, t) - u(x, t - \Delta t)]$,

$$u_{x_i}(x, t) = \frac{1}{\Delta x}[u(x_1, \ldots, x_i + \Delta x, \ldots, x_n, t) - u(x_1, \ldots, x_i, \ldots, x_n t)],$$

$$u_{\bar{x}_i}(x, t) = \frac{1}{\Delta x}[u(x_1, \ldots, x_i, \ldots, x_n, t) - u(x_1, \ldots, x_i - \Delta x, \ldots, x_n, t)],$$

$$f^\Delta(x, t) = \frac{1}{(\Delta x)^n \Delta t}\int_{x_1 - \Delta x}^{x_1} \cdots \int_{x_n - \Delta x}^{x_n} \int_{t - \Delta t}^{t} f(\xi_1, \ldots, \xi_n, \tau)\, d\xi\, d\tau,$$

and the $a_{ij}^\Delta(x, t)$ are defined in the same way with respect to the $a_{ij}(x, t)$. If $a_{ij}(x, t)$ or $f(x, t)$ are continuous functions of x, t, then in place of $a_{ij}^\Delta(x, t)$ or $f^\Delta(x, t)$ one can take their value at the lattice point (x, t). To relations (16.17) we add the equalities

$$u|_{S_k^\Delta} = 0, \quad k = 1, \ldots, m, \text{ and } u(x, 0) = \psi_0^\Delta(x), \tag{16.18}$$

where $\psi_0^\Delta(x) = (\Delta x)^{-n}\int_{x_1-\Delta x}^{x_1} \cdots \int_{x_n-\Delta x}^{x_n} \psi_0(\xi_1, \cdots, \xi_n)\, d\xi_1 \cdots d\xi_n$ at the interior points of Ω_0 and $\psi_0^\Delta(x) = 0$ at the points of S_0.

Relations (16.17) and (16.18) permit us to calculate an approximate solution u^Δ at all lattice points of Q_T^Δ successively layer by layer: $t = k\Delta t$, $k = 1, 2, \cdots$ $\cdots, m$. These values of u^Δ are determined uniquely (in the case of a general equation they will be unique for all Δt not exceeding some number τ_0; in our particular example they will be unique for all Δt).

Indeed, for determining u^Δ on the kth layer (on Ω_k) we have a system of linear algebraic equations containing as many equations as unknowns (their number is equal to the number of lattice points of $\Omega_k^\Delta \backslash S_k^\Delta$). The corresponding homogeneous system has the form

$$\frac{1}{\Delta t} u^{\Delta}(x, t) - \left(a^{\Delta}_{ij}(x, t) u^{\Delta}_{x_j}(x, t)\right)_{\bar{x}_i} = 0, \quad (x, t) \in \Omega^{\Delta}_k - S^{\Delta}_k. \tag{16.19}$$

We multiply this equality by $(\Delta x)^n u^{\Delta}(x, t)$ and sum over all points of Ω^{Δ}_k, taking u^{Δ} here and below to be equal to zero on S^{Δ}_k and outside of Ω^{Δ}_k. The second term is transformed by the formula for "summation by parts":

$$(\Delta x)^n \sum_{\Omega^{\Delta}_k} v_{\bar{x}_i} u = -(\Delta x)^n \sum_{\Omega^{\Delta}_k} v u_{x_i}, \tag{16.20}$$

which is valid for any functions u and v, one of which is equal to zero on S^{Δ}_k and outside of Ω^{Δ}_k. The derivation of formula (16.20) is given in [65c]; it is easily done on one's own, however, using the formula for calculating the divided difference of a product:

$$u(x)\, v(x))_{x_i} = u(x)\, v_{x_i}(x) + u_{x_i}(x)\, v(x_1, \ldots, x_i + \Delta x_i, \ldots, x_n). \tag{16.21}$$

Thus, owing to (16.20) we have from (16.19)

$$(\Delta x)^n \sum_{\Omega^{\Delta}_k} \left[\frac{1}{\Delta t}(u^{\Delta})^2 + a^{\Delta}_{ij} u^{\Delta}_{x_j} u^{\Delta}_{x_i}\right] = 0,$$

and consequently $u^{\Delta} \equiv 0$ on Ω^{Δ}_k, i.e. the homogeneous systems corresponding to the systems (16.17) have only trivial solutions on each layer Ω_k, and hence the complete nonhomogeneous systems are uniquely solvable for any right sides. In order to obtain a uniform estimate for all u^{Δ} for various Δx and Δt we multiply (16.17) by $2(\Delta x)^n \Delta t\, u^{\Delta}(x, t)$, sum over all interior points of $Q^{\Delta}_{k_0 \Delta t}$ and transform the first term by formula (16.10), while the second is transformed by formula (16.20). This gives

$$(\Delta x)^n \sum_{\Omega^{\Delta}_{k_0}} (u^{\Delta})^2 - (\Delta x)^n \sum_{\Omega^{\Delta}_0} \psi_0^2 + (\Delta t)^2 (\Delta x)^n \sum_{k=1}^{k_0} \sum_{\Omega^{\Delta}_k} (u^{\Delta}_{\bar{t}})^2$$
$$+ 2\Delta t (\Delta x)^n \sum_{\Omega^{\Delta}_k} a^{\Delta}_{ij} u^{\Delta}_{x_j} u^{\Delta}_{x_i} = 2\Delta t (\Delta x)^n \sum_{k=1}^{k_0} \sum_{\Omega^{\Delta}_k} f^{\Delta} u^{\Delta}. \tag{16.22}$$

From here it is not difficult to obtain the basic estimate

$$\max_{0 \leqslant k_0 \leqslant m} (\Delta x)^n \sum_{\Omega^{\Delta}_{k_0}} (u^{\Delta})^2 + (\Delta t)^2 (\Delta x)^n \sum_{k=1}^{m} \sum_{\Omega^{\Delta}_k} (u^{\Delta}_t)^2$$
$$+ \Delta t (\Delta x)^n \sum_{k=1}^{m} \sum_{\Omega^{\Delta}_k} (u^{\Delta}_x)^2 \leqslant c \tag{16.23}$$

with the constant c depending only on $\|f\|_{2,Q_T}$, $\|\psi_0\|_{2,\Omega}$, the numbers ν and μ and the height T, and thereby a general estimate for the whole family of solutions $\{u^\Delta\}$ for various Δx and Δt. Estimate (16.23) makes it possible to pass to the limit as $\Delta x \to 0$, $\Delta t \to 0$ and prove that the limit of $\{u^\Delta\}$ is a generalized solution of the problem. A detailed description of this with the necessary auxiliary propositions concerning the interpolation of functions on nets and embedding theorems for net functions is given in §6 of Chapter I, §2 of Chapter III and §2 of Chapter IV in the book [65c]. A rough picture of it is as follows: with respect to each of the u^Δ a continuous function $u^{\Delta'}(x, t)$ is constructed in $\overline{Q}_T^\Delta$ that coincides with u^Δ at the vertices of the net, is linear in each argument when the others are fixed, and is equal to zero for $x \overline{\in} \Omega_k^\Delta$, $t \overline{\in} [0, m\Delta t]$. The functions $u^{\Delta'}(x, t)$ are elements of $\overset{\circ}{W}{}_2^{1,1}(Q_T)$. By virtue of (16.23) they have uniformly bounded norms

$$|u^{\Delta'}|_{Q_T} \leqslant c. \tag{16.24}$$

Because of this one can choose from $\{u^{\Delta'}\}$ a subsequence $\{u^{\Delta'_k}\}$ with $\Delta_k x$ and $\Delta_k t \to 0$, which converges weakly in $W_2^{1,0}(Q_T)$ to some function $u(x, t)$ from $\overset{\circ}{W}{}_2^{1,0}(Q_T)$. This function u is a generalized solution of problem (16.5). In order to prove this we introduce a different interpolation of the functions u^Δ, giving step functions $\overline{u^{\Delta_k}}(x, t)$, that are equal on each elementary cell of the net to their values at one of the vertices of the cell (suppose for definiteness at the vertex having the smallest coordinates). It has been proved that the $\overline{u^{\Delta_k}}(x, t)$ converge weakly in $L_2(Q_T)$ to the same function $u(x, t)$ as the $u^{\Delta'_k}(x, t)$, while the $(u_{x_i}^{\Delta_k})$, $i = 1, \cdots, n$, converge weakly in $L_2(Q_T)$ to its derivatives $\partial u/\partial x_i$. We take an arbitrary smooth function $\eta(x, t)$ that is equal to zero both near S_T and for $t \geq T - \epsilon$, $\epsilon > 0$, and multiply both sides of equations (16.17) by $\Delta t(\Delta x)^n \eta(x, t)$. Then we sum the obtained equalities over all interior lattice points (x, t) of Q_T^Δ and in the first and second terms we carry out a summation by parts in accordance with formulas (16.15) and (16.20). This gives us the identity

$$\begin{aligned} -\Delta t\,(\Delta x)^n \sum_{k=0}^{m} \sum_{\Omega_k^\Delta} u^\Delta \eta_t - (\Delta x)^n \sum_{\Omega_0^\Delta} \psi_0 \eta(x, 0) \\ + \Delta t\,(\Delta x)^n \sum_{k=1}^{m} \sum_{\Omega_k^\Delta} a_{ij}^\Delta u_{x_j}^\Delta \eta_{x_i} = \Delta t\,(\Delta x)^n \sum_{k=1}^{m} \sum_{\Omega_k^\Delta} f^\Delta \eta, \end{aligned} \tag{16.25}$$

which, upon using the above-mentioned step interpolation, can be written in the form

$$\int_{Q_T} -\overline{u^\Delta}(\overline{\eta_t})\,dx\,dt - \int_\Omega \overline{\psi}_0\overline{\eta}(x,0)\,dx + \int_{\Delta t}^{T}\int_\Omega \overline{a^\Delta_{ij}}(\overline{u^\Delta_{x_j}})(\overline{\eta_{x_i}})\,dx\,dt = \int_{\Delta t}^{T}\int_\Omega \overline{f^\Delta}\overline{\eta}\,dx\,dt. \tag{16.26}$$

In this identity we can pass to the limit with respect to the above-chosen subsequence Δ_k, since the $\overline{u^{\Delta_k}}$ and $(\overline{u^{\Delta_k}_{x_i}})$ converge weakly in $L_2(Q_T)$ to u and $\partial u/\partial x_i$ respectively, $\overline{\eta}$, $(\overline{\eta_{x_i}})$ and $(\overline{\eta_t})$ converge uniformly in $\overline{Q}_T$ to η, $\partial\eta/\partial x_i$ and $\partial\eta/\partial t$, the $\overline{a^\Delta_{ij}}$ (more precisely, some subsequence of them), remaining uniformly bounded, converge almost everywhere to a_{ij}, the $\overline{\psi_0^\Delta}$ converge in $L_2(\Omega)$ to ψ_0, while the $\overline{f^\Delta}$ converge in $L_2(Q_T)$ to f. In the limit we arrive at identity (16.14) for u and thereby see that u is a generalized solution of problem (16.5). By the uniqueness theorem for generalized solutions of problem (16.5) in the class $W_2^{1,0}(Q_T)$ we can assert that the whole set of approximate solutions $\{u^\Delta\}$ has as its limit for $\Delta x \to 0$ and $\Delta t \to 0$ the function u.

The difference scheme considered by us, as well as the method of proof of its convergence, can be applied to general parabolic equations. It is well known that the method of finite differences permits one to observe an improvement in the differential properties of solutions in relation to an improvement in the differential properties of the known functions defining the problem. The dependences established in this way are not exact but are not far from it. As to how it is done, the reader can become acquainted with it in the book [65c].

§17. ON FOURIER'S METHOD

Let us consider the problem

$$\mathscr{L}u \equiv u_t - \mathscr{M}u = 0, \quad u|_{S_t} = 0, \quad u|_{t=0} = \psi_0(x), \tag{17.1}$$

where $\mathfrak{M}u = (a_{ij}(x)\,u_{x_j})_{x_i} - a(x)\,u$.

As is well known, its solution can be formally represented in the form of a Fourier series

$$u(x,t) = \sum_{k=1}^{\infty} a_k e^{-\lambda_k t}\psi_k(x), \tag{17.2}$$

in which the $a_k = (\psi_0, \psi_k) \equiv \int_\Omega \psi_0 \psi_k dx$, while the $\psi_k(x)$ and the $-\lambda_k$ are the orthonormal functions and their corresponding eigenvalues for the elliptic problem

$$\mathcal{M}\psi_k = -\lambda_k \psi_k, \quad \psi_k|_S = 0. \tag{17.3}$$

Each term of the series (17.2) is a solution of equation (17.1) satisfying the boundary condition, while the whole series is for $t = 0$ an expansion of the function $\psi_0(x)$ in the functions $\psi_k(x)$. For a justification of Fourier's method it is necessary to investigate the convergence of the series (17.2).

Depending on the character of the convergence of the series (17.2) its sum $u(x, t)$ will be a generalized solution from this or that function space. Thus, for example, if the series (17.2) converges in $L_2(Q_T)$, then $u(x, t)$ will be a generalized solution of problem (17.1) from $L_2(Q_T)$; if it converges in $W_2^{1,0}(Q_T)$, then $u(x, t)$ will be a generalized solution from $W_2^{1,0}(Q_T)$, and so on. An investigation of the convergence of the series (17.2) reduces to a study of the expansions

$$\varphi(x) = \sum_{k=1}^{\infty} a_k \psi_k(x), \quad a_k = (\varphi, \psi_k), \tag{17.4}$$

of arbitrary functions $\phi(x)$ in the eigenfunctions ψ_k.

The convergence of (17.4) in $L_2(\Omega)$ for an arbitrary function $\phi(x)$ from $L_2(\Omega)$ is a consequence of general facts from the theory of selfadjoint operators and a theorem proved by K. Friedrichs in the middle thirties on the possibility of extending an operator $\mathcal{M}$, prescribed on a set of smooth functions equal to zero on S, to a selfadjoint operator. From these same propositions it is also easily deduced that the series (17.4) converge in $W_2^1(\Omega)$ for any function $\phi(x)$ from $\overset{\circ}{W}{}_2^1(\Omega)$. These propositions are in turn a development of investigations on the convergence in $L_2(\Omega)$ of expansions in eigenfunctions of the Laplace operator and the one-dimensional differential operator $(d/dx)(p(x)\, d/dx) + q(x)$ (see [15b]).

The convergence of the series (17.4) in the spaces $W_2^m(\Omega)$ for any integer m was investigated in [65c]. In particular, the following theorem is proved in Chapter II of that work.

Theorem 17. 1. *If $\phi(x)$ is an element of $W_2^m(\Omega)$, $m \geq 0$, satisfying the boundary conditions $\phi|_S = \mathfrak{M}\phi|_S = \cdots = \mathfrak{M}^{[(m-1)/2]}\phi|_S = 0$ (for $m = 0$ all of the conditions drop out, while for $m = 1$ the first is retained: $\phi|_S = 0$), then the series (17.4) converges to it in the norm of $W_2^m(\Omega)$. In this connection the boundary S and the coefficients $a_{ij}(x)$ and $a(x)$ are assumed to possess a certain smoothness.*

It is easy to see that this result is exact and convertible (i.e., from the convergence of the series (17.4) in the space $W_2^m(\Omega)$ it follows that its sum $\phi(x)$ possesses the properties indicated in Theorem 17.1) This follows from two conjec tures proved in [65c]: 1) for the sum of any series $\Sigma_{k=1}^{\infty} a_k \psi_k(x)$ one has the estimates

$$\left\| \sum_{k=1}^{\infty} a_k \psi_k(x) \right\|_{2,\Omega}^{(m)} \leqslant c_m \left[\sum_{k=1}^{\infty} a_k^2 (|\lambda_k|^m + 1) \right]^{\frac{1}{2}} \tag{17.5}$$

and 2) for any function $\phi(x)$ satisfying the conditions of Theorem 17.1 the coefficients $a_k = (\phi, \psi_k)$ are subject to the inequalities

$$|a_k| \leqslant c'_m a_k |\lambda_k|^{-\frac{m}{2}} \text{ c } \sqrt{\sum_{k=1}^{\infty} a_k^2} \leqslant \|\varphi\|_{2,\Omega}^{(m)} \leqslant c''_m. \tag{17.6}$$

The constants c_m, c'_m, c''_m depend only on m, S and the coefficients of $\mathfrak{M}$.

The numbers λ_k, as is well known, tend to $+\infty$ for $k \to \infty$.

In view of this and the estimates (17.5), (17.6) the series (17.2) will converge for $t > 0$ in the norm of any $W_2^m(\Omega)$, $m \geq 0$ (if only S and the a_{ij}, a are sufficiently smooth) as long as $\psi_0 \in L_2(\Omega)$. And its convergence for $t = 0$ (and thereby for $t \geq 0$) is determined by the properties of the function $\psi_0(x)$ as formulated in Theorem 17.1.

Remark 17.1. We note that if one confines oneself to the convergence of the series $\Sigma_{k=1}^{\infty} a_k \psi_k(x)$ in $W_2^m(\Omega')$ for any strictly interior subdomain Ω' of the domain Ω, then it is not necessary to impose any smoothness conditions on the boundary S. Indeed, in deriving inequalities (17.6) only the formula for integration by parts is used, while inequalities (17.5) can be replaced by the inequalities

$$\left\| \sum_{k=1}^{\infty} a_k \psi_k(x) \zeta(x) \right\|_{2,\Omega}^{(m)} \leqslant c_m (|\zeta_x|) \left[\sum_{k=1}^{\infty} a_k^2 (|\lambda_k|^m + 1) \right]^{\frac{1}{2}}, \tag{17.7}$$

where $\zeta(x)$ is an arbitrary sufficiently smooth function that is equal to zero near all of the boundary S, and these inequalities are derived without making use of any properties of S at all (in connection with inequalities (17.7) see, for example, §7 of Chapter III in the book [65q]).

The paper [48] is devoted to an investigation of the convergence of the series $\Sigma_{k=1}^{\infty} a_k \psi_k(x)$ in domains with poorly behaved boundaries for the case when $\mathfrak{M}$ is the Laplace operator. The question of convergence of the series (17.2) in $L_2(\Omega)$ for the case $\mathfrak{M}u = (a_{ij}(x) u_{x_j})_{x_i} + b_i(x) u_{x_i} + a(x) u$ is investigated in the papers [72]. For an investigation of the convergence of the series (17.2) in the spaces $W_2^m(\Omega)$ it is necessary to take into account the inequalities

$$\|u(x)\|_{2,\Omega}^{(2m)} \leqslant c_m [\|\mathcal{M}^m u\|_{2,\Omega} + \|u\|_{2,\Omega}], \tag{17.7'}$$

$$\|u(x)\|_{2,\Omega}^{(2m+1)} \leqslant c_m [|(\mathcal{M}^{m+1}u, \mathcal{M}^m u)| + \|u\|_{2,\Omega}], \tag{17.8}$$

proved for nonselfadjoint $\mathfrak{M}$ and for various homogeneous boundary conditions (see Chapter II of [65c]). In connection with the series $\sum_{k=1}^\infty a_k\psi_k(x)$ they give an estimate of $\|\sum_{k=1}^\infty a_k\psi_k\|_{2,\Omega}^{(2m)}$ in terms of $\|\sum_{k=1}^\infty a_k\lambda_k^m\psi_k\|_{2,\Omega}$ and $\|\sum_{k=1}^\infty a_k\psi_k\|_{2,\Omega}$ and an estimate of $\|\sum_{k=1}^\infty a_k\psi_k\|_{2,\Omega}^{(2m+1)}$ in terms of $|(\sum_{k=1}^\infty a_k\lambda_k^{m+1}\phi_k, \sum_{k=1}^\infty a_k\lambda_k^m\psi_k)|$ and $\|\sum_{k=1}^\infty a_k\psi_k\|_{2,\Omega}$.

Fourier's method also yields a good representation for a solution of the inhomogeneous equation

$$u_t - \mathcal{M}u = f(x, t), \tag{17.9}$$

as long as $\mathfrak{M}$, as in equation (17.1), is a formally selfadjoint operator with coefficients not depending on t. A solution of this equation that is equal to zero on Γ_T is given by the series

$$u(x, t) = \sum_{k=1}^\infty \psi_k(x) \int_0^t e^{-\lambda_k(t-\tau)} f_k(\tau)\, d\tau, \tag{17.10}$$

where $\psi_k(x)$ and λ_k have their former meaning, while $f_k(t) = (f(x,t), \psi_k(x))$. The convergence of such series in $W_2^m(\Omega)$ can be investigated by using the propositions formulated above for the series $\sum_{k=1}^\infty a_k\psi_k(x)$. Analogous investigations are carried out in [65c] for the boundary conditions $\partial u/\partial N + \sigma u|_S = a_{ij}(x)u_{x_j}\cos(\mathbf{n}, x_i) + \sigma u|_S = 0$, in which $\mathbf{n}$ is the normal to S

§18. ON THE LAPLACE TRANSFORM METHOD

For solving the Cauchy problem and the boundary value problems for the equations

$$u_t - \mathcal{M}u \equiv u_t - (a_{ij}(x)u_{x_j} + a_i(x)u)_{x_i} + b_i(x)u_{x_i} + a(x)u = f(x, t) \tag{18.1}$$

with coefficients not depending on t, one often makes use of the Laplace transform in t, which reduces them to corresponding problems of elliptic type with a complex parameter λ. Let us describe the basic plan of attack for this by using as an example the first boundary value problem

$$u_t - \mathcal{M}u = f(x, t), \quad u|_{S_\infty} = 0, \quad u|_{t=0} = \psi_0(x) \tag{18.2}$$

in the semi-infinite cylinder $Q = \Omega \times (0, \infty)$.

We multiply both sides of the equation by $e^{-\lambda t}$, where $\lambda = \lambda_1 + i\lambda_2$ is a complex number with $\lambda_1 > 0$, and then integrate over t from 0 to ∞. Assuming that all of the integrals converge, we transform the first of them as follows:

$$\int_0^\infty u_t e^{-\lambda t}\, dt = ue^{-\lambda t}\Big|_0^\infty + \lambda \int_0^\infty ue^{-\lambda t}\, dt. \tag{18.3}$$

We denote the Laplace transforms of u and f by

$$\tilde{u}(x, \lambda) = \int_0^\infty u(x, t)e^{-\lambda t} dt \text{ and } \tilde{f}(x, \lambda) = \int_0^\infty f(x, t)e^{-\lambda t}\, dt$$

and take into account that by virtue of the assumed convergence of the integrals we have $ue^{-\lambda t} \to 0$ for $t \to \infty$, while by virtue of the initial condition $ue^{-\lambda t} \to \psi_0$ for $t \to 0$. In this way we see that $\tilde{u}$ must satisfy the equation

$$\lambda\tilde{u} - \mathcal{M}\tilde{u} = \tilde{f} + \psi_0. \tag{18.4}$$

In addition, since $u(x, t)|_{S_\infty} = 0$ it must satisfy the boundary condition

$$\tilde{u}|_S = 0. \tag{18.5}$$

It is well known that problem (18.4), (18.5) is uniquely solvable for all values of λ on the complex λ plane except the spectral values, which in the case of a finite domain Ω (and reasonable properties of the coefficients) form a countable set going off to $-\infty$. In other words, the operator $\mathfrak{M} - \lambda$ has an inverse in $L_2(\Omega)$ for all nonspectral λ. This inverse operator $(\mathfrak{M} - \lambda)^{-1}$ is called the resolvent of the operator $\mathfrak{M}$, and with its use the solution of problem (18.4), (18.5) can be represented in the form

$$\tilde{u}(x, \lambda) = -(\mathcal{M} - \lambda)^{-1}\tilde{f} - (\mathcal{M} - \lambda)^{-1}\psi_0. \tag{18.6}$$

If $\tilde{u}(x, \lambda)$ decreases sufficiently fast as $\lambda_2 \to \pm\infty$ when $\lambda_1 \geq 0$ is fixed, then $u(x, t)$ can be recovered from it by means of the inverse Laplace transform:

$$u(x, t) = \frac{1}{2\pi i}\int_{\lambda_1 - i\infty}^{\lambda_1 + i\infty} \tilde{u}(x, \lambda)e^{\lambda t}\, d\lambda. \tag{18.6$'$}$$

This formula gives the desired solution of problem (18.2). Such is the formal plan of attack for the Laplace transform method. In it the functions $u(x, t)$ and also $f(x, t)$ are actually continued by zero into the domain of negative values of t. This permits one to take $\lambda_1 > 0$ and achieve convergence of the integral $\int_{-\infty}^{\infty} ue^{-\lambda t}dt$ even in those cases when the solution $u(x, t)$ increases as $t \to +\infty$. But such a continuation is not very good when $\psi_0(x) \not\equiv 0$, since it leads to a function that is discontinuous at $t = 0$, while the Laplace transforms of nonsmooth functions decrease quite slowly as $\lambda_2 \to \pm\infty$ (and conversely, a slow decrease in $\tilde{u}$ generates a nonsmoothness in u). In order to avoid "spoiling" problem (18.2) in this way it is better to first reduce it to one with zero initial condition (and, if possible, with f and its derivatives with respect to t equal to zero at $t = 0$) and then, after continuing u and f by zero into the domain $t < 0$, apply the Laplace transform.

A justification of Laplace's method consists essentially in a study of the smoothness properties of the solutions $\tilde{u}(x, \lambda)$ of problem (18.4), (18.5) with respect to x and the manner in which $\tilde{u}(x, \lambda)$ (as an element of some function space of functions of x) decreases for $\lambda_2 \to \pm\infty$ and fixed $\lambda_1 \geq 0$. This was done for equations of parabolic and hyperbolic types, and also for strongly parabolic and strongly hyperbolic systems, by one of the present authors (a detailed discussion of the hyperbolic case is given in [65c]) and for the parabolic case by the authors of [67]. This method is also analyzed in many papers on the abstract equations (14.1) with operators S not depending on t (see [43; 73; 76; 115a,b 127 et al.]).

CHAPTER IV

LINEAR EQUATIONS WITH SMOOTH COEFFICIENTS

In this chapter we consider the Cauchy problem and the basic boundary value problems for parabolic equations of second order with, on the whole, smooth coefficients.

These problems are considered here by means of two distinct methods. In §§1–9 we present a method dealing with exact estimates of a solution in terms of the data of the problem and involving the construction of a special operator, which is an analogue of the regularizer in the theory of elliptic systems [11a; 23] and permits one to solve the Cauchy problem or a boundary value problem "in the small" with respect to t. It is then possible, by making repeated use of a theorem on the solvability of a problem "in the small," to construct a solution in any finite interval $0 < t < T$ without loss in the exactness of the result.

By means of this method we study the Cauchy problem, the first boundary value problem and the problem with directional derivative in the classes $H^{l,l/2}$ and $W_q^{2m,m}$. The construction of the regularizor is based on the idea of Korn and Schauder of "freezing" the coefficients of an operator L, and hence this method demands a detailed study of the so-called model problems: the Cauchy problem and boundary value problems in a half space for equations with constant coefficients without minor terms. Such problems are reduced by a linear transformation of the coordinates to analogous problems, for the heat equation, which are considered in §1. This section has an auxiliary character. In it are cited the well-known representations in terms of heat potentials of the solutions of the Cauchy problem and of the first boundary value problem. In addition, an analogous representation, first mentioned in [92], is deduced in §1 for the problem with directional derivative. In §§2 and 3 exact estimates are proved for all of these potentials in the norms of $H^{l,l/2}$ and $W_q^{2m,m}$. Some definitions and auxiliary propositions that are necessary for what follows are contained in §4. The basic results on the solvability of the Cauchy problem and of the boundary value problems for parabolic equations of second order in classes $H^{l,l/2}$ are formulated in §5 and proved in §§6–8. In §9, using as an example the first boundary value problem, it is shown how analogous results in the spaces $W_q^{2m,m}$ are proved by means of a general plan set forth in the preceding sections. §10 is devoted to exact estimates of a solution in subdomains $Q' \subset Q_T$, both purely interior and adjacent to the lower base or lateral surface of the cylinder Q_T.

It should be emphasized that the indicated method is used to prove the existence of solutions of the Cauchy problem and of the boundary value problems in classes of functions all of whose derivatives entering into the equation, or even derivatives of higher order, are Hölder continuous or summable with power $q > 1$ in the whole domain Q_T.

The other method, which is presented in §§11–17, is the method of potential theory for equations with variable coefficients. The central feature in it is the construction of a fundamental solution $Z(x, \xi, t, \tau)$ of a parabolic equation of form (2.2) of Chapter I. This was first done in the papers of F. Dressel [21], who constructed a fundamental solution by means of the classical method of E. Levi [69]. The essence of this method consists in the fact that a fundamental solution is looked for in the form of a sum of two terms: a fundamental solution Z_0 of the equation without minor terms and with frozen leading coefficients (it is written out in explicit form and is the principal term with respect to the order of the singularity at the point $x = \xi$, $t = \tau$) and an additional term in the form of an integral operator with kernel Z_0 and a certain density Q' determined from an integral equation. In the parabolic case this integral equation turns out to be of Volterra type with respect to the variable t, and consequently is solved by the method of successive approximations.

The restrictions that must be imposed on the coefficients in order to construct a fundamental solution depend in much on the choice of the principal term Z_0. For example, if one chooses for Z_0 a fundamental solution of an equation whose coefficients are frozen at the point (x, t), then it is necessary to require that the leading coefficients a_{ij} of the operator L have derivatives $\partial^2 a_{ij}/\partial x_k \partial x_l$ and $\partial a_{ij}/\partial t$ satisfying a Hölder condition. Indeed, under these conditions a fundamental solution Z is constructed in [21] and shown to have the same singularity at the point $z = \xi$, $t = \tau$ as the function Z_0. If one chooses for Z_0 a fundamental solution of an equation whose coefficients are frozen at the point (ξ, τ), then all of the subsequent constructions are possible for Hölder-continuous a_{ij}. This is done in the papers of W. Pogorzelski [94a,c] (for elliptic equations such a method was proposed still earlier by L. Lichtenstein; see in this connection his paper [71] and also the monograph of Carlo Miranda, *Equazioni alle derivate parzioli di tipo ellittico*). By freezing the coefficients only with respect to x, one can waive the Hölder condition with respect to t for the a_{ij} [25b, Part II]. A further significant weakening of the conditions on the a_{ij} is achieved in [79].

§11 of the present chapter is devoted to the construction of a fundamental solution. For the sake of simplicity in the presentation we do not reduce the restrictions on the a_{ij} to a minimum and we assume that these functions satisfy a Hölder condition with respect to all of their arguments. Under these conditions we prove exact estimates for the derivatives $D_x Z$, $D_x^2 Z$, $D_t Z$ [25a] and their differences with respect to x and t. As pointed out in [79], exact estimates of differences were first obtained in an unpublished work of S. D. Ivasišen.

In §14 the function Z is used to investigate the Cauchy problem in the class of bounded functions and a still different method is employed to prove exact estimates of its solution in the norms of $H^{l,l/2}$.

In §§15 and 16 a potential theory is presented for the first and second boundary value problems in domains with bounded boundary S. This theory is a generalization of the classical theory of heat potentials [93b; 86; 119b]. It permits one to prove the classical solvability of boundary value problems, assuming only continuity of the boundary values. However, in using a double-layer potential for solving the first boundary value problem one must impose additional conditions on the leading coefficients of the equation: the differentiability of these coefficients with respect to the x_k and the Hölder-continuity of their derivatives. W. Pogorzelski [94d] got rid of these additional restrictions by constructing a Green's function for the first boundary value problem with the use of a single-layer potential. We present his method in §16.

Finally, in §17 we present estimates of S. N. Bernšteĭn for the first derivatives u_{x_k} of solutions of the first boundary value problem.

§1. THE HEAT EQUATION AND HEAT POTENTIALS

In this section we examine some model problems for the heat equation

$$\frac{\partial u(x, t)}{\partial t} - \Delta u(x, t) = f(x, t). \tag{1.1}$$

Let us begin with an examination of the Cauchy problem

$$\frac{\partial u}{\partial t} - \Delta u = f \quad (t > 0, -\infty < x_i < +\infty, \ i = 1, \dots, n),$$
$$u|_{t=0} = \varphi(x). \tag{1.2}$$

Under certain natural restrictions on the functions f and ϕ the solution of this problem is explicitly written out in terms of potentials, the densities of which are simply expressed in terms of the functions f and ϕ, while the kernel is the fundamental solution of the heat equation

$$\Gamma(x, t) = \begin{cases} \dfrac{1}{(4\pi t)^{\frac{n}{2}}} e^{-\frac{x^2}{4t}} \left(x^2 = \sum\limits_{i=1}^{n} x_i^2 \right) & \text{for } t > 0, \\ 0 & \text{for } t < 0, \end{cases} \tag{1.3}$$

which for the heat equation plays the same important role as the function $|x|^{2-n}$ plays for the Laplace equation. The potentials formed from the function (1.3) are called heat potentials. We consider at once those heat potentials that solve the Cauchy problem (1.2). They are the volume heat potential

$$(\Gamma * f) = \int_{-\infty}^{t} d\tau \int_{E_n} \Gamma(x - y, t - \tau) f(y, \tau) \, dy, \tag{1.4}$$

playing the role of a newtonian potential for equation (1.1), and the potential

$$(\Gamma *_1 \varphi) = \int_{E_n} \Gamma(x - y, t) \varphi(y) \, dy. \tag{1.5}$$

The integrals (1.4) and (1.5) converge if the functions f and ϕ satisfy certain conditions at infinity: they do not increase too rapidly for $|x| \to \infty$ (not faster than e^{cx^2}) and the function f decreases as a power of t for $|t| \to -\infty$. Since the results of this and the following two sections will be applied below mainly to potentials with densities having compact supports, we will not consider increasing f and ϕ.

The potential (1.5) is for $t > 0$ a solution of the homogeneous equation

$$\left(\frac{\partial}{\partial t} - \Delta \right) (\Gamma *_1 \varphi) = 0.$$

In addition, if the function ϕ is continuous at the point x, then

$$\lim_{t \to 0} (\Gamma *_1 \varphi) = \varphi(x). \tag{1.6}$$

In order to prove this, we make use of the following easily verified property of the function Γ: for $t > 0$

$$\int_{E_n} D_t^r D_x^s \Gamma(x, t)\, dx = \begin{cases} 1, & \text{if } r = 0,\ s = 0, \\ 0, & \text{if } r + s \geqslant 1. \end{cases} \tag{1.7}$$

From this property it follows that

$$\lim_{t\to 0} (\Gamma *_1 \varphi) = \lim_{t\to 0} \int_{E_n} \Gamma(x - y,\ t)\,[\varphi(y) - \varphi(x)]\, dy + \varphi(x)$$

$$= (4\pi)^{-\frac{n}{2}} \lim_{t\to 0} \int_{E_n} e^{-\frac{z^2}{4}} \left[\varphi\left(x - z\sqrt{t}\right) - \varphi(x)\right] dz + \varphi(x) = \varphi(x).$$

Let us now consider the volume potential

$$g(x,\ t) = (\Gamma * f).$$

The kernels $\Gamma(x - y,\ t - \tau)$ and $\partial\Gamma(x - y,\ t - \tau)/\partial x_i$ have weak (locally integrable) singularities at the point $y = x$, $t = \tau$ and therefore

$$\frac{\partial g}{\partial x_i} = \left(\frac{\partial \Gamma}{\partial x_i} * f\right).$$

It is impossible, however, to differentiate the volume potential twice with respect to x or once with respect to t in this manner. Let us calculate these derivatives, assuming that $f(x, t)$ satisfies a Hölder condition in the variables x and is continuous in t. We consider the sequence of functions

$$g_h(x,\ t) = \int_{-\infty}^{t-h} d\tau \int_{E_n} \Gamma(x - y,\ t - \tau)\, f(y,\ \tau)\, dy.$$

It is evident that $\lim_{h\to 0} g_h(x, t) = g(x, t)$. Let us prove the existence of a limit for the sequence $\partial g_h/\partial t$. We have

$$\frac{\partial g_h}{\partial t} = \int_{-\infty}^{t-h} d\tau \int_{E_n} \frac{\partial \Gamma(x - y,\ t - \tau)}{\partial t} f(y,\ \tau)\, dy$$

$$+ \int_{E_n} \Gamma(x - y,\ h)\, f(y,\ t - h)\, dy$$

$$= \int_{-\infty}^{t-h} d\tau \int_{E_n} \frac{\partial \Gamma(x - y,\ t - \tau)}{\partial t} [f(y,\ \tau) - f(x,\ \tau)]\, dy$$

$$+ \int_{E_n} \Gamma(x - y,\ h)\, f(y,\ t - h)\, dy.$$

Taking into consideration (1.6), we conclude that

$$\frac{\partial g}{\partial t}=\lim_{h\to 0}\frac{\partial g_h}{\partial t}$$
$$=\int_{-\infty}^{t}d\tau\int_{E_n}\frac{\partial\Gamma(x-y,\,t-\tau)}{\partial t}[f(y,\,\tau)-f(x,\,\tau)]\,dy+f(x,\,t). \tag{1.8}$$

Analogously

$$\frac{\partial^2 g}{\partial x_i\,\partial x_j}=\lim_{h\to 0}\frac{\partial^2 g_h}{\partial x_i\,\partial x_j}$$
$$=\int_{-\infty}^{t}d\tau\int_{E_n}\frac{\partial^2\Gamma(x-y,\,t-\tau)}{\partial x_i\,\partial x_j}[f(y,\,\tau)-f(x,\,\tau)]\,dy. \tag{1.9}$$

From the last two relations it follows that

$$\left(\frac{\partial}{\partial t}-\Delta\right)(\Gamma * f)=f. \tag{1.10}$$

If the function f satisfies a Hölder condition in the variable t, then $\partial g/\partial t$ has the representation

$$\frac{\partial g}{\partial t}=\int_{-\infty}^{t}d\tau\int_{E_n}\frac{\partial\Gamma(x-y,\,t-\tau)}{\partial t}[f(y,\,\tau)-f(y,\,t)]\,dy. \tag{1.11}$$

Let us express the solution $u(x, t)$ of the Cauchy problem in terms of the potentials (1.4), (1.5). We assume that the function f satisfies a Hölder condition, while ϕ is continuous. We extend $f(x, t)$ in some manner into the domain $t < 0$. By virtue of (1.10) the function

$$v(x,\,t)=u(x,\,t)-g(x,\,t)$$

must be a solution of the problem

$$\frac{\partial v}{\partial t}-\Delta v=0,$$
$$v|_{t=0}=\varphi(x)-g(x,\,0)\equiv\psi(x).$$

Consequently,

$$v(x,\,t)=(\Gamma *_1\psi)$$

and thus

$$u(x,\,t)=(\Gamma * f)+(\Gamma *_1\psi),\quad \psi=\varphi-(\Gamma * f)|_{t=0}. \tag{1.12}$$

We will normally make use of this formula by setting the function f equal to zero in the domain $t < 0$. In this case

$$g(x, t) = \int_0^t d\tau \int_{E_n} \Gamma(x-y, t-\tau) f(y, \tau)\, dy,$$

and

$$\psi(x) = \varphi(x)$$

$$u(x, t) = \int_0^t d\tau \int_{E_n} \Gamma(x-y, t-\tau) f(y, \tau)\, dy + (\Gamma *_1 \varphi). \tag{1.13}$$

Besides the Cauchy problem, we examine in this section two boundary value problems for the heat equation in the domain $x_n > 0$, $t > 0$ of the space E_{n+1}: the first boundary value problem

$$\frac{\partial u}{\partial t} - \Delta u = 0,$$

$$u|_{t=0} = 0, \quad u|_{x_n=0} = \Phi_1(x', t) \tag{1.14}$$

and the problem with directional derivative

$$\frac{\partial v}{\partial t} - \Delta v = 0,$$

$$v|_{t=0} = 0, \quad \sum_{i=1}^n b_i \frac{\partial v}{\partial x_i}\bigg|_{x_n=0} = \Phi_2(x', t). \tag{1.15}$$

Here $x' = (x_1, \cdots, x_{n-1})$, while the b_i are certain real constants with $b_n \neq 0$. We will suppose that $b_n > 0$.

Both problems are formulated for the homogeneous heat equation and homogeneous initial conditions since inhomogeneities in the equation and initial conditions are removed by means of the potentials (1.4), (1.5).

Let us introduce the notation

$$(G *_2 \Phi^0) = \int_{-\infty}^{\infty} d\tau \int_{E_{n-1}} G(x'-y', x_n, t-\tau)\, \Phi^0(y', \tau)\, dy',$$

$$\Phi^0 = \begin{cases} \Phi & \text{for } t > 0, \\ 0 & \text{for } t < 0. \end{cases}$$

The solution of problem (1.14) is the double-layer heat potential

$$u = -2\left(\frac{\partial \Gamma}{\partial x_n} * {}_2\Phi_1^0\right)$$
$$= -2\int_0^t d\tau \int_{E_{n-1}} \frac{\partial \Gamma(x'-y', x_n, t-\tau)}{\partial x_n} \Phi_1(y', \tau)\, dy'. \qquad (1.16)$$

In order to verify this it is sufficient to demonstrate that it satisfies the necessary boundary conditions for $x_n = 0$. Since

$$-2\int_0^\infty d\tau \int_{E_{n-1}} \frac{\partial \Gamma(x', x_n, \tau)}{\partial x_n} dx'$$
$$= \frac{x_n}{(4\pi)^{\frac{n}{2}}} \int_0^\infty e^{-\frac{x_n^2}{4\tau}} \frac{d\tau}{\tau^{\frac{n+2}{2}}} \int_{E_{n-1}} e^{-\frac{x'^2}{4\tau}} dx' = 1, \qquad (1.17)$$

it follows that any function $\Phi(x', t)$ which is continuous at the point (x', t) satisfies the relation

$$-2\lim_{x_n\to 0}\left(\frac{\partial \Gamma}{\partial x_n} * {}_2\Phi\right)$$
$$= \Phi(x', t) - 2\lim_{x_n\to 0}\int_0^\infty d\tau \int_{E_{n-1}} \frac{\partial \Gamma(y', x_n, \tau)}{\partial x_n}$$
$$\times [\Phi(x'-y', t-\tau) - \Phi(x', t)]\, dy' = \Phi(x', t). \qquad (1.18)$$

Consequently the potential (1.16) satisfies the necessary boundary condition for any continuous Φ_1.

The solution of problem (1.15) is expressed in the form of a potential

$$v = (G * {}_2\Phi_2^0).$$

In order to calculate the kernel G of this potential we take in (1.15) the Fourier transform with respect to the variables x' and the Laplace transform with respect to t. The functions transformed in this manner will be denoted by the symbol $\sim$ placed over them:

$$\tilde{f}(\xi, p) = \frac{1}{(2\pi)^{\frac{n-1}{2}}} \int_0^\infty e^{-pt} dt \int_{E_{n-1}} e^{-ix'\xi} f(x', t)\, dx' \qquad (1.19)$$

(here and below $x'\zeta = \sum_{\alpha=1}^{n-1} x_\alpha \zeta_\alpha$).

Under such a transformation problem (1.15) goes over into the following problem for an ordinary differential equation in the variable x_n:

$$\frac{d^2\tilde{v}}{dx_n^2} - (p + \zeta^2)\tilde{v} = 0, \tag{1.20}$$

$$b_n \frac{d\tilde{v}}{dx_n} + i \sum_{\alpha=1}^{n-1} b_\alpha \zeta_\alpha \tilde{v} = \tilde{\Phi}_2^0. \tag{1.21}$$

In this problem we are for the moment short one condition for the function $\tilde{v}$, which can be formulated on the basis of the following considerations. The function v is expressed in terms of $\tilde{v}$ by means of the inverse integral transform with respect to (1.19), i.e.,

$$v(x, t) = \frac{1}{(2\pi)^{\frac{n-1}{2}} 2\pi i} \int_{E_{n-1}} e^{ix'\zeta} d\zeta \int_{\sigma - i\infty}^{\sigma + i\infty} \tilde{v}(\zeta, p, x_n) e^{pt} dp, \tag{1.22}$$

if the integral in the right side exists. Equation (1.20) has two linearly independent solutions: $e^{-x_n\sqrt{p+\zeta^2}}$ and $e^{x_n\sqrt{p+\zeta^2}}$, where by $\sqrt{z}$ is meant the value of the root corresponding to a variation of $\arg z$ within the limits $-\pi < \arg z \leq \pi$; the function $\tilde{v}$ is a linear combination of these two solutions. In order for the integral in the right side of (1.22) to converge for any $x_n > 0$ we must require that the function $\tilde{v}$ be expressed in terms of only the first of these solutions. This can be achieved by imposing the condition

$$\tilde{v} \xrightarrow[x_n \to +\infty]{} 0. \tag{1.23}$$

This is the missing condition for the function $\tilde{v}$.

Thus we must solve the problem (1.20), (1.21), (1.23). It is evident that

$$\tilde{v}(\zeta, p, x_n) = F(\zeta, p) e^{-x_n\sqrt{p+\zeta^2}}.$$

Calculating the function F from condition (1.21), we obtain

$$\tilde{v}(\zeta, p, x_n) = \frac{e^{-x_n\sqrt{p+\zeta^2}}}{i\sum_{\alpha=1}^{n-1} b_\alpha\zeta_\alpha - b_n\sqrt{p+\zeta^2}} \tilde{\Phi}_2^0(\zeta, p)$$

and consequently

$$v(x,\ t)=\frac{1}{(2\pi)^{\frac{n-1}{2}}2\pi i}\int\limits_{E_{n-1}} e^{ix'\zeta}\,d\zeta\int\limits_{\sigma-i\infty}^{\sigma+i\infty} e^{pt}$$

$$\times\frac{e^{-x_n\sqrt{p+\zeta^2}}}{i\sum\limits_{\alpha=1}^{n-1}b_\alpha\zeta_\alpha-b_n\sqrt{p+\zeta^2}}\tilde{\Phi}_2^0(\zeta,\ p)\,dp.$$

By virtue of the well-known formula for the Fourier transform of the convolution of two functions we conclude that

$$v(x,\ t)=(G * {}_2\Phi_2^0), \tag{1.24}$$

where

$$G(x,\ t)=\frac{1}{(2\pi)^{n-1}2\pi i}\int\limits_{E_{n-1}} e^{ix'\zeta}\,d\zeta\int\limits_{\sigma-i\infty}^{\sigma+i\infty} e^{pt}\frac{e^{-x_n\sqrt{p+\zeta^2}}}{ib'\zeta-b_n\sqrt{p+\zeta^2}}\,dp \tag{1.25}$$

(we have introduced the notation $b'=(b_1,\cdots,b_{n-1})$).

We choose the number σ in such a way that the path of integration lies in the domain of analyticity of the integrand. This will be the case for $\sigma=-\zeta^2+a^2$, $a>0$.

From the analyticity of the integrand with respect to p in the half plane $\operatorname{Re}p>-\zeta^2$ it is not difficult to see that $G(x,t)=0$ for negative values of t. One can calculate the integral (1.25) for positive t. Let us introduce the new variable of integration $s=s_1+is_2$ connected with p by the relation

$$\sqrt{p+\zeta^2}=s.$$

Then

$$\int\limits_{a^2-\zeta^2-i\infty}^{a^2-\zeta^2+i\infty} e^{pt}\frac{e^{-x_n\sqrt{p+\zeta^2}}}{ib'\zeta-b_n\sqrt{p+\zeta^2}}\,dp=2e^{-\zeta^2t}\int\limits_{H(a)} e^{s^2t-sx_n}\frac{s\,ds}{ib'\zeta-b_ns},$$

where $H(a)$ is the right branch of the hyperbola $s_1^2-s_2^2=a^2$. It is evident that the value of the integral in the right side is not changed if the path $H(a)$ is replaced by the path $\operatorname{Re}s=a$. Let us make this substitution and put $\beta_\alpha=b_\alpha/b_n$, $\beta=(\beta_1,\cdots,\beta_{n-1})$. We get

$$G(x,\ t)=-\frac{1}{(2\pi)^{n-1}\pi ib_n}\int\limits_{E_{n-1}} e^{ix'\zeta-\zeta^2t}\,d\zeta\int\limits_{a-i\infty}^{a+i\infty} e^{s^2t-sx_n}\frac{s\,ds}{s-i\beta\zeta}.$$

Let us again introduce a new variable of integration $v = s - i\beta\zeta$ and interchange the order of integration with respect to ζ and v. We will have

$$G(x, t) = -\frac{1}{(2\pi)^{n-1}\pi i b_n} \int_{a-i\infty}^{a+i\infty} e^{v^2 t - vx_n} \frac{dv}{v}$$

$$\times \int_{E_{n-1}} \exp[ix'\zeta - \zeta^2 t - (\beta\zeta)^2 t + 2ivt\beta\zeta - ix_n\beta\zeta](v + i\beta\zeta)\, d\zeta$$

$$= -\frac{1}{(2\pi)^{n-1}\pi i b_n}\left[\int_{a-i\infty}^{a+i\infty} e^{v^2 t - vx_n} J\, dv\right.$$

$$\left. + \sum_{\alpha=1}^{n-1} \beta_\alpha \frac{\partial}{\partial x_\alpha} \int_{a-i\infty}^{a+i\infty} e^{v^2 t - vx_n} J \frac{dv}{v}\right], \tag{1.26}$$

where

$$J = \int_{E_{n-1}} \exp\{-[\zeta^2 + (\beta\zeta)^2]\, t + i(x' + 2vt\beta - x_n\beta)\zeta\}\, d\zeta.$$

We assume that $\beta \neq 0$ (if $\beta = 0$, then it is easily seen that $J = (\pi/t)^{(n-1)/2} e^{-x'^2/4t}$).

Let B be an orthogonal matrix of order $n - 1$ with elements β_{kj}, where

$$\beta_{j1} = \frac{\beta_j}{|\beta|} \quad (j = 1, \ldots, n-1),$$

and the remaining elements are chosen arbitrarily. Let us introduce in the integral J the new variables of integration η_j $(j = 1, \cdots, n - 1)$ connected with the ζ_j by the relation $\zeta = B\eta$. Then

$$\beta\zeta = |\beta|\,\eta_1,$$

$$\zeta^2 + (\beta\zeta)^2 = \eta^2 + \beta^2\eta_1^2,$$

$$x'\zeta = y'\eta,$$

where

$$y'_i = \sum_{k=1}^{n-1} \beta_{ki} x'_k, \tag{1.27}$$

and consequently

$$J=\prod_{j=2}^{n-1}\int_{-\infty}^{\infty}e^{-\eta_j^2t+iy_j\eta_j}\,d\eta_j$$

$$\times\int_{-\infty}^{\infty}e^{-\left(1+\beta^2\right)\eta_1^2t+i\left(y_1+2v|\beta|t-x_n|\beta|\right)\eta_1}\,d\eta_1$$

$$=\left(\frac{\pi}{t}\right)^{\frac{n-1}{2}}\frac{1}{\sqrt{1+\beta^2}}\exp\left[-\frac{1}{4t}\sum_{j=2}^{n-1}y_j^2-\frac{(y_1+2vt|\beta|-x_n|\beta|)^2}{4t(1+\beta^2)}\right]$$

$$=\left(\frac{\pi}{t}\right)^{\frac{n-1}{2}}\frac{1}{\sqrt{1+\beta^2}}\exp\left[-\frac{1}{4t}\sum_{j=2}^{n-1}y_j^2-\frac{(y_1-x_n|\beta|)^2}{4t(1+\beta^2)}-\frac{v^2t\beta^2}{1+\beta^2}\right.$$

$$\left.-v|\beta|\frac{y_1-x_n|\beta|}{1+\beta^2}\right] \tag{1.28}$$

and by virtue of (1.26)

$$G(x,\ t)=-\frac{1}{(4\pi t)^{\frac{n-1}{2}}\pi ib_n\sqrt{1+\beta^2}}$$

$$\times\left\{\exp\left[-\frac{1}{4t}\sum_{j=2}^{n-1}y_j^2-\frac{(y_1-x_n|\beta|)^2}{4t(1+\beta^2)}\right]\right.$$

$$\times\int_{a-i\infty}^{a+i\infty}\exp\left(\frac{v^2t}{1+\beta^2}-\frac{vx_n}{1+\beta^2}-\frac{vy_1|\beta|}{1+\beta^2}\right)dv$$

$$+\sum_{\alpha=1}^{n-1}\beta_\alpha\frac{\partial}{\partial x_\alpha}\exp\left[-\frac{1}{4t}\sum_{j=2}^{n-1}y_j^2-\frac{(y_1-x_n|\beta|)^2}{4t(1+\beta^2)}\right]$$

$$\left.\times\int_{a-i\infty}^{a+i\infty}\exp\left(\frac{v^2t}{1+\beta^2}-\frac{vx_n}{1+\beta^2}-\frac{vy_1|\beta|}{1+\beta^2}\right)\frac{dv}{v}\right\}. \tag{1.29}$$

The first integral in the right side of (1.29) is easily calculated:

$$I=\int_{a-i\infty}^{a+i\infty}\exp\left(\frac{v^2t}{1+\beta^2}-\frac{vx_n}{1+\beta^2}-\frac{|\beta|y_1v}{1+\beta^2}\right)dv=$$

$$= i\int_{-\infty}^{\infty}\exp\left[-\frac{w^2t+iw(x_n+|\beta|y_1)}{1+\beta^2}\right]dw$$

$$= i\sqrt{\frac{\pi(1+\beta^2)}{t}}\,e^{-\frac{(x_n+|\beta|y_1)^2}{4t(1+\beta^2)}}. \qquad (1.30)$$

The second integral, which we denote by K, is most simply calculated by means of the obvious relation

$$\frac{\partial K}{\partial x_n} = -\frac{1}{1+\beta^2}I.$$

Since

$$|K| \leqslant \exp\left(\frac{a^2t-ax_n-ay_1|\beta|}{1+\beta^2}\right)\int_{-\infty}^{+\infty}\exp\left(-\frac{w^2t}{1+\beta^2}\right)\frac{dw}{\sqrt{a^2+w^2}},$$

it follows that $\lim_{x_n\to+\infty}K=0$, and consequently

$$K = \frac{1}{1+\beta^2}\int_{x_n}^{\infty}I\,dz = i\sqrt{\frac{\pi}{t(1+\beta^2)}}\int_{x_n}^{\infty}\exp\left[-\frac{(z+|\beta|y_1)^2}{4t(1+\beta^2)}\right]dz$$

$$= 2i\sqrt{\pi}\int_{\frac{x_n+|\beta|y_1}{2\sqrt{t(1+\beta^2)}}}^{\infty} e^{-z^2}\,dz. \qquad (1.31)$$

Substituting (1.30) and (1.31) into (1.29) and making use of the fact that by virtue of (1.27)

$$y_1 = \frac{\beta x'}{|\beta|}, \qquad y'^2 = x'^2$$

and consequently

$$-\sum_{j=2}^{n-1}y_j^2 - \frac{(y_1-x_n|\beta|)^2}{1+\beta^2} - \frac{(x_n+|\beta|y_1)^2}{1+\beta^2} = -y'^2 - x_n^2 = -x^2,$$

$$-\sum_{j=2}^{n-1}y_j^2 - \frac{(y_1-x_n|\beta|)^2}{1+\beta^2} =$$

$$= -x^2 + \frac{(x_n+|\beta|y_1)^2}{1+\beta^2} = -x^2 + \frac{(x_n+\beta x')^2}{1+\beta^2},$$

we obtain

$$G(x, t) = -\frac{2}{(4\pi t)^{\frac{n}{2}} b_n}\left\{e^{-\frac{x^2}{4t}} + \frac{2\sqrt{t}}{\sqrt{1+\beta^2}}\sum_{\alpha=1}^{n-1}\beta_\alpha \frac{\partial}{\partial x_\alpha} e^{-\frac{x^2}{4t}+\frac{(x_n+\beta x)^2}{4t(1+\beta^2)}}\int\limits_{\frac{x_n+\beta x'}{2\sqrt{t(1+\beta^2)}}}^{\infty} e^{-z^2}dz\right\}$$

$$= -\frac{2}{(4\pi t)^{\frac{n}{2}} b_n}\left\{e^{-\frac{x^2}{4t}} + \frac{2\sqrt{t}}{|b|}\sum_{\alpha=1}^{n-1} b_\alpha \frac{\partial}{\partial x_\alpha} e^{-\frac{x^2}{4t}+\frac{(bx)^2}{4tb^2}}\int\limits_{\frac{bx}{2\sqrt{t}|b|}}^{\infty} e^{-z^2}dz\right\} \tag{1.32}$$

Here $b = (b_1, \cdots, b_n)$ and $bx = \sum_{i=1}^n b_i x_i$. Carrying out the differentiation in the second term, we get

$$G(x, t) = -\frac{2}{(4\pi t)^{\frac{n}{2}} b^2}\left\{b_n e^{-\frac{x^2}{4t}} + \frac{b^2 x_n - b_n(bx)}{|b|\sqrt{t}} \times e^{-\frac{x^2}{4t}+\frac{(bx)^2}{4tb^2}}\int\limits_{\frac{bx}{2\sqrt{t}|b|}}^{\infty} e^{-z^2}dz\right\}. \tag{1.33}$$

This function, being a solution of the homogeneous heat equation, has the property

$$\sum_{i=1}^{n} b_i \frac{\partial G(x, t)}{\partial x_i} = \frac{x_n}{(4\pi t)^{\frac{n}{2}} t} e^{-\frac{x^2}{4t}} = -2\frac{\partial \Gamma(x, t)}{\partial x_n}, \tag{1.34}$$

and therefore by virtue of (1.18) it is clear that the function (1.24) is really the solution of problem (1.15).

A special case of problem (1.15) is the second boundary value problem for which $b_1 = \cdots = b_{n-1} = 0$, $b_n = 1$. Its solution is a single-layer heat potential:

$$v(x, t) = -2(\Gamma *_2 \Phi_2^0) = -2\int_0^t d\tau \int_{E_{n-1}} \Gamma(x' - y', x_n, t - \tau)\Phi_2(y', \tau)\, dy'.$$

§2. ESTIMATES OF THE HEAT POTENTIALS IN HÖLDER NORMS

In this and in the following sections exact estimates will be obtained for the basic potentials $(\Gamma * f)$, $(\Gamma *_1 \phi)$, $(G *_2 \Phi)$, introduced in §1, in terms of their densities.

It is convenient to introduce the standard notation for the domains and spaces in which the functions f, ϕ, Φ and the potentials are defined. We will make use of the following notation (some of which we have encountered above):

E_{n+1} is the euclidean space of points $(x_1, \cdots, x_n, t)$,

E_n is the space of points $(x_1, \cdots, x_n)$,

$\widetilde{E}_n$ is the subspace $x_n = 0$ in E_{n+1},

E_{n-1} is the space of points $x' = (x_1, \cdots, x_{n-1})$,

D_{n+1} is the half-space $t > 0$ in E_{n+1},

$\widetilde{D}_{n+1}$ is the half-space $x_n > 0$ in E_{n+1},

D_n is the half-space $x_n > 0$ in E_n,

$\widetilde{D}_n$ is the half-space $t > 0$ in $\widetilde{E}_n$,

R is the domain $x_n > 0$, $t > 0$ in E_{n+1}.

We will denote by $B^{(T)}$ $(T > 0)$ the set of points of the domain B that are situated in the domain $t < T$, with B standing for any of the domains introduced above, except E_n and D_n. Thus, for example, $D_{n+1}^{(T)} = E_n \times (0, T)$, $E_{n+1}^{(T)} = E_n \times (-\infty, T)$, etc.

The following inequalities will be proved below:

$$\langle(\Gamma * f)\rangle_{E_{n+1}^{(T)}}^{(l+2)} \leqslant c_1 \langle f\rangle_{E_{n+1}^{(T)}}^{(l)}, \tag{2.1}$$

$$\langle(\Gamma *_1 \varphi)\rangle_{D_{n+1}^{(T)}}^{(l)} \leqslant c_2 \langle\varphi\rangle_{E_n}^{(l)}, \tag{2.2}$$

$$\langle(G *_2 \Phi)\rangle_{\widetilde{D}_{n+1}^{(T)}}^{(l+1)} \leqslant c_3 \langle\Phi\rangle_{\widetilde{E}_n^{(T)}}^{(l)}. \tag{2.3}$$

In all three inequalities l is an arbitrary nonintegral positive number, T is any positive number, including $+\infty$ (in this case it is natural to take $B^{(\infty)} = B$), while $\langle v\rangle^{(l)}$ is the norm defined in §1 of Chapter I.[1] The constants are everywhere independent of T.

Inequality (2.3) for $b_n = 1$, $b_\alpha = 0$ $(\alpha < n)$ gives us an estimate of a single-layer potential. In addition, by virtue of (1.34)

$$-2\left(\frac{\partial\Gamma}{\partial x_n} * {}_2\Phi\right) = \sum_{i=1}^{n} b_i \frac{\partial}{\partial x_i}(G * {}_2\Phi),$$

and hence (2.3) also implies an estimate of a double-layer potential:

$$\left\langle\left(\frac{\partial\Gamma}{\partial x_n} * {}_2\Phi\right)\right\rangle^{(l)}_{\widetilde{D}^{(T)}_{n+1}} \leqslant c\,\langle\Phi\rangle^{(l)}_{\widetilde{E}^{(T)}_n}. \tag{2.4}$$

For a proof of inequalities (2.1)–(2.3) we will make use of the following estimates of the kernels Γ and G:

$$|D_t^r D_x^s \Gamma(x, t)| \leqslant c_{r,s} t^{-\frac{n}{2}-r-\frac{s}{2}} \exp\left(-C\frac{x^2}{t}\right), \tag{2.5}$$

$$|D_t^r D_x^s G(x, t)| \leqslant c'_{r,s} t^{-\frac{n}{2}-r-\frac{s}{2}} \exp\left(-C'\frac{x^2}{t}\right). \tag{2.6}$$

Estimate (2.5) is obvious for $C < \frac{1}{4}$. For a proof of estimate (2.6) it is sufficient to demonstrate that

$$e^{-\frac{x^2}{4t}+\frac{(bx)^2}{4tb^2}} \int_{\frac{bx}{2\sqrt{t}\,|b|}}^{\infty} e^{-z^2}\,dz \leqslant c \exp\left(-C''\frac{x^2}{4t}\right) \tag{2.7}$$

for any $x \in D_n$. If $bx \geq 0$, then

$$\int_{\frac{bx}{2\sqrt{t}\,|b|}}^{\infty} e^{-z^2}\,dz \leqslant e^{-\frac{(bx)^2}{8tb^2}} \int_0^{\infty} e^{-\frac{z^2}{2}}\,dz = \sqrt{\frac{\pi}{2}}\, e^{-\frac{(bx)^2}{8tb^2}}$$

and

$$e^{-\frac{x^2}{4t}+\frac{(bx)^2}{4tb^2}} \int_{\frac{bx}{2\sqrt{t}\,|b|}}^{\infty} e^{-z^2}\,dz \leqslant \sqrt{\frac{\pi}{2}}\, e^{-\frac{x^2}{4t}+\frac{(bx)^2}{8tb^2}} \leqslant \sqrt{\frac{\pi}{2}}\, e^{-\frac{x^2}{8t}},$$

since $x^2 - (bx)^2/b^2 \geq 0$. But if $bx < 0$, then

$$x^2 - \frac{(bx)^2}{b^2} \geqslant x^2\,[1 - \max\cos^2\vartheta],$$

[1] Here we suppose $\rho_0 = \rho_1 = +\infty$ in (1.11) and (1.12) of Chapter I.

where ϑ is the angle between the vectors b and x and the maximum is taken with respect to all $x \in D_n$ such that $bx < 0$ (i.e., $\cos\vartheta < 0$). The maximum value is achieved when $x_n = 0$ and the vectors $x' = (x_1, \cdots, x_{n-1})$ and $b' = (b_1, \cdots, b_{n-1})$ have opposite directions, with

$$\max\cos^2\vartheta = \frac{b'^2}{b^2} = 1 - \frac{b_n^2}{b^2}.$$

Therefore if $bx < 0$, then

$$x^2 - \frac{(bx)^2}{b^2} \geqslant \frac{b_n^2}{b^2}x^2.$$

As a result we obtain inequality (2.7) and consequently (2.6). The constant C' depends on b_n^2/b^2.

Besides estimates (2.5) and (2.6) we will make use of the fact that the potential $(\Gamma * f)$ satisfies the inhomogeneous heat equation (1.10), while the remaining potentials satisfy the homogeneous heat equation. Finally, we will make use of equalities (1.7) and (1.17).

Let us proceed to the proof of inequality (2.1). In accordance with the definition of the norm $\langle v\rangle^{(l)}_{E^{(T)}_{n+1}}$ (see (1.10) ff. of Chapter I) it reduces to a proof of two estimates:

$$\langle D_t^r D_x^s(\Gamma * f)\rangle^{(l-l')}_{x, E^{(T)}_{n+1}} \leqslant c\langle f\rangle^{(l)}_{E^{(T)}_{n+1}}, \tag{2.8}$$

where $2r + s = l' + 2$, $l' = [l]$, and

$$\langle D_t^r D_x^s(\Gamma * f)\rangle^{\left(\frac{l+2-2r-s}{2}\right)}_{t, E^{(T)}_{n+1}} \leqslant c\langle f\rangle^{(l)}_{E^{(T)}_{n+1}}, \tag{2.9}$$

where r and s satisfy the condition $0 < l + 2 - 2r - s < 2$.

By virtue of equation (1.10) both estimates can be proved by restricting oneself to the case $r = 0$ and estimating only the derivatives $D_x^s(\Gamma * f)$.

Let us prove (2.8) for $r = 0$ and $s = l' + 2$. Since

$$D_x^s(\Gamma * f) = D_x^2(\Gamma * D_x^{l'} f), \tag{2.10}$$

it is sufficient to prove estimate (2.8) for $r = 0$, $l' = 0$ (i.e., for $l = \alpha < 1$). Suppose, as in §1, $(\Gamma * f) = g(x, t)$. By virtue of formula (1.9) we have

$$D_x^2 g(x, t) - D_{x'}^2 g(x, t)$$
$$= \int_{-\infty}^{t} d\tau \int_{|x-y|\leqslant 2|x-x'|} D_x^2\Gamma(x-y, t-\tau)[f(y, \tau)-f(x, \tau)]\, dy$$
$$- \int_{-\infty}^{t} d\tau \int_{|x-y|\leqslant 2|x-x'|} D_{x'}^2\Gamma(x'-y, t-\tau)$$
$$\times [f(y, \tau)-f(x', \tau)]\, dy$$
$$+ \int_{-\infty}^{t} d\tau \int_{|x-y|\geqslant 2|x-x'|} \left[D_x^2\Gamma(x-y, t-\tau)\right.$$
$$\left. - D_{x'}^2\Gamma(x'-y, t-\tau)\right][f(y, \tau)-f(x, \tau)]\, dy$$
$$+ \int_{-\infty}^{t} [f(x', \tau)-f(x, \tau)]\, d\tau \times$$
$$\times \int_{|x-y|\geqslant 2|x-x'|} D_x^2\Gamma(x-y, t-\tau)\, dy = \sum_{i=1}^{4} I_i. \tag{2.11}$$

We estimate the moduli of each of the four terms of the right side with the use of (2.5). The first and second terms are estimated in precisely the same way, for example,

$$|I_1| \leqslant c \int_{|x-y|\leqslant 2|x-x'|} |x-y|^{\alpha}\, dy$$
$$\times \int_{-\infty}^{t} (t-\tau)^{-\frac{n}{2}-1} \exp\left(-C\frac{|x-y|^2}{t-\tau}\right) d\tau \langle f\rangle_{x, E_{n+1}^{(T)}}^{(\alpha)}$$
$$= c \int_{|x-y|\leqslant 2|x-x'|} |x-y|^{-n+\alpha}\, dy \langle f\rangle_{x, E_{n+1}^{(T)}}^{(\alpha)}$$
$$= c|x-x'|^{\alpha} \langle f\rangle_{x, E_{n+1}^{(T)}}^{(\alpha)}.$$

In the third term, for $|x-y| \geq 2|x-x'|$ we have

$$|D_x^2\Gamma(x-y, t-\tau) - D_{x'}^2\Gamma(x'-y, t-\tau)|$$
$$\leqslant c|x-x'|(t-\tau)^{-\frac{n}{2}-\frac{3}{2}} \exp\left(-C\frac{|x-y|^2}{t-\tau}\right)$$

and hence

$$|I_3| \leqslant c \langle f\rangle_{x, E_{n+1}^{(T)}}^{(\alpha)} |x-x'| \int_{|x-y|\geqslant 2|x-x'|} |x-y|^{\alpha}\, dy \times$$

$$\times \int_{-\infty}^{t} (t-\tau)^{-\frac{n+3}{2}} \exp\left(-C\frac{|x-y|^2}{(t-\tau)}\right) d\tau$$

$$\leqslant c\,|x-x'|^{\alpha} \langle f\rangle^{(\alpha)}_{x,E^{(T)}_{n+1}}.$$

Finally, I_4 can be written in the form

$$I_4 = -\int_{-\infty}^{t} [f(x',\tau) - f(x,\tau)]\,d\tau \times \int_{|x-y|=2|x-x'|} \frac{\partial\Gamma(x-y,t-\tau)}{\partial x_k} n_j\,dS_y,$$

and since

$$\left|\frac{\partial\Gamma(x,t)}{\partial x_k}\right| = \frac{|x_k|}{2(4\pi)^{\frac{n}{2}} t^{\frac{n+2}{2}}} e^{-\frac{x^2}{4t}},$$

it follows that

$$|I_4| \leqslant |x-x'|^{1+\alpha} \langle f\rangle^{(\alpha)}_{x,E^{(T)}_{n+1}} \int_{|x-y|=2|x-x'|} dS_y$$

$$\times \int_{-\infty}^{t} (t-\tau)^{-\frac{n+2}{2}} e^{-\frac{|x-x'|^2}{4(t-\tau)}} d\tau = c\,|x-x'|^{\alpha} \langle f\rangle^{(\alpha)}_{x,E^{(T)}_{n+1}}.$$

Consequently,

$$\langle g\rangle^{(\alpha+2)}_{x,E^{(T)}_{n+1}} \leqslant c\,\langle f\rangle^{(\alpha)}_{E^{(T)}_{n+1}}.$$

Applying this estimate to the potential standing in the right side of (2.10) for $\alpha = l - l'$, we obtain inequality (2.8).

Let us pass to the proof of inequality (2.9) for $r = 0$, $0 < l + 2 - s < 2$. From the latter inequality it is seen that $s > l$ and consequently $D_x^s(\Gamma * f) = D_x^{s_1}(\Gamma * D_x^{l'} f)$, where $s_1 = s - l'$ satisfies the condition $l - l' < s_1 < l - l' + 2$, i.e. s_1 can assume two values: $s_1 = 1$ and $s_1 = 2$. Thus the proof of inequality (2.9), in the same way as the proof of inequality (2.8), reduces to the special case $l = \alpha < 1$, $r = 0$ and $s = 1, 2$. Let us prove inequality (2.9) for these values of the indices.

Using formula (1.9) for the case $s = 2$, and equality (1.7) for the case $s = 1$, we see that for $s = 1, 2$

$$D_x^s g(x,t) = \int_{-\infty}^{t} d\tau \int_{E_n} D_x^s\Gamma(x-y,t-\tau)[f(y,\tau) - f(x,\tau)]\,dy.$$

Therefore, assuming for definiteness that $t' < t \leq T$, we have

$$D_x^s g(x, t) - D_x^s g(x, t') = \int_{2t'-t}^{t} d\tau \int_{E_n} D_x^s \Gamma(x-y, t-\tau)$$

$$\times [f(y, \tau) - f(x, \tau)]\, dy - \int_{2t'-t}^{t'} d\tau \int_{E_n} D_x^s \Gamma(x-y, t'-\tau)$$

$$\times [f(y, \tau) - f(x, \tau)]\, dy + \int_{-\infty}^{2t'-t} d\tau \int_{E_n} [D_x^s \Gamma(x-y, t-\tau)$$

$$- D_x^s \Gamma(x-y, t'-\tau)] [f(y, \tau) - f(x, \tau)]\, dy.$$

Estimating the right side of this representation in the same way as the right side of (2.11), we obtain

$$| D_x^s g(x, t) - D_x^s g(x, t') | \leqslant c(t-t')^{\frac{\alpha+2-s}{2}} \langle f \rangle_{x, E_{n+1}^{(T)}}^{(\alpha)},$$

which proves inequality (2.9) for $r = 0$, $l = \alpha < 1$. As a result we obtain inequality (2.9) in the general case.

Thus inequalities (2.8) and (2.9), and consequently inequality (2.1), are proved.

Let us pass to the proof of inequality (2.2). It also reduces to the estimates

$$\langle D_t^r D_x^s (\Gamma *_1 \varphi) \rangle_{x, D_{n+1}^{(T)}}^{(l-l')} \leqslant c \langle \varphi \rangle_{E_n}^{(l)} \quad (2r + s = l') \tag{2.12}$$

and

$$\langle D_t^r D_x^s (\Gamma *_1 \varphi) \rangle_{t, D_{n+1}^{(T)}}^{\left(\frac{l-2r-s}{2}\right)} \leqslant c \langle \varphi \rangle_{E_n}^{(l)} \quad (0 < l - 2r - s < 2). \tag{2.13}$$

Since $(\Gamma *_1 \phi)$ satisfies the homogeneous heat equation it is sufficient to prove both inequalities for $r = 0$. A fact to be taken advantage of here is

$$D_x^s (\Gamma *_1 \varphi) = (\Gamma *_1 D_x^s \varphi).$$

This permits us to restrict the proof of inequality (2.12) to the case $r = 0$, $s = 0$, $l = \alpha < 1$, and the proof of inequality (2.13) to the case $r = 0$, $s = 0$, $0 < l < 2$.

Let $(\Gamma *_1 \phi) = h(x, t)$.

For the indicated values of the parameters inequality (2.12) is proved in the following way: since $\Gamma > 0$ it follows by virtue of (1.7) that

$$|h(x,t)-h(x',t)| = \left|\int_{E_n} \Gamma(y,t)[\varphi(x-y)-\varphi(x'-y)]\,dy\right|$$
$$\leqslant \int_{E_n} \Gamma(y,t)\,|\varphi(x-y)-\varphi(x'-y)|\,dy$$
$$\leqslant |x-x'|^{\alpha}\langle\varphi\rangle_{E_n}^{(\alpha)}\int_{E_n}\Gamma(y,t)\,dy = |x-x'|^{\alpha}\langle\varphi\rangle_{E_n}^{(\alpha)}.$$

Let us prove (2.13) for $r=0$, $s=0$, $l=\rho\in(0,2)$. Since the function Γ is even with respect to all of the variables x it follows that

$$h(x,t) = \int_{E_n}\Gamma(y,t)\,\varphi(x-y)\,dy = \int_{E_n}\Gamma(y,t)\,\varphi(x+y)\,dy$$
$$= \frac{1}{2}\int_{E_n}\Gamma(y,t)\,[\varphi(x-y)+\varphi(x+y)]\,dy$$
$$= \varphi(x) + \frac{1}{2}\int_{E_n}\Gamma(y,t)\,[\varphi(x-y)-2\varphi(x)+\varphi(x+y)]\,dy$$

(we have also taken advantage of (1.7)) and

$$h(x,t)-h(x,t') = \frac{1}{2}\int_{E_n}[\Gamma(y,t)-\Gamma(y,t')]$$
$$\times[\varphi(x-y)-2\varphi(x)+\varphi(x+y)]\,dy$$
$$= \frac{1}{2}\int_{t'}^{t}d\tau\int_{E_n}\frac{\partial\Gamma(y,\tau)}{\partial\tau}[\varphi(x-y)-2\varphi(x)+\varphi(x+y)]\,dy. \tag{2.14}$$

We now note that

$$|\varphi(x-y)-2\varphi(x)+\varphi(x+y)| \leqslant c\,\langle\varphi\rangle_{E_n}^{(\rho)}\,|y|^{\rho}. \tag{2.15}$$

For $\rho<1$ this follows from

$$\varphi(x-y)-2\varphi(x)+\varphi(x+y)$$
$$= [\varphi(x-y)-\varphi(x)]+[\varphi(x+y)-\varphi(x)],$$

and for $1<\rho<2$ it follows from

$$\varphi(x-y)-2\varphi(x)+\varphi(x+y)$$

$$=\int_0^1 \frac{\partial}{\partial t}[\varphi(x+ty)+\varphi(x-ty)]\,dt$$

$$=\sum_{l=1}^{n} y_l \int_0^1 [\varphi_{x_l}(x+ty)-\varphi_{x_l}(x-ty)]\,dt.$$

Estimating the right side of (2.14) by means of (2.15) and (2.5), we obtain

$$|h(x,t)-h(x,t')|\leqslant c\langle\varphi\rangle_{E_n}^{(\rho)}\int_{t'}^{t} d\tau\int_{E_n} |y|^{\rho}\left|\frac{\partial\Gamma(y,\tau)}{\partial\tau}\right|dy$$

$$\leqslant c\langle\varphi\rangle_{E_n}^{(\rho)}\int_{t'}^{t}\tau^{-1+\frac{\rho}{2}}\,d\tau\leqslant c(t-t')^{\frac{\rho}{2}}\langle\varphi\rangle_{E_n}^{(\rho)}.$$

Thus we have proved inequalities (2.12), (2.13) and, consequently, (2.2).

Let us pass to the proof of estimate (2.3). We will establish the following inequalities:

$$\langle(D_xG*_2\Phi)\rangle_{x,\widetilde{D}_{n+1}^{(T)}}^{(\alpha)}\leqslant c\langle\Phi\rangle_{\widetilde{E}_n^{(T)}}^{(\alpha)},\tag{2.16}$$

$$\langle(D_tG*_2\Phi)\rangle_{x,\widetilde{D}_{n+1}^{(T)}}^{(\alpha)}\leqslant c\langle\Phi\rangle_{t,\widetilde{E}_n^{(T)}}^{\left(\frac{\alpha+1}{2}\right)},\tag{2.17}$$

$$\langle(D_xG*_2\Phi)\rangle_{t,\widetilde{D}_{n+1}^{(T)}}^{(\alpha)}\leqslant c\langle\Phi\rangle_{\widetilde{E}_n^{(T)}}^{(2\alpha)},\quad\alpha\neq\frac{1}{2}\tag{2.18}$$

$$\langle(D_tG*_2\Phi)\rangle_{t,\widetilde{D}_{n+1}^{(T)}}^{\left(\frac{\alpha}{2}\right)}\leqslant c\langle\Phi\rangle_{\widetilde{E}_n^{(T)}}^{(\alpha+1)},\tag{2.19}$$

$$\langle(G*_2\Phi)\rangle_{t,\widetilde{D}_{n+1}^{(T)}}^{\left(\frac{\alpha+1}{2}\right)}\leqslant c\langle\Phi\rangle_{\widetilde{E}_n^{(T)}}^{(\alpha)},\tag{2.20}$$

in which $0<\alpha<1$.

In order to prove (2.16) we will make use of the representation

$$(D_xG*_2\Phi)=\int_0^\infty d\tau\int_{E_{n-1}} D_xG(x'-y',x_n,\tau)[\Phi(y',t-\tau)$$

$$-\Phi(y',t)]\,dy'+\int_0^\infty d\tau\int_{E_{n-1}} D_xG(x'-y',x_n,\tau)[\Phi(y',t)$$

$$-\Phi(x',t)]\,dy'+\Phi(x',t)\int_0^\infty d\tau\int_{E_{n-1}} D_xG(x',x_n,\tau)\,dx'.\tag{2.21}$$

Let us demonstrate that for all $x_n > 0$

$$\int_0^\infty d\tau \int_{E_{n-1}} \frac{\partial}{\partial x_i} G(x', x_n, \tau)\, dx' = \frac{\delta_{in}}{b_n}. \tag{2.22}$$

It is clear that the left side of (2.22) is equal to zero for $i < n$. On the other hand the derivative $\partial G/\partial x_n$ is expressed by means of (1.34) in the following manner:

$$\frac{\partial G}{\partial x_n} = -\frac{2}{b_n}\frac{\partial \Gamma}{\partial x_n} - \sum_{i=1}^{n-1} \frac{b_i}{b_n}\frac{\partial G}{\partial x_i}. \tag{2.23}$$

Therefore

$$\int_0^\infty d\tau \int_{E_{n-1}} \frac{\partial G(x', x_n, \tau)}{\partial x_n}\, dx'$$

$$= -\frac{2}{b_n}\int_0^\infty d\tau \int_{E_{n-1}} \frac{\partial \Gamma(x', x_n, \tau)}{\partial x_n}\, dx' = \frac{1}{b_n}.$$

Let $x, z \in D_n$. Let K be the intersection of the ball with center x and radius $2|x - z|$ in the space E_n with the hyperplane E_{n-1} (it can be empty).

From (2.21) and (2.22) one easily obtains the equality

$$\left(\frac{\partial G}{\partial x_i} *_2 \Phi\right) - \left(\frac{\partial G}{\partial z_i} *_2 \Phi\right)$$

$$= \int_0^{|x-z|^2} d\tau \int_{E_{n-1}} \frac{\partial}{\partial x_i} G(x'-y', x_n, \tau)[\Phi(y', t-\tau) - \Phi(y', t)]\, dy'$$

$$- \int_0^{|x-z|^2} d\tau \int_{E_{n-1}} \frac{\partial}{\partial z_i} G(z'-y', z_n, \tau)[\Phi(y', t-\tau) - \Phi(y', t)]\, dy'$$

$$+ \int_{|x-z|^2}^\infty d\tau \int_{E_{n-1}} \left[\frac{\partial}{\partial x_i} G(x'-y', x_n, \tau)\right.$$

$$\left. - \frac{\partial}{\partial z_i} G(z'-y', z_n, \tau)\right][\Phi(y', t-\tau) - \Phi(y', t)]\, dy'$$

$$+ \int_K [\Phi(y', t) - \Phi(x', t)]\, dy' \int_0^\infty \frac{\partial G(x'-y', x_n, \tau)}{\partial x_i}\, d\tau -$$

$$- \int_K [\Phi(y', t) - \Phi(z', t)]\, dy' \int_0^\infty \frac{\partial G(z' - y', z_n, \tau)}{\partial z_i}\, d\tau$$

$$+ \int_{E_{n-1}\setminus K} dy' \int_0^\infty \Big[\frac{\partial G(x' - y', x_n, \tau)}{\partial x_i}$$

$$- \frac{\partial G(z' - y', z_n, \tau)}{\partial z_i}\Big] [\Phi(y', t) - \Phi(z', t)]\, d\tau$$

$$+ [\Phi(z', t) - \Phi(x', t)] \int_0^\infty d\tau \int_{E_{n-1}\setminus K} \frac{\partial G(x' - y', x_n, \tau)}{\partial x_i}\, dy'$$

$$+ \frac{\delta_{in}}{b_n} [\Phi(x', t) - \Phi(z', t)]. \qquad (2.24)$$

We will estimate each term of the right side of (2.24) with the use of (2.6). We note that for $i \neq n$

$$\int_0^\infty d\tau \int_{E_{n-1}\setminus K} \frac{\partial G(x' - y', x_n, \tau)}{\partial x_i}\, dy'$$

$$= \int_0^\infty d\tau \int_\Sigma G(x' - y', x_n, \tau)\, n\, dS_{y'},$$

where Σ is the boundary of K, which for nonempty K is a sphere in the space E_{n-1} with center x', such that if $y' \in \Sigma$, then $\sqrt{|x' - y'|^2 + x_n^2} = 2|x - z|$. Let G' and G'' be the two summands in the right side of (1.33), whose sum is G. It is easy to see that

$$\int_\Sigma G'(x' - y', x_n, \tau)\, n_i\, dS_{y'} = 0$$

and

$$|G''(\xi, \tau)| \leqslant \frac{c\,|\xi|}{t^{\frac{n+1}{2}}} e^{-C'\frac{\xi^2}{\tau}},$$

so that for $i < n$

$$\left| \int_0^\infty d\tau \int_{E_{n-1}\setminus K} \frac{\partial G(x' - y', x_n, \tau)}{\partial x_i}\, dy' \right|$$

$$\leqslant \int_0^\infty d\tau \int_\Sigma |G''(x' - y', x_n, \tau)|\, dS_y \leqslant$$

$$\leqslant c|x-z|^{n-1}\int_0^\infty e^{-C'\frac{|x-z|^2}{t}}\frac{d\tau}{\tau^{\frac{n+1}{2}}}=c.$$

This inequality is also valid for $i=n$ on account of (2.23) and the fact that

$$\left|\int_0^\infty d\tau\int_{E_{n-1}\setminus K}\frac{\partial\Gamma}{\partial x_n}\,dy'\right|=\int_0^\infty d\tau\int_{E_{n-1}\setminus K}\left|\frac{\partial\Gamma}{\partial x_n}\right|dy'\leqslant\frac{1}{2}.$$

Taking into consideration this inequality and estimate (2.6), we obtain

$$\left|\left(\frac{\partial G}{\partial x_i}*_2\Phi\right)-\left(\frac{\partial G}{\partial z_i}*_2\Phi\right)\right|$$

$$\leqslant\langle\Phi\rangle^{\left(\frac{\alpha}{2}\right)}_{t,\widetilde{E}_n^{(T)}}\Bigg[\int_0^{|x-z|^2}\tau^{\frac{\alpha}{2}}\,d\tau\int_{E_{n-1}}\left(\left|\frac{\partial G(x'-y',x_n,\tau)}{\partial x_i}\right|\right.$$

$$\left.+\left|\frac{\partial G(z'-y',z_n,\tau)}{\partial z_i}\right|\right)dy'$$

$$+\sum_{j=1}^n|x_j-z_j|\int_0^1 d\lambda\int_{|x-z|^2}^\infty\tau^{\frac{\alpha}{2}}\,d\tau$$

$$\times\int_{E_{n-1}}\left|\frac{\partial^2G(u'-y',u_n,\tau)}{\partial u_i\,\partial u_j}\right|\Bigg|_{u=\lambda x+(1-\lambda)z}dy'\Bigg]$$

$$+\langle\Phi\rangle^{(\alpha)}_{x,\widetilde{E}_n^{(T)}}\Bigg[\int_K dy'\int_0^\infty\left(|x'-y'|^\alpha\left|\frac{\partial G(x'-y',x_n,\tau)}{\partial x_i}\right|\right.$$

$$\left.+|z'-y'|^\alpha\left|\frac{\partial G(z'-y',z_n,\tau)}{\partial z_i}\right|\right)d\tau$$

$$+\int_{E_{n-1}\setminus K}|y'-z'|^\alpha\,dy'\int_0^\infty\left|\frac{\partial G(x'-y',x_n,\tau)}{\partial x_i}\right.$$

$$\left.-\frac{\partial G(z'-y',z_n,\tau)}{\partial z_i}\right|d\tau+c|x'-z'|^\alpha\Bigg]$$

$$\leqslant c|x-z|^\alpha\left(\langle\Phi\rangle^{\left(\frac{\alpha}{2}\right)}_{t,\widetilde{E}_n^{(T)}}+\langle\Phi\rangle^{(\alpha)}_{x,\widetilde{E}_n^{(T)}}\right)=c|x-z|^\alpha\langle\Phi\rangle^{(\alpha)}_{\widetilde{E}_n^{(T)}},$$

which proves (2.16).

Inequality (2.17) is proved analogously. Since

$$(D_tG *_2\Phi) = \int_{E_{n-1}} dy' \int_0^\infty D_tG(x'-y', x_n, \tau)[\Phi(y', t-\tau)-\Phi(y', t)]\,d\tau,$$

which is analogous to the representation (2.21), but without the second and third terms of the right side, it follows that the difference of the potential $(D_tG *_2\Phi)$ at the points x and z satisfies an equality, in the right side of which are found the first three terms of the right side of (2.24) with kernel $\partial G/\partial t$ in place of $\partial G/\partial x_i$, and inequality (2.17) is proved by using exactly the same estimates as were used in proving inequality (2.16).

Let us pass to the proof of estimates (2.18)–(2.20). We denote the kernels D_xG, D_tG, G of the potentials being estimated by the common symbol $\mathfrak{G}(x, t)$. We have

$$(\mathfrak{G} *_2\Phi) = \int_{-\infty}^{t} d\tau \int_{E_{n-1}} \mathfrak{G}(x'-y', x_n, t-\tau)[\Phi(y', \tau)-\Phi(y', t)]\,dy'$$
$$+ \int_{E_{n-1}} \Phi(y', t)\,dy' \int_0^\infty \mathfrak{G}(x'-y', x_n, \tau)\,d\tau$$

and, assuming for definiteness that $t > t'$, we obtain

$$\begin{aligned}
&(\mathfrak{G}(x, t) *_2\Phi) - (\mathfrak{G}(x, t') *_2\Phi)\\
&= \int_{2t'-t}^{t} d\tau \int_{E_{n-1}} \mathfrak{G}(x'-y', x_n, t-\tau)[\Phi(y', \tau)-\Phi(y', t)]\,dy'\\
&- \int_{2t'-t}^{t'} d\tau \int_{E_{n-1}} \mathfrak{G}(x'-y', x_n, t'-\tau)[\Phi(y', \tau)-\Phi(y', t')]\,dy'\\
&\quad + \int_{-\infty}^{2t'-t} d\tau \int_{E_{n-1}} [\mathfrak{G}(x'-y', x_n, t-\tau)\\
&- \mathfrak{G}(x'-y', x_n, t'-\tau)][\Phi(y', \tau)-\Phi(y', t')]\,dy'\\
&+ \int_{E_{n-1}} [\Phi(y', t)-\Phi(y', t')]\,dy' \int_0^{2(t-t')} \mathfrak{G}(x'-y', x_n, \tau)\,d\tau. \qquad (2.25)
\end{aligned}$$

The first three terms of the right side are easily estimated with the use of inequality (2.6). Only the last term is estimated in a manner that depends on what is meant by $\mathfrak{G}(x, t)$. If $\mathfrak{G} = G$, then

$$\left|\int_{E_{n-1}} [\Phi(y', t)-\Phi(y', t')]\,dy' \int_0^{2(t-t')} G(x'-y', x_n, \tau)\,d\tau\right|$$
$$\leqslant (t-t')^{\frac{\alpha}{2}} \langle\Phi\rangle^{\left(\frac{\alpha}{2}\right)}_{t, \widetilde{E}_n^{(T)}} \int_0^{2(t-t')} d\tau \int_{E_{n-1}} |G(y', x_n, \tau)|\,dy'$$
$$\leqslant c(t-t')^{\frac{\alpha+1}{2}} \langle\Phi\rangle^{\left(\frac{\alpha}{2}\right)}_{t, \widetilde{E}_n^{(T)}}.$$

For $\mathfrak{G} = \partial G/\partial t$ we have

$$\left|\int_{E_{n-1}} [\Phi(y', t)-\Phi(y', t')]\,dy' \int_0^{2(t-t')} \frac{\partial G(x'-y', x_n, \tau)}{\partial\tau}\,d\tau\right|$$
$$=\left|\int_{E_{n-1}} G(x'-y', x_n, 2(t-t'))\,[\Phi(y', t)-\Phi(y', t')]\,dy'\right|$$
$$\leqslant (t-t')^{\frac{\alpha+1}{2}} \langle\Phi\rangle^{\left(\frac{\alpha+1}{2}\right)}_{t, \widetilde{E}_n^{(T)}} \int_{E_{n-1}} |G(x'-y', x_n, 2(t-t'))|\,dy'$$
$$\leqslant c(t-t')^{\frac{\alpha}{2}} \langle\Phi\rangle^{\left(\frac{\alpha+1}{2}\right)}_{t, \widetilde{E}_n^{(T)}}.$$

Suppose $\mathfrak{G} = \partial G/\partial x_i$, $\alpha < \frac{1}{2}$. Then

$$\int_{E_{n-1}} [\Phi(y', t)-\Phi(y', t')]\,dy' \int_0^{2(t-t')} \frac{\partial G(x'-y', x_n, \tau)}{\partial x_i}\,d\tau$$
$$= \int_0^{2(t-t')} d\tau \int_{E_{n-1}} [\Phi(y', t)-\Phi(x', t)]\,\frac{\partial G(x'-y', x_n, \tau)}{\partial x_i}\,dy'$$
$$- \int_0^{2(t-t')} d\tau \int_{E_{n-1}} [\Phi(y', t')-\Phi(x', t')]\,\frac{\partial G(x'-y', x_n, \tau)}{\partial x_i}\,dy'$$
$$+ [\Phi(x', t)-\Phi(x', t')] \int_0^{2(t-t')} d\tau \int_{E_{n-1}} \frac{\partial G(x'-y', x_n, \tau)}{\partial x_i}\,dy'.$$

Since

$$\int_0^{2(t-t')} d\tau \int_{E_{n-1}} \frac{\partial}{\partial x_i} G(x'-y',\ x_n,\ \tau)\,dy'$$
$$= -\frac{2}{b_n}\delta_{in} \int_0^{(2t-t')} d\tau \int_{E_{n-1}} \frac{\partial}{\partial x_n} \Gamma(x'-y',\ x_n,\ \tau)\,dy',$$

it follows that

$$\left| \int_{E_{n-1}} [\Phi(y',t)-\Phi(y',t')]\,dy' \int_0^{2(t-t')} \frac{\partial}{\partial x_i} G(x'-y',x_n,\tau)\,d\tau \right|$$
$$\leqslant 2\langle\Phi\rangle^{(2\alpha)}_{x,\widetilde{E}^{(T)}_n} \int_0^{2(t-t')} d\tau \int_{E_{n-1}} |x-y'|^{2\alpha} \left| \frac{\partial G(x'-y',\ x_n,\ \tau)}{\partial x_i} \right| dy'$$
$$+ \frac{1}{|b_n|} \langle\Phi\rangle^{(\alpha)}_{t,\widetilde{E}^{(T)}_n} (t-t')^\alpha \leqslant c\,(t-t')^\alpha \langle\Phi\rangle^{(2\alpha)}_{\widetilde{E}^{(T)}_n}.$$

It remains for us to prove inequality (2.18) in the case $\alpha > ½$. We note that if $k \neq n$, then

$$\left(\frac{\partial G}{\partial x_k} *_2 \Phi\right) = \left(G *_2 \frac{\partial \Phi}{\partial x_k}\right)$$

and by virtue of (2.20)

$$\left\langle\left(\frac{\partial G}{\partial x_k} *_2 \Phi\right)\right\rangle^{(\alpha)}_{t,\widetilde{D}^{(T)}_{n+1}}$$
$$= \left\langle\left(G *_2 \frac{\partial \Phi}{\partial x_k}\right)\right\rangle^{(\alpha)}_{t,\widetilde{D}^{(T)}_{n+1}} \leqslant c\left\langle\frac{\partial \Phi}{\partial x_k}\right\rangle^{(2\alpha-1)}_{\widetilde{E}^{(T)}_n} \leqslant c\,\langle\Phi\rangle^{(2\alpha)}_{\widetilde{E}^{(T)}_n}.$$

Further, from (1.34) we obtain

$$\left(\frac{\partial G}{\partial x_n} *_2 \Phi\right) = -\frac{2}{b_n}\left(\frac{\partial \Gamma}{\partial x_n} *_2 \Phi\right) - \sum_{k=1}^{n-1} \frac{b_k}{b_n}\left(\frac{\partial G}{\partial x_k} *_2 \Phi\right).$$

The terms under the summation sign have just been estimated, while the first term of the right side is estimated in the following manner:

$$\left| \int_0^\infty d\tau \int_{E_{n-1}} \frac{\partial \Gamma(y',x_n,\tau)}{\partial x_n} [\Phi(x'-y',t-\tau)-\Phi(x'-y',t'-\tau)]\,dy' \right|$$
$$\leqslant (t-t')^\alpha \langle\Phi\rangle^{(\alpha)}_{t,\widetilde{E}^{(T)}_n} \int_0^\infty d\tau \int_{E_{n-1}} \left| \frac{\partial \Gamma(y',\ x_n,\ \tau)}{\partial x_n} \right| dy'$$
$$\leqslant \frac{1}{2}(t-t')^\alpha \langle\Phi\rangle^{(\alpha)}_{t,\widetilde{E}^{(T)}_n}.$$

Consequently

$$\left\langle\left(\frac{\partial G}{\partial x_n} *_2 \Phi\right)\right\rangle^{(\alpha)}_{t,\,\tilde{D}^{(T)}_{n+1}} \leqslant c\,\langle\Phi\rangle^{(2\alpha)}_{\tilde{E}^{(T)}_n},$$

and inequality (2.18) is proved for any $\alpha < 1$, $\alpha \neq \frac{1}{2}$.

Thus inequalities (2.16)–(2.20) are proved.

Let us now use these inequalities to prove estimate (2.3). Like estimates (2.1) and (2.2) it reduces to a proof of the fact that for $2r + s = l' + 1$

$$\langle D^r_t D^s_x (G *_2 \Phi)\rangle^{(l-l')}_{x,\,\tilde{D}^{(T)}_{n+1}} \leqslant c\,\langle\Phi\rangle^{(l)}_{\tilde{E}^{(T)}_n} \tag{2.26}$$

and for $0 < l + 1 - 2r - s < 2$

$$\langle D^r_t D^s_x (G *_2 \Phi)\rangle^{\left(\frac{l+1-2r-s}{2}\right)}_{t,\,\tilde{D}^{(T)}_{n+1}} \leqslant c\,\langle\Phi\rangle^{(l)}_{\tilde{E}^{(T)}_n}. \tag{2.27}$$

Since $(G *_2 \Phi)$ satisfies the heat equation, we need only prove inequalities (2.26), (2.27) for the case when $D^s_x = D^s_{x'}$ and $D^s_x = D^{s-1}_{x'} D_{x_n}$, i.e. we must estimate the derivatives $D^r_t D^{s-k}_{x'} D^k_{x_n} (G *_2 \Phi)$ for $k \leq 1$.

Let us prove inequality (2.26) for such derivatives. We choose numbers $r_1 \leq r$ and $s_1 \leq s - k$ so that the sum $2r_1 + s_1$ has a maximum value not greater than l'. Since $k \leq 1$ it follows that $2r_1 + s_1 \geq l' - 1$.

Let $r_2 = r - r_1$, $s_2 = s - k - s_1$. We have

$$D^r_t D^{s-k}_{x'} D^k_{x_n} (G *_2 \Phi) = D^{r_2}_t D^{s_2}_{x'} D^k_{x_n} \left(G *_2 D^{r_1}_t D^{s_1}_{x'} \Phi\right). \tag{2.28}$$

Suppose that r_1 and s_1 can be chosen so that $2r_1 + s_1 = l'$. Since $2r + s = l' + 1$, we have $2r_2 + s_2 + k = 1$, from which it follows that $r_2 = 0$, $s_2 + k = 1$.

If on the other hand the numbers r_1 and s_1 cannot be chosen so that $2r_1 + s_1 = l'$, but can be chosen so that $2r_1 + s_1 = l' - 1$, then in this case $2r_2 + k = 2$, $s_2 = 0$ (otherwise one could increase s_1 by unity and achieve the result that $2r_1 + s_1 = l'$), and consequently $k = 0$, $r_2 = 1$.

Enlisting inequality (2.16) or (2.17), depending on the value of $2r_1 + s_1$, for an estimate of the right side of (2.28), we obtain (2.26).

In (2.27) the sum $2r + s$ can have two possible values: $2r + s = l' + 1$ and $2r + s = l'$.

In the case $2r + s = l' + 1$ estimate (2.27) is established in exactly the same way as (2.26), with the use of inequalities (2.18) and (2.19).

Consider the case $2r + s = l'$. If $2r_1 + s_1 = l'$, then $r_2 = s_2 = k = 0$, while if $2r_1 + s_1 = l' - 1$, then $r_2 = 0$, $s_2 = 0$, $k = 1$, and hence for the proof of inequality (2.27) when $2r + s = l'$ one should make use of estimates (2.20) or (2.18).

Thus inequalities (2.1)–(2.3) are proved.

§3. ESTIMATES OF THE HEAT POTENTIALS IN THE NORMS OF $W_q^{2m,m}$

Let us prove that the potentials considered in the preceding sections satisfy the inequalities

$$\langle\langle(\Gamma * f)\rangle\rangle_{q, E_{n+1}}^{(2m+2)} \leqslant c\, \langle\langle f\rangle\rangle_{q, E_{n+1}}^{(2m)}, \tag{3.1}$$

$$\langle\langle(\Gamma *_1 \varphi)\rangle\rangle_{q, D_{n+1}}^{(2m+2)} \leqslant c\, \langle\langle\varphi\rangle\rangle_{q, E_n}^{\left(2m+2-\frac{2}{q}\right)}, \tag{3.2}$$

$$\langle\langle(G *_2 \Phi)\rangle\rangle_{q, \widetilde{D}_{n+1}}^{(2m+2)} \leqslant c\, \langle\langle\Phi\rangle\rangle_{q, \widetilde{E}_n}^{\left(2m+1-\frac{1}{q}\right)}, \tag{3.3}$$

$$\left\langle\left\langle\left(\frac{\partial\Gamma}{\partial x_n} *_2 \Phi\right)\right\rangle\right\rangle_{q, \widetilde{D}_{n+1}}^{(2m+2)} \leqslant c\, \langle\langle\Phi\rangle\rangle_{q, \widetilde{E}_n}^{\left(2m+2-\frac{1}{q}\right)}, \tag{3.4}$$

where $q > 1$ and m is a nonnegative integer. The norms in inequalities (3.1)–(3.4) are defined in §1 of Chapter I.

It is easy to demonstrate that inequalities (3.1)–(3.4) for any $m > 0$ readily reduce to the same inequalities with $m = 0$. For inequality (3.1) this is implied by the following property of the volume potential $(\Gamma * f)$, which we have already made use of above:

$$D_t^r D_x^s (\Gamma * f) = (\Gamma * D_t^r D_x^s f).$$

The other potentials satisfy the heat equation. Thus, for example, instead of (3.2) it is sufficient to establish the inequality

$$\sum_{(s=2m+2)} \langle\langle D_x^s (\Gamma *_1 \varphi)\rangle\rangle_{q, D_{n+1}}^{(2m+2)} \leqslant c\, \langle\langle\varphi\rangle\rangle_{q, E_n}^{\left(2m+2-\frac{2}{q}\right)},$$

and instead of (3.3), the inequality

$$\sum_{\substack{2r+s=2m+2\\ k\leqslant 1}} \langle\langle D_t^r D_{x'}^{s-k} D_{x_n}^k (G *_2 \Phi)\rangle\rangle_{q, \widetilde{D}_{n+1}^{(T)}}^{(2m+2)} \leqslant c\, \langle\langle\Phi\rangle\rangle_{q, \widetilde{E}_n^{(T)}}^{\left(2m+1-\frac{1}{q}\right)}.$$

Since

$$D_x^s(\Gamma *_1 \varphi) = (\Gamma *_1 D_x^s\varphi),$$

$$D_t^r D_{x'}^{s'}(G *_2 \Phi) = (G *_2 D_t^r D_{x'}^{s'}\Phi),$$

it is evident that inequalities (3.2) and (3.3) with $m > 0$ follow from the same inequalities with $m = 0$. The same applies to (3.4).

We will therefore prove inequalities (3.1)–(3.4) under the assumption that $m = 0$. In proving (3.1) we will use a theorem concerning multiplicators in Fourier integrals [82].

We denote the Fourier transform of a function $f(x, t)$ with respect to the variables x and t by $\tilde{f}(\xi, \xi_0)$:

$$\tilde{f}(\xi, \xi_0) = \frac{1}{(2\pi)^{\frac{n+1}{2}}} \int_{E_n} e^{-ix\xi}\, dx \int_{-\infty}^{\infty} e^{-it\xi_0} f(x, t)\, dt,$$

$$\xi = (\xi_1, \ldots, \xi_n).$$

It is not difficult to calculate that

$$\tilde{\Gamma} = \frac{1}{(2\pi)^{\frac{n+1}{2}} (i\xi_0 + \xi^2)}.$$

According to the well-known theorem on the Fourier transform of the convolution of two functions,

$$\widetilde{(\Gamma * f)} = (2\pi)^{\frac{n+1}{2}} \tilde{\Gamma}\tilde{f} = \frac{\tilde{f}}{i\xi_0 + \xi^2},$$

and consequently

$$\widetilde{\frac{\partial}{\partial t}(\Gamma * f)} = \frac{i\xi_0}{i\xi_0 + \xi^2}\tilde{f}, \quad \widetilde{\frac{\partial^2}{\partial x_k \partial x_j}(\Gamma * f)} = -\frac{\xi_k\xi_j}{i\xi_0 + \xi^2}\tilde{f}.$$

The functions $\Lambda_0(\xi_0, \xi) = i\xi_0/(i\xi_0 + \xi^2)$ and $\Lambda_{jk}(\xi_0, \xi) = -\xi_k\xi_j/(i\xi_0 + \xi^2)$ satisfy the conditions

$$\left|\xi_{i_1} \cdots \xi_{i_s} \frac{\partial^s \Lambda}{\partial \xi_{i_1} \cdots \partial \xi_{i_s}}\right| \leqslant c \tag{3.5}$$

for $s = 1, \cdots, n+1$, $i_j = 0, \cdots, n$ and $i_k \neq i_j$ $(k \neq j)$. Hence they are multiplicators of class (L_q, L_q)[1)] and we have the estimate

$$\left\| \frac{\partial}{\partial t}(\Gamma * f) \right\|_{q, E_{n+1}} + \sum_{k, j=1}^{n} \left\| \frac{\partial^2}{\partial x_k \partial x_j}(\Gamma * f) \right\|_{q, E_{n+1}} = \langle\langle(\Gamma * f)\rangle\rangle_{q, E_{n+1}}^{(2)} \leqslant c \| f \|_{q, E_{n+1}},$$

representing inequality (3.1) with $m = 0$. Thus inequality (3.1) is proved.

Let us pass to the proof of (3.2).

Let $\mathfrak{G}(x, t)$ be a symbol denoting $D_t\Gamma$ or $D_x^2\Gamma$. The kernel $\mathfrak{G}(x, t)$ is even with respect to all of the variables x, and in addition, for any $t > 0$,

$$\int_{E_n} \mathfrak{G}(x, t)\, dx = 0.$$

Therefore

$$(\mathfrak{G} *_1 \varphi) = \frac{1}{2} \int_{E_n} \mathfrak{G}(y, t)\, [\varphi(x - y) - 2\varphi(x) + \varphi(x + y)]\, dy$$

and

$$\| (\mathfrak{G} *_1 \varphi) \|_{q, E_n} \leqslant \frac{1}{2} \int_{E_n} | \mathfrak{G}(y, t) | N(y)\, dy \leqslant \frac{c}{t^{\frac{n+2}{2}}} \int_{E_n} e^{-\frac{y^2}{8t}} N(y)\, dy,$$

where

$$N(y) = \| \varphi(x - y) - 2\varphi(x) + \varphi(x + y) \|_{q, E_n}.$$

Applying Hölder's inequality, we get

$$\| (\mathfrak{G} *_1 \varphi) \|_{q, E_n} \leqslant \frac{c}{t^{\frac{n+2}{2}}} \left(\int_{E_n} e^{-\frac{y^2}{8t}} N^q(y)\, dy \right)^{\frac{1}{q}} \left(\int_{E_n} e^{-\frac{y^2}{8t}}\, dy \right)^{1 - \frac{1}{q}} = \frac{c}{t^{1 + \frac{n}{2q}}} \left(\int_{E_n} e^{-\frac{y^2}{8t}} N^q(y)\, dy \right)^{\frac{1}{q}}$$

and

---

1) Although the formulation of the basic theorem [82] being used here contains a somewhat stronger condition than inequality (3.5), it is completely clear from the proof that the theorem is also valid for condition (3.5).

$$\|(\mathfrak{G} *_1 \varphi)\|_{q, D_{n+1}} \leqslant c \left(\int_0^\infty \frac{dt}{t^{q+\frac{n}{2}}} \int_{E_n} e^{-\frac{y^2}{8t}} N^q(y)\, dy \right)^{\frac{1}{q}}$$

$$= c \left(\int_{E_n} \int_{E_n} \frac{|\varphi(x-y) - 2\varphi(x) + \varphi(x-y)|^q}{|y|^{n-2+2q}} dx\, dy \right)^{\frac{1}{q}}$$

$$\leqslant c \langle\langle \varphi \rangle\rangle_{q, E_n}^{\left(2-\frac{2}{q}\right)},$$

Q.E.D. The latter in this chain of inequalities is proved for $q \neq 2$ in exactly the same way as (2.15), and for $q = 2$ is implied by Parseval's equality. Indeed,

$$\phi(x-y) - 2\phi(x) + \phi(x+y)$$

$$= -\frac{4}{(2\pi)^{n/2}} \int_{E_n} e^{ix\xi} \sin^2 \frac{y\xi}{2} \tilde{\phi}(\xi)\, d\xi,$$

where $\tilde{\phi}(\xi)$ is the Fourier transform of the function $\phi(x)$ with respect to all of the variables x. Hence

$$\int_{E_n} \frac{dy}{|y|^{n+2}} \int_{E_n} |\phi(x-y) - 2\phi(x) + \phi(x+y)|^2 dx$$

$$= 16 \int_{E_n} |\tilde{\phi}(\xi)|^2 d\xi \int_{E_n} \sin^4 \frac{y\xi}{2} \frac{dy}{|y|^{n+2}} = c \int_{E_n} |\tilde{\phi}(\xi)|^2 |\xi|^2 d\xi$$

$$= c \left[\langle\langle \Phi \rangle\rangle_{2,E_n}^{(1)} \right]^2.$$

For $m = 0$ estimate (3.3) reduces to the inequalities

$$\left\| \left(\frac{\partial G}{\partial t} *_2 \Phi \right) \right\|_{q, \tilde{D}_{n+1}} \leqslant c \langle\langle \Phi \rangle\rangle_{q, \tilde{E}_n}^{\left(1-\frac{1}{q}\right)}, \tag{3.6}$$

$$\left\| (D_{x'}^{2-k} D_{x_n}^k G *_2 \Phi) \right\|_{q, \tilde{D}_{n+1}} \leqslant c \langle\langle \Phi \rangle\rangle_{q, \tilde{E}_n}^{\left(1-\frac{1}{q}\right)}, \quad k = 0, 1, \tag{3.7}$$

and in the case $n = 1$ it reduces to just (3.6).

Let us prove (3.6). We note that

$$\left(\frac{\partial G}{\partial t} * {}_2\Phi\right) = \int_0^\infty d\tau \int_{E_{n-1}} \frac{\partial G(y', x_n, \tau)}{\partial \tau} [\Phi(x' - y', t-\tau) - \Phi(x'-y', t)]\, dy',$$

so that by virtue of (2.6)

$$\left\| \left(\frac{\partial G}{\partial t} * {}_2\Phi\right)\right\|_{q, \tilde{E}_n} \leqslant \int_0^\infty d\tau \int_{E_{n-1}} \left| \frac{\partial G(y', x_n, \tau)}{\partial \tau} \right| N_1(\tau)\, dy'$$

$$\leqslant c \int_0^\infty e^{-C' \frac{x_n^2}{\tau}} N_1(\tau) \frac{d\tau}{\tau^{\frac{3}{2}}},$$

where

$$N_1(\tau) = \| \Phi(x', t-\tau) - \Phi(x', t) \|_{q, \tilde{E}_n}.$$

Applying Hölder's inequality, we get

$$\left\| \left(\frac{\partial G}{\partial t} *_2 \Phi\right)\right\|_{q, \tilde{E}_n} \leqslant c \left(\int_0^\infty e^{-C' \frac{x_n^2}{\tau}} N_1^q(\tau) \frac{d\tau}{\tau^{\frac{3}{4}+\frac{q}{2}}} \right)^{\frac{1}{q}}$$

$$\times \left(\int_0^\infty e^{-\frac{x_n^2}{8\tau}} \frac{d\tau}{\tau^{1+\frac{q'}{4q}}} \right)^{\frac{1}{q'}} = \frac{c}{x_n^{\frac{1}{2q}}} \left(\int_0^\infty e^{-C' \frac{x_n^2}{\tau}} N_1^q(\tau) \frac{d\tau}{\tau^{\frac{3}{4}+\frac{q}{2}}} \right)^{\frac{1}{q}}$$

$$\left(\frac{1}{q'} = 1 - \frac{1}{q} \right).$$

Let us now form the L_q norm of both sides with respect to the variable x_n on the semiaxis $x_n > 0$. We will have

$$\left\| \left(\frac{\partial G}{\partial t} * {}_2\Phi\right)\right\|_{q, \tilde{D}_{n+1}} \leqslant c \left(\int_0^\infty N_1^q(\tau) \frac{d\tau}{\tau^{\frac{3}{4}+\frac{q}{2}}} \int_0^\infty e^{-C' \frac{x_n^2}{\tau}} \frac{dx_n}{x_n^{\frac{1}{2}}} \right)^{\frac{1}{q}}$$

$$= c \left(\int_0^\infty N_1^q(\tau) \frac{d\tau}{\tau^{\frac{1}{2}+\frac{q}{2}}} \right)^{\frac{1}{q}} = c \langle\langle \Phi \rangle\rangle_{q, t, \tilde{E}_n}^{\left(\frac{1}{2} - \frac{1}{2q}\right)}.$$

and (3.6) is proved.

In proving (3.7) it should be assumed that $n > 1$. We use the fact that

$$(D_{x'}^{2-k} D_{x_n}^k G * {}_2\Phi) = \int_0^\infty d\tau \int_{E_{n-1}} D_{y'}^{2-k} D_{x_n}^k G(y', x_n, \tau)$$

$$\times [\Phi(x' - y', t-\tau) - \Phi(x', t-\tau)]\, dy'.$$

Consequently,

$$\left\| \left(D_{x'}^{2-k} D_{x_n}^k G *_2 \Phi\right) \right\|_{q,\tilde{E}_n} \leqslant c \int_0^\infty \frac{d\tau}{\tau^{\frac{n+2}{2}}} \int_{E_{n-1}} e^{-C\frac{y'^2+x_n^2}{\tau}} N_2(y')\,dy'$$

$$= c' \int_{E_{n-1}} \frac{N_2(y')\,dy'}{(y'^2+x_n^2)^{\frac{n}{2}}} \leqslant c' \left(\int_{E_{n-1}} \frac{N_2^q(y')\,dy'}{(y'^2+x_n^2)^{\frac{n-1}{2}+\frac{q}{2}-\frac{1}{4}}} \right)^{\frac{1}{q}}$$

$$\times \left(\int_{E_{n-1}} \frac{dy'}{(y'^2+x_n^2)^{\frac{n-1}{2}+\frac{q'}{4q}}} \right)^{\frac{1}{q'}}$$

$$= c'' \frac{1}{x_n^{\frac{1}{2q}}} \left(\int_{E_{n-1}} \frac{N_2^q(y')\,dy'}{(y'^2+x_n^2)^{\frac{n-1}{2}+\frac{q}{2}-\frac{1}{4}}} \right)^{\frac{1}{q}},$$

where

$$N_2(y') = \|\Phi(x'-y',\ t) - \Phi(x',\ t)\|_{q,\tilde{E}_n},$$

and

$$\left\| \left(D_{x'}^{2-k} D_{x_n}^k G *_2 \Phi\right) \right\|_{q,\tilde{D}_{n+1}}$$

$$\leqslant c \left(\int_{E_{n-1}} N_2^q(y')\,dy' \int_0^\infty \frac{dx_n}{x_n^{\frac{1}{2}} (y'^2+x_n^2)^{\frac{n-1}{2}+\frac{q}{2}-\frac{1}{4}}} \right)^{\frac{1}{q}}$$

$$= c \int_{E_{n-1}} \frac{N_2^q(y')\,dy'}{|y'|^{n-2+q}} = c \langle\langle \Phi \rangle\rangle_{q,x,\tilde{E}_n}^{\left(1-\frac{1}{q}\right)}.$$

Thus we have proved estimate (3.3).

It remains to prove estimate (3.4), which, like (3.3), reduces to the inequalities

$$\left\| \left(\frac{\partial^2 \Gamma}{\partial t\, \partial x_n} *_2 \Phi \right) \right\|_{q,\tilde{D}_{n+1}} \leqslant c \langle\langle \Phi \rangle\rangle_{q,\tilde{E}_n}^{\left(2-\frac{1}{q}\right)}, \tag{3.8}$$

$$\left\| D_{x'}^{2-k} D_{x_n}^k \left(\frac{\partial \Gamma}{\partial x_n} *_2 \Phi \right) \right\|_{q,\tilde{D}_{n+1}} \leqslant c \langle\langle \Phi \rangle\rangle_{q,\tilde{E}_n}^{\left(2-\frac{1}{q}\right)}, \quad k = 0,\ 1. \tag{3.9}$$

Since

$$\left(\frac{\partial^2\Gamma}{\partial t\,\partial x_n} *_2\Phi\right) = \int_0^\infty d\tau \int_{E_{n-1}} \frac{\partial^2\Gamma(y', x_n, \tau)}{\partial x_n\,\partial\tau}[\Phi(x'-y', t-\tau) - \Phi(x'-y', t)]\,dy',$$

(3.8) is proved in the same way as (3.6). As to inequality (3.9), it can be reduced to the already proved estimate (3.3) with $m = 0$ if one takes advantage of equation (1.34). We have

$$D_{x'}^{2-k}D_{x_n}^k\left(\frac{\partial\Gamma}{\partial x_n} *_2\Phi\right) = D_{x'}^{1-k}D_{x_n}^k\left(\frac{\partial\Gamma}{\partial x_n} *_2 D_{x'}\Phi\right)$$
$$= -\frac{1}{2}\sum_{i=1}^n b_i\frac{\partial}{\partial x_i}D_{x'}^{1-k}D_{x_n}^k(G *_2 D_{x'}\Phi)\quad (k=0,\ 1).$$

Estimating the second derivatives of the potential $(G *_2 D_{x'}\Phi)$ with the use of (3.3), we obtain

$$\left\| D_{x'}^{2-k}D_{x_n}^k\left(\frac{\partial\Gamma}{\partial x_n} *_2\Phi\right)\right\|_{q,\widetilde{D}_{n+1}}$$
$$\leqslant c\sum_{i=1}^{n-1}\left\langle\!\left\langle\frac{\partial\Phi}{\partial x_i}\right\rangle\!\right\rangle_{q,\widetilde{E}_n}^{\left(1-\frac{1}{q}\right)} \leqslant c\langle\!\langle\Phi\rangle\!\rangle_{q,\widetilde{E}_n}^{\left(2-\frac{1}{q}\right)}.$$

Estimates (3.1)–(3.4) are proved.

§4. DOMAINS. SOME AUXILIARY PROPOSITIONS

In this section we gather definitions and auxiliary propositions used only in the present chapter and partially in §10 of Chapter VII.

We will study below boundary value problems for equations with variable coefficients in domains of cylindrical form $Q_T = \Omega \times (0, T) \subset E_{n+1}$, where $T > 0$ is a finite number and Ω is a domain in the space E_n.

We will formulate at once the restrictions imposed on the domain.

The domain Ω and its boundary S can be both finite and infinite; the boundary S must be sufficiently smooth. At each point of the boundary there must exist a tangent plane. Let $\mathbf{n}(\xi)$ be the unit vector of the outer normal to S at the point ξ. A Cartesian coordinate system $\{y\}$ with origin at ξ, the y_n axis of which is directed along $\mathbf{n}(\xi)$, is, as usual, called a local coordinate system. It is assumed that there exists a number $d > 0$ such that, in a sphere of radius d with center at any point $\xi \in S$, the surface S is given in a local system at the point ξ by the equation

$$y_n = F(y') \quad (y' = (y_1, \ldots, y_{n-1})), \tag{4.1}$$

where F is a single-valued function. We will say that $S \in H^l$ ($S \in C^m$) if $F(y') \in H^l(K_0)$ ($F(y') \in C^m(K_0)$) for any $\xi \in S$, where K_0 is the ball $|y'| \leq d/2$, and if the norms $|F|_{K_0}^{(l)}$ ($|F|_{K_0}^{(m)}$) are bounded by a common constant. We will consider that at least $S \in H^{1+\alpha}$; $0 < \alpha \leq 1$. Under this assumption, in a neighborhood of ξ we have the inequality

$$\left|\frac{\partial F}{\partial y_k}\right| \leqslant c\,|y'|^{\alpha}. \tag{4.2}$$

We will assume that it is possible to construct in the domain Ω for any $\lambda > 0$, no matter how small, a finite or countable (for unbounded Ω) number of subdomains $\omega^{(k)}$ and $\Omega^{(k)}$ possessing the following properties:

1. $$\omega^{(k)} \subset \Omega^{(k)} \subset \Omega, \quad \bigcup_k \omega^{(k)} = \bigcup_k \Omega^{(k)} = \Omega.$$

2. For any point $x \in \Omega$ there exists an $\omega^{(k)}$ such that $x \in \omega^{(k)}$ and the distance from x to $\Omega^{(k)} \backslash \omega^{(k)}$ is not less than $d\lambda$.

3. There exists an N_0 not depending on λ such that the intersection of any $N_0 + 1$ distinct $\Omega^{(k)}$ (and consequently any $N_0 + 1$ distinct $\omega^{(k)}$) is empty.

4. The sets $\omega^{(k)}$ and $\Omega^{(k)}$ that are separated from the boundary S by a positive distance (we denote the set of their indices k by $\mathfrak{M}$) are n-dimensional cubes with common center $\xi^{(k)} \in \Omega$ whose linear dimensions are equal to $\beta\lambda$ and $2\beta\lambda$ ($\beta > 0$) respectively, while the sets $\omega^{(k)}$ and $\Omega^{(k)}$ adjacent to the boundary (the set of their indices is denoted by $\mathfrak{N}$) are defined in local coordinates at a point $\xi^{(k)} \in S$ by the inequalities

$$|y_\alpha| < \frac{\lambda}{2} \ (\alpha = 1, \ldots, n-1); \ 0 < y_n - F(y') < \lambda,$$
$$|y_\alpha| < \lambda \ (\alpha = 1, \ldots, n-1); \ 0 < y_n - F(y') < 2\lambda.$$

The set of domains $\omega^{(k)}$ will be denoted by $\mathfrak{k}^{(\lambda)}$, while the subdomains $\Omega^{(k)}$ are denoted by $\mathfrak{K}^{(\lambda)}$.

It is possible to show (we will not do this) that in domains with smooth boundaries satisfying the conditions formulated above one can construct systems of subdomains $\mathfrak{k}^{(\lambda)}$ and $\mathfrak{K}^{(\lambda)}$ for any $\lambda > 0$.

The change of variables

$$z_\alpha = y_\alpha \quad (\alpha < n), \qquad z_n = y_n - F(y') \tag{4.3}$$

takes a domain $\Omega^{(k)}$ $(k \in \mathfrak{N})$ into the cube

$$|z_\alpha| < \lambda, \quad 0 < z_n < 2\lambda,$$

which is denoted by $\mathfrak{R}$. Its face $|z_\alpha| < \lambda$, $z_n = 0$ is denoted by σ. We also set

$$s^{(k)} = \overline{\omega}^{(k)} \cap S, \quad S^{(k)} = \overline{\Omega}^{(k)} \cap S \qquad (k \in \mathfrak{N}).$$

And we introduce the functions $\zeta^{(k)}(x)$ having the properties

$$0 \leqslant \zeta^{(k)} \leqslant 1, \ |D_x^s \zeta^{(k)}| \leqslant \frac{c_s}{\lambda^s}, \quad \zeta^{(k)}(x) = \begin{cases} 1 & \text{for } x \in \omega^{(k)}, \\ 0 & \text{for } x \in \Omega \setminus \Omega^{(k)}. \end{cases}$$

By virtue of property 3 of the domains $\Omega^{(k)}$

$$1 \leqslant \sum_k \zeta^{(k)^2}(x) \leqslant N_0,$$

and hence the functions

$$\eta^{(k)}(x) = \frac{\zeta^{(k)}(x)}{\sum_j \zeta^{(j)^2}(x)}$$

possess the following properties: $\eta^{(k)} = 0$ in $\Omega \setminus \Omega^{(k)}$,

$$|D^s \eta^{(k)}(x)| \leqslant \frac{c_s}{\lambda^s}$$

and moreover,

$$\sum_k \eta^{(k)}(x) \zeta^{(k)}(x) = 1. \tag{4.4}$$

Let us pass to the proof of the necessary auxiliary propositions for the functions from Hölder classes $H^{l,l/2}(\overline{Q}_T)$ and from classes $W_q^{2m,m}(\overline{Q}_T)$.

We begin by considering the problem of extending functions from these classes.

Suppose V and V_1 are two domains in some euclidean space, with $V \subset V_1$, and suppose a function u is given in V and belongs to some function space $\mathfrak{B}(V)$ with norm $\|u\|_V$. We say that it is extended with preservation of class into the domain V_1 if a function u^* is constructed in V_1 that coincides with u in V and belongs to the space $\mathfrak{B}(V_1)$, with

$$\|u^*\|_{V_1} \leqslant c \|u\|_V$$

where the constant is independent of u.

We first consider the case when the domains V and $V_1 \setminus V$ are symmetric with respect to some plane. Suppose for definiteness that V is the half space D_{n+1}, while V_1 is the whole space E_{n+1}, and one wishes to extend into E_{n+1} a function $u(x, t)$ given in D_{n+1}, i.e. for $t > 0$. We extend it into the domain $t < 0$

by putting for $t < 0$

$$u(x, t) = \sum_{l=1}^{N} \lambda_l u\left(x, -\frac{t}{l}\right), \tag{4.5}$$

where the λ_i are numbers determined from the linear system of equations

$$\sum_{l=1}^{N} \lambda_l \left(-\frac{1}{l}\right)^s = 1 \qquad (s = 0, \ldots, N-1).$$

It is easy to verify by means of this system that every sufficiently smooth function extended in the indicated manner will have at least $N - 1$ continuous derivatives $\partial^k u/\partial t^k$ $(k = 0, \cdots, N-1)$ on the plane $t = 0$.

The above construction for extending functions is due to Hestenes and Whitney and is described, for example, in [28].

It is possible to show by means of quite elementary estimates that the extension of functions in the indicated manner is an extension with preservation of any of the classes used by us. In particular, the inequalities

$$\langle u \rangle_{E_{n+1}}^{(l)} \leqslant c \langle u \rangle_{D_{n+1}}^{(l)}, \quad | u |_{E_{n+1}}^{(l)} \leqslant c | u |_{D_{n+1}}^{(l)}, \tag{4.6}$$

and also

$$\langle\langle u \rangle\rangle_{q, E_{n+1}}^{(l)} \leqslant c \langle\langle u \rangle\rangle_{q, D_{n+1}}^{(l)}, \quad \| u \|_{q, E_{n+1}}^{(l)} \leqslant c \| u \|_{q, D_{n+1}}^{(l)} \tag{4.7}$$

hold for any $l \geq 0$, both integral and nonintegral, as long as $N \geq [l/2]$ (N is the number of summands in (4.5)).

It is clear that the described method for extending functions preserves class not only under an extension from a half plane onto the whole plane but also under an extension from the domain $\Omega \times (0, T)$ into $\Omega \times (-T, T)$, as well as under an extension with respect to the space variables x (below we will extend functions in this way from the domain $R^{(T)}$ into $D_{n+1}^{(T)}$ with respect to the variable x_n).

This method can also be used to extend functions from domains with smooth boundaries. One constructs in the given domain Ω a sufficiently fine "partition of unity"

$$1 = \sum_k \psi_k$$

(the partition of unity (4.4) can be used). Then an extension of a function u reduces to an extension of the function $u_k = u\psi_k$ for those k for which the boundary of the support of the function ψ_k has a common part S_k with the boundary S of the domain Ω. And this can be done by the above-mentioned method after a "straightening" of S_k.

In this way one can prove the following theorem used below.

Theorem 4.1. *Suppose $\Omega \subset E_n$ is a finite or infinite domain satisfying the conditions formulated at the beginning of the section, and its boundary S belongs to the class H^l. Then every function $u(x) \in H^l(\Omega)$ can be extended with preservation of class onto the whole space E_n.*

An analogous theorem is valid for the spaces $W_q^l(\Omega)$ (it is proved for integral l in [5a]).

In addition to an extension with preservation of class from a smaller domain into a larger domain of the same dimension we will make use of an extension of functions given in the space E_n into the domain $D_{n+1}^{(T)}$ of larger dimension. We will do this by solving the Cauchy problem (1.2), the solution of which can be considered as an extension of the function $\phi(x)$ defining the initial condition into the domain $D_{n+1}^{(T)}$. First we note that the estimates of §2 imply the following result.

Theorem 4.2. *If $f \in H^{l,l/2}(\overline{D}_{n+1})$, $\phi \in H^{l+2}(\overline{E}_n)$, then the solution (1.12) of problem (1.2) satisfies the inequality*

$$\langle u\rangle_{D_{n+1}}^{(l+2)} \leqslant c\left(\langle f\rangle_{D_{n+1}}^{(l)} + \langle \varphi\rangle_{E_n}^{(l+2)}\right)$$

with the constant c depending only on n and l.

Proof. We extend the function f from D_{n+1} into E_{n+1} by formula (4.5) with $N = [l/2]$; then according to (4.6)

$$\langle f\rangle_{E_{n+1}}^{(l)} \leqslant c_l \langle f\rangle_{D_{n+1}}^{(l)}.$$

By virtue of (2.1) and (2.2) with $T = \infty$ the function (1.12) satisfies the estimate

$$\langle u\rangle_{D_{n+1}}^{(l+2)} \leqslant c\left(\langle f\rangle_{E_{n+1}}^{(l)} + \langle \psi\rangle_{E_n}^{(l+2)}\right) \leqslant c\left(\langle f\rangle_{D_{n+1}}^{(l)} + \langle \varphi\rangle_{E_n}^{(l+2)} + \langle (\Gamma * f)\rangle_{E_n}^{(l+2)}\right) \leqslant c\left(\langle f\rangle_{D_{n+1}}^{(l)} + \langle \varphi\rangle_{E_n}^{(l+2)}\right),$$

and the theorem is proved.

This theorem permits one to construct in the domain $D_{n+1}^{(T)}$ a function from class $H^{l,l/2}(D_{n+1}^{(T)})$ if the values of it and certain of its derivatives are given for $t = 0$.

Theorem 4.3. *Let there be given in the space E_n functions*

$\phi_j(x) \in H^{l-2j}(\overline{E}_n)$ $(j = 0, \cdots, [l/2])$, *with* l *being an arbitrary positive nonintegral number. It is possible to construct a function* $u(x, t) \in H^{l, l/2}(D_{n+1}^{(T)})$ *such that*

$$\left.\frac{\partial^j u(x, t)}{\partial t^j}\right|_{t=0} = \varphi_j(x), \tag{4.8}$$

$$|u|_{D_{n+1}^{(T)}}^{(l)} \leqslant c \sum_{j=0}^{\left[\frac{l}{2}\right]} |\varphi_j|_{E_n}^{(l-2j)} \tag{4.9}$$

for any finite T *and*

$$\langle u\rangle_{D_{n+1}}^{(l)} \leqslant c \sum_{j=0}^{\left[\frac{l}{2}\right]} \langle \varphi_j\rangle_{E_n}^{(l-2j)}. \tag{4.10}$$

Proof. Let $l > 2$. We introduce the functions

$$\psi_j(x) = \sum_{s=0}^{j} \binom{j}{s} (-1)^s \Delta^s \varphi_{j-s}(x) \quad \left(j = 0, \ldots, \left[\frac{l}{2}\right]\right). \tag{4.11}$$

Every function $u(x, t)$ that is given for $t > 0$ and satisfies condition (4.8) also satisfies the condition

$$\left.\left(\frac{\partial}{\partial t} - \Delta\right)^j u(x, t)\right|_{t=0} = \psi_j(x) \tag{4.12}$$

and conversely. Let us verify that (4.12) implies (4.8). Conditions (4.12) can be written in the form

$$\sum_{s=0}^{j} \binom{j}{s} (-1)^s \Delta^s \left[\left.\frac{\partial^{j-s} u(x, t)}{\partial t^{j-s}}\right|_{t=0} - \varphi_{j-s}(x)\right] = 0$$

$$\left(j = 0, \ldots, \left[\frac{l}{2}\right]\right). \tag{4.13}$$

From the equation corresponding to $j = 0$ we obtain

$$u|_{t=0} = \phi_0(x).$$

Using this equality, we get from the equation corresponding to $j = 1$ the condition

$$\left.\frac{\partial u}{\partial t}\right|_{t=0} = \varphi_1(x)$$

and so on.

We define $u(x, t)$ from the Cauchy problem

$$\frac{\partial u}{\partial t} - \Delta u = u^{(1)}(x, t), \quad u|_{t=0} = \psi_0(x) = \varphi_0(x).$$

the function $u^{(1)}$ from the problem

$$\frac{\partial u^{(1)}}{\partial t} - \Delta u^{(1)} = u^{(2)}, \quad u^{(1)}|_{t=0} = \psi_1(x)$$

and so on, while the function $u^{([l/2])}$ is defined from the problem

$$\left(\frac{\partial}{\partial t} - \Delta\right) u^{\left(\left[\frac{l}{2}\right]\right)} = 0, \quad u^{\left(\left[\frac{l}{2}\right]\right)}\Big|_{t=0} = \psi_{\left[\frac{l}{2}\right]}(x).$$

Since $u^{(j)} = (\partial/\partial t - \Delta)^j u$, the function $u(x, t)$ satisfies conditions (4.12) and consequently (4.8). In addition, by virtue of Theorem 4.3

$$\langle u\rangle^{(l)}_{D_{n+1}} \leqslant c_1\left(\langle u^{(1)}\rangle^{(l-2)}_{D_{n+1}} + \langle\psi_0\rangle^{(l)}_{E_n}\right)$$

$$\leqslant c_2\left(\langle u^{(2)}\rangle^{(l-4)}_{D_{n+1}} + \langle\psi_0\rangle^{(l)}_{E_n} + \langle\psi_1\rangle^{(l-2)}_{E_n}\right) \leqslant \cdots \leqslant c\sum_{j=0}^{\left[\frac{l}{2}\right]}\langle\psi_j\rangle^{(l-2j)}_{E_n}.$$

But (4.11) implies

$$\langle\psi_j\rangle^{(l-2j)}_{E_n} \leqslant c\sum_{s=0}^{j}\langle\varphi_{j-s}\rangle^{(l-2j+2s)}_{E_n},$$

so that

$$\sum_{j=0}^{\left[\frac{l}{2}\right]}\langle\psi_j\rangle^{(l-2j)}_{E_n} \leqslant c\sum_{j=0}^{\left[\frac{l}{2}\right]}\langle\varphi_j\rangle^{(l-2j)}_{E_n},$$

and inequality (4.10) is proved.

We have assumed that $l > 2$. For $l < 2$ it is possible to take $u(x, t) = \phi_0(x)$.

It remains to prove that inequality (4.9) is fulfilled for any finite T. Here we make use of the identity

$$g(x, t) = \sum_{i=0}^{k}\frac{t^i}{i!}\frac{\partial^i g(x, \tau)}{\partial\tau^i}\Big|_{t=0}$$

$$+ \frac{1}{(k-1)!}\int_0^t (t-\tau)^{k-1}\left[\frac{\partial^k g(x, \tau)}{\partial\tau^k} - \frac{\partial^k g(x, \tau)}{\partial\tau^k}\Big|_{\tau=0}\right]d\tau.$$

This identity implies that for any $\alpha < 1$

$$|g|^{(0)}_{D^{(\tau)}_{n+1}} \leqslant c_1(T)\sum_{i=0}^{k}\left|\left.\frac{\partial^i g(\cdot, \tau)}{\partial\tau^i}\right|_{\epsilon=0}\right|^{(0)}_{E_n} + c_2(T)\langle g\rangle^{(k+\alpha)}_{t, D^{(\tau)}_{n+1}}. \tag{4.14}$$

We take the function $D_t^r D_x^s u$ with $2r + s < l$ as g and put

$$k = \left[\frac{l-2r-s}{2}\right], \quad \alpha = \frac{l-2r-s}{2} - \left[\frac{l-2r-s}{2}\right].$$

Estimating the second term of the right side of (4.14) with the already proved inequality (4.10), we obtain estimate (4.9).

Thus the theorem is completely proved.

An analogous result for the spaces $W_q^{2m,m}(D_{n+1}^{(T)})$ is formulated in the following manner.

Theorem 4.4. *Let there be given functions* $\phi_j \in W_q^{2m-2j-2/q}(E_n)$ $(j = 0, \cdots, m-1;\ m \geq 1)$. *It is possible to construct in the domain* D_{n+1} *a function* $u(x, t)$ *such that for any* $T > 0$

$$u \in W_q^{2m,\,m}(D_{n+1}^{(T)}), \quad \left.\frac{\partial^j u}{\partial t^j}\right|_{t=0} = \varphi_j(x),$$

$$\|u\|_{q,\,D_{n+1}^{(T)}}^{(2m)} \leqslant c \sum_{j=0}^{m-1} \|\varphi_j\|_{q,\,E_n}^{\left(2m-2j-\frac{2}{q}\right)}$$

and

$$\langle\langle u \rangle\rangle_{q,\,D_{n+1}}^{(2m)} \leqslant c \sum_{j=0}^{m-1} \langle\langle \varphi_j \rangle\rangle_{q,\,E_n}^{\left(2m-2j-\frac{2}{q}\right)}.$$

This result is well known [9; 49c]. Theorem 4.3 is also known, but in the theory of functions it is normally considered as a special case of corresponding results for spaces $H_q^{l,l/2}$ of S. M. Nikol'skiĭ [89], which we will not use.

Let us now formulate an important definition for what follows.

The set of functions from class $H^{l,l/2}(\overline{Q}_T)$ satisfying the zero initial conditions

$$\left.\frac{\partial^k u}{\partial t^k}\right|_{t=0} = 0 \quad \left(k = 0, \ldots, \left[\frac{l}{2}\right]\right) \tag{4.15}$$

is called the space $\overset{\circ}{H}{}^{l,l/2}(\overline{Q}_T)$.

Analogously, the set of functions from class $W_q^{2m,m}(Q_T)$ satisfying the zero initial conditions

$$\left.\frac{\partial^k u}{\partial t^k}\right|_{t=0} = 0 \qquad (k = 0, \ldots, m-1)$$

is called the space $\overset{\circ}{W}{}_q^{2m,m}(Q_T)$.

These spaces possess the following property: the zero extension of functions from these spaces into the domain $t < 0$ is an extension with preservation of class.

The following assertion is valid for functions from $\overset{\circ}{H}{}^{l,l/2}(\overline{Q}_T)$.

Lemma 4.1. *Let $\Omega' \subseteq \Omega$ and $Q'_\tau = \Omega' \times (0, \tau)$. If $u \in \overset{\circ}{H}{}^{l,l/2}(Q'_\tau)$, then for all integral $j < l$*

$$\langle u \rangle^{(j)}_{Q'_\tau} \leqslant c\tau^{\frac{l-j}{2}} \langle u \rangle^{\left(\frac{l}{2}\right)}_{t, Q'_\tau}.$$

This lemma is obvious.

Let us take an arbitrary small λ and consider the system of subsets $\mathfrak{K}^{(\lambda)}$. Let $Q^{(k)}_\tau = \Omega^{(k)} \times (0, \tau)$. The following norm is obviously finite for every function $u \in H^{l,l/2}(\overline{Q}_\tau)$:

$$\{u\}^{(l)}_{Q_\tau} = \sup_k \langle u \rangle^{(l)}_{Q^{(k)}_\tau}. \tag{4.16}$$

Lemma 4.2. *Suppose*

$$\tau = \lambda^2 \kappa, \tag{4.17}$$

with $\kappa \leq 1$, $\lambda^2 \kappa \leq T$. Then in the class $\overset{\circ}{H}{}^{l,l/2}(\overline{Q}_T)$ the norms $\langle u \rangle^{(l)}_{Q_\tau}$ and $\{u\}^{(l)}_{Q_\tau}$ are equivalent:

$$\{u\}^{(l)}_{Q_\tau} \leqslant \langle u \rangle^{(l)}_{Q_\tau} \leqslant c\, \{u\}^{(l)}_{Q_\tau} \tag{4.18}$$

and the constant c does not depend on λ and κ.

Proof. The left inequality of (4.18) is completely obvious. It is also obvious that

$$\langle u \rangle^{\left(\frac{l}{2}\right)}_{t, Q_\tau} \leqslant c \sup_k \langle u \rangle^{\left(\frac{l}{2}\right)}_{t, Q^{(k)}_\tau} \leqslant c\, \{u\}^{(l)}_{Q_\tau}.$$

It remains to show that $\langle u \rangle^{(l)}_{x, Q_\tau} < c\, \{u\}^{(l)}_{Q_\tau}$. Let $2r + s = l'$; we consider the function

$$\frac{1}{|x - x'|^{l-l'}} \left| D^r_t D^s_x u(x, t) - D^r_t D^s_{x'} u(x', t) \right|.$$

Let x_0, x'_0 and t_0 be points such that

$$\frac{1}{|x_0 - x'_0|^{l-l'}} \left| D^r_t D^s_x u(x, t) \Big|_{\substack{x = x_0 \\ t = t_0}} - D^r_t D^s_x u(x, t) \Big|_{\substack{x = x'_0 \\ t = t_0}} \right|$$
$$\geqslant \frac{1}{2} \sup_{x, x', t} \frac{1}{|x - x'|^{l-l'}} \left| D^r_t D^s_x u(x, t) - D^r_t D^s_{x'} u(x', t) \right|.$$

If $|x_0 - x_0'| \geq d\lambda$, then

$$\frac{1}{|x_0 - x_0'|^{l-l'}} \left| D_t^r D_x^s u(x, t)\Big|_{\substack{x=x_0\\ t=t_0}} - D_t^r D_x^s u(x, t)\Big|_{\substack{x=x_0'\\ t=t_0}} \right|$$
$$\leqslant \frac{2}{d^{l-l'}\lambda^{l-l'}} \sup_{Q_\tau} |D_t^r D_x^s u| = \frac{1}{d^{l-l'}\lambda^{l-l'}} \sup_k \sup_{Q_\tau^{(k)}} |D_t^r D_x^s u|.$$

Now by virtue of Lemma 4.1

$$\sup_{Q_\tau^{(k)}} |D_t^r D_x^s u(x, t)| \leqslant c\lambda^{l-l'} \varkappa^{\frac{l-l'}{2}} \langle u\rangle_{t, Q_\tau^{(k)}}^{\left(\frac{l}{2}\right)}$$

and, consequently, for $|x_0 - x_0'| \geq d\lambda$

$$\langle u\rangle_{x, Q_\tau}^{(l)} \leqslant c\, \frac{\varkappa^{\frac{l-l'}{2}}}{d^{l-l'}} \{u\}_{Q_\tau}^{(l)} \leqslant \frac{c}{d^{l-l'}} \{u\}_{Q_\tau}^{(l)}.$$

But if $|x_0 - x_0'| \leq d\lambda$, then by virtue of property 2 of the sets $\Omega^{(k)}$ the points x_0 and x_0' belong to some one set $\Omega^{(k_0)}$, and therefore

$$\langle u\rangle_{x, Q_\tau}^{(l)} \leqslant 2\, \{u\}_{Q_\tau}^{(l)}.$$

The lemma is proved.

Lemma 4.3. *Suppose a function $\psi_k(x)$ defined in $\Omega^{(k)}$ possesses the property*

$$|D_x^s \psi_k(x)| \leqslant \frac{c_s}{\lambda^s} \quad (s \leqslant [l] + 1), \tag{4.19}$$

while the numbers τ and λ are connected by relation (4.17). *Then, for any function $u \in \overset{\circ}{H}{}^{l, l/2}(Q_\tau^{(k)})$,*

$$\langle \psi_k u\rangle_{Q_\tau^{(k)}}^{(l)} \leqslant c\, \langle u\rangle_{Q_\tau^{(k)}}^{(l)}, \tag{4.20}$$

in which the constant c does not depend on λ and τ.

Proof. It is easy to see that

$$\langle \psi_k u\rangle_{x, Q_\tau^{(k)}}^{(l)} \leqslant c \sum_{j=0}^{[l]} \left(\langle \psi_k\rangle_{\Omega^{(k)}}^{(j)} \langle u\rangle_{x, Q_\tau^{(k)}}^{(l-j)} + \langle u\rangle_{Q_\tau^{(k)}}^{(j)} \langle \psi_k\rangle_{\Omega^{(k)}}^{(l-j)} \right),$$

from which, by virtue of (4.19), we get

$$\langle \psi_k u\rangle_{x, Q_\tau^{(k)}}^{(l)} \leqslant c \sum_{j=0}^{[l]} \left(\frac{1}{\lambda^j} \langle u\rangle_{x, Q_\tau^{(k)}}^{(l-j)} + \frac{1}{\lambda^{l-j}} \langle u\rangle_{Q_\tau^{(k)}}^{(j)} \right)$$
$$\leqslant c \left(\langle u\rangle_{x, Q_\tau^{(k)}}^{(l)} + \sum_{j=0}^{[l]} \frac{1}{\lambda^{l-j}} \langle u\rangle_{Q_\tau^{(k)}}^{(j)} \right).$$

We have made use of the fact that for $j > 0$

$$\langle u\rangle^{(l-j)}_{x,\,Q^{(k)}_\tau} \leqslant c\lambda^{1-(l-[l])}\langle u\rangle^{[l]+1-j}_{Q^{(k)}_\tau}.$$

One estimates $\langle\psi_k u\rangle^{(l/2)}_{t,Q^{(k)}_T}$ in an analogous manner. Using Lemma 4.1 for an estimate of $\langle u\rangle^{(j)}_{Q^{(k)}_T}$ and taking into consideration (4.17), we obtain the required inequality (4.20). The lemma is proved.

In what follows we will frequently have to estimate functions of the form

$$w(x,\ t) = \sum_k w^{(k)}(x,\ t), \tag{4.21}$$

where $w^{(k)} = 0$ for $x \in \Omega\setminus\Omega^{(k)}$.

Lemma 4.4.

$$|w|^{(l)}_{Q_\tau} \leqslant N_0 \sup_k \langle w^{(k)}\rangle^{(l)}_{Q^{(k)}_\tau}. \tag{4.22}$$

Proof. Since

$$\langle w\rangle^{(l)}_{Q^{(k)}_\tau} \leqslant \sum_{k'} \langle w^{(k')}\rangle^{(l)}_{Q^{(k)}_\tau}$$

$$= \sum_{s=1}^{N_0} \langle w^{(k_s)}\rangle^{(l)}_{Q^{(k)}_\tau} \leqslant \sum_{s=1}^{N_0} \langle w^{(k_s)}\rangle^{(l)}_{Q^{(k_s)}_\tau} \leqslant N_0 \sup_{k'} \langle w^{(k')}\rangle^{(l)}_{Q^{(k')}_\tau},$$

it follows that

$$|w|^{(l)}_{Q_\tau} = \sup_k \langle w\rangle^{(l)}_{Q^{(k)}_\tau} \leqslant N_0 \sup_k \langle w^{(k)}\rangle^{(l)}_{Q^{(k)}_\tau}$$

and (4.22) is proved.

Corollary. *Suppose* (4.17) *holds and*

$$\tilde{w}(x,\ t) = \sum_k \psi^{(k)}(x)\,\tilde{w}^{(k)}(x,\ t), \tag{4.23}$$

where $\tilde{w}^{(k)} \in \overset{\circ}{H}{}^{l,l/2}(\overline{Q^{(k)}_T})$, *while the* $\psi^{(k)}$ *are functions, each of which is different from zero only in* $\Omega^{(k)}$ *and satisfies inequality* (4.19) *Then*

$$|\tilde{w}|^{(l)}_{Q_\tau} \leqslant c \sup_k \langle \tilde{w}^{(k)}\rangle^{(l)}_{Q^{(k)}_\tau},$$

with the constant c *being independent of* λ *and* τ.

Let us turn to a consideration of the space $\overset{\circ}{W}{}^{2m,m}_q(Q_T)$. We first note that by virtue of property 3 of the domains $\Omega^{(k)}$ the norm $\|u\|_{q,Q_T}$ is equivalent to the norm

$$\{u\}_{q,Q_\tau}=\Big(\sum_k \|u\|^q_{q,Q^{(k)}_\tau}\Big)^{\frac{1}{q}},$$

since

$$\|u\|_{q,Q_\tau}\leqslant\{u\}_{q,Q_\tau}\leqslant N_0^{\frac{1}{q}}\|u\|_{q,Q_\tau}.$$

Consequently the norm $\langle\langle u\rangle\rangle^{(2m)}_{q,Q_T}$ is equivalent to the norm

$$\{u\}^{(2m)}_{q,Q_\tau}=\Big[\sum_k\Big(\langle\langle u\rangle\rangle^{(2m)}_{q,Q^{(k)}_\tau}\Big)^q\Big]^{\frac{1}{q}},$$

i.e.

$$\langle\langle u\rangle\rangle^{(2m)}_{q,Q_\tau}\leqslant\{u\}^{(2m)}_{q,Q_\tau}\leqslant c\,\langle\langle u\rangle\rangle^{(2m)}_{q,Q_\tau}.$$

In the space $\overset{\circ}{W}{}^{2m,m}_q(Q_T)$ the norms $\langle\langle u\rangle\rangle^{(2m)}_{q,Q_T}$ and $\|u\|^{(2m)}_{q,Q_T}$ are equivalent. This is implied by the following lemma, which is analogous to Lemma 4.1.

Lemma 4.5. *If* (4.17) *holds, then for any* $u\in\overset{\circ}{W}{}^{2m,m}_q(Q^{(k)}_T)$

$$\langle\langle u\rangle\rangle^{(j)}_{q,Q^{(k)}_\tau}\leqslant c\tau^{m-\frac{j}{2}}\langle\langle u\rangle\rangle^{(2m)}_{q,Q^{(k)}_\tau}\tag{4.24}$$

and for any $u\in\overset{\circ}{W}{}^{2m,m}_q(Q_T)$

$$\langle\langle u\rangle\rangle^{(j)}_{q,Q_\tau}\leqslant c\tau^{m-\frac{j}{2}}\langle\langle u\rangle\rangle^{(2m)}_{q,Q_\tau},\tag{4.24'}$$

where $j<2m$.

Proof. Let us prove (4.24). If $j=2m'<2m$ the proof is obvious. For $j=2m'+1$ one must use inequality (3.15) of Chapter II, which in the present case is valid for $\delta=\tau^{1/2}$ by virtue of (4.17). As a result we obtain

$$\langle\langle u\rangle\rangle^{(j)}_{q,Q^{(k)}_\tau}\leqslant\tau^{m-\frac{j}{2}}\langle\langle u\rangle\rangle^{(2m)}_{q,Q^{(k)}_\tau}+c\tau^{-\frac{j}{2}}\|u\|_{q,Q^{(k)}_\tau},$$

and since for $u\in\overset{\circ}{W}{}^{2m,m}_q(Q^{(k)}_T)$

$$\|u\|_{q,Q^{(k)}_\tau}\leqslant c\tau^m\Big\|\frac{\partial^m u}{\partial t^m}\Big\|_{q,Q^{(k)}_\tau}\leqslant c\tau^m\langle\langle u\rangle\rangle^{(2m)}_{q,Q^{(k)}_\tau},$$

it follows that (4.24) is proved. Inequality (4.24′) is proved in exactly the same way.

Lemma 4.6. *If $\psi_k(x)$ possesses property (4.19) for $s \le 2m$ and (4.17) holds, then*

$$\langle\langle \psi_k u\rangle\rangle^{(2m)}_{q,\, Q^{(k)}_\tau} \leqslant c\, \langle\langle u\rangle\rangle^{(2m)}_{q,\, Q^{(k)}_\tau}$$

with the constant c being independent of λ and τ.

Indeed, with the use of Lemma 4.5 and inequality (4.17) we obtain

$$\langle\langle \psi_k u\rangle\rangle^{(2m)}_{q,\, Q^{(k)}_\tau} \leqslant c \sum_{j=0}^{2m} \frac{1}{\lambda^j} \langle\langle u\rangle\rangle^{(2m-j)}_{q,\, Q^{(k)}_\tau} \leqslant c\, \langle\langle u\rangle\rangle^{(2m)}_{q,\, Q^{(k)}_\tau}.$$

Lemma 4.7. *If (4.17) holds, then, for a function $w(x, t)$ of (4.21),*

$$\{w\}^{(2m)}_{q,\, Q_\tau} \leqslant c \left[\sum_k \left(\langle\langle w^{(k)}\rangle\rangle^{(2m)}_{q,\, Q^{(k)}_\tau}\right)^q\right]^{\frac{1}{q}}$$

with the constant c depending only on N_0.

Proof. We have

$$\begin{aligned}
\{w\}^{(2m)}_{q,\, Q_\tau} &= \left[\sum_k \left(\langle\langle w\rangle\rangle^{(2m)}_{q,\, Q^{(k)}_\tau}\right)^q\right]^{\frac{1}{q}} \\
&\leqslant \left[\sum_k \left(\sum_{k'} \langle\langle w^{(k')}\rangle\rangle^{(2m)}_{q,\, Q^{(k)}_\tau}\right)^q\right]^{\frac{1}{q}} \\
&= \left[\sum_k \left(\sum_{s=1}^{N_0} \langle\langle w^{(k_s)}\rangle\rangle^{(2m)}_{q,\, Q^{(k)}_\tau}\right)^q\right]^{\frac{1}{q}} \\
&\leqslant c(N_0) \left[\sum_k \sum_{s=1}^{N_0} \left(\langle\langle w^{(k_s)}\rangle\rangle^{(2m)}_{q,\, Q^{(k)}_\tau}\right)^q\right]^{\frac{1}{q}} \\
&= c(N_0) \left[\sum_k \sum_{k'} \left(\langle\langle w^{(k')}\rangle\rangle^{(2m)}_{q,\, Q^{(k)}_\tau}\right)^q\right]^{\frac{1}{q}} \\
&= c(N_0) \left[\sum_{k'} \sum_{s=1}^{N_0} \left(\langle\langle w^{(k')}\rangle\rangle^{(2m)}_{q,\, Q^{(k_s)}_\tau}\right)^q\right]^{\frac{1}{q}} \\
&\leqslant N_0^{\frac{1}{q}} c(N_0) \left[\sum_{k'} \left(\langle\langle w^{(k')}\rangle\rangle^{(2m)}_{q,\, Q^{(k')}_\tau}\right)^q\right]^{\frac{1}{q}},
\end{aligned}$$

Q.E.D.

Lemmas 4.6 and 4.7 imply

Corollary. *If (4.17) holds, then the function (4.23) is subject to the inequality*

$$\{\tilde{w}\}^{(2m)}_{q,\, Q_\tau} \leqslant c \left[\sum_k \left(\langle\langle \tilde{w}^{(k)}\rangle\rangle^{(2m)}_{q,\, Q^{(k)}_\tau}\right)^q\right]^{\frac{1}{q}}. \tag{4.25}$$

in which the constant c *does not depend on* λ *and* τ.

Lemma 4.8. *Let* $S \in O^{2m}$. *Then every function* $u(x, t) \in \overset{\circ}{W}{}_q^{2m,m}(Q_\tau)$ *can be approximated in the norm* $\|u\|_{q,Q_\tau}^{(2m)}$ *by infinitely differentiable functions that are equal to zero for small* t.

Proof. We extend the function $u(x, t)$ with preservation of class into the domain $D_{n+1}^{(\tau)}$, and then into $E_{n+1}^{(\tau)}$, putting $u(x, t) = 0$ for $t < 0$. Finally, we extend u with preservation of class from $E_{n+1}^{(\tau)}$ onto the whole space E_{n+1}. Under such an extension

$$\|u\|_{q,E_{n+1}}^{(2m)} \leqslant c\|u\|_{q,E_{n+1}^{(\tau)}}^{(2m)} = c\|u\|_{q,D_{n+1}^{(\tau)}}^{(2m)} \leqslant c\|u\|_{q,Q_\tau}^{(2m)}.$$

The sequence of functions

$$u_n(x, t) = u\left(x, t - \frac{1}{n}\right),$$

each of which is equal to zero for $t < 1/n$, converges to $u(x, t)$ in the norm $\|u\|_{q,E_{n+1}}^{(2m)}$. Now each function u_n can be approximated by averagings of it, viz., infinitely differentiable functions that vanish for small t. The assertion of the lemma follows.

If $u(x, t) \in W_q^{2m,m}(Q_\tau)$, then, as is seen from Lemma 3.4 of Chapter II, the values of u and certain of its derivatives on S_τ belong to the classes $W_q^{s-1/q, \frac{1}{2}(s-1/q)}(S_\tau)$, where s is an integer, and inequality (3.20) of Chapter II holds. If the function $u(x, t)$ is given in the domain $\widetilde{D}_{n+1}^{(T)}$ and $u \in W_q^{2m,m}(\widetilde{D}_{n+1}^{(T)})$, then, besides inequality (3.20), an analogous inequality is fulfilled between the principal parts of the norms in (3.20). Namely, we have

Theorem 4.5. *For any function* $u \in W_q^{2m,m}(\widetilde{D}_{n+1}^{(T)})$

$$\left\langle\left\langle D_t^r D_x^s u\,\big|_{x_n=0}\right\rangle\right\rangle_{q,\widetilde{E}_n^{(T)}}^{\left(2m-2r-s-\frac{1}{q}\right)} \leqslant c\,\langle\langle u\rangle\rangle_{q,\widetilde{D}_{n+1}^{(T)}}^{(2m)},$$

in which $2r + s < 2m$ *and the constant* c *does not depend on* T.

The norms in the spaces $H^{l,l/2}(S_\tau)$ and $W_q^{l,l/2}(S_\tau)$ (l a nonintegral number), as was indicated in §3 of Chapter II, are usually determined by means of a parametrization of the surface S_τ, viz., a partitioning of it into sufficiently fine parts and a mapping of each part into the space $\widetilde{E}_n$. In this chapter we will always make use of a parametrization connected with a partitioning of the surface S into subsets $S^{(k)} = \overline{\Omega}^{(k)} \cap S$ (we denote the systems of sets $s^{(k)} = \overline{\omega}^{(k)} \cap S$ by $\mathfrak{f}^{(\lambda)}$, while the sets $S^{(k)}$ are denoted by $\mathfrak{S}^{(\lambda)}$), and consequently of the surface S_τ into subsets $S_\tau^{(k)} = S^{(k)} \times (0, \tau)$. They are mapped into the space $\widetilde{E}_n$ by the transformation of space coordinates (4.3). Under this transformation $S^{(k)}$ goes over into the $(n-1)$-dimensional cube σ

$$|z_\alpha| < \lambda \quad (\alpha = 1, \ldots, n-1)$$

and $S_\tau^{(k)}$ goes into $\sigma_\tau = \sigma \times (0, \tau)$.

The space of functions from $H^{l,l/2}(S_\tau)$ satisfying the initial conditions (4.15) is denoted by $\overset{\circ}{H}{}^{l,l/2}(S_\tau)$. In this space we have the equivalent norms

$$|u|_{S_\tau}^{(l)} = \sup_{k \in \mathfrak{N}} |u^{(k)}(z', t)|_{\sigma_\tau}^{(l)}$$

and

$$\{u\}_{S_\tau}^{(l)} = \sup_{k \in \mathfrak{N}} \langle u^{(k)}(z', t)\rangle_{\sigma_\tau}^{(l)},$$

in which $u^{(k)}(z', t)$ $(z' \in \sigma, k \in \mathfrak{N})$ is the function $u(x, t)$ given on $S_\tau^{(k)}$ in the coordinates of (4.3), i.e. in local coordinates at the point $\xi^{(k)} \in S^{(k)}$. Also, Lemmas 4.3 and 4.4 are valid in the spaces $\overset{\circ}{H}{}^{l,l/2}(S_\tau)$.

Let us now introduce the spaces $\overset{\circ}{W}{}_q^{l,l/2}(\widetilde{D}_n^{(\tau)})$ with nonintegral $l > 0$. We base the introduction of these spaces on the following natural principle: if a function $u \in \overset{\circ}{W}{}_q^{l,l/2}(\widetilde{D}_n^{(\tau)})$ is extended by zero into the domain $t < 0$, then the resultant function u^0 must belong to the class $W_q^{l,l/2}(\widetilde{E}_n^{(\tau)})$. Let us determine what conditions must be satisfied by the function u for this to be true. We first consider the case when $\lambda = l/2 < 1$. We calculate the norm

$$[u^0]_{q, E_n^{(\tau)}}^{(l)} = \left(\sum_{i=1}^{n-1} \int_{-\infty}^{\tau} dt \int_{E_{n-1}} dx' \int_{E_{n-1}} \left| \frac{\partial^{[l]} u^0(x', t)}{\partial x_i^{[l]}} - \frac{\partial^{[l]} u^0(y', t)}{\partial y_i^{[l]}} \right|^q \frac{dy'}{|x'-y'|^{n-1+q(l-[l])}} \right.$$
$$\left. + \int_{E_{n-1}} dx' \int_{-\infty}^{\tau} dt \int_{-\infty}^{\tau} |u^0(x', t) - u^0(x', t')|^q \frac{dt'}{|t-t'|^{1+q\lambda}} \right)^{\frac{1}{q}}$$

This norm is equivalent to the norm $\langle\langle u^0\rangle\rangle_{q, \widetilde{E}_n^{(\tau)}}^{(l)}$. For $l < 1$ this is obvious; for $l > 1$ the second norm contains the terms

$$\int\limits_{E_{n-1}} dx' \int\limits_{-\infty}^{T} dt \int\limits_{-\infty}^{T} \left| \frac{\partial u^0(x', t)}{\partial x_i} - \frac{\partial u^0(x', t')}{\partial x_i} \right|^2 \frac{dt'}{|t-t'|^{1+q\frac{l-1}{2}}} \qquad (i = 1, \cdots, n-1),$$

which are absent in the norm $[u^0]^{(l)}_{q, \widetilde{E}_n^{(T)}}$. They are estimated in terms of this norm with the use of embedding theorems for the spaces $W_q^{l, l/2}$ with nonintegral l [9].

We do not present these theorems here; we merely note that a limiting special case of them ($q \to \infty$) is given by Lemma 3.1 of Chapter II.

The first summand of the norm $[u^0]^{(l)}_{q, \widetilde{E}_n^{(T)}}$ is obviously equal to

$$\sum_{i=1}^{n-1} \int\limits_0^{\tau} dt \int\limits_{E_{n-1}} dx' \int\limits_{E_{n-1}} \left| \frac{\partial^{[l]} u(x', t)}{\partial x_i^{[l]}} - \frac{\partial^{[l]} u(y', t)}{\partial y_i^{[l]}} \right|^q \frac{dy'}{|x'-y'|^{n-1+q(l-[l])}},$$

while the second is transformed in the following manner:

$$\begin{aligned} &\int\limits_{E_{n-1}} dx' \int\limits_{-\infty}^{\tau} dt \int\limits_{-\infty}^{\tau} |u^0(x', t) - u^0(x', t')|^q \frac{dt'}{|t-t'|^{1+q\lambda}} \\ &\quad = \int\limits_{E_{n-1}} dx' \int\limits_0^{\tau} dt \int\limits_0^{\tau} |u(x', t) - u(x', t')|^q \frac{dt'}{|t-t'|^{1+q\lambda}} \\ &\quad + \int\limits_{E_{n-1}} dx' \int\limits_0^{\tau} |u(x', t)|^q dt \int\limits_{-\infty}^{0} \frac{dt'}{|t-t'|^{1+q\lambda}} \\ &\quad + \int\limits_{E_{n-1}} dx' \int\limits_0^{\tau} |u(x', t')|^q dt' \int\limits_{-\infty}^{0} \frac{dt}{|t-t'|^{1+q\lambda}} \\ &= \int\limits_{E_{n-1}} dx' \int\limits_0^{\tau} dt \int\limits_0^{\tau} |u(x', t) - u(x', t')|^q \frac{dt'}{|t-t'|^{1+q\lambda}} \\ &\qquad + \frac{2}{q\lambda} \int\limits_{E_{n-1}} dx' \int\limits_0^{\tau} \frac{|u(x', t)|^q \, dt}{t^{q\lambda}}. \end{aligned}$$

Thus

$$\left([u^0]^{(l)}_{q,\,\widetilde{E}^{(\tau)}_n}\right)^q=\left([u]^{(l)}_{q,\,\widetilde{D}^{(\tau)}_n}\right)^q+\frac{2}{q\lambda}\int\limits_{E_{n-1}}dx'\int\limits_0^{\tau}\frac{|u(x',t)|^q\,dt}{t^{q\lambda}}. \tag{4.26}$$

Concerning the latter term of the right side of (4.26) one can show the following. If $q\lambda<1$, then

$$\int\limits_{E_{n-1}}dx'\int\limits_0^{\tau}\frac{|u(x',t)|^q\,dt}{t^{q\lambda}}\leqslant c_1\int\limits_{E_{n-1}}dx'\int\limits_0^{\tau}dt\int\limits_0^{\tau}\frac{|u(x',t)-u(x',t')|^q}{|t-t'|^{1+q\lambda}}dt'+\frac{c_2}{\tau^{q\lambda}}\int\limits_{E_{n-1}}dx'\int\limits_0^{\tau}|u(x',t)|^q\,dt. \tag{4.27}$$

If $q\lambda>1$, then this integral is bounded only if $u(x,0)=0$ (such a limiting value of the function $u(x,t)$ is meaningful for $q\lambda>1$), in which case

$$\int\limits_{E_{n-1}}dx'\int\limits_0^{\tau}\frac{|u(x',t)|^q\,dt}{t^{q\lambda}}\leqslant c\int\limits_{E_{n-1}}dx'\int\limits_0^{\tau}dt\int\limits_0^{\tau}\frac{|u(x',t)-u(x',t')|^q}{|t-t'|^{1+q\lambda}}dt'. \tag{4.28}$$

But if $q\lambda=1$, then an estimate in the form of (4.27) or (4.28) does not hold (see in this connection [116b]).

In view of all that has been said it is natural to define the space $\overset{\circ}{W}{}^{l,l/2}_q(\widetilde{D}^{(\tau)}_n)$ for $l<2$ as the set of functions with finite norm

$$|u|^{(l)}_{q,\,\widetilde{D}^{(\tau)}_n}=\langle\langle u\rangle\rangle^{(l)}_{q,\,\widetilde{D}^{(\tau)}_n}+\left(\int\limits_0^{\tau}\frac{dt}{t^{\frac{q}{2}l}}\int\limits_{E_{n-1}}|u(x',t)|^q\,dx'\right)^{\frac{1}{q}}.$$

For $l/2>1/q$ this means that the condition $u(x,0)=0$ is fulfilled.

For $l>2$ the space $\overset{\circ}{W}{}^{l,l/2}_q(D^{(\tau)}_n)$ is naturally defined as the set of functions satisfying the zero initial conditions

$$\left.\frac{\partial^k u}{\partial t^k}\right|_{t=0}=0\ \left(k=0,\ \ldots,\ \left[\frac{l}{2}\right]-1\right) \tag{4.29}$$

and, in addition, having the finite norm

$$|u|^{(l)}_{q,\,\widetilde{D}^{(\tau)}_n}=\langle\langle u\rangle\rangle^{(l)}_{q,\,\widetilde{D}^{(\tau)}_n}+\left(\int_0^\tau\frac{dt}{t^{q\lambda}}\int_{E_{n-1}}\left|\frac{\partial^{\left[\frac{l}{2}\right]}u(x',t)}{\partial t^{\left[\frac{l}{2}\right]}}\right|^q dx'\right)^{\frac{1}{q}},$$

where $\lambda=l/2-[l/2]$ (if $\lambda>1/q$ this norm is finite only for $D_t^{[l/2]}u|_{t=0}=0$).

Theorem 4.6. *For any function* $u\in\overset{\circ}{W}{}^{2m,m}_q(R^{(\tau)})$

$$\left|D_t^rD_x^su\,|_{x_n=0}\right|^{\left(2m-2r-s-\frac{1}{q}\right)}_{q,\,\widetilde{D}^{(\tau)}_n}\leqslant c\,\langle\langle u\rangle\rangle^{(2m)}_{q,\,R^{(\tau)}} \tag{4.30}$$

$(2r+s<2m)$.

Indeed, if the function $u(x,t)$ is extended into the domain $t<0$ by zero, we will have, on the one hand,

$$\langle\langle u\rangle\rangle^{(2m)}_{q,\,\widetilde{D}^{(\tau)}_{n+1}}=\langle\langle u\rangle\rangle^{(2m)}_{q,\,R^{(\tau)}}$$

and, on the other hand, by virtue of (4.26)

$$\left|D_t^rD_x^su|_{x_n=0}\right|^{\left(2m-2r-s-\frac{1}{q}\right)}_{q,\,\widetilde{D}^{(\tau)}_n}\leqslant c\left\langle\left\langle D_t^rD_x^su|_{x_n=0}\right\rangle\right\rangle^{\left(2m-2r-s-\frac{1}{q}\right)}_{q,\,\widetilde{E}^{(\tau)}_n}.$$

It only remains to apply Theorem 4.5.

In exactly the same way as Lemma 4.7, one proves

Lemma 4.9. *Every function* $u\in\overset{\circ}{W}{}^{l,l/2}(\widetilde{D}^{(\tau)}_n)$ *can be approximated in the norm* $|u|^{(l)}_{q,\widetilde{D}^{(\tau)}_n}$ *by infinitely differentiable functions that are equal to zero for small* t.

The space $\overset{\circ}{W}{}^{l,l/2}_q(S_T)$ for nonintegral $l>0$ is now defined as the set of functions satisfying for $l>2$ conditions (4.29) and having the finite norm

$$|u|^{(l)}_{q,\,S_\tau}=\left[\sum_{k\in\mathfrak{N}}\left(|u^{(k)}(z',t)|^{(l)}_{q,\,\sigma_\tau}\right)^q\right]^{\frac{1}{q}},$$

where

$$|u^{(k)}|^{(l)}_{q,\,\sigma_\tau}=\langle\langle u^{(k)}\rangle\rangle^{(l)}_{q,\,\sigma_\tau}+\left(\int_0^\tau\frac{dt}{t^{q\lambda}}\int_\sigma\left|\frac{\partial^{\left[\frac{l}{2}\right]}u^{(k)}(z',t)}{\partial t^{\left[\frac{l}{2}\right]}}\right|^q dz'\right)^{\frac{1}{q}}.$$

with $\langle\langle u\rangle\rangle^{(l)}_{q,\sigma_T}$ being defined by equality (3.17) of Chapter II.

It is not difficult to show on the basis of Theorem 4.6 that the inequality

$$\{D_t^r D_x^s u\}^{(2m-2r-s-1/q)}_{q,S_T} \leq c\ \langle\langle u\rangle\rangle^{(2m)}_{q,Q_T} \tag{4.31}$$

which is analogous to (4.30), is valid for any $u \in \overset{\circ}{W}{}^{2m,m}_q$. For $m-r-s/2-1/2q = k+1/q$, where k is an integer, the norm on the left side is stronger than the norm $\|D_t^r D_x^s u\|^{(2m-2r-s-1/q)}_{q,S_T}$.

It is possible to show that the characteristic given by (4.31) of the properties of the boundary values of functions from the classes $\overset{\circ}{W}{}^{2m,m}_q(Q_T)$ and of their derivatives on the surface S_T is exact in the following sense. It is possible to find a function $u \in \overset{\circ}{W}{}^{2m,m}_q(Q_T)$, whose norm can be estimated in terms of corresponding norms of a certain number of its normal derivatives on S_T of the order r, $0 \leq r \leq 2m-1$, which are given. One such possible extension of a function from S_T into Q_T is obtained by solving a boundary value problem in Q_T for a homogeneous parabolic equation, which will be done below in §9.

Let us at once establish the necessary properties of the surfaces S_T and Γ_0 (we recall that Γ_0 is the set of points $x \in \Omega$, $t=0$) of a function from $W^{2m,m}_q(Q_T)$ that does not satisfy zero initial conditions. We confine ourselves to the case $m=1$. Let $\Re_T = \Re \times (0,T)$, where $\Re$ is a cube in E_n: $\Re = \sigma \times (0, 2\lambda)$; $\sigma \subset E_{n-1}$ is the cube $|x_\alpha| < \lambda$ $(\alpha = 1,\cdots, n-1)$, $0 < x_n < 2\lambda$.

Theorem 4.7. *For any function* $u \in W^{2,1}_q(\Re_T)$ *for* $q<3$

$$\int_\sigma dx' \int_0^T dt \int_0^{2\lambda} \frac{|u(x',0,t)-u(x',x_n,0)|^q}{(t+x_n^2)^q}\, dx_n \leq c\left(\|u\|^{(2)}_{q,\Re_T}\right)^q; \tag{4.32}$$

furthermore, when $q>2$,

$$\int_\sigma dx' \int_0^T dt \int_0^{2\lambda} \frac{|u_{x_n}(x',0,t)-u_{x_n}(x',x_n,0)|^q}{(t+x_n^2)^{q/2}}\, dx_n \leq c\left(\|u\|^{(2)}_{q,\Re_T}\right)^q. \tag{4.33}$$

Proof. We make use of the identity

$$u(x', 0, t) - u(x', x_n, 0) = \left[u(x', 0, t) - 2u\left(x', \frac{x_n}{2}, t\right) + u(x', x_n, t)\right]$$

$$+ 2\left[u\left(x', \frac{x_n}{2}, t\right) - u\left(x', \frac{x_n}{2}, 0\right)\right] - [u(x', x_n, t) - u(x', x_n, 0)]$$

$$- 2\left[u(x', x_n, 0) - u\left(x', \frac{x_n}{2}, 0\right)\right] = \int_{-\frac{x_n}{2}}^{\frac{x_n}{2}} u_{\eta\eta}\left(x', \frac{x_n}{2} + \eta, t\right)\left(\frac{x_n}{2} - |\eta|\right) d\eta$$

$$+ 2\int_0^t u_\xi\left(x', \frac{x_n}{2}, \xi\right) d\xi - \int_0^t u_\xi(x', x_n, \xi)\, d\xi - 2\left[u(x', x_n, 0) - u\left(x', \frac{x_n}{2}, 0\right)\right].$$

The first three terms of the right side are estimated by means of Hölder's inequality in the following manner:

$$\left|\int_{-\frac{x_n}{2}}^{\frac{x_n}{2}} u_{\eta\eta}\left(x', \frac{x_n}{2} + \eta, t\right)\left(\frac{x_n}{2} - |\eta|\right) d\eta\right| \le \frac{x_n}{2}\int_0^{x_n} |u_{y_n y_n}(x', y_n, t)|\, dy_n$$

$$\le \frac{x_n}{2}\left[\int_0^{x_n} \frac{dy_n}{y_n^{\epsilon q'}}\right]^{\frac{1}{q'}}\left[\int_0^{x_n} |u_{y_n y_n}(x', y_n, t)|^q y_n^{\epsilon q}\, dy_n\right]^{\frac{1}{q}}$$

$$= c x_n^{1+\frac{1}{q'}-\epsilon}\left[\int_0^{x_n} |u_{y_n y_n}(x', y_n, t)|^q y_n^{\epsilon q} dy_n\right]^{\frac{1}{q}},$$

$$\left|\int_0^t u_\xi\left(x', \frac{x_n}{2}, \xi\right) d\xi\right| \le c t^{\frac{1}{q'}-\epsilon}\left[\int_0^t |u_\xi(x', x_n, \xi)|^q \xi^{\epsilon q}\, d\xi\right]^{\frac{1}{q}} \tag{4.34}$$

with the third term being estimated in exactly the same way as the second. Here and everywhere ϵ is a sufficiently small positive number ($\epsilon < 1/q'$).

From here with the use of elementary estimates one obtains an inequality that differs from (4.32) only by the presence in the right side of an additional term

$$c\int_\sigma dx' \int_0^{2\lambda} \left| u(x', x_n, 0) - u\left(x', \frac{x_n}{2}, 0\right)\right|^q \frac{dx_n}{x_n^{2(q-1)}}.$$

Let us estimate this quantity in terms of the norm $\langle\langle u(x, 0)\rangle\rangle_{q,\Re}^{(2-2/q)}$. Let $K_\rho(x)$ be the ball with center at $x = (x', x_n) \in \Re$ and radius ρ. We have

$$u(x', x_n, 0) - u\left(x', \frac{x_n}{2}, 0\right) = \frac{1}{\operatorname{mes}(K_{x_n}(x) \cap \Re)}\left\{\int_{K_{x_n}(x)\cap\Re} [u(x', x_n, 0) - u(y', y_n, 0)]\, dy\right.$$

$$\left. + \int_{K_{x_n}(x)\cap\Re} \left[u(y', y_n, 0) - u\left(x', \frac{x_n}{2}, 0\right)\right] dy\right\},$$

whence

$$\left| u(x', x_n, 0) - u\left(x', \frac{x_n}{2}, 0\right)\right|$$

$$\le \frac{c}{x_n^{\frac{n}{q}}}\left\{\left(\int_{K_{x_n}(x)\cap\Re} |u(x, 0) - u(y, 0)|^q dy\right)^{\frac{1}{q}} + \left(\int_{K_{x_n}(x)\cap\Re} \left|u(y', y_n, 0) - u\left(x', \frac{x_n}{2}, 0\right)\right|^q dy\right)^{\frac{1}{q}}\right\}$$

and

$$\int_\sigma dx' \int_0^{2\lambda} \left| u(x', x_n, 0) - u\left(x', \frac{x_n}{2}, 0\right)\right|^q \frac{dx_n}{x_n^{2(q-1)}} \le$$

$$\leq c\left[\int_{\Re}\frac{dx}{x_n^{n+2q-2}}\int_{K_{x_n}(x)\cap\Re}|u(x,0)-u(y,0)|^q dy\right.$$

$$\left.+\int_{\Re}\frac{dx}{x_n^{n+2q-2}}\int_{K_{x_n}(x)\cap\Re}\left|u\left(x',\frac{x_n}{2},0\right)-u(y,0)\right|^q dy\right]$$

$$\leq c\int_{\Re}dx\int_{\Re}|u(x,0)-u(y,0)|^q\frac{dy}{|x-y|^{n+2q-2}}=c\left[\langle\langle u(x,0)\rangle\rangle_{q,\Re}^{\left(2-\frac{2}{q}\right)}\right]^q. \qquad (4.35)$$

The norm in the right side is estimated in terms of $\|u\|_{q,\Re_T}^{(2)}$ by means of the embedding theorem (Lemma 3.4 of Chapter II). Inequality (4.32) is proved.

Estimate (4.33) is proved in an analogous way with the use of the identity

$$u_{x_n}(x',0,t)-u_{x_n}(x',x_n,0)=-\int_0^{x_n}u_{\xi\xi}(x',\xi,t)\,d\xi+[u_{x_n}(x,t)-u_{x_n}(x,0)]. \qquad (4.36)$$

For the expression in square brackets we have the estimate

$$\int_{\Re}dx\int_0^T|u_{x_n}(x,t)-u_{x_n}(x,0)|^q\frac{dt}{t^{\frac{q-1}{2}}}\leq c\left\{\|u\|_{q,\Re_T}^{(2)}\right\}^q. \qquad (4.37)$$

From it and from (4.36) we easily obtain (4.33).

Estimate (4.37) is proved in exactly the same way as (4.32), with the use of the identity

$$u_{x_n}(x,t)-u_{x_n}(x,0)=\frac{1}{\sqrt{t}}\int_0^{\sqrt{t}}[u_{x_n}(x',x_n+\xi,t)-u_{x_n}(x',x_n+\xi,0)]\,d\xi$$

$$+\frac{1}{\sqrt{t}}\int_0^{\sqrt{t}}[u_{x_n}(x',x_n+\xi,0)-u_{x_n}(x',x_n,0)]\,d\xi-$$

$$-\frac{1}{\sqrt{t}}\int_0^{\sqrt{t}}[u_{x_n}(x', x_n+\xi, t)-u_{x_n}(x', x_n, t)]\,d\xi=\frac{1}{\sqrt{t}}\int_0^t u_\eta(x', x_n+\sqrt{t}, \eta)\,d\eta$$

$$-\frac{1}{\sqrt{t}}\int_0^t u_\eta(x', x_n, \eta)\,d\eta-\frac{1}{\sqrt{t}}\int_0^{\sqrt{t}}d\xi\int_0^{\xi}u_{x_nx_n}(x', x_n+\eta, t)\,d\eta$$

$$+\frac{1}{\sqrt{t}}\int_0^{\sqrt{t}}[u_{x_n}(x', x_n+\eta, 0)-u_{x_n}(x', x_n, 0)]\,d\eta.$$

In this identity the arguments of the derivatives of u in the right side can fall outside the limits of the domain $\Re_\tau$ when $(x, t)\in\Re_\tau$, but if $t\le 4\lambda^2$, then they are confined to the domain $\widetilde{\Re}_\tau=\widetilde{\Re}\times(0, \tau)$, where $\widetilde{\Re}=\sigma\times(0, 4\lambda)$.

We extend the function $u(x, t)$ from $\Re_\tau$ into $\widetilde{\Re}_\tau$ by the method of Hestenes and Whitney described at the beginning of the section. The elementary estimates, which we will not repeat, lead to the inequality

$$\int_{\Re}dx\int_0^{\min(\tau, 4\lambda^2)}|u_{x_n}(x, t)-u_{x_n}(x, 0)|^q\frac{dt}{t^{\frac{q-1}{2}}}\le c\left[\langle\langle u\rangle\rangle^{(2)}_{q,\widetilde{\Re}_\tau}\right]^q\le c\left[\langle\langle u\rangle\rangle^{(2)}_{q,\Re_\tau}\right]^q.$$

In the case $\tau>4\lambda^2$ we also have

$$\int_{\Re}dx\int_{4\lambda^2}^{\tau}|u_{x_n}(x, t)-u_{x_n}(x, 0)|^q\frac{dt}{t^{\frac{q-1}{2}}}\le c\left[\|u_{x_n}\|^q_{q,\Re_\tau}+\|u_{x_n}(x, 0)\|^q_{q,\Re}\right]\le c\left[\|u\|^{(2)}_{q,\Re_\tau}\right]^q.$$

Estimate (4.37), and with it (4.33), is proved.

Inequalities (4.32) for $q\ne 3/2$ and (4.33) for $q\ne 3$ can be deduced from Lemma 3.4, since under these conditions the integrals in the left sides of (4.32) and (4.33) are estimated in terms of the sum of the norms $\|u(x', 0, t)\|^{(2-1/q)}_{q,\sigma_\tau}$ (or $\|u_{x_n}(x', 0, t)\|^{(1-1/q)}_{q,\sigma_\tau}$) and $\|u(x, 0)\|^{(2-2/q)}_{q,\Re}$.

We now introduce, for functions given on a surface Γ_τ, namely the sum of the lower base and lateral surface of a cylindrical domain Q_τ, the norm

$$\{\{u\}\}^{(l)}_{q,\Gamma_T} = \|u\|^{(l)}_{q,S_T} + \|u(x, 0)\|^{\left(l-\frac{1}{q}\right)}_{q,\Omega}$$

$$+\left[\sum_{k\in\mathfrak{N}} \int_\sigma dz' \int_0^{2\lambda} dz_n \int_0^T \frac{|u^{(k)}(z', 0, t) - u^{(k)}(z', z_n, 0)|^q}{(t + z_n^2)^{\frac{1}{2}(1+ql)}} dt\right]^{\frac{1}{q}} \tag{4.38}$$

for $1/q < l < 1 + 1/q$. If $u(x, 0) = 0$, then it is easily seen that this norm is equivalent to the norm $\{u\}^{(l)}_{q,S_T}$ introduced earlier:

$$c_1\{u\}^{(l)}_{q,S_T} \le \{\{u\}\}^{(l)}_{q,\Gamma_T} \le c_2\{u\}^{(l)}_{q,S_T}. \tag{4.39}$$

From Theorem 4.7 we obtain the following refinement of inequality (4.30) in the case $m = 1$, $q = 3/2$ and $q = 3$:

$$\{\{u\}\}^{\left(\frac{4}{3}\right)}_{\frac{3}{2},\Gamma_T} \le c\,\|u\|^{(2)}_{\frac{3}{2},Q_T} \tag{4.40}$$

and

$$\left\{\left\{\sum_{i=1}^n b_i(x, t)\frac{\partial u}{\partial x_i}\right\}\right\}^{\left(\frac{2}{3}\right)}_{3,\Gamma_T} \le c\,\|u\|^{(2)}_{3,Q_T} \tag{4.41}$$

Here the b_i are smooth functions given on Γ_T and having on S_T the property $\sum_{i=1}^n b_i(x, t)n_i(x) \neq 0$.

§5. FORMULATION OF BASIC RESULTS ON THE SOLVABILITY OF THE CAUCHY PROBLEM AND BOUNDARY VALUE PROBLEMS FOR EQUATIONS WITH VARIABLE COEFFICIENTS IN HÖLDER FUNCTION CLASSES

The basic result of this chapter is a theorem on the unique solvability of the Cauchy problem and two boundary value problems for a parabolic equation of second order in a cylindrical domain $Q_T = \Omega \times (0, T)$.

Let us formulate these problems.

We denote by $\mathfrak{L}(x, t, \partial/\partial x, \partial/\partial t)$ the linear parabolic differential operator with real coefficients

$$\mathcal{L}\left(x, t, \frac{\partial}{\partial x}, \frac{\partial}{\partial t}\right) u = \frac{\partial u}{\partial t} - \sum_{i, j=1}^{n} a_{ij}(x, t) \frac{\partial^2 u}{\partial x_i \partial x_j}$$

$$+ \sum_{i=1}^{n} a_i(x, t) \frac{\partial u}{\partial x_i} + a(x, t) u. \tag{5.1}$$

We assume that this operator is uniformly parabolic, i.e., in the domains where the above-mentioned problems are solved, inequality (2.5) of Chapter I is fulfilled for any real $\xi_1, \dots, \xi_n$.

We assume that the coefficients of the operator of (5.1) are defined in a layer $D_{n+1}^{(T)} = E_n \times (0, T)$. We will consider in this domain the Cauchy problem

$$\left.\begin{aligned} &\mathcal{L}\left(x, t, \frac{\partial}{\partial x}, \frac{\partial}{\partial t}\right) u(x, t) = f(x, t), \\ &u|_{t=0} = \varphi(x). \end{aligned}\right\} \tag{5.2}$$

Suppose, further, that Ω is a domain satisfying the conditions mentioned in §4. In the cylindrical domain $Q_T = \Omega \times (0, T)$ with lateral surface $S_T = S \times (0, T)$ we will consider the first boundary value problem and the problem with directional derivative:

$$\left.\begin{aligned} &\mathcal{L}\left(x, t, \frac{\partial}{\partial x}, \frac{\partial}{\partial t}\right) u(x, t) = f(x, t), \\ &u|_{t=0} = \varphi(x), \quad u|_{S_T} = \Phi(x, t), \end{aligned}\right\} \tag{5.3}$$

$$\left.\begin{aligned} &\mathcal{L}\left(x, t, \frac{\partial}{\partial x}, \frac{\partial}{\partial t}\right) u(x, t) = f(x, t), \\ &u|_{t=0} = \varphi(x), \\ &\mathcal{B}\left(x, t, \frac{\partial}{\partial x}\right) u\Big|_{S_T} \\ &\equiv \sum_{i=1}^{n} b_i(x, t) \frac{\partial u}{\partial x_i} + b(x, t)\, u|_{S_T} = \Phi(x, t). \end{aligned}\right\} \tag{5.4}$$

We assume that the functions $b_i(x, t)$ satisfy everywhere on S_T the condition

$$\left|\sum_{i=1}^{n} b_i(x, t)\, n_i(x)\right| \geqslant \delta > 0. \tag{5.5}$$

This condition can be written in the form $|b \cdot n| \geqslant \delta > 0$, where $\mathbf{b} = (b_1, \dots, b_n)$, and it means that the vector $\mathbf{b}$ does not at any point lie in the tangent plane to S. The boundary condition of (5.4) can be written in the form

$$|\mathbf{b}(x, t)|\frac{\partial u}{\partial \mathbf{l}} + bu\Big|_{S_T} = \Phi,$$

where $\mathbf{l} = \mathbf{b}/|\mathbf{b}|$.

In problem (5.2), and in the case of unbounded Ω problems (5.3) and (5.4), it is necessary to restrict the growth of the solution for $|x| \to \infty$. We will consider these problems only in the function classes $H^{l, l/2}$, the elements of which are bounded, and in the classes $W_q^{2m, m}$, the elements of which tend to zero in a certain sense for $|x| \to \infty$.

We will assume that the functions f, ϕ, Φ in (5.3) and (5.4) satisfy the compatibility conditions for $x \in S$, $t = 0$. These conditions consist in the fact that the derivatives $\partial^k u/\partial t^k|_{t=0}$, which can be determined for $t = 0$ by means of the equation and initial condition, must satisfy for $x \in S$ the boundary conditions of (5.3) or (5.4). We introduce the notation

$$u^{(k)}(x) = \frac{\partial^k u(x, t)}{\partial t^k}\Big|_{t=0},$$

$$\mathcal{A}\left(x, t, \frac{\partial}{\partial x}\right)u = \sum_{i, j=1}^{n} a_{ij}(x, t)\frac{\partial^2 u}{\partial x_i \partial x_j} - \sum_{i=1}^{n} a_i \frac{\partial u}{\partial x_i} - au. \tag{5.6}$$

It is obvious that the functions $u^{(k)}(x)$ $(k = 0, 1)$ are determined in the following manner:

$$u^{(0)}(x) = \varphi(x), \quad u^{(1)}(x) = \mathcal{A}\left(x, 0, \frac{\partial}{\partial x}\right)\varphi(x) + f(x, 0), \tag{5.7}$$

while the remaining functions are found from the recursion relations

$$\begin{aligned} u^{(k+1)}(x) &= \frac{\partial^k}{\partial t^k}\mathcal{A}\left(x, t, \frac{\partial}{\partial x}\right)u(x, t) + \frac{\partial^k f(x, t)}{\partial t^k}\Big|_{t=0} \\ &= \sum_{j=0}^{k}\binom{k}{j}\mathcal{A}^{(j)}\left(x, 0, \frac{\partial}{\partial x}\right)u^{(k-j)}(x) + f^{(k)}(x), \end{aligned} \tag{5.8}$$

where $\mathcal{A}^{(j)}(x, t, \partial/\partial x)$ is the operator resulting from the operator $\mathcal{A}$ of (5.6) upon differentiating the coefficients j times with respect to t.

We will say that the compatibility conditions of order $m \geq 0$ are fulfilled for problem (5.3) or (5.4) if

$$u^{(k)}(x)\big|_{x \in S} = \frac{\partial^k \Phi}{\partial t^k}\Big|_{t=0} = \Phi^{(k)}(x) \qquad (k = 0, \ldots, m)$$

or correspondingly

$$\frac{\partial^k}{\partial t^k}\left[\sum_{i=1}^{n} b_i(x,t)\frac{\partial u}{\partial x_i}+b(x,t)\,u\right]\Bigg|_{\substack{t=0\\ x\in S}} =\Phi^{(k)}(x)\qquad (k=0,\ldots,m),$$

i.e.

$$\sum_{j=0}^{k}\binom{k}{j}\left[\sum_{i=1}^{n} b_i^{(k-j)}(x)\frac{\partial u^{(j)}(x)}{\partial x_i}+b^{(k-j)}(x)\,u^{(j)}(x)\right]\Bigg|_{x\in S} =\Phi^{(k)}(x)\qquad (k=0,\ldots,m).$$

We will first consider problems (5.2)–(5.4) in the Hölder function classes $H^{l,l/2}$. We formulate directly the basic results on the solvability of these problems in the form of theorems which will be proved in the sections immediately following.

Theorem 5.1. *Suppose $l>0$ is a nonintegral number and the coefficients of the operator $\mathfrak{L}$ belong to the class $H^{l,l/2}(\overline{D_{n+1}^{(T)}})$. Then for any $f\in H^{l,l/2}(\overline{D_{n+1}^{(T)}})$, $\phi\in H^{l+2}(\overline{E}_n)$ problem (5.2) has a unique solution from the class $H^{l+2,l/2+1}(\overline{D_{n+1}^{(T)}})$. It satisfies the inequality*

$$|u|_{D_{n+1}^{(T)}}^{(l+2)}\leqslant c\left(|f|_{D_{n+1}^{(T)}}^{(l)}+|\varphi|_{E_n}^{(l+2)}\right) \tag{5.9}$$

with the constant not depending on f and ϕ.

Theorem 5.2. *Suppose $l>0$ is a nonintegral number, the coefficients of the operator $\mathfrak{L}$ belong to the class $H^{l,l/2}(\overline{Q}_T)$, and the boundary S belongs to the class H^{l+2}. Then for any $f\in H^{l,l/2}(\overline{Q}_T)$, $\phi\in H^{l+2}(\overline{\Omega})$, $\Phi\in H^{l+2,l/2+1}(\overline{S}_T)$ satisfying the compatibility conditions of order $[l/2]+1$ problem (5.3) has a unique solution from the class $H^{l+2,l/2+1}(\overline{Q}_T)$. It satisfies the inequality*

$$|u|_Q^{(l+2)}\leqslant c\left(|f|_Q^{(l)}+|\varphi|_\Omega^{(l+2)}+|\Phi|_{S_T}^{(l+2)}\right). \tag{5.10}$$

Theorem 5.3. *Suppose $l>0$ is a nonintegral number, $S\in H^{l+2}$, the coefficients of the operator $\mathfrak{L}$ belong to the class $H^{l,l/2}(\overline{Q}_T)$ and, finally, b_i, $b\in H^{l+1,l/2+1/2}(\overline{S}_T)$. Then for any $f\in H^{l,l/2}(\overline{Q}_T)$, $\phi\in H^{l+2}(\overline{\Omega})$, $\Phi\in H^{l+1,(l+1)/2}(\overline{S}_T)$ satisfying the compatibility conditions of order $[(l+1)/2]$, problem (5.4) has a* *unique solution from the class* $H^{l+2,l/2+1}(\overline{Q}_T)$, *with*

$$|u|_Q^{(l+2)} \leqslant c\left(|f|_Q^{(l)} + |\varphi|_\Omega^{(l+2)} + |\Phi|_{S_T}^{(l+1)}\right). \tag{5.11}$$

Since the proofs of these theorems are almost completely identical, we will confine ourselves to a detailed proof of Theorem 5.3. The calculations in the proof of this theorem are somewhat more complicated than those in the proof of Theorems 5.1 and 5.2.

Theorems 5.1–5.3 are first proved under special assumptions concerning f, ϕ and Φ, which amount to the following: $\phi = 0$, while f and Φ satisfy the zero initial conditions

$$\left.\frac{\partial^i f}{\partial t^i}\right|_{t=0} = 0 \qquad \left(i = 0, \ldots, \left[\frac{l}{2}\right]\right)$$

and

$$\left.\frac{\partial^i \Phi}{\partial t^i}\right|_{t=0} = 0,$$

where $i = 0, \cdots, [l/2] + 1$ for problem (5.3) and $i = 0, \cdots, [(l+1)/2]$ for problem (5.4). It is easily verified that under these conditions the functions f and Φ satisfy the compatibility conditions of the order indicated in Theorems 5.2 and 5.3. In this connection $u^{(k)}(x) = 0$ for $k = 0, \cdots, [l/2] + 1$.

If the functions f, ϕ, Φ satisfy the indicated conditions, then we will call the corresponding problems (5.2)–(5.4) problems with zero initial data. By making use of the notion of the classes $\overset{\circ}{H}{}^{l,l/2}(\overline{Q}_T)$ introduced in §4, we can formulate these problems in the following manner.

1. Find a function $u \in \overset{\circ}{H}{}^{l+2,l/2+1}(\overline{D^{(T)}_{n+1}})$ such that

$$\mathscr{L}\left(x, t, \frac{\partial}{\partial x}, \frac{\partial}{\partial t}\right) u = f, \tag{5.2'}$$

where

$$f \in \overset{\circ}{H}{}^{l,\frac{l}{2}}(\overline{D^{(T)}_{n+1}}).$$

2. Find a function $u \in \overset{\circ}{H}{}^{l+2,l/2+1}(\overline{Q}_T)$ such that

$$\left.\begin{aligned} \mathscr{L}\left(x, t, \frac{\partial}{\partial x}, \frac{\partial}{\partial t}\right) u = f, \\ u|_{S_T} = \Phi, \end{aligned}\right\} \tag{5.3'}$$

where

$$f \in \overset{\circ}{H}{}^{l, \frac{l}{2}}(\overline{Q}_T), \quad \Phi \in \overset{\circ}{H}{}^{l+2, \frac{l}{2}+1}(\overline{S}_T).$$

3. Find a function $u \in \overset{\circ}{H}{}^{l+2,l/2+1}(\overline{Q}_T)$ for which

$$\left.\begin{aligned} \mathscr{L}\left(x, t, \frac{\partial}{\partial x}, \frac{\partial}{\partial t}\right) u &= f, \\ \mathscr{B}\left(x, t, \frac{\partial}{\partial x}\right) u \Big|_{S_T} &= \Phi, \end{aligned}\right\} \tag{5.4'}$$

where

$$f \in \overset{\circ}{H}{}^{l, \frac{l}{2}}(\overline{Q}_T), \quad \Phi \in \overset{\circ}{H}{}^{l+1, \frac{l+1}{2}}(\overline{S}_T).$$

The chief step in the proof of Theorems 5.1–5.3 is the proof of the following assertion.

Theorem 5.4. *Under the conditions of Theorems 5.1–5.3 problems (5.3′) and (5.4′) have unique solutions in the cylinder $Q_\tau = \Omega \times (0, \tau)$, while problem (5.2′) has a unique solution in $D_{n+1}^{(\tau)}$, for $\tau \le \tau_0$, where τ_0 is a certain positive number depending on the coefficients of the operator $\mathfrak{L}$, on b_i, b (in the case of problem (5.4′)) and on the various characteristics of the domain Ω and boundary S, but not on f and Φ. The solutions are subject to the inequalities*

$$|u|_{D_{n+1}^{(\tau)}}^{(l+2)} \leqslant c\,|f|_{D_{n+1}^{(\tau)}}^{(l)}, \tag{5.12}$$

$$|u|_{Q_\tau}^{(l+2)} \leqslant c\left(|f|_{Q_\tau}^{(l)} + |\Phi|_{S_\tau}^{(l+2)}\right), \tag{5.13}$$

$$|u|_{Q_\tau}^{(l+2)} \leqslant c\left(|f|_{Q_\tau}^{(l)} + |\Phi|_{S_\tau}^{(l+1)}\right), \tag{5.14}$$

in which the constants remain bounded for $\tau \to 0$.

With the use of this theorem one can easily prove Theorems 5.1–5.3 for a cylindrical domain Q_T of arbitrary finite height T.

As will be shown in §10 of Chapter VII, Theorems 5.1–5.3 carry over to parabolic systems (see Theorem 3.1 of Chapter VII). Here we formulate a theorem on the solvability of the first boundary value problem for a parabolic system of second order, the principal part of which splits into separate parabolic equations:

$$\frac{\partial u^k}{\partial t} - \sum_{i=1}^{n} a_{ij}(x, t)\frac{\partial^2 u^k}{\partial x_i \partial x_j} + \sum_{l=1}^{m}\sum_{i=1}^{n} a_i^{(k,l)}(x, t)\frac{\partial u^l}{\partial x_i} + \sum_{l=1}^{m} a^{(k,l)}(x, t) u^l = f^k(x, t),$$

$$\begin{aligned} u^k|_{t=0} &= \varphi^k(x), \\ u^k|_{S_T} &= \Phi^k(x, t) \end{aligned} \tag{5.15}$$

$$(k = 1, \ldots, m).$$

Theorem 5.5. *Suppose $S \in H^{l+2}$ and the coefficients a_{ij}, $a_i^{(k,r)}$, $a^{(k,r)}$ belong to the class $H^{l,l/2}(\overline{Q}_T)$. Then, for any $f^k \in H^{l,l/2}(\overline{Q}_T)$, $\phi^k \in H^{l+2}(\overline{\Omega})$, $\Phi^k \in H^{l+2,\, l/2+1}(\overline{S}_T)$ satisfying the compatibility conditions of order $[l/2]+1$, problem (5.15) has a unique solution $u = (u^1, \cdots, u^m)$ with $u^j \in H^{l+2,l/2+1}(\overline{Q}_T)$ and*

$$\sum_{k=1}^m |u^k|_{Q_T}^{(l+2)} \leqslant c\left(\sum_{k=1}^m |f^k|_{Q_T}^{(l)} + \sum_{k=1}^m |\varphi^k|_{\Omega}^{(l+2)} + \sum_{k=1}^m |\Phi^k|_{S_T}^{(l+2)}\right).$$

The compatibility conditions for problem (5.15) are formulated in exactly the same way as for problem (5.3), while the proof of Theorem 5.5 coincides verbatim with the proof of Theorem 5.2. It is only necessary in all of the calculations to take u as a vector and the a_i and a as matrices with elements $a_i^{(k,l)}$ and $a^{(k,l)}$.

We selected problem (5.15) here because it will be the object of a special study in Chapter VII.

§6. MODEL PROBLEMS IN A HALF SPACE

We consider the Cauchy problem and boundary value problems with zero initial data in the domain $R^{(T)}$ for equations having constant coefficients and containing only the leading terms. In the Cauchy problem it is required that one find a function $w \in \overset{\circ}{H}{}^{l+2,l/2+1}(\overline{D_{n+1}^{(T)}})$ satisfying in $D_{n+1}^{(T)}$ the equation

$$\frac{\partial w}{\partial t} - \sum_{i,j=1}^n a_{ij} \frac{\partial^2 w}{\partial x_i \partial x_j} = f, \tag{6.1}$$

where $f \in \overset{\circ}{H}{}^{l,l/2}(\overline{D_{n+1}^{(T)}})$ and the a_{ij} are constants satisfying condition (2.5) of Chapter I.

In addition, we consider in the domain $R^{(T)}$ the boundary value problems

$$\left.\begin{aligned} &\frac{\partial v}{\partial t}-\sum_{i,j=1}^{n} a_{ij}\frac{\partial^2 v}{\partial x_i \partial x_j}=f,\\ &v\big|_{x_n=0}=\Phi_1(x',t),\end{aligned}\right\} \tag{6.2}$$

$$\left.\begin{aligned} &\frac{\partial u}{\partial t}-\sum_{i,j=1}^{n} a_{ij}\frac{\partial^2 u}{\partial x_i \partial x_j}=f,\\ &\sum_{i=1}^{n} b_i \frac{\partial u}{\partial x_i}\bigg|_{x_n=0}=\Phi_2(x',t)\end{aligned}\right\} \tag{6.3}$$

(the b_i are constants with $b_n \neq 0$).

It is assumed that $f \in \overset{\circ}{H}{}^{l,l/2}(\overline{R^{(T)}})$, $\Phi_1 \in \overset{\circ}{H}{}^{l+2,l/2+1}(\overline{\widetilde{D}_n^{(T)}})$, $\Phi_2 \in \overset{\circ}{H}{}^{l+1,(l+1)/2}(\overline{\widetilde{D}_n^{(T)}})$. The solution of problems (6.2) and (6.3) must belong to the class $\overset{\circ}{H}{}^{l+2,l/2+1}(\overline{R^{(T)}})$.

Theorem 6.1. *Problem* (6.1) *is uniquely solvable in the class* $\overset{\circ}{H}{}^{l+2,l/2+1}(\overline{D_{n+1}^{(T)}})$ *while problems* (6.2), (6.3) *are uniquely solvable in the class* $\overset{\circ}{H}{}^{l+2,l/2+1}(\overline{R^{(T)}})$ *for any* $T > 0$. *The solutions of these problems are subject to the inequalities*

$$\langle w\rangle^{(l+2)}_{D^{(T)}_{n+1}} \leqslant c\,\langle f\rangle^{(l)}_{D^{(T)}_{n+1}}, \tag{6.4}$$

$$\langle v\rangle^{(l+2)}_{R^{(T)}} \leqslant c\left(\langle f\rangle^{(l)}_{R^{(T)}} + \langle \Phi_1\rangle^{(l+2)}_{\widetilde{D}^{(T)}_n}\right), \tag{6.5}$$

$$\langle u\rangle^{(l+2)}_{R^{(T)}} \leqslant c\left(\langle f\rangle^{(l)}_{R^{(T)}} + \langle \Phi_2\rangle^{(l+1)}_{\widetilde{D}^{(T)}_n}\right), \tag{6.6}$$

in which the constants do not depend on T.

Proof. We first prove this theorem for the heat equation ($a_{ij} = \delta_{ij}$). As can be seen from the results of §1, in this case the solutions of problems (6.1)–(6.3) are given by

$$w=\int_0^t d\tau \int_{E_n} \Gamma(x-y,\,t-\tau)\,f(y,\,\tau)\,dy=(\Gamma * f^0), \tag{6.7}$$

where

$$f^0(x,\,t)=\begin{cases} f(x,\,t) & \text{for } t \geqslant 0,\\ 0 & \text{for } t<0;\end{cases} \tag{6.8}$$

$$v(x,\,t)=(\Gamma * f^{*0})-2\left(\frac{\partial \Gamma}{\partial x_n} *_2 \omega_1^0\right). \tag{6.9}$$

where f^* is an extension with respect to the variable x_n of the function $f(x, t)$, given in $R^{(T)}$, into the domain $D_{n+1}^{(T)}$ by means of a formula of type (4.5), so that

$$\langle f^* \rangle_{D_{n+1}^{(T)}}^{(l)} \leqslant c \langle f \rangle_{R^{(T)}}^{(l)},$$

f^{*0} is the extension of f^* into the domain $t < 0$ determined by formula (6.8), and

$$\omega_1 = \Phi_1 - (\Gamma * f^{*0})|_{x_n = 0}; \tag{6.10}$$

and, finally,

$$u(x, t) = (\Gamma * f^{*0}) + (G *_2 \omega_2^0), \tag{6.11}$$

where

$$\omega_2 = \Phi_2 - \sum_{i=1}^{n} b_i \frac{\partial}{\partial x_i} (\Gamma * f^{*0})|_{x_n = 0}.$$

Estimates (6.4)–(6.6) are consequences of estimates (2.1), (2.3), (2.4) and the fact that, under the zero extension (6.8) of functions from the classes $\overset{\circ}{H}{}^{l, l/2}$, the norm is preserved.

For example, inequality (6.6) is proved by means of (2.1) and (2.3) in the following manner:

$$\langle u \rangle_{R^{(T)}}^{(l+2)} \leqslant \langle (\Gamma * f^{*0}) \rangle_{E_{n+1}^{(T)}}^{(l+2)} + \langle (G *_2 \omega_2^0) \rangle_{\tilde{D}_{n+1}^{(T)}}^{(l+2)} \leqslant \langle (\Gamma * f^{*0}) \rangle_{E_{n+1}^{(T)}}^{(l+2)}$$
$$+ c \langle \omega_2^0 \rangle_{\tilde{E}_n^{(T)}}^{(l+1)} \leqslant \langle (\Gamma * f^{*0}) \rangle_{E_{n+1}^{(T)}}^{(l+2)} + c \left\{ \langle \Phi_2^0 \rangle_{\tilde{E}_n^{(T)}}^{(l+1)} + \langle (\Gamma * f^{*0}) \rangle_{E_{n+1}^{(T)}}^{(l+2)} \right\}$$
$$\leqslant c \left\{ \langle \Phi_2 \rangle_{\tilde{D}_n^{(T)}}^{(l+1)} + \langle f^{*0} \rangle_{E_{n+1}^{(T)}}^{(l)} \right\} \leqslant c \left\{ \langle \Phi_2 \rangle_{\tilde{D}_n^{(T)}}^{(l+1)} + \langle f \rangle_{R^{(T)}}^{(l)} \right\}.$$

The uniqueness of these solutions follows from the results of §2 of Chapter I (see Theorems 2.6 and 2.7 of Chapter I). The conditions of Theorem 2.7 are fulfilled since one can take as $\phi(x)$ in the half space $x_n > 0$ the function

$$\varphi(x_n) = \frac{1}{2} + \frac{1}{4mx_n + 2}.$$

Thus the theorem is proved for the case when $a_{ij} = \delta_{ij}$, i.e. when problems (6.1)–(6.3) are considered for the heat equation. The general case reduces to this one, since problems (6.1)–(6.3), as is known, can be written as problems for the heat equation. For this purpose it is necessary to carry out certain linear transformations of the coordinates. First we make an orthogonal transformation of coordinates

$$y_k = \sum_{l=1}^{n} \beta_{kl} x_l,$$

reducing the matrix A with elements a_{ij}, which without loss of generality can be regarded as symmetric, to one of diagonal form, i.e.

$$\sum_{i,j=1}^{n} \beta_{ki} a_{ij} \beta_{lj} = \alpha_l \delta_{kl}.$$

It is necessary to take as the β_{lj} the components of an lth eigenvector of the matrix A; the numbers α_l are the eigenvalues of this matrix. Equation (6.1) is written in the form

$$\frac{\partial w}{\partial t} - \sum_{l=1}^{n} \alpha_l \frac{\partial^2 w}{\partial y_l^2} = f(y, t).$$

If one introduces the new variables

$$z_l = \frac{y_l}{\sqrt{\alpha_l}},$$

then this equation goes over into the heat equation

$$\frac{\partial w(z, t)}{\partial t} - \Delta w(z, t) = f(z, t).$$

In problems (6.2) and (6.3) these transformations rotate the plane $x_n = 0$ in some way. Therefore in considering these problems it is convenient to make an additional orthogonal transformation of coordinates

$$\xi_l = \sum_{k=1}^{n} \gamma_{lk} z_k,$$

which will take this plane into the plane $\xi_n = 0$. The Laplace operator does not vary under this transformation. Thus it is possible to transform problems (6.1)–(6.3) by means of simply a linear transformation of coordinates into the same problems for the heat equation. In this connection the boundary condition of (6.3) goes over into

$$\sum_{l=1}^{n} b_l' \left.\frac{\partial u(\xi, t)}{\partial \xi_l}\right|_{\xi_n=0} = \Phi_2(\xi', t),$$

where

$$b'_l = \sum_{l,k=1}^{n} \frac{b_l \beta_{kl} \gamma_{lk}}{\sqrt{\alpha_k}}.$$

Let us verify that the necessary condition

$$b'_n = \sum_{l,k=1}^{n} \frac{\gamma_{nk}}{\sqrt{\alpha_k}} \beta_{kl} b_l \neq 0 \tag{6.12}$$

is fulfilled. The equation of the plane $x_n = 0$ has the following form in the coordinates $\{y\}$ and $\{z\}$:

$$\sum_{k=1}^{n} \beta_{kn} y_k = 0,$$

$$\sum_{k=1}^{n} \beta_{kn} \sqrt{\alpha_k}\, z_k = 0.$$

On the other hand it can also be written in the form

$$\sum_{k=1}^{n} \gamma_{nk} z_k = 0.$$

Consequently,

$$\beta_{kn} \sqrt{\alpha_k} = \gamma_{nk} d, \tag{6.13}$$

from which it follows that

$$d^2 = \sum_{k=1}^{n} \alpha_k \beta_{kn}^2,$$

i.e.

$$\sqrt{\nu} \leqslant |d| \leqslant \sqrt{\mu},$$

where μ and ν are the largest and smallest eigenvalues of the matrix A (the constants in inequality (2.5) of Chapter I).

From (6.12) and (6.13) it follows that

$$b'_n = \frac{1}{d} \sum_{l,k=1}^{n} \beta_{kn} \beta_{kl} b_l = \frac{b_n}{d}$$

and

$$\frac{1}{\sqrt{\mu}} |b_n| \leqslant |b'_n| \leqslant \frac{1}{\sqrt{\nu}} |b_n|.$$

Thus the condition $b'_n \neq 0$ is proved, and the proof of the theorem is complete.

§7. ON THE SOLVABILITY OF PROBLEM (5.4′)

In this section Theorem 5.4 will be proved for problem (5.4′). The basic idea of the proof consists in the following. Let A be the linear operator defined in the space $\overset{\circ}{H}{}^{l+2,l/2+1}(\overline{Q}_T)$ that associates with each element of this space the pair of functions $(\mathfrak{L}u, \mathfrak{B}u|_{S_T})$. Under the restrictions on the coefficients of the operators $\mathfrak{L}$ and $\mathfrak{B}$ imposed in the condition of Theorem 5.4 one has for each $u \in \overset{\circ}{H}{}^{l+2,l/2+1}(\overline{Q}_T)$ the inequality

$$|\mathscr{L}u|_{Q_\tau}^{(l)} + |\mathscr{B}u|_{S_\tau}^{(l+1)} \leqslant c\,|u|_{Q_\tau}^{(l+2)}. \tag{7.1}$$

Let

$$\mathfrak{H}^{(l)} = \overset{\circ}{H}{}^{l,\frac{l}{2}}(\overline{Q}_\tau) \times \overset{\circ}{H}{}^{l+1,\frac{l+1}{2}}(\overline{S}_\tau)$$

be the Banach space whose elements are the pairs of functions $h = (f, \Phi)$, where $f \in \overset{\circ}{H}{}^{l,l/2}(\overline{Q}_T)$, $\Phi \in \overset{\circ}{H}{}^{l+1,(l+1)/2}(\overline{S}_T)$, and whose norm is defined as the sum of the norms $|f|_{Q_T}^{(l)}$ and $|\Phi|_{S_T}^{(l+1)}$. Inequality (7.1) expresses the fact that the operator A is a bounded operator acting from $\overset{\circ}{H}{}^{l+2,l/2+1}(\overline{Q}_T)$ into $\mathfrak{H}^{(l)}$. Problem (5.4′) can be interpreted as the problem of solving the equation

$$Au = h, \tag{7.2}$$

where $h \in \mathfrak{H}^{(l)}$, while Theorem 5.4 can be interpreted as a theorem on the existence of a bounded inverse operator A^{-1}.

In order to prove Theorem 5.4 we will construct in this section a bounded operator R acting from $\mathfrak{H}^{(l)}$ into $\overset{\circ}{H}{}^{l+2,l/2+1}(\overline{Q}_T)$ and such that for any $h \in \mathfrak{H}^{(l)}$ and $v \in \overset{\circ}{H}{}^{l+2,l/2+1}(\overline{Q}_T)$

$$ARh = h + Th, \tag{7.3}$$

$$RAv = v + Wv, \tag{7.4}$$

where T and W are bounded operators in the spaces $\mathfrak{H}^{(l)}$ and $\overset{\circ}{H}{}^{l+2,l/2+1}(\overline{Q}_T)$ respectively, the norms of which are small if the height T of the cylinder Q_T is small.

Equalities (7.3) and (7.4) permit one to prove Theorem 5.4 with the use of very elementary methods of functional analysis.

Let us fix the number T in such a way that

$$\|T\| < 1, \quad \|W\| < 1.$$

On the basis of the contraction mapping principle we can now conclude that the operators $I + T$ and $I + W$ have bounded inverses $(I + T)^{-1}$ and $(I + W)^{-1}$ defined in the spaces $\mathfrak{H}^{(l)}$ and $\overset{\circ}{H}{}^{l+2,l/2+1}(\overline{Q}_\tau)$ respectively. By applying the operator $(I + W)^{-1}$ to both sides of equality (7.4) and replacing h by $(I + T)^{-1}h$ in (7.3), we show that for any $h \in \mathfrak{H}^{(l)}$ and $v \in \overset{\circ}{H}{}^{l+2,l/2+1}(\overline{Q}_\tau)$

$$AR(I+T)^{-1}h = h,$$
$$(I+W)^{-1}RAv = v.$$

This means that the operator A has bounded right and left inverse operators equal to $R(I + T)^{-1}$ and $(I + W)^{-1}R$ respectively. These operators, as is known, coincide:

$$R(I+T)^{-1} = (I+W)^{-1}R = A^{-1}. \tag{7.5}$$

Consequently the operator A establishes a one-to-one correspondence between the spaces $\overset{\circ}{H}{}^{l+2,l/2+1}(\overline{Q}_\tau)$ and $\mathfrak{H}^{(l)}$, i.e. equation (7.2), or, what is the same thing, problem (5.4′), is uniquely solvable for any $h \in \mathfrak{H}^{(l)}$. Estimate (5.14) is equivalent to the assertion that the operator A^{-1} is bounded. But this follows from equality (7.5):

$$\|A^{-1}\| \leqslant \|(I+T)^{-1}\| \|R\|,$$
$$\|A^{-1}\| \leqslant \|(I+W)^{-1}\| \|R\|.$$

If one chooses τ so small that, for example, $\|T\| < \frac{1}{2}$, then $\|(I + T)^{-1}\| \leq 2$ and $\|A^{-1}\| \leq 2\|R\|$. The norm of the operator R will be estimated below (Theorem 7.1).

Thus everything reduces to a construction of the operator R, a proof of (7.3), (7.4) and estimates of the norms $\|R\|$, $\|T\|$, $\|W\|$.

Let us proceed to the construction of the operator R. We fix in an arbitrary manner a small $\lambda > 0$ and introduce the sets $\Omega^{(k)} \subset \mathfrak{R}^{(\lambda)}$ from §4. We introduce in addition the following notation. We denote by Z_k the operator, defined on the functions $u(z)$ given in the domain $\mathfrak{R}$, that with each such function associates the same function under transformation from the coordinates $\{z\}$, connected with the local coordinates at the point $\xi^{(k)}$ by formulas (4.3), to the original coordinates $\{x\}$. Further, let

$$\mathscr{L}_0\left(x, t, \frac{\partial}{\partial x}, \frac{\partial}{\partial t}\right)u = \frac{\partial u}{\partial t} - \sum_{i,j=1}^{n} a_{ij}(x, t)\frac{\partial^2 u}{\partial x_i \partial x_j},$$

and let $\mathfrak{L}_0^{(k)}$ be the operator $\mathfrak{L}_0$ in the local system of coordinates $\{y\}$ at the point $\xi^{(k)}$, i.e.

$$\mathscr{L}_0^{(k)} = \frac{\partial}{\partial t} - \sum_{i,j=1}^{n} a_{ij}^{(k)} \frac{\partial^2}{\partial y_i \partial y_j},$$

where

$$a_{rs}^{(k)} = \sum_{i,j=1}^{n} a_{ij} \beta_{ri}^{(k)} \beta_{sj}^{(k)},$$

with the $\beta_{ri}^{(k)}$ being the elements of an orthogonal matrix connecting the coordinates $\{x\}$ and $\{y\}$:

$$y_r = \sum_{i=1}^{n} \beta_{ri}^{(k)} (x_i - \xi_i^{(k)}).$$

It is obvious that the operator $\mathfrak{L}_0^{(k)}$ also satisfies the condition of parabolicity (2.5) of Chapter I with the very same μ and ν as $\mathfrak{L}_0$.

Analogously, let

$$\mathscr{B}_0\left(x, t, \frac{\partial}{\partial x}\right) = \sum_{i=1}^{n} b_i(x, t) \frac{\partial}{\partial x_i}$$

and

$$\mathscr{B}_0^{(k)}\left(y, t, \frac{\partial}{\partial y}\right) = \sum_{i=1}^{n} b_i^{(k)} \frac{\partial}{\partial y_i}$$

where

$$b_i^{(k)} = \sum_{j=1}^{n} b_j \beta_{ij}^{(k)}.$$

Let $f^{(k)}(x, t)$ $(k \in \mathfrak{M})$ be a function from $\overset{\circ}{H}{}^{l, l/2}(\overline{D_{n+1}^{(\tau)}})$. We denote by $R^{(k)}$ the operator that associates with this function the solution of the Cauchy problem with zero initial data

$$\mathscr{L}_0\left(\xi^{(k)}, 0, \frac{\partial}{\partial x}, \frac{\partial}{\partial t}\right) v^{(k)}(x, t) = f^{(k)}(x, t),$$

$$v^{(k)} \in \overset{\circ}{H}{}^{l+2, \frac{l}{2}+1}(\overline{D_{n+1}^{(\tau)}}), \tag{7.6}$$

which, as was shown in §6, exists and is unique. Further, for $k \in \mathfrak{N}$ let $f^{(k)}(z, t) \in \overset{\circ}{H}{}^{l, l/2}(\overline{R^{(\tau)}})$, $\Phi^{(k)}(z', t) \in \overset{\circ}{H}{}^{l+1, (l+1)/2}(\overline{\tilde{D}_n^{(\tau)}})$ and $R^{(k)}$ denote the operator that associates with the pair of functions $h^{(k)} = (f^{(k)}, \Phi^{(k)})$ the solution of the boundary value problem with zero initial data

$$\mathscr{L}_0^{(k)}\left(\xi_0^{(k)}, 0, \frac{\partial}{\partial z}, \frac{\partial}{\partial t}\right) v^{(k)}(z, t) = f^{(k)}(z, t),$$

$$\mathfrak{B}_0^{(k)}\left(\xi^{(k)}, 0, \frac{\partial}{\partial z}\right) v^{(k)}\Big|_{z_n=0} = \Phi^{(k)}(z', t). \tag{7.7}$$

The solution of this problem, as was shown in §6, also exists and is unique in the class $\overset{\circ}{H}{}^{l+2,l/2+1}(\overline{R^{(\tau)}})$.

Let $\zeta^{(k)}(x)$ and $\eta^{(k)}(x)$ denote the functions introduced in §4. We define an operator R by the formula

$$Rh = \sum_k \eta^{(k)}(x)\, v^{(k)}(x, t), \tag{7.8}$$

where

$$v^{(k)}(x, t) = \begin{cases} R^{(k)}\zeta^{(k)} f & \text{for } k \in \mathfrak{M}, \\ Z_k R^{(k)}(Z_k^{-1}\zeta^{(k)} f,\ Z_k^{-1}\zeta^{(k)}\Phi) & \text{for } k \in \mathfrak{N}. \end{cases} \tag{7.9}$$

By virtue of (6.4) and (6.6)

$$\langle R^k f^{(k)}\rangle_{D_{n+1}^{(\tau)}}^{(l+2)} \leqslant c\, \langle f^{(k)}\rangle_{D_{n+1}^{(\tau)}}^{(l)} \quad (k \in \mathfrak{M}), \tag{7.10}$$

$$\langle R^{(k)}(f^{(k)}, \Phi^{(k)})\rangle_{R^{(\tau)}}^{(l+2)} \leqslant c\left(\langle f^{(k)}\rangle_{R^{(\tau)}}^{(l)} + \langle \Phi^{(k)}\rangle_{\widetilde{D}_n^{(\tau)}}^{(l+1)}\right) \tag{7.11}$$

$$(k \in \mathfrak{N}),$$

with the constants in both inequalities being independent of τ and k by virtue of the uniform parabolicity of the equation and condition (5.5).

We introduce in the space $\overset{\circ}{H}{}^{l+2,l/2+1}(\overline{Q}_\tau)$ the norm $\{v\}_{Q_\tau}^{(l+2)}$ of (4.16), and in the space $\mathfrak{H}^{(l)}$ the norm $\|h\|_{\mathfrak{H}^{(l)}} = \{f\}_{Q_\tau}^{(l)} + \{\Phi\}_{S_\tau}^{(l+1)}$. By virtue of Lemmas 4.1 and 4.2 this norm is equivalent to the norm $|f|_{Q_\tau}^{(l)} + |\Phi|_{S_\tau}^{(l+1)}$.

Let the numbers τ and λ be connected with each other by relation (4.17).

Theorem 7.1. *The operator R is a bounded operator acting from the space $\mathfrak{H}^{(l)}$ into the space $\overset{\circ}{H}{}^{l+2,l/2+1}(\overline{Q}_\tau)$, i.e.,*

$$\{Rh\}_{Q_\tau}^{(l+2)} \leqslant c\, \|h\|_{\mathfrak{H}^{(l)}}, \tag{7.12}$$

with the constant c being independent of λ and τ.

Proof. Let us estimate the vector $v = Rh$ by means of the Corollary of Lemma 4.4. We will have

$$\{v\}_{Q_\tau}^{(l+2)} \leqslant c \sup_k \langle v^{(k)} \rangle_{Q_\tau^{(k)}}^{(l+2)}.$$

Using inequalities (7.10) or (7.11) for an estimate of $v^{(k)}$ and the obvious property of boundedness of the operators Z_k and Z_k^{-1}, we get

$$\langle v^{(k)} \rangle_{Q_\tau^{(k)}}^{(l+2)} \leqslant c \langle \zeta^{(k)} f \rangle_{Q_\tau^{(k)}}^{(l)} \tag{7.13}$$

for $k \in \mathfrak{M}$ and

$$\langle v^{(k)} \rangle_{Q_\tau^{(k)}}^{(l+2)} \leqslant c \left(\langle \zeta^{(k)} f \rangle_{Q_\tau^k}^{(l)} + \langle Z_k^{-1} \zeta^{(k)} \Phi \rangle_{\sigma_\tau}^{(l+1)} \right) \tag{7.14}$$

for $k \in \mathfrak{N}$. From here with the use of Lemma 4.3 we deduce

$$\{v\}_{Q_\tau}^{(l+2)} \leqslant c \left(\sup_k \langle \zeta^{(k)} f \rangle_{Q_\tau^{(k)}}^{(l)} + \sup_{k \in \mathfrak{N}} \langle Z_k^{-1} \zeta^{(k)} \Phi \rangle_{\sigma_\tau}^{(l+1)} \right)$$
$$\leqslant c \left(\sup_k \langle f \rangle_{Q_\tau^k}^{(l)} + \sup_{k \in \mathfrak{N}} \langle Z_k^{-1} \Phi \rangle_{\sigma_\tau}^{(l+1)} \right) = c \| h \|_{\mathfrak{H}^{(l)}},$$

and the theorem is proved.

Theorem 7.2. *For any $h \in \mathfrak{H}^{(l)}$ and $v \in \overset{\circ}{H}{}^{l+2,l/2+1}(\overline{Q}_\tau)$ relations* (7.3) *and* (7.4) *are valid, in which T and W are bounded operators in the spaces $\mathfrak{H}^{(l)}$ and $\overset{\circ}{H}{}^{l+1,l/2+1}$ respectively and their norms are small for small τ.*

Proof. Let us introduce for the minor terms of the operators $\mathfrak{L}$ and $\mathfrak{B}$ the notation

$$\mathscr{L}_1\left(x, t, \frac{\partial}{\partial x}\right) u = \sum_{i=1}^n a_i(x, t) \frac{\partial u}{\partial x_i} + a(x, t) u,$$

$$\mathscr{B}_1(x, t) u = b(x, t) u.$$

Also, let

$$A_0 v = (\mathscr{L}_0 v, \mathscr{B}_0 v |_{S_\tau}), \quad A_1 v = (\mathscr{L}_1 v, \mathscr{B}_1 v |_{S_\tau}).$$

We have

$$\mathscr{L}_0 Rh = \sum_k \mathscr{L}_0 \eta^{(k)} v^{(k)} = \sum_k (\mathscr{L}_0 \eta^{(k)} v^{(k)} - \eta^{(k)} \mathscr{L}_0 v^{(k)}) +$$

$$+\sum_k \eta^{(k)}\left[\mathscr{L}_0\left(x,\, t,\, \frac{\partial}{\partial x},\, \frac{\partial}{\partial t}\right)-\mathscr{L}_0\left(\xi^{(k)},\, 0,\, \frac{\partial}{\partial x},\, \frac{\partial}{\partial t}\right)\right] v^k$$

$$+\sum_k \eta^{(k)}\mathscr{L}_0\left(\xi^{(k)},\, 0,\, \frac{\partial}{\partial x},\, \frac{\partial}{\partial t}\right) v^{(k)}.$$

If $k \in \mathfrak{M}$, then

$$\mathscr{L}_0\left(\xi^{(k)},\, 0,\, \frac{\partial}{\partial x},\, \frac{\partial}{\partial t}\right) v^{(k)}$$

$$=\mathscr{L}_0\left(\xi^{(k)},\, 0,\, \frac{\partial}{\partial x},\, \frac{\partial}{\partial t}\right) R^{(k)}\zeta^{(k)} f=\zeta^{(k)} f;$$

while if $k \in \mathfrak{N}$, then

$$\mathscr{L}_0\left(\xi^{(k)},\, 0,\, \frac{\partial}{\partial x},\, \frac{\partial}{\partial t}\right) v^{(k)}$$

$$=\mathscr{L}_0\left(\xi^{(k)},\, 0,\, \frac{\partial}{\partial x},\, \frac{\partial}{\partial t}\right) Z_k R^{(k)}\left(Z_k^{-1}\zeta^{(k)} f,\; Z_k^{-1}\zeta^{(k)}\Phi\right)$$

$$=Z_k\mathscr{L}_0^{(k)}\left(\xi^{(k)},\, 0,\, \frac{\partial}{\partial z}-\operatorname{grad} F\,\frac{\partial}{\partial z_n},\, \frac{\partial}{\partial t}\right) R^{(k)}\left(Z_k^{-1}\zeta^{(k)} f,\, Z_k^{-1}\zeta^{(k)}\Phi\right)$$

$$=Z_k\left[\mathscr{L}_0^{(k)}\left(\xi^{(k)},\, 0,\, \frac{\partial}{\partial z}-\operatorname{grad} F\,\frac{\partial}{\partial z_n},\, \frac{\partial}{\partial t}\right)-\right.$$

$$\left.-\mathscr{L}_0^{(k)}\left(\xi^{(k)},\, 0,\, \frac{\partial}{\partial z},\, \frac{\partial}{\partial t}\right)\right] R^{(k)}\left(Z_k^{-1}\zeta^{(k)} f,\; Z_k^{-1}\zeta^{(k)}\Phi\right)+\zeta^{(k)} f,$$

since, by the definition of $R^{(k)}$, for $z \in \mathfrak{N}$ we have

$$\mathscr{L}_0^{(k)}\left(\xi^{(k)},\, 0,\, \frac{\partial}{\partial z},\, \frac{\partial}{\partial t}\right) R^{(k)}\left(Z_k^{-1}\zeta^{(k)} f,\; Z_k^{-1}\zeta^{(k)}\Phi\right)=Z_k^{-1}\zeta^{(k)} f.$$

We will always denote the n-dimensional vector with components $\partial F/\partial z_1, \cdots, \partial F/\partial z_{n-1}$, 0 by $\operatorname{grad} F$. Under the change of variables (4.3)

$$\frac{\partial}{\partial y_\alpha}=\frac{\partial}{\partial z_\alpha}-\frac{\partial F}{\partial z_\alpha}\frac{\partial}{\partial z_n}\quad (\alpha \neq n), \qquad \frac{\partial}{\partial y_n}=\frac{\partial}{\partial z_n},$$

i.e.

$$\frac{\partial}{\partial y}=\frac{\partial}{\partial z}-\operatorname{grad} F\,\frac{\partial}{\partial z_n}.$$

Now using the fact that $\sum_k \eta^{(k)}\zeta^{(k)}=1$, and, for the sake of conciseness, letting

$$\tilde{v}^{(k)}=R^{(k)}\left(Z_k^{-1}\zeta^{(k)} f,\; Z_k^{-1}\zeta^{(k)}\Phi\right) \quad (k \in \mathfrak{N}),$$

we obtain

$$\mathscr{L}Rh=h+T_1 h,$$

where

$$T_1h = \mathscr{L}_1 Rh + \sum_k (\mathscr{L}_0\eta^{(k)}v^{(k)} - \eta^{(k)}\mathscr{L}_0 v^{(k)})$$
$$+ \sum_k \eta^{(k)}\left[\mathscr{L}_0\left(x, t, \frac{\partial}{\partial x}, \frac{\partial}{\partial t}\right) - \mathscr{L}_0\left(\xi^{(k)}, 0, \frac{\partial}{\partial x}, \frac{\partial}{\partial t}\right)\right] v^{(k)}$$
$$+ \sum_{k\in\mathfrak{N}} \eta^{(k)} Z_k \left[\mathscr{L}_0^{(k)}\left(\xi^{(k)}, 0, \frac{\partial}{\partial z} - \operatorname{grad} F \frac{\partial}{\partial z_n}, \frac{\partial}{\partial t}\right)\right.$$
$$\left. - \mathscr{L}_0^{(k)}\left(\xi^{(k)}, 0, \frac{\partial}{\partial z}, \frac{\partial}{\partial t}\right)\right] \tilde{v}^{(k)}(z, t). \tag{7.15}$$

In a completely analogous way we establish that

$$\mathscr{B} Rh|_{S_\tau} = h + T_2 h,$$

where

$$T_2h = \mathscr{B}_1 Rh|_{S_\tau} + \sum_{k\in\mathfrak{N}} (\mathscr{B}_0\eta^{(k)}v^{(k)} - \eta^{(k)}\mathscr{B}_0 v^{(k)})|_{S_\tau}$$
$$+ \sum_{k\in\mathfrak{N}} \eta^{(k)}\left[\mathscr{B}_0\left(x, t, \frac{\partial}{\partial x}\right) - \mathscr{B}_0\left(\xi^{(k)}, 0, \frac{\partial}{\partial x}\right)\right] v^{(k)}\Big|_{S_\tau}$$
$$+ \sum_{k\in\mathfrak{N}} \eta^{(k)} Z_k \left[\mathscr{B}_0^{(k)}\left(\xi^{(k)}, 0, \frac{\partial}{\partial z} - \operatorname{grad} F \frac{\partial}{\partial z_n}\right)\right.$$
$$\left. - \mathscr{B}_0^{(k)}\left(\xi^{(k)}, 0, \frac{\partial}{\partial z}\right)\right] \tilde{v}^{(k)}.$$

Thus

$$ARh = (\mathscr{L}Rh, \mathscr{B}Rh|_{S_\tau}) = (f, \Phi) + (T_1h, T_2h) = h + Th.$$

In order to estimate the norm of the operator T we express the operators T_1 and T_2 in terms of the coefficients of the operators $\mathfrak{L}$ and $\mathfrak{B}$. We have

$$T_1h = \left(\sum_{i=1}^n a_i(x, t)\frac{\partial}{\partial x_i} + a(x, t)\right) Rh$$
$$- \sum_k \sum_{i, j=1}^n \left(2a_{ij}(x, t)\frac{\partial\eta^{(k)}}{\partial x_i}\frac{\partial v^{(k)}}{\partial x_j} + v^{(k)} a_{ij}(x, t)\frac{\partial^2\eta^{(k)}}{\partial x_i\,\partial x_j}\right)$$
$$- \sum_k \sum_{i, j=1} \eta^{(k)}[a_{ij}(x, t) - a_{ij}(\xi^{(k)}, 0)]\frac{\partial^2 v^{(k)}}{\partial x_i\,\partial x_j}$$
$$+ \sum_{k\in\mathfrak{N}}^n \eta^{(k)} Z_k \left[\sum_{\alpha, \beta=1}^{n-1}\left(2a_{\alpha\beta}^{(k)}(z, t)\frac{\partial F}{\partial z_\alpha}\frac{\partial^2\tilde{v}^{(k)}}{\partial z_\beta\,\partial z_n} - \right.\right.$$

$$- a^{(k)}_{\alpha\beta}(z, t)\frac{\partial F}{\partial z_\alpha}\frac{\partial F}{\partial z_\beta}\frac{\partial^2 \tilde{v}^{(k)}}{\partial z_n^2} + a^{(k)}_{\alpha\beta}\frac{\partial^2 F}{\partial z_\alpha \partial z_\beta}\frac{\partial v^{(k)}}{\partial z_n}\Bigg)$$

$$+ 2\sum_{\alpha=1}^{n-1} a^{(k)}_{\alpha n}(z, t)\frac{\partial F}{\partial z_\alpha}\frac{\partial^2 \tilde{v}^{(k)}}{\partial z_n^2}\Bigg],$$

$$T_2 h = b(x, t) Rh|_{S_\tau} + \sum_{k\in\mathfrak{N}}\sum_{i=1}^{n} b_i(x, t)\frac{\partial \eta^{(k)}}{\partial x_i} v^{(k)}\Big|_{S_\tau}$$

$$+ \sum_{k\in\mathfrak{N}}\sum_{i=1}^{n} \eta^{(k)}\left[b_i(x, t) - b_i(\xi^{(k)}, 0)\right]\frac{\partial v^{(k)}}{\partial x_i}\Big|_{S_\tau}$$

$$- \sum_{k\in\mathfrak{N}} \eta^{(k)} Z_k \sum_{\alpha=1}^{n-1} b^{(k)}_\alpha(z, t)\frac{\partial F}{\partial z_\alpha}\frac{\partial \tilde{v}^{(k)}}{\partial z_n}\Big|_{z_n=0}.$$

By taking into consideration the fact that for $x \in \Omega^{(k)}$

$$|a_{ij}(x, t) - a_{ij}(\xi^{(k)}, 0)|$$

$$\leqslant |a_{ij}(x, t) - a_{ij}(\xi^{(k)}, t)| + |a_{ij}(\xi^{(k)}, t) - a_{ij}(\xi^{(k)}, 0)|$$

$$\leqslant \left\{\begin{matrix} c\lambda & \text{for} & l > 1 \\ c\lambda^l & \text{for} & l < 1 \end{matrix}\right\} \leqslant c\lambda^{l-[l]},$$

$$|b_i(x, t) - b_i(\xi^{(k)}, 0)| \leqslant c\lambda$$

and for $z' \in \sigma$

$$\left|\frac{\partial F}{\partial z_\alpha}\right| \leqslant c\lambda, \tag{7.16}$$

and also making use of Lemmas 4.1 and 4.4, one can estimate the function $T_1 h$ in the following manner:

$$\{T_1 h\}^{(l)}_{Q_\tau} \leqslant c\Bigg(\{Rh\}^{(l+1)}_{Q_\tau} + \frac{1}{\lambda}\sup_k \langle v^{(k)}\rangle^{(l+1)}_{Q^{(k)}_\tau} + \frac{1}{\lambda^2}\sup_k \langle v^{(k)}\rangle^{(l)}_{Q^{(k)}_\tau}$$

$$+ \lambda^{l-[l]}\sup_k \langle v^{(k)}\rangle^{(l+2)}_{Q^{(k)}_\tau}\Bigg)$$

$$\leqslant c\left[\tau^{\frac{1}{2}}\{Rh\}^{(l+2)}_{Q_\tau} + \left(\frac{\tau^{\frac{1}{2}}}{\lambda} + \frac{\tau}{\lambda^2} + \lambda^{l-[l]}\right)\sup_k \langle v^{(k)}\rangle^{(l+2)}_{Q^{(k)}_\tau}\right],$$

from which, by virtue of (7.12)–(7.14) and Lemma 4.3, we obtain

$$\{T_1h\}^{(l)}_{Q_\tau} \leqslant c\left(\varkappa^{\frac{1}{2}}\lambda + \varkappa^{\frac{1}{2}} + \varkappa + \lambda^{l-[l]}\right)\|h\|_{\mathfrak{H}(l)}$$
$$\leqslant c\left(\varkappa^{\frac{1}{2}} + \lambda^{l-[l]}\right)\|h\|_{\mathfrak{H}(l)}.$$

In a completely analogous way one can show that

$$\{T_2h\}^{(l+1)}_{S_\tau} \leqslant c\left(\varkappa^{\frac{1}{2}} + \lambda\right)\|h\|_{\mathfrak{H}(l)}.$$

Consequently,

$$\|Th\|_{\mathfrak{H}(l)} \leqslant c\left(\varkappa^{\frac{1}{2}} + \lambda^{l-[l]}\right)\|h\|_{\mathfrak{H}(l)}, \tag{7.17}$$

from which it is seen that for small $\varkappa$ and λ the norm of the operator T will indeed be small. Since the parameter λ can be taken arbitrarily small, it can be considered that we have obtained the desired estimate for the norm of the operator T.

Let us pass to the proof of (7.4). We have

$$RAv = RA_0v + RA_1v.$$

For $k \in \mathfrak{M}$

$$\begin{aligned} R^{(k)}\zeta^{(k)}\mathscr{L}_0v = {} & R^{(k)}(\zeta^{(k)}\mathscr{L}_0v - \mathscr{L}_0\zeta^{(k)}v) \\ & + R^{(k)}\left(\mathscr{L}_0\left(x,\ t,\ \frac{\partial}{\partial x},\ \frac{\partial}{\partial t}\right)\right. \\ & \left. - \mathscr{L}_0\left(\xi^{(k)},\ 0,\ \frac{\partial}{\partial x},\ \frac{\partial}{\partial t}\right)\right)\zeta^{(k)}v + \zeta^{(k)}v. \end{aligned} \tag{7.18}$$

Since, by the uniqueness of the solution of problem (7.6),

$$R^{(k)}\mathscr{L}_0\left(\xi^{(k)},\ 0,\ \frac{\partial}{\partial x},\ \frac{\partial}{\partial t}\right)\zeta^{(k)}v = \zeta^{(k)}v.$$

Further, for $k \in \mathfrak{N}$

$$\begin{aligned} & R^{(k)}(Z_k^{-1}\zeta^{(k)}\mathscr{L}_0v,\ Z_k^{-1}\zeta^{(k)}\mathscr{B}_0v|_{S_\tau}) \\ & = R^{(k)}[Z_k^{-1}(\zeta^{(k)}\mathscr{L}_0v - \mathscr{L}_0\zeta^{(k)}v),\ Z_k^{-1}(\zeta^{(k)}\mathscr{B}_0v - \mathscr{B}_0\zeta^{(k)}v)|_{S_\tau}] \\ & + R^{(k)}\left[Z_k^{-1}\left(\mathscr{L}_0\left(x,\ t,\ \frac{\partial}{\partial x},\ \frac{\partial}{\partial t}\right) - \mathscr{L}_0\left(\xi^{(k)},\ 0,\ \frac{\partial}{\partial x},\ \frac{\partial}{\partial t}\right)\right)\zeta^{(k)}v,\right. \\ & \left. Z_k^{-1}\left(\mathscr{B}_0\left(x,\ t,\ \frac{\partial}{\partial x}\right) - \mathscr{B}_0\left(\xi^{(k)},\ 0,\ \frac{\partial}{\partial x}\right)\right)\zeta^{(k)}v\Big|_{S_\tau}\right] \\ & \quad + R^{(k)}\left[\left(\mathscr{L}_0^{(k)}\left(\xi^{(k)},\ 0,\ \frac{\partial}{\partial z} - \operatorname{grad} F\frac{\partial}{\partial z_n},\ \frac{\partial}{\partial t}\right)\right.\right. \\ & \qquad \left.\left. - \mathscr{L}_0^{(k)}\left(\xi^{(k)},\ 0,\ \frac{\partial}{\partial z},\ \frac{\partial}{\partial t}\right)\right)Z_k^{-1}\zeta^{(k)}v,\right. \end{aligned}$$

$$\left(\mathscr{B}_0^{(k)}\left(\xi^{(k)}, 0, \frac{\partial}{\partial z} - \operatorname{grad} F \frac{\partial}{\partial z_n}\right) - \mathscr{B}_0^{(k)}\left(\xi^{(k)}, 0, \frac{\partial}{\partial z}\right)\right) Z_k^{-1}\zeta^{(k)}v\Big|_{z_n=0}\Big] + Z_k^{-1}\zeta^{(k)}v, \tag{7.19}$$

since, by the uniqueness of the solution of problem (7.7) (Theorem 6.1),

$$R^{(k)}\left[\mathscr{L}_0^{(k)}\left(\xi^{(k)}, 0, \frac{\partial}{\partial z}, \frac{\partial}{\partial t}\right) Z_k^{-1}\zeta^{(k)}v,\; \mathscr{B}_0^{(k)}\left(\xi^{(k)}, 0 \frac{\partial}{\partial z}\right) Z_k^{-1}\zeta^{(k)}v\big|_{z_n=0}\right] = Z_k^{-1}\zeta^{(k)}v.$$

From (7.18), (7.19), (7.8) and (7.9) it follows that equality (7.4) holds, and

$$\begin{aligned}
Wv = {} & RA_1v + \sum_{k\in\mathfrak{M}} \eta^{(k)}R^{(k)}(\zeta^{(k)}\mathscr{L}_0v - \mathscr{L}_0\zeta^{(k)}v) \\
& + \sum_{k\in\mathfrak{M}} \eta^{(k)}R^{(k)}\left[\left(\mathscr{L}_0\left(x, t, \frac{\partial}{\partial x}, \frac{\partial}{\partial t}\right) - \mathscr{L}_0\left(\xi^{(k)}, 0, \frac{\partial}{\partial x}, \frac{\partial}{\partial t}\right)\right)\zeta^{(k)}v\right] \\
& + \sum_{k\in\mathfrak{N}} \eta^{(k)}Z_kR^{(k)}[Z_k^{-1}(\zeta^{(k)}\mathscr{L}_0v - \mathscr{L}_0\zeta^{(k)}v), \\
& \qquad Z_k^{-1}(\zeta^{(k)}\mathscr{B}_0v - \mathscr{B}_0\zeta^{(k)}v)|_{S_\tau}] \\
& + \sum_{k\in\mathfrak{N}} \eta^{(k)}Z_kR^{(k)}\left[Z_k^{-1}\left(\mathscr{L}_0\left(x, t, \frac{\partial}{\partial x}, \frac{\partial}{\partial t}\right) - \mathscr{L}_0\left(\xi^{(k)}, 0, \frac{\partial}{\partial x}, \frac{\partial}{\partial t}\right)\right)\zeta^{(k)}v,\right. \\
& \qquad \left. Z_k^{-1}\left(\mathscr{B}_0\left(x, t, \frac{\partial}{\partial x}\right) - \mathscr{B}_0\left(\xi^{(k)}, 0, \frac{\partial}{\partial x}\right)\right)\zeta^{(k)}v|_{S_\tau}\right] \\
& + \sum_{k\in\mathfrak{N}} \eta^{(k)}Z_kR^{(k)}\left[\left(\mathscr{L}_0^{(k)}\left(\xi^{(k)}, 0, \frac{\partial}{\partial z} - \operatorname{grad} F \frac{\partial}{\partial z_n}\right) - \mathscr{L}_0^{(k)}\left(\xi^{(k)}, 0, \frac{\partial}{\partial z}, \frac{\partial}{\partial t}\right)\right) Z_k^{-1}\zeta^{(k)}v,\right. \\
& \qquad \left. \left(\mathscr{B}_0^{(k)}\left(\xi^{(k)}, 0, \frac{\partial}{\partial z} - \operatorname{grad} F \frac{\partial}{\partial z_n}\right) - \mathscr{B}_0^{(k)}\left(\xi^{(k)}, 0, \frac{\partial}{\partial z}\right)\right) Z_k^{-1}\zeta^{(k)}v\big|_{z_n=0}\right].
\end{aligned}$$

An estimate of the norm $\langle Wv\rangle_{Q_T}^{(l+2)}$ is obtained with the use of this representation in exactly the same way as estimate (7.17). By inequality (7.12) and Lemma 4.1 we have

$$\{RA_1v\}_{Q_\tau}^{(l+2)} \leqslant c\,\|A_1v\|_{\mathfrak{G}^{(l)}} \leqslant c\,\{v\}_{Q_\tau}^{(l+1)} \leqslant c\tau^{\frac{1}{2}}\,\{v\}_{Q_\tau}^{(l+2)}.$$

Making use once more of Lemmas 4.3 and 4.4 and inequalities (7.10) and (7.11), we obtain the estimate

$$
\{Wv\}_{Q_\tau}^{(l+2)} \leqslant c\left(\tau^{\frac{1}{2}} \{v\}_{Q_\tau}^{(l+2)} + \frac{1}{\lambda} \sup_k \langle v\rangle_{Q_\tau^{(k)}}^{(l+1)} + \frac{1}{\lambda^2} \sup_k \langle v\rangle_{Q_\tau^{(k)}}^{(l)} + \lambda^{l-[l]} \sup_k \langle v\rangle_{Q_\tau^{(k)}}^{(l+2)}\right) \leqslant c\left(\lambda^{l-[l]} + \varkappa^{\frac{1}{2}}\right) \{v\}_{Q_\tau}^{(l+2)},
$$

which completes the proof of the theorem.

Thus we have constructed the operator R, proved equalities (7.3) and (7.4) and obtained estimates of the norms $\|R\|$, $\|T\|$, $\|W\|$. As was shown at the beginning of the section, these results imply the unique solvability of problem (5.4′) and the following estimate for its solution:

$$
\{u\}_{Q_\tau}^{(l+2)} \leqslant \| A^{-1} \| \| h \|_{\mathfrak{H}^{(l)}} \leqslant 2c \| h \|_{\mathfrak{H}^{(l)}}.
$$

Here c is the constant in (7.12). By virtue of Lemmas 4.1 and 4.2 this estimate is equivalent to (5.14). Theorem 5.4 is proved for problem (5.4′).

As we have already noted, this theorem is proved in a completely analogous way for problems (5.2′) and (5.3′); moreover, the calculations for the proof are simplified, since there are no boundary conditions in problem (5.2′), while in problem (5.3′) they have a simpler form.

We again emphasize that in exactly the same way it is possible to consider problem (5.4′) for the very same equation in a cylindrical domain whose lower base lies in the plane $t = t_0 < T$. The height of the cylinder in which the existence of a solution is guaranteed can be taken to be the same for any t_0; this is a consequence of the uniform parabolicity of the operator $\mathfrak{L}$ and inequality (5.5). We denote this height by τ_0.

§8. ON THE SOLVABILITY OF PROBLEM (5.4)

In this section we deduce Theorem 5.3 from the just-proved Theorem 5.4. It is obvious that this requires reducing problem (5.4) to the problem with zero initial data (5.4′). For this purpose it is necessary to calculate from the equation and initial condition the functions $u^{(k)}(x)$, $k = 0, \cdots, 1 + [l/2]$ (this is done with the use of (5.7) and (5.8)), and then to construct a function $v(x, t)$ which possesses the property that

$$
\left.\frac{\partial^k v(x, t)}{\partial t^k}\right|_{t=0} = u^{(k)}(x) \quad \left(k = 0, \ldots, 1 + \left[\frac{l}{2}\right]\right), \tag{8.1}
$$

and to introduce a new unknown function

$$
u'(x, t) = u(x, t) - v(x, t). \tag{8.2}
$$

It must be a solution of the problem with zero initial data

$$\left.\begin{aligned} &\mathscr{L}\left(x, t, \frac{\partial}{\partial x}, \frac{\partial}{\partial t}\right)u' = f', \\ &\mathscr{B}\left(x, t, \frac{\partial}{\partial x}\right)u'\Big|_{S_T} = \Phi', \end{aligned}\right\} \tag{8.3}$$

where

$$f' = f - \mathfrak{L}v \in \overset{\circ}{H}{}^{l, \frac{l}{2}}(\bar{Q}_T), \quad \Phi' = \Phi - \mathfrak{B}v|_{S_T} \in \overset{\circ}{H}{}^{l+1, \frac{l+1}{2}}(S_T).$$

Let us verify that if the conditions of Theorem 5.3 are fulfilled, then $u^{(k)}(x) \in H^{l+2-2k}(\bar{\Omega})$ for $k < 1 + l/2$ and

$$|u^{(k)}|_{\Omega}^{(l+2-2k)} \leqslant c\left(|f|_{Q_t}^{(l)} + |\varphi|_{\Omega}^{(l+2)}\right) \quad (0 < t \leqslant T). \tag{8.4}$$

This is easily done with the use of (5.7) and (5.8). Since the coefficients of the operator $A^{(j)}$, being derivatives of the coefficients of the operator A, belong to the class $H^{l-2j, l/2-j}(\bar{Q}_T)$, it follows from (5.8) that for $k < 1 + l/2$

$$\begin{aligned} |u^{(k)}|_{\Omega}^{(l+2-2k)} &\leqslant |f^{(k-1)}|_{\Omega}^{(l+2-2k)} + c\sum_{j=0}^{k-1}|u^{(j)}|_{\Omega}^{(l+2-2(k-1))} \\ &\leqslant |f|_{Q_t}^{(l)} + c\sum_{j=0}^{k-1}|u^{(j)}|_{\Omega}^{(l+2-2j)}. \end{aligned} \tag{8.5}$$

In this way an estimate of $|u^{(k)}|_{\Omega}^{(l+2-2k)}$ reduces to an estimate of the norm $|u^{(j)}|_{\Omega}^{(l+2-2j)}$ for $j < k$. Since $u^{(0)} = \phi$ it is obvious that (8.5) implies (8.4).

Thus it is shown that $u^{(k)}(x) \in H^{l+2-2k}(\bar{\Omega})$. By means of Theorem 4.1 we can now extend the functions $u^{(k)}(x)$ onto the whole space E_n in such a way that

$$|u^{(k)}|_{E_n}^{(l+2-2k)} \leqslant c\,|u^{(k)}|_{\Omega}^{(l+2-2k)} \tag{8.6}$$

By virtue of Theorem 4.3 it is possible to construct a function $v(x, t) \in H^{l+2, l/2+1}(\overline{D_{n+1}^{(T)}})$ satisfying the initial conditions (8.1) and the inequality

$$\begin{aligned} |v|_{D_{n+1}^{(T)}}^{(l+2)} &\leqslant c\sum_{k=0}^{1+\left[\frac{l}{2}\right]}|u^{(k)}|_{E_n}^{(l+2-2k)} \leqslant c\sum_{k=0}^{1+\left[\frac{l}{2}\right]}|u^{(k)}|_{\Omega}^{(l+2-2k)} \\ &\leqslant c\left(|f|_{Q}^{(l)} + |\varphi|_{\Omega}^{(l+2)}\right) \end{aligned} \tag{8.7}$$

(we have also made use of (8.4) and (8.6)). Therefore the functions f' and Φ' in (8.3) possess the following differential properties: $f' \in \overset{\circ}{H}{}^{l, l/2}(\bar{Q}_T)$, $\Phi' \in \overset{\circ}{H}{}^{l+1, (l+1)/2}(S_T)$.

According to Theorem 5.4 there exists in the cylinder Q_{τ_0} a unique solution of problem (8.3) from the class $\overset{\circ}{H}{}^{l+2,\, l/2+1}(\overline{Q}_{\tau_0})$, that satisfies the inequality

$$|u'|_{Q_{\tau_0}}^{(l+2)} \leqslant c\left(|f'|_{Q_{\tau_0}}^{(l)} + |\Phi'|_{S_{\tau_0}}^{(l+1)}\right). \tag{8.8}$$

This means that problem (5.4) also has a unique solution in Q_{τ_0}, with (8.8) and (8.7) implying that

$$|u|_{Q_{\tau_0}}^{(l+2)} \leqslant |u'|_{Q_{\tau_0}}^{(l+2)} + |v|_{Q_{\tau_0}}^{(l+2)} \leqslant c\left(|f|_{Q_{\tau_0}}^{(l)} + |\varphi|_{\Omega}^{(l+2)} + |\Phi|_{S_{\tau_0}}^{(l+1)}\right).$$

Let us show that a solution of this problem exists in a cylinder of large height. For this purpose we fix some point in the interval $[0, \tau_0]$, for example $\tau_0/2$, and calculate $(\partial^k u/\partial t^k)|_{t=\tau_0/2}$. Let

$$\left.\frac{\partial^k u}{\partial t^k}\right|_{t=\frac{\tau_0}{2}} = \varphi^{(k)}(x).$$

Since $u \in H^{l+2,\, l/2+1}(Q_\tau)$, we have $\phi^{(k)} \in H^{l+2-2k}(\overline{\Omega})$. We construct a function $v'(x, t) \in H^{l+2,\, l/2+1}(\overline{Q}')$, where $Q' = \Omega \times (\tau_0/2, T)$, in such a way that for $k < 1 + l/2$

$$\left.\frac{\partial^k v'}{\partial t^k}\right|_{t=\frac{\tau_0}{2}} = \varphi^{(k)}(x)$$

and

$$|v'|_{Q'}^{(l+2)} \leqslant c \sum_{k=0}^{1+\left[\frac{l}{2}\right]} |\varphi^{(k)}|_{\Omega}^{(l+2-2k)}. \tag{8.9}$$

This can be done by exactly the same method as was used above in constructing the function v.

The function $u'' = u - v'$ is a solution of the following problem with zero initial data posed for $t = \tau_0/2$:

$$\mathscr{L}u'' = f - \mathscr{L}v' = f'',$$
$$\mathscr{B}u''|_S = \Phi - \mathscr{B}v'|_S = \Phi''.$$

We have established that a solution of this problem exists for $\tau_0/2 \leq t \leq \tau_0$, while by virtue of Theorem 5.4 it is possible to assert that it exists for $\tau_0/2 \leq t \leq 3\tau_0/2$ and satisfies in the cylinder $Q'' = \Omega \times (\tau_0/2, 3\tau_0/2)$ the inequality

$$|u''|_{Q''}^{(l+2)} \leqslant c\left(|f''|_{Q''}^{(l)} + |\Phi''|_{\tilde S_{\tau_0}}^{(l+1)}\right),$$
$$\left(\tilde S_{\tau_0} = S \times \left(\frac{\tau_0}{2}, \frac{3\tau_0}{2}\right)\right). \tag{8.10}$$

Hence, a solution of problem (5.4) exists for $0 \le t \le 3\tau_0/2$; it is easily verified that (8.9), (8.10) imply

$$|u|_{Q_{\frac{3\tau_0}{2}}}^{(l+2)} \leqslant c\left(|f|_{Q_{\frac{3\tau_0}{2}}}^{(l)} + |\Phi|_{S_{\frac{3\tau_0}{2}}}^{(l+1)} + |\varphi|_{\Omega}^{(l+2)}\right).$$

By continuing to argue in the same way, we exhaust the whole interval $[0, T]$ and prove the existence of a solution of problem (5.4) in the cylinder Q_T, and also estimate (5.11).

The uniqueness of the solution of problem (5.4) or of the equivalent problem (8.3) in a cylinder Q_T is an obvious consequence of the uniqueness of the solution of problem (8.3) in a cylinder of small height Q_{τ_0} established in Theorem 5.4.

In this way Theorem 5.3 is completely proved.

§9. THE FIRST BOUNDARY VALUE PROBLEM IN CLASSES $W_q^{2,1}(Q_T)$

Problems (5.2)–(5.4) can also be solved in the spaces $W_q^{2m,m}(Q_T)$ ($q > 1$, m a positive integer). The general plan for investigating problems (5.2)–(5.4) in these classes does not undergo any changes in comparison with the plan presented in §§6–8 for application to the spaces $H^{l,l/2}(\overline{Q}_T)$. We will confine ourselves here to an illustration of this plan by using as an example the first boundary value problem (5.3), which we will consider in the classes $W_q^{2,1}(Q_T)$.

As a preliminary, we introduce the norm

$$\|f\|_{r,Q_T}^{(\mathrm{loc})} = \sup_{q_T} \|f\|_{r,q_T}, \tag{9.1}$$

where the supremum is taken over with respect to all cylinders $q_T = \omega \times (0, T)$, the bases ω of which are the intersection of a domain Ω with some domain of unit measure, for example a cube. If the domain Ω is bounded, then the set of functions with finite norm $\|f\|_{r,Q_T}^{(\mathrm{loc})}$ coincides with $L_r(Q_T)$, but for unbounded Ω it is wider than $L_r(Q_T)$.

Theorem 9.1. *Let $q > 1$. Suppose that the coefficients a_{ij} of the operator $\mathfrak{L}$ are bounded continuous functions in Q_T, while the coefficients a_i and a have finite norms $\|a_i\|_{r,Q_T}^{(\mathrm{loc})}$ and $\|a\|_{s,Q_T}^{(\mathrm{loc})}$ respectively, with*

$$r=\begin{cases}\max(q,\ n+2) & \text{for} \quad q\neq n+2,\\ n+2+\varepsilon & \text{for} \quad q=n+2,\end{cases}$$

$$s=\begin{cases}\max\left(q,\ \dfrac{n+2}{2}\right) & \text{for} \quad q\neq \dfrac{n+2}{2},\\ \dfrac{n+2}{2}+\varepsilon & \text{for} \quad q=\dfrac{n+2}{2}\end{cases}$$

and ϵ being an arbitrarily small positive number. Suppose, further, that the quantities $\|a_i\|^{(\mathrm{loc})}_{r,Q_{t,t+\tau}}$ and $\|a\|^{(\mathrm{loc})}_{s,Q_{t,t+\tau}}$ tend to zero for $\tau\to 0$. Let $S\in O^2$. Then, for any $f\in L_q(Q_T)$, $\phi\in W_q^{2-2/q}(\Omega)$ and $\Phi\in W_q^{2-1/q,1-1/2q}(S_T)$, with $q\neq 3/2$, satisfying in the case $q>3/2$ the compatibility condition of zero order

$$\phi|_S=\Phi|_{t=0}, \tag{9.2}$$

problem (5.3) has a unique solution $u\in W_q^{2,1}(Q_T)$. It satisfies the estimate

$$\|u\|^{(2)}_{q,Q_T}\leqslant c\left(\|f\|_{q,Q_T}+\|\varphi\|^{\left(2-\frac{2}{q}\right)}_{q,\Omega}+\|\Phi\|^{\left(2-\frac{1}{q}\right)}_{q,S_T}+T^{-1+\frac{1}{2q}}\|\Phi\|_{q,S_T}\right). \tag{9.3}$$

(the last term can be omitted in the case $q>\frac{3}{2}$). The constant $c=c(T)$ remains bounded for finite values of T.

For any $f\in L_{3/2}(Q_T)$ and any $\varphi\in W_{3/2}^{2/3}(\Omega)$, $\Phi\in W_{3/2}^{4/3,2/3}(S_T)$, such that the function

$$\Psi=\begin{cases}\varphi(x) & x\in\Omega,\\ \Phi(x,t) & x\in S,\end{cases}$$

given on Γ_T, has the finite norm $\{\{\Psi\}\}^{(4/3)}_{3/2,\Gamma_T}$ of (4.38), problem (5.3) has a unique solution $u\in W_{3/2}^{2,1}(Q_T)$. It satisfies the inequality

$$\|u\|^{(2)}_{\frac{3}{2},Q_T}\leq c\left[\|f\|_{\frac{3}{2},Q_T}+\{\{\Psi\}\}^{\left[\frac{4}{3}\right]}_{\frac{3}{2},\Gamma_T}\right]. \tag{9.3'}$$

From Theorem 9.1 and Lemma 3.3 of Chapter II we have

Corollary. *If the conditions of Theorem 9.1 are fulfilled for $q>(n+2)/2$, then the solution of problem (5.3) satisfies a Hölder condition in x and t; when $q>n+2$ the derivatives of the solution of problem (5.4) with respect to the x_i also satisfy a Hölder condition in x and t.*

Indeed, it follows from Lemma 3.3 of Chapter II that for $q>(n+2)/2$, $q\neq n+2$,

$$|u|_Q^{\left(2-\frac{n+2}{q}\right)} \leqslant c\|u\|_{q,Q}^{(2)}.$$

And if $q > n + 2$, then $2 - (n + 2)/q > 1$.

In the proof of Theorem 9.1, just as in the proof of Theorems 5.1–5.3, a basic role is played by the investigation of a problem with zero initial data, which is formulated in the following manner: find a function $u \in \overset{\circ}{W}{}_q^{2,1}(Q_\tau)$ such that

$$\left.\begin{aligned} \mathscr{L}u &= f, \\ u|_{S_T} &= \Phi, \end{aligned}\right\} \tag{9.4}$$

where

$$f \in L_q(Q_\tau), \quad \Phi \in \overset{\circ}{W}{}_q^{2-\frac{1}{q},\, 1-\frac{1}{2q}}(S_\tau).$$

Theorem 9.2. *Under the conditions of Theorem 9.1, problem (9.4) has a unique solution* $u \in \overset{\circ}{W}{}_q^{2,1}(Q_\tau)$ *for any* $f \in L_q(Q_\tau)$, $\Phi \in \overset{\circ}{W}{}_q^{2-1/q,\,1-1/2q}(S_\tau)$ *if* $\tau \leq \tau_0'$, *where* τ_0' *is some number not depending on* f, Φ. *In this connection*

$$\|u\|_{q,Q_\tau}^{(2)} \leqslant c\left(\|f\|_{q,Q_\tau} + \{\Phi\}_{q,S_\tau}^{\left(2-\frac{1}{q}\right)}\right). \tag{9.5}$$

The index q *can have any value greater than* 1, *in particular*, 3/2.

The proof of this theorem requires the consideration of a model problem in the domain $R^{(T)}$:

$$\left.\begin{aligned} \frac{\partial v}{\partial t} - \sum_{i,j=1}^{n} a_{ij}\frac{\partial^2 v}{\partial x_i \partial x_j} &= f(x, t), \\ v|_{x_n=0} &= \Phi(x', t). \end{aligned}\right\} \tag{9.6}$$

It is assumed that $f \in L_q(R^{(T)})$, $\Phi \in \overset{\circ}{W}{}_q^{2-1/q,\,1-1/2q}(\widetilde{D}_n^{(T)})$; the solution must be found in the class $\overset{\circ}{W}{}_q^{2,1}(R^{(T)})$. Let us show that problem (9.6) has a unique solution in this class satisfying the inequality

$$\langle\langle v\rangle\rangle_{q,R^{(T)}}^{(2)} \leqslant c\left(\|f\|_{q,R^{(T)}} + |\Phi|_{q,\widetilde{D}_n^{(T)}}^{\left(2-\frac{1}{q}\right)}\right). \tag{9.7}$$

In exactly the same way as in the proof of Theorem 6.1 it is sufficient to confine oneself to a consideration of the heat equation. It is shown in §6 that for $f \in \overset{\circ}{H}{}^{l,l/2}(R^{(T)})$, and $\Phi \in \overset{\circ}{H}{}^{l+2,l/2+1}(\widetilde{D}_n^{(T)})$ $(l > 0)$ problem (9.6) has a solution in the class $\overset{\circ}{H}^{l+2,l/2+1}(\overline{R^{(T)}})$ that is expressed by formula (6.9). Let us show that estimate (9.7) is valid for the function (6.9). We extend the function f^{*0} of (6.9) from the domain $E^{(T)}_{n+1}$ into the space E_{n+1}, and the function ω^0_1 from $\widetilde{E}^{(T)}_n$ into $\widetilde{E}_n$, in such a way that

$$\| f^{*0} \|_{q, E_{n+1}} \leqslant c \| f^{*0} \|_{q, E^{(T)}_{n+1}} \leqslant c \| f \|_{q, R^{(T)}}, \tag{9.8}$$

$$\langle\langle \omega^0_1 \rangle\rangle^{\left(2-\frac{1}{q}\right)}_{q, \widetilde{E}_n} \leqslant c \langle\langle \omega^0_1 \rangle\rangle^{\left(2-\frac{1}{q}\right)}_{q, \widetilde{E}^{(T)}_n}. \tag{9.9}$$

As was explained in §4, such an extension is possible. Using inequalities (9.8) and (9.9), and also estimates (3.1) and (3.4), we obtain

$$\left.\begin{aligned} &\langle\langle\langle \Gamma * f^{*0} \rangle\rangle\rangle^{(2)}_{q, R^{(T)}} \leqslant \langle\langle\langle \Gamma * f^{*0} \rangle\rangle\rangle^{(2)}_{q, E_{n+1}} \leqslant \\ &\qquad\qquad \leqslant c \| f^{*0} \|_{q, E_{n+1}} \leqslant c \| f \|_{q, R^{(T)}}, \\ &\left\langle\left\langle\left(\frac{\partial \Gamma}{\partial x_n} *_2 \omega^0_1 \right)\right\rangle\right\rangle^{(2)}_{q, R^{(T)}} \leqslant c \langle\langle \omega^0_1 \rangle\rangle^{\left(2-\frac{1}{q}\right)}_{q, \widetilde{E}^{(T)}_n}. \end{aligned}\right\} \tag{9.10}$$

In order to estimate the norm of the function ω^0_1 we make use of the definition (6.10) of this function. Taking into consideration (4.26) and an embedding theorem (Theorem 4.5), and also estimate (9.10), we obtain

$$\begin{aligned} \langle\langle \omega^0_1 \rangle\rangle^{\left(2-\frac{1}{q}\right)}_{q, \widetilde{E}^{(T)}_n} &\leqslant \langle\langle \Phi^0 \rangle\rangle^{\left(2-\frac{1}{q}\right)}_{q, \widetilde{E}^{(T)}_n} + \left\langle\left\langle (\Gamma * f^{*0})\big|_{x_n=0} \right\rangle\right\rangle^{\left(2-\frac{1}{q}\right)}_{q, \widetilde{E}^{(T)}_n} \\ &\leqslant c\left(| \Phi |^{\left(2-\frac{1}{q}\right)}_{q, \widetilde{D}^{(T)}_n} + \langle\langle\langle \Gamma * f^{*0} \rangle\rangle\rangle^{(2)}_{q, E_{n+1}} \right) \\ &\leqslant c\left(| \Phi |^{\left(2-\frac{1}{q}\right)}_{q, \widetilde{D}^{(T)}_n} + \| f \|_{q, R^{(T)}} \right). \end{aligned} \tag{9.11}$$

Combining estimates (9.10) and (9.11), we obtain inequality (9.7) for the function (6.9).

Now suppose $f \in L_q(R^{(T)})$, $\Phi \in \overset{\circ}{W}{}^{2-1/q, 1-1/2q}_q(\widetilde{D}^{(T)}_n)$. We approximate these functions by functions $f^{(n)} \in \overset{\circ}{H}^{l,l/2}(\overline{R^{(T)}})$, $\Phi^{(n)} \in \overset{\circ}{H}^{l+2,l/2+1}(\overline{\widetilde{D}^{(T)}_n})$ $(l>0)$. Let $v^{(n)} \in \overset{\circ}{H}^{l+2,l/2+1}(\overline{R^{(T)}})$ be the solution of the problem with the functions $f^{(n)}$, $\Phi^{(n)}$ in place of f, Φ in the right sides of the equation and boundary condition. The function $v^{(n)} - v^{(m)}$ is estimated in terms of $f^{(n)} - f^{(m)}$ and $\Phi^{(n)} - \Phi^{(m)}$ by inequality (9.7), from which it follows that the sequence $\{v^{(n)}\}$ converges in the norm $\langle\langle v\rangle\rangle^{(2)}_{q,R^{(T)}}$ to some limit $v \in \overset{\circ}{W}{}_q^{2,1}(R^{(T)})$, which is the desired solution of the problem. This solution is subject to inequality (9.7).

The uniqueness of the solution follows from the fact that inequality (9.7) holds for any solution of problem (9.6) from the class $\overset{\circ}{W}{}_q^{2,1}(R^{(T)})$, which is easily demonstrated by approximating this solution with the functions $v^{(n)} \in \overset{\circ}{H}{}^{l+2,l/2+1}(R^{(T)})$, and using the fact that the inequality (9.7) is valid for the functions $v^{(n)}$.

Thus we have proved the unique solvability of problem (9.6) in the classes $\overset{\circ}{W}{}_q^{2,1}(R^{(T)})$.

The Cauchy problem with zero initial data for the equation

$$\frac{\partial w}{\partial t} - \sum_{i,j=1}^{n} a_{ij}\frac{\partial^2 w}{\partial x_i \partial x_j} = f \tag{9.12}$$

with fixed a_{ij} is considered in exactly the same way. For any $f \in L_q(D_{n+1}^{(T)})$ this problem has a unique solution $w \in \overset{\circ}{W}{}_q^{2,1}(D_{n+1}^{(T)})$, with

$$\langle\langle w\rangle\rangle^{(2)}_{q,D_{n+1}^{(T)}} \leqslant c\,\|f\|_{q,D_{n+1}^{(T)}}. \tag{9.13}$$

Let us pass to the proof of Theorem 9.2. In the same way as in §7, we write problem (9.4) in the form

$$A'u = h, \tag{9.14}$$

where A' is the operator associating with each function $u \in \overset{\circ}{W}{}_q^{2,1}(Q_T)$ the pair of functions $A'u = (\mathfrak{L}u, u|_{S_T})$. By virtue of Lemma 3.4 of Chapter II the operator A' is a bounded operator acting from the space $\overset{\circ}{W}{}_q^{2,1}(Q_T)$ into the space

$$\mathfrak{B}_q = L_q(Q_\tau)\times \overset{\circ}{W}{}_q^{2-\frac{1}{q},\,1-\frac{1}{2q}}(S_\tau).$$

The solvability of equation (9.14) for any $h \in \mathfrak{B}_q$, or, what is the same thing, of problem (9.4), can be proved for small τ by constructing an operator R' that is completely analogous to the operator R of §7. We define it by formulas (7.8) and (7.9), but in this section the $R^{(k)}$ will be operators that differ somewhat from the corresponding operators of §7. Namely, we will now denote by $R^{(k)}$ for $k \in \mathfrak{M}$ the operator associating with a function $f^{(k)} \in L_q(D_{n+1}^{(\tau)})$ the solution of the Cauchy problem with zero initial data

$$\mathscr{L}_0\left(\xi^{(k)},\ 0,\ \frac{\partial}{\partial x},\ \frac{\partial}{\partial t}\right)v^{(k)} = f^{(k)},$$
$$v^{(k)} \in \overset{\circ}{W}{}^{2,1}_q(D^{(\tau)}_{n+1}), \tag{9.15}$$

and for $k \in \mathfrak{N}$ the operator associating with a pair of functions $h^{(k)} = (f^{(k)}, \Phi^{(k)})$, where $f^{(k)} \in L_q(R^{(\tau)})$, $\Phi^{(k)} \in \overset{\circ}{W}{}^{2-1/q,\,1-1/2q}_q(\widetilde{D}^{(\tau)}_n)$, the solution of the first boundary value problem with zero initial data

$$\mathscr{L}^{(k)}_0\left(\xi^{(k)},\ 0,\ \frac{\partial}{\partial z},\ \frac{\partial}{\partial t}\right)v^{(k)}(z,\ t) = f^{(k)}(z,\ t),$$
$$v^{(k)}\Big|_{z_n=0} = \Phi^{(k)}(z',\ t), \tag{9.16}$$
$$v^{(k)} \in \overset{\circ}{W}{}^{2,1}_q(R^{(\tau)}).$$

It is not difficult to verify with the use of estimates (9.7) and (9.13) that if in (9.15) the function $f^{(k)}(x, t)$ vanishes outside of $Q^{(k)}_\tau$, while in (9.16) the function $f^{(k)}(z, t)$ vanishes outside of the domain $\mathfrak{R}$ and $\Phi^{(k)}$ vanishes outside of the domain $\sigma'_\tau = \sigma' \times (0, \tau)$, where $\sigma' \subset \sigma$ is the $(n-1)$-dimensional cube

$$|z_\alpha| < \frac{3}{4} \qquad (\alpha = 1,\ \ldots,\ n-1),$$

then for $k \in \mathfrak{M}$

$$\langle\langle R^{(k)} f^{(k)}\rangle\rangle^{(2)}_{q,\ Q^{(k)}_\tau} \leqslant c\, \| f^{(k)} \|_{q,\ Q^{(k)}_\tau} \tag{9.17}$$

and for $k \in \mathfrak{N}$

$$\langle\langle R^{(k)} h^{(k)}\rangle\rangle^{(2)}_{q,\ \mathfrak{R}_\tau} \leqslant c\left(\| f^{(k)} \|_{q,\ \mathfrak{R}_\tau} + | \Phi^{(k)} |^{\left(2-\frac{1}{q}\right)}_{q,\ \sigma_\tau}\right) \qquad (\mathfrak{R}_\tau = \mathfrak{R} \times (0,\ \tau)). \tag{9.18}$$

The definition of the domains $Q^{(k)}_\tau$, $\mathfrak{R}$, σ was given in §4.

We introduce in the space $\overset{\circ}{W}{}^{2,1}_q(Q_\tau)$ the norm $\{v\}^{(2)}_{q,Q_\tau}$, and in the space $\mathfrak{B}_q$ the norm

$$\| h \|_{\mathfrak{B}_q} = \{f\}_{q,\ Q_\tau} + \{\Phi\}^{\left(2-\frac{1}{q}\right)}_{q,\ S_\tau},$$

where $\{f\}_{q,Q_\tau}$, $\{v\}^{(2)}_{q,Q_\tau}$ and $\{\Phi\}^{(2-1/q)}_{q,Q_\tau}$ are norms defined in §4. Using (9.17) and (9.18), it is easy to show that for any $h \in \mathfrak{B}_q$

$$|R'h|^{(2)}_{q,Q_\tau} \leqslant c\,\|h\|_{\mathfrak{B}_q}. \tag{9.19}$$

This inequality is proved in exactly the same way as (7.8), but in place of Lemma 4.4 it is necessary to use Lemma 4.7. It should only be noted that inequality (9.18) can be applied in the proof of (9.19) under the condition that the functions $Z_k^{-1}\zeta^{(k)}\Phi$ in (7.9) are equal to zero outside of σ'_τ for any $k \in \mathfrak{N}$. This will be the case if $Z_k^{-1}\zeta^{(k)} = 0$ when $z' \in E_{n-1}\backslash\sigma'$ for any $k \in \mathfrak{N}$. It is obvious that this latter condition can be fulfilled without damage to the other properties of the functions $\zeta^{(k)}$, which are formulated in §4 and made use of throughout.

Let us prove, finally, that for any $h \in \mathfrak{B}_q$ and $v \in \overset{\circ}{W}{}_q^{2,1}(Q_\tau)$ one has the relations

$$A'R'h = h + T'h, \tag{9.20}$$

$$R'A'v = v + W'v \tag{9.21}$$

with bounded operators T' and W', the norms of which are small for small τ. Let

$$A_0'v = (\mathscr{L}_0 v,\ v|_S),\quad A_1'v = (\mathscr{L}_1 v,\ 0).$$

By repeating the arguments of Theorem 7.2 we easily see that relations (9.20) and (9.21) indeed hold, with

$$T'h = (T_1 h,\ 0),$$

where T_1 is the operator defined by equality (7.15), in which R is replaced by R' and

$$\begin{aligned}
W'v = {} & R'A_1'v + \sum_{k\in\mathfrak{M}} \eta^{(k)} R^{(k)} (\zeta^{(k)}\mathscr{L}_0 v - \mathscr{L}_0\zeta^{(k)} v) \\
& + \sum_{k\in\mathfrak{M}} \eta^{(k)} R^{(k)} \Big[\Big(\mathscr{L}_0\Big(x,\ t,\ \frac{\partial}{\partial x},\ \frac{\partial}{\partial t}\Big) \\
& - \mathscr{L}_0\Big(\xi^{(k)},\ 0,\ \frac{\partial}{\partial x},\ \frac{\partial}{\partial t}\Big)\Big]\zeta^{(k)} v \\
& + \sum_{k\in\mathfrak{N}} \eta^{(k)} Z_k R^{(k)} [Z_k^{-1}(\zeta^{(k)}\mathscr{L}_0 v - \mathscr{L}_0\zeta^{(k)} v),\ 0] \\
& + \sum_{k\in\mathfrak{N}} \eta^{(k)} Z_k R^{(k)} \Big[Z_k^{-1}\Big(\mathscr{L}_0\Big(x,\ t,\ \frac{\partial}{\partial x},\ \frac{\partial}{\partial t}\Big) \\
& - \mathscr{L}_0\Big(\xi^{(k)},\ 0,\ \frac{\partial}{\partial x},\ \frac{\partial}{\partial t}\Big)\Big)\zeta^{(k)} v,\ 0\Big] \\
& + \sum_{k\in\mathfrak{N}} \eta^{(k)} Z_k R^{(k)} \Big[\Big(\mathscr{L}_0^{(k)}\Big(\xi^{(k)},\ 0,\ \frac{\partial}{\partial z} - \operatorname{grad} F\frac{\partial}{\partial z_n},\ \frac{\partial}{\partial t}\Big) \\
& - \mathscr{L}_0^{(k)}\Big(\xi^{(k)},\ 0,\ \frac{\partial}{\partial z},\ \frac{\partial}{\partial t}\Big)\Big) Z_k^{-1}\zeta^{(k)} v,\ 0\Big].
\end{aligned} \tag{9.22}$$

Let us proceed to an estimate of $\|T'h\|_{\mathfrak{B}_q}$ and $\{W'v\}^{(2)}_{q,Q_T}$. We will show that for any $v \in \overset{\circ}{W}{}^{2,1}_q(Q_\tau)$

$$\{\mathscr{L}_1 v\}_{q,Q_\tau} \leqslant \eta(\tau)\,\{v\}^{(2)}_{q,Q_\tau}, \tag{9.23}$$

where $\eta(\tau) \to 0$ for $\tau \to 0$. We have

$$\begin{aligned}\{\mathscr{L}_1 v\}_{q,Q_\tau} &\leqslant \sum_{i=1}^n \left\{a_i \frac{\partial v}{\partial x_i}\right\}_{q,Q_\tau} + \{av\}_{q,Q_\tau} \\ &= \sum_{i=1}^n \Big[\sum_k \Big\| a_i \frac{\partial v}{\partial x_i}\Big\|^q_{q,Q_\tau^{(k)}}\Big]^{\frac{1}{q}} + \Big[\sum_k \|av\|^q_{q,Q_\tau^{(k)}}\Big]^{\frac{1}{q}}.\end{aligned}$$

Let us estimate $\|a_i\,\partial v/\partial x_i\|_{q,Q_\tau^{(k)}}$. Applying Hölder's inequality, we get

$$\Big\| a_i \frac{\partial v}{\partial x_i}\Big\|_{q,Q_\tau^{(k)}} \leqslant \|a_i\|_{r,Q_\tau^{(k)}} \Big\| \frac{\partial v}{\partial x_i}\Big\|_{\frac{rq}{r-q},\,Q_\tau^{(k)}}. \tag{9.24}$$

Here $r \geq q$ is the index defined in the formulation of Theorem 9.1. If $r = q$, we as usual take

$$\Big\| \frac{\partial v}{\partial x_i}\Big\|_{\frac{rq}{r-q},\,Q_\tau^{(k)}} = \Big\| \frac{\partial v}{\partial x_i}\Big\|_{\infty,\,Q_\tau^{(k)}} = \operatorname*{vrai\,max}_{Q_\tau^{(k)}} \Big| \frac{\partial v}{\partial x_i}\Big|;$$

in this case inequality (9.24) remains valid.

Using Lemma 3.3 of Chapter II and Lemma 4.5, we get

$$\Big\| \frac{\partial v}{\partial x_i}\Big\|_{\frac{rq}{r-q},\,Q_\tau^{(k)}} \leqslant c\,\langle\langle v\rangle\rangle^{(2)}_{q,Q_\tau^{(k)}}, \tag{9.25}$$

and since by condition the functions a_i have a finite norm $\|a_i\|^{(\mathrm{loc})}_{r,Q_T}$ it follows from (9.24) and (9.25) that

$$\Big\| a_i \frac{\partial v}{\partial x_i}\Big\|_{q,Q_\tau^{(k)}} \leqslant c\,\|a_i\|^{(\mathrm{loc})}_{r,Q_\tau}\,\langle\langle v\rangle\rangle^{(2)}_{q,Q_\tau^{(k)}}.$$

In a completely analogous way

$$\|av\|_{q,Q_\tau^{(k)}} \leqslant c\,\|a\|^{(\mathrm{loc})}_{s,Q_\tau}\,\langle\langle v\rangle\rangle^{(2)}_{q,Q_\tau^{(k)}},$$

so that

$$\{\mathscr{L}_1 v\}_{q,Q_\tau} \leqslant c\Big(\sum_{i=1}^n \|a_i\|^{(\mathrm{loc})}_{r,Q_\tau} + \|a\|^{(\mathrm{loc})}_{s,Q_\tau}\Big)\Big[\sum_k \Big(\langle\langle v\rangle\rangle^{(2)}_{q,Q_\tau^{(k)}}\Big)^q\Big]^{\frac{1}{q}}$$

and, consequently, (9.23) holds with

$$\eta(\tau)=c\left(\sum_{i=1}^{n}\|a_i\|_{r,Q_\tau}^{(\mathrm{loc})}+\|a\|_{s,Q_\tau}^{(\mathrm{loc})}\right).$$

By virtue of a condition of the theorem this quantity is small for small τ.

Estimate (9.23) together with inequality (9.19) permits us to estimate the first terms of the right sides of (7.15) and (9.22). The remaining terms are estimated in the same way as in §7, with the use of Lemmas 4.6 and 4.7, inequalities (9.17)–(9.19), (7.16), and also the inequalities

$$|a_{ij}(x,t)|\leqslant c,$$
$$|a_{ij}(x,t)-a_{ij}(\xi^{(k)},0)|\leqslant\eta_1(\lambda)\qquad ((x,t)\in Q_\tau^{(k)}),$$

where $\eta_1(\lambda)\to 0$ together with λ. As a result we obtain the estimates

$$\|T'h\|_{\mathfrak{W}_q}\leqslant c\left(\eta(\tau)+\eta_1(\lambda)+\lambda+\varkappa^{\frac{1}{2}}\right)\|h\|_{\mathfrak{W}_q},$$
$$\{W'v\}_{q,Q_\tau}^{(2)}\leqslant c\left(\eta(\tau)+\eta_1(\lambda)+\lambda+\varkappa^{\frac{1}{2}}\right)\{v\}_{q,Q_\tau}^{(2)}.$$

From here it is clear that, by choosing sufficiently small numbers λ and κ, we can make the norms of the operators T' and W' arbitrarily small. This proves Theorem 9.2 in the same way as in the Hölder-continuous case.

Theorem 9.1 is proved from Theorem 9.2 in the same manner as Theorem 5.3 in §8. In order to prove the existence of the solution of problem (5.3) and estimate (9.3) we reduce problem (5.3) to a problem with zero initial data for a function $u'=u-v$, where $v(x,t)$ is a function given in Q_T and satisfying the initial condition

$$v(x,0)=\varphi(x),$$

with

$$\|v\|_{q,Q}^{(2)}\leqslant c\|\varphi\|_{q,\Omega}^{\left(2-\frac{2}{q}\right)}. \tag{9.26}$$

This function can be constructed by extending $\varphi(x)$ with preservation of class onto the whole space E_n and putting $v=(\Gamma *_1\varphi)$. Then estimate (9.26) will be a consequence of (3.2).

The function u' must be a solution of the problem

$$\mathscr{L}\left(x,t,\frac{\partial}{\partial x},\frac{\partial}{\partial t}\right)u'=f',\qquad u'|_{S_T}=\Phi',\qquad u'|_{t=0}=0. \tag{9.27}$$

where $f' = f - \mathfrak{L}v$ and $\Phi' = \Phi - v|_S$. In this connection, on account of Lemma 3.4 of Chapter II, we have

$$\|f'\|_{q,Q_T} \le \|f\|_{q,Q_T} + c\,\|v\|^{(2)}_{q,Q_T} \le \|f\|_{q,Q_T} + c\,\|\varphi\|^{(2-2/q)}_{q,\Omega},$$

$$\|\Phi'\|^{\left(2-\frac{1}{q}\right)}_{q,S_T} \leqslant \|\Phi\|^{\left(2-\frac{1}{q}\right)}_{q,S_T} + \|v\|^{\left(2-\frac{1}{q}\right)}_{q,S_T} \leqslant \|\Phi\|^{\left(2-\frac{1}{q}\right)}_{q,S_T} \tag{9.28}$$

$$+ c\,\|v\|^{(2)}_{q,Q_T} \leqslant \|\Phi\|^{\left(2-\frac{1}{q}\right)}_{q,S_T} + c\,\|\varphi\|^{\left(2-\frac{2}{q}\right)}_{q,\Omega},$$

while by virtue of the compatibility condition (9.2) the function Φ' for $q > 3/2$ satisfies the zero initial condition $\Phi'(x, 0) = 0$.

Suppose to begin with that $q \neq 3/2$; then $1 - 1/2q \neq 1/q$ and as was shown in §4 (see (4.27), (4.28)),

$$\{\Phi'\}^{\left(2-\frac{1}{q}\right)}_{q,S_T} \leqslant c\,\|\Phi'\|^{\left(2-\frac{1}{q}\right)}_{q,S_T} + \frac{C}{T^{1-\frac{1}{2q}}}\|\Phi'\|_{q,S_T}. \tag{9.29}$$

(the last term can be omitted in the case $q > 3/2$).

Applying Theorem 9.2 successively for the intervals $t \in (k\tau_0', (k+1)\tau_0')$, $k = 0, ½, 1, \cdots$ and using (9.29), we prove, in exactly the same way as in §8, the solvability of problem (9.27) and the estimate

$$\left(\|u'\|^{(2)}_{q,Q} \leqslant c\ \|f\|_{q,Q} + \|\Phi'\|^{\left(2-\frac{1}{q}\right)}_{q,S_T} + T^{-1+\frac{1}{2q}}\|\Phi\|_{q,S_T}\right) \tag{9.30}$$

This result and estimate (9.28) imply Theorem 9.1 in the case $q \neq 3/2$.

Now consider the case $q = 3/2$. Let us estimate the norm $\{\Phi'\}^{(4/3)}_{3/2,S_T}$ of the function Φ' of (9.27). Let

$$\Phi_0' = \begin{cases} 0 & x \in \Omega,\ t = 0 \\ \phi'(x, t) & (x, t) \in S_T \end{cases}.$$

Taking into account (4.39), (4.40) and (9.26), we obtain

$$\{\Phi'\}^{\left(\frac{4}{3}\right)}_{\frac{3}{2},S_T} \le c\,\{\{\Phi_0'\}\}^{\left(\frac{4}{3}\right)}_{\frac{3}{2},\Gamma_T} = c\,\{\{\Psi - v\}\}^{\left(\frac{4}{3}\right)}_{\frac{3}{2},\Gamma_T} \le c\left[\{\{\Psi\}\}^{\left(\frac{4}{3}\right)}_{\frac{3}{2},\Gamma_T} + \{\{v\}\}^{\left(\frac{4}{3}\right)}_{\frac{3}{2},\Gamma_T}\right] \tag{9.31}$$

$$\le c\left[\{\{\Psi\}\}^{\left(\frac{4}{3}\right)}_{\frac{3}{2},\Gamma_T} + \|v\|^{(2)}_{\frac{3}{2},Q_T}\right] \le c\,\{\{\Psi\}\}^{\left(\frac{4}{3}\right)}_{\frac{3}{2},\Gamma_T}.$$

Thus $\Phi' \in \overset{\circ}{W}{}^{4/3,2/3}_{3/2}(S_T)$. The solvability of problem (9.27) in the class $\overset{\circ}{W}{}^{2,1}_{3/2}(Q_T)$ is established by means of Theorem 9.2 in exactly the same way as in the classes $\overset{\circ}{W}{}^{2,1}_{q}(Q_T)$ with $q \neq 3/2$.

The solution of this problem is subject to the inequality

$$\|u'\|^{(2)}_{\frac{3}{2},Q_T} \leq c\left[\|f'\|_{\frac{3}{2},Q_T} + \{\Phi'\}^{\left(\frac{4}{3}\right)}_{\frac{3}{2},S_T}\right]$$

which together with (9.26), (9.28) and (9.31) proves estimate (9.3′).

Thus Theorem 9.1 is completely proved.

It carries over word for word to problem (5.15).

It should be emphasized that the estimates (9.3) and (9.3′), proved for all indices q, of the solution of problem (5.3) are exact in the sense that their right sides are estimated in terms of their left sides multiplied by certain constants which do not depend on the functions f, φ and Φ.

An analogous theorem is valid for problem (5.4), but the singular index for it will be $q = 3$. In proving this theorem it is necessary to use (4.41) instead of (4.40).

§10. LOCAL ESTIMATES OF THE SOLUTIONS OF PROBLEMS (5.3) AND (5.4)

This section is devoted to estimates of the solutions of problems (5.4) and (5.5) in the norms $|u|^{(l)}_{Q'}$, where $Q' = \Omega' \times (T_1, T_2)$, $\Omega' \subset \Omega$, $0 \leq T_1 < T_2 \leq T$. In the right side of these estimates, besides the norms of the known functions, will be some weak norm of the solution, for example $\max|u|$, but in a certain larger domain Q''.

We introduce the following notation. The intersection $S \cap \overline{\Omega}'$ is denoted by S'. Let $\Omega \supset \Omega'' \supset \Omega'$, with the distance from Ω' to $\Omega \backslash \Omega''$ being positive, and let $S'' = S \cap \overline{\Omega}''$. If the set S' is empty, then the domain Ω'' will be chosen so that S'' will also be empty. Let $Q'' = \Omega'' \times (T_0, T_2)$, with $0 < T_0 < T_1$ if $T_1 > 0$, and $T_0 = 0$ if $T_1 = 0$. Finally, let $S'_{T_1T_2} = S' \times (T_1, T_2)$ and $S''_{T_1T_2} = S'' \times (T_1, T_2)$.

Theorem 10.1. *Suppose* $u \in H^{l+2,l/2+1}(\overline{Q}'')$ *and suppose that in* Q''

$$\mathscr{L}\left(x,\, t,\, \frac{\partial}{\partial x},\, \frac{\partial}{\partial t}\right) u = f; \tag{10.1}$$

in addition, if $T_0 = 0$, *let there be fulfilled the initial condition*

$$u\,|_{t=0} = \varphi(x); \quad x \in \Omega'' \tag{10.2}$$

and, finally, if S'' *is not empty, let there be fulfilled the boundary condition*

$$u\,|_{S''_{T_1 T_2}} = \Phi(x,\, t) \tag{10.3}$$

or

$$\mathscr{B} u\,|_{S''_{T_1 T_2}} = \Phi(x,\, t). \tag{10.4}$$

Suppose the coefficients of the operator $\mathfrak{L}$ *belong to the class* $H^{l,l/2}(\overline{Q}'')$ *and* $S'' \in H^{l+2}$. *If condition* (10.3) *is fulfilled, then*

$$|u|_{Q'}^{(l+2)} \leqslant c_1\left(|f|_{Q''}^{(l)} + |\varphi|_{\Omega''}^{(l+2)} + |\Phi|_{S''_{T_0 T_2}}^{(l+2)}\right) + c_2 |u|_{Q''}; \tag{10.5}$$

but if the boundary condition (10.4) *is fulfilled and the coefficients* b_i *and* b *of the operator* $\mathfrak{B}$ *belong to the class* $H^{l+1,(l+1)/2}(\overline{S''_{T_0 T_2}})$, *then*

$$|u|_{Q'}^{(l+2)} \leqslant c\left(|f|_{Q''}^{(l)} + |\varphi|_{\Omega''}^{(l+2)} + |\Phi|_{S''_{T_0 T_2}}^{(l+1)}\right) + c_2 |u|_{Q''}. \tag{10.6}$$

If S'' *is empty, then the right sides of* (10.5) *and* (10.6) *will not contain the norms of the function* Φ, *and for* $T_0 > 0$ *they will not contain* $|\varphi|_{\Omega''}^{(l+2)}$.

Proof. For the sake of definiteness we will assume that S'' is not empty, that the boundary condition has the form (10.3), and that $T_0 = T_1 = 0$, and we will prove inequality (10.5). We note that it is sufficient to prove it for domains Ω'' and Ω' of canonical form, i.e. for those entering into $\mathfrak{k}^{(\lambda)}$ and $\mathfrak{K}^{(\lambda)}$.

Let the domain Ω'' be given in a local coordinate system at some point $\xi \in S$ by the inequalities

$$-\lambda_1 \leqslant y_\alpha \leqslant \lambda_1 \quad (\alpha = 1,\, \ldots,\, n-1), \quad 0 < y_n - F(y') < 2\lambda_1,$$

where λ_1 is a small positive number. Along with Ω'' we consider the domains $\Omega^{(\lambda)}$ $(0 \leq \lambda \leq \lambda_1)$ defined by the inequalities

$$\begin{gathered} -(\lambda_1 - \lambda) \leqslant y_\alpha \leqslant (\lambda_1 - \lambda) \quad (\alpha = 1,\, \ldots,\, n-1), \\ 0 \leqslant y_n - F(y') \leqslant 2(\lambda_1 - \lambda). \end{gathered}$$

We introduce the notation $Q^{(\lambda)} = \Omega^{(\lambda)} \times (0,\, T_2)$, $S^{(\lambda)} = \overline{\Omega}^{(\lambda)} \cap S$, $S_{T_2}^{(\lambda)} = S^{(\lambda)} \times (0,\, T_2)$.

Let $\zeta^{(\lambda)}(x)$ be a function having the properties

$$\zeta^{(\lambda)}=\begin{cases}1 & \text{for} \quad x\in\Omega^{(\lambda)},\\ 0 & \text{for} \quad x\in\Omega\setminus\Omega^{\left(\frac{\lambda}{2}\right)},\end{cases} \qquad |D_x^s\zeta^{(\lambda)}|\leqslant\frac{c_s}{\lambda^s}.$$

The function $v=\zeta^{(\lambda)}u$ is defined in Q_{T_2} (it is equal to zero outside of Q'') and is a solution of the first boundary value problem

$$\mathscr{L}v=g,$$
$$v|_{t=0}=\varphi\zeta^{(\lambda)},\quad v|_{S_{T_2}}=\Phi\zeta^{(\lambda)},$$

where

$$g=f\zeta^{(\lambda)}+(\mathscr{L}u\zeta^{(\lambda)}-\zeta^{(\lambda)}\mathscr{L}u).$$

On the basis of Theorem 5.2

$$\begin{aligned}|v|_{Q_{T_2}}^{(l+2)}&\leqslant c\left(|g|_{Q_{T_2}}^{(l)}+|\varphi\zeta^{(\lambda)}|_{\Omega}^{(l+2)}+|\Phi\zeta^{(\lambda)}|_{S_{T_2}}^{(l+2)}\right)\\ &\leqslant c\left\{|f\zeta^{(\lambda)}|_{Q^{\left(\frac{\lambda}{2}\right)}}^{(l)}+|\mathscr{L}u\zeta^{(\lambda)}-\zeta^{(\lambda)}\mathscr{L}u|_{Q^{\left(\frac{\lambda}{2}\right)}}^{(l)}+|\varphi\zeta^{(\lambda)}|_{\Omega^{\left(\frac{\lambda}{2}\right)}}^{(l+2)}\right.\\ &\qquad\left.+|\Phi\zeta^{(\lambda)}|_{S_{T_2}^{\left(\frac{\lambda}{2}\right)}}^{(l+2)}\right\}.\end{aligned} \tag{10.7}$$

We have

$$\mathscr{L}u\zeta^{(\lambda)}-\zeta^{(\lambda)}\mathscr{L}u=-2\sum_{i,j=1}^{n}a_{ij}\frac{\partial u}{\partial x_i}\frac{\partial\zeta^{(\lambda)}}{\partial x_j}-\sum_{i,j=1}^{n}a_{ij}u\frac{\partial^2\zeta^{(\lambda)}}{\partial x_i\,\partial x_j}+\sum_{i=1}^{n}a_iu\frac{\partial\zeta^{(\lambda)}}{\partial x_i}$$

and therefore

$$\begin{aligned}&|\mathscr{L}u\zeta^{(\lambda)}-\zeta^{(\lambda)}\mathscr{L}u|_{Q^{\left(\frac{\lambda}{2}\right)}}^{(l)}\\ &\qquad\leqslant c\left\{\sum_{k=1}^{[l]+2}\frac{1}{\lambda^k}\langle u\rangle_{Q^{\left(\frac{\lambda}{2}\right)}}^{(l+2-k)}+\sum_{k=0}^{[l]+1}\frac{1}{\lambda^{l+2-k}}\langle u\rangle_{Q^{\left(\frac{\lambda}{2}\right)}}^{(k)}\right\}.\end{aligned} \tag{10.8}$$

In the norms of the functions $f\zeta^{(\lambda)}$, $\varphi\zeta^{(\lambda)}$, $\Phi\zeta^{(\lambda)}$ we isolate $\langle f\rangle^{(l)}$, $\langle\varphi\rangle^{(l+2)}$, $\langle\Phi\rangle^{(l+2)}$, while the remaining terms are estimated in terms of the norms of the function $u(x,t)$ by means of relations (10.1)–(10.3). For example,

$$|f\zeta^{(\lambda)}|_{Q^{\left(\frac{\lambda}{2}\right)}}^{(l)}=\langle f\zeta^{(\lambda)}\rangle_{Q^{\left(\frac{\lambda}{2}\right)}}^{(l)}+\sum_{k=0}^{[l]}\langle f\zeta^{(\lambda)}\rangle_{Q^{\left(\frac{\lambda}{2}\right)}}^{(k)}\leqslant\langle f\rangle_{Q^{\left(\frac{\lambda}{2}\right)}}^{(l)}+$$

$$+c\left(\sum_{k=1}^{[l]}\frac{1}{\lambda^k}\langle f\rangle^{(l-k)}_{Q\left(\frac{\lambda}{2}\right)}+\sum_{k=0}^{[l]}\frac{1}{\lambda^{l-k}}\langle f\rangle^{(k)}_{Q\left(\frac{\lambda}{2}\right)}\right)\leqslant\langle f\rangle^{(l)}_{Q\left(\frac{\lambda}{2}\right)}$$

$$+c\left(\sum_{k=1}^{[l]+2}\frac{1}{\lambda^k}\langle u\rangle^{(l+2-k)}_{Q\left(\frac{\lambda}{2}\right)}+\sum_{k=0}^{[l]+2}\frac{1}{\lambda^{l+2-k}}\langle u\rangle^{(k)}_{Q\left(\frac{\lambda}{2}\right)}\right).$$

Treating the remaining terms of the right side of (10.7) in the same way and taking into consideration (10.8), we get

$$|u|^{(l+2)}_{Q(\lambda)}\leqslant|v|^{(l+2)}_{Q_{T_1}}\leqslant c_1\left\{\langle f\rangle^{(l)}_{Q\left(\frac{\lambda}{2}\right)}+\langle\varphi\rangle^{(l+2)}_{\Omega\left(\frac{\lambda}{2}\right)}+\langle\Phi\rangle^{(l+2)}_{S_{T_1}\left(\frac{\lambda}{2}\right)}\right\}$$

$$+c_2\left\{\sum_{k=1}^{[l]+2}\frac{1}{\lambda^k}\langle u\rangle^{(l+2-k)}_{Q\left(\frac{\lambda}{2}\right)}+\sum_{k=0}^{[l]+2}\frac{1}{\lambda^{l+2-k}}\langle u\rangle^{(k)}_{Q\left(\frac{\lambda}{2}\right)}\right\}. \tag{10.9}$$

The norms of the function $u(x, t)$ that appear in the right side are estimated with the use of Lemma 3.2 of Chapter II. By virtue of this lemma

$$\left.\begin{aligned}\langle u\rangle^{(l+2-k)}_{Q\left(\frac{\lambda}{2}\right)}&\leqslant\delta^k\langle u\rangle^{(l+2)}_{Q\left(\frac{\lambda}{2}\right)}+c\delta^{-(l+2-k)}|u|^{(0)}_{Q\left(\frac{\lambda}{2}\right)},\\ \langle u\rangle^{(k)}_{Q\left(\frac{\lambda}{2}\right)}&\leqslant\delta^{l+2-k}\langle u\rangle^{(l+2)}_{Q\left(\frac{\lambda}{2}\right)}+c\delta^{-k}|u|^{(0)}_{Q\left(\frac{\lambda}{2}\right)}.\end{aligned}\right\} \tag{10.10}$$

These inequalities are valid for $\delta\le c_0(\lambda_1-\lambda/2)$, where c_0 is some constant not depending on λ. One can therefore take $\delta=\epsilon\lambda$ with ϵ small. If we estimate the right side of (10.9) with the use of (10.10), we will have

$$|u|^{(l+2)}_{Q(\lambda)}\leqslant\varepsilon|u|^{(l+2)}_{Q\left(\frac{\lambda}{2}\right)}+c\left\{\langle f\rangle^{(l)}_{Q^{\prime\prime}}+\langle\varphi\rangle^{(l+2)}_{\Omega^{\prime\prime}}+\langle\Phi\rangle^{(l+2)}_{S^{\prime\prime}_{T_1}}\right\}$$

$$+\frac{c_\varepsilon}{\lambda^{l+2}}|u|^{(0)}_{Q\left(\frac{\lambda}{2}\right)} \tag{10.11}$$

Let

$$N_0=\langle f\rangle^{(l)}_{Q^{\prime\prime}}+\langle\varphi\rangle^{(l+2)}_{\Omega^{\prime\prime}}+\langle\Phi\rangle^{(l+2)}_{S^{\prime\prime}_{T_1}},$$

$$N_1=|u|^{(0)}_{Q^{\prime\prime}}$$

$$f(\lambda)=\lambda^{l+2}|u|^{(l+2)}_{Q(\lambda)},\quad K(\lambda)=c\lambda^{l+2}N_0+c_\varepsilon N_1$$

and

$$\varepsilon = \frac{\mu}{2^{l+3}}.$$

The number $\mu \leq 1$ is fixed so that $\delta = \epsilon\lambda \leq c_0(\lambda_1 - \lambda/2)$; for this purpose it is sufficient to take $\mu = \min(1, 2^{l+2}c_0)$. Multiplying (10.11) by λ^{l+2}, we get

$$f(\lambda) \leqslant \frac{\mu}{2} f\left(\frac{\lambda}{2}\right) + K(\lambda) \leqslant \frac{1}{2} f\left(\frac{\lambda}{2}\right) + K(\lambda).$$

Since the function $f(\lambda)$ is bounded for $0 \leq \lambda \leq \lambda_1$ it follows from this inequality that

$$f(\lambda) \leqslant \sum_{j=0}^{\infty} \frac{1}{2^j} K\left(\frac{\lambda}{2^j}\right).$$

The function $K(\lambda)$ is monotonically increasing, and hence

$$f(\lambda) \leqslant \sum_{j=0}^{\infty} \frac{1}{2^j} K(\lambda) = 2K(\lambda).$$

Dividing this inequality by λ^{l+2}, we obtain the desired estimate

$$| u |_{Q^{(\lambda)}}^{(l+2)} \leqslant c_1 \left\{ \langle f \rangle_{Q'}^{(l)} + \langle \varphi \rangle_{\Omega'}^{(l+2)} + \langle \Phi \rangle_{S'_{T_1}}^{(l+2)} \right\} + \frac{c_2}{\lambda^{l+2}} | u |_{Q''}^{(0)},$$

which is valid for any $\lambda < \lambda_1$. Inequality (10.5) is proved.

We have assumed that $T_0 = T_1 = 0$. For $T_1 > T_0 > 0$ estimate (10.5) is proved analogously; it is merely necessary to take as the cutting function a function depending on x and t and vanishing for $t \leq T_0$.

We note, further, that in place of $|u|_{Q''}^{(0)}$ in the right sides of (10.5) and (10.6) we can put the weaker norm $\|u\|_{r,Q''}$ $r \geqslant 1$. We can also obtain estimates analogous to estimates (10.5), (10.6) in the norms $\langle \langle u \rangle \rangle_{Q''}^{(2m)}$. For example, for a function $u \in W_q^{2,1}(Q'')$ satisfying relations (10.1)–(10.3) we have

$$\| u \|_{q,Q'}^{(2)} \leqslant \left(\| f \|_{q,Q'} + \| \varphi \|_{q,\Omega'}^{\left(2-\frac{2}{q}\right)} + \| \Phi \|_{q,S_{T_0T_1}}^{\left(2-\frac{1}{q}\right)} \right) + c_2 \| u \|_{\tilde{q},Q''}, \tag{10.12}$$

where $1 \leq \tilde{q} \leq q$. In the case $T_0 = T_1 = 0$ this inequality has another form for $q = 3/2$. It is proved in exactly the same way as (10.5), with the use of estimate (9.3), it being assumed that the coefficients of the operator $\mathfrak{L}$ satisfy the conditions of Theorem 9.1 in Q''.

§11. A FUNDAMENTAL SOLUTION OF THE PARABOLIC EQUATION OF SECOND ORDER

In this section we commence the presentation of another method for investigating the Cauchy problem and the basic boundary value problems for parabolic equations of second order, namely the method of potential theory. We begin with the construction of a fundamental solution for a parabolic operator of second order.

Let there be given in the domain $D_{n+1}^{(T)} = E_n \times (0, T)$ a parabolic operator $\mathfrak{L}(x, t, \partial/\partial x, \partial/\partial t)$ with coefficients belonging to the class $H^{\alpha, \alpha/2}(\overline{D_{n+1}^{(T)}})$, $\alpha < 1$. We will construct a fundamental solution $Z(x, \xi, t, \tau)$ that satisfies the equation

$$\mathscr{L}\left(x, t, \frac{\partial}{\partial x}, \frac{\partial}{\partial t}\right) Z(x, \xi, t, \tau) = \delta(x - \xi)\delta(t - \tau) \tag{11.1}$$

and is bounded for $|x| \to \infty$.

The function Z plays the same important role for the operator $\mathfrak{L}$ as the function Γ plays for the heat operator. The function Z can be constructed by the method of E. Levi [69], which consists in looking for Z in the form of a sum of two terms: a principal term, having the desired singularity at $x = \xi$, $t = \tau$, and an additional term, where for the principal term one selects a function $Z_0(x - \xi, \xi, t, \tau)$, namely a fundamental solution of the equation with parabolic operator $\mathfrak{L}_0(\xi, \tau, \partial/\partial x, \partial/\partial t)$, which is obtained from the original operator $\mathfrak{L}$ by discarding the minor terms and "freezing" the coefficients at the point (ξ, τ). The second term is looked for in the form of an integral operator with kernel Z_0. The density of this operator is determined from an integral equation.

The function Z_0, which is obviously obtained from the function Γ by a certain linear change of the space variables x, has for $t > \tau$ the form

$$Z_0(x - \xi, \xi, t, \tau) = \frac{1}{[4\pi(t - \tau)]^{\frac{n}{2}} (\det A(\xi, \tau))^{1/2}} \times \exp\left(-\frac{1}{4(t - \tau)} \sum_{i, j=1}^{n} A^{(i, j)}(\xi, \tau)(x_i - \xi_i)(x_j - \xi_j)\right). \tag{11.2}$$

where $A(\xi, \tau)$ is the matrix composed of the leading coefficients $a_{ij}(\xi, \tau)$ of the operator $\mathfrak{L}_0$, while the $A^{(i,j)}(\xi, \tau)$ are the elements of the inverse matrix $A^{-1}(\xi, \tau)$. For $t < \tau$ we set $Z_0(x - \xi, \xi, t, \tau) = 0$.

The function $Z_0(x - \xi, \xi, t, \tau)$ possesses the following properties. As a function of the arguments x and t it is infinitely differentiable for $t > \tau$ and

$$|D_t^r D_x^s Z_0(x-\xi, \xi, t, \tau)| \leqslant c(t-\tau)^{-\frac{n}{2}-r-\frac{s}{2}} \exp\left(-C\frac{|x-\xi|^2}{t-\tau}\right) \tag{11.3}$$

$(C, c>0)$. In addition, the function $Z_0(z, \xi, t, \tau)$ and its derivatives with respect to z and t satisfy a Hölder condition in the arguments ξ and τ. In particular,

$$|D_t^r D_z^s Z_0(z, \xi, t, \tau) - D_t^r D_z^s Z_0(z, \xi', t, \tau)| \leqslant c|\xi-\xi'|^\beta (t-\tau)^{-\frac{n}{2}-r-\frac{s}{2}} \exp\left(-C\frac{z^2}{t-\tau}\right), \quad 0<\beta\leqslant\alpha. \tag{11.4}$$

Finally, we will also make use of the following property of the function Z_0: for any fixed ξ and for $t-\tau>0$

$$\int_{E_n} Z_0(z, \xi, t, \tau)\,dz = 1;$$

consequently, for $2r+s>0$,

$$\int_{E_n} D_t^r D_z^s Z_0(z, \xi, t, \tau)\,dz = 0. \tag{11.5}$$

It will frequently be necessary for us to consider the integral

$$\int_{E_n} D_t^r D_x^s Z_0(x-\xi, \xi, t, \tau)\,d\xi \qquad (2r+s>0).$$

By virtue of (11.5)

$$\int_{E_n} D_t^r D_x^s Z_0(x-\xi, \xi, t, \tau)\,d\xi = \int_{E_n} \left[D_t^r D_x^s Z_0(x-\xi, \xi, t, \tau) - D_t^r D_x^s Z_0(x-\xi, z, t, \tau)|_{z=x}\right] d\xi,$$

from which, with the use of (11.4), we obtain

$$\left|\int_{E_n} D_t^r D_x^s Z_0(x-\xi, \xi, t, \tau)\,d\xi\right| \leqslant c(t-\tau)^{-\frac{n+2r+s}{2}} \int_{E_n} |x-\xi|^\alpha \exp\left(-C\frac{|x-\xi|^2}{t-\tau}\right) d\xi = c(t-\tau)^{-\frac{2r+s-\alpha}{2}} \tag{11.6}$$

Let us consider the volume potential

$$u(x, t, \tau) = \int_{\tau}^{t} d\lambda \int_{E_n} Z_0(x - y, y, t, \lambda) f(y, \lambda)\, dy, \tag{11.7}$$

where f is a function satisfying a Hölder condition in all its arguments. The derivatives of this potential are expressed by the formulas

$$\frac{\partial u}{\partial x_i} = \int_{\tau}^{t} d\lambda \int_{E_n} \frac{\partial Z_0(x-y, y, t, \lambda)}{\partial x_i} f(y, \lambda)\, dy,$$

$$\frac{\partial^2 u}{\partial x_i \partial x_j} = \int_{\tau}^{t} d\lambda \int_{E_n} \frac{\partial^2 Z_0(x-y, y, t, \lambda)}{\partial x_i \partial x_j} [f(y, \lambda) - f(x, \lambda)]\, dy$$

$$+ \int_{\tau}^{t} f(x, \lambda)\, d\lambda \int_{E_n} \frac{\partial^2 Z_0(x-y, y, t, \lambda)}{\partial x_i \partial x_j} dy, \tag{11.8}$$

$$\frac{\partial u}{\partial t} = f(x, t) +$$

$$+ \int_{\tau}^{t} d\lambda \int_{E_n} \frac{\partial Z_0(x-y, y, t, \lambda)}{\partial t} [f(y, \lambda) - f(x, \lambda)]\, dy$$

$$+ \int_{\tau}^{t} f(x, \lambda)\, d\lambda \int_{E_n} \frac{\partial Z_0(x-y, y, t, \lambda)}{\partial t} dy. \tag{11.9}$$

It is easily verified with the use of estimates (11.3) and (11.6) that the integrals in the right sides of these equalities converge.

Equalities (11.8) and (11.9) are the easiest to prove, by considering a sequence of functions

$$u_h(x, t, \tau) = \int_{\tau}^{t-h} d\lambda \int_{E_n} Z_0(x-y, y, t, \lambda) f(y, \lambda)\, dy$$

and arguing in exactly the same way as in the proof of (1.8) and (1.9). Here one makes use of the relation

$$\lim_{h \to 0} \int_{E_n} Z_0(x-y, y, t, t-h)\, \varphi(y)\, dy = \varphi(x), \tag{11.10}$$

which is valid for any continuous $\phi(x)$ that does not increase too rapidly at infinity. Indeed,

$$\int_{E_n} Z_0(x-y,\ y,\ t,\ t-h)\,\varphi(y)\,dy$$
$$= \int_{E_n} Z_0(x-y,\ x,\ t,\ t-h)\,\varphi(y)\,dy$$
$$+ \int_{E_n} [Z_0(x-y,\ y,\ t,\ t-h) - Z_0(x-y,\ x,\ t,\ t-h)]\,\varphi(y)\,dy$$
$$= I_1 + I_2.$$

The first term I_1 of the right side tends to $\phi(x)$ for $h \to 0$ (this is established in exactly the same way as (1.7)). As to the second term, by virtue of (11.4) we have the estimate

$$|I_2| \leqslant Ch^{-\frac{n}{2}} \int_{E_n} |x-y|^{\alpha} \exp\left(-C\frac{|x-y|^2}{h}\right) dy = ch^{\frac{\alpha}{2}},$$

which implies that it tends to zero for $h \to 0$.

In this way one proves relation (11.10) and, consequently, equalities (11.8) and (11.9).

It is also not difficult to calculate that

$$\mathscr{L}\left(x,\ t,\ \frac{\partial}{\partial x},\ \frac{\partial}{\partial t}\right) u(x,\ t,\ \tau) = \lim_{h\to 0} \mathscr{L}\left(x,\ t,\ \frac{\partial}{\partial x},\ \frac{\partial}{\partial t}\right) u_h(x,\ t,\ \tau)$$
$$= f(x,\ t) + \int_{\tau}^{t} d\lambda \int_{E_n} K(x,\ y,\ t,\ \lambda) f(y,\ \lambda)\,dy, \tag{11.11}$$

where

$$K(x,\ y,\ t,\ \lambda) = \sum_{i,\ j=1}^{n} [a_{ij}(y,\ \lambda) - a_{ij}(x,\ t)] \frac{\partial^2 Z_0(x-y,\ y,\ t,\ \lambda)}{\partial x_i\,\partial x_j}$$
$$+ \mathscr{L}_1\left(x,\ t,\ \frac{\partial}{\partial x}\right) Z_0(x-y,\ y,\ t,\ \lambda),$$
$$\mathscr{L}_1\left(x,\ t,\ \frac{\partial}{\partial x}\right) = \sum_{i=1}^{n} a_i(x,\ t) \frac{\partial}{\partial x_i} + a_0(x,\ t). \tag{11.12}$$

Let us proceed to the construction of the fundamental solution Z. We will look for it in the form

$$Z(x,\ \xi,\ t,\ \tau) = Z_0(x-\xi,\ \xi,\ t,\ \tau)$$
$$+ \int_{\tau}^{t} d\lambda \int_{E_n} Z_0(x-y,\ y,\ t,\ \lambda)\,Q(y,\ \xi,\ \lambda,\ \tau)\,dy, \tag{11.13}$$

where Q is a function to be chosen in such a way that the equation (11.1) holds. Applying to the potential

$$Z'(x, \xi, t, \tau) = \int_{\tau}^{t} d\lambda \int_{E_n} Z_0(x-y, y, t, \lambda) Q(y, \xi, \lambda, \tau)\, dy \tag{11.14}$$

formula (11.11) and noting that

$$\mathscr{L}\left(x, t, \frac{\partial}{\partial x}, \frac{\partial}{\partial t}\right) Z_0(x-\xi, \xi, t, \tau) = \delta(x-\xi)\,\delta(t-\tau) + K(x, \xi, t, \tau),$$

we see that (11.1) will hold when

$$Q(x, \xi, t, \tau) + \int_{\tau}^{t} d\lambda \int_{E_n} K(x, y, t, \lambda) Q(y, \xi, \lambda, \tau)\, dy + K(x, \xi, t, \tau) = 0. \tag{11.15}$$

This equality is a Volterra integral equation for Q, whose kernel K has a weak singularity since

$$|a_{ij}(y, \tau) - a_{ij}(x, t)| \leqslant c\left(|x-y|^{\alpha} + |t-\tau|^{\frac{\alpha}{2}}\right) \tag{11.16}$$

and consequently

$$|K(x, y, t, \tau)| \leqslant c\,(t-\tau)^{-\frac{n+2-\alpha}{2}} \exp\left(-C\,\frac{|x-y|^2}{t-\tau}\right). \tag{11.17}$$

Let us also estimate the difference $K(x, \xi, t, \tau) - K(x', \xi, t, \tau)$. Let x'' be that one of the two points x and x' which is nearest to ξ. Suppose for the sake of definiteness that $x'' = x'$. We have

$$\begin{aligned}
&|K(x, \xi, t, \tau) - K(x', \xi, t, \tau)| \\
&\quad \leqslant \sum_{i, j=1}^{n} |a_{ij}(x, t) - a_{ij}(x', t)| \left|\frac{\partial^2 Z_0(x-\xi, \xi, t, \tau)}{\partial x_i\, \partial x_j}\right| \\
&\qquad + \sum_{i, j=1}^{n} |a_{ij}(\xi, \tau) - a_{ij}(x', t)| \\
&\qquad \times \left|\frac{\partial^2 Z_0(x-\xi, \xi, t, \tau)}{\partial x_i\, \partial x_j} - \frac{\partial^2 Z_0(x'-\xi, \xi, t, \tau)}{\partial x'_i\, \partial x'_j}\right| \\
&\quad + \left|\left[\mathscr{L}_1\left(x, t, \frac{\partial}{\partial x}\right) - \mathscr{L}_1\left(x', t, \frac{\partial}{\partial x}\right)\right] Z_0(x-\xi, \xi, t, \tau)\right| \\
&\qquad + \left|\mathscr{L}_1\left(x', t, \frac{\partial}{\partial x}\right) Z_0(x-\xi, \xi, t, \tau)\right. \\
&\qquad \left. - \mathscr{L}_1\left(x', t, \frac{\partial}{\partial x'}\right) Z_0(x'-\xi, \xi, t, \tau)\right|.
\end{aligned} \tag{11.18}$$

One easily verifies that

$$\left|\frac{\partial^2 Z_0(x-\xi, \xi, t, \tau)}{\partial x_i \partial x_j} - \frac{\partial^2 Z_0(x'-\xi, \xi, t, \tau)}{\partial x'_i \partial x'_j}\right| \leqslant c|x-x'|(t-\tau)^{-\frac{n+3}{2}} \exp\left(-C\frac{|x''-\xi|^2}{t-\tau}\right) \tag{11.19}$$

(the elementary proof of this inequality is omitted).

From (11.19) and (11.3) it follows that for any $\gamma \in [0, 1]$

$$\left|\frac{\partial^2 Z_0(x-\xi, \xi, t, \tau)}{\partial x_i \partial x_j} - \frac{\partial^2 Z_0(x'-\xi, \xi, t, \tau)}{\partial x'_i \partial x'_j}\right| \leqslant \left|\frac{\partial^2 Z_0(x-\xi, \xi, t, \tau)}{\partial x_i \partial x_j} - \frac{\partial^2 Z_0(x'-\xi, \xi, t, \tau)}{\partial x'_i \partial x'_j}\right|^\gamma \left(\left|\frac{\partial^2 Z_0}{\partial x_i \partial x_j}\right| + \left|\frac{\partial^2 Z_0}{\partial x'_i \partial x'_j}\right|\right)^{1-\gamma}$$
$$\leqslant c|x-x'|^\gamma (t-\tau)^{-\frac{n+2+\gamma}{2}} \exp\left(-C\frac{|x''-\xi|^2}{t-\tau}\right). \tag{11.20}$$

Using inequalities (11.16) and (11.20) for $\gamma = \alpha$ and (11.3) for an estimate of the first two terms of the right side of (11.18), and making use of analogous inequalities for an estimate of the remaining terms, we obtain

$$|K(x, \xi, t, \tau) - K(x', \xi, t, \tau)| \leqslant c|x-x'|^\alpha (t-\tau)^{-\frac{n+2}{2}} \exp\left(-C\frac{|x''-\xi|^2}{t-\tau}\right). \tag{11.21}$$

From this estimate and from estimate (11.17) one easily obtains the more general inequality (cf. (11.20))

$$|K(x, \xi, t, \tau) - K(x', \xi, t, \tau)| \leqslant c|x-x'|^\beta (t-\tau)^{-\frac{n+2-(\alpha-\beta)}{2}} \exp\left(-C\frac{|x''-\xi|^2}{t-\tau}\right), \tag{11.22}$$

in which β is any number satisfying the condition $0 \leq \beta \leq \alpha$.

The integral equation (11.15) is solved by the method of successive approximations, and its solution Q is expressed in the form of a series

$$Q(x, \xi, t, \tau) = \sum_{m=1}^{\infty} (-1)^m K_m(x, \xi, t, \tau), \tag{11.23}$$

where K_m is an iterated kernel:

$$K_m(x, \xi, t, \tau) = \int_\tau^t d\lambda \int_{E_n} K(x, y, t, \lambda) K_{m-1}(y, \xi, \lambda, \tau)\, dy. \qquad (11.24)$$

With the use of induction one can prove the estimate

$$|K_m(x, \xi, t, \tau)|$$
$$\leqslant c^m \left(\frac{\pi}{C}\right)^{\frac{n(m-1)}{2}} \frac{\Gamma^m\left(\frac{\alpha}{2}\right)}{\Gamma\left(\frac{m\alpha}{2}\right)} (t-\tau)^{\frac{m\alpha - n - 2}{2}} \exp\left(-C\frac{|x-\xi|^2}{t-\tau}\right). \qquad (11.25)$$

This estimate holds for $m = 1$ (see (11.17)), and in order to prove it for $m > 1$ it is necessary to estimate the integral

$$|K_{m+1}(x, \xi, t, \tau)|$$
$$= \left| \int_\tau^t d\lambda \int_{E_n} K(x, y, t, \lambda) K_m(y, \xi, \lambda, \tau)\, dy \right|$$
$$\leqslant c^{m+1} \left(\frac{\pi}{C}\right)^{\frac{n(m-1)}{2}} \frac{\Gamma^m\left(\frac{\alpha}{2}\right)}{\Gamma\left(\frac{m\alpha}{2}\right)} \int_\tau^t (t-\lambda)^{\frac{\alpha-n-2}{2}} (\lambda-\tau)^{\frac{m\alpha-n-2}{2}} d\lambda$$
$$\times \int_{E_n} \exp\left(-C\frac{|x-y|^2}{t-\lambda} - C\frac{|y-\xi|^2}{\lambda-\tau}\right) dy.$$

Since

$$\int_{E_n} \exp\left(-C\frac{|x-y|^2}{t-\lambda} - C\frac{|y-\xi|^2}{\lambda-\tau}\right) dy$$
$$= \left(\frac{\pi}{C}\right)^{\frac{n}{2}} \left(\frac{(t-\lambda)(\lambda-\tau)}{t-\tau}\right)^{\frac{n}{2}} \exp\left(-C\frac{|x-\xi|^2}{t-\tau}\right)$$

and

$$\int_\tau^t (t-\tau)^{\frac{\alpha-2}{2}} (\lambda-\tau)^{\frac{m\alpha-2}{2}} d\lambda = \frac{\Gamma\left(\frac{\alpha}{2}\right)\Gamma\left(\frac{m\alpha}{2}\right)}{\Gamma\left(\frac{(m+1)\alpha}{2}\right)} (t-\tau)^{\frac{\alpha+m\alpha-2}{2}},$$

it follows that

$$|K_{m+1}| \leqslant c^{m+1} \left(\frac{\pi}{C}\right)^{\frac{nm}{2}} \frac{\Gamma^{m+1}\left(\frac{\alpha}{2}\right)}{\Gamma\left(\frac{(m+1)\alpha}{2}\right)}$$
$$\times (t-\tau)^{\frac{(m+1)\alpha-n-2}{2}} \exp\left(-C\frac{|x-\xi|^2}{t-\tau}\right),$$

and estimate (11.25) is proved. Inequality (11.25) guarantees the uniform convergence of the series (11.23) for $t - \tau > 0$ and yields the estimate

$$|Q(x, \xi, t, \tau)| \leqslant c(t-\tau)^{-\frac{n+2-\alpha}{2}} \exp\left(-C\frac{|x-\xi|^2}{t-\tau}\right). \tag{11.26}$$

for Q. This estimate ensures the convergence of the integrals in (11.14) and (11.15).

Thus we have constructed a solution of the integral equation (11.15) and proved estimate (11.26) for it. Let us now prove that the solution $Q(x, \xi, t, \tau)$ satisfies a Hölder condition in its first argument.

From inequality (11.26) and inequality (11.22), in which $\beta = \alpha' < \alpha$ and the exponential factor in the right side is replaced by

$$\exp\left(-C\frac{|x-\xi|^2}{t-\tau}\right) + \exp\left(-C\frac{|x'-\xi|^2}{t-\tau}\right)$$

it follows that

$$\begin{aligned} &\left|\int_\tau^t d\lambda \int_{E_n} [K(x, y, t, \lambda) - K(x', y, t, \lambda)]\, Q(y, \xi, \lambda, \tau)\, dy\right| \\ &\leqslant c|x-x'|^{\alpha'} \int_\tau^t (t-\lambda)^{-\frac{n+2-(\alpha-\alpha')}{2}} (\lambda-\tau)^{-\frac{n+2-\alpha}{2}} d\lambda \\ &\quad \times \int_{E_n} \left[\exp\left(-C\frac{|x-y|^2}{t-\lambda}\right) + \exp\left(-C\frac{|x'-y|^2}{t-\lambda}\right)\right] \exp\left(-C\frac{|y-\xi|^2}{\lambda-\tau}\right) dy \\ &\leqslant c|x-x'|^{\alpha'} (t-\tau)^{-\frac{n+2-2\alpha+\alpha'}{2}} \exp\left(-C\frac{|x''-\xi|^2}{t-\tau}\right). \end{aligned} \tag{11.27}$$

Now estimating $Q(x, \xi, t, \tau) - Q(x', \xi, t, \tau)$ with the use of equation (11.15), we see that

$$\begin{aligned} &|Q(x, \xi, t, \tau) - Q(x', \xi, t, \tau)| \\ &\leqslant c|x-x'|^{\alpha'} (t-\tau)^{-\frac{n+2-(\alpha-\alpha')}{2}} \exp\left(-C\frac{|x''-\xi|^2}{t-\tau}\right), \quad \alpha' < \alpha. \end{aligned} \tag{11.28}$$

One should now substitute the obtained solution Q into (11.13) and convince oneself of the fact that the function (11.13) really satisfies equation (11.1). But this has in fact already been done above in the derivation of equation (11.15); one need only note that the application of formula (11.11) to the potential (11.14) is valid by virtue of estimate (11.28).

The construction of the fundamental solution Z is complete.

§12. SOME AUXILIARY INEQUALITIES FOR THE FUNCTION Q

In this section we will obtain certain inequalities for the function Q which will be needed by us for exact estimates of the fundamental solution Z.

First we will improve estimate (11.28) by proving it for $\alpha' = \alpha$:

$$|Q(x, \xi, t, \tau) - Q(x', \xi, t, \tau)| \leqslant c|x - x'|^{\alpha} (t - \tau)^{-\frac{n+2}{2}} \exp\left(-C \frac{|x'' - \xi|^2}{t - \tau}\right); \tag{12.1}$$

x'', as in §11, is the closest to ξ of the points x and x'. Secondly, we will show that for $t > t' > \tau$

$$|Q(x, \xi, t, \tau) - Q(x, \xi, t', \tau)| \leqslant c(t - t')^{\frac{\alpha}{2}} (t' - \tau)^{-\frac{n+2}{2}} \exp\left(-C \frac{|x - \xi|^2}{t - \tau}\right). \tag{12.2}$$

Finally, we will consider the function

$$q(x, t) = \int_0^t d\tau \int_{E_n} Q(x, \xi, t, \tau) \, d\xi$$

and show that it belongs to the class $H^{\alpha, \alpha/2}(\overline{D_{n+1}^{(T)}})$:

$$|q|^{(\alpha)}_{D^{(T)}_{n+1}} \leqslant c. \tag{12.3}$$

Let us proceed to the proof of estimate (12.1). Let

$$K_1(x, \xi, t, \tau) = \sum_{i, j=1}^{n} [a_{ij}(\xi, \tau) - a_{ij}(x, t)] \frac{\partial^2 Z_0(x - \xi, \xi, t, \tau)}{\partial x_i \partial x_j},$$

$$K_2(x, \xi, t, \tau) = \mathscr{L}_1\left(x, t, \frac{\partial}{\partial x}\right) Z_0(x - \xi, \xi, t, \tau).$$

The kernel K_2 has a weaker singularity than K_1, since it does not contain second derivatives of the function Z_0. In particular, in place of (11.21), for the kernel K_2 we have

$$|K_2(x, \xi, t, \tau) - K_2(x', \xi, t, \tau)| \leqslant c|x - x'|^{\alpha} (t - \tau)^{-\frac{n+1+\alpha}{2}} \exp\left(-C \frac{|x'' - \xi|^2}{t - \tau}\right) \tag{12.4}$$

and therefore in place of (11.27) we will have

$$\left|\int_\tau^t d\lambda \int_{E_n} [K_2(x, y, t, \lambda) - K_2(x', y, t, \lambda)]\, Q(y, \xi, \lambda, \tau)\, dy\right| \leqslant c\,|x - x'|^\alpha (t-\tau)^{-\frac{n+1}{2}} \exp\left(-C\frac{|x'' - \xi|^2}{t-\tau}\right). \tag{12.5}$$

Let us now estimate

$$\begin{aligned}\int_\tau^t d\lambda \int_{E_n} & [K_1(x, y, t, \lambda) - K_1(x', y, t, \lambda)]\, Q(y, \xi, \lambda, \tau)\, dy \\ &= \int_\tau^t d\lambda \int_\sigma K_1(x, y, t, \lambda) Q(y, \xi, \lambda, \tau)\, dy \\ &\quad - \int_\tau^t d\lambda \int_\sigma K_1(x', y, t, \lambda) Q(y, \xi, \lambda, \tau)\, dy \\ &+ \int_\tau^t d\lambda \int_{E_n\setminus\sigma} [K_1(x, y, t, \lambda) - K_1(x', y, t, \lambda)]\, Q(y, \xi, \lambda, \tau)\, dy. \end{aligned}\tag{12.6}$$

Here $\sigma = \sigma_1 \setminus \sigma_2$, σ_1 is the ball with center x'' and radius $2|x - x'|$, and σ_2 is the ball with center ξ and radius $|x'' - \xi|/2$.

In order to estimate the first two integrals of the right side of (12.6) it is necessary to take into consideration the fact that if $y \in \sigma$ then $y \in E_n \setminus \sigma_2$, and consequently

$$|y - \xi| \geqslant \frac{|x'' - \xi|}{2}.$$

We have

$$\begin{aligned}\left|\int_\tau^t d\lambda \int_\sigma K_1(x, y, t, \lambda) Q(y, \xi, \lambda, \tau)\, dy\right| &\leqslant c \int_\tau^t (t-\lambda)^{-\frac{n+2-\alpha}{2}} (\lambda-\tau)^{-\frac{n+2-\alpha}{2}} d\lambda \\ &\times \int_\sigma \exp\left(-C\frac{|y-\xi|^2}{\lambda-\tau}\right) \exp\left(-C\frac{|x-y|^2}{t-\lambda}\right) dy.\end{aligned}$$

We divide the integral with respect to λ into two parts: an integral with limits $(\tau, (t+\tau)/2)$ and one with limits $((t+\tau)/2, t)$. In the first integral we apply the inequalities

$$t-\lambda \geqslant \frac{1}{2}(t-\tau), \quad \exp\left(-C\frac{|x-y|^2}{t-\lambda}\right) \leqslant 1,$$

$$\exp\left(-C\frac{|y-\xi|^2}{\lambda-\tau}\right) \leqslant \exp\left(-\frac{C}{2}\frac{|y-\xi|^2}{\lambda-\tau}\right)\exp\left(-\frac{C}{4}\frac{|x''-\xi|^2}{t-\tau}\right),$$

and in the second, the inequalities

$$\lambda-\tau \geqslant \frac{1}{2}(t-\tau), \quad \exp\left(-C\frac{|y-\xi|^2}{\lambda-\tau}\right) \leqslant \exp\left(-\frac{C}{4}\frac{|x''-\xi|^2}{t-\tau}\right).$$

As a result we obtain

$$\begin{aligned}
&\left|\int_\tau^t d\lambda \int_\sigma K_1(x, y, t, \lambda) Q(y, \xi, \lambda, \tau)\, dy\right| \\
&\leqslant c\exp\left(-\frac{C}{4}\frac{|x''-\xi|^2}{t-\tau}\right)(t-\tau)^{-\frac{n+2-\alpha}{2}} \\
&\times\left[\int_\sigma dy \int_\tau^{\frac{t+\tau}{2}} (\lambda-\tau)^{-\frac{n+2-\alpha}{2}} \exp\left(-\frac{C}{2}\frac{|y-\xi|^2}{\lambda-\tau}\right) d\lambda\right. \\
&\left.+\int_\sigma dy \int_{\frac{t+\tau}{2}}^t (t-\lambda)^{-\frac{n+2-\alpha}{2}} \exp\left(-C\frac{|x-y|^2}{t-\lambda}\right) d\lambda\right] \\
&\leqslant c|x-x'|^\alpha (t-\tau)^{-\frac{n+2-\alpha}{2}} \exp\left(-\frac{C}{4}\frac{|x''-\xi|^2}{t-\tau}\right).
\end{aligned} \tag{12.7}$$

In completely the same way we prove the inequality

$$\begin{aligned}
&\left|\int_\tau^t d\lambda \int_\sigma K_1(x', y, t, \lambda) Q(y, \xi, \lambda, \tau)\, dy\right| \\
&\qquad \leqslant c|x-x'|^\alpha (t-\tau)^{-\frac{n+2-\alpha}{2}} \exp\left(-\frac{C}{4}\frac{|x''-\xi|^2}{t-\tau}\right).
\end{aligned}$$

It remains to estimate the last integral in (12.6). We assume for the sake of definiteness that $x'' = x'$. We have

$$\begin{aligned}
&\int_\tau^t d\lambda \int_{E_n \setminus \sigma} [K_1(x, y, t, \lambda) - K_1(x', y, t, \lambda)]\, Q(y, \xi, \lambda, \tau)\, dy \\
&\qquad = \sum_{i,j=1}^n [a_{ij}(x', t) - a_{ij}(x, t)] \times
\end{aligned}$$

$$\times \int_{\tau}^{t} d\lambda \int_{E_n \setminus \sigma} \frac{\partial^2 Z_0(x-y, y, t, \lambda)}{\partial x_i \partial x_j} Q(y, \xi, \lambda, \tau)\, dy$$

$$+ \sum_{i,j=1}^{n} \int_{\tau}^{t} d\lambda \int_{E_n \setminus \sigma} [a_{ij}(y, \lambda) - a_{ij}(x', t)] \Big[\frac{\partial^2 Z_0(x-y, y, t, \lambda)}{\partial x_i \partial x_j}$$

$$- \frac{\partial^2 Z_0(x'-y, y, t, \lambda)}{\partial x_i' \partial x_j'}\Big] Q(y, \xi, \lambda, \tau)\, dy. \tag{12.8}$$

Let us consider the latter term. We note that $E_n \setminus \sigma = (E_n \setminus \sigma_1) \cup (\sigma_1 \cap \sigma_2)$, and estimate the integral over $E_n \setminus \sigma_1$.

If $y \in E_n \setminus \sigma_1$, then it is easily seen that

$$\frac{1}{2} |x'-y| \leqslant |x-y| \leqslant \frac{3}{2} |x'-y|,$$

and so, by virtue of (11.16), (11.20) and (11.26),

$$\Big| \int_{\tau}^{t} d\lambda \int_{E_n \setminus \sigma_1} [a_{ij}(y, \lambda) - a_{ij}(x', t)] \Big[\frac{\partial^2 Z_0(x-y, y, t, \lambda)}{\partial x_i \partial x_j}$$

$$- \frac{\partial^2 Z_0(x'-y, y, t, \lambda)}{\partial x_i' \partial x_j'}\Big] Q(y, \xi, \lambda, \tau)\, dy \Big|$$

$$\leqslant c \Big\{ |x-x'|^{\alpha} \int_{\tau}^{\frac{t+\tau}{2}} (t-\lambda)^{-\frac{n+2+\alpha}{2}} (\lambda-\tau)^{-\frac{n+2-\alpha}{2}} d\lambda$$

$$\times \int_{E_n \setminus \sigma_1} \exp\Big(-C \frac{|x'-y|^2}{t-\lambda}\Big) \exp\Big(-C \frac{|y-\xi|^2}{\lambda-\tau}\Big) \Big[|x'-y|^{\alpha}$$

$$+ (t-\lambda)^{\frac{\alpha}{2}}\Big] dy + |x-x'| \int_{\frac{t+\tau}{2}}^{t} (t-\lambda)^{-\frac{n+3}{2}} (\lambda-\tau)^{-\frac{n+2-\alpha}{2}} d\lambda$$

$$\times \int_{E_n \setminus \sigma_1} \exp\Big(-C \frac{|x'-y|^2}{t-\lambda}\Big) \exp\Big(-C \frac{|y-\xi|^2}{\lambda-\tau}\Big)$$

$$\times \Big[|x'-y|^{\alpha} + (t-\lambda)^{\frac{\alpha}{2}}\Big] dy \Big\}. \tag{12.9}$$

Here we make use of the inequalities

$$|x'-y|^\alpha \exp\left(-C\frac{|x'-y|^2}{t-\lambda}\right) \leqslant c(t-\lambda)^{\frac{\alpha}{2}} \exp\left(-C'\frac{|x'-y|^2}{t-\lambda}\right), \quad (C'<C),$$

$$\exp\left(-C'\frac{|x'-y|^2}{t-\lambda}\right)\exp\left(-C'\frac{|y-\xi|^2}{\lambda-\tau}\right) \leqslant \exp\left[-\frac{C'}{t-\tau}(|x'-y|^2+|y-\xi|^2)\right] \leqslant \exp\left(-\frac{C'}{2}\frac{|x'-\xi|^2}{t-\tau}\right). \tag{12.10}$$

Denoting the left side of (12.9) by I_{ij}, we obtain

$$\begin{aligned} I_{ij} \leqslant c\Big\{ & |x-x'|^\alpha (t-\tau)^{-\frac{n+2}{2}} \exp\left(-C_1\frac{|x'-\xi|^2}{t-\tau}\right) \\ & \times \int_\tau^{\frac{t+\tau}{2}} (\lambda-\tau)^{-\frac{2-\alpha}{2}} d\lambda + |x-x'|(t-\tau)^{-\frac{n+2-\alpha}{2}} \\ & \times \exp\left(-C_1\frac{|x'-\xi|^2}{t-\tau}\right) \int_{E_n\setminus\sigma_1} dy \int_{\frac{t+\tau}{2}}^t (t-\lambda)^{-\frac{n+3-\alpha}{2}} \\ & \times \exp\left(-C_2\frac{|x'-y|^2}{t-\lambda}\right) d\lambda\Big\} \\ & \leqslant c|x-x'|^\alpha (t-\tau)^{-\frac{n+2-\alpha}{2}} \exp\left(-C_1\frac{|x'-\xi|^2}{t-\tau}\right). \end{aligned}$$

Let us now consider the integral

$$I'_{ij} = \int_\tau^t d\lambda \int_{\sigma_1\cap\sigma_2} [a_{ij}(y,\lambda)-a_{ij}(x',t)]\left[\frac{\partial^2 Z_0(x-y,y,t,\lambda)}{\partial x_i\,\partial x_j} - \frac{\partial^2 Z_0(x'-y,y,t,\lambda)}{\partial x'_i\,\partial x'_j}\right] Q(y,\xi,\lambda,\tau)\,dy.$$

If $y \in \sigma_2$, then $|x'-y| \le |x'-\xi| + |\xi-y| \le 3|x'-\xi|/2$ and $|x-y| \ge |x-\xi| - |\xi-y| \ge |x-\xi| - |x'-\xi|/2 \ge |x'-\xi|/2 \ge |x'-y|/3$. Consequently

$$|I'_{ij}| \leqslant c\int_\tau^t (t-\lambda)^{-\frac{n+2}{2}} (\lambda-\tau)^{-\frac{n+2-\alpha}{2}} d\lambda \int_{\sigma_1\cap\sigma_2} \left[|x'-y|^\alpha + (t-\lambda)^{\frac{\alpha}{2}}\right] \exp\left(-C_3\frac{|x'-y|^2}{t-\lambda}\right)\exp\left(-C_3\frac{|y-\xi|^2}{\lambda-\tau}\right) dy \le$$

$$\leqslant c\exp\left(-C_4\frac{|x'-\xi|^2}{t-\tau}\right)\left[(t-\tau)^{-\frac{n+2-\alpha}{2}}\int\limits_{\sigma_1\cap\sigma_2}dy\int\limits_{\tau}^{\frac{t+\tau}{2}}(\lambda-\tau)^{-\frac{n+2-\alpha}{2}}\right.$$

$$\times\exp\left(-C_5\frac{|y-\xi|^2}{\lambda-\tau}\right)dy+$$

$$\left.+(t-\tau)^{-\frac{n+2-\alpha}{2}}\int\limits_{\sigma_1\cap\sigma_2}dy\int\limits_{\frac{\tau+t}{2}}^{t}(t-\lambda)^{-\frac{n+2-\alpha}{2}}\exp\left(-C_5\frac{|x'-y|^2}{t-\lambda}\right)d\lambda\right]$$

$$\leqslant c|x-x'|^\alpha(t-\tau)^{-\frac{n+2-\alpha}{2}}\exp\left(-C_4\frac{|x'-\xi|^2}{t-\tau}\right).$$

Thus we have shown that

$$\left|\int\limits_{\tau}^{t}d\lambda\int\limits_{E_n\setminus\sigma}[a_{ij}(y,\ \lambda)-a_{ij}(x',\ t)]\right.$$

$$\left.\times\left[\frac{\partial^2 Z_0(x-y,\ y,\ t,\ \lambda)}{\partial x_i\,\partial x_j}-\frac{\partial^2 Z_0(x'-y,\ y,\ t,\ \lambda)}{\partial x_i'\,\partial x_j'}\right]Q(y,\ \xi,\ \lambda,\ \tau)\,dy\right|$$

$$\leqslant c|x-x'|^\alpha(t-\tau)^{-\frac{n+2-\alpha}{2}}\exp\left(-C\frac{|x'-\xi|^2}{t-\tau}\right). \tag{12.11}$$

Finally, let us show that

$$\left|\int\limits_{\tau}^{t}d\lambda\int\limits_{E_n\setminus\sigma}\frac{\partial^2 Z_0(x-y,\ y,\ t,\ \lambda)}{\partial x_i\,\partial x_j}Q(y,\ \xi,\ \lambda,\ \tau)\,dy\right|$$

$$\leqslant c(t-\tau)^{-\frac{n+2-\alpha}{2}}\exp\left(-C\frac{|x-\xi|^2}{t-\tau}\right). \tag{12.12}$$

We have

$$\left|\int\limits_{\tau}^{\frac{t+\tau}{2}}d\lambda\int\limits_{E_n\setminus\sigma}\frac{\partial^2 Z_0}{\partial x_i\,\partial x_j}Q\,dy\right|$$

$$\leqslant c\int\limits_{\tau}^{\frac{t+\tau}{2}}(t-\lambda)^{-\frac{n+2}{2}}(\lambda-\tau)^{-\frac{n+2-\alpha}{2}}d\lambda$$

$$\times\int\limits_{E_n}\exp\left(-C\frac{|x-y|^2}{t-\lambda}-C\frac{|y-\xi|^2}{\lambda-\tau}\right)dy$$

$$\leqslant c(t-\tau)^{-\frac{n+2-\alpha}{2}}\exp\left(-C\frac{|x-\xi|^2}{t-\tau}\right).$$

Further,

$$\int_{\frac{t+\tau}{2}}^{t} d\lambda \int_{E_n\setminus\sigma} \frac{\partial^2 Z_0}{\partial x_i \partial x_j} Q\, dy$$

$$= \int_{\frac{t+\tau}{2}}^{t} d\lambda \int_{E_n\setminus\sigma} \frac{\partial^2 Z_0(x-y, y, t, \lambda)}{\partial x_i \partial x_j} [Q(y, \xi, \lambda, \tau) - Q(x, \xi, \lambda, \tau)]\, dy$$

$$+ \int_{\frac{t+\tau}{2}}^{t} Q(x, \xi, \lambda, \tau)\, d\lambda \int_{E_n\setminus\sigma} \Big[\frac{\partial^2 Z_0(x-y, y, t, \lambda)}{\partial x_i \partial x_j}$$

$$- \frac{\partial^2 Z_0(x-y, x, t, \lambda)}{\partial y_i \partial y_j}\Big] dy - \int_{\frac{t+\tau}{2}}^{t} Q(x, \xi, \lambda, \tau)\, d\lambda$$

$$\times \int_{\omega} \frac{\partial Z_0(x-y, x, ', \lambda)}{\partial y_i} n_j(y)\, d\omega_y = J_1 + J_2 + J_3,$$

where ω is the boundary of σ. Making use of estimates (11.3) and (11.28), and repeating calculations that have already been carried out several times above, we obtain

$$|J_1| \leqslant c \int_{\frac{t+\tau}{2}}^{t} (t-\lambda)^{-\frac{n+2}{2}} (\lambda - \tau)^{-\frac{n+2-(\alpha-\alpha')}{2}} d\lambda$$

$$\times \int_{E_n} |x-y|^{\alpha'} \exp\Big(-C \frac{|x-y|^2}{t-\lambda}\Big) \Big[\exp\Big(-C \frac{|x-\xi|^2}{\lambda-\tau}\Big)$$

$$+ \exp\Big(-C \frac{|y-\xi|^2}{\lambda-\tau}\Big)\Big] dy \leqslant c(t-\tau)^{-\frac{n+2-\alpha}{2}} \exp\Big(-C' \frac{|x-\xi|^2}{t-\tau}\Big),$$

$$|J_2| \leqslant c \int_{\frac{\tau+t}{2}}^{t} (t-\lambda)^{-\frac{2-\alpha}{2}} (\lambda-\tau)^{-\frac{n+2-\alpha}{2}} \exp\Big(-C \frac{|x-\xi|^2}{\lambda-\tau}\Big) d\lambda$$

$$\leqslant c(t-\tau)^{-\frac{n+2-2\alpha}{2}} \exp\Big(-C \frac{|x-\xi|^2}{t-\tau}\Big).$$

In order to estimate J_3 we use the inequality

$$\Big|\frac{\partial Z_0(x-y, x, t, \lambda)}{\partial y_i}\Big| \leqslant c|x-y|(t-\lambda)^{-\frac{n+2}{2}} \exp\Big(-C \frac{|x-y|^2}{t-\lambda}\Big) \qquad (12.13)$$

implied by (11.2).

We also note that the set ω coincides with the boundary ω_1 of the ball σ_1 if $|x' - \xi| \geq 4|x - x'|$; while if $|x' - \xi| < 4|x - x'|$, then $\omega = \omega'_1 \cup \omega'_2$ where ω'_1 is the set of points of ω_1 the distance of which to ξ is not less than $|x' - \xi|/2$, and ω'_2 is that part of the boundary ω_2 of the ball σ_2 lying in σ_1.

If $y \in \omega'_1$, then

$$|x - x'| \leqslant |x - y| \leqslant 3|x - x'|;$$

while if $y \in \omega'_2$ (in case $|x' - \xi| < 4|x - x'|$), then

$$\frac{1}{2}|x - \xi| \leqslant |x - y| \leqslant \frac{3}{2}|x - \xi|.$$

Therefore

$$\left| \int_{\omega'_1} \frac{\partial Z_0(x - y, x, t, \lambda)}{\partial y_l} n_j(y)\, d\omega_y \right| \leqslant c|x - x'|^n (t - \lambda)^{-\frac{n+2}{2}} \exp\left(-C\frac{|x - x'|^2}{t - \lambda}\right),$$

$$\left| \int_{\omega'_2} \frac{\partial Z_0(x - y, x, t, \lambda)}{\partial y_l} n_j(y)\, d\omega_y \right| \leqslant c|x - \xi|^n (t - \lambda)^{-\frac{n+2}{2}} \exp\left(-C\frac{|x - \xi|^2}{t - \lambda}\right)$$

and

$$\begin{aligned} |J_3| \leqslant{}& c(t - \tau)^{-\frac{n+2-\alpha}{2}} \exp\left(-C\frac{|x - \xi|^2}{t - \tau}\right) \\ &\times \left[|x - x'|^n \int_{\frac{t+\tau}{2}}^{t} (t - \lambda)^{-\frac{n+2}{2}} \exp\left(-C\frac{|x - x'|^2}{t - \lambda}\right) d\lambda \right. \\ &\left. + |x - \xi|^n \int_{\frac{t+\tau}{2}}^{t} (t - \lambda)^{-\frac{n+2}{2}} \exp\left(-C\frac{|x - \xi|^2}{t - \lambda}\right) d\lambda \right] \\ \leqslant{}& c(t - \tau)^{-\frac{n+2-\alpha}{2}} \exp\left(-C\frac{|x - \xi|^2}{t - \tau}\right). \end{aligned}$$

Thus inequality (12.12) is proved. From (12.8), (12.11) and (12.12) we obtain

$$\left|\int_\tau^t d\lambda \int_{E_n\setminus\sigma} [K_1(x, y, t, \lambda) - K_1(x', y, t, \lambda)]\, Q(y, \xi, \lambda, \tau)\, dy\right|$$
$$\leqslant c\,|x - x'|^\alpha (t-\tau)^{-\frac{n+2-\alpha}{2}} \exp\left(-C\frac{|x'' - \xi|^2}{t-\tau}\right). \tag{12.14}$$

Combining estimates (12.5), (12.7) and (12.14), we find that

$$\left|\int_\tau^t d\lambda \int_{E_n} [K(x, y, t, \lambda) - K(x', y, t, \lambda)]\, Q(y, \xi, \lambda, \tau)\, dy\right|$$
$$\leqslant c\,|x - x'|^\alpha (t-\tau)^{-\frac{n+2-\alpha}{2}} \exp\left(-C\frac{|x'' - \xi|^2}{t-\tau}\right). \tag{12.15}$$

This inequality and (11.21) imply estimate (12.1).

Let us turn to inequality (12.2). We assume that $t > t' > \tau$ and $t - t' < (t-\tau)/2$. First we note that

$$|K(x, \xi, t, \tau) - K(x, \xi, t', \tau)|$$
$$\leqslant c(t-t')^{\frac{\alpha}{2}} (t'-\tau)^{-\frac{n+2}{2}} \exp\left(-C\frac{|x-\xi|^2}{t-\tau}\right), \tag{12.16}$$

while for K_2 we have

$$|K_2(x, \xi, t, \tau) - K_2(x, \xi, t', \tau)|$$
$$\leqslant c(t-t')^{\frac{\alpha}{2}} (t'-\tau)^{-\frac{n+1+\alpha}{2}} \exp\left(-C\frac{|x-\xi|^2}{t-\tau}\right). \tag{12.17}$$

We have

$$\int_\tau^t d\lambda \int_{E_n} K(x, y, t, \lambda)\, Q(y, \xi, \lambda, \tau)\, dy$$
$$- \int_\tau^{t'} d\lambda \int_{E_n} K(x, y, t', \lambda)\, Q(y, \xi, \lambda, \tau)\, dy$$
$$= \int_{2t'-t}^t d\lambda \int_{E_n} K(x, y, t, \lambda)\, Q(y, \xi, \lambda, \tau)\, dy$$
$$- \int_{2t'-t}^{t'} d\lambda \int_{E_n} K(x, y, t', \lambda)\, Q(y, \xi, \lambda, \tau)\, dy +$$

$$+\int_{\tau}^{2t'-t} d\lambda \int_{E_n} [K_2(x, y, t, \lambda)-K_2(x, y, t', \lambda)]\, Q(y, \xi, \lambda, \tau)\, dy$$

$$+\sum_{i, j=1}^{n} [a_{ij}(x, t')-a_{ij}(x, t)] \int_{\tau}^{2t'-t} d\lambda$$

$$\times \int_{E_n} \frac{\partial^2 Z_0(x-y, y, t, \lambda)}{\partial x_i\, \partial x_j} Q(y, \xi, \lambda, \tau)\, dy$$

$$+\sum_{i, j=1}^{n} \int_{\tau}^{2t'-t} d\lambda \int_{E_n} [a_{ij}(y, \lambda)-a_{ij}(x, t')]$$

$$\times \left[\frac{\partial^2 Z_0(x-y, y, t, \lambda)}{\partial x_i\, \partial x_j}-\frac{\partial^2 Z_0(x-y, y, t', \lambda)}{\partial x_i\, \partial x_j}\right] Q(y, \xi, \lambda, \tau)\, dy. \qquad (12.18)$$

Applying estimates (11.17) and (11.26), we obtain

$$\left|\int_{2t'-t}^{t} d\lambda \int_{E_n} KQ\, dy\right| \leqslant c \int_{2t'-t}^{t} (t-\lambda)^{-\frac{n+2-\alpha}{2}} (\lambda-\tau)^{-\frac{n+2-\alpha}{2}} d\lambda$$

$$\times \int_{E_n} \exp\left(-C\frac{|x-y|^2}{t-\lambda}\right) \exp\left(-C\frac{|y-\xi|^2}{\lambda-\tau}\right) dy$$

$$\leqslant c\,(t-t')^{\frac{\alpha}{2}} (t-\tau)^{-\frac{n+2-\alpha}{2}} \exp\left(-C\frac{|x-\xi|^2}{t-\tau}\right).$$

The second term of the right side of (12.18) is estimated analogously. The remaining terms are estimated with the use of inequalities (12.17), (11.16) and (11.3). In the process one proves the inequality

$$\left|\int_{\tau}^{2t'-t} d\lambda \int_{E_n} \frac{\partial^2 Z_0(x-y, y, t, \lambda)}{\partial x_i\, \partial x_j} Q(y, \xi, \lambda, \tau)\, dy\right|$$

$$\leqslant c\,(t-\tau)^{-\frac{n+2-\alpha}{2}} \exp\left(-C\frac{|x-\xi|^2}{t-\tau}\right) \qquad (12.19)$$

which is analogous to (12.12).

The proof makes use of the representation

$$\int_{\tau}^{2t'-t} d\lambda \int_{E_n} \frac{\partial^2 Z_0}{\partial x_i \partial x_j} Q\, dy$$

$$= \int_{\frac{t+\tau}{2}}^{2t'-t} d\lambda \int_{E_n} \frac{\partial^2 Z_0}{\partial x_i \partial x_j} [Q(y, \xi, \lambda, \tau) - Q(x, \xi, \lambda, \tau)]\, dy$$

$$+ \int_{\frac{t+\tau}{2}}^{2t'-t} Q(x, \xi, \lambda, \tau)\, d\lambda \int_{E_n} \frac{\partial^2 Z_0(x-y, y, t, \lambda)}{\partial x_i \partial x_j} dy$$

$$+ \int_{\tau}^{\frac{t+\tau}{2}} d\lambda \int_{E_n} \frac{\partial^2 Z_0(x-y, y, t, \lambda)}{\partial x_i \partial x_j} Q(y, \xi, \lambda, \tau)\, dy.$$

As a result of estimating all of the terms of the right side of (12.18) we obtain inequality (12.2).

We have assumed that $t - t' < (t' - \tau)/3$. But inequality (12.2) also holds for $t - t' \geq (t' - \tau)/3$. In this case it follows from (11.26) since

$$(t' - \tau)^{-\frac{n+2-\alpha}{2}} \leqslant (t - t')^{\frac{\alpha}{2}} (t' - \tau)^{-\frac{n+2}{2}}$$

Finally, let us consider the function

$$q(x, t) = \int_0^t d\tau \int_{E_n} Q(x, y, t, \tau)\, dy.$$

Let σ_1 be the ball $|x' - y| \leq 2|x - x'|$ We have

$$q(x, t) - q(x', t) = \int_{\sigma_1} dy \int_0^t Q(x, y, t, \tau)\, d\tau$$

$$- \int_{\sigma_1} dy \int_0^t Q(x', y, t, \tau)\, d\tau$$

$$+ \int_{E_n \setminus \sigma_1} dy \int_0^t [Q(x, y, t, \tau) - Q(x', y, t, \tau)]\, d\tau.$$

With the use of (11.26) it is very easy to prove that the first two integrals can be estimated in terms of $c|x-x'|^\alpha$. In order to estimate the third term we represent Q in the form

$$Q = -K_1 - Q_1,$$

where

$$Q_1 = K_2 + \int_\tau^t d\lambda \int_{E_n} K(x, y, t, \lambda) Q(y, \xi, \lambda, \tau)\, dy.$$

From (12.4) and (12.15) we obtain

$$|Q_1(x, \xi, t, \tau) - Q_1(x', \xi, t, \tau)| \leqslant c|x-x'|^\alpha (t-\tau)^{-\frac{n+2}{2}+\frac{\beta}{2}} \exp\left(-C\frac{|x''-\xi|^2}{t-\tau}\right),$$

where $\beta = \min(\alpha, 1-\alpha)$. Therefore

$$\int_0^t d\tau \int_{E_n \setminus \sigma_1} |Q_1(x, y, t, \tau) - Q_1(x', y, t, \tau)|\, dy \leqslant c|x-x'|^\alpha.$$

Finally,

$$\int_{E_n \setminus \sigma_1} dy \int_0^t [K_1(x, y, t, \tau) - K_1(x', y, t, \tau)]\, d\tau$$

$$= \sum_{i,j=1}^n [a_{ij}(x', t) - a_{ij}(x, t)] \int_0^t d\tau \int_{E_n \setminus \sigma_1} \frac{\partial^2 Z_0(x-y, y, t, \tau)}{\partial x_i \partial x_j}\, dy$$

$$+ \sum_{i,j=1}^n \int_0^t d\tau \int_{E_n \setminus \sigma_1} [a_{ij}(y, \tau) - a_{ij}(x', t)] \times \left[\frac{\partial^2 Z_0(x-y, y, t, \tau)}{\partial x_i \partial x_j} - \frac{\partial^2 Z_0(x'-y, y, t, \tau)}{\partial x'_i \partial x'_j}\right] dy,$$

from which, by repeating arguments already carried out above, we obtain

$$\left|\int_0^t d\tau \int_{E_n \setminus \sigma_1} [K_1(x, y, t, \tau) - K_1(x', y, t, \tau)]\, dy\right| \leqslant c|x-x'|^\alpha.$$

Thus $\langle q \rangle_{x, Q_T}^{(\alpha)} \le c$. Analogously one can prove that $\langle q \rangle_{t, Q_T}^{(\alpha/2)} \le c$.

It is necessary in this regard to use the representation

$$q(x, t)-q(x, t')=\int_{\max(0, 2t'-t)}^{t} d\tau \int_{E_n} Q(x, y, t, \tau)\, dy$$
$$-\int_{\max(0, 2t'-t)}^{t'} d\tau \int_{E_n} Q(x, y, t', \tau)\, dy$$
$$+\int_{0}^{\max(0, 2t'-t)} d\tau \int_{E_n} [Q_1(x, y, t, \tau)-Q_1(x, y, t', \tau)]\, dy$$
$$+\sum_{i, j=1}^{n} [a_{ij}(x, t')-a_{ij}(x, t)]$$
$$\times \int_{0}^{\max(0, 2t'-t)} d\tau \int_{E_n} \frac{\partial^2 Z_0(x-y, y, t, \tau)}{\partial x_i \partial x_j}\, dy$$
$$+\sum_{i, j=1}^{n} \int_{0}^{\max(0, 2t'-t)} d\tau \int_{E_n} [a_{ij}(y, \tau)-a_{ij}(x, t')]$$
$$\times\left[\frac{\partial^2 Z_0(x-y, y, t, \tau)}{\partial x_i \partial x_j}-\frac{\partial^2 Z_0(x-y, y, t', \tau)}{\partial x_i \partial x_j}\right] dy.$$

The explicit calculations are omitted.

Thus inequalities (12.1)–(12.3) are established.

§ 13. ESTIMATES OF THE FUNDAMENTAL SOLUTION

In this section we obtain the following estimates of the fundamental solution $Z(x, \xi, t, \tau)$, which are necessary for establishing exact estimates of a volume potential with kernel Z:

$$|D_t^r D_x^s Z(x, \xi, t, \tau)| \leqslant c(t-\tau)^{-\frac{n+2r+s}{2}} \exp\left(-C\frac{|x-\xi|^2}{t-\tau}\right), \tag{13.1}$$

where $2r+s \leq 2$, $t > \tau$,

$$|D_t^r D_x^s Z(x, \xi, t, \tau)-D_t^r D_{x'}^s Z(x', \xi, t, \tau)|$$
$$\leqslant c\left[|x-x'|^{\gamma}(t-\tau)^{-\frac{n+2+\gamma}{2}}+|x-x'|^{\beta}(t-\tau)^{-\frac{n+2-\alpha+\beta}{2}}\right]$$
$$\times \exp\left(-C\frac{|x''-\xi|^2}{t-\tau}\right), \tag{13.2}$$

where $2r+s=2$ (i.e. $r=0$, $s=2$ and $r=1$, $s=0$), $0 \leq \gamma \leq 1$, $0 \leq \beta \leq \alpha$, $t > \tau$,

and

$$|D_t^r D_x^s Z(x, \xi, t, \tau) - D_{t'}^r D_x^s Z(x, \xi, t', \tau)|$$
$$\leqslant c\left[(t-t')(t'-\tau)^{-\frac{n+2r+s+2}{2}} + (t-t')^{\frac{2-2r-s+\alpha}{2}}(t'-\tau)^{-\frac{n+2}{2}}\right]$$
$$\times \exp\left(-C\frac{|x-\xi|^2}{t-\tau}\right), \tag{13.3}$$

where $2r + s = 1, 2$ and $t > t' > \tau$.

When applied to the function Z_0 estimates (13.1)–(13.3) follow from (11.3), and hence it is sufficient to establish them for the function Z' of (11.14). If $s = 0$ or $s = 1$, then by virtue of (11.3) and (11.26) we have

$$|D_x^s Z'(x, \xi, t, \tau)|$$
$$\leqslant \int_\tau^t d\lambda \int_{E_n} |D_x^s Z_0(x-y, y, t, \lambda)||Q(y, \xi, \lambda, \tau)|dy$$
$$\leqslant C\int_\tau^t (t-\lambda)^{-\frac{n+s}{2}}(\lambda-\tau)^{-\frac{n+2-\alpha}{2}} d\lambda$$
$$\times \int_{E_n} \exp\left(-C\frac{|x-y|^2}{t-\lambda}\right)\exp\left(-C\frac{|y-\xi|^2}{\lambda-\tau}\right)dy$$
$$\leqslant c(t-\tau)^{-\frac{n}{2}}\int_\tau^t (t-\lambda)^{-\frac{s}{2}}(\lambda-\tau)^{-\frac{2-\alpha}{2}} d\lambda \exp\left(-C\frac{|x-\xi|^2}{t-\tau}\right)$$
$$= c(t-\tau)^{-\frac{n+s-\alpha}{2}}\exp\left(-C\frac{|x-\xi|^2}{t-\tau}\right).$$

In order to estimate $D_x^2 Z'$ we take advantage of the formula

$$\frac{\partial^2 Z'}{\partial x_i \partial x_j} = \int_\tau^{\frac{t+\tau}{2}} d\lambda \int_{E_n} \frac{\partial^2 Z_0(x-y, y, t, \lambda)}{\partial x_i \partial x_j} Q(y, \xi, \lambda, \tau)dy$$
$$+ \int_{\frac{t+\tau}{2}}^t d\lambda \int_{E_n} \frac{\partial^2 Z_0(x-y, y, t, \lambda)}{\partial x_i \partial x_j}[Q(y, \xi, \lambda, \tau) - Q(x, \xi, \lambda, \tau)]dy$$
$$+ \int_{\frac{t+\tau}{2}}^t Q(x, \xi, \lambda, \tau)d\lambda \int_{E_n} \frac{\partial^2 Z_0(x-y, y, t, \lambda)}{\partial x_i \partial x_j}dy, \tag{13.4}$$

which follows from (11.8). With the use of inequalities (11.3), (11.26), (11.6) and (12.1) we obtain

$$\left|\frac{\partial^2 Z'}{\partial x_i \partial x_j}\right| \leqslant c \left\{ \int_\tau^{\frac{t+\tau}{2}} (t-\lambda)^{-\frac{n+2}{2}} (\lambda-\tau)^{-\frac{n+2-\alpha}{2}} d\lambda \right.$$

$$\times \int_{E_n} \exp\left(-C\frac{|x-y|^2}{t-\lambda}\right) \exp\left(-C\frac{|y-\xi|^2}{\lambda-\tau}\right) dy$$

$$+ \int_{\frac{t+\tau}{2}}^{t} (t-\lambda)^{-\frac{n+2}{2}} (\lambda-\tau)^{-\frac{n+2}{2}} d\lambda \int_{E_n} \exp\left(-C\frac{|x-y|^2}{t-\lambda}\right)$$

$$\times |x-y|^\alpha \left[\exp\left(-C\frac{|y-\xi|^2}{\lambda-\tau}\right) + \exp\left(-C\frac{|x-\xi|^2}{\lambda-\tau}\right)\right] dy$$

$$\left. + \int_{\frac{t+\tau}{2}}^{t} (t-\lambda)^{-1+\frac{\alpha}{2}} (\lambda-\tau)^{-\frac{n+2-\alpha}{2}} \exp\left(-C\frac{|x-\xi|^2}{\lambda-\tau}\right) d\lambda \right\}$$

$$\leqslant c(t-\tau)^{-\frac{n+2-\alpha}{2}} \exp\left(-C\frac{|x-\xi|^2}{t-\tau}\right)$$

Precisely the same estimate is also valid for $|\partial Z'/\partial t|$ Thus for $2r + s \leq 2$ we have

$$|D_t^r D_x^s Z'| \leqslant c(t-\tau)^{-\frac{n+2r+s-\alpha}{2}} \exp\left(-C\frac{|x-\xi|^2}{t-\tau}\right), \tag{13.5}$$

and inequality (13.1) is proved.

Let us estimate

$$\frac{\partial^2 Z'(x, \xi, t, \lambda)}{\partial x_i \partial x_j} - \frac{\partial^2 Z'(x', \xi, t, \lambda)}{\partial x_i' \partial x_j'}$$

For this purpose we use formula (13.4). By virtue of (11.20) and (11.26)

$$\left| \int_\tau^{\frac{t+\tau}{2}} d\lambda \int_{E_n} \left[\frac{\partial^2 Z_0(x-y, y, t, \lambda)}{\partial x_i \partial x_j} - \frac{\partial^2 Z_0(x'-y, y, t, \lambda)}{\partial x_i' \partial x_j'}\right] \right.$$

$$\left. \times Q(y, \xi, \lambda, \tau)\, dy \right| \leqslant c|x-x'|^\alpha \int_\tau^{\frac{t+\tau}{2}} (t-\lambda)^{-\frac{n+2+\alpha}{2}} \times$$

$$\times(\lambda-\tau)^{-\frac{n+2-\alpha}{2}}d\lambda\int\limits_{E_n}\Big[\exp\Big(-C\frac{|x-y|^2}{t-\lambda}\Big)$$

$$+\exp\Big(-C\frac{|x'-y|^2}{t-\lambda}\Big)\Big]\exp\Big(-C\frac{|y-\xi|^2}{\lambda-\tau}\Big)dy$$

$$\leqslant C|x-x'|^\alpha(t-\tau)^{-\frac{n+2+\alpha}{2}}\int\limits_{\tau}^{\frac{t+\tau}{2}}(\lambda-\tau)^{-\frac{2-\alpha}{2}}d\lambda$$

$$\times\Big[\exp\Big(-C\frac{|x-\xi|^2}{t-\tau}\Big)+\exp\Big(-C\frac{|x'-\xi|^2}{t-\tau}\Big)\Big]$$

$$\leqslant c|x-x'|^\alpha(t-\tau)^{-\frac{n+2}{2}}\exp\Big(-C\frac{|x''-\xi|^2}{t-\tau}\Big) \qquad (13.6)$$

We denote by $H_{ij}(x, \xi, t, \tau)$ the sum of the last two terms in (13.4), and by σ_1, as above, the ball with center x'' and radius $2|x-x'|$.

It is not difficult to verify that

$$H_{ij}(x,\xi,t,\tau)-H_{ij}(x',\xi,t,\tau)$$

$$=\int\limits_{\frac{t+\tau}{2}}^{t}d\lambda\int\limits_{\sigma_1}\frac{\partial^2 Z_0(x-y,y,t,\lambda)}{\partial x_i\,\partial x_j}[Q(y,\xi,\lambda,\tau)-Q(x,\xi,\lambda,\tau)]\,dy$$

$$-\int\limits_{\frac{t+\tau}{2}}^{t}d\lambda\int\limits_{\sigma_1}\frac{\partial^2 Z_0(x'-y,y,t,\lambda)}{\partial x'_i\,\partial x'_j}[Q(y,\xi,\lambda,\tau)-Q(x',\xi,\lambda,\tau)]\,dy$$

$$+\int\limits_{\frac{t+\tau}{2}}^{t}d\lambda\int\limits_{E_n\setminus\sigma_1}\Big[\frac{\partial^2 Z_0(x-y,y,t,\lambda)}{\partial x_i\,\partial x_j}-\frac{\partial^2 Z_0(x'-y,y,t,\lambda)}{\partial x'_i\,\partial x'_j}\Big]$$

$$\times[Q(y,\xi,\lambda,\tau)-Q(x',\xi,\lambda,\tau)]\,dy$$

$$+\int\limits_{\frac{t+\tau}{2}}^{t}[Q(x,\xi,\lambda,\tau)-Q(x',\xi,\lambda,\tau)]\,d\lambda\int\limits_{\sigma_1}\frac{\partial^2 Z_0(x-y,y,t,\lambda)}{\partial x_i\,\partial x_j}dy$$

$$+\int\limits_{\frac{t+\tau}{2}}^{t}Q(x',\xi,\lambda,\tau)\,d\lambda\int\limits_{E_n}\Big[\frac{\partial^2 Z_0(x-y,y,t,\lambda)}{\partial x_i\,\partial x_j}$$

$$-\frac{\partial^2 Z_0(x'-y,y,t,\lambda)}{\partial x'_i\,\partial x'_j}\Big]dy=\sum_{l=1}^{5}M_l.$$

Using (12.10), we obtain

$$|M_1| \leqslant c \int\limits_{\frac{t+\tau}{2}}^{t} (t-\lambda)^{-\frac{n+2}{2}} (\lambda-\tau)^{-\frac{n+2}{2}} d\lambda$$

$$\times \int\limits_{\sigma_1} |x-y|^{\alpha} \exp\left(-C\frac{|x-y|^2}{t-\lambda}\right)\left[\exp\left(-C\frac{|y-\xi|^2}{\lambda-\tau}\right)\right.$$

$$\left. + \exp\left(-C\frac{|x-\xi|^2}{\lambda-\tau}\right)\right] dy \leqslant c\,(t-\tau)^{-\frac{n+2}{2}} \exp\left(-C'\frac{|x-\xi|^2}{t-\tau}\right)$$

$$\times \int\limits_{\sigma_1} |x-y|^{\alpha}\, dy \int\limits_{\frac{t+\tau}{2}}^{t} (t-\lambda)^{-\frac{n+2}{2}} \exp\left(-C'\frac{|x-y|^2}{t-\lambda}\right) d\lambda$$

$$\leqslant c\,|x-x'|^{\alpha} (t-\tau)^{-\frac{n+2}{2}} \exp\left(-C'\frac{|x-\xi|^2}{t-\tau}\right).$$

In a completely analogous way

$$|M_2| \leqslant c\,|x-x'|^{\alpha} (t-\tau)^{-\frac{n+2}{2}} \exp\left(-C\frac{|x-\xi|^2}{t-\tau}\right).$$

If $y \in E_n \setminus \sigma_1$ then $c_1|x'-y| \le |x-y| \le c_2|x'-y|$. Therefore

$$|M_3| \leqslant c\,|x-x'| \int\limits_{\frac{t+\tau}{2}}^{t} (t-\lambda)^{-\frac{n+3}{2}} (\lambda-\tau)^{-\frac{n+2}{2}} d\lambda$$

$$\times \int\limits_{E_n \setminus \sigma_1} |x''-y|^{\alpha} \exp\left(-C\frac{|x''-y|^2}{t-\lambda}\right)$$

$$\times \left[\exp\left(-C\frac{|y-\xi|^2}{\lambda-\tau}\right) + \exp\left(-C\frac{|x'-\xi|^2}{\lambda-\tau}\right)\right] dy$$

$$\leqslant c\,|x-x'|\,(t-\tau)^{-\frac{n+2}{2}} \exp\left(-\frac{C}{2}\frac{|x''-\xi|^2}{t-\tau}\right)$$

$$\times \int\limits_{E_n \setminus \sigma_1} |x''-y|^{\alpha}\, dy \int\limits_{\frac{t+\tau}{2}}^{t} (t-\lambda)^{-\frac{n+3}{2}} \exp\left(-C'\frac{|x''-y|^2}{t-\lambda}\right) d\lambda$$

$$\leqslant c\,|x-x'|^{\alpha} (t-\tau)^{-\frac{n+2}{2}} \exp\left(-\frac{C}{2}\frac{|x''-\xi|^2}{t-\tau}\right).$$

Before estimating M_4 we note that

$$\int_{\sigma_1} \frac{\partial^2 Z_0(x-y, y, t, \lambda)}{\partial x_i \partial x_j} dy = \int_{\sigma_1} \Big[\frac{\partial^2 Z_0(x-y, y, t, \lambda)}{\partial x_i \partial x_j}$$

$$- \frac{\partial^2 Z_0(x-y, x, t, \lambda)}{\partial y_i \partial y_j} \Big] dy + \int_{\omega_1} \frac{\partial Z_0(x-y, x, t, \lambda)}{\partial y_i} n_j d\omega_y. \tag{13.7}$$

In consequence of (11.4)

$$\Big| \int_{\sigma_1} \Big[\frac{\partial^2 Z_0(x-y, y, t, \lambda)}{\partial x_i \partial x_j} - \frac{\partial^2 Z_0(x-y, x, t, \lambda)}{\partial y_i \partial y_j} \Big] dy \Big|$$

$$\leqslant c(t-\lambda)^{-\frac{2+n}{2}} \int_{\sigma_1} |x-y|^\alpha \exp\left(-C\frac{|x-y|^2}{t-\lambda}\right) dy, \tag{13.8}$$

whence

$$\Big| \int_{\sigma_1} \Big[\frac{\partial^2 Z_0(x-y, y, t, \lambda)}{\partial x_i \partial x_j} - \frac{\partial^2 Z_0(x-y, x, t, \lambda)}{\partial y_i \partial y_j} \Big] dy \Big|$$

$$\leqslant c(t-\lambda)^{-\frac{2-\alpha}{2}}.$$

If we estimate the second term of the right side of (13.7) with the use of (12.13), we will have

$$\Big| \int_{\sigma_1} \frac{\partial^2 Z_0(x-y, y, t, \lambda)}{\partial x_i \partial x_j} dy \Big| \leqslant c \Big[|t-\lambda|^{-\frac{2-\alpha}{2}}$$

$$+ |x-x'|^n (t-\lambda)^{-\frac{n+2}{2}} \exp\left(-C\frac{|x-x'|^2}{t-\lambda}\right) \Big]. \tag{13.9}$$

From this inequality and from estimate (12.1) it follows that

$$|M_4| \leqslant c|x-x'|^\alpha (t-\tau)^{-\frac{n+2}{2}}$$

$$\times \exp\left(-C\frac{|x''-\xi|^2}{t-\tau}\right) \Big[\int_{\frac{t+\tau}{2}}^{t} (t-\lambda)^{-\frac{2-\alpha}{2}} d\lambda$$

$$+ |x-x'|^n \int_{\frac{t+\tau}{2}}^{t} (t-\lambda)^{-\frac{n+2}{2}} \exp\left(-C\frac{|x-x'|^2}{t-\lambda}\right) d\lambda \Big]$$

$$\leqslant c|x-x'|^\alpha (t-\tau)^{-\frac{n+2}{2}} \exp\left(-C\frac{|x''-\xi|^2}{t-\tau}\right).$$

Finally, let us estimate M_5. We have

$$\int_{E_n} \left[\frac{\partial^2 Z_0}{\partial x_i \partial x_j} - \frac{\partial^2 Z_0}{\partial x_i' \partial x_j'} \right] dy$$

$$= \int_{E_n \setminus \sigma_1} \left[\frac{\partial^2 Z_0 (x - y, y, t, \lambda)}{\partial x_i \partial x_j} - \frac{\partial^2 Z_0 (x' - y, y, t, \lambda)}{\partial x_i' \partial x_j'} \right.$$

$$\left. - \frac{\partial^2 Z_0 (x - y, x, t, \lambda)}{\partial y_i \partial y_j} + \frac{\partial^2 Z_0 (x' - y, x', t, \lambda)}{\partial y_i \partial y_j} \right] dy$$

$$+ \int_{\sigma_1} \left[\frac{\partial^2 Z_0 (x - y, y, t, \lambda)}{\partial x_i \partial x_j} - \frac{\partial^2 Z_0 (x - y, x, t, \lambda)}{\partial y_i \partial y_j} \right] dy$$

$$- \int_{\sigma_1} \left[\frac{\partial^2 Z_0 (x' - y, y, t, \lambda)}{\partial x_i' \partial x_j'} - \frac{\partial^2 Z_0 (x' - y, x', t, \lambda)}{\partial y_i \partial y_j} \right] dy.$$

The second term of the right side is estimated by inequality (13.8); for the third term we have the analogous inequality

$$\left| \int_{\sigma_1} \left[\frac{\partial^2 Z_0 (x' - y, y, t, \lambda)}{\partial x_i \partial x_j} - \frac{\partial^2 Z_0 (x' - y, x', t, \lambda)}{\partial y_i \partial y_j} \right] dy \right|$$

$$\leqslant c (t - \lambda)^{-\frac{n+2}{2}} \int_{\sigma_1} | x' - y |^{\alpha} \exp \left(- C \frac{| x' - y |^2}{t - \lambda} \right) dy.$$

We represent the integral I over $E_n \setminus \sigma_1$ (the first term) in the form

$$I = \int_{E_n \setminus \sigma_1} \left[\frac{\partial^2 Z_0 (x - y, y, t, \lambda)}{\partial x_i \partial x_j} - \frac{\partial^2 Z_0 (x' - y, y, t, \lambda)}{\partial x_i' \partial x_j'} \right.$$

$$\left. - \frac{\partial^2 Z_0 (x - y, x, t, \lambda)}{\partial y_i \partial y_j} + \frac{\partial^2 Z_0 (x' - y, x, t, \lambda)}{\partial y_i \partial y_j} \right] dy$$

$$- \int_{\omega_1} \left[\frac{\partial^2 Z_0 (x' - y, x', t, \lambda)}{\partial y_i} \right.$$

$$\left. - \frac{\partial Z_0 (x' - y, x, t, \lambda)}{\partial y_i} \right] n_j (y) \, d\omega_y = I_1 + I_2.$$

By representing the difference of the function $\partial^2 Z_0 / \partial x_i \partial x_j'$ at the points x and x' in terms of the derivative $\partial^3 Z_0 / \partial l \, \partial z_i \partial z_j$ where z is a point on the segment between the points x and x', and l is the direction of the vector $x - x'$, we obtain, with the use of (11.4),

$$|I_1| \leqslant \left| \int_{x}^{x'} dl_z \int_{E_n \setminus \sigma_1} \left[\frac{\partial^3 Z_0(z-y, y, t, \lambda)}{\partial l\, \partial z_i\, \partial z_j} - \frac{\partial^3 Z_0(z-y, x, t, \lambda)}{\partial l\, \partial z_i\, \partial z_j} \right] dy \right| \leqslant c\,|x-x'|\,(t-\lambda)^{-\frac{n+3}{2}} \times \int_{E_n \setminus \sigma_1} |x''-y|^{\alpha} \exp\left(-C\frac{|x''-y|^2}{t-\lambda}\right) dy. \tag{13.10}$$

In order to estimate I_2 we take advantage of the inequality

$$\left| \frac{\partial Z_0(x'-y, x', t, \lambda)}{\partial y_i} - \frac{\partial Z_0(x'-y, x, t, \lambda)}{\partial y_i} \right| \leqslant c\,|x-x'|^{\alpha}\,|x'-y|\,(t-\lambda)^{-\frac{n+2}{2}} \exp\left(-C\frac{|x'-y|^2}{t-\lambda}\right),$$

which is obtained from (11.2) with the use of elementary estimates. By virtue of this inequality

$$|I_2| \leqslant c\,|x-x'|^{n+\alpha}(t-\lambda)^{-\frac{n+2}{2}} \exp\left(-C\frac{|x'-x|^2}{t-\lambda}\right). \tag{13.11}$$

Combining inequalities (13.8), (13.10) and (13.11), we obtain

$$\int_{-\infty}^{t} d\lambda \left| \int_{E_n} \left[\frac{\partial^2 Z_0}{\partial x_i\, \partial x_j} - \frac{\partial^2 Z_0}{\partial x'_i\, \partial x'_j} \right] dy \right| \leqslant c \left\{ \int_{\sigma_1} |x-y|^{\alpha} dy \int_{-\infty}^{t} (t-\lambda)^{-\frac{n+2}{2}} \exp\left(-C\frac{|x-y|^2}{t-\lambda}\right) d\lambda \right.$$
$$+ \int_{\sigma_1} |x'-y|^{\alpha} dy \int_{-\infty}^{t} (t-\lambda)^{-\frac{n+2}{2}} \exp\left(-C\frac{|x'-y|^2}{t-\lambda}\right) d\lambda$$
$$+ |x-x'| \int_{E_n \setminus \sigma_1} |x''-y|^{\alpha} dy \int_{-\infty}^{t} (t-\lambda)^{-\frac{n+3}{2}}$$
$$\times \exp\left(-C\frac{|x''-y|^2}{t-\lambda}\right) d\lambda + |x-x'|^{n+\alpha} \int_{-\infty}^{t} (t-\lambda)^{-\frac{n+2}{2}}$$
$$\left. \times \exp\left(-C\frac{|x-x'|^2}{t-\lambda}\right) d\lambda \right\} \leqslant c\,|x-x'|^{\alpha}. \tag{13.12}$$

From this inequality and from estimate (11.26) of the function Q it follows that

$$|M_5| \leqslant c|x-x'|^\alpha (t-\tau)^{-\frac{n+2-\alpha}{2}} \exp\left(-C\frac{|x'-\xi|^2}{t-\tau}\right)$$
$$\leqslant c|x-x'|^\alpha (t-\tau)^{-\frac{n+2-\alpha}{2}} \exp\left(-C\frac{|x''-\xi|^2}{t-\tau}\right).$$

Thus all of the M_i are estimated and

$$|H_{ij}(x, \xi, t, \tau)-H_{ij}(x', \xi, t, \tau)|$$
$$\leqslant c|x-x'|^\alpha (t-\tau)^{-\frac{n+2}{2}} \exp\left(-C\frac{|x''-\xi|^2}{t-\tau}\right).$$

This inequality together with (13.6) proves the estimate

$$\left|\frac{\partial^2 Z'(x, \xi, t, \tau)}{\partial x_i \partial x_j}-\frac{\partial^2 Z'(x', \xi, t, \tau)}{\partial x_i' \partial x_j'}\right|$$
$$\leqslant c|x-x'|^\alpha (t-\tau)^{-\frac{n+2}{2}} \exp\left(-C\frac{|x''-\xi|^2}{t-\tau}\right). \tag{13.13}$$

In order to obtain an analogous estimate for $\partial Z'/\partial t$ we make use of the fact that, as was determined in §11, the function Z' satisfies the equation

$$\mathscr{L}\left(x, t, \frac{\partial}{\partial x}, \frac{\partial}{\partial t}\right) Z'(x, \xi, t, \tau)$$
$$= Q(x, \xi, t, \tau)+\int_\tau^t d\lambda \int_{E_n} K(x, y, t, \lambda) Q(y, \xi, \lambda, \tau)\, d\xi$$
$$= -K(x, \xi, t, \tau). \tag{13.14}$$

Transferring to the right side of this equation all of the terms except $\partial Z'/\partial t$ and making use of the already proved estimates (13.5) and (13.13), we obtain

$$\left|\frac{\partial Z'(x, \xi, t, \lambda)}{\partial t}-\frac{\partial Z'(x', \xi, t, \tau)}{\partial t}\right|$$
$$\leqslant c|x-x'|^\alpha (t-\tau)^{-\frac{n+2}{2}} \exp\left(-C\frac{|x''-\xi|^2}{t-\tau}\right).$$

Thus for $2r+s=2$

$$|D_t^r D_x^s Z'(x, \xi, t, \tau)-D_t^r D_{x'}^s Z'(x', \xi, t, \tau)|$$
$$\leqslant c|x-x'|^\alpha (t-\tau)^{-\frac{n+2}{2}} \exp\left(-C\frac{|x''-\xi|^2}{t-\tau}\right).$$

From here and from (13.5) it follows that

$$|D_t^r D_x^s Z'(x, \xi, t, \tau) - D_t^r D_{x'}^s Z'(x', \xi, t, \tau)|$$
$$\leqslant c|x - x'|^{\beta}(t - \tau)^{-\frac{n+2-\alpha+\beta}{2}} \exp\left(-C\frac{|x'' - \xi|^2}{t - \tau}\right) \qquad (13.15)$$

for $0 \le \beta \le \alpha$. This completes the proof of inequality (13.2).

Let us proceed to an estimate of the difference

$$D_x^s Z'(x, \xi, t, \tau) - D_x^s Z'(x, \xi, t', \tau)$$

for $s = 1, 2$, $t > t' > \tau$, $t - t' < \frac{1}{4}(t' - \tau)$. We make use of the representation

$$D_x^s Z'(x, \xi, t, \tau)$$
$$= \int_{\tau}^{\frac{t'+\tau}{2}} d\lambda \int_{E_n} D_x^s Z_0(x - y, y, t, \lambda) Q(y, \xi, \lambda, \tau)\, dy$$
$$+ \int_{\frac{t'+\tau}{2}}^{t} d\lambda \int_{E_n} D_x^s Z_0(x - y, y, t, \lambda)$$
$$\times [Q(y, \xi, \lambda, \tau) - Q(x, \xi, \lambda, \tau)]\, dy$$
$$+ \int_{\frac{t'+\tau}{2}}^{t} Q(x, \xi, \lambda, \tau)\, d\lambda \int_{E_n} D_x^s Z_0(x - y, y, t, \lambda)\, dy \qquad (13.16)$$

and of the same representation for $D_x^s Z'(x, \xi, t', \tau)$, obtained from (13.16) by replacing t everywhere by t'. Forming the difference of these two functions, we obtain

$$D_x^s Z'(x, \xi, t, \tau) - D_x^s Z'(x, \xi, t', \tau)$$
$$= \int_{\tau}^{\frac{t'+\tau}{2}} d\lambda \int_{E_n} [D_x^s Z_0(x-y, y, t, \lambda) - D_x^s Z_0(x-y, y, t', \lambda)] Q\, dy$$
$$+ \int_{\frac{t'+\tau}{2}}^{2t'-t} d\lambda \int_{E_n} [D_x^s Z_0(x - y, y, t, \lambda) - D_x^s Z_0(x - y, y, t', \lambda)]$$
$$\times [Q(y, \xi, \lambda, \tau) - Q(x, \xi, \lambda, \tau)]\, dy + \int_{\frac{t'+\tau}{2}}^{2t'-t} Q(x, \xi, \lambda, \tau)\, d\tau \times$$

$$\times \int_{E_n} [D_x^s Z_0(x-y,\ y,\ t,\ \lambda) - D_x^s Z_0(x-y,\ y,\ t',\ \lambda)]\, dy$$

$$+ \int_{2t'-t}^{t} d\lambda \int_{E_n} D_x^s Z_0(x-y,\ y,\ t,\ \lambda)[Q(y,\ \xi,\ \lambda,\ \tau) - Q(x,\ \xi,\ \lambda,\ \tau)]\, dy$$

$$- \int_{2t'-t}^{t'} d\lambda \int_{E_n} D_x^s Z_0(x-y,\ y,\ t',\ \lambda)$$

$$\times [Q(y,\ \xi,\ \lambda,\ \tau) - Q(x,\ \xi,\ \lambda,\ \tau)]\, dy$$

$$+ \int_{2t'-t}^{t} Q(x,\ \xi,\ \lambda,\ \tau)\, d\lambda \int_{E_n} D_x^s Z_0(x-y,\ y,\ t,\ \lambda)\, dy$$

$$- \int_{2t'-t}^{t'} Q(x,\ \xi,\ \lambda,\ \tau)\, d\lambda \int_{E_n} D_x^s Z_0(x-y,\ y,\ t',\ \lambda)\, dy \equiv \sum_{i=1}^{7} N_i.$$

Each term of the right side of this representation is estimated with the use of inequalities (11.3), (11.6), (11.26) and (12.1). The assumptions $t > t' > \tau$, $t - t' < (t' - \tau)/4$ are also used:

$$|N_1| \leqslant c(t-t') \int_{\tau}^{\frac{t'+\tau}{2}} (t'-\lambda)^{-\frac{n+s+2}{2}} (\lambda-\tau)^{-\frac{n+2-\alpha}{2}}\, d\lambda$$

$$\times \int_{E_n} \exp\left(-C\frac{|x-y|^2}{t-\lambda}\right) \exp\left(-C\frac{|y-\xi|^2}{\lambda-\tau}\right) dy$$

$$\leqslant c(t-t')(t'-\tau)^{-\frac{n+s+2-\alpha}{2}} \exp\left(-C\frac{|x-\xi|^2}{t-\tau}\right)$$

$$\leqslant c(t-t')^{\frac{2-s+\alpha}{2}} (t'-\tau)^{-\frac{n+2}{2}} \exp\left(-C\frac{|x-\xi|^2}{t-\lambda}\right),$$

$$|N_2| \leqslant c(t-t') \int_{\frac{t'+\tau}{2}}^{2t'-t} (t'-\lambda)^{-\frac{n+s+2}{2}} (\lambda-\tau)^{-\frac{n+2}{2}}\, d\lambda$$

$$\times \int_{E_n} |x-y|^{\alpha} \left[\exp\left(-C\frac{|y-\xi|^2}{\lambda-\tau}\right) + \exp\left(-C\frac{|x-\xi|^2}{\lambda-\tau}\right)\right]$$

$$\times \exp\left(-C\frac{|x-y|^2}{t-\lambda}\right) dy \leqslant c(t-t')(t'-\tau)^{-\frac{n+2}{2}} \times$$

$$\times \exp\left(-C\frac{|x-\xi|^2}{t-\tau}\right)\int_{\frac{t'+\tau}{2}}^{2t'-t}(t'-\lambda)^{-\frac{s+2-\alpha}{2}}\,d\lambda$$

$$\leqslant c(t-t')^{\frac{2-s+\alpha}{2}}(t'-\tau)^{-\frac{n+2}{2}}\exp\left(-C\frac{|x-\xi|^2}{t-\tau}\right).$$

The fourth and fifth, and also the sixth and seventh, terms are estimated in the same way; for example,

$$|N_4|\leqslant c\int_{2t'-t}^{t}(t-\lambda)^{-\frac{n+s}{2}}(\lambda-\tau)^{-\frac{n+2}{2}}\,d\lambda$$

$$\times\int_{E_n}\exp\left(-C\frac{|x-y|^2}{t-\lambda}\right)\left[\exp\left(-C\frac{|y-\xi|^2}{\lambda-\tau}\right)\right.$$

$$\left.+\exp\left(-C\frac{|x-\xi|^2}{\lambda-\tau}\right)\right]|x-y|^\alpha\,dy$$

$$\leqslant c(t-\tau)^{-\frac{n+2}{2}}\exp\left(-C\frac{|x-\xi|^2}{t-\tau}\right)\int_{2t'-t}^{t}(t-\lambda)^{-\frac{s-\alpha}{2}}\,d\lambda$$

$$=c(t-t')^{\frac{2-s+\alpha}{2}}(t-\tau)^{-\frac{n+2}{2}}\exp\left(-C\frac{|x-\xi|^2}{t-\tau}\right),$$

and by (11.6)

$$|N_6|\leqslant c\int_{2t-t'}^{t}|Q(x,\xi,\lambda,\tau)|\,(t-\lambda)^{-\frac{s-\alpha}{2}}\,d\lambda$$

$$\leqslant c(t-t')^{\frac{2-s+\alpha}{2}}(t-\tau)^{-\frac{n+2-\alpha}{2}}\exp\left(-C\frac{|x-\xi|^2}{t-\tau}\right).$$

It remains for us to estimate the third term N_3

$$\int_{E_n}\left[D_x^sZ_0(x-y,y,t,\lambda)-D_x^sZ_0(x-y,y,t',\lambda)\right]dy$$

$$=\int_{t'}^{t}dt''\int_{E_n}D_{t''}D_x^sZ_0(x-y,y,t'',\lambda)\,dy$$

and since, by (11.6),

$$\left|\int_{E_n} [D_x^s Z_0(x-y,\, y,\, t,\, \lambda) - D_x^s Z_0(x-y,\, y,\, t',\, \lambda)]\, dy\right| \leqslant c\int_{t'}^{t} (t''-\lambda)^{-\frac{2+s-\alpha}{2}}\, dt'', \tag{13.17}$$

it follows that

$$|N_3| \leqslant c(t'-\tau)^{-\frac{n+2-\alpha}{2}} \exp\left(-C\frac{|x-\xi|^2}{t-\tau}\right) \times \int_{t'}^{t} dt'' \int_{\frac{t'+\tau}{2}}^{2t'-t} (t''-\lambda)^{-\frac{2+s-\alpha}{2}}\, d\lambda$$
$$\leqslant c(t-t')^{\frac{2-s+\alpha}{2}} (t'-\tau)^{-\frac{n+2-\alpha}{2}} \exp\left(-C\frac{|x-\xi|^2}{t-\tau}\right).$$

Combining the estimates, we obtain for $s = 1, 2$

$$|D_x^s Z'(x,\, \xi,\, t,\, \tau) - D_x^s Z'(x,\, \xi,\, t',\, \tau)| \leqslant c(t-t')^{\frac{2-s+\alpha}{2}} (t'-\tau)^{-\frac{n+2}{2}} \exp\left(-C\frac{|x-\xi|^2}{t-\tau}\right). \tag{13.18}$$

This estimate is proved under the assumption that $t - t' < (t' - \tau)/4$. But it also holds for $t - t' \geq (t' - \tau)/4$, since in this case it follows from (13.5).

Let us estimate the difference

$$D_t Z'(x,\, \xi,\, t,\, \tau) - D_{t'} Z'(x,\, \xi,\, t',\, \tau).$$

This is easily done with the use of the already proved estimate (13.18) if one takes advantage of equation (13.14). From this equation, from (13.18), (13.1) and from the easily verified inequality

$$|K(x,\, \xi,\, t,\, \tau) - K(x,\, \xi,\, t',\, \tau)| \leqslant c(t-t')^{\alpha/2} (t'-\tau)^{-\frac{n+2}{2}} \exp\left(-C\frac{|x-\xi|^2}{t-\tau}\right)$$

it follows that

$$|D_t Z'(x,\, \xi,\, t,\, \tau) - D_{t'} Z'(x,\, \xi,\, t',\, \tau)| \leqslant c(t-t')^{\alpha/2} (t'-\tau)^{-\frac{n+2}{2}} \exp\left(-C\frac{|x-\xi|^2}{t-\tau}\right) \tag{13.19}$$

(we assume, as above, that $t > t' > \tau$).

Estimates (13.18) and (13.19) together with (11.3) prove inequality (13.3).

§14. SOLUTION OF THE CAUCHY PROBLEM

Let us consider in the domain $D_{n+1}^{(T)}$ the Cauchy problem

$$\mathscr{L}\left(x,\ t,\ \frac{\partial}{\partial x},\ \frac{\partial}{\partial t}\right) v(x,\ t) = f(x,\ t),$$
$$v|_{t=0} = \varphi(x). \tag{14.1}$$

We assume that the function f satisfies a Hölder condition in all of its arguments, and that φ is continuous. In this case the solution of problem (14.1) can be written in the form of a sum of two potentials with kernel Z:

$$v(x,\ t) = \int_0^t d\tau \int_{E_n} Z(x,\ \xi,\ t,\ \tau) f(\xi,\ \tau)\, d\xi$$
$$+ \int_{E_n} Z(x,\ \xi,\ t,\ 0)\varphi(\xi)\, d\xi. \tag{14.2}$$

In order to ensure the convergence of the integrals in the right side of (14.2) it is necessary to require that the functions f and φ do not increase too rapidly as $|x| \to \infty$; it is sufficient to require that they increase no faster than a function e^{ax^2}, where a is some positive constant depending on T and on the constant C in inequality (13.1). These restrictions on f and φ are formulated more precisely in [25a], where the potentials of (14.2) are studied in detail in classes of functions increasing at infinity. We will consider problem (14.1) only in a class of bounded functions.

In order to verify that the function (14.2) is really a solution of problem (14.1) one should consider separately the functions

$$v_1(x,\ t) = \int_0^t d\tau \int_{E_n} Z(x,\ \xi,\ t,\ \tau) f(\xi,\ \tau)\, d\tau,$$
$$v_2(x,\ t) = \int_{E_n} Z(x,\ \xi,\ t,\ 0)\varphi(\xi)\, d\xi.$$

They are solutions of the problems

$$\mathscr{L}v_1 = f,\quad v_1|_{t=0} = 0,$$
$$\mathscr{L}v_2 = 0,\quad v_2|_{t=0} = \varphi. \tag{14.3}$$

Of these equalities only the equality $v_2|_{t=0} = \varphi$ requires a verification, since the construction of a fundamental solution in § 11 was based on the relation $\mathfrak{L}v_1 = f$, while the remaining equalities are obvious. Since by virtue of (13.5) for

$r = s = 0$

$$\lim_{t\to 0}\int_{E_n} Z'(x, \xi, t, 0)\varphi(\xi)\,d\xi = 0,$$

we see, upon using relation (11.10), that

$$\lim_{t\to 0} v_2(x, t) = \lim_{t\to 0}\int_{E_n} Z_0(x-\xi, \xi, t, 0)\varphi(\xi)\,d\xi = \varphi(x),$$

Q.E.D.

Let us now prove, using estimates of the preceding sections, that if $f \in H^{\alpha, \alpha/2}(\overline{D^{(T)}_{n+1}})$ and $\varphi \in H^{2+\alpha}(\overline{E}_n)$, then

$$|v_1|^{(2+\alpha)}_{D^{(T)}_{n+1}} \leqslant c\,|f|^{(\alpha)}_{D^{(T)}_{n+1}}, \tag{14.4}$$

$$|v_2|^{(2+\alpha)}_{D^{(T)}_{n+1}} \leqslant c\,|\varphi|^{(2+\alpha)}_{E_n} \tag{14.5}$$

and, consequently,

$$|v|^{(2+\alpha)}_{D^{(T)}_{n+1}} \leqslant c\left(|f|^{(\alpha)}_{D^{(T)}_{n+1}} + |\varphi|^{(2+\alpha)}_{E_n}\right). \tag{14.6}$$

We will thereby, in a different way, have proved Theorem 5.1 for $l = \alpha < 1$. In order to prove inequality (14.4) one must establish that

$$\langle v_1\rangle^{(2+\alpha)}_{x, D^{(T)}_{n+1}} \leqslant c\,|f|^{(\alpha)}_{D^{(T)}_{n+1}} \tag{14.7}$$

$$\langle D^s_x v_1\rangle^{\left(\frac{2-s+\alpha}{2}\right)}_{t, D^{(T)}_{n+1}} \leqslant c\,|f|^{(\alpha)}_{D^{(T)}_{n+1}} \qquad (s = 1, 2), \tag{14.8}$$

$$\langle D_t v_1\rangle^{\left(\frac{\alpha}{2}\right)}_{t, D^{(T)}_{n+1}} \leqslant c\,|f|^{(\alpha)}_{D^{(T)}_{n+1}}, \tag{14.9}$$

and also

$$\langle v_1\rangle^{(s)}_{D^{(T)}_{n+1}} \leqslant c\,|f|^{(\alpha)}_{D^{(T)}_{n+1}} \qquad (s = 0, 1, 2). \tag{14.10}$$

Since these estimates are proved in approximately the same way as estimates (2.8), (2.9), we will not verify them in every detail. We make use of the equality

$$D^s_x v_1(x, t) = \int_0^t d\tau \int_{E_n} D^s_x Z(x, y, t, \tau)\,[f(y, \tau) - f(x, \tau)]\,dy + \int_0^t f(x, \tau)\,d\tau \int_{E_n} D^s_x Z(x, y, t, \tau)\,dy. \tag{14.11}$$

Since by virtue of (11.6) and (13.5)

$$\left|\int_{E_n} D_x^s Z(x, y, t, \tau)\, dy\right| \leqslant c(t-\tau)^{-\frac{s-\alpha}{2}} \qquad (s=1, 2),$$

$$\left|\int_{E_n} Z\, dy\right| \leqslant c, \tag{14.12}$$

it follows from representation (14.11) that

$$|D_x^s v_1| \leqslant ct^{\frac{2-s+\alpha}{2}} |f|^{(\alpha)}_{D^{(T)}_{n+1}} \qquad (s=1, 2),$$

$$|v_1| \leqslant ct|f|^{(\alpha)}_{D^{(T)}_{n+1}} \tag{14.13}$$

Using the equation $\mathfrak{L}v_1 = f$ for an estimate of $D_t v_1$, we obtain

$$|D_t v_1| \leqslant ct^{\frac{\alpha}{2}} |f|^{(\alpha)}_{D^{(T)}_{n+1}}$$

The latter three inequalities prove estimate (14.10), which, however, is completely obvious for $s = 0, 1$.

Let us pass to the proof of estimates (14.7)–(14.9). We begin with (14.8). In order to prove it one must estimate the difference

$$D_x^s v_1(x, t) - D_x^s v_1(x, t') \qquad (s=1, 2).$$

For the sake of definiteness let $t' < t \le T$. If $t > 2t'$, i.e., $t - t' > t'$, then by virtue of (14.13)

$$|D_x^s v_1(x, t) - D_x^s v_1(x, t')| \leqslant c\,(t^{\frac{2-s+\alpha}{2}}$$
$$+ t'^{\frac{2-s+\alpha}{2}})|f|^{(\alpha)}_{D^{(T)}_{n+1}} \leqslant c(t-t')^{\frac{2-s+\alpha}{2}} |f|^{(\alpha)}_{D^{(T)}_{n+1}}. \tag{14.14}$$

The same inequality also holds for $t - t' \le t'$, but in this case the proof requires more subtle arguments. With the use of (14.11) one can prove that

$$D_x^s v_1(x, t) - D_x^s v_1(x, t')$$
$$= \int_{2t'-t}^{t} d\tau \int_{E_n} D_x^s Z(x, y, t, \tau)\,[f(y, \tau) - f(x, \tau)]\, dy -$$

$$-\int_{2t'-t}^{t'} d\tau \int_{E_n} D_x^s Z(x, y, t', \tau)[f(y, \tau) - f(x, \tau)]\, dy$$

$$+\int_0^{2t'-t} d\tau \int_{E_n} [D_x^s Z(x, y, t, \tau)$$

$$- D_x^s Z(x, y, t', \tau)][f(y, \tau) - f(x, \tau)]\, dy$$

$$+\int_{2t'-t}^{t} f(x, \tau)\, d\tau \int_{E_n} D_x^s Z(x, y, t, \tau)\, dy$$

$$-\int_{2t'-t}^{t'} f(x, \tau)\, d\tau \int_{E_n} D_x^s Z(x, y, t', \tau)\, dy$$

$$+\int_0^{2t'-t} [f(x, \tau) - f(x, t)]\, d\tau \int_{E_n} [D_x^s Z(x, y, t, \tau)$$

$$-D_x^s Z(x, y, t', \tau)]\, dy + f(x, t) \int_0^{2t'-t} d\tau \int_{E_n} [D_x^s Z(x, y, t, \tau)$$

$$- D_x^s Z(x, y, t', \tau)]\, dy = \sum_{j=1}^{7} J_j. \tag{14.15}$$

The first six terms of the right side are easily estimated with the use of inequalities (13.1), (13.3), (13.5) and (11.6):

$$\sum_{j=1}^{6} |J_j| \leqslant c(t-t')^{\frac{2-s+\alpha}{2}} |f|_{D_{n+1}^{(T)}}^{(\alpha)}. \tag{14.16}$$

The J_7 term is transformed in the following manner:

$$\int_0^{2t'-t} d\tau \int_{E_n} [D_x^s Z(x, y, t, \tau) - D_x^s Z(x, y, t', \tau)]\, dy$$

$$= \int_0^{2t'-t} d\tau \int_{E_n} [D_x^s Z_0(x-y, y, t, \tau) - D_x^s Z_0(x-y, y, t', \tau)]\, dy$$

$$-\int_{2t'-t}^{t} d\tau \int_{E_n} D_x^s Z'(x, y, t, \tau)\, dy +$$

$$+\int_{2t'-t}^{t'} d\tau \int_{E_n} D_x^s Z'(x, y, t', \tau)\, dy$$

$$+\left[\int_0^t d\tau \int_{E_n} D_x^s Z'(x, y, t, \tau)\, dy - \int_0^{t'} d\tau \int_{E_n} D_x^s Z'(x, y, t', \tau)\, dy\right].$$

By virtue of (13.17) and (13.5) the first three integrals of the right side do not exceed $c(t-t')^{(2-s+\alpha)/2}$. The fourth term can be written in the form

$$\int_0^t d\tau \int_{E_n} D_x^s Z'(x, y, t, \tau)\, dy - \int_0^{t'} d\tau \int_{E_n} D_x^s Z'(x, y, t', \tau)\, dy$$

$$= D_x^s\left[\int_0^t d\tau \int_{E_n} Z_0(x-y, y, t, \tau)\, q(y, \tau)\, dy\right.$$

$$\left. - \int_0^{t'} d\tau \int_{E_n} Z_0(x-y, y, t', \tau)\, q(y, \tau)\, dy\right], \tag{14.17}$$

where

$$q(y, \tau) = \int_0^\tau d\lambda \int_{E_n} Q(y, \xi, \tau, \lambda)\, d\xi$$

is a function that, as was shown in §12, belongs to the class $H^{\alpha, \alpha/2}(\overline{D_{n+1}^{(T)}})$ of (12.3).

If from the very beginning we had considered instead of v_1 the volume potential

$$v_1^0 = \int_0^t d\tau \int_{E_n} Z_0(x-y, y, t, \tau)\, f(y, \tau)\, dy,$$

then for this potential we would have had representation (14.15), in which the kernel Z is replaced everywhere by Z_0. For the right side of this representation we would have had estimate (14.16) and also, by virtue of (13.17), the estimate

$$|J_7| \leqslant c(t-t')^{\frac{2-s+\alpha}{2}} |f|_{D_{n+1}^{(T)}}^{(0)}. \tag{14.18}$$

Consequently inequality (14.8) holds for the potential v_1^0 and the right side of (14.17) is estimated in terms of $c(t-t')^{(2-s+\alpha)/2}$ (we also make use of (12.3) here). Hence the J_7 term of representation (14.15) is also estimated by inequality (14.18), and estimate (14.8) is thus proved.

Inequality (14.7) is obtained by means of the same arguments from the representation

$$\frac{\partial^2 v_1(x,t)}{\partial x_i \partial x_j} - \frac{\partial^2 v_1(x',t)}{\partial x'_i \partial x'_j} = \int_0^t d\tau \int_{\sigma_1} \frac{\partial^2 Z(x,y,t,\tau)}{\partial x_i \partial x_j}$$

$$\times \{[f(y,\tau) - f(x,\tau)] + [f(x,\tau) - f(x',t)]\}\, dy$$

$$- \int_0^t d\tau \int_{\sigma_1} \frac{\partial^2 Z(x',y,t,\tau)}{\partial x'_i \partial x'_j} \{[f(y,\tau) - f(x',\tau)]$$

$$+ [f(x',\tau) - f(x',t)]\}\, dy + \int_0^t d\tau \int_{E_n \setminus \sigma_1} \left[\frac{\partial^2 Z(x,y,t,\tau)}{\partial x_i \partial x_j}\right.$$

$$\left. - \frac{\partial^2 Z(x',y,t,\tau)}{\partial x'_i \partial x'_j}\right][f(y,\tau) - f(x',t)]\, dy$$

$$+ f(x',t) \int_0^t d\tau \int_{E_n} \left[\frac{\partial^2 Z(x,y,t,\tau)}{\partial x_i \partial x_j} - \frac{\partial^2 Z(x',y,t,\tau)}{\partial x'_i \partial x'_j}\right] dy,$$

which is a consequence of formula (14.11). Here σ_1 is the ball with center x and radius $2|x - x'|$. In estimating the right side one should use inequalities (13.1), (13.2) and (13.9).

Finally, inequality (14.9) is a consequence of the already proved inequality (14.8) and the equation $\mathfrak{L} v_1 = f$.

Thus we have proved inequality (14.4).

Inequality (14.5) can be proved by writing the function v_2 in terms of a volume potential. Since this function is a solution of problem (14.3), the function

$$v_3(x,t) = v_2(x,t) - \varphi(x)$$

is a solution of the problem

$$\mathscr{L} v_3 = -\mathscr{L}\varphi(x), \quad v_3|_{t=0} = 0.$$

Consequently,

$$v_3 = -\int_0^t d\tau \int_{E_n} Z(x,y,t,\tau)\, \mathscr{L}\varphi(y)\, dy$$

and

$$v_2(x, t) = \varphi(x) - \int_0^t d\tau \int_{E_n} Z(x, y, t, \tau) \mathscr{L}\varphi(y)\, dy.$$

If we now apply inequality (14.4) to the right side, we obtain (14.5).

Estimate (14.6) is proved.

§15. THE SINGLE-LAYER POTENTIAL

Let $Q = \Omega \times (0, T)$, where Ω is a bounded domain in the space E_n with boundary $S \in H^{1+\beta}$, $\beta < 1$. Consider in $D_{n+1}^{(T)}$ a parabolic operator $\mathfrak{L}(x, t, \partial/\partial x, \partial/\partial t)$ with coefficients from $H^{\alpha, \alpha/2}(D_{n+1}^{(T)})$ and let Z be its fundamental solution.

The potential

$$V(x, t) = \int_0^t d\tau \int_S Z(x, \xi, t, \tau)\varphi(\xi, \tau)\, dS_\xi$$

is called a single-layer potential. This potential possesses many of the properties of an ordinary electrostatic or heat single-layer potential.

By means of estimate (13.1) it is not difficult to show that the function $V(x, t)$ is for any bounded φ a continuous function of x and t. Let us calculate the conormal derivative of the potential $V(x, t)$ on S:

$$\lim_{x \to x_0 \in S} \sum_{i, j=1}^{n} a_{ij}(x, t) \frac{\partial V(x, t)}{\partial x_j} n_i(x_0).$$

Let

$$V(x, t) = V_0(x, t) + V'(x, t),$$

where

$$V_0 = \int_0^t d\tau \int_S Z_0 \varphi\, dS_\xi, \quad V' = \int_0^t d\tau \int_S Z' \varphi\, dS_\xi.$$

Estimate (13.5) for $r = 0$, $s = 1$ implies

$$\lim_{x \to x_0} \sum_{i, j=1}^{n} a_{ij}(x, t) \frac{\partial V'(x, t)}{\partial x_j} n_i(x_0) = \sum_{i, j=1}^{n} a_{ij}(x_0, t) \frac{\partial V'(x, t)}{\partial x_j}\Bigg|_{x=x_0} n_i(x_0). \tag{15.1}$$

Let us calculate the conormal derivative of the function V_0. We first consider the directly defined value of the conormal derivative on S

$$W_0(x_0, t) = \int_0^t d\tau \int_S H_0(x_0, \xi, t, \tau)\varphi(\xi, \tau)\, dS_\xi,$$

where

$$H_0(x_0, \xi, t, \tau) = \sum_{i, j=1}^n a_{ij}(x_0, t) \frac{\partial Z_0}{\partial x_j}\bigg|_{x=x_0} n_i(x_0)$$

$$= -\frac{1}{2(t-\tau)} \sum_{i, j, k=1}^n a_{ij}(x_0, t) A^{(j, k)}(\xi, \tau)$$

$$\times (x_{0_k} - \xi_k) n_i(x_0) Z_0(x_0 - \xi, \xi, t, \tau).$$

Since

$$\left|\sum_{i, j, k=1}^n a_{ij}(\xi, \tau) A^{(j, k)}(\xi, \tau)(x_{0_k} - \xi_k) n_i(x_0)\right|$$

$$= \left|\sum_{i=1}^n (x_{0_i} - \xi_i) n_i(x_0)\right| \leqslant c\,|x_0 - \xi|^{1+\beta}$$

and

$$|a_{ij}(x_0, t) - a_{ij}(\xi, \tau)| \leqslant c(|x_0 - \xi|^\alpha + (t-\tau)^{\frac{\alpha}{2}}),$$

it follows that

$$|H_0| \leqslant c(|x_0 - \xi|^\beta + |x_0 - \xi|^\alpha + (t-\tau)^{\frac{\alpha}{2}})(t-\tau)^{-\frac{n+1}{2}}$$

$$\times \exp\left(-C\frac{|x_0 - \xi|^2}{t-\tau}\right) \leqslant c(t-\tau)^{-\frac{n+1-\delta}{2}} \exp\left(-C\frac{|x_0 - \xi|^2}{t-\tau}\right), \quad (15.2)$$

where $\delta = \min(\alpha, \beta)$. Consequently, the kernel H_0 has a weak (integrable) singularity.

We assume that the point x approaches x_0 along the normal to x_0. Consider the difference

$$\sum_{i, j=1}^n a_{ij}(x, t) \frac{\partial V_0(x, t)}{\partial x_j} n_i(x_0) - W_0(x_0, t).$$

One easily verifies the identity

$$\sum_{i, j=1}^n a_{ij}(x, t) \frac{\partial V_0(x, t)}{\partial x_j} n_i(x_0) - W_0(x_0, t)$$

$$= \left[\int_0^t d\tau \int_S \sum_{i, j=1}^n a_{ij}(x, t) \frac{\partial Z_0(x-\xi, \xi, t, \tau)}{\partial x_j} n_i(x_0) \times\right.$$

$$\times (\varphi(\xi, \tau) - \varphi(x_0, \tau))\, dS_\xi - \int_0^t d\tau \int_S H_0(x_0, \xi, t, \tau)$$

$$\times (\varphi(\xi, \tau) - \varphi(x_0, \tau))\, dS_\xi \Bigg] + \Bigg[\int_0^t (\varphi(x_0, \tau) - \varphi(x_0, t))\, d\tau$$

$$\times \int_S \sum_{i, j=1}^n a_{ij}(x, t) \frac{\partial Z_0(x-\xi, \xi, t, \tau)}{\partial x_j} n_i(x_0)\, dS_\xi$$

$$- \int_0^t (\varphi(x_0, \tau) - \varphi(x_0, t))\, d\tau \int_S H_0(x_0, \xi, t, \tau)\, dS_\xi \Bigg]$$

$$+ \varphi(x_0, t) \Bigg[\int_0^t d\tau \int_S \sum_{i, j=1}^n a_{ij}(x, t)$$

$$\times \left(\frac{\partial Z_0(x-\xi, \xi, t, \tau)}{\partial x_j} - \frac{\partial Z_0(x-\xi, z, t, \tau)}{\partial x_j} \bigg|_{z=x} \right) n_i(x_0)\, dS_\xi$$

$$- \int_0^t d\tau \int_S \sum_{i, j=1}^n a_{ij}(x_0, t) \left(\frac{\partial Z_0(x-\xi, \xi, t, \tau)}{\partial x_j} \bigg|_{x=x_0} \right.$$

$$\left. - \frac{\partial Z_0(x-\xi, z, t, \tau)}{\partial x_j} \bigg|_{\substack{z=x_0 \\ x=x_0}} \right) n_i(x_0)\, dS_\xi \Bigg]$$

$$+ \varphi(x_0, t) \Bigg[\int_0^t d\tau \int_S \sum_{i, j=1}^n a_{ij}(x, t) \frac{\partial Z_0(x-\xi, z, t, \tau)}{\partial x_j} \bigg|_{z=x}$$

$$\times (n_i(x_0) - n_i(\xi))\, dS_\xi - \int_0^t d\tau \int_S \sum_{i, j=1}^n a_{ij}(x_0, t)$$

$$\times \frac{\partial Z_0(x-\xi, z, t, \tau)}{\partial x_j} \bigg|_{\substack{z=x_0 \\ x=x_0}} (n_i(x_0) - n_i(\xi))\, dS_\xi \Bigg]$$

$$+ \varphi(x_0, t) \Bigg\{ \int_0^t d\tau \int_S \sum_{i, j=1}^n a_{ij}(x, t) \frac{\partial Z_0(x-\xi, z, t, \tau)}{\partial x_j} \bigg|_{z=x}$$

$$\times n_i(\xi)\, dS_\xi - \int_0^t d\tau \int_S \sum_{i, j=1}^n a_{ij}(x_0, t)$$

$$\times \frac{\partial Z_0(x-\xi, z, t, \tau)}{\partial x_j} \bigg|_{\substack{z=x_0 \\ x=x_0}} n_i(\xi)\, dS_\xi \Bigg\}. \tag{15.3}$$

If the function ϕ is continuous, then all of the expressions in square brackets in the right side tend to zero as $x \to x_0$. Consider the first expression. We fix an arbitrary small positive ϵ. Let δ be a number such that for $|\xi - x_0| \le \delta$ we have $\varphi(\xi, \tau) - \varphi(x_0, \tau) \le \epsilon$. Let Σ be a part of the surface S lying inside the ball with center x_0 and radius δ, and let σ be a part of S lying inside the ball with center x_0 and radius $2|x - x_0|$. We assume that $\delta/2 > |x - x_0|$. We have

$$\left| \int_0^t d\tau \int_\sigma \sum_{i,j=1}^n a_{ij}(x, t) \frac{\partial Z_0}{\partial x_j} n_i [\varphi(\xi, \tau) - \varphi(x_0, \tau)]\, dS_\xi \right|$$
$$\leqslant c\varepsilon \int_\sigma dS_\xi \int_0^t (t-\tau)^{-\frac{n+1}{2}} \exp\left(-C \frac{|x-\xi|^2}{t-\tau}\right) d\tau$$
$$\leqslant c\varepsilon \int_\sigma \frac{dS_\xi}{|x-\xi|^{n-1}} \leqslant \frac{c\varepsilon}{|x-x_0|^{n-1}} \int_\sigma dS_\xi \leqslant c\varepsilon,$$

since x lies on the normal to x_0 and, consequently, x_0 is the nearest point to x of the surface S.

By virtue of (15.2)

$$\left| \int_0^t d\tau \int_\sigma H_0(x_0, \xi, t, \tau) [\varphi(\xi, \tau) - \varphi(x_0, \tau)]\, dS_\xi \right|$$
$$\leqslant c\varepsilon \int_\sigma dS_\xi \int_0^t (t-\tau)^{-\frac{n+1-\delta}{2}} \exp\left(-C \frac{|x-\xi|^2}{t-\tau}\right) d\tau \leqslant c\varepsilon.$$

In order to estimate the remaining integrals we make use of the inequality

$$\left| \sum_{i,j=1}^n a_{ij}(x, t) \frac{\partial Z_0(x-\xi, \xi, t, \tau)}{\partial x_j} n_i(x_0) - H_0(x_0, \xi, t, \tau) \right|$$
$$\leqslant c \left\{ |x - x_0| (t-\tau)^{-\frac{n+2}{2}} + |x-x_0|^\alpha (t-\tau)^{-\frac{n+1}{2}} \right\}$$
$$\times \left\{ \exp\left(-C \frac{|x-\xi|^2}{t-\tau}\right) + \exp\left(-C \frac{|x_0-\xi|^2}{t-\tau}\right) \right\}, \tag{15.4}$$

which follows from (11.3) and from

$$|a_{ij}(x, t) - a_{ij}(x_0, t)| \leqslant c |x - x_0|^\alpha. \tag{15.5}$$

Since for $\xi \in S \setminus \sigma$ we have

$$\frac{1}{2} |x_0 - \xi| \leqslant |x - \xi| \leqslant \frac{3}{2} |x_0 - \xi|.$$

it follows that

$$\left| \int_0^t d\tau \int_{\Sigma \setminus \sigma} \left[\sum_{i,j=1}^n a_{ij}(x, t) \frac{\partial Z_0}{\partial x_j} n_i(x_0) - H_0 \right] \right.$$

$$\left. \times [\varphi(\xi, \tau) - \varphi(x_0, \tau)] \, dS_\xi \right|$$

$$\leqslant c\varepsilon \int_0^t \left\{ |x - x_0| (t - \tau)^{-\frac{n+2}{2}} \right.$$

$$\left. + |x - x_0|^\alpha (t - \tau)^{-\frac{n+1}{2}} \right\} d\tau \int_{\Sigma \setminus \sigma} \exp\left(-C \frac{|x_0 - \xi|^2}{t - \tau}\right) dS_\xi$$

$$\leqslant c\varepsilon \left[|x - x_0| \int_{\Sigma \setminus \sigma} \frac{dS_\xi}{|x_0 - \xi|^n} + |x - x_0|^\alpha \int_{\Sigma \setminus \sigma} \frac{dS_\xi}{|x_0 - \xi|^{n-1}} \right]$$

$$\leqslant c\varepsilon [1 + |x - x_0|^\alpha |\log |x - x_0||] \leqslant c\varepsilon$$

and

$$\left| \int_0^t d\tau \int_{S \setminus \Sigma} \left[\sum_{i,j=1}^n a_{ij}(x, t) \frac{\partial Z_0}{\partial x_j} n_i(x_0) - H_0 \right] \right.$$

$$\left. \times [\varphi(\xi, \tau) - \varphi(x_0, \tau)] \, dS_\xi \right| \leqslant c(\delta) |x - x_0|^\alpha.$$

Combining all of these estimates, we see that for small $|x - x_0|$ the first term of (15.3) does not exceed $c\epsilon$, and consequently it vanishes for $|x - x_0| \to 0$.

Let us consider the second term. Let η be a number such that for $|t - \tau| \leq \eta$ we have

$$|\varphi(x, t) - \varphi(x, \tau)| \leqslant \varepsilon.$$

Repeating the calculations just carried out, we obtain

$$\left| \int_{t-\eta}^t [\varphi(x_0, \tau) - \varphi(x_0, t)] \, d\tau \int_\sigma \left[\sum_{i,j=1}^n a_{ij} \frac{\partial Z_0}{\partial x_j} n_i - H_0 \right] dS_\xi \right|$$

$$\leqslant \varepsilon \int_{t-\eta}^t d\tau \left[\int_\sigma \sum_{i,j=1}^n \left| a_{ij} \frac{\partial Z_0}{\partial x_j} n_i \right| dS_\xi + \int_\sigma |H_0| \, dS_\xi \right] \leqslant$$

$$\leqslant c\varepsilon \int_{\sigma} dS_\xi \left[\int_{t-\eta}^{t} (t-\tau)^{-\frac{n+1}{2}} \exp\left(-C\frac{|x_0-\xi|^2}{t-\tau}\right) d\tau \right.$$

$$\left. + \int_{t-\eta}^{t} (t-\tau)^{-\frac{n+1-\delta}{2}} \exp\left(-C\frac{|x_0-\xi|^2}{t-\tau}\right) d\tau \right]$$

$$\leqslant c\varepsilon \left[\int_{\sigma} \frac{dS_\xi}{|x-\xi|^{n-1}} + \int_{\sigma} \frac{dS_\xi}{|x_0-\xi|^{n-1-\delta}} \right] \leqslant c\varepsilon,$$

and, further,

$$\left| \int_{t-\eta}^{t} [\varphi(x_0, \tau) - \varphi(x_0, t)]\, d\tau \int_{S\setminus\sigma} \left[\sum_{i,j=1}^{n} a_{ij} \frac{\partial Z_0}{\partial x_j} n_i - H_0 \right] dS_\xi \right|$$

$$\leqslant c\varepsilon \int_{S\setminus\sigma} dS_\xi \int_{t-\eta}^{t} \left[|x-x_0| (t-\tau)^{-\frac{n+2}{2}} + |x-x_0|^{\alpha} (t-\tau)^{-\frac{n+1}{2}} \right]$$

$$\times \left[\exp\left(-C\frac{|x-\xi|^2}{t-\tau}\right) + \exp\left(-C\frac{|x_0-\xi|^2}{t-\tau}\right) \right] d\tau$$

$$\leqslant c\varepsilon\, [1 + |x-x_0|^{\alpha}\, |\log|x-x_0||] \leqslant c\varepsilon$$

and, finally,

$$\left| \int_{0}^{t-\eta} [\varphi(x_0, \tau) - \varphi(x_0, t)]\, d\tau \int_{S} \left[\sum_{i,j=1}^{n} a_{ij} \frac{\partial Z_0}{\partial x_j} - H_0 \right] dS_\xi \right|$$

$$\leqslant c(\eta)\, |x-x_0|^{\alpha}.$$

Combining these estimates, we see that the second term of the right side also tends to zero for $x \to x_0$.

The third and fourth terms also vanish for $x \to x_0$, as is easily established with the use of estimates (11.3), (15.5) and

$$|n_i(x_0) - n_i(\xi)| \leqslant c\, |x_0 - \xi|^{\beta}. \tag{15.6}$$

In these terms one can pass to the limit under the integral sign for $x \to x_0$, since the kernels of the potentials in these terms have a weak singularity.

Let us consider the expression in braces in (15.3). We have

$$\int_0^t d\tau \int_S \sum_{i,j=1}^n a_{ij}(x,t) \frac{\partial Z_0(x-\xi, z, t, \tau)}{\partial x_j}\Big|_{z=x} n_i(\xi)\, dS_\xi$$

$$= -\lim_{\varepsilon\to 0} \int_0^{t-\varepsilon} d\tau \int_S \sum_{i,j=1}^n a_{ij}(x,t) \frac{\partial Z_0(x-\xi, x, t, \tau)}{\partial \xi_j} n_i(\xi)\, dS_\xi$$

$$= -\lim_{\varepsilon\to 0} \int_0^{t-\varepsilon} d\tau \int_\Omega \sum_{i,j=1}^n a_{ij}(x,t) \frac{\partial^2 Z_0(x-\xi, x, t, \tau)}{\partial \xi_i \partial \xi_j}\, d\xi$$

$$= -\lim_{\varepsilon\to 0} \int_0^{t-\varepsilon} d\tau \int_\Omega \frac{\partial Z_0(x-\xi, x, t, \tau)}{\partial t}\, d\xi$$

$$+ \sum_{i,j=1}^n \int_0^t d\tau \int_\Omega [a_{ij}(x,\tau) - a_{ij}(x,t)] \frac{\partial^2 Z_0(x-\xi, x, t, \tau)}{\partial \xi_i \partial \xi_j}\, d\xi.$$

The function Z_0 depends on τ through the difference $t-\tau$ and through the functions $A^{(i,j)}(\xi, \tau)$ and $\det A(\xi, \tau)$. Let

$$\tilde{Z}(x-\xi, x, t-\tau, \tau) = Z_0(x-\xi, x, t, \tau).$$

The last argument τ of $\tilde{Z}$ coincides with the argument of $A^{(i,j)}$ and $\det A$.

It is obvious that

$$\frac{\partial Z_0}{\partial t} = -\frac{\partial \tilde{Z}(x-\xi, x, t-\tau, \lambda)}{\partial \tau}\Big|_{\lambda=\tau}$$
$$= -\frac{\partial \tilde{Z}(x-\xi, x, t-\tau, t)}{\partial \tau} + \mathfrak{Z}(x, \xi, t, \tau),$$

where

$$\mathfrak{Z} = \left(\frac{\partial \tilde{Z}(x-\xi, x, t-\tau, t)}{\partial \tau} - \frac{\partial \tilde{Z}(x-\xi, x, t-\tau, \lambda)}{\partial \tau}\Big|_{\lambda=\tau}\right).$$

One easily sees that

$$|\mathfrak{Z}(x, \xi, t, \tau)| \leqslant c(t-\tau)^{-\frac{n+2-\alpha}{2}} \exp\left(-C\frac{|x-\xi|^2}{t-\tau}\right)$$

and consequently

$$
-\lim_{\varepsilon\to 0}\int_0^{t-\varepsilon} d\tau \int_\Omega \frac{\partial Z_0(x-\xi, x, t, \tau)}{\partial t}\, d\xi
$$

$$
= \lim_{\varepsilon\to 0}\int_\Omega \tilde{Z}(x-\xi, x, \varepsilon, t)\, d\xi - \int_\Omega \tilde{Z}(x-\xi, x, t, t)\, d\xi
$$

$$
- \int_0^t d\tau \int_\Omega \mathcal{Z}(x, \xi, t, \tau)\, d\xi.
$$

By transforming in a similar way the second term of the expression in braces in the right side of (15.3), we see that all of this expression is equal to

$$
\lim_{\varepsilon\to 0}\int_\Omega \tilde{Z}(x-\xi, x, \varepsilon, t)\, d\xi - \lim_{\varepsilon\to 0}\int_\Omega \tilde{Z}(x_0-\xi, x_0, \varepsilon, t)\, d\xi
$$

$$
-\left[\int_\Omega \tilde{Z}(x-\xi, x, t, t)\, d\xi - \int_\Omega \tilde{Z}(x_0-\xi, x_0, t, t)\, d\xi\right]
$$

$$
-\left[\int_0^t d\tau \int_\Omega \mathcal{Z}(x, \xi, t, \tau)\, d\xi - \int_0^t d\tau \int_\Omega \mathcal{Z}(x_0, \xi, t, \tau)\, d\xi\right]
$$

$$
+\left[\sum_{i, j=1}^n \int_0^t d\tau \int_\Omega (a_{ij}(x, \tau) - a_{ij}(x, t)) \frac{\partial^2 Z_0(x-\xi, x, t, \tau)}{\partial \xi_i\, \partial \xi_j}\, d\xi\right.
$$

$$
- \sum_{i, j=1}^n \int_0^t d\tau \int_\Omega (a_{ij}(x_0, \tau) - a_{ij}(x_0, t))
$$

$$
\left.\times \frac{\partial^2 Z_0(x_0-\xi, x_0, t, \tau)}{\partial \xi_i\, \partial \xi_j}\, d\xi\right].
$$

It is obvious that the expressions in square brackets here also tend to zero for $x \to x_0$. Therefore

$$
\lim_{x\to x_0}\left[\sum_{i, j=1}^n a_{ij}(x, t)\frac{\partial V_0}{\partial x_j} n_i(x_0) - W_0(x_0, t)\right]
$$

$$
= \varphi(x_0, t)\left\{\lim_{\varepsilon\to 0}\int_\Omega \tilde{Z}(x-\xi, x, \varepsilon, t)\, d\xi\right.
$$

$$
\left. - \lim_{\varepsilon\to 0}\int_\Omega \tilde{Z}(x_0-\xi, x_0, \varepsilon, t)\, d\xi\right\}.
$$

One easily calculates, introducing the new variables of integration $z = (x-\xi)/\sqrt{\epsilon}$, that

$$\lim_{\varepsilon\to 0}\int_\Omega \tilde{Z}(x-\xi,\ x,\ \varepsilon,\ t)\,d\xi = \begin{cases} 1, & \text{if} \quad x\in\Omega, \\ 0, & \text{if} \quad x\in E_n\setminus\bar{\Omega}, \end{cases}$$

and

$$\lim_{\varepsilon\to 0}\int_\Omega \tilde{Z}(x_0-\xi,\ x_0,\ \varepsilon,\ t)\,d\xi = \int_{\mathfrak{D}} \tilde{Z}(z,\ x_0,\ 1,\ t)\,dz$$

$$= \frac{1}{(4\pi)^{\frac{n}{2}}\,(\det A_0(x_0,\ t))^{\frac{1}{2}}}$$

$$\times\int_{\mathfrak{D}} \exp\left(-\frac{1}{4}\sum_{i,\,j=1}^{n} A^{(i,\,j)}(x_0,\ t)\,z_iz_j\right)dz,$$

where $\mathfrak{D}$ is a certain half space whose boundary plane passes through the origin. It is easily seen that the latter integral is equal to one half the integral of the same integrand taken over the whole space E_n, since this function is even with respect to all of the variables z. Thus

$$\lim_{\varepsilon\to 0}\int_\Omega \tilde{Z}(x_0-\xi,\ x_0,\ \varepsilon,\ t)\,d\xi = \frac{1}{2}\int_{E_n} \tilde{Z}(z,\ x_0,\ 1,\ t)\,dz = \frac{1}{2}$$

and

$$\lim_{x\to x_0}\sum_{i,\,j=1}^{n} a_{ij}\frac{\partial V_0}{\partial x_j}n_i(x_0) = W_0(x_0,\ t) \pm \frac{1}{2}\varphi(x_0,\ t), \tag{15.7}$$

where the plus sign in the second term corresponds to the case when the point x tends to x_0 from within Ω, while the minus sign corresponds to the case when $x \to x_0$ from without.

In proving relation (15.7) we assumed that the point x tends to x_0 while remaining on the normal to S at the point x_0, but, just as in classical potential theory, this relation is also valid under more general assumptions. Suppose x is in a neighborhood of x_0 and $\bar{x}$ is the nearest point to x of the surface S, so that x lies on the normal to S at the point $\bar{x}$. For definiteness let $x\in\Omega$. We have

$$\left| \sum_{i,j=1}^{n} a_{ij}(x,t) \frac{\partial V_0(x,t)}{\partial x_j} n_i(x_0) - W_0(x_0,t) - \frac{1}{2}\varphi(x_0,t) \right|$$
$$\leqslant \left| \sum_{i,j=1}^{n} a_{ij}(x,t) \frac{\partial V_0(x,t)}{\partial x_j} n_i(\bar{x}) - W_0(\bar{x},t) - \frac{1}{2}\varphi(\bar{x},t) \right|$$
$$+ | W_0(\bar{x},t) - W_0(x_0,t) | + \frac{1}{2} | \varphi(\bar{x},t) - \varphi(x_0,t) |$$
$$+ \left| \sum_{i,j=1}^{n} a_{ij}(x,t) \frac{\partial V_0(x,t)}{\partial x_j} [n_i(x_0) - n_i(\bar{x})] \right|.$$

It is easy to see that the first three terms of the right side tend to zero as $x \to x_0$. In order to estimate the fourth term we make use of inequalities (15.6) and the fact that

$$\left| \frac{\partial V_0(x,t)}{\partial x_j} \right|$$
$$\leqslant c \max|\varphi| \int_S dS_\xi \int_0^t \exp\left(-C \frac{|x-\xi|^2}{t-\tau}\right)(t-\tau)^{-\frac{n+1}{2}} d\tau$$
$$\leqslant c_1 \max|\varphi| \int_S \frac{dS_\xi}{|x-\xi|^{n-1}} \leqslant c_2 |\log|x-\bar{x}||.$$

Thus the fourth term also tends to zero if x tends to x_0 in such a way that

$$|x_0 - \bar{x}|^\beta \log|x - \bar{x}| \to 0. \tag{15.8}$$

This is always the case if x remains within a certain cone with apex at x_0 and axis directed along $\mathbf{n}(x_0)$.

From (15.1) and (15.7) it follows that under condition (15.8) we have

$$\lim_{x \to x_0} \sum_{i,j=1}^{n} a_{ij}(x,t) \frac{\partial V(x,t)}{\partial x_j} n_i(x_0) = W(x_0,t) \pm \frac{1}{2}\varphi(x_0,t), \tag{15.9}$$

where

$$W(x_0,t) = \int_0^t d\tau \int_S H(x_0,\xi,t,\tau)\varphi(\xi,\tau)\, dS_\xi,$$

$$H(x_0,\xi,t,\tau) = \sum_{i,j=1}^{n} a_{ij}(x_0,t) \frac{\partial Z(x,\xi,t,\tau)}{\partial x_j}\Big|_{x=x_0} n_i(x_0). \tag{15.10}$$

This permits us to use a single-layer potential to solve the second boundary value problem for the equation $\mathfrak{L}u = 0$, both in the interior domain Q_T and in the exterior domain $D_{n+1}^{(T)}\setminus\overline{Q}_T$. Consider, for example, the interior problem

$$\mathscr{L}\left(x,\ t,\ \frac{\partial}{\partial x},\ \frac{\partial}{\partial t}\right)u = 0 \qquad (x \in \Omega),$$

$$\sum_{i,\ j=1}^{n} a_{ij}\frac{\partial u}{\partial x_j}n_i(x)\bigg|_{S_T} = \Phi(x,\ t), \quad u|_{t=0} = 0. \tag{15.11}$$

Its solution is sought in the form of a single-layer potential

$$u(x,\ t) = \int_0^t d\tau \int_S Z(x,\ \xi,\ t,\ \tau)\,\varphi(\xi,\ \tau)\,d\xi. \tag{15.12}$$

The potential is continuous in the closure of the domain and satisfies inside Q the homogeneous equation of (15.11) and also the zero initial condition. Requiring that its conormal derivative be equal to Φ, we arrive by virtue of (15.9) at an integral equation for the density φ:

$$\varphi(\xi,\ t) = -2\int_0^t d\tau \int_S H(\xi,\ \eta,\ t,\ \tau)\,\varphi(\eta,\ \tau)\,dS_\eta + 2\Phi(\xi,\ t).$$

Since the kernel H has a weak singularity, this equation is solved by the method of successive approximations. Its solution will be a continuous function if Φ is continuous. It is also easy to show that if Φ satisfies a Hölder condition in its arguments, then φ will also satisfy a Hölder condition.

The solution of the exterior problem

$$\mathscr{L}u = 0 \qquad (x \in E_n \setminus \overline{\Omega}),$$

$$\sum_{i,\ j=1}^{n} a_{ij}\frac{\partial u}{\partial x_j}n_i(x)\bigg|_{S_T} = \Phi, \quad u|_{t=0} = 0 \tag{15.13}$$

can also be sought in the form (15.12), and the equation for the density φ will be

$$\varphi(\xi,\ t) = 2\int_0^t d\tau \int_S H(\xi,\ \eta,\ t,\ \tau)\,\varphi(\eta,\ \tau)\,dS_\eta - 2\Phi(\xi,\ t).$$

Thus we have proved

Theorem 15.1. *Let the coefficients of the operator* $\mathfrak{L}$ *be defined in* $D_{n+1}^{(T)}$ *and belong to the class* $H^{\alpha,\,\alpha/2}(\overline{D_{n+1}^{(T)}})$. *Let* Ω *be a bounded domain with boundary*

$S \in H^{1+\beta}$. *For any continuous function* Φ *on* S_T *problems* (15.11) *and* (15.13) *have classical solutions that* (i) *are continuous in the closed domains* $\overline{Q}_T$ *and* $D_{n+1}^{(T)} \setminus Q_T$ *respectively,* (ii) *have continuous derivatives, namely, those entering into the equations, inside these domains and* (iii) *satisfy the boundary conditions when the boundary is approached along any curve on which relation* (15.8) *is fulfilled. The solution of the exterior problem is bounded at infinity.*

§16. SOLUTION OF THE FIRST BOUNDARY VALUE PROBLEM

In this section we obtain a solution of the first boundary value problem

$$\left.\begin{aligned} \mathscr{L}\left(x, t, \frac{\partial}{\partial x}, \frac{\partial}{\partial t}\right)u = f, \\ u\,|_S = \Phi(x, t), \quad u\,|_{t=0} = \varphi(x). \end{aligned}\right\} \tag{16.1}$$

For $f = 0$, $\varphi = 0$ the solution of this problem can be sought in the form of a double-layer potential

$$u(x, t) = \int_0^t d\tau \int_S \sum_{i,j=1}^n a_{ij}(\xi, \tau) \frac{\partial Z(x, \xi, t, \tau)}{\partial \xi_j} n_i(\xi) \mu(\xi, \tau)\, dS_\xi. \tag{16.2}$$

However, it is necessary in this connection to impose additional restrictions on the leading coefficients a_{ij} of the operator $\mathfrak{L}$; these restrictions are connected with the need to differentiate the function Z with respect to the variables ξ_j. If one assumes that the functions a_{ij} satisfy a Hölder condition in t with exponent $\alpha/2$ and have derivatives $\partial a_{ij}/\partial x_k$ satisfying a Hölder condition with exponent α in the variables x, then one can show that the function Z has derivatives $\partial Z/\partial \xi_j$, $\partial^2 Z/\partial x_k \partial \xi_j$, $\partial^3 Z/\partial x_i \partial x_k \partial \xi_j$, $\partial^2 Z/\partial t \partial \xi_j$ that are continuous functions for $x \neq \xi$ and are subject to the inequalities

$$\left| D_t^r D_x^s \frac{\partial Z(x, \xi, t, \tau)}{\partial \xi_j} \right| \leqslant c(t-\tau)^{-\frac{n+2r+s+1}{2}} \exp\left(-C\frac{|x-\xi|^2}{t-\tau}\right) \tag{16.3}$$

for $2r + s \leq 2$, so that the double-layer potential has continuous derivatives $D_t^r D_x^s u(x, t)$ $(2r + s \leq 2)$ inside Q_T and $D_{n+1}^{(T)} \setminus \overline{Q}_T$. For the function Z' one has the estimates

$$\left| D_x^s \frac{\partial Z'}{\partial \xi_j} \right| \leqslant c(t-\tau)^{-\frac{n+s}{2}} \exp\left(-C\frac{|x-\xi|^2}{t-\tau}\right) \quad (s \leqslant 2). \tag{16.4}$$

The proof of estimates (16.3), (16.4) and the continuity of all of the above-mentioned derivatives of the function Z will not be carried out; it can be found in [25a] or [31b].

The limiting value of the potential (16.2) for $x \to x_0 \in S$ is calculated in the same way as the limiting value of the conormal derivative of the single-layer potential in §15. It is expressed by the formula

$$\lim_{x \to x_0 \in S} u(x, t) = \int_0^t d\tau \int_S M(x_0, \xi, t, \tau) \mu(\xi, \tau) dS_\xi \mp \frac{1}{2} \mu(x_0, t), \tag{16.5}$$

where

$$M(x_0, \xi, t, \tau) = \sum_{i, j=1}^{n} a_{ij}(\xi, \tau) \frac{\partial Z(x_0, \xi, t, \tau)}{\partial \xi_j} n_i(\xi)$$

is the kernel, which satisfies for $x_0 \in S$ the inequality

$$|M_0(x_0, \xi, t, \tau)| \leqslant c(t-\tau)^{-\frac{n+1-\delta}{2}} \exp\left(-C\frac{|x_0-\xi|^2}{t-\tau}\right), \quad \delta > 0, \tag{16.6}$$

and hence has an integrable singularity at $x_0 = \xi$, $t = \tau$. The minus sign in formula (16.5) is taken when $x \to x_0$ from within Ω, while the plus sign is taken when $x \to x_0$ from without. Relation (16.5) is first established in the case when x tends to x_0 along the normal to S at the point x_0. But after that, in exactly the same way as in classical potential theory (see, for example, [36], Russian p. 71, and also the end of §15), one can show that (16.5) holds no matter how x tends to x_0, even without condition (15.8). From this it follows, in particular, that for a continuous μ the function $u(x, t)$ is uniformly continuous both in $\overline{Q}_T$ and in $D_{n+1}^{(T)} \setminus Q_T$.

Relation (16.5) permits us to reduce problem (16.1) with $f = 0$ to an integral equation both in the interior (with respect to S) domain Q_T and in the exterior domain $D_{n+1}^{(T)} \setminus \overline{Q}_T$. The solution of the interior problem is expressed in the form of the potential (16.2), the density $\mu(\xi, t)$ of which satisfies the integral equation

$$\mu(\xi, t) = 2\int_0^t d\tau \int_S M(\xi, \eta, t, \tau) \mu(\eta, \tau) dS_\eta - 2\Phi(\xi, t),$$

while in the case of the exterior problem the integral equation has the form

$$\mu(\xi, t) = -2 \int_0^t d\tau \int_S M(\xi, \eta, t, \tau)\mu(\eta, \tau)\, dS_\eta + 2\Phi(\xi, t).$$

Both equations are Volterra equations with a weakly polar kernel and are solved by the method of successive approximations. Their solutions are as smooth as the function $\Phi(\xi, t)$.

Thus we have

Theorem 16.1. *If the coefficients of the operator $\mathfrak{L}$ belong to the class $H^{\alpha, \alpha/2}(D_{n+1}^{(T)})$, and in addition the leading coefficients a_{ij} have derivatives $\partial a_{ij}/\partial x_k$ satisfying a Hölder condition in the variables x with exponent α, while the surface S satisfies Ljapunov's conditions, then problem* (16.1) *for $f = 0, \varphi = 0$ and any continuous Φ has a classical solution both in the interior domain Q_T and in the exterior domain $D_{n+1}^{(T)} \setminus \overline{Q}_T$. This solution has continuous derivatives, namely those entering into the equation, at interior points of the corresponding domain, and it itself is continuous up to the boundary.*

We now proceed to solve the first boundary value problem (16.1) with the use of a Green's function.

The Green's function for problem (16.1) in an interior domain is a function $G(x, y, t, \tau)$ possessing the following properties for all $(y, \tau) \in Q_T$:

$$\mathscr{L}\left(x, t, \frac{\partial}{\partial x}, \frac{\partial}{\partial t}\right) G(x, y, t, \tau) = \delta(x - y)\delta(t - \tau),$$
$$G(x, y, \tau, \tau) = 0, \quad G(x, y, t, \tau)|_{x \in S} = 0.$$

It is connected with the fundamental solution Z by the relation

$$G(x, y, t, \tau) = Z(x, y, t, \tau) - g(x, y, t, \tau), \tag{16.7}$$

where the function $g(x, y, t, \tau)$ is the solution of the problem

$$\mathscr{L}\left(x, t, \frac{\partial}{\partial x}, \frac{\partial}{\partial t}\right) g(x, y, t, \tau) = 0,$$
$$g(x, y, \tau, \tau) = 0,$$
$$g(x, y, t, \tau)|_{x \in S} = Z(x, y, t, \tau)|_{x \in S}. \tag{16.8}$$

If one requires that the coefficients of the operator $\mathfrak{L}$ satisfy the conditions of Theorem 16.1, then one can express the function g in the form of a double-layer potential. W. Pogorzelski [94d] proposed the ingenious method set forth below, which permits one to obtain the solution of problem (16.8) in the form of a single-layer potential for $a_{ij} \in H^{\alpha, \alpha/2}(\overline{D_{n+1}^{(T)}})$.

Let $(y, \tau) \in Q_T$, and for $x \in E_n \setminus \overline{\Omega}$ and $\tau \leq t \leq T$ let

$$g(x, y, t, \tau) = Z(x, y, t, \tau). \tag{16.9}$$

This function is obviously the solution of the following second boundary value problem in the exterior domain $D_{n+1}^{(T)} \setminus \overline{Q}_T$:

$$\mathscr{L}g = 0,$$
$$g(x, y, \tau, \tau) = 0,$$
$$\sum_{i,j=1}^{n} a_{ij} \frac{\partial g}{\partial x_j} n_i(x) \Big|_{x \in S} = H(x, y, t, \tau),$$

where $H(x, y, t, \tau)$ is the function (15.10).

From the results of §15 it follows that the function g is representable in the form of a single-layer potential

$$g(x, y, t, \tau) = \int_\tau^t d\lambda \int_S Z(x, \xi, t, \lambda)\, \omega(\xi, y, \lambda, \tau)\, dS_\xi, \tag{16.10}$$

where the density ω is determined from the integral equation

$$\omega(\xi, y, t, \tau) = 2 \int_\tau^t d\lambda \int_S H(\xi, \eta, t, \lambda)\, \omega(\eta, y, \lambda, \tau)\, dS_\xi - 2H(\xi, y, t, \tau), \tag{16.11}$$

which is solved by the method of successive approximations.

Let us now define the function g in Q_T by formula (16.10) and show that it satisfies the relations of (16.8). Of these relations only the boundary condition requires verification. It follows from the continuity of the single-layer potential in the domain $D_{n+1}^{(T)}$. Indeed, if the point x tends to a point $x_0 \in S$ while remaining in the domain $E_n \setminus \overline{\Omega}$, then by virtue of (16.9)

$$\lim_{x \to x_0} g(x, y, t, \tau) = Z(x, y, t, \tau). \tag{16.12}$$

But since the limiting value of the single-layer potential does not depend on from which side of the surface S the point x approaches x_0, it follows that relation (16.12) implies the necessary boundary condition.

Thus the function g is expressed by formula (16.10).

It is seen from this formula that $g(x, y, t, \tau)$ does not have any singularities when the points x and y are not on the surface S.

Let us estimate g with the use of relations (16.10) and (16.11). We first note that for $y \in \Omega$ the function $H(\xi, y, t, \tau)$ possesses the following properties:

$$|H(\xi, y, t, \tau)| \leqslant c(t-\tau)^{-\frac{n+1}{2}} \exp\left(-C\frac{|\xi-y|^2}{t-\tau}\right),$$

$$\int_\tau^t d\lambda \int_S |H(\xi, y, \lambda, \tau)|\, dS_\xi \leqslant c. \tag{16.13}$$

In order to prove the latter estimate we represent H in the form

$$H = H_0 + H',$$

where

$$H' = \sum_{i,j=1}^n a_{ij}(\xi, t)\, \frac{\partial Z'(\xi, y, t, \tau)}{\partial \xi_j}\, n_i(\xi)$$

is a function for which the estimate

$$|H'(\xi, y, t, \tau)| \leqslant c(t-\tau)^{-\frac{n+1-\alpha}{2}} \exp\left(-C\frac{|\xi-y|^2}{t-\tau}\right)$$

holds by virtue of (13.5), and hence

$$\int_\tau^t d\lambda \int_S |H'(\xi, y, \lambda, \tau)|\, dS_\xi \leqslant c(t-\tau)^{\frac{\alpha}{2}} \leqslant c_1.$$

For the function H_0 we have the formula

$$\begin{aligned} H_0(\xi, y, t, \tau) &= \sum_{i,j=1}^n a_{ij}(\xi, t)\, \frac{\partial Z_0(\xi-y, y, t, \tau)}{\partial \xi_j}\, n_i(\xi) \\ &= \sum_{i,j=1}^n [a_{ij}(\xi, t) - a_{ij}(y, \tau)]\, \frac{\partial Z_0(\xi-y, y, t, \tau)}{\partial \xi_j}\, n_i(\xi) \\ &\quad - \frac{1}{2(t-\tau)} \sum_{i=1}^n (\xi_i - y_i)\, n_i(\xi)\, Z_0(\xi-y, y, t, \tau), \end{aligned}$$

so that

$$\int_\tau^t d\lambda \int_S |H_0(\xi, y, \lambda, \tau)|\, dS_\xi$$
$$\leqslant c\left((t-\tau)^{\frac{\alpha}{2}} + \int_S \left|\sum_{i=1}^n (\xi_i - y_i)\, n_i(\xi)\right| \frac{dS_\xi}{|\xi - y|^n}\right) \leqslant c.$$

The boundedness of the latter integral is proved in classical potential theory (see, for example, [36], Russian pages 275–276, for $n = 3$).

Thus $H(\xi, y, t, \tau)$ is subject to inequalities (16.13). By using the representation of the function ω in the form of the series obtained in the solution of equation (16.11) by the method of successive approximations, it is not difficult to show that the same estimates hold for ω:

$$|\omega(\xi, y, t, \tau)| \leqslant c\,(t-\tau)^{-\frac{n+1}{2}} \exp\left(-C\frac{|\xi - y|^2}{t-\tau}\right),$$
$$\int_\tau^t d\lambda \int_S |\omega(\xi, y, \lambda, \tau)|\, dS_\xi \leqslant c. \tag{16.14}$$

We now estimate the function g with the use of (16.10). Let $\rho(x)$ be the distance from the point x to the surface S; then for $0 \le 2r + s \le 2$ we have

$$|D_t^r D_x^s g(x, y, t, \tau)|$$
$$\leqslant \sup_{\substack{\xi \in S \\ \tau \leqslant \lambda \leqslant t}} |D_t^r D_x^s Z(x, \xi, t, \lambda)| \int_\tau^t d\lambda \int_S |\omega(\xi, y, \lambda, \tau)|\, dS_\xi$$
$$\leqslant c \sup_{\substack{\xi \in S \\ \tau \leqslant \lambda \leqslant t}} (t-\lambda)^{-\frac{n+2r+s}{2}} \exp\left(-C\frac{|x-\xi|^2}{t-\lambda}\right)$$
$$\leqslant \frac{c}{\rho^{n+2r+s}(x)} \exp\left(-C'\frac{\rho^2(x)}{t-\tau}\right). \tag{16.15}$$

In addition, we can obtain an estimate for g that does not depend on $\rho(x)$. Let σ be a part of the surface S lying in the ball with center x and radius $|x - y|/2$. We have

$$|g(x, y, t, \tau)| \leqslant c \Bigg\{ (t-\tau)^{-\frac{n+1}{2}} \int_{\frac{t+\tau}{2}}^{t} (t-\lambda)^{-\frac{n}{2}} d\lambda$$

$$\times \int_S \exp\left(-C \frac{|x-\xi|^2}{t-\lambda}\right) \exp\left(-C \frac{|\xi-y|^2}{\lambda-\tau}\right) dS_\xi$$

$$+ (t-\tau)^{-\frac{n}{2}} \exp\left(-C \frac{|x-y|^2}{t-\tau}\right) \int_\tau^{\frac{t+\tau}{2}} d\lambda \int_{S \setminus \sigma} |\omega(\xi, y, \lambda, \tau)| dS_\xi$$

$$+ (t-\tau)^{-\frac{n}{2}} \exp\left(-\frac{C}{2} \frac{|x-y|^2}{t-\tau}\right)$$

$$\times \int_\tau^{\frac{t+\tau}{2}} (\lambda-\tau)^{-\frac{n+1}{2}} \exp\left(-\frac{C}{2} \frac{|x-y|^2}{\lambda-\tau}\right) d\lambda \int_\sigma dS_\xi \Bigg\}$$

$$\leqslant c(t-\tau)^{-\frac{n}{2}} \exp\left(-C_1 \frac{|x-y|^2}{t-\tau}\right).$$

It follows from this estimate that the Green's function has the same singularity at the point $x = y$, $t = r$ as the fundamental solution:

$$|G(x, y, t, \tau)| \leqslant c(t-\tau)^{-\frac{n}{2}} \exp\left(-C \frac{|x-y|^2}{t-\tau}\right). \tag{16.16}$$

Let us consider the function

$$u(x, t) = \int_0^t d\tau \int_\Omega G(x, y, t, \tau) f(y, \tau)\, dy$$

$$+ \int_\Omega G(x, y, t, 0) \varphi(y)\, dy = u_1(x, t) + u_2(x, t) \tag{16.17}$$

for $f \in H^{\alpha, \alpha/2}(\overline{Q}_T)$ and $\varphi \in C(\overline{\Omega})$, and show that it is the solution of problem (16.1) for $\Phi = 0$. On account of (16.7) the equation $\mathfrak{L}u = f$ is fulfilled in the domain Q_T. Further, by virtue of (16.16) and the results of §14 we have

$$\lim_{t \to 0} u(x, t) = \lim_{t \to 0} u_2(x, t) = \varphi(x).$$

Finally, let us verify that the function u satisfies the zero boundary condition. It is obvious that for $t > 0$

$$u_2|_S = 0.$$

We fix a small $\delta > 0$. On account of (16.16) we have

$$\left| \int_{t-\delta}^{t} d\tau \int_{\Omega} G(x, y, t, \tau) f(y, \tau) dy \right| \leqslant c \max|f|\delta.$$

Since the function $G(x, y, t, \tau)$ does not have any singularities for $t > \tau$, for any preassigned $\eta > 0$ we can find a $d > 0$ such that, when the distance from the point x to the boundary is less than d,

$$\left| \int_{0}^{t-\delta} d\tau \int_{\Omega} G(x, y, t, \tau) f(y, \tau) dy \right| \leqslant \eta.$$

Thus for a point x that is sufficiently close to the boundary we have $|u_1(x, t)| \leq c\delta \max|f| + \eta$, i.e. $u_1|_S = 0$.

We have thus proved that the function (16.10) is the classical solution of problem (16.1) for $\Phi = 0$.

We have considered problem (16.1) in an interior domain, but all of the arguments with obvious changes carry over to the exterior domain $D_{n+1}^{(T)} \setminus \overline{Q}_T$. Thus we have

Theorem 16.2. *Suppose the coefficients of the operator* $\mathfrak{L}$ *belong to the class* $H^{\alpha, \alpha/2}(D_{n+1}^{(T)})$. *Then problem* (16.1) *with* $\Phi = 0$ *has a classical solution in both interior and exterior domains for any bounded* f *and* φ *if* f *satisfies a Hölder condition and* φ *is continuous.*

Theorem 10.1 and inequality (16.16) permit us to obtain exact estimates for the derivatives of the Green's function. We present them without proof.

Theorem 16.3. *If the coefficients of the operator* $\mathfrak{L}$ *belong to the class* $H^{\alpha, \alpha/2}$, *then the following inequalities are valid:*

$$|D_t^r D_x^s G(x, y, t, \tau)| \leqslant c(t-\tau)^{-\frac{n+2r+s}{2}} \exp\left(-C\frac{|x-y|^2}{t-\tau}\right),$$

$$|D_t^r D_x^s G(x, y, t, \tau) - D_{t'}^r D_x^s G(x, y, t', \tau)|$$
$$\leqslant c(t-t')^{\frac{\alpha}{2}+1-r-\frac{s}{2}}(t'-\tau)^{-\frac{n+2+\alpha}{2}} \exp\left(-C\frac{|x-y|^2}{t-\tau}\right),$$

where $2r + s = 1, 2$ *and* $\tau < t' < t$, *and*

$$|D_t^r D_x^s G(x, y, t, \tau) - D_t^r D_{x'}^s G(x', y, t, \tau)|$$
$$\leqslant c|x-x'|^{\alpha}(t-\tau)^{-\frac{n+2+\alpha}{2}} \exp\left(-C\frac{|x''-y|^2}{t-\tau}\right),$$

where $2r + s = 2$ *and* x'' *is that one of the points* x *and* x' *which is closest to* y.

§17. ON THE ESTIMATES OF S. N. BERNSTEIN

In the paper [8a] S. N. Bernstein proposed a method for estimating the maximums of the moduli of derivatives of any order of solutions of linear parabolic equations under the assumption that the solution itself and all of the known functions in the equation are sufficiently smooth. Local (interior) estimates are given in [8a]. In combination with the estimates of derivatives on a boundary given by him earlier (see [8b] and §3 of Chapter VI) they form a sufficiently complete collection of estimates even in the closure of a domain (with the use, of course, of the boundary and initial conditions). The fact that the estimates on a boundary are carried out in [8b] only for elliptic equations, and in two-dimensional domains, is nonessential: they go through without any changes for equations of both types, elliptic and parabolic, in any multidimensional domain. These estimates, as we will show below, are simple and beautiful. But they cannot be considered as exact estimates (concerning this see §§3 and 4 of Chapter I) and therefore cannot be as effectively used in proving theorems on the solvability of various boundary value problems as the estimates in Hölder spaces and in the spaces $W_q^{2m,m}$ given above.

Let us present Bernstein's method for $|u_x|$. Suppose u is a classical solution of the equation

$$\mathscr{L}u = \frac{\partial u}{\partial t} - a_{ij}\frac{\partial^2 u}{\partial x_i \partial x_j} + a_i \frac{\partial u}{\partial x_i} = -au + f \tag{17.1}$$

with continuous coefficients and free term f and with the a_{ij} satisfying condition (2.5) of Chapter I.

If its values are known on Γ_T: $u|_{\Gamma_T} = \psi$, with S and ψ being sufficiently smooth, then $\max_{\Gamma_T}|u_x|$ is estimated in terms of known quantities in the way done in Lemma 3.1 of Chapter VI (with natural simplifications for the given linear case). In order to estimate $|u_x|$ one must know that the derivatives u_t and $u_{x_ix_j}$ and all of the coefficients and the free term of equation (17.1) are differentiable with respect to x at each point of Q_T. Suppose that $\max_{Q_T}|u| = M$. Consider the function $v(x, t) = u_x^2(x, t) + \lambda u^2(x, t) - \mu t$, where λ and μ are two numbers which will be chosen sufficiently large later on. Let us calculate $\mathfrak{L}v$:

$$\mathcal{L}v = 2u_{x_k}\left(u_{x_k t} - a_{ij}u_{x_k x_i x_j} + a_i u_{x_k x_i}\right)$$
$$+ 2\lambda u\left(u_t - a_{ij}u_{x_i x_j} + a_i u_{x_i}\right) - 2a_{ij}u_{x_k x_i}u_{x_k x_j}$$
$$- 2\lambda a_{ij}u_{x_i}u_{x_j} - \mu.$$

By virtue of equation (17.1)

$$\mathcal{L}v = 2u_{x_k}(-au_{x_k} - a_{x_k}u + f_{x_k} + a_{ijx_k}u_{x_i x_j} - a_{ix_k}u_{x_i})$$
$$+ 2\lambda u(-au + f) - 2a_{ij}u_{x_k x_i}u_{x_k x_j} - 2\lambda a_{ij}u_{x_i}u_{x_j} - \mu.$$

We estimate $\mathcal{L}v$ from above, using Cauchy's inequality (1.2) of Chapter II and condition (2.5) of Chapter I, as follows:

$$\mathcal{L}v \leqslant c\left[\left(1 + \frac{1}{\varepsilon}\right)u_x^2 + (1 + \lambda)M^2 + f_x^2\right]$$
$$+ \varepsilon u_{xx}^2 + \lambda f^2 - 2\nu u_{xx}^2 - 2\nu\lambda u_x^2 - \mu,$$

where the constant c depends only on $\max_{Q_T}|a_{ij}, a_i, a_{ijx_k}, a_{ix_k}, a, a_{x_k}|$. Take $\epsilon = 2\nu$, $\lambda = (1/2\nu)c(1 + 1/2\nu)$ and $\mu = c(1+\lambda)M^2 + c\max_{Q_T} f_x^2 + \lambda \max_{Q_T} f^2 + 1$. Then $\mathcal{L}v \le -1 < 0$ at all points of Q_T and hence the maximum of v is on Γ_T, i.e.

$$\max_{Q_T} v = \max_{Q_T}\left(u_x^2 + \lambda u^2 - \mu t\right) \leqslant \max_{\Gamma_T}\left(u_x^2 + \lambda u^2 - \mu t\right). \tag{17.2}$$

And this is the desired estimate for $\max_{Q_T}|u_x|$ in terms of M and the already estimated $\max_{\Gamma_T}|u_x|$.

In order to obtain a local (interior) estimate of $|u_x|$ we consider the auxiliary function

$$w(x, t) = \zeta^2(x, t)\,u_x^2(x, t) + \lambda u^2(x, t) - \mu t,$$

where $\zeta(x, t)$ is a cutting function that is equal to zero in the vicinity of Γ_T. It is easily seen that $\mathcal{L}w$ will be negative for sufficiently large λ and μ, and hence $\max_{Q_T} w$ will occur on Γ_T, i.e.

$$w(x, t) \leqslant \max_{\Gamma_T} w \leqslant \lambda M^2.$$

This inequality also gives an estimate of $\max_{Q'}|u_x|$ for any domain Q' that is separated from Γ_T by a positive distance.

In order to estimate second-order derivatives one must consider the function

$$v(x, t) = u_{xx}^2(x, t) + \lambda u_x^2(x, t) - \mu t$$

or the function

$$w(x, t) = \zeta^2(x, t)\, u_{xx}^2(x, t) + \lambda u_x^2(x, t) - \mu t,$$

depending on whether a total or a local estimate of $|u_{xx}|$ is desired, and choose λ and μ so large that $\mathfrak{L}v$ or $\mathfrak{L}w$ will be negative in Q_T. (A detailed derivation of this can be found in [65h] and in [123a].) The derivatives of u of any order are estimated in essentially the same way.

CHAPTER V

QUASI-LINEAR EQUATIONS WITH PRINCIPAL PART IN DIVERGENCE FORM

This chapter is devoted to quasi-linear equations with principal part in divergence form, i.e. to equations of the form

$$\mathscr{L}u \equiv u_t - \frac{d}{dx_i} a_i(x, t, u, u_x) + a(x, t, u, u_x) = 0. \tag{0.1}$$

They do not include all of the quasi-linear equations. But all of the widely known equations of mechanics that are of parabolic type belong to this class. Certainly the equations

$$u_t - a_{ij}(x, t, u) u_{x_i x_j} + a(x, t, u, u_x) = 0$$

and

$$u_t - a_{11}(x_1, t, u, u_{x_1}) u_{x_1 x_1} + a(x_1, t, u, u_{x_1}) = 0$$

can be reduced to it in an obvious way. We will investigate the generalized and classical solutions of the equations (0.1) and prove the unique solvability of the basic boundary value problems. The restrictions under which this is done are essential and, as the examples of Chapter I show, they cannot be weakened (in the terms of the adopted spaces) for the whole class of equations (0.1).

The solvability of the boundary value problems, as was noted in §4 of Chapter I, is proved by us on the basis of the Leray-Schauder theorem and a priori estimates of the norms $|u|^{(1+\alpha)}_{Q_T}$ of all possible solutions of them. We begin the present chapter by obtaining these estimates. In §§1 and 2 we give estimates of the norms $\max_{Q_T} |u|$, $|u|^{(\alpha)}_{Q'}$ and $|u|^{(\alpha)}_{Q_T}$, then (in §3) of the norms $|u_x|^{(\alpha)}_{Q'}$, and after this of $\max_{Q_T} |u_x|$ (§4) and $|u_x|^{(\alpha)}_{Q_T}$ (§5) for the whole domain Q_T. In this connection we essentially use the results of §§7 and 8 of Chapter II on functions of the classes $\mathfrak{B}_2$, which contain the solution u themselves and their derivatives u_x. Estimates in the vicinity of S_T, and thereby in the whole domain, are carried out in these sections for the first boundary conditions. They and Theorem 5.2 of Chapter IV on linear equations with smooth coefficients permit one to verify the fulfillability of all of the conditions of the Leray-Schauder theorem and to establish with its use the solvability of the first boundary value problem. This is done in §6. In §7 analogous arguments are carried out for other (nonlinear) boundary conditions. In §8 the Cauchy problem is briefly analyzed. In §9 the Stefan problem is considered. Its specific peculiarity is the fact that the common boundary of different phases (or the boundary of the domain of definition in the case of a one-phase problem) is unknown. Finally, in §10 we give another method for estimating the oscillations of the solutions of equations (0.1), which does not require one to turn to the classes $\mathfrak{B}_2$ but uses essentially the same analytic facts as the first method.

At the end of §6 we give a proof, which is based only on the energy estimate, of the generalized solvability of the boundary value problems.

§1. BOUNDED GENERALIZED SOLUTIONS. HÖLDER CONTINUITY

Suppose the functions $a_i(x, t, u, p)$ and $a(x, t, u, p)$ making up equation (0.1) are defined for $(x, t) \in \overline{Q}_T$ and arbitrary values of the arguments u and $p = (p_1, \cdots, p_n)$, are continuous in u and p, and satisfy the conditions

$$a_i(x, t, u, p)\, p_i \geqslant \nu(|u|)\, p^2 - \varphi_0(x, t), \tag{1.1}$$

$$|a_i(x, t, u, p)| \leqslant \mu(|u|)\,|p| + \varphi_1(x, t), \tag{1.2}$$

$$|a(x, t, u, p)| \leqslant \mu_1(|u|)\, p^2 + \varphi_2(x, t), \tag{1.3}$$

in which $\nu(\xi)$ and the $\mu_i(\xi)$ are, as elsewhere, positive continuous functions of $\xi \geq 0$, with $\nu(\xi)$ monotonically decreasing, the $\mu_i(\xi)$ monotonically increasing, and the functions $\phi_i(x, t)$ nonnegative and having the finite norms

$$\|\varphi_0,\ \varphi_2\|_{q, r, Q_T}, \quad \|\varphi_1\|_{2q, 2r, Q_T} \leqslant \mu_2, \tag{1.4}$$

where q and r are arbitrary positive numbers satisfying the condition

$$\frac{1}{r} + \frac{n}{2q} = 1 - \varkappa_1,$$

with

$$\left.\begin{array}{l} q\in\left[\frac{n}{2(1-\varkappa_1)},\ \infty\right],\quad r\in\left[\frac{1}{1-\varkappa_1},\ \infty\right],\quad 0<\varkappa_1<1 \\ \qquad\qquad\qquad\qquad\qquad\qquad\text{for}\quad n\geqslant 2, \\ q\in[1,\ \infty],\quad r\in\left[\frac{1}{1-\varkappa_1},\ \frac{2}{1-2\varkappa_1}\right],\quad 0<\varkappa_1<\frac{1}{2} \\ \qquad\qquad\qquad\qquad\qquad\qquad\text{for}\quad n=1. \end{array}\right\} \tag{1.5}$$

By a *bounded generalized solution* (bo. gen. solution) of equation (0.1) we will mean a function $u(x, t)$ from $V_2^{1,0}(Q_T)$ with $\operatorname{vrai\,max}_{Q_T}|u| < \infty$ that satisfies the integral identity

$$\int_\Omega u(x, t)\eta(x, t)\,dx\Big|_0^t + \int_0^t\int_\Omega [-u\eta_t + a_i(x, t, u, u_x)\eta_{x_i} + a(x, t, u, u_x)\eta]\,dx\,dt = 0 \tag{1.6}$$

for any function $\eta(x, t)$ from $\overset{\circ}{W}{}_2^{1,1}(Q_T)$ with $\operatorname{vrai\,max}_{Q_T}|\eta| < \infty$ and any t from $[0, T]$.

As a consequence of assumptions (1.1)–(1.5) all of the terms of (1.6) are finite for arbitrary u and η from the indicated classes, and from (1.6) it follows that

$$\begin{aligned} &\int_\Omega u(x, t)\eta(x, t)\,dx\Big|_{t_0}^t \\ &\quad + \int_{t_0}^t\int_\Omega [-u\eta_t + a_i(x, t, u, u_x)\eta_{x_i}]\,dx\,dt \\ &\leqslant \int_{t_0}^t\int_\Omega (\mu_1 u_x^2 + \varphi_2)|\eta|\,dx\,dt, \qquad 0\leqslant t_0\leqslant t\leqslant T, \end{aligned} \tag{1.7}$$

where $\mu_1 = \mu_1(\operatorname{vrai\,max}_{Q_T}|u|)$. Let $\operatorname{vrai\,max}_{Q_T}|u| = M$, $\mu = \mu(M)$, $\nu = \nu(M)$ and $\mu_1 = \mu_1(M)$. The following theorem holds:

Theorem 1.1. *Suppose the functions $a_i(x, t, u, p)$ and $a(x, t, u, p)$ possess properties* (1.1)–(1.5), *while $u(x, t)$ is a bo. gen. solution of equation* (0.1) *with* $\operatorname{vrai\,max}_{Q_T}|u| = M$. *Then $u(x, t)$ belongs to $H^{\alpha,\alpha/2}(Q_T)$ with some positive α depending only on n, ν, μ, q, r and $M\mu_1/\nu$. The quantity $|u|_{Q'}^{(\alpha)}$, for any $Q'\subset Q_T$ separated from Γ_T by a positive distance d, is estimated from above by a constant depending only on n, M, ν, μ, μ_1, μ_2, q, r and the distance d.*

If u is Hölder continuous (*of exponent β with respect to x and $\beta/2$ with respect to t*) *on some part Γ' of the boundary Γ_T satisfying condition* (A), *then*

$u(x, t)$ will belong to $H^{\alpha,\alpha/2}(Q_T \cup \Gamma')$ with some $\alpha > 0$ depending only on n, ν, μ, q, r, $M\mu_1/\nu$ and β. The norm $|u|^{(\alpha)}_{Q'}$, for any Q' belonging to Q_T and separated from $\Gamma_T \setminus \Gamma'$ by a positive distance d, is estimated from above by a constant determined by n, M, ν, μ, μ_1, μ_2, q, r, β, $|u|^{(\beta)}_{\Gamma'}$, the constants θ_0 and a_0 of condition (A) and the distance d. In particular, when $\Gamma' = \Gamma_T$, S satisfies condition (A), $u|_{\Gamma_T} \in H^{\beta,\beta/2}(\Gamma_T)$, the solution $u(x, t) \in H^{\alpha,\alpha/2}(\overline{Q}_T)$ and $|u|^{(\alpha)}_{Q_T}$ is estimated from above by a constant depending only on n, M, ν, μ, μ_1, μ_2, q, r, β, the constants θ_0 and a_0 of condition (A) and $|u|^{(\beta)}_{\Gamma_T}$.

Remark 1.1. The parameters q and r in (1.4) can be different for different ϕ_i, and in place of the $\phi_i(x, t)$ in the right sides of (1.1)–(1.3) there can be finite sums of different ϕ, it being only necessary that the parameters q and r corresponding to them satisfy the requirements (1.5).

In order to prove the theorem it is sufficient to demonstrate that $u(x, t)$ belongs to the class $\mathfrak{B}_2(\overline{Q}_T, M, \gamma, m, \delta, \kappa)$ with parameters γ, m, δ and κ determined only by M and the numbers entering into conditions (1.1)–(1.5).

To this end we first note that identity (1.6) implies the identities

$$\int_{t_0}^{t_1}\int_\Omega [u_{ht}\hat{\eta} + a_i(x, t, u, u_x)_{\bar{h}}\hat{\eta}_{x_i} + a(x, t, u, u_x)_{\bar{h}}\hat{\eta}]\,dx\,dt = 0 \tag{1.8}$$

for $h \le t_0 \le t_1 \le T - h$ with any function $\hat{\eta}(x, t)$ from $\overset{\circ}{V}{}^{1,0}_2(Q_{T-h})$. The symbols $(\ )_h$ and $(\ )_{\bar{h}}$ in them denote the Steklov averagings over future and past t defined by formulas (2.7) and (2.10) of Chapter III. Relations (1.8) are deduced from (1.6) in the same way as (2.11) was deduced from (1.16) in Chapter III. In (1.8) we put $\hat{\eta}(x, t) = \zeta^2(x, t)\max\{u_h(x, t) - k;\ 0\} \equiv \zeta^2 u_h^{(k)}$, where $\zeta(x, t)$ is an arbitrary nonnegative continuous piecewise-smooth function that is equal to zero on S_T, and we transform the first term as follows:

$$\int_{t_0}^{t_1}\int_\Omega u_{ht}u_h^{(k)}\zeta^2\,dx\,dt = \frac{1}{2}\int_\Omega (u_h^{(k)}(x, t))^2\zeta^2(x, t)\,dx\Big|_{t_0}^{t_1} - \int_{t_0}^{t_1}\int_\Omega (u_h^{(k)})^2\zeta\zeta_t\,dx\,dt.$$

We then pass to the limit in all of the terms as $h \to 0$. As a result we obtain

$$\frac{1}{2}\|u^{(k)}(x, t)\zeta(x, t)\|^2_{2,\Omega}\Big|_{t_0}^{t_1}+\int_{t_0}^{t_1}\int_{\Omega}[-(u^{(k)})^2\zeta\zeta_t$$
$$+a_i(u^{(k)}\zeta^2)_{x_i}+au^{(k)}\zeta^2]\,dx\,dt=0. \tag{1.9}$$

The validity of the passage to the limit in all of the terms as $h \longrightarrow 0$ follows from Lemmas 4.5–4.7 of Chapter II and from assumptions (1.1)–(1.5) on the functions $a_i(x, t, u, p)$ and $a(x, t, u, p)$, and also from the membership of u in $V_2^{1,0}(Q_T)$. In (1.9) t_0 and t_1 can be taken arbitrarily from $(0, T)$ and, since all of the terms of (1.9) are continuous functions of t_0 and t_1, also from $[0, T]$.

We will assume $\zeta(x, t)$ to be different from zero only for $x \in K_\rho$. By virtue of assumptions (1.1)–(1.3) it follows from (1.9) that

$$\frac{1}{2}\|u^{(k)}(x, t)\zeta(x, t)\|^2_{2,K_\rho}\Big|_{t_0}^{t}+\nu\int_{t_0}^{t}\int_{K_\rho}[u_x^{(k)}]^2\zeta^2\,dx\,dt$$
$$\leqslant\int_{t_0}^{t}\int_{A_{k,\rho}}[\varphi_0\zeta^2+(\mu|u_x|+\varphi_1)2(u-k)\zeta|\zeta_x|$$
$$+(\mu_1u_x^2+\varphi_2)(u-k)\zeta^2+(u-k)^2\zeta|\zeta_t|]\,dx\,dt. \tag{1.10}$$

Estimating the terms of the right side by Cauchy's inequality (1.2) of Chapter II, we obtain the inequality

$$\frac{1}{2}\|u^{(k)}(x, t)\zeta(x, t)\|^2_{2,K_\rho}\Big|_{t_0}^{t}+\nu\int_{Q(\rho,t-t_0)}[u_x^{(k)}]^2\zeta^2\,dx\,dt$$
$$\leqslant\int_{t_0}^{t}\int_{A_{k,\rho}}\Big[\varphi_0\zeta^2+\varepsilon_1\mu u_x^2\zeta^2+\frac{1}{\varepsilon_1}\mu(u-k)^2\zeta_x^2+\varphi_1^2\zeta^2$$
$$+(u-k)^2\zeta_x^2+(\mu_1u_x^2+\varphi_2)\zeta^2\Big(\max_{Q(\rho,t-t_0)}u(x, t)-k\Big)$$
$$+(u-k)^2\zeta|\zeta_t|\Big]\,dx\,dt, \tag{1.11}$$

in which $Q(\rho, t-t_0)=K_\rho\times(t_0, t)$ and ϵ_1 is an arbitrary positive number.

This inequality has the same character as inequality (10.1) of Chapter III. The only difference consists in the fact that the right side of (1.11) contains the term $(\max_{Q(\rho,t-t_0)}u(x, t)-k)\mu_1u_x^2$, the coefficient $\mu_1(\max_{Q(\rho,t-t_0)}u(x, t)-k)$ of which is in general not small and which accordingly cannot be cancelled by the second term of the left side. In order for this to be possible we impose a restriction on the choice of the level k. Namely, we take only those k for which

$$\max_{Q(\rho,\tau)} u(x,t)-k\leqslant\delta=\frac{\nu}{4\mu_1}. \tag{1.12}$$

Then (1.11) with $\epsilon_1=\nu/4\mu$ for $t\in[t_0,t_0+\tau]$ implies

$$\begin{aligned}\|u^{(k)}(x,t)\zeta(x,t)\|^2_{2,K_\rho}&+\nu\int_{Q(\rho,t-t_0)}[u_x^{(k)}]^2\zeta^2\,dx\,dt\\ \leqslant\|u^{(k)}(x,t_0)\zeta(x,t_0)\|^2_{2,K_\rho}&+c\int_{Q(\rho,t-t_0)}[u^{(k)}]^2(\zeta_x^2+\zeta|\zeta_t|)\,dx\,dt\\ &+\int_{t_0}^{t}\int_{A_{k,\rho}(t)}\mathscr{D}_1(x,t)\,dx\,dt,\end{aligned} \tag{1.13}$$

where $c=2(4\mu^2/\nu+1)$ and $\mathfrak{D}_1(x,t)=2(\phi_0+\phi_1^2+\delta\phi_2)\zeta^2$. This inequality coincides with inequality (10.2) of Chapter III for the case $M=0$. Moreover, the assumptions (1.4)–(1.5) on the ϕ_i are such that the $\mathfrak{D}_1$ from (1.13) and the $\mathfrak{D}_1$ from (10.2) belong to one and the same space $L_{q,r}(Q_T)$. In view of this (1.13) implies inequalities (10.4) and (10.6) of Chapter III, in which M is put equal to zero, or equivalently, inequalities (7.5) of Chapter II. Analogous inequalities hold for $-u(x,t)$. They imply inequalities (7.1) and (7.2) of Chapter II and thereby the fact that the solution u is an element of

$$\mathfrak{B}_2(Q_T,M,\gamma,\hat r,\delta,\varkappa)\qquad\text{with}\quad\gamma=\frac{\max\{c;\|\mathscr{D}_1\|_{q,r,Q_T}\}}{\min\{1;\nu\}},$$

$$\varkappa=\frac{2\varkappa_1}{n},\quad\hat r=\frac{2r}{r-1}(1+\varkappa),\quad\delta=\frac{\nu}{4\mu_1},$$

and an element of

$$\mathfrak{B}_2(Q(\rho_0,\rho_0^2),M,\gamma,\hat r,\delta,\varkappa)\ \text{with}\ \ \rho_0=\left[c\|\mathscr{D}_1\|^{-1}_{q,r,Q_T}\right]^{\frac{1}{n\varkappa}}\varkappa_n^{\frac{1}{\hat r}-\frac{n}{4}},$$

$$\gamma=\frac{c}{\min\{1;\nu\}},\quad\varkappa=\frac{\varkappa_1}{n},\quad\hat r=\frac{2r}{r-1}(1+2\varkappa),\quad\delta=\frac{\nu}{4\mu_1}.$$

But, as was proved in §7 of Chapter II, this guarantees the membership of u in $H^{\alpha,\alpha/2}(Q_T)$ with some positive α, which is a continuous function of the arguments $\gamma=(2/\min\{1;\nu\})(4\mu^2/\nu+1)$, q, r and $M/\delta=4M\mu_1/\nu$ in an admissible domain of variation of these arguments (for M/δ this domain is the semiaxis $[0,\infty)$), and the estimate of $\langle u\rangle^{(\alpha)}_{Q}$.

The other two assertions of the theorem follow from Theorem 8.1 of Chapter II and the fact that the solution u belongs to $\mathfrak{B}_2(\overline{Q}_T, M, \gamma, \hat{r}, \delta, \kappa)$ and $\mathfrak{B}_2(\overline{Q}(\rho_0, \rho_0^2), M, \gamma, \hat{r}, \delta, \kappa)$ with the same values of the parameters $M, \gamma, \hat{r}, \delta$ and κ as above.

Remark 1.2. It is easy to see that the membership of the function $u(x, t)$ to the classes $\mathfrak{B}_2$ was actually deduced by us not from the fact that $u(x, t)$ satisfies the identity (1.6) but from the fact that inequalities (1.7) are valid for it. Therefore all of the assertions of Theorem 1.1 are valid for an arbitrary function $u(x, t)$ from $V_2^{1,0}(Q_T)$ with finite vrai $\max_{Q_T} |u|$ that satisfies inequalities (1.7) for all of the indicated η, t_0 and t.

Inequalities (1.13) for u (but not for $-u$!), and hence inequalities (7.1) and (7.2) of Chapter II for u, underlying the definitions of the classes $\mathfrak{B}_2$, were deduced by us from inequalities (1.7), in which η takes only nonnegative values.

§2. ON THE BOUNDEDNESS OF GENERALIZED SOLUTIONS

We consider here arbitrary (not necessarily bounded) generalized solutions of equation (0.1) that belong to the space $V_2^{1,0}(Q_T)$. It is natural to define such solutions $u(x, t)$ as elements of $V_2^{1,0}(Q_T)$ satisfying the identity (1.6) for any function $\eta(x, t)$ from $\overset{\circ}{W}{}_2^{1,1}(Q_T)$. In order for this definition to make sense it is necessary to impose restrictions on the functions $a_i(x, t, u, p)$ and $a(x, t, u, p)$ that are somewhat more rigid than (1.2)–(1.5). If one formulates them in the form

$$\begin{gathered} |a_i(x, t, u, p)| \leqslant c|p| + c|u|^{\alpha} + \varphi_1(x, t), \\ |a(x, t, u, p)| \leqslant c|p|^{2-\beta} + c|u|^{\gamma} + \varphi_2(x, t), \end{gathered} \tag{2.1}$$

then the function $\phi_1(x, t)$, $\phi_2(x, t)$ and the parameters α, β, γ must satisfy the conditions

$$\left.\begin{aligned} & 0 \leqslant \alpha \leqslant \frac{n+2}{n}; \quad \frac{n}{n+2} \leqslant \beta \leqslant 2; \\ & 0 \leqslant \gamma \leqslant 1 + \frac{4}{n}; \quad \varphi_1 \in L_2(Q_T), \quad \varphi_2 \in L_{q_0, r_0}(Q_T); \\ & \frac{1}{r_0} + \frac{n}{2q_0} = 1 + \frac{n}{4}, \quad q_0 \geqslant 1, \quad r_0 \geqslant 1. \end{aligned}\right\} \tag{2.2}$$

Conditions (2.1) and (2.2), as can easily be demonstrated by means of Hölder's inequality and inequality (3.4) of Chapter II, guarantee the finiteness of the integrals $\int_{Q_T} a_i(x, t, u, u_x)\eta_{x_i}\,dx\,dt$ and $\int_{Q_T} a(x, t, u, u_x)\eta\,dx\,dt$ for any functions u and η from $V_2^{1,0}(Q_T)$. Let us show that any generalized solution $u(x, t)$ from

$V_2^{1,0}(Q_T)$ of equation (0.1) will be bounded and that it is possible to estimate $\operatorname{vrai\,max}_{Q_T}|u|$ from above in terms of $\|u\|_{q_1,r_1,Q_T}$ and known quantities if for $|u| \geq k_0$ the functions $a_i(x, t, u, p)$ and $a(x, t, u, p)$ also satisfy the conditions

$$a_i(x, t, u, p)\,p_i \geqslant \nu|p|^2 - \mu|u|^\delta - u^2\varphi(x, t), \quad \nu > 0, \tag{2.3}$$

and

$$-a(x, t, u, p)\,u \leqslant \mu_0|p|^2 + \mu|u|^\delta + u^2\varphi(x, t), \tag{2.4}$$

in which

$$\left.\begin{array}{l} (\delta-2)\left(\frac{1}{r_1}+\frac{n}{2q_1}\right) = 1-\varkappa_1, \quad \text{where} \quad \delta > 2, \\ \frac{q_1}{\delta-2} \in \left[\frac{n}{2(1-\varkappa_1)}, \infty\right], \ \frac{r_1}{\delta-2} \in \left[\frac{1}{1-\varkappa_1}, \infty\right], \\ \qquad 0 < \varkappa_1 < 1, \quad \text{for} \quad n \geqslant 2 \text{ and} \\ \frac{q_1}{\delta-2} \in [1, \infty], \ \frac{r_1}{\delta-2} \in \left[\frac{1}{1-\varkappa_1}, \frac{2}{1-2\varkappa_1}\right], \\ \qquad 0 < \varkappa_1 < \frac{1}{2}, \quad \text{for} \quad n = 1, \end{array}\right\} \tag{2.5}$$

while $\phi(x, t)$ is an element of $L_{q_2,r_2}(Q_T)$ with q_2 and r_2 satisfying the conditions

$$\left.\begin{array}{l} \frac{1}{r_2}+\frac{n}{2q_2} = 1-\varkappa_2, \\ q_2 \in \left[\frac{n}{2(1-\varkappa_2)}, \infty\right], \ r_2 \in \left[\frac{1}{1-\varkappa_2}, \infty\right], \ 0 < \varkappa_2 < 1, \\ \qquad \text{for} \quad n \geqslant 2, \\ q_2 \in [1, \infty], \ r_2 \in \left[\frac{1}{1-\varkappa_2}, \frac{2}{1-2\varkappa_2}\right], \ 0 < \varkappa_2 < \frac{1}{2} \\ \qquad \text{for} \quad n = 1. \end{array}\right\} \tag{2.6}$$

The numbers q_1 and r_1 can be any positive numbers, even less than unity, and $\mu_0 < \nu$. The number δ is assumed to be greater than 2; for $\delta = 2$ the term $|u|^\delta$ can be assumed to be included in $u^2\phi(x, t)$. We confine ourselves here to a total estimate of $|u|$, i.e. to an estimate of $\operatorname{vrai\,max}_{Q_T}|u|$, assuming in this connection, of course, that $\operatorname{vrai\,max}_{\Gamma_T}|u| < \infty$. From the present section and §8 of Chapter III it can be seen how one varies the reasoning to obtain local estimates of $\operatorname{vrai\,max}|u|$ (without assuming that $\operatorname{vrai\,max}_{\Gamma_T}|u|$ is bounded). If $\mu = 0$ in (2.3) and (2.4), then $\operatorname{vrai\,max}_{Q_T}|u|$ is estimated only in terms of known quantities.

If it is known a priori that the solution $u(x, t)$ is bounded, then the condition $\mu_0 < \nu$ can be dropped. We will prove all of these assertions.

Theorem 2.1. *Suppose the functions* $a_i(x, t, u, p)$ *and* $a(x, t, u, p)$ *satisfy conditions* (2.1)–(2.6) *and* $\mu_0 < \nu$. *Then any generalized solution* $u(x, t)$ *from* $V_2^{1,0}(Q_T)$ *of equation* (0.1) *with* $\operatorname{vrai\,max}_{\Gamma_T} |u| \equiv M_0 < \infty$ *is a bounded function and* $\operatorname{vrai\,max}_{Q_T} |u|$ *is estimated from above by a constant* c *determined only by the quantities* $\nu - \mu_0$, μ, $\|\phi\|_{q_2, r_2, Q_T}$, q_2, r_2, q_1, r_1, δ, k_0, $\|u\|_{q_1, r_1, Q_T}$, T *and* mes Ω. *The parameters* q_1 *and* r_1 *can be any positive numbers. When* $\mu = 0$ *the constant* c *does not depend on* $\|u\|_{q_1, r_1, Q_T}$.

In order to prove Theorem 2.1 we consider relation (1.9). For $k \geq M_0$ the function $\zeta(x, t)$ in it can be put equal to 1 since the function $\hat{\eta} = u_h^{(k)} \zeta^2$ will be equal to zero on Γ_T. From (1.9) with $\zeta \equiv 1$, $t_0 = 0$ and $k \geq \max\{M_0; k_0\}$ it follows by virtue of assumptions (2.3) and (2.4) that

$$\frac{1}{2} \| u^{(k)} \|^2_{2, A_k(t_1)} + \int_{Q_{t_1}(k)} (\nu u_x^2 - \mu |u|^\delta - u^2 \varphi)\, dx\, dt$$

$$\leqslant - \int_{Q_{t_1}(k)} au \frac{u^{(k)}}{u}\, dx\, dt \leqslant \int_{Q_{t_1}(k)} (\mu_0 u_x^2 + \mu |u|^\delta + u^2 \varphi)\, dx\, dt, \qquad (2.7)$$

where, as elsewhere, $u^{(k)}(x, t) = \max\{u(x, t) - k; 0\}$, $A_k(t)$ is the set of points $x \in \Omega$ at which $u(x, t) > k$, and $Q_{t_1}(k)$ is the set of points $(x, t) \in Q_{t_1} = \Omega \times (0, t_1)$ at which $u(x, t) > k$. Since by assumption $\mu_0 < \nu$, from (2.7) we have the inequality

$$| u^{(k)} \|^2_{2, A_k(t_1)} + \| u_x^{(k)} \|^2_{Q_{t_1}(k)} \leqslant c \int_{Q_{t_1}(k)} (\mu |u|^\delta + u^2 \varphi)\, dx\, dt, \qquad (2.8)$$

in which c depends only on $\nu - \mu_0$ from conditions (2.3) and (2.4). Let us estimate the integrals $j_1 = \int_{Q_{t_1}(k)} |u|^\delta dx\, dt$ and $j_2 = \int_{Q_{t_1}(k)} u^2 \phi\, dx\, dt$, assuming that $\delta > 2$ and the quantity $\|u\|_{q_1, r_1, Q_T}$ is known. By virtue of conditions (2.5) and Hölder's inequality we get

$$j_1 = \int_{Q_{t_1}(k)} |u|^{\delta - 2} u^2\, dx\, dt$$

$$\leqslant \| u^{\delta - 2} \|_{\frac{q_1}{\delta - 2}, \frac{r_1}{\delta - 2}, Q_{t_1}(k)} \| u^2 \|_{\frac{q_1}{q_1 - (\delta - 2)}, \frac{r_1}{r_1 - (\delta - 2)}, Q_{t_1}(k)} =$$

$$\equiv \|u\|^{\delta-2}_{q_1, r_1, Q_{t_1}(k)} \|u\|^2_{\bar q_1, \bar r_1, Q_{t_1}(k)}$$

$$\leqslant 2\|u\|^{\delta-2}_{q_1, r_1, Q_T}\left[\|u-k\|^2_{\bar q_1, \bar r_1, Q_{t_1}(k)} + k^2\left(\int_0^{t_1} \operatorname{mes}^{\frac{\bar r_1}{\bar q_1}} A_k(t)\,dt\right)^{\frac{2}{\bar r_1}}\right], \tag{2.9}$$

where $\bar q_1 = 2q_1/q_1 - (\delta - 2)$ and $\bar r_1 = 2r_1/(r_1 - (\delta - 2))$. Conditions (2.5) imply $1/\bar r_1 + n/2\bar q_1 = n(1 + \hat\kappa_1)/4$, where $\hat\kappa_1 = 2\kappa_1/n$ and

$$\bar q_1 \in \left[2, \frac{2n}{n-2+2\varkappa_1}\right], \quad \bar r_1 \in \left[2, \frac{2}{\varkappa_1}\right] \quad \text{for} \quad n \geqslant 2,$$

$$\bar q_1 \in [2, \infty], \quad \bar r_1 \in \left[\frac{4}{1+2\varkappa_1}, \frac{2}{\varkappa_1}\right] \quad \text{for} \quad n = 1.$$

We take the numbers $\hat q_1 = \bar q_1(1 + \hat\kappa_1)$, $\hat r_1 = \bar r_1(1 + \hat\kappa_1)$. They satisfy relations (3.3) of Chapter II; thus the embedding theorem (3.4) of Chapter II is valid. In view of this and inequality (3.5) of Chapter II,

$$j_1 \leqslant 2\|u\|^{\delta-2}_{q_1, r_1, Q_T}\left[\beta^2 |u^{(k)}|^2_{Q_{t_1}(k)} \left(\int_0^{t_1} \operatorname{mes}^{\frac{\hat r_1}{\hat q_1}} A_k(t)\,dt\right)^{\frac{2}{\bar r_1}-\frac{2}{\hat r_1}} + k^2\left(\int_0^{t_1} \operatorname{mes}^{\frac{\hat r_1}{\hat q_1}} A_k(t)\,dt\right)^{2\frac{1+\varkappa_1}{\hat r_1}}\right]$$

$$\leqslant 2\|u\|^{\delta-2}_{q_1, r_1, Q_T}\left[\beta^2 |u^{(k)}|^2_{Q_{t_1}(k)} \left(t_1^{\frac{1}{\hat r_1}} \operatorname{mes}^{\frac{1}{\hat q_1}} \Omega\right)^{2\hat\varkappa_1} + k^2\left(\int_0^{t_1} \operatorname{mes}^{\frac{\hat r_1}{\hat q_1}} A_k(t)\,dt\right)^{2\frac{1+\hat\varkappa_1}{\hat r_1}}\right]. \tag{2.10}$$

For $q_1 = \delta - 2$ (this is possible only in the case $n = 1$) $\bar q_1$ goes to ∞ and in place of (2.10) we will have

$$j_1 \leqslant 2\|u\|^{\delta-2}_{q_1, r_1, Q_T}\left[\beta^2 |u^{(k)}|^2_{Q_{t_1}(k)} \mu^{\frac{2\hat\varkappa_1}{\hat r_1}}(k, t_1) + k^2 \mu^{2\frac{1+\hat\varkappa_1}{\hat r_1}}(k, t_1)\right], \tag{2.11}$$

where $\mu(k, t_1)$ is the measure of the set of those points t in the interval $(0, t_1)$ at which mes $A_k(t) > 0$.

The integral j_2 is estimated analogously:

$$\begin{aligned} j_2 &\leqslant \|\varphi\|_{q_2, r_2, Q_{t_1}(k)} \|u\|^2_{\bar{q}_2, \bar{r}_2, Q_{t_1}(k)} \\ &\leqslant 2\|\varphi\|_{q_2, r_2, Q_T}\left[\|u-k\|^2_{\bar{q}_2, \bar{r}_2, Q_{t_1}(k)} + k^2\|1\|^2_{\bar{q}_2, \bar{r}_2, Q_{t_1}(k)}\right] \\ &\leqslant 2\|\varphi\|_{q_2, r_2, Q_T}\left[\beta^2|u^{(k)}|^2_{Q_{t_1}(k)}\left(t_1^{\frac{1}{\hat{r}_2}} \operatorname{mes}^{\frac{1}{\hat{q}_2}}\Omega\right)^{2\hat{\varkappa}_2}\right. \\ &\qquad\left. + k^2\left(\int_0^{t_1} \operatorname{mes}^{\frac{\hat{r}_2}{\hat{q}_2}} A_k(t)\,dt\right)^{2\frac{1+\hat{\varkappa}_2}{\hat{r}_2}}\right], \end{aligned} \tag{2.12}$$

where $\bar{q}_2 = 2q_2/(q_2 - 1)$, $\bar{r}_2 = 2r_2/(r_2 - 1)$, $\hat{\kappa}_2 = 2\kappa_2/n$, $\hat{q}_2 = \bar{q}_2(1 + \hat{\kappa}_2)$, $\hat{r}_2 = \bar{r}_2(1 + \hat{\kappa}_2)$, with the numbers $\hat{q}_2$ and $\hat{r}_2$ satisfying relations (3.3) of Chapter II. For $q_2 = 1$ the index $\bar{q}_2 = \infty$ and instead of (2.12) we get

$$\begin{aligned} j_2 \leqslant 2\|\varphi\|_{q_2, r_2, Q_T}&\left[\beta^2|u^{(k)}|^2_{Q_{t_1}(k)}\mu^{\frac{2\hat{\varkappa}_2}{\hat{r}_2}}(k, t_1)\right. \\ &\left. + k^2\mu^{2\frac{1+\hat{\varkappa}_2}{\hat{r}_2}}(k, t_1)\right], \end{aligned} \tag{2.13}$$

where $\mu(k, t_1)$ has the same meaning as in (2.10). Let us substitute our estimates for j_1 and j_2 into inequality (2.8). Then for values of t_1 so small that

$$\begin{aligned} 4c\beta^2&\left[\mu\|u\|^{\delta-2}_{q_1, r_1; Q_T}\left(t_1^{\frac{1}{\hat{r}_1}} \operatorname{mes}^{\frac{1}{\hat{q}_1}}\Omega\right)^{2\hat{\varkappa}_1}\right. \\ &\left. + \|\varphi\|_{q_2, r_2, Q_T}\left(t_1^{\frac{1}{\hat{r}_2}} \operatorname{mes}^{\frac{1}{\hat{q}_2}}\Omega\right)^{2\hat{\varkappa}_2}\right] \leqslant \frac{1}{2}, \end{aligned}$$

we will have

$$\begin{aligned} |u^{(k)}|_{Q_{t_1}(k)} \leqslant c_1 k&\left[\left(\int_0^{t_1} \operatorname{mes}^{\frac{\hat{r}_1}{\hat{q}_1}} A_k(t)\,dt\right)^{\frac{1+\hat{\varkappa}_1}{\hat{r}_1}}\right. \\ &\left. + \left(\int_0^{t_1} \operatorname{mes}^{\frac{\hat{r}_2}{\hat{q}_2}} A_k(t)\,dt\right)^{\frac{1+\hat{\varkappa}_2}{\hat{r}_2}}\right] \end{aligned} \tag{2.14}$$

for $k \geq \max\{M_0, k_0\}$, where for $\hat{q}_i = \infty$ it is necessary to take $\mu(k, t_1)$ in place of the integral $\int_0^{t_1} \operatorname{mes}^{\hat{r}_i/\hat{q}_i} A_k(t)dt$. The constant c_1 is determined only by known quantities. For $\mu = 0$ (i.e. for $\delta = 2$) it does not depend on $\|u\|_{q_1, r_1, Q_T}$. From inequality (2.14) we conclude on the basis of Theorem 6.1 of Chapter II (more precisely, Remark 6.1 to it) that $u(x, t)$ is bounded in Q_{t_1} from above and $\operatorname{vrai}\max_{Q_{t_1}} u$ does not exceed a certain number determined only by M_0, k_0, the constants c_1, $\hat{r}_i$, $\hat{q}_i$, $\hat{\kappa}_i$, and mes Ω and T, i.e. by the quantities indicated in Theorem 2.1. By considering $u(x, t)$ successively in the cylinders $\Omega \times (t_1, 2t_1)$, $\Omega \times (2t_1, 3t_1)$ and so on, we estimate in a finite number of steps vrai max u from above in all of Q_T. In order to estimate $u(x, t)$ from below it is necessary to apply the result just obtained to the function $\tilde{u}(x, t) = -u(x, t)$, which satisfies an equation of the same type as $u(x, t)$, namely

$$\tilde{u}_t - \frac{d}{dx_i}\tilde{a}_i(x, t, \tilde{u}, \tilde{u}_x) + \tilde{a}(x, t, \tilde{u}, \tilde{u}_x) = 0$$

with

$$\tilde{a}_i(x, t, \tilde{u}, \tilde{u}_x) = -a_i(x, t, -\tilde{u}, -\tilde{u}_x)$$

and

$$\tilde{a}(x, t, \tilde{u}, \tilde{u}_x) = -a(x, t, -\tilde{u}, -u_x),$$

subject to conditions (2.3)–(2.6). Theorem 2.1 is proved.

Remark 2.1. We note that the estimate of $\operatorname{vrai}\max_{Q_T} u$ from above is actually deduced not from the identities (1.6), which the generalized solution $u(x, t)$ of equation (0.1) satisfies, but from the corresponding inequalities

$$\int_\Omega u(x, t)\eta(x, t)\,dx\Big|_0^t + \int_0^t\int_\Omega [-u\eta_t + a_i(x, t, u, u_x)\eta_{x_i} + a(x, t, u, u_x)\eta]\,dx\,dt \leq 0 \qquad (1.6')$$

with a nonnegative function η from $\overset{\circ}{W}{}_2^{1,1}(Q_T)$. Analogously, only inequalities (1.6′) with a nonpositive function η are needed for an estimate of $\operatorname{vrai}\min_{Q_T} u$ from below.

As indicated, conditions (2.1)–(2.2) are imposed only in order that the integrals in (1.6) will have a meaning for any of the solutions in question. If one assumes that the solution u is better behaved than in Theorem 2.1, then the requirements (2.1)–(2.2) can be weakened. In addition, it is possible to get rid of the restriction $\mu_0 < \nu$, which was imposed above. Namely, the following theorem holds.

Theorem 2.2. *If the functions* $a_i(x, t, u, p)$ *and* $a(x, t, u, p)$ *for* $(x, t) \in \overline{Q}_T$, $|u|, |p| < \infty$ *are subject to restrictions* (1.2)–(1.3) *with* ϕ_1^2, ϕ_2 *from* $L_1(Q_T)$, *and satisfy for* $k \geq k_0$ *and the same* x, t, u, p *conditions* (2.3)–(2.6) *with an arbitrary constant* μ_0 *in* (2.4), *then for any bounded generalized solution* $u \in V_2^{1,0}(Q_T)$ *the quantity* vrai $\max_{Q_T} |u|$ *is estimated from above by a constant* M *depending only on* $k_0, \nu, \mu, \mu_0, \delta, q_1, r_1, \|\phi\|_{q_2, r_2, Q_T}, q_2$ *and* r_2 *from conditions* (2.3)–(2.6), *and on* $M_0 =$ vrai $\max_{\Gamma_T} |u|$, *the norm* $\|u\|_{q_1, r_1, Q_T}$, mes Ω *and* T.

We consider the function $v = |u|^m$, $m > 1$, and the function $v_{(h)} = |u_h|^m$, corresponding to it, and in identity (1.8) we put

$$\hat{\eta}(x, t) = |u_h(x, t)|^{m-1} v_{(h)}^{(k)}(x, t) \operatorname{sign} u_h(x, t), \qquad k \geqslant M_0^m.$$

Here u_h is the Steklov averaging of u over t defined in §2 of Chapter III, while $v_{(h)}^{(k)}(x, t) = \max\{v_{(h)}(x, t) - k; 0\}$.

This leads us to the inequality

$$\frac{1}{2m} \int_\Omega [v_{(h)}^{(k)}(x, t)]^2 dx \Big|_{t_0}^{t_1} + \int_{t_0}^{t_1} \int_\Omega [a_l(x, t, u, u_x)(|u_h|^{m-1} \operatorname{sign} u_h v_{(h)}^{(k)})_{\bar{h} x_l} + a(x, t, u, u_x)(|u_h|^{m-1} \operatorname{sign} u_h v_{(h)}^{(k)})_{\bar{h}}] \, dx \, dt = 0.$$

from which, passing to the limit as $h \to 0$ and setting $t_0 = 0$, we deduce

$$\frac{1}{2m} \int_\Omega [v^{(k)}(x, t)]^2 dx + \int_0^{t_1} \int_\Omega [a_l(x, t, u, u_x)(|u|^{m-2} u v_{x_l}^{(k)} + (m-1)|u|^{m-2} u_{x_l} v^{(k)}) + a(x, t, u, u_x)|u|^{m-2} u v^{(k)}] \, dx \, dt = 0; \tag{2.15}$$

for $k \geq \max\{M_0^m; k_0^m\}$ we obtain from (2.15), (2.3) and (2.4) the inequality

$$\frac{1}{2m} \int_\Omega [v^{(k)}(x, t_1)]^2 dx + \int_{Q_{t_1}(k)} (\nu u_x^2 - \mu |u|^\delta - u^2 \varphi)(m|u|^{2m-2} + (m-1)|u|^{m-2} v^{(k)}) \, dx \, dt \leqslant \int_{Q_{t_1}(k)} (\mu_0 u_x^2 + \mu |u|^\delta + u^2 \varphi)|u|^{m-2} v^{(k)}) \, dx \, dt,$$

where $Q_{t_1}(k)$ is the set of points (x, t) from Q_{t_1} at which $v(x, t) > k$.

Hence, choosing $m > \mu_0/\nu$ and taking into account that $|u|^m = v$ and $m|u|^{m-1}|u_x| = |v_x|$, we see that

$$\int_\Omega [v^{(k)}(x, t_1)]^2\, dx + \int_{Q_{t_1}(k)} v_x^2\, dx\, dt \leqslant c \int_{Q_{t_1}(k)} \left(v^{\frac{\delta-2}{m}+2} + v^2\varphi\right) dx\, dt. \tag{2.16}$$

Here the function $\phi(x, t)$ is the same as in (2.8), the exponent $\tilde{\delta} = (\delta - 2)/m + 2$ is subject to the restriction

$$(\tilde{\delta} - 2)\left(\frac{m}{r_1} + \frac{mn}{2q_1}\right) = 1 - \varkappa_1,$$

and in regard to $v(x, t)$ it is known that

$$\| v \|^{\frac{1}{m}}_{\frac{q_1}{m}, \frac{r_1}{m}, Q_T} = \left\| v^{\frac{1}{m}} \right\|_{q_1, r_1, Q_T} = \| u \|_{q_1, r_1, Q_T}.$$

Thus inequalities (2.16) have exactly the same form as (2.8), with the same restrictions on ϕ, $\tilde{\delta} = (\delta - 2)/m + 2$ and v that were imposed on ϕ, δ and u respectively in inequalities (2.8). This permits us to draw the desired conclusion concerning the estimate of $\operatorname{vrai}\max_{Q_T} v$ and hence of $\operatorname{vrai}\max_{Q_T}|u|$ from above in terms of the constants indicated in the theorem.

In connection with the above discussion we mention the paper [53], in which the validity of the strong maximum principle is established for generalized solutions from $W_2^{1,1/2}(Q_T)$ of equation (0.1) if in it $a_i = a_{ij}(x, t)\, u_{x_j}$ and $a = a(u, u_x) \geq M|u_x|$.

§3. ESTIMATES OF $\max_{Q'}|u_x|$ AND $\langle u_x \rangle^{(\alpha)}_{Q'}$

We assume that the properties of the functions $a_i(x, t, u, p)$ and $a(x, t, u, p)$ are somewhat better than in §1, namely, that the partial derivatives $\partial a_i(x, t, u, p)/\partial p_j$, $\partial a_i/\partial u$ and $\partial a_i/\partial x_j$ exist and that for $(x, t) \in Q_T$, $|u| \leq M$ and arbitrary p the following estimates are valid:

$$\nu \sum_{i=1}^{n} \xi_i^2 \leqslant \frac{\partial a_i(x, t, u, p)}{\partial p_j} \xi_i \xi_j \leqslant \mu \sum_{i=1}^{n} \xi_i^2, \tag{3.1}$$

$$|a_i(x, t, u, p)| + \left|\frac{\partial a_i(x, t, u, p)}{\partial u}\right| \leqslant \mu|p| + \varphi_1(x, t), \tag{3.2}$$

$$\left|\frac{\partial a_i(x, t, u, p)}{\partial x_j}\right| \leqslant \mu p^2 + \varphi_2(x, t), \tag{3.3}$$

$$|a(x, t, u, p)| \leqslant \mu p^2 + \varphi_3(x, t), \tag{3.4}$$

in which

$$\|\varphi_1, \varphi_2, \varphi_3\|_{2q, 2r, Q_T} \leqslant \mu_1, \tag{3.5}$$

with

$$\left.\begin{array}{l} \frac{1}{r} + \frac{n}{2q} = 1 - \varkappa_1 \quad \text{and} \\ q \in \left[\frac{n}{2(1-\varkappa_1)}, \infty\right], \quad r \in \left[\frac{1}{1-\varkappa_1}, \infty\right], \\ \qquad 0 < \varkappa_1 < 1 \quad \text{for} \quad n \geqslant 2, \\ q \in [1, \infty], \qquad r \in \left[\frac{1}{1-\varkappa_1}, \frac{2}{1-2\varkappa_1}\right], \\ \qquad 0 < \varkappa_1 < \frac{1}{2} \quad \text{for} \quad n = 1. \end{array}\right\} \tag{3.6}$$

We will prove that under the fulfilment of these conditions the quantities $\max_{Q'}|u_x|$ and $\langle u_x\rangle^{(\alpha)}_{Q'}$ for the solutions $u(x, t)$ of equations (0.1) and some $\alpha > 0$ are estimated from above by a constant depending only on the numerical parameters in conditions (3.1)–(3.6), $\max_{Q_T}|u|$ and the distance from Q' to Γ_T. For the sake of simplicity we will assume that $u \in O^{2,1}(Q_T)$. As to how to obtain these estimates for generalized solutions we refer the reader to Chapter IV of the book [65q], in which this is shown for equations of elliptic type. The indices q and r in (3.5) can be different for different functions. Thus, suppose $u(x, t) \in O^{2,1}(Q_T)$ and $\max_{Q_T}|u| = M$. We will first estimate $\|u_x\zeta\|_{L_2(Q_T)}$, where $\zeta(x, t)$ is a smooth function with values between 0 and 1 that is equal to zero on Γ_T. To this end we consider the equality

$$\int_0^t \int_\Omega \mathscr{L}u e^{\lambda u}\zeta^2 dx\, dt = 0, \quad \lambda = \text{const} \gg 1,$$

and transform it by means of an integration by parts to the form

$$\frac{1}{\lambda}\int_\Omega e^{\lambda u(x,t)}\zeta^2(x, t)dx\Big|_0^t + \int_0^t \int_\Omega e^{\lambda u}\Big[-\frac{1}{\lambda}2\zeta\zeta_t + a_i(x, t, u, u_x)(\lambda u_{x_i} + 2\zeta\zeta_{x_i}) + a\zeta^2\Big]dx\, dt = 0. \tag{3.7}$$

By virtue of assumptions (3.1) and (3.2) we have

$$a_i(x,t,u,p)p_i = p_i p_j \int_0^1 \frac{\partial a_i(x,t,u,\tilde{p})}{\partial \tilde{p}_j}\bigg|_{\tilde{p}=\tau p} d\tau + p_i a_i(x,t,u,0) \geqslant \nu p^2 - \sum_{i=1}^n |p_i|\varphi_1(x,t) \geqslant \frac{\nu}{2}p^2 - \frac{n}{2\nu}\varphi_1^2.$$

Therefore

$$\int_0^t\int_\Omega \frac{\lambda\nu}{2}u_x^2 e^{\lambda u}\zeta^2 dx\,dt \leqslant \int_0^t\int_\Omega e^{\lambda u}\Big[\frac{2}{\lambda}|\zeta\zeta_t| + \frac{\lambda n}{2\nu}\varphi_1^2 + (\mu|u_x| + \varphi_1)2\zeta|\zeta_x| + (\mu u_x^2 + \varphi_3)\zeta^2\Big]dx\,dt$$
$$\leqslant \int_0^t\int_\Omega e^{\lambda u}\Big(\frac{2}{\lambda}|\zeta\zeta_t| + \frac{\lambda n}{2\nu}\varphi_1^2 + \mu u_x^2\zeta^2 + \mu\zeta_x^2 + 2\varphi_1\zeta|\zeta_x| + \mu u_x^2\zeta^2 + \varphi_3\zeta^2\Big)dx\,dt. \tag{3.8}$$

We choose $\lambda = 8\mu/\nu$ and take into account the fact that $\|\phi_1, \phi_1^2, \phi_3\|_{1,Q_T} \leq c(\mu_1)$; then from (3.8) it is seen that

$$\int_{Q_T} u_x^2\zeta^2 dx\,dt \leqslant c, \tag{3.9}$$

where c is a constant depending only on n, $1/\nu$, μ, μ_1, M, $\max_{Q_T}(|\zeta_x|, |\zeta_t|)$ and mes Q_T (c is an increasing function of the indicated arguments).

Let us now show that it is possible to estimate the integrals $\int_{K_\rho}|u_x|^{2s}dx$ for any s (this need only be done for s not exceeding a certain fixed number s_0 determined by the parameters q and r of (3.6)) in terms of known parameters, as long as K_ρ and the concentric ball $K_{2\rho}$ belong to Ω and ρ does not exceed a certain number ρ_0. The number ρ_0 is also determined only by known parameters and s_0. It is easy to verify that the assumptions of Theorem 1.1 follow from conditions (3.1)–(3.6). We arbitrarily fix an interior cylinder Q' and carry out all considerations within it. By virtue of Theorem 1.1, for any $K_\rho \subset \overline{\Omega}'$ we have

$$\operatorname{osc}^2\{u, K_\rho\} \leqslant c\rho^{\alpha_1}. \tag{3.10}$$

Let us take a cylinder $K_{2\rho}\times(0, t_1)$ and a smooth function $\zeta(x,t)$ with values between 0 and 1 that is equal to zero in the vicinity of the lower base and lateral surface of this cylinder.

We consider the equality

$$-\int_0^t \int_{K_{2\rho}} \sum_{k=1}^n \mathscr{L}u \frac{d}{dx_k}(|u_x|^{2s} u_{x_k}\zeta^2)\,dx\,dt = 0,$$
$$t \leqslant t_1, \quad s = 0,\ 1,\ \ldots, \tag{3.11}$$

and transform it by means of a double integration by parts to the form

$$\frac{1}{2s+2}\int_{K_{2\rho}} |u_x(x,\ t)|^{2s+2}\zeta^2(x,\ t)\,dx + \int_0^t \int_{K_{2\rho}} \Big[\frac{-1}{2s+2}|u_x|^{2s+2}\,2\zeta\zeta_t$$
$$+\frac{da_l}{dx_k}\frac{d}{dx_l}(|u|^{2s}u_{x_k}\zeta^2) - a\frac{d}{dx_k}(|u_{x_k}|^{2s}u_{x_k}\zeta^2)\Big]\,dx\,dt = 0. \tag{3.12}$$

It is not hard to see that the passage from (3.11) to (3.12) is legitimate in spite of the fact that the derivatives u_{tx} and $D_x^3 u$ were used in the intervening transformations (see page 211).

We leave the trivially nonnegative terms in the left side, while the remaining terms are carried over to the right side and estimated from above, using inequality (1.2) of Chapter II and (3.1)–(3.6):

$$\frac{1}{2s+2}\int_{K_{2\rho}} |u_x(x,\ t)|^{2s+2}\zeta^2(x,\ t)\,dx$$
$$+\int_0^t \int_{K_{2\rho}} \Big[\frac{\partial a_l}{\partial u_{x_j}}|u_x|^{2s}u_{x_kx_l}u_{x_kx_j}\zeta^2$$
$$+\ 2s\frac{\partial a_l}{\partial u_{x_j}}|u_x|^{2s-2}u_{x_k}u_{x_kx_j}u_{x_l}u_{x_lx_i}\zeta^2\Big]\,dx\,dt$$
$$=\int_0^t \int_{K_{2\rho}} \Big[\frac{\partial a_l}{\partial u_{x_j}}u_{x_kx_j}|u_x|^{2s}u_{x_k}2\zeta\zeta_{x_l}$$
$$+\Big(\frac{\partial a_l}{\partial u}u_{x_k} + \frac{\partial a_l}{\partial x_k} - a\delta_k^l\Big)\frac{d}{dx_l}(|u_x|^{2s}u_{x_k}\zeta^2)$$
$$+\ \frac{1}{s+1}|u_x|^{2s+2}\zeta\zeta_t\Big]\,dx\,dt \leqslant \int_0^t \int_{K_{2\rho}} \Big[\varepsilon|u_x|^{2s}u_{xx}^2\zeta^2 + \frac{\mu^2}{\varepsilon}|u_x|^{2s+2}\zeta_x^2 +$$

$$+ (\mu u_x^2 + \varphi_1 |u_x| + \mu u_x^2 + \varphi_2 + \mu u_x^2 + \varphi_3) \sum_{l,k=1}^{n} \frac{d}{dx_l} (|u_x|^{2s} u_{x_k} \zeta^2)$$
$$+ \frac{1}{s+1} |u_x|^{2s+2} \zeta |\zeta_t| \Big] \, dx \, dt. \tag{3.13}$$

Hence

$$j \equiv \frac{1}{2s+2} \int_{K_{2\rho}} |u_x(x, t)|^{2s+2} \zeta^2(x, t) \, dx + \int_0^t \int_{K_{2\rho}} \Big[\frac{\nu}{2} |u_x|^{2s} u_{xx}^2 \zeta^2$$
$$+ \nu s |u_x|^{2s-2} \sum_{l=1}^{n} \Big(\sum_{k=1}^{n} u_{x_k} u_{x_k x_l} \Big)^2 \zeta^2 \Big] \, dx \, dt$$
$$\leqslant \int_0^t \int_{K_{2\rho}} \Big[\varepsilon u_{xx}^2 |u_x|^{2s} \zeta^2 + \frac{\mu^2}{\varepsilon} |u_x|^{2s+2} \zeta_x^2 + \varepsilon u_{xx}^2 |u_x|^{2s} \zeta^2$$
$$+ \frac{9n^2}{4\varepsilon} \mu^2 |u_x|^{2s+4} \zeta^2 + \varepsilon_1 s \sum_{l=1}^{n} \Big(\sum_{k=1}^{n} u_{x_k} u_{x_k x_l} \Big)^2 |u_x|^{2s-2} \zeta^2$$
$$+ \frac{9sn^2}{\varepsilon_1} \mu^2 |u_x|^{2s+4} \zeta^2 + 3n^2 \mu |u_x|^{2s+4} \zeta^2 + 3n^2 \mu |u_x|^{2s+2} \zeta_x^2$$
$$+ |u_x|^{2s+2} \zeta |\zeta_t| + (\varphi_1 |u_x| + \varphi_2 + \varphi_3) \sum_{l,k=1}^{n} \Big| (|u_x|^{2s} u_{x_k} \zeta^2)_{x_l} \Big| \Big] \, dx \, dt.$$

We choose $\epsilon = \nu/8$ and $\epsilon_1 = \nu/2$, and we transfer the terms containing second derivatives of u with respect to x to the left side and reduce similar terms. This gives us

$$j \leqslant c\Big(n, \frac{1}{\nu}, \mu\Big) \int_0^t \int_{K_{2\rho}} \Big[(s+1) |u_x|^{2s+4} \zeta^2$$
$$+ |u_x|^{2s+2} (\zeta_x^2 + |\zeta_t|) + (\varphi_1 |u_x| + \varphi_2 + \varphi_3)$$
$$\times \sum_{l,k=1}^{n} \big(|u_x|^{2s} u_{x_k} \zeta^2 \big)_{x_l} \Big] \, dx \, dt. \tag{3.14}$$

In order to estimate the first term of the right side we make use of the inequality

$$\int_{K_{2\rho}} |u_x|^{2s+4}\zeta^2\,dx \leqslant c_1\rho^{\alpha_1}\int_{K_{2\rho}} [(s+1)^2|u_x|^{2s}u_{xx}^2\zeta^2+|u_x|^{2s+2}\zeta_x^2]\,dx, \tag{3.15}$$

which by virtue of (3.10) follows from inequality (5.8) of Chapter II. The constant c_1 in it depends only on n and c from (3.10). Let us take ρ_0 such that

$$c_1 c\left(n, \frac{1}{\nu}, \mu\right)(s_0+1)^3\rho_0^{\alpha_1}=\frac{\nu}{4}. \tag{3.16}$$

Then for $\rho \le \rho_0$ and $s = s_0$ it follows from (3.14) and (3.15) that

$$\begin{aligned}\frac{1}{2s+2}\int_{K_{2\rho}}|u_x|^{2s+2}\zeta^2\,dx+\int_0^t\int_{K_{2\rho}}\Bigg[\frac{\nu}{4}|u_x|^{2s}u_{xx}^2\zeta^2 \\ +\nu s|u_x|^{2s-2}\sum_{l=1}^n\left(\sum_{k=1}^n u_{x_k}u_{x_kx_l}\right)^2\zeta^2\Bigg]\,dx\,dt \\ \leqslant c\left(n,\frac{1}{\nu},\mu\right)\int_0^t\int_{K_\rho}\{[c_1\rho_0^{\alpha_1}(s+1)+1]|u_x|^{2s+2}\zeta_x^2 \\ +|u_x|^{2s+2}\zeta_t+(\varphi_1+\varphi_2+\varphi_3)(1+|u_x|) \\ \times\sum_{i,k}\left|\left(|u_x|^{2s}u_{x_k}\zeta^2\right)_{x_i}\right|\}\,dx\,dt.\end{aligned} \tag{3.17}$$

Inequalities (3.17) have the same character as inequalities (11.7) of Chapter III. The differences in the constants and domains of integration are nonessential. The role of the function $\mathfrak{D}$ is played by the sum $\phi_1+\phi_2+\phi_3$, which, as can be seen from a comparison of conditions (3.2)–(3.6) with conditions (11.1) of Chapter III, has the same finite norm $\|\mathfrak{D}\|_{2q,2r,Q_T}$.

In the same way as in Chapter III, we obtain from (3.17) and (3.15) for $s = 0$ and from inequality (3.9) the estimate

$$\max_{0\leqslant t\leqslant T}\|u_x\zeta\|^2_{2,K_{2\rho}}+\int_0^T\int_{K_{2\rho}}[u_{xx}^2+|u_x|^4]\zeta^2\,dx\,dt\leqslant c,$$

and from it and equation (1.1) itself the estimate

$$\int_0^T\int_{K_{2\rho}}u_t^2\zeta^2\,dx\,dt\leqslant c,$$

where the c's are constants determined only by known quantities.

In addition, the following integral estimates were deduced from inequalities (11.7):

$$\max_{\varepsilon \leqslant t \leqslant T} \int_{K_\rho} |u_x(x, t)|^{2s+2}\,dx + \int_\varepsilon^T \int_{K_\rho} |u_x|^{2s+4}\,dx\,dt \leqslant c_{s_0}, \tag{3.18}$$
$$\varepsilon > 0,\ 0 \leqslant s \leqslant s_0,\quad \rho \leqslant \rho_0,$$

with the constants c_{s_0} depending only on n, M, s_0, the parameters from conditions (3.1)–(3.6) and $\epsilon > 0$. Consequently, estimates (3.18) are also valid in the present case as long as ρ_0 and s_0 are subject to condition (3.16). In the case when $u(x, 0)$ is summed over $K_{2\rho}$ with power $2s_0 + 2$ one can take the function ζ to be independent of t and thereby obtain estimates (3.18) also for $\epsilon = 0$.

Let us now proceed to an estimate of $\max_{Q'} |u_x|$ and $\langle u_x \rangle_{Q'}^{(\alpha)}$. Each of the derivatives u_{x_k} can be considered as a generalized solution of a certain linear equation whose coefficients satisfy all of the conditions used in §10 of Chapter III for estimating the Hölder norm of any generalized solution of such a linear equation in terms of its norm in $L_2(Q_T)$. To this end we transform the identity

$$-\int_0^T \int_{K_\rho} \mathscr{L} u \eta_{x_k}\,dx\,dt = 0 \tag{3.19}$$

by means of an integration by parts as follows:

$$\begin{aligned} &\int_{K_\rho} u_{x_k}(x, t)\,\eta(x, t)\,dx \\ &\quad + \int_0^t \int_{K_\rho} \Big[-u_{x_k}\eta_t + \Big(\frac{\partial a_l(x, t, u, u_x)}{\partial u_{x_j}} u_{x_k x_j} \\ &\quad + \frac{\partial a_l}{\partial u} u_{x_k} + \frac{\partial a_l}{\partial x_k} \Big) \eta_{x_l} - a\eta_{x_k} \Big]\,dx\,dt = 0; \end{aligned} \tag{3.20}$$

here we assume that $\eta(x, t)$ is equal to zero on the lower base and lateral surface of the cylinder $Q_1 = K_\rho \times (0, t_1)$. The passage from (3.19) to (3.20) is legitimate, although the intervening calculations contain the derivatives $u_{x_k t}$ and $D_x^3 u$ (see the explanation with regard to a similar situation on page 211). By virtue of (3.20) the function $u_k = u_{x_k}$ can be considered as a generalized solution from $V_2^{1,0}(Q_1)$ of the linear equation

$$\frac{\partial u_k}{\partial t} - \frac{\partial}{\partial x_l}\left(a_{lj}(x, t)\,u_{kx_j} + f_l^k(x, t)\right) = 0, \tag{3.21}$$

in which

$$a_{ij}(x, t)=\frac{\partial a_i(x, t, u(x, t), u_x(x, t))}{\partial u_{x_j}},$$

while

$$f_i^k(x, t)=\frac{\partial a_i(x, t, u(x, t), u_x(x, t))}{\partial u} u_{x_k}(x, t) + \frac{\partial a_i(x, t, u(x, t), u_x(x, t))}{\partial x_k} - \delta_k^i a(x, t, u(x, t), u_x(x, t)).$$

The functions $a_{ij}(x, t)$ and $f_i(x, t)$ making up this equation, by virtue of our assumptions on the solution $u(x, t)$ and the functions $a_i(x, t, u, p)$ and $a(x, t, u, p)$, possess the following properties:

$$\nu(M)\xi^2 \leqslant a_{ij}(x, t)\xi_i\xi_j \leqslant \mu(M)\xi^2$$

and

$$\begin{aligned} |f_i^k(x, t)| &\leqslant (\mu|u_x(x, t)| + \varphi_1(x, t))|u_x(x, t)| \\ &+ \mu u_x^2(x, t) + \varphi_2(x, t) + \mu u_x^2(x, t) + \varphi_3(x, t) \\ &\leqslant 3\mu u_x^2(x, t) + \varphi_2 + \varphi_3 + \frac{1}{1+\delta}\varphi_1^{1+\delta} + \frac{\delta}{1+\delta}|u_x|^{\frac{1+\delta}{\delta}}, \end{aligned}$$

where δ is an arbitrary positive number. We take δ so small that the numbers $q_1 = q/(1+\delta)$ and $r_1 = r/(1+\delta)$ satisfy a condition of type (3.6); namely,

$$\frac{1}{r_1}+\frac{n}{2q_1}=\frac{1+\delta}{r}+\frac{(1+\delta)n}{q}=(1-\varkappa_1)(1+\delta)<1. \tag{3.22}$$

For such a δ the function $\phi_2 + \phi_3 + (1/(1+\delta))\phi_1^{1+\delta_1}$ will have a finite norm in L_{q_1,r_1}. In order for $3\mu u_x^2 + (\delta/(1+\delta))|u_x|^{(1+\delta)/\delta}$ to belong to L_{q_1,r_1} it is sufficient, in consequence of (3.18), to take $s_0 = [mq_1]$, where $m = \max\{1, (1+\delta)/2\delta\}$, and then f_i^k will belong to $L_{q_1,r_1}(Q(\rho_0, \tau))$ in an arbitrary cylinder $Q(\rho_0, \tau) \subset Q_T$ that is separated from Γ_T (or from S_T if $u(x, 0) \in L_{2s_0+2}(\Omega)$) by a positive distance, with ρ_0 determined by condition (3.16).

In view of this Theorems 8.1 and 10.1 of Chapter III apply to $u_k = u_{x_k}$ as the solution of the linear equation (3.21). They and considerations analogous to those developed in §6 of Chapter III guarantee the validity of the following proposition.

Theorem 3.1. *Suppose $u(x, t)$ is a solution from $O^{2,1}(Q_T)$ of equation (0.1)* *with* $\max_{Q_T}|u| = M$. *If the functions* $a_i(x, t, u, p)$ *and* $a(x, t, u, p)$ *in equation* (0.1) *satisfy conditions* (3.1)–(3.6), *then for any domain* $Q' \subset Q_T$ *separated from* Γ_T *by a positive distance* d *the quantity* $|u_x|^{(\alpha)}_{Q'}$ *with some* $\alpha > 0$ *and the norms* $\|u_{xx}, u_t\|_{2,Q'}$ *are estimated from above by a constant depending only on* n, M, *the parameters* ν, μ, μ_1, q *and* r *from conditions* (3.1)–(3.6), *and the distance* d. *The number* $\alpha > 0$ *is determined by* n, ν, μ, q *and* r. *If furthermore it is known that* $\max_{x\in\Omega}|u_x(x, 0)| < \infty$ (*or* $|u_x(x, 0)|^{(\beta)}_{\Omega} < \infty$), *then* $\max_{Q'}|u_x|$ (*respectively* $|u_x|^{(\alpha)}_{Q'}$) *and the norms* $\|u_{xx}, u_t\|_{2,Q'}$, *for any domain* $Q' \subset Q_T$ *separated from* S_T *by a positive distance* d, *are finite and are estimated from above by a constant determined only by* $n, M, \nu, \mu, \mu_1, q, r, d$ *and* $\max_{x\in\Omega}|u_x(x, 0)|$ (*respectively* $|u_x(x, 0)|^{(\beta)}_{\Omega}$ *and* β).

§4. AN ESTIMATE OF $\max|u_x|$ IN THE WHOLE DOMAIN

Let us now concern ourselves with an estimate of $\max|u_x|$ in the vicinity of S_T. Suppose the functions $a_i(x, t, u, p)$ and $a(x, t, u, p)$ satisfy conditions (3.1)–(3.6), while the solution $u(x, t)$ of equation (0.1) is an element of $O^{2,1}(\overline{Q}_T)$. We first assume that $u|_{S_T} = 0$ and S_T satisfies condition (A). From the equality

$$\int_0^t \int_\Omega \mathscr{L}u\,(e^{\lambda u} - 1)\,dx\,dt = 0, \quad t \leqslant T,$$

where λ is a sufficiently large number, we obtain the estimate

$$\int_{Q_T} u_x^2\,dx\,dt \leqslant c, \tag{4.1}$$

where c is a constant depending only on n, $M = \operatorname{vrai}\max_{Q_T}|u|$, ν, μ, μ_1 from conditions (3.1)–(3.6), T and $\operatorname{mes}\Omega$ (c is an increasing function of these arguments). This is deduced in literally the same way as inequality (3.9) was deduced from (3.7). We will now obtain estimates of the integrals $\int|u_x|^{2s+2}\,dx$ for any s and for domains adjacent to S. This is done in basically the same way as in §11 of Chapter III and in §3 of the present chapter.

Let us assume that

$$\max|u_x|\,|_{S_T} = M_2. \tag{4.2}$$

We consider the function

$$b(x, t)=\begin{cases} 0 & \text{for } |u_x(x, t)|^2 \leqslant M_2^2 \equiv \hat{M}, \\ |u_x|^2-\hat{M} & \text{for } \hat{M} \leqslant |u_x|^2 \leqslant \hat{M}+1, \\ 1 & \text{for } |u_x|^2 \geqslant \hat{M}+1, \end{cases}$$

which is equal to zero on S_T.

We form with its help the following integral identity:

$$-\int_0^t \int_{\Omega_{2\rho}} \sum_{k=1}^n \mathscr{L}u \frac{\partial}{\partial x_k}(u_{x_k} b\zeta^2)\, dx\, dt = 0, \tag{4.3}$$

where $\Omega_{2\rho} = K_{2\rho} \cap \Omega$, $K_{2\rho}$ is an arbitrary ball with center on S and ρ not exceeding some number ρ_0, while $\zeta(x)$ is a cutting function for $K_{2\rho}$.

Let us denote u_x^2 by v. We transform the term containing u_t as follows:

$$\begin{aligned} \frac{1}{2}\int_0^t \int_{\Omega_{2\rho}} v_t b\zeta^2\, dx\, dt &= \frac{1}{2}\int_{\Omega_{2\rho}} vb\zeta^2\, dx \Big|_0^t - \frac{1}{2}\int_0^t \int_{\Omega_{2\rho}} v(b\zeta^2)_t\, dx\, dt \\ &= \frac{1}{2}\int_{\Omega_{2\rho}} vb\zeta^2\, dx \Big|_0^t - \frac{1}{2}\int_0^t \int_{\Omega_{2\rho}} (b+\hat{M})\, b_t\zeta^2\, dx\, dt \\ &= \frac{1}{2}\int_{\Omega_{2\rho}} vb\zeta^2\, dx \Big|_0^t - \frac{1}{4}\int_{\Omega_{2\rho}} (b+\hat{M})^2\zeta^2\, dx \Big|_0^t \\ &= \frac{1}{2}\int_{\Omega_{2\rho}} \left(vb - \frac{b^2}{2} - \hat{M}b\right)\zeta^2\, dx \Big|_0^t. \end{aligned}$$

The remaining terms are transformed by means of a double integration by parts. As a result, from (4.3) we obtain

$$\begin{aligned} \frac{1}{2}\int_{\Omega_{2\rho}} \left(vb - \frac{b^2}{2} - \hat{M}b\right)\zeta^2\, dx \Big|_0^t &+ \int_0^t \int_{\Omega_{2\rho}} \bigg[\bigg(\frac{\partial a_l}{\partial u_{x_j}} u_{x_k x_j} \\ &+ \frac{\partial a_l}{\partial u} u_{x_k} + \frac{\partial a_l}{\partial x_k}\bigg)(u_{x_k x_l} b\zeta^2 + u_{x_k} b_{x_l}\zeta^2 + 2u_{x_k} b\zeta\zeta_{x_l}) \\ &+ a\left(\Delta u b\zeta^2 + u_{x_k} b_{x_k}\zeta^2 + 2u_{x_k} b\zeta\zeta_{x_k}\right)\bigg]\, dx\, dt = 0. \end{aligned} \tag{4.4}$$

The passage from (4.3) to (4.4) is legitimate in spite of the fact that the derivatives u_{xt} and $D_x^3 u$ were used in the intervening calculations. In the left side of equality (4.4) we leave the positive integrals

$$\frac{1}{2}\int_{\Omega_{2\rho}} vb\zeta^2\,dx\Big|^t, \quad \int_0^t\int_{\Omega_{2\rho}} \frac{\partial a_l}{\partial u_{x_j}} u_{x_k x_j} u_{x_k x_l} b\zeta^2\,dx\,dt,$$

$$\int_0^t\int_{\Omega_{2\rho}} \frac{\partial a_l}{\partial u_{x_j}} u_{x_k x_j} u_{x_k} b_{x_l}\zeta^2\,dx\,dt = \frac{1}{2}\int_0^t\int_{\Omega_{2\rho}} \frac{\partial a_l}{\partial u_{x_j}} b_{x_l} b_{x_j}\zeta^2\,dx\,dt$$

and carry out estimates of the moduli of the remaining terms from above, using conditions (3.1)–(3.4) and Cauchy's inequality. As a result of these estimates and an estimate of the positive terms of the left side from below we obtain

$$\begin{aligned}\frac{1}{2}\int_{\Omega_{2\rho}} vb\zeta^2 dx\Big|^t &+ \nu\int_0^t\int_{\Omega_{2\rho}}\left(u_{xx}^2 b + \frac{1}{2}b_x^2\right)\zeta^2\,dx\,dt\\ &\leqslant \frac{1}{2}\int_{\Omega_{2\rho}} vb\zeta^2\,dx\Big|^{t=0} + \frac{1}{2}\int_{\Omega_{2\rho}}\left(\frac{b^2}{2}+\hat{M}b\right)\zeta^2\,dx\Big|_0^t\\ &+ c\int_0^t\int_{\Omega_{2\rho}}\sum_{i=1}^3 \varphi_i^2\,dx\,dt + c\int_0^t\int_{\Omega_{2\rho}} vb\zeta_x^2\,dx\,dt\\ &+ c\int_0^t\int_{\Omega_{2\rho}} \varphi_1^2 u_x^2 b^2\zeta^2\,dx\,dt + c\int_0^t\int_{\Omega_{2\rho}} v^2 b\zeta^2\,dx\,dt. \end{aligned} \tag{4.5}$$

One can easily see that the left side of (4.5) is estimated from below in terms of

$$c'\,|w|^2_{\Omega_{2\rho}\times(0,t)} - c\int_0^t\int_{\Omega_{2\rho}} vb\zeta_x^2\,dx\,dt, \text{ where } w = |u_x|\,b\zeta,\ c'>0.$$

The first four integrals in the right side of (4.5) are bounded by known quantities, since $0 \leq b(x, t) \leq 1$ and estimate (4.1) is already known for the integral $\int_{Q_T} v\,dx\,dt$. The latter integrals can be estimated in the following manner by means of inequality (3.6) of Chapter II, taking into account conditions (3.5) and (3.6), as well as the boundary condition $u|_S = 0$ and the Hölder continuity of u in x:

$$\int_0^t\int_{\Omega_{2\rho}} \varphi_1^2 w^2\,dx \leqslant c_1\rho^{\varepsilon}\,|w|^2_{\Omega_{2\rho}\times(0,t)},$$

$$\varepsilon = \frac{n(q-1)\varkappa_1}{q(2\varkappa_1+n)},$$

$$\int\limits_{\Omega_{2\rho}} |u_x|^4 b\zeta^2\,dx = \int\limits_{\Omega_{2\rho}} \sum_{k=1}^{n} u_{x_k}^2 vb\zeta^2\,dx$$
$$= -\int\limits_{\Omega_{2\rho}} u\,[\Delta ub\zeta^2 + 2u_{x_k}u_{x_l}u_{x_lx_k}b\zeta^2$$
$$+ u_{x_k}vb_{x_k}\zeta^2 + 2u_{x_k}vb\zeta\zeta_{x_k}]\,dx$$
$$\leqslant c_1\rho^\alpha \int\limits_{\Omega_{2\rho}} \left(u_{xx}^2 b\zeta^2 + \frac{1}{2}b_x^2\zeta^2\right)dx + c_1\rho^\alpha \int\limits_{\Omega_{2\rho}} v^2 b\zeta^2\,dx + c_1. \tag{4.6}$$

Choosing

$$\rho^\varepsilon + \rho^\alpha \leqslant \frac{1}{4c_1(c+1)} \min\{1;\ \nu;\ c'\},$$

from (4.6) and (4.5) we get

$$\max_{0\leqslant t\leqslant T} \int\limits_{\Omega_{2\rho}} vb\zeta^2\,dx + \nu\int\limits_0^T \int\limits_{\Omega_{2\rho}} (u_{xx}^2 b + b_x^2)\zeta^2\,dx\,dt$$
$$+ \int\limits_0^T \int\limits_{\Omega_{2\rho}} v^2 b\zeta^2\,dx\,dt \leqslant \text{const.}$$

These estimates together with equation (1.1) and analogous estimates for the interior balls $K_{2\rho}$ obtained in §3, guarantee the inequalities

$$\left.\begin{aligned} &\max_{0\leqslant t\leqslant T} \int\limits_{\Omega} u_x^2(x,\ t)\,dx + \int\limits_{Q_T} |u_x|^4\,dx\,dt \leqslant c_2, \\ &\int\limits_0^T \int\limits_{\Omega'} (u_{xx}^2 + u_t^2)\,dx\,dt \leqslant c_2', \end{aligned}\right\} \tag{4.7}$$

in which the constant c_2 depends on c from (4.1), M_2, $\|u_x(x,\ 0)\|_{2,\Omega}$, $|u|_{Q_T}^{(\alpha)}$ and α, while c_2' also depends on the distance from Ω' to S.

The above arguments are insufficient for obtaining an estimate of $\int_0^t\int_\Omega u_{xx}^2\,dx\,dt$ since b vanishes on the boundary of the domain. To this end special considerations are required in the vicinity of the boundary, which we will now develop under the assumption that $S \in O^2$. Suppose S_1 is a not large portion of the boundary S, while $\zeta(x)$ is a cutting function for a ball K_ρ with center on S_1 and not intersecting $S\setminus S_1$. Without loss in generality one can assume that the equation for S_1 has the form $x_n = 0$. In the same way as in the interior estimates we consider equality (3.11) with $\zeta = 0$ and transform it by means of an integration by parts to a form analogous to (3.12). Since $\zeta(x)$ does not equal zero on S_1 it follows from a comparison with the case of interior estimates in (3.12) that there will exist additional integrals over S_1:

$$\int_0^t \int_{S_1} a_i [u_{x_k x_k} \operatorname{Cos}(n, x_i) - u_{x_k x_i} \operatorname{Cos}(n, x_k)] \zeta^2 \, dx \, dt$$

$$+ \int_0^t \int_{S_1} a_i u_{x_k} [2\zeta \zeta_{x_k} \operatorname{Cos}(n, x_i) - 2\zeta \zeta_{x_i} \operatorname{Cos}(n, x_k)] \, dx \, dt \equiv I_1 + I_2.$$

The second of these integrals is obviously estimated in terms of M_2 and known constants. The first, on account of the boundary condition $u|_{S_T} = 0$, takes the form

$$I_1 = \int_0^t \int_{S_1} (a_n u_{x_n x_n} - a_i u_{x_n x_i}) \zeta^2 \, dx \, dt$$

$$= \int_0^t \int_{S_1} \sum_{i=1}^{n-1} a_i u_{x_n x_i} \zeta^2 \, dx \, dt.$$

Introducing the notation $b_i(x, t, u, p) = \int_0^{p_n} a_i(x, t, u, p_1, \cdots, p_{n-1}, \tau) \, d\tau$, one can transform I_1 as follows:

$$I_1 = \int_0^t \int_{S_1} \sum_{i=1}^{n-1} \left[\frac{db_i(x, t, u, u_x)}{dx_i} - \frac{\partial b_i(x, t, u, u_x)}{\partial x_i} \right] \zeta^2 \, dx \, dt$$

$$= - \int_0^t \int_{S_1} \sum_{i=1}^{n-1} (2b_i \zeta \zeta_{x_i} + b_{ix_i} \zeta^2) \, dx \, dt.$$

Hence we see that the integral I_1 is estimated in the same way as I_2 in terms of M_2 and known constants. In other respects the estimates of $\int_0^t \int_\Omega u_{xx}^2 \zeta^2 \, dx \, dt$ in the vicinity of the boundary do not differ from the interior estimates. In this way we estimate $\int_0^T \int_\Omega u_{xx}^2 \zeta^2 \, dx \, dt$. Using equation (1.1) again and the second of the inequalities (4.7) we arrive at the estimate

$$\int_0^T \int_\Omega (u_{xx}^2 + u_t^2) \, dx \, dt \leq c_3, \tag{4.8}$$

where c_3 depends only on the quantities determining c_2 in (4.7) and on the norms in O^2 of the functions defining the boundary S.

In order to obtain estimates of the integrals of higher powers of u_x it is necessary to employ the relations

$$-\int_0^t \int_{A_{2\rho,\hat{M}}(t)} \mathscr{L}u \frac{d}{dx_k}\left[(v-\hat{M})^s u_{x_k}\zeta^2\right] dx\,dt = 0,$$
$$s = 1, 2, \ldots,$$

where, as above, $v = u_x^2$, $\hat{M} = M_2^2$, $A_{2\rho,\hat{M}}(t)$ is the set of points $x \in \Omega_{2\rho} = K_{2\rho} \cap \Omega$ at which $v(x, t) > \hat{M}$, while $K_{2\rho}$ is a ball with center on S, and with them to develop arguments analogous to the derivation of estimates (3.18) from (3.11). This leads to the inequalities

$$\max_{0\leqslant t\leqslant T} \int_{\Omega_\rho} |u_x|^{2s+2} dx + \int_0^T \int_{\Omega_\rho} |u_x|^{2s+4} dx\,dt \leqslant c_s \tag{4.9}$$

for $1 \le s \le s_0$ and $\rho \le \rho_0$, where s_0 and ρ_0 are connected by a condition of type (3.16). Inequalities (4.7) and (4.9) together with the same estimates (3.18) for interior cylinders permit one to conclude that for any $s > 0$

$$\max_{0\leqslant t\leqslant T} \int_{\Omega} |u_x|^{2s+2} dx + \int_0^T \int_{\Omega} |u_x|^{2s+4} dx\,dt \leqslant c'_s, \tag{4.10}$$

where c'_s is some constant depending only on s, M_2, n, the constants ν, μ, μ_1, q, r from conditions (3.1)–(3.6), $\operatorname{mes}\Omega$, T, $\|u_x(x, 0)\|_{2s+2,\Omega}$ and $|u|_{Q_T}^{(\alpha)}$. By virtue of Theorem 1.1 the norm $|u|_{Q_T}^{(\alpha)}$ is estimated in terms of $|u|_{\Gamma_T}^{(\beta)}$, β, $\max_{Q_T}|u|$ and the constants in conditions (A) and (3.1)–(3.6). Further, by arguing in the same way as at the end of the preceding section, we show that $u_k \equiv u_{x_k}$ is a solution of the linear equation (3.21) with the a_{ij} and f_i satisfying the conditions of Theorems 8.1 and 10.1 of Chapter III; and hence their conclusions hold for u_{x_k}. Thus the following theorem is proved.

Theorem 4.1. *Suppose $u(x, t) \in O^{2,1}(\bar{Q}_T)$ satisfies equation* (0.1) *and vanishes on S_T, which is assumed to satisfy condition* (A). *If the functions $a_i(x, t, u, p)$ and $a(x, t, u, p)$ satisfy conditions* (3.1)–(3.6), *then $\max_{Q_T}|u_x|$ is estimated from above by a constant depending only on the parameters in conditions* (3.1)–(3.6), *n, $\max_{Q_T}|u|$, $\max_{\Gamma_T}|u_x|$, $\operatorname{mes}\Omega$, T and a_0 and θ_0 in condition* (A), *while $|u_x|_{Q_T}^{(\alpha)}$ is estimated from above by a constant depending on the same parameters, $|u_x|_{\Gamma_T}^{(\beta)}$ and β. The number α is determined here by n, μ, ν, μ_1, q, r and* *β. If furthermore* $s \in O^2$, *then* $\|u_{xx}, u_t\|_{2,Q_T}$ *is estimated from above by a constant depending only on the parameters from conditions* (3.1)–(3.6), n, $\max_{Q_T}|u|$, $\max_{\Gamma_T}|u_x|$, mes Ω, T *and the boundary* S.

This theorem given an estimate of $\max_{Q_T}|u_x|$, $\|u_{xx}, u_t\|_{2,Q_T}$ and $|u_x|^{(\alpha)}_{Q_T}$ in terms of the quantities $\max_{\Gamma_T}|u_x|$ and $|u_x|^{(\beta)}_{\Gamma_T}$, which are not known from the conditions of the boundary value problems. In Chapter VI (see Lemma 3.1) $\max_{\Gamma_T}|u_x|$ will be estimated for solutions of the first boundary value problem in terms of known quantities only, and this will be done directly for solutions of general quasi-linear equations without referring to any results of the present chapter. An estimate of $|u_x|^{(\alpha)}_{Q_T}$ will be given in the following section in terms of only known quantities and $\max_{Q_T}|u_x|$.

Theorem 4.1 is easily generalized to the case of a nonhomogeneous boundary condition. To this end it is necessary to introduce in place of $u(x, t)$ a new unknown function $v(x, t) = u(x, t) - \hat{\psi}(x, t)$, where $\hat{\psi}(x, t)$ is some function from $O^{2,1}(\overline{Q}_T)$ that is equal to u on S_T, and to note that Theorem 4.1 is applicable to it. This leads to the following proposition.

Theorem 4.2. *Theorem* 4.1 *is also valid for the case when* u *coincides on* S_T *with the values of some function* $\hat{\psi}(x, t)$ *from* $O^{2,1}(\overline{Q}_T)$. *The estimates of* $\max_{Q_T}|u_x|$, $\|u_{xx}, u_t\|_{2,Q_T}$ *and* $|u_x|^{(\alpha)}_{Q_T}$ *will depend in this connection on the norm of* $\hat{\psi}$ *in* $O^{2,1}(\overline{Q}_T)$, *i.e. on* $\max_{Q_T}|\psi, \psi_x, \psi_{xx}, \psi_t|$.

§5. ESTIMATE OF $\langle u_x\rangle^{(\alpha)}_{Q'}$ AND HIGHER DERIVATIVES IN AN ARBITRARY SUBDOMAIN OF THE DOMAIN Q_T

In this section we give estimates of the quantities $\langle u_x\rangle^{(\alpha)}_{Q'}$ and $|u_t, u_{xx}|^{(\alpha)}_{Q'}$ for solutions $u(x, t)$ in an arbitrary (not necessarily strictly interior) subdomain Q' of the domain Q_T. To this end we first prove a theorem guaranteeing an estimate of $\max_{Q'}|u_t|$.

Theorem 5.1. *Suppose that a solution* $u(x, t)$ *of equation* (0.1) *belongs to* $C^{2,1}(\overline{Q}_T)$ *and has* $\max_{Q_T}|u| = M$, $\max_{Q_T}|u_x| = M_1$ *and suppose that for* $(x, t) \in \overline{Q}_T$, $|u| \le M$, $|p| \le M_1$ *the functions* $a_i(x, t, u, p)$ *and* $a(x, t, u, p)$ *satisfy a Lipschitz condition in* t, *are differentiable with respect to* u *and* p *and are subject to the inequalities*

$$\nu\xi^2 \leqslant \frac{\partial a_i(x, t, u, p)}{\partial p_j}\xi_i\xi_j \leqslant \mu\xi^2, \tag{5.1}$$

$$\left| \frac{\partial a_i(x, t, u, p)}{\partial u}, \frac{a_i(x, t+h, u, p) - a_i(x, t, u, p)}{h}, \right.$$
$$\left. \frac{\partial a}{\partial p}, \frac{\partial a}{\partial u}, \frac{a(x, t+h, u, p) - a(x, t, u, p)}{h} \right| \leqslant \varphi(x, t), \tag{5.2}$$

where $\|\phi\|_{q,r,Q_T} \le \mu_1$, *and the numbers* q *and* r *satisfy condition* (3.6). *Then* $\max_{Q_T}|u_t|$ *is estimated from above by a constant depending only on* n, ν, μ_1, q, r *and* $\max_{\Gamma_T}|u_t|$. *If* Γ' *is an arbitrary portion of the surface* Γ, *then* $\max_{Q'}|u_t|$, *where* Q' *is any part of the cylinder* Q_T *that does not intersect with* $\Gamma_T \setminus \Gamma'$, *is estimated in terms of* n, ν, μ, μ_1, q, r, $\max_{\Gamma'}|u_t|$ *and the distance from* Q' *to* $\Gamma_T \setminus \Gamma'$.

For a proof we consider in the cylinder Q_{T-h} the function

$$v^h(x, t) = \frac{1}{\Delta t}[u(x, t+\Delta t) - u(x, t)], \quad \Delta t = h > 0,$$

and show that it is a generalized solution of a certain linear equation. We take the divided difference in t of both sides of equation (0.1):

$$\frac{\Delta \mathscr{L} u}{\Delta t} \equiv \frac{\partial v^h}{\partial t} - \frac{d}{dx_i}\frac{\Delta a_i}{\Delta t} + \frac{\Delta a}{\Delta t} = 0$$

and represent $\Delta a_i/\Delta t$, $\Delta a/\Delta t$ in the following form:

$$\begin{aligned}
\frac{\Delta a_i}{\Delta t} &= \frac{1}{\Delta t}\{a_i[x, t+\Delta t, u(x, t+\Delta t), u_x(x, t+\Delta t)] \\
&\qquad - a_i[x, t, u(x, t), u_x(x, t)]\} \\
&= \frac{1}{\Delta t}\int_0^1 \frac{\partial}{\partial \tau} a_i[x, t+\Delta t, \tau u(x, t+\Delta t) + (1-\tau)u(x, t), \\
&\qquad \tau u_x(x, t+\Delta t) + (1-\tau)u_x(x, t)]\, d\tau \\
&+ \frac{1}{\Delta t}\{a_i[x, t+\Delta t, u(x, t), u_x(x, t)] - \\
&\qquad - a_i[x, t, u(x, t), u_x(x, t)]\} \\
&= v^h_{x_j}\int_0^1 \frac{\partial a_i[\dots]}{\partial u_{x_j}}\, d\tau + v^h\int_0^1 \frac{\partial a_i[\dots]}{\partial u}\, d\tau + \frac{1}{h}\{\dots\} \\
&\equiv a_{ij}v^h_{x_j} + b_i v^h + b,
\end{aligned} \tag{5.3}$$

and analogously

$$\frac{\Delta a}{\Delta t} = c_i v^h_{x_i} + c v^h + d.$$

This gives the following relation for v^h:

$$\frac{\partial v^h}{\partial t} - \frac{d}{dx_i}[a_{ij}v^h_{x_j} + b_i v^h + b] + c_i v^h_{x_i} + cv^h + d = 0, \tag{5.4}$$

which can be considered as a linear parabolic equation for v^h with coefficients and free terms satisfying the conditions of Theorems 7.1 and 10.1 of Chapter III.

By virtue of Theorem 7.1

$$\max_{Q_{T-h}} |v^h| \leq 2c \max \{\max_{\Gamma_{T-h}} |v^h|;\ 1\}, \tag{5.5}$$

where c is determined by n, ν, μ_1, q and r and does not depend on h, and hence an analogous estimate is valid for $\lim_{h\to 0} v^h = u_t$. The first part of Theorem 5.1 is proved.

For a proof of the second of its assertions, on local estimates of $\max_{Q'} |u_t|$, we make use of Theorem 8.1 of Chapter III, which implies the inequality

$$\max_{Q'} |v^h| \leq c'$$

for any part Q' of the domain $\overline{Q}_T$. If Q' is an interior subdomain of Q_T, then the quantity c' is determined by the constants determining c in (5.5) and the distance from Q' to Γ_{T-h}, while for cylinders adjacent to a portion Γ' of the surface Γ_T, the quantity c' depends only on n, ν, μ, μ_1, q, r, $\max_{\Gamma'} |v^h|$ and the distance from Q' to $\Gamma_T \setminus \Gamma'$. It is clear that analogous estimates will also hold for u_t.

Remark 5.1. On the basis of Theorem 10.1 of Chapter III, for v^h, and hence also for u_t, it is possible to estimate $|v^h|^{(\alpha)}_{Q'}$ in terms of the same quantities as $\max_{Q'} |v^h|$ and the distance from Q' to Γ_T if the latter is positive. If S satisfies condition (A) and $|u_t|^{(\beta)}_{\Gamma_T} < \infty$, then it is also possible to estimate $|u_t|^{(\alpha)}_{Q_T}$.

Remark 5.2. It is not difficult to see that the assertion of Theorem 5.1 remains in force for solutions $u(x, t)$ from $C^{2,1}(\overline{Q}_T \setminus S_0) \cap C(\overline{Q}_T)$ with $\operatorname{vrai\,max}_{\Gamma_T} |u_t| < \infty$ where $S_0 = \{x \in S,\ t = 0\}$, it being only necessary to understand $\max_{\Gamma_T} |u_t|$ as

$$\lim_{\Delta t \to +0} \sup_{(x, t) \in \Gamma_{T-h}} \left| \frac{u(x, t+\Delta t) - u(x, t)}{\Delta t} \right|.$$

Let us now proceed to an estimate of $\langle u_x \rangle^{(\alpha)}_{Q_T}$, assuming that S belongs to O^2, while the functions $a_i(x, t, u, p)$, $a(x, t, u, p)$ satisfy in addition to conditions (5.1), (5.2) the following conditions for $|u| \leq M$, $|p| \leq M_1$:

$$\sup_{t\in[0,T]}\left\|\frac{\partial a_i(x,t,u,p)}{\partial u},\ \frac{\partial a_i}{\partial x_j},\ a\right\|_{q_2,\Omega}\leqslant\mu_2,\ q_2>n. \tag{5.6}$$

To this end we consider (0.1) for fixed t as an elliptic equation

$$-\frac{d}{dx_i}(a_i(x,t,u,u_x))+f(x,t)=0 \tag{5.7}$$

with free term

$$f(x,t)=u_t(x,t)+a(x,t,u(x,t),u_x(x,t)).$$

Conditions (5.6) and the estimates of $\max_{Q'}|u_t|$ established in Theorem 5.1 imply the boundedness of the norms

$$\|f(x,t)\|_{q_2,\Omega'}\leqslant\mu_3(\Omega'),\quad q_2>n. \tag{5.8}$$

For such equations the following assertion is valid.

Theorem 5.2. *Suppose $u(x)$ belongs to $W_2^2(\Omega)$, has* vrai $\max_\Omega|u_x|\le M_1$ *and satisfies almost everywhere in Ω equation (5.7) (in which t is fixed: $t=t_0$). If the conditions*

$$\nu\xi^2\leqslant\frac{\partial a_i(x,t_0,u(x),u_x(x))}{\partial u_{x_j}}\xi_i\xi_j\leqslant\mu\xi^2,\quad\nu>0,$$

$$\left\|\frac{\partial a_i(x,t_0,u(x),u_x(x))}{\partial u},\ \frac{\partial a_i}{\partial x_j},\ f\right\|_{q_2,\Omega}\leqslant\mu_2,\quad q_2>n,$$

are fulfilled, then for any subdomain $\Omega'\subset\Omega$ separated from $S\setminus S'$ by a positive distance d the quantity $\langle u_x(x)\rangle_{\Omega'}^{(\alpha)}$ is estimated from above by a constant depending only on n, ν, μ, μ_2, q_2, M_1, $\|u\|_{q_3,S'}^{(2-1/q_3)}$, $q_3>n$, S' and the distance d. The constant $\alpha>0$ is determined by the same quantities except for d. The piece of boundary S' (it can be the empty set) must be a surface of class O^2 (or at least $W_{q_3}^2$, $q_3>n$).

This theorem is proved in §6 of Chapter IV of the book [65q] (see Theorem 6.1 and the remark to it), only there the dependence on the boundary values of u is expressed in terms of the norms $\|\psi\|_{q_3,\Omega}^{(2)}$, where $\psi(x)$ is some function coinciding with u on S. But this difference is nonessential since by virtue of Theorem 2.3 of Chapter II the norms $\|\hat\psi\|_{q_3,S}^{(2-1/q_3)}$ and $\inf\|\psi\|_{q_3,\Omega}^{(2)}$, where the infimum is taken over all functions ψ from $W_{q_3}^{(2)}(\Omega)$ that are equal to $\hat\psi$ on S, are equivalent.

On the basis of Theorem 5.2 and the assumptions stated above, for the solution

$u(x, t)$ in question, we obtain an estimate of $\langle u_x(x, t)\rangle^{(\alpha)}_{x,\Omega'}$ for arbitrary domains $\Omega' \subset \Omega$ in terms of known quantities.

Suppose $\max_{0\le t\le T}\langle u_x(x, t)\rangle^{(\alpha)}_{x,\Omega} \le c$. By virtue of Lemma 3.1 of Chapter II, from this inequality and the inequality $\max_{Q_T}|u_t| \le c_1$ there follows the inequality $\langle u_x\rangle^{(\alpha/(1+\alpha))}_{t,Q_T} \le c_2$ with c_2 depending only on c, c_1, α and the boundary S. Analogously, using local estimates of $\max_{Q'}|u_t|$ and $\langle u_x\rangle^{(\alpha)}_{x,\Omega'}$, we obtain an estimate of $\langle u_x\rangle^{(\alpha/(1+\alpha))}_{t,Q'}$ for any $Q' \subset Q_T$. In this way the following proposition is proved.

Theorem 5.3. *Suppose the functions $a_i(x, t, u, p)$ and $a(x, t, u, p)$ satisfy conditions* (5.1), (5.2) *and* (5.6), *while $S \in O^2$. Then for any solution $u \in C^{2,1}(\overline{Q}_T)$ of equation* (0.1) *with $\max_{Q_T}|u| = M$ and $\max_{Q_T}|u_x| = M_1$ the norm $|u_x|^{(\alpha)}_{Q_T}$ is estimated from above by a constant depending only on n, ν, μ, μ_1, q, r, μ_2, q_2, M, M_1, $\max_{\Gamma_T}|u_t|$, S, $\max_{0\le t\le T}\|u(x, t)\|^{(2-1/q_3)}_{q_3,S}$ and q_3, with $q_3 > n$. For any part Q' of the cylinder Q_T intersecting with a portion Γ' of the surface Γ_T and separated from $\Gamma_T\setminus\Gamma'$ by a positive distance d, the norm $|u_x|^{(\alpha)}_{Q'}$ is estimated from above by a constant depending only on n, ν, μ, μ_1, q, r, μ_2, q_2, M, M_1, d, Γ', $\max_{\Gamma'}|u_t|$ and the quantities $\max_{t_1\le t\le t_2}\|u(x, t)\|^{(2-1/q_3)}_{q_3,S'}$ and $q_3 > n$ if $\Gamma' = S' \times [t_1, t_2]$ and $t_1 > 0$, and $\max_{0\le t\le t_2}\|u(x, t)\|^{(2-1/q_3)}_{q_3,S'}$, $\|u(x, 0)\|^{(2)}_{q_3,\Omega'}$, and $q_3 > n$ if Γ' is the union of the surface $S' \times [0, t_2]$ and the part $\{(x, 0): x \in \Omega'\}$ of the lower base of Q_T. The index α is determined in both cases by the same quantities as $|u_x|^{(\alpha)}$ except for d.*

Remark 5.3. The assertion of Theorem 5.3 remains valid for the solutions described in Remark 5.2.

After estimating the norms $|u_x|^{(\alpha)}_{Q_T}$ and $|u_x|^{(\alpha)}_{Q'}$ one can consider equation (0.1) as a linear equation with smooth coefficients (if, of course, $a_i(x, t, u, p)$ and $a(x, t, u, p)$ are smooth functions of their arguments) and make use of the corresponding theorems of Chapter IV on linear equations for estimates and other qualitative investigations of its solutions.

Let us formulate a proposition that follows from Theorem 5.2 of Chapter IV and Theorem 5.3 of this section.

Theorem 5.4. *Suppose that on the set $\mathfrak{M} = \{(x, t, u, p): (x, t) \in \overline{Q}_T, |u| \le M, |p| \le M_1\}$ the functions $a_i(x, t, u, p)$ and $a(x, t, u, p)$ satisfy conditions* (5.1) *and* (5.2), *while the functions a_i, $\partial a_i/\partial p_j$, $\partial a_i/\partial u$, $\partial a_i/\partial x_j$ and a are Hölder continuous in the arguments x, t, u, p with exponents β, $\beta/2$, β and β respectively. If S belongs to the class $H^{2+\beta}$, then for any solution u from $H^{2+\beta, 1+\beta/2}(\overline{Q}_T)$*

of equation (0.1) *having* $\max_{Q_T}|u| \le M$ *and* $\max_{Q_T}|u_x| \le M_1$ *the norm* $|u|^{(2+\beta)}_{Q_T}$ *is estimated from above by a constant depending only on* M, M_1, β, *the quantities* ν, μ, q, r, μ_1 *from conditions* (5.1), (5.2), *the norms of* a_i, $\partial a_i/\partial p_j$, $\partial a_i/\partial u$, $\partial a_i/\partial x_j$ *and* a *in* $H^{\beta, \beta/2, \beta, \beta}(\mathfrak{M})$, S *and the norm* $|u|^{(2+\beta)}_{\Gamma_T}$.

Let Γ'_T *be an arbitrary portion of the surface* Γ_T. *For any subdomain* Q' *of the cylinder* $\overline{Q}_T$ *that is separated from* $\Gamma_T \setminus \Gamma'$ *by a distance* $d > 0$, *the norm* $|u|^{(2+\beta)}_{Q'}$ *is estimated from above in terms of* d *and the same quantities as the norm* $|u|^{(2+\beta)}_{Q_T}$ *except for the quantity* $|u|^{(2+\beta)}_{\Gamma_T}$ *which is replaced by the norm* $|u|^{(2+\beta)}_{\Gamma'}$.

Theorem 5.4 together with the theorems of §§1–4 of this chapter and Theorem 5.2 of Chapter IV for linear equations permits one to prove the solvability in the large of the first boundary value problem for equations (0.1).

§6. THE SOLVABILITY OF THE FIRST BOUNDARY VALUE PROBLEM

Let us consider in Q_T the first boundary value problem for equation (0.1), i.e. the problem of determining in $\overline{Q}_T$ a function $u(x, t)$ satisfying in Q_T the equation

$$\mathscr{L}u \equiv u_t - \frac{d}{dx_i} a_i(x, t, u, u_x) + a(x, t, u, u_x) = 0, \tag{6.1}$$

and coinciding on Γ_T with some known function ψ:

$$u|_{\Gamma_T} = \psi|_{\Gamma_T} \tag{6.2}$$

Equality (6.2) contains in itself the initial and boundary conditions. We will assume that ψ is the value on Γ_T of a function $\psi(x, t)$ defined in $\overline{Q}_T$ and possessing a certain smoothness, which will be specified below. If $\psi(x, t) \in C(\overline{Q}_T)$, then the compatibility condition of zero order (for the initial and boundary conditions) is said to be fulfilled. If $\psi(x, t) \in C^{2,1}(\overline{Q}_T)$ and on the manifold $S_0 \equiv \{x \in S,\ t = 0\}$

$$\psi_t - \frac{d}{dx_i} a_i(x, t, \psi, \psi_x) + a(x, t, \psi, \psi_x) = 0, \tag{6.3}$$

then the compatibility condition of first order (for the initial and boundary conditions and the equation) is said to be fulfilled. It is clear that the compatibility condition of zero order must be fulfilled for a solution $u(x, t)$ of problem (6.1), (6.2) belonging to $C(\overline{Q}_T)$, while the compatibility condition of first order must be fulfilled for a solution from $C^{2,1}(\overline{Q}_T)$ that satisfies (6.1) in $\overline{Q}_T$.

The solvability of problem (6.1), (6.2) can be proved by making use of theorems from the theory of abstract nonlinear equations, bearing in mind that it has a uniqueness theorem for the classical solution. We will apply the Leray-Schauder principle (its formulation can be found in [68b] or in §8 of Chapter IV of [65q]).

We write equation (6.1) in the form

$$\mathscr{L}u \equiv u_t - \frac{\partial a_i(x, t, u, u_x)}{\partial u_{x_j}} u_{x_i x_j} + A(x, t, u, u_x) = 0, \tag{6.4}$$

where

$$A(x, t, u, u_x) = a(x, t, u, u_x) - \frac{\partial a_i(x, t, u, u_x)}{\partial u} u_{x_i} - \frac{\partial a_i(x, t, u, u_x)}{\partial x_i},$$

and consider the family of linear problems

$$\left.\begin{aligned} v_t - \left[\tau \frac{\partial a_i(x, t, w, w_x)}{\partial w_{x_j}} + (1-\tau)\delta_i^j\right] v_{x_i x_j} \\ + \tau A(x, t, w, w_x) - (1-\tau)[\psi_t - \Delta\psi] = 0, \\ v|_{\Gamma_T} = \psi|_{\Gamma_T}, \quad 0 \leqslant \tau \leqslant 1, \end{aligned}\right\} \tag{6.5}$$

on determining the function v; the function $w(x, t)$ in this connection is assumed to be given. We introduce a linear Banach space B_δ of functions $w(x, t)$ that are continuous together with their derivatives with respect to x in $\overline{Q}_T$ and have the finite norm

$$|w|_{B_\delta} \equiv |w|_{Q_T}^{(\delta)} + |w_x|_{Q_T}^{(\delta)}.$$

Under certain restrictions on the functions a_i, a, ψ and S, problem (6.5) defines an operator Φ in B_δ which associates each function w of B_δ with a solution v of the linear problem (6.5):

$$v = \Phi(w;\ \tau). \tag{6.6}$$

This operator is nonlinear and depends on τ. Its fixed points for $\tau = 1$ are solutions of problem (6.1), (6.2). In searching for the fixed points of $\Phi(w;\ \tau)$ we will apply the Leray-Schauder principle.

Let u^τ be one of the fixed points of the transformation $\Phi(w;\ \tau)$, i.e. let

$$u^\tau = \Phi(u^\tau;\ \tau).$$

This means that u^τ is a solution of the nonlinear problem

$$\left.\begin{aligned} \mathcal{L}_\tau u &\equiv \tau\mathcal{L}u + (1-\tau)(u_t - \Delta u - \psi_t + \Delta\psi) \\ &\equiv u_t - \frac{d}{dx_i}\left[\tau a_i(x, t, u, u_x) + (1-\tau)u_{x_i}\right] \\ &\quad + \tau a(x, t, u, u_x) - (1-\tau)[\psi_t - \Delta\psi] = 0, \\ u|_{\Gamma_T} &= \psi|_{\Gamma_T}, \qquad 0 \leqslant \tau \leqslant 1. \end{aligned}\right\} \tag{6.7}$$

The operators $\mathfrak{L}_\tau$ possess the property that if the conditions of some theorem of the preceding sections, in which an estimate is obtained of one of the norms considered in the book, are valid for the operator $\mathfrak{L}$, then the same conditions are also valid for $\mathfrak{L}_\tau$ when $\tau \in [0, 1]$. In fact, suppose conditions (2.29), (2.30) or (2.29), (2.32), (2.33) of Chapter I are fulfilled for $\mathfrak{L}$. Then, as is easily seen, analogous conditions are fulfilled for $\mathfrak{L}_\tau$, and estimate (2.31) or (2.34) of Chapter I holds for all possible classical solutions u^τ of problems (6.7), so that

$$\max_{Q_T} |u^\tau| \leqslant M, \qquad \tau \in [0, 1]. \tag{6.8}$$

We assume further that for $(x, t) \in \overline{Q}_T$, $|u| \leq M$ and arbitrary p the functions $a_i(x, t, u, p)$ and $a(x, t, u, p)$ satisfy the inequalities

$$\left.\begin{aligned} &\nu\xi^2 \leqslant \frac{\partial a_i(x, t, u, p)}{\partial p_j}\xi_i\xi_j \leqslant \mu\xi^2, \qquad \nu > 0, \\ &\sum_{i=1}^n \left(|a_i| + \left|\frac{\partial a_i}{\partial u}\right|\right)(1 + |p|) + \sum_{i,j=1}^n \left|\frac{\partial a_i}{\partial x_j}\right| + |a| \\ &\qquad\qquad \leqslant \mu(1 + |p|)^2. \end{aligned}\right\} \tag{6.9}$$

Obviously, the coefficients of the operators $\mathfrak{L}_\tau$ will also satisfy conditions of the same type, where the constants ν_τ and μ_τ for them can be chosen independently of all τ from $[0, 1]$. We retain the notation ν and μ for them. From conditions (6.9) follow the conditions of Theorems 1.1 and 4.1 of the present chapter and the conditions of Lemma 3.1 of Chapter VI, and hence the following estimates hold for all solutions u^τ from $C^{2,1}(\overline{Q}_T)$ of problems (6.7):

$$|u^\tau|_{Q_T}^{(\alpha)} \leqslant M'; \quad \max_{Q_T} |u_x^\tau| \leqslant M_1, \qquad 0 \leqslant \tau \leqslant 1, \tag{6.10}$$

with the constants M' and M_1 being the same for all u^τ and depending only on M, the constants ν and μ from conditions (6.9), $|\psi|_{S_T}^{(2)}$, $\max_{x \in \overline{\Omega}} |\psi_x(x, 0)|$ and the boundary S, which we assume belongs to the class O^2.

Further restrictions on the $a_i(x, t, u, p)$ and $a(x, t, u, p)$ will be imposed only for $(x, t) \in \overline{Q}_T$, $|u| \leq M$ and $|p| \leq M_1$. Thus, in order to obtain a priori estimates of $\max_{Q_T}|u_t^\tau|$ and $|u_x^\tau|_{Q_T}^{(\alpha)}$ it is necessary to add the requirements (5.2) to (6.9). They will be fulfilled for all $\mathfrak{L}_\tau$ with the same function $\phi(x, t)$ and consequently, for the whole class of solutions u^τ,

$$\max_{Q_T}|u_t^\tau|, \quad |u_x^\tau|_{Q_T}^{(\alpha)} \leqslant M_2, \qquad 0 \leqslant \tau \leqslant 1, \tag{6.11}$$

with one and the same α and one and same constant M_2 depending only on the known parameters from our conditions, $S \in O^2$ and $|\psi|_{\Gamma_T}^{(2)}$. Finally, the estimate

$$|u^\tau|_{Q_T}^{(2+\beta)} \leqslant M_3 \tag{6.12}$$

is given by Theorem 5.2 of Chapter IV on the solutions of linear equations with smooth coefficients if one assumes in addition that the functions $a_i(x, t, u, p)$, $\partial a_i/\partial p_j$, $\partial a_i/\partial u$, $\partial a_i/\partial x_i$ and a are continuous functions of (x, t, u, p) satisfying a Hölder condition in x, t, u and p with exponents β, $\beta/2$, β and β, that $S \in H^{2+\beta}$, $\psi \in H^{2+\beta, 1+\beta/2}(\overline{Q}_T)$ and that ψ satisfies the compatibility condition (6.3). In this way one establishes the uniform boundedness in the norm of $H^{2+\beta, 1+\beta/2}(\overline{Q}_T)$ of all possible solutions u^τ of problems (6.7) or, what is the same, of all fixed points of the transformation $\Phi(w; \tau)$. The uniform boundedness of the norms of the u^τ (in the space in which the transformation $\Phi(w; \tau)$ is considered) is one of the conditions of applicability of the Leray-Schauder Theorem, and the most difficult to verify.

Let us now formulate the first existence theorem for the classical solution.

Theorem 6.1. *Suppose that the following conditions hold.*

a) *For* $(x, t) \in \overline{Q}_T$ *and arbitrary* u *conditions* (2.29), (2.30) *or* (2.29), (2.32), (2.33) *of Chapter* I *are fulfilled, which when applied to equation* (0.1) *have the form*

$$\frac{\partial a_i(x, t, u, p)}{\partial p_j}\xi_i\xi_j\Big|_{p=0} \geqslant 0, \quad A(x, t, u, 0)\,u \geqslant -b_1u^2 - b_2$$

or

$$A(x, t, u, 0)\,u \geqslant -\Phi(|u|)|u| - b_2, \quad \int_0^\infty \frac{d\tau}{\Phi(\tau)} = \infty, \quad \Phi > 0.$$

b) *For* $(x, t) \in \overline{Q}_T$, $|u| \leq M$ *(*M *is taken from* (6.8)*) and arbitrary* p *the functions* $a_i(x, t, u, p)$ *and* $a(x, t, u, p)$ *are continuous, the* $a_i(x, t, u, p)$ *are differentiable with respect to* x, u *and* p *and inequalities* (6.9) *are fulfilled.*

c) *For* $(x, t) \in \overline{Q}_T$, $|u| \leq M$ *and* $|p| \leq M_1$ *(*M_1 *is taken from* (6.10)*) the* *functions* a_i, a, $\partial a_i/\partial p_j$, $\partial a_i/\partial u$, *and* $\partial a_i/\partial x_i$ *are continuous functions satisfying a Hölder condition in* x, t, u *and* p *with exponents* β, $\beta/2$, β *and* β *respectively.*

d) *For* $(x, t) \in \overline{Q}_T$, $|u| \leq M$ *and* $|p| \leq M_1$ *the function* $a(x, t, u, p)$ *has the partial derivatives* $\partial a/\partial p_i$ *and* $\partial a/\partial u$ *and the functions* a_i *and* a *satisfy condition* (5.2).

e) $\psi(x, t) \in H^{2+\beta, 1+\beta/2}(\overline{Q}_T)$ *and satisfies condition* (6.3).

f) $S \in H^{2+\beta}$

Under these conditions there exists a unique solution of problem (6.1), (6.2) *from the class* $H^{2+\beta, 1+\beta/2}(\overline{Q}_T)$. *Moreover, this solution has derivatives* u_{xt} *from* $L_2(Q_T)$.

Proof. As the basic space in which we will consider the transformation Φ defined by (6.6), we take the space B_δ with sufficiently small δ (δ will be stipulated below). As was shown above, the theorems of the preceding sections guarantee the uniform boundedness

$$\max_{Q_T} |u^\tau| \leqslant M, \ \max_{Q_T} |u^\tau_x| \leqslant M_1, \ |u^\tau|_{B_\alpha} \equiv |u^\tau|^{(\alpha)}_{Q_T} + |u^\tau_x|^{(\alpha)}_{Q_T} \leqslant M_3 \quad (6.12')$$

with some $0 < \alpha < 1$ for all solutions u^τ from $C^{2,1}(\overline{Q}_T)$ of problems (6.7). But each such solution u^τ is obviously a fixed point of the transformation $\Phi(w; \tau)$ and, conversely, any fixed point $u^\tau \in B_\delta$ of the transformation $\Phi(w; \tau)$ is a classical solution of problem (6.7), more precisely, a solution from $H^{2+\beta\delta, 1+\beta\delta/2}(\overline{Q}_T)$. In fact, from the membership of u^τ in B_δ and condition c) it follows that the coefficients of the $v_{x_ix_j}$ and the free term in (6.5) are elements of $H^{\beta\delta, \beta\delta/2}(\overline{Q}_T)$ for $w = u^\tau$, and therefore by virtue of Theorem 5.2 of Chapter IV the solution $v = \Phi(u^\tau; \tau)$ of problem (6.5) will belong to $H^{2+\beta\delta, 1+\beta\delta/2}(\overline{Q}_T)$.

We modify, if necessary, the functions $a_i(x, t, u, p)$ and $a(x, t, u, p)$ outside of the domain of variation of the arguments $\{(x, t) \in \overline{Q}_T, |u| \leq M, |p| \leq M_1\}$ so that they satisfy properties a)–d) of Theorem 6.1 on the set $\{(x, t) \in \overline{Q}_T$, $|u| \leq M + \epsilon$, $M \leq M_1 + \epsilon\}$, where ϵ is some positive number. We take $\delta = \alpha < 1$ from (6.12′) and consider the operator $\Phi(w; \tau)$ on a convex bounded set $\mathfrak{M}$ of the space B_δ, consisting of all elements $w \in B_\delta$ satisfying the inequalities

$$\max_{Q_T}|w| \leqslant M + \varepsilon, \ \max_{Q_T}|w_x| \leqslant M_1 + \varepsilon, \ |w|_{B_\delta} \leqslant M_3 + \varepsilon. \quad (6.13)$$

By virtue of (6.12′) all of the fixed points u^τ of the transformation $\Phi(w;\tau)$ lie strictly in the interior of this set $\mathfrak{M}$.

Let us verify that $\Phi(w;\tau)$ on $\mathfrak{M}\times[0,1]$ is a family of operators that are equicontinuous in w and τ and uniformly compact. For all w with $|w|_{B_\delta}\le M_3+\epsilon$ and $\tau\in[0,1]$ the functions $v=\Phi(w;\tau)$, as solutions of problem (6.5), have, by virtue of what has been said above, the uniformly bounded norms

$$|v,\, v_x,\, v_t,\, v_{xx}|_{Q_T}^{(\alpha\beta)}\leqslant c. \tag{6.14}$$

By the uniform boundedness of $|v_{xx}|$ and $|v_t|$, as was proved in Lemma 3.1 of Chapter II, the norms $|v_x|_{Q_T}^{(1)}$ for the functions v are uniformly bounded, while the set of such v is compact in B_α, since $\alpha<1$ and, consequently, the operators Φ, $0\le\tau\le 1$, take the whole set $\mathfrak{M}$ into a set that is compact in B_α. In order to verify the uniform continuity in w of $\Phi(w;\tau)$ on $\mathfrak{M}\times[0,1]$ we take two close elements w' and w'' from $\mathfrak{M}$ and, corresponding to them, $v'=\Phi(w';\tau)$ and $v''=\Phi(w'';\tau)$. We subtract equation (6.5) for v'' from equation (6.5) for v' and consider the result as a linear equation for $v=v'-v''$ of the form

$$\begin{aligned} v_t-\left[\tau\frac{\partial a_i(x,t,w',w'_x)}{\partial w_{x_j}}+(1-\tau)\delta_i^j\right]v_{x_ix_j} &= \tau\left[\frac{\partial a_i(x,t,w',w'_x)}{\partial w_{x_j}}-\frac{\partial a_i(x,t,w'',w''_x)}{\partial w_{x_j}}\right]v''_{x_ix_j}\\ &\quad-\tau\left[A(x,t,w',w'_x)-A(x,t,w'',w''_x)\right]. \end{aligned}\tag{6.15}$$

Also,

$$v|_{\Gamma_T}=0. \tag{6.16}$$

We consider the right side of (6.15) as a free term. It is not difficult to demonstrate that when the quantity $|w'-w''|_{B_\alpha}$ is small the right side of (6.15) is small uniformly in $\tau\in[0,1]$ as a function of (x,t) in the norm of the space $H^{\alpha\beta,\,\alpha\beta/2}(\bar Q_T)$. But then by Theorem 5.2 of Chapter IV the quantities $|v,\, v_x,\, v_{xx},\, v_t|_{Q_T}^{(\alpha\beta)}$, and hence also the norm $|v|_{B_\alpha}$, will be small.

The uniform continuity in τ of $\Phi(w;\tau)$ on $\mathfrak{M}\times[0,1]$, and thereby the uniform continuity of $\Phi(w;\tau)$ in $(w;\tau)$ on $\mathfrak{M}\times[0,1]$, is proved analogously. The applicability of the Leray-Schauder Theorem will follow if we establish that the transformation $\Phi(w;0)$ for $\tau=0$ has a unique fixed point inside $\mathfrak{M}$ and that the transformation $w - \Phi(w; 0)$ is invertible in a neighborhood of this point. But this is indeed the case since $\Phi(w; 0)$ takes all of $\mathfrak{M}$ into the single element $\hat{v}(x, t) \equiv \psi(x, t)$ which is the solution of the problem

$$\hat{v}_t - \Delta\hat{v} = \psi_t - \Delta\psi, \quad \hat{v}|_{\Gamma_T} = \psi|_{\Gamma_T}.$$

Thus for each τ from $[0, 1]$ there exists at least one fixed point $u^\tau(x, t)$ for Φ, which, as explained above, will be a solution of problem (6.7) from $H^{2+\alpha\beta, 1+\alpha\beta/2}(\overline{Q}_T)$ (by virtue of (6.12) the variation of the functions a_i and a beyond the limits of the set $\{(x, t) \in \overline{Q}_T, |u| \le M, |p| \le M_1\}$ has no influence on the u^τ). Let us verify that the u^τ will actually belong to $H^{2+\beta, 1+\beta/2}(\overline{Q}_T)$. Indeed, the membership of the u^τ in $H^{2+\alpha\beta, 1+\alpha\beta/2}(\overline{Q}_T)$ implies by virtue of Lemma 3.1 of Chapter II the membership of the u^τ_x in $H^{1,1/2}(\overline{Q}_T)$, and from this and condition c) of the theorem it follows that the coefficients of equation (6.5), if one substitutes $u^\tau(x, t)$ in them for $w(x, t)$, will as functions of (x, t) be elements of $H^{\beta, \beta/2}(\overline{Q}_T)$. But a solution u^τ of problem (6.7) can be regarded as a solution v of problem (6.5) with $w = u^\tau$, and therefore (see Theorem 5.2 of Chapter IV) $v = u^\tau$ is an element of $H^{2+\beta, 1+\beta/2}(\overline{Q}_T)$. Thus it is proved that problems (6.7) for all τ from $[0, 1]$, and in particular problem (6.1), (6.2), have solutions from $H^{2+\beta, 1+\beta/2}(\overline{Q}_T)$.

The uniqueness of such a solution is proved in the usual way. Let u' and u'' be two solutions of problem (6.1), (6.2). They satisfy the integral identities

$$\int_0^t \int_\Omega [u_t\eta + a_i(x, t, u, u_x)\eta_{x_i} + a(x, t, u, u_x)\eta]\, dx\, dt = 0 \tag{6.17}$$

with an arbitrary continuously differentiable function $\eta(x, t)$ that is equal to zero on S_T. We subtract one identity from the other and write the increments of the a_i and a, following (5.3), in the form

$$\begin{aligned}\Delta a_i &\equiv a_i(x, t, u', u'_x) - a_i(x, t, u'', u''_x) \\ &= v_{x_j} \int_0^1 \frac{\partial a_i[x, t, \tau u' + (1-\tau)u'', \tau u'_x + (1-\tau)u''_x]}{\partial u_{x_j}}\, d\tau \\ &\quad + v \int_0^1 \frac{\partial a_i[\ldots]}{\partial u}\, d\tau \equiv \hat{a}_{ij} v_{x_j} + \hat{b}_i v; \quad \Delta a = \hat{c}_i v_{x_i} + \hat{c} v\end{aligned}$$

where $v = u' - u''$. As a result we obtain the identity

$$\int_0^t \int_\Omega [v_t\eta + (\tilde{a}_{ij}v_{x_j} + \tilde{b}_i v)\eta_{x_i} + (\tilde{c}_i v_{x_i} + \tilde{c}v)\eta]\, dx\, dt = 0,$$

from which by virtue of the theorems in §3 of Chapter III on linear equations it follows that $v \equiv 0$.

For a proof of the existence of the derivatives $\partial^2 u/\partial t \partial x$ from $L_2(Q_T)$ of the solution u we revert to the proof of Theorem 5.1. By virtue of our assumptions on the a_i and a, relation (5.4) can be regarded as a linear equation for

$$\frac{\Delta u}{\Delta t} = \frac{1}{\Delta t}[u(x, t + \Delta t) - u(x, t)]$$

with such properties of the coefficients as ensure the uniform (in Δt) boundedness of the integrals. $\int_0^{T-\Delta t} \int_\Omega (\Delta u_x/\Delta t)^2\, dx\, dt$. From this latter fact we obtain the existence of u_{tx} and the finiteness of $\|u_{tx}\|_{2,Q_T}$. The theorem is proved.

If one assumes that the data possess a greater smoothness, then the solutions will also be smoother. In fact, if it is known that the solution $u(x, t)$ of problem (6.1), (6.2) belongs, for example, to $H^{2+\beta, 1+\beta/2}(\overline{Q}_T)$, then, considering $u(x, t)$ as a solution of the linear problem (6.5) with $w = u$, we obtain on the basis of Theorem 5.2 of Chapter IV additional information on it, for example, its membership in $H^{3+\beta, (3+\beta)/2}(Q_T)$, if concerning the $a_i(x, t, u, p)$ and $a(x, t, u, p)$ it is known that they, the partial derivatives of the a_i of first and second order and the partial derivatives of a of first order satisfy a Hölder condition with exponents β, $\beta/2$, β and β respectively. In general, this improvement is already clear from the theorems of Chapter IV on linear equations and does not require an analysis of the nonlinear problem itself. We will not enumerate here all of the implications which one can extract from Theorem 6.1 and the theorems of Chapter IV, assuming that the reader can do this independently.

From a comparison of the conditions of Theorem 6.1 of the present chapter with Theorem 5.2 of Chapter IV it would seem that condition d) in Theorem 6.1 on the differentiability of a with respect to u and p and the Lipschitz continuity of the a_i and a in t is redundant (overstated). From the point of view of the uniqueness theorem the differentiability of a with respect to u and p (or, what is roughly the same, the Lipschitz continuity of a in u and p) is necessary. But it is possible to prove the existence theorem without these assumptions, using the just proved Theorem 6.1 and various theorems of Chapters III–V on a priori estimates for quasi-linear and linear equations. The differential properties of the solutions guaranteed in this connection will be determined by the smoothness conditions for the functions a_i and a, and also by the smoothness of $\psi|_{\Gamma_T}$ and S and by the order of compatibility of the boundary and initial conditions, while the solutions themselves will be obtained as limits of solutions $u^m(x, t)$ of boundary problems for equations $\mathfrak{L}^m u = 0$ with smoothed functions a_i^m and a^m and smoothed S^m and ψ^m. Among the possible consequences of Theorem 6.1 and the a priori estimates of the preceding sections we establish, for example, the following propositions.

Theorem 6.2. *Suppose conditions* a)–c) *of Theorem* 6.1 *are fulfilled. Suppose further that the following conditions are fulfilled:*

d) $\psi|_{S_T} \in O^{2,1}(S_T);\ \max_{x \in \Omega} |\psi_x(x, 0)| < \infty;\ \psi \in H^{\gamma, \frac{\gamma}{2}}(\bar{Q}_T);$

e) $S \in O^2.$

Then there exists at least one solution $u(x, t)$ *of problem* (6.1), (6.2) *belonging to* $H^{\alpha, \alpha/2}(\bar{Q}_T)$ *and having* u_x *that are bounded in* $\bar{Q}_T$ *and derivatives* u_t, u_{xx} *that belong to* $H^{\beta,\beta/2}(Q_T)$. *For the uniqueness of such a solution it is sufficient that the function* $a(x, t, u, p)$ *satisfy a Lipschitz condition in* u *and* p *uniformly on any compactum of the form* $\{(x, t) \in \bar{Q}_T,\ |u| \le c,\ |p| \le c\}$.

We take an infinite sequence of embedded cylinders $Q_T^m = \Omega_m \times (0, T)$, $m = 1, 2, \cdots$, such that $Q_T^m \supset Q_T^{m+1} \supset Q_T$ for all m and such that the boundaries Ω_m (and thereby also the lateral surfaces S_T^m of the cylinders Q_T^m) belong to $H^{2+\beta}$ and have uniformly bounded norms in Q^2, and we will assume that the function $\psi(x, t)$ is extended onto the domain $\bar{Q}_T^1$, while the functions $a_i(x, t, u, p)$ and $a(x, t, u, p)$ are extended onto the domain $\{(x, t) \in \bar{Q}_T^1,\ |u| < \infty,\ |p| < \infty\}$, so that $\psi \in O^{2,1}(\bar{Q}_T^1 \setminus \Omega \times [0, T])$, while for the a_i and a the smoothness properties and inequalities enumerated in conditions a)–c) of Theorem 6.1 are fulfilled.

We construct with respect to the $a_i(x, t, u, p)$, $a(x, t, u, p)$ and $\psi(x, t)$ averagings $a_i^m(x, t, u, p)$, $a^m(x, t, u, p)$ and $\psi^m(x, t)$ over x, t, u and p with averaging radius $\rho_m = d/m$, where d is the distance from S_T^1 to S_T^2, with a Steklov averaging to the right being taken over t.

We consider now the family of boundary value problems

$$\mathscr{L}^m u \equiv u_t - \frac{d}{dx_i} a_i^m(x, t, u, u_x) + A^m(x, t, u, u_x) = 0,$$
$$u|_{\Gamma_T^m} = \psi^m|_{\Gamma_T^m}, \quad m = 2, 3, \ldots, \tag{6.18}$$

in the domains Q_T^m, in which the $a_i^m(x, t, u, p)$ and $\psi^m(x, t)$ are the averagings constructed above of the a_i and ψ respectively, while $A^m(x, t, u, p)$ equals $a^m(x, t, u, p)$ for $\{(x, t) \in \overline{Q}_T, |u| < \infty, |p| < \infty\}$, equals

$$-\psi_t^m(x, 0) + \frac{d}{dx_i} a_i(x, 0, \psi^m(x, 0), \psi_x^m(x, 0))$$

for $\{(x, t) \in S_0^m, u = \psi^m(x, 0), p = \psi_x^m(x, 0)\}$, [1] and for the remaining values of the arguments is defined so that $A^m(x, t, u, p)$ is a smooth function of its arguments and the functions $A^m(x, t, u, p)$, $m = 1, 2, \cdots$, satisfy conditions (6.9) uniformly in m, i.e.

$$|A^m(x, t, u, p)| \leqslant \mu(|u|)(1 + |p|)^2.$$

Problem (6.18) satisfies all of the conditions of Theorem 6.1 in Q_T^m and therefore has a solution $u^m(x, t)$ possessing the properties indicated in Theorem 6.1. On the basis of Theorems 1.1 and 4.1 of the present chapter, inequalities (2.31) and (2.34) of Chapter I and Lemma 3.1 of Chapter IV, we have the uniform (in m) estimates

$$|u^m|_{Q_T^m}^{(\alpha)} \leqslant c, \quad \max_{Q_T^m} |u_x^m| \leqslant c, \tag{6.19}$$

with some $\alpha \leq \gamma$, and

$$|u^m|_{Q'}^{(2+\beta)} \leqslant c(d), \tag{6.20}$$

where Q' is a subdomain of Q_T that is separated from Γ_T by a positive distance $d > 0$, while $c(d)$ is a constant depending only on the constants of conditions a)–c), $\max_{Q_T^1} |\psi|$ and the distance d. In view of (6.19) it is possible to choose a subsequence from the sequence $\{u^m\}$ that uniformly converges in $\overline{Q}_T$ to $u(x, t)$, which by virtue of (6.19) and (6.20) is a solution of problem (6.1), (6.2). The uniqueness is proved in the same way as in Theorem 6.1.

If condition c) in Theorem 6.2 is discarded, then problem (6.1), (6.2) has a solution u belonging to $H^{\alpha, \alpha/2}(\overline{Q}_T)$, having derivatives u_{x_i} that are bounded in Q_T and Hölder continuous inside Q_T, and having u_t and u_{xx} belonging to $L_2(Q'_T)$, $Q'_T = \Omega' \times [0, T]$, $\overline{\Omega}' \subset \Omega$. This assertion is proved in the same way as

---

1) We recall that according to our notation Γ_T^m is the totality of points (x, t) belonging to the lateral surface and lower base of the cylinder Q_T^m, while S_0^m is the totality of points $(x, 0)$ with x belonging to the boundary S^m of the domain Ω_m.

Theorem 6.2, only in place of (6.20) we will have according to Theorems 3.1 and 4.2 the uniform boundedness of weaker norms of u^m, namely

$$|u_x^m|_{Q'}^{(\gamma)} + \|u_t^m\|_{2, Q_T'} + \|u_{xx}^m\|_{2, Q_T'} \leqslant c(d), \tag{6.21}$$

where Q' is any subdomain of Q_T that is separated from Γ_T by a distance $d > 0$. Thus we have

Theorem 6.3. *Suppose conditions* a) *and* b) *of Theorem* 6.1 *and conditions* d) *and* e) *of Theorem* 6.2 *are fulfilled. Then problem* (6.1), (6.2) *has a solution* $u(x, t)$ *from* $H^{\alpha, \alpha/2}(\overline{Q}_T) \cap W_2^{2,1}(Q_T')$ *with* u_x *from* $H^{\gamma, \gamma/2}(Q_T)$ *and with finite* $\sup_{Q_T}|u_x|$. *If furthermore* $a(x, t, u, p)$ *satisfies a Lipschitz condition in* u *and* p *(uniformly on any compact set), then the solution is unique in the indicated class.*

The uniqueness is proved in the same way as in Theorem 6.1, but it is better to verify the fact that $u(x, t)$ satisfies equation (6.1) for almost all $(x, t) \in Q_T$ in the following way: write in place of equations (6.18) the identities

$$\int_{Q_T^m} [u_t^m \eta + a_i^m(x, t, u^m, u_x^m) \eta_{x_i} + a^m(x, t, u^m, u_x^m) \eta] \, dx \, dt = 0 \tag{6.22}$$

with a smooth function η that is finite in Q_T and pass to the limit in them with respect to some subsequence $\{u^{m_k}\}$, $k = 1, 2, \cdots$, converging to u uniformly in $\overline{Q}_T$ and such that the $u_x^{m_k}$ converge uniformly to u_x in each interior subdomain $Q' \subset Q_T$, while the $u_t^{m_k}$ and $u_{xx}^{m_k}$ converge to u_t and u_{xx} weakly in $L_2(Q_T')$. Since the functions $a_i(x, t, u, p)$ and $a(x, t, u, p)$ are continuous in (x, t, u, p), their averagings converge to them uniformly.

By what we have said, it is possible in (6.22) to pass to the limit with respect to $k \to \infty$ and obtain an analogous identity for u, which is equivalent to equation (6.1) since u has, besides continuous u_x, generalized derivatives u_t and u_{xx}.

Under a further weakening of the conditions on ψ and S we lose the boundedness of $\max_{Q_T}|u_x|$. Thus, if in Theorem 6.3 we require instead of condition b) that the $a_i(x, t, u, p)$ and $a(x, t, u, p)$ be continuous in u and p and satisfy inequalities (3.1)–(3.6), and we replace conditions d) and e) by the membership of $\psi|_{\Gamma_T}$ in $H^{\beta}(\Gamma_T)$ and condition (A), then instead of (6.19)–(6.21) we will have the lesser information

and

$$\left.\begin{aligned} &|u^m|^{(\alpha)}_{Q_T^m} \leqslant c \\ &|u^m_x|^{(\gamma)}_{Q'} + \|u^m_t\|_{2,\,Q'} + \|u^m_{xx}\|_{2,\,Q'} \leqslant c\,(d). \end{aligned}\right\} \tag{6.23}$$

These uniform estimates permit us to conclude that there exists a limit function u for the u^m that belongs to $H^{\alpha,\,\alpha/2}(\overline{Q}_T) \cap W_2^{2,1}(Q')$, has u_x from $H^{\gamma,\,\gamma/2}(Q_T)$, satisfies (6.2) at each point of Γ_T, and equation (6.1) almost everywhere.

In this way, with the use of the approximation described above, we prove Theorem 6.4.

Theorem 6.4. *Suppose that the following conditions hold.*

a) *Condition* a) *of Theorem* 6.1 *is fulfilled.*

b) *The* $a_i(x, t, u, p)$ *and* $a(x, t, u, p)$ *are continuous in* (u, p), *the* $a_i(x, t, u, p)$ *are differentiable with respect to* x, u *and* p, *and inequalities* (3.1)-(3.6) *are valid for them.*

c) $\psi|_{\Gamma_T} \in H^{\beta}(\Gamma_T)$.

d) S *satisfies condition* (A).

Then problem (6.1), (6.2) *has a solution* u *from* $H^{\alpha,\,\alpha/2}(\overline{Q}_T) \cap W_2^{2,1}(Q')$ *with* u_x *from* $H^{\gamma,\,\gamma/2}(Q_T)$, *where* Q' *is any interior subdomain of* Q_T.

Condition a) in Theorems 6.1–6.4 allows one to obtain an a priori estimate of $\max_{Q_T}|u|$ for all possible classical solutions of problems (6.1) by making use of the so-called "maximum principle". It can be replaced by conditions of integral type, which permit one to estimate $\max_{Q_T}|u|$ on the basis of Theorems 2.1 or 2.2. Since these estimates are applied to "good" (classical) solutions, the restrictions on the $a_i(x, t, u, p)$ and $a(x, t, u, p)$ reduce to the fact that these functions are integrable over Q_T for any finite u and p and conditions (2.3)–(2.6) are fulfilled. Thus, for example, from Theorem 2.2 and Theorem 6.1 follows

Theorem 6.5. *The assertions of Theorem* 6.1 *remain in force if its condition* a) *is replaced by the following condition.*

a′) *For* $(x, t) \in \overline{Q}_T$ *and arbitrary* u *and* p *the inequality*

$$\sum |a_i(x, t, u, p)| + |a(x, t, u, p)| \leqslant \mu(|u|, |p|)\,\varphi_1(x, t)$$

is valid with any continuous function $\mu(\xi, r)$ *and* $\phi_1(x, t) \in L_1(Q_T)$, *while for* $|u|$ *exceeding some number* k_0 *the assumptions* (2.3)–(2.6) *are fulfilled, where, if* $\mu > 0$ *in them, then the uniform boundedness of* $\|u^r\|_{q_1, r_1, Q_T}$ *must be known for all possible solutions* u^τ *(in Theorem* 6.1 *this must be known simply for the solutions from* $H^{2+\beta, 1+\beta/2}(\overline{Q}_T)$ *of problem* (6.7)).

Condition a′) guarantees by virtue of Theorem 2.2 the uniform boundedness of $\max_{Q_T}|u^\tau|$, i.e. what we had extracted from condition a) of Theorem 6.1. We recall that the numbers q_1 and r_1 in conditions (2.3)–(2.6) can be any numbers greater than zero and that for $\mu = 0$ none of the a priori requirements on the boundedness of any of the norms of the solutions u^τ are imposed.

It is possible to give different sufficient conditions when any weak norms of the solutions are estimated in terms of known quantities. For example, if for $(x, t) \in \overline{Q}_T$ and arbitrary u and p

$$a_i(x, t, u, p)p_i + a(x, t, u, p)u \geqslant \nu p^2 - \varphi_2(x, t)|u|^\alpha - \varphi_1(x, t), \tag{6.24}$$

where $\nu > 0$, $\phi_1(x, t) \in L_1(Q_T)$, $\alpha \in (0, 2)$, $\phi_2(x, t) \in L_{q_2, r_2}(Q_T)$, and the numbers q_2 and r_2 are subject to the conditions $1/r_2 + n/2q_2 = 1 + (n/4)(2-\alpha)$, r_2, $q_2 \geq 1$, then the norms $|u^\tau|_{Q_T}$ for the u^τ when $u^\tau|_{\Gamma_T} = \psi|_{\Gamma_T} = 0$ are uniformly bounded. Indeed, we take the equality

$$\int_{Q_t} \left[u_t^\tau u^\tau + \tau a_i(x, t, u^\tau, u_x^\tau) u_{x_i}^\tau + (1-\tau)(u_x^\tau)^2 + \tau a(x, t, u^\tau, u_x^\tau)\right] dx\, dt = 0, \tag{6.25}$$

which is obviously valid for any solution u^τ of problem (6.7), and, making use of assumption (6.24), we derive from it the inequality

$$\begin{aligned}
\frac{1}{2}\int_\Omega (u^\tau)^2 dx \Big|_0^t &+ (\tau\nu + 1 - \tau)\int_{Q_T} (u_x^\tau)^2 dx\, dt \\
&\leqslant \tau \int_{Q_T} (\varphi_2|u^\tau|^\alpha + \varphi_1)\, dx\, dt \\
&\leqslant \tau \|\varphi_2\|_{q_2, r_2, Q_T} \|u^\tau\|^\alpha_{\frac{\alpha q_2}{q_2-1}, \frac{\alpha r_2}{r_2-1}, Q_T} + \tau\|\varphi_1\|_{1, Q_T} \\
&\leqslant \tau\beta^{(\alpha)} \|\varphi_2\|_{q_2, r_2, Q_T} |u^\tau|^\alpha_{Q_T} + \tau\|\varphi_1\|_{1, Q_T}.
\end{aligned} \tag{6.26}$$

The desired estimate of $|u^\tau|_{Q_T}$ follows from (6.26) since

$$\tau\nu + 1 - \tau \geqslant \min\{\nu;\ 1\} \text{ and } \alpha < 2. \tag{6.27}$$

If for $(x, t) \in Q_T$ and arbitrary u and p

$$a_i(x, t, u, p) p_i + a(x, t, u, p) u \geqslant \nu |u|^{q_1} - \varphi_1(x, t) \tag{6.28}$$

with some $\nu > 0$, $q_1 \geq 0$, $\phi_1 \in L_1(Q_T)$, then from the same relation (6.25) we obtain a uniform estimate of

$$\| u^\tau \|_{2, \infty Q_T} + \| u^\tau \|_{q, Q_T}, \quad q = \min\left\{q_1, 2\frac{n+2}{n}\right\}.$$

It is possible to obtain criteria similar to the criteria just indicated by starting not from (6.25) but from the equality

$$\int_{Q_t} \left[\frac{1}{q+1}\frac{\partial (u^\tau)^{q+1}}{\partial t} + q\tau a_i(x, t, u^\tau, u^\tau_x) u^\tau_{x_i} (u^\tau)^{q-1} \right.$$
$$\left. + q(1-\tau)(u^\tau_x)^2 (\tau^\tau)^{q-1} + \tau a(x, t, u^\tau, u^\tau_x)(u^\tau)^q\right] dx\, dt = 0, \tag{6.29}$$

which is obviously valid for the solutions u^τ of problem (6.7) when $\psi|_{\Gamma_T} = 0$.

The replacement of condition a) by a condition of type a′) can also be done in other theorems of the present section. Another admissible modification of the conditions of Theorem 6.1 consists in the replacement of condition e) on ψ by the condition of simple continuity of ψ on Γ_T. We will do this in Theorem 4.4 of Chapter VI directly for equations of general form. It is possible to go further in weakening the conditions of the theorems and arrive at generalized solutions of equations (6.1) having only derivatives of first order. Let us cite one of the propositions analogous to Theorem 8.8 of Chapter IV of the book [65q].

Theorem 6.6. *Suppose the following conditions are fulfilled.*

a) *For* $(x, t) \in \overline{Q}_T$ *and arbitrary* u *and* p *inequality* (6.24) *is valid and*

$$|a_i(x, t, u, p)|(1+|p|) \leqslant \mu(1+|p|)^2 + (1+|u|^\alpha)\varphi_3(x, t) \tag{6.30}$$

and

$$|a(x, t, u, p)| \leqslant \mu(1+|p|)^2 + (1+|u|^\alpha)\varphi_3(x, t) \tag{6.31}$$

with

$$\alpha < 2 \text{ and } \varphi_3 \in L_{q_3, r_3}(Q_T), \quad \frac{1}{r_3} + \frac{n}{2q_3} < 1; \; r_3, q_3 \geqslant 1.$$

b) *For* $(x, t) \in \overline{Q}_T$, *arbitrary* p, *and* $|u|$ *exceeding some constant* R

$$a_i(x, t, u, p) p_i \geqslant \nu |p|^2 - \mu |u|^\beta - u^2\varphi_3(x, t)$$

and

$$-a(x, t, u, p)u \leqslant \mu|p|^2 + \mu|u|^\beta + u^2\varphi_3(x, t)$$

with $\beta < 2 + 4/n$ *and with function* ϕ_3 *being of the same type as in* (6.30).

c) *For* (x, t) *and* $(x', t') \in \overline{Q}_T$ *and arbitrary* u, v, p *and* q

$$(p_i - q_i)[a_i(x, t, u, p) - a_i(x, t, u, q)] \geqslant \nu(|u|)|p - q|^2, \tag{6.32}$$

$$\begin{aligned} |a_i(x, t, u, p) - a_i(x, t, u, q)| &\leqslant \mu(|u|)|p - q|, \\ |a_i(x, t, u, p) - a_i(x', t', v, p)| &\leqslant \varepsilon(|x - x'| \\ &+ |t - t'| + |u - v|)[|p| + \mu(|u| + |v|) \\ &+ \varphi_4(x, t) + \varphi_4(x', t')] \end{aligned} \tag{6.33}$$

and

$$\begin{aligned} |a(x, t, u, p) - a(x', t', v, q)| &\leqslant \varepsilon(|x - x'| \\ &+ |t - t'| + |u - v| + |p - q|)[\mu(|u| + |v| \\ &+ |p| + |q|) + \varphi_1(x, t) + \varphi_1(x', t')], \end{aligned} \tag{6.34}$$

where $\nu(r)$ *and* $\mu(r)$ *are continuous positive functions of* $r \geq 0$, $\epsilon(r)$ *is a continuous function of* $r \geq 0$ *that is equal to zero for* $r = 0$, $\phi_4 \in L_2(Q_T)$, *and* $\phi_1(x, t) \in L_1(Q_T)$.

Then problem (6.1), (6.2) *with* $\psi|_{S_T} = 0$ *has at least one solution from* $H^{\gamma, \gamma/2}(\overline{Q}_T) \cap W_2^{1,0}(Q_T)$ *with some* $\gamma > 0$ *for any function* $\psi \in H^{\delta, \delta/2}(\Gamma_T)$. *The boundary* S *in this connection must satisfy condition* (A). *If furthermore the functions* $a_i(x, t, u, p)$ *are differentiable with respect to* u *and* p *and*

$$\left| \frac{\partial a_i(x, t, u, p)}{\partial u} \right| \leqslant \mu(|u|)[|p| + \varphi_4(x, t)], \quad \varphi_4 \in L_2(Q_T), \tag{6.35}$$

while the function $a(x, t, u, p)$ *is independent of* u *and* p, *then the solution is unique in the class of bounded functions from* $W_2^{1,0}(Q_T)$.

This theorem is proved with the use of an approximation of the same type as in the proof of Theorem 8.8 in Chapter IV of [65q]. Namely, the solutions $u(x, t)$ of problem (6.1), (6.2) are obtained as limit points in the strong topology of the spaces $H^{\gamma, \gamma/2}(\overline{Q}_T)$ and $W_2^{1,0}(Q_T)$ for the set of solutions $u^\rho(x, t)$, $\rho \to +0$, of the auxiliary problems

$$u_t - \frac{d}{dx_i} a_i^\rho(x, t, u, u_x) + a^\rho(x, t, u, u_x) = 0, \quad u|_{S_T} = 0, \quad u|_{t=0} = \psi(x, 0), \tag{6.36}$$

in which the $a_i^\rho(x, t, u, p)$ and $a^\rho(x, t, u, p)$ are averagings of the functions $a_i(x, t, u, p)$ and $a(x, t, u, p)$ with an infinitely differentiable nonnegative kernel $\omega(\tau)$ of the form

$$
\begin{aligned}
a^\rho(x, t, u, p) = & \int\limits_{|x-x'|<\rho} \omega\left(\frac{|x-x'|}{\rho}\right)\frac{dx'}{\rho^n} \\
\times & \int\limits_{|t-t'|<\rho} \omega\left(\frac{|t-t'|}{\rho}\right)\frac{dt'}{\rho} \int\limits_{|u-u'|<\rho} \omega\left(\frac{|u-u'|}{\rho}\right)\frac{du'}{\rho} \\
\times & \int\limits_{|p-p'|\le\rho} \omega\left[\frac{|p-p'|}{\rho}\right] a(x', t', u', p')\frac{dp'}{\rho^n}
\end{aligned}
$$

(in this connection it is necessary as a preliminary to extend the functions $a_i(x, t, u, p)$ and $a(x, t, u, p)$ from $\overline{Q}_T$ onto a somewhat larger domain with preservation of the properties of them enumerated in the conditions of the theorem).

Assumptions a), b) and (6.32) guarantee the fulfillability of all of the conditions of Theorem 6.4 (with condition a) replaced by condition a'); see Theorem 6.5) for problems (6.36) and consequently also the solvability of these problems. Moreover, they guarantee the uniform boundedness of the norms $|u^\rho|_{Q_T}^{(\hat{\gamma})}$ and $|u^\rho|_{Q_T}$ (with some $\hat{\gamma} > 0$) of all solutions of (6.36). This fact, conditions a)–c) and relations (6.36) permit one to choose a subsequence from $\{u^\rho\}$ that converges in $H^{\gamma,\gamma/2}(\overline{Q}_T)$, $\gamma < \hat{\gamma}$, and in $W_2^{1,0}(Q_T)$, and prove that the limit functions u for them will satisfy the integral identity

$$
\int\limits_{Q_T} [-u\eta_t + a_i(x, t, u, u_x)\eta_{x_i} + a(x, t, u, u_x)\eta]\, dx\, dt = 0 \tag{6.37}
$$

for all smooth η that are equal to zero on the boundary of Q_T, and consequently will be the desired generalized solutions of problem (6.1), (6.2). The question of uniqueness of the solution guaranteed by Theorem 6.6 when to its conditions are added only conditions of the type

$$
|a(x, t, u, p) - a(x, t, v, p)| \leqslant \mu|u-v|[|p|^2 + \varphi_1(x, t)]
$$

and

$$
\begin{aligned}
&|a(x, t, u, p) - a(x, t, u, q)| \\
&\qquad \leqslant \mu|p-q|[|p| + |q| + \varphi_4(x, t)],
\end{aligned}
$$

where $\phi_1 \in L_1(Q_T)$, while $\phi_4 \in L_2(Q_T)$, has not been resolved in the general case. But for the case noted in Theorem 6.6 uniqueness follows from Theorem 3.4

of Chapter III. In fact, for the difference $v(x, t) = u'(x, t) - u''(x, t)$ of two possible solutions u' and u'' of problem (6.1), (6.2) there follows from (6.37) the identity

$$\int_{Q_T} [-v\eta_t + (\hat{a}_{ij}(x, t) v_{x_j} + \hat{a}_i(x, t) v) \eta_{x_i}] \, dx \, dt = 0,$$

in which

$$\hat{a}_{ij}(x, t) = \int_0^1 \frac{\partial}{\partial p_j} a_i [x, t, \tau u'(x, t) + (1-\tau) u''(x, t), \tau u'_x(x, t) + (1-\tau) u''_x(x, t)] \, d\tau,$$

while $\hat{a}_i(x, t) = \int_0^1 (\partial a_i[\cdots]/\partial u) \, d\tau$. The conditions of Theorem 3.4 of Chapter III, and hence its assertion on the fact that $v \equiv 0$, are realized by virtue of conditions (6.32), (6.33) and (6.35). Theorem 3.4 makes it possible to strengthen somewhat the theorem of uniqueness just proved.

The existence of generalized solutions u of the boundary value problems for equations (6.1) in the class of functions having only derivatives u_x can be proved in another way without using Theorems 6.1–6.5 on their classical solvability. Namely, such solutions can be obtained as limits of approximate solutions u^N computed by Galerkin's method. The u^N are defined in the same way as for linear equations (see §4 of Chapter III). The special character of the quasi-linear case consists in the need to carry out passages to the limit in functions $b(x, t, u^N, u^N_x)$ having in general a nonlinear dependence on the u^N and u^N_x. We do this below by making use of an idea of Minty [84a], concerning weak passages to the limit through convex functions and permitting one to pass to the limit in the leading terms $a_i(x, t, u, u_x)$ of an equation, and also one of the techniques in the paper [68a] of Leray and Lions, which is useful for passing to the limit in the minor term $a(x, t, u, u_x)$ (in connection with the topic see also [84, 11, 134, 22]).

We introduce the space $\overset{\circ}{V}_{m,2}(Q_T)$, $m \geq 1$, consisting of all measurable functions $u(x, t)$ that are equal to zero on S_T and have the finite norm

$$|u|_{m, 2, Q_T} = \operatorname*{vrai\,max}_{0<t<T} \|u\|_{2, \Omega} + \|u_x\|_{m, Q_T},$$

and the space $\overset{\circ}{V}{}^{1,0}_{m,2}(Q_T)$ obtained by a completion in this norm of all smooth functions that are equal to zero on S_T.

They are embedded in the spaces $L_{q,r}(Q_T)$ with q and r satisfying the conditions

$$\left.\begin{aligned} &\frac{1}{r}+\frac{2n}{mn+2m-2n}\frac{1}{q}=\frac{n}{mn+2m-2n}, \\ &\text{where} \\ &q\in\left[2, \frac{nm}{n-m}\right], \quad r\in[m, \infty], \text{ for } n>m\geqslant\frac{2n}{n+2}, \\ &q\in[2, \infty), \; r\in\left(m-2+2\frac{m}{n}, \infty\right], \text{ for } 1<n\leqslant m, \\ &q\in[2, \infty], \; r\in[3m-2, \infty], \text{ for } m\geqslant n=1, \end{aligned}\right\} \tag{6.38}$$

so that under such q and r the inequality

$$\|u\|_{q,r,Q_T}\leqslant\beta|u|_{m,2,Q_T}. \tag{6.39}$$

is valid for any function u from $\overset{\circ}{V}_{m,2}(Q_T)$. One can easily verify that if $m>2n/(n+2)$ for $n\geq 2$, and $m\geq 1$ for $n=1$, then the indices $r=q=m(n+2)/n$ obey conditions (6.38).

Inequality (6.39) is deduced from inequality (2.9) of Chapter II for arbitrary $m\geq 1$ in the same way as for $m=2$, that is, in the same way as inequality (3.4) of Chapter II. We will establish the following theorem.

Theorem 6.7. *The problem*

$$u_t-\frac{d}{dx_i}a_i(x, t, u, u_x)+a(x, t, u, u_x)=0, \quad u|_{S_T}=0, \quad u|_{t=0}=\psi_0(x), \tag{6.40}$$

for any $\psi_0\in L_2(\Omega)$ *has at least one generalized solution* u *from* $\overset{\circ}{V}{}^{1,0}_{m,2}(Q_T)$, *with* $\int_0^T h^{-2}\|\Delta_h u\|^2_{2,Q_{T-h}}dh<\infty$, *where* $\Delta_h u(x, t)\equiv u(x, t+h)-u(x, t)$, *if the following conditions are fulfilled.*

1) *For* $(x, t, u, p)\in\{\bar{\Omega}\times[0, T]\times E_1\times E_n\}$ *the functions* $a_i(x, t, u, p)$ *and* $a(x, t, u, p)$ *are measurable in* (x, t, u, p) *and continuous in* (u, p) *for almost all* (x, t) *from* Q_T; *the functions* $a_i(x, t, u, p)$ *satisfy the inequalities*

$$|a_i(x, t, u, p)|\leq\phi_1(x, t)+c|u|^{\frac{q^*}{m'}}+c|p|^{m-1}, \quad \phi_1\in L_{m'}(Q_T), \tag{6.41}$$

where $q^*<q=m(n+2)/n$, $m'=m/(m-1)$, *with* $m>2n/(n+2)$ *for* $n\geq 2$ *and* $m>1$ *for* $n=1$; *and the function* $a(x, t, u, p)$ *satisfies the inequality*

$$|a(x, t, u, p)| \le \phi_2(x, t) + c|u|^{\frac{q^*}{q'}} + c|p|^{\frac{m^*}{q'}}, \quad \phi_2 \in L_{q'}(Q_T), \tag{6.42}$$

where $q' = q/(q-1)$ *and* $m^* < m$.

2) *For any function* $u(x)$ *from* $\overset{\circ}{W}{}^1_m(\Omega)$

$$\int_\Omega [a_i(x, t, u, u_x) u_{x_i} + a(x, t, u, u_x) u] \, dx$$

$$\geqslant \nu \int_\Omega |u_x|^m dx - c(t) \int_\Omega (1 + u^2) \, dx, \quad \nu > 0, \quad \int_0^T c(t) \, dt \leqslant c. \tag{6.43}$$

3) *A monotonicity condition of the form*

$$[a_i(x, t, u, p) - a_i(x, t, u, q)](p_i - q_i) \geqslant \nu(|p - q|, u, q, x, t), \tag{6.44}$$

where $\nu(\tau, u, q, x, t)$ *is continuous on* $E_1^+ \times E_1 \times E_n \times \bar{\Omega} \times [0, T]$, *positive for* $\tau > 0$ *and equal to zero for* $\tau = 0$.

By a generalized solution u of problem (6.40) we mean a function of the class indicated in the theorem that satisfies the integral identity

$$\int_{Q_t} [-u\phi_t + a_i(x, t, u, u_x)\phi_{x_i} + a(x, t, u, u_x)\phi] \, dx \, dt$$

$$+ \int_\Omega u(x, t)\phi(x, t) dx - \int_\Omega \psi_0(x)\phi(x, 0) dx = 0$$

for any smooth $\phi(x, t)$ that is equal to zero on S_T.

In order to prove the solvability of problem (6.40) we make use of Galerkin's method. We take a fundamental system $\{\psi_k(x)\}$ in the space $\overset{\circ}{W}{}^1_m(\Omega)$ such that $(\psi_k, \psi_l) = \delta_k^l$ and $\max_\Omega |\psi_k, \psi_{kx}| = c_k < \infty$. An approximate solution $u^N(x, t)$ will be sought in the usual form $u^N = \sum_{k=1}^N c_k^N(t)\psi_k(x)$ where the $c_k^N(t)$ are determined from the system of ordinary differential equations

$$(u_t^N, \psi_k) + (a_i(x, t, u^N, u_x^N), \psi_{kx_i}) + (a(x, t, u^N, u_x^N), \psi_k) = 0, \tag{6.45}$$

$k = 1, 2, \cdots, N$ and initial conditions

$$c_k^N(0) = (\psi_0, \psi_k), \quad k = 1, 2, \cdots, N. \tag{6.46}$$

From assumptions (6.41), (6.42) and the indicated properties of the functions ψ_k

it follows that the second and third terms in (6.45) are summable functions of t on $[0, T]$ with summability power q' for $|c_k^N| \leq \text{const}$, and are continuous in the c_k^N, $k = 1, \cdots, N$ (we note that $q' \leq m'$). Thus for the existence of at least one solution of problem (6.45), (6.46) on the whole interval $[0, T]$ it is sufficient to know that all of its possible solutions c_k^N are uniformly bounded on $[0, T]$.

This type of boundedness follows from the a priori estimate

$$|u^N|_{m,2,Q_T} \leq c, \tag{6.47}$$

in which c does not depend on N, since

$$\max_{0 \leqslant t \leqslant T} \sum_{k=1}^{N} (c_k^N(t))^2 = \max_{0 \leqslant t \leqslant T} \|u^N(t)\|_{2,\Omega}^2 \leqslant |u^N|_{m,2,Q_T}^2 .$$

Estimate (6.47) is derived from (6.45) in the following manner. We multiply the kth equation of (6.45) by c_k^N and then sum all of the equations over k from 1 to N. From the resultant equality integrated with respect to t we obtain by virtue of assumption (6.43)

$$\frac{1}{2}\| u^N(x, t)\|_{2,\Omega}^2 + \nu \int_{Q_t} |u_x^N|^m \, dx\, dt \leqslant \int_0^t c(t) \int_\Omega [1 + (u^N)^2] \, dx\, dt,$$

from which, as we proved in Lemma 5.5 of Chapter II, follows (6.47). From (6.47) and (6.39) we also have

$$\| u^N \|_{q, Q_T} \leqslant c\beta, \quad q = m\frac{n+2}{n}. \tag{6.48}$$

We now let N tend to ∞. Owing to the uniform estimates (6.47) and (6.48) it is possible to choose a subsequence from $\{u^N\}$ that is weakly convergent in $L_q(Q_T)$ to some function u and is such that the derivatives $u_{x_i}^N$ of its elements converge weakly in $L_m(Q_T)$ to u_{x_i}. Let us agree for the sake of conciseness not to take note of the necessity of distinguishing a convergent subsequence from a compact (in the strong or weak sense) sequence but to speak of the convergence of the sequence itself. In accordance with this we assume that the u^N, $N = 1, 2, \cdots$, converge weakly in $L_q(Q_T)$ to u, and the $u_{x_i}^N$, $N = 1, 2, \cdots$, converge weakly in $L_m(Q_T)$ to u_{x_i}.

Because of assumptions (6.41) and (6.42) the following estimates are valid for any u and v from $\overset{\circ}{V}_{m,2}(Q_T)$:

$$\|a_i(x, t, u(x, t), v_x(x, t))\|_{m',Q'} \le \|\phi_1\|_{m',Q}$$
$$+ c\|u\|_{q,Q'}^{\frac{q^*}{m'}} (\text{mes } Q')^{\frac{1}{m'}(1-\frac{q^*}{q})} + c\|v_x\|_{m,Q'}^{m-1}$$
$$\equiv \mu_1(\text{mes } Q', \|u\|_{q,Q'}, \|v_x\|_{m,Q'}), \qquad (6.49)$$

where Q' is any measurable subset of Q_T, and

$$\|a(x, t, u(x, t), v_x(x, t))\|_{q',Q'} \le \|\phi_2\|_{q',Q'}$$
$$+ c\|u\|_{q,Q'}^{\frac{q^*}{q'}} (\text{mes } Q')^{\frac{1}{q'}(1-\frac{q^*}{q})} + c\|v_x\|_{m,Q'}^{\frac{m^*}{q'}} (\text{mes } Q')^{\frac{1}{q'}(1-\frac{m^*}{m})}$$
$$\equiv \mu_2(\text{mes } Q', \|u\|_{q,Q'}, \|v_x\|_{m,Q'}), \qquad (6.50)$$

in which the function $\mu_1(\tau_1, \tau_2, \tau_3) \longrightarrow 0$ for $\tau_1 + \tau_3 \longrightarrow 0$, and $\mu_2(\tau_1, \tau_2, \tau_3) \longrightarrow 0$ for $\tau_1 \longrightarrow 0$, uniformly on any bounded set of the space (τ_1, τ_2, τ_3).

Let us consider the functions $l_{N,k}(t) = (u^N(x, t), \psi_k(x))$, $N, k = 1, 2, \cdots$. They are continuous in t and equicontinuous for any fixed k. To see this we integrate (6.45) from t to $t + h$ and exploit estimates (6.47)–(6.50) in the following manner:

$$|(u^N(x, t+h) - u^N(x, t), \psi_k(x))|$$
$$\leqslant \int_t^{t+h} |(a_i(x, t, u^N, u_x^N), \psi_{kx_i}) + (a(x, t, u^N, u_x^N), \psi_k)|\, dx\, dt$$
$$\leqslant c_k \int_t^{t+h} \int_\Omega \Big(\sum_{i=1}^n |a_i(x, t, u^N, u_x^N)| + |a(x, t, u^N, u_x^N)|\Big) dx\, dt$$
$$\leqslant c_k \sum_{k=1}^n \|a_i(x, t, u^N, u_x^N)\|_{m', Q_{t,t+h}} (h \cdot \text{mes}\, \Omega)^{\frac{1}{m}}$$
$$+ c_k \|a(x, t, u^N, u_x^N)\|_{q', Q_{t,t+h}} (h \cdot \text{mes}\, \Omega)^{\frac{1}{q}} \leqslant c_k' h^{\frac{1}{q}}.$$

The constants c_k and c'_k depend on k but not on N for $N \ge k$. In view of this the families of functions $l_{N,k}(t)$, $N = 1, 2, \cdots$, are indeed equicontinuous in $t \in [0, T]$ for any fixed k. Furthermore, from (6.47) it follows that $\max_{0 \le t \le T} |l_{N,k}(t)| \le c$ for all $N, k = 1, 2, \cdots$. This, as was shown in the proof of Theorem 4.1 of Chapter III, guarantees the weak convergence in $L_2(\Omega)$, which is uniform in $t \in [0, T]$, of the u^N to u. The limit function u will thereby be an element of $\overset{\circ}{V}_{m,2}(Q_T)$, and estimates (6.47) and (6.48) are valid for it.

Let us prove that u is a generalized solution of problem (6.40). From (6.45) follows the identity

$$\int_{Q_t}\left[-u^N\phi_t+a_i(x,t,u^N,u^N_x)\phi_{x_i}+a(x,t,u^N,u^N_x)\phi\right]dx\,dt + \int_\Omega u^N\varphi\,dx\Big|_{t=0}^{t=t}=0, \tag{6.51}$$

which is valid for any function $\phi=\sum_{k=1}^N d_k(t)\psi_k(x)$, where the $d_k(t)$ are continuous functions of t having generalized derivatives $d'_k(t)$ from $L_{q'}[0,T]$ on $[0,T]$. We denote the totality of such ϕ by $\mathcal{P}_N$. The function u^N belongs to $\mathcal{P}_N$.

Let us attempt to pass to the limit as $N\to\infty$ in (6.51) for a fixed ϕ. The functions $A_i^N(x,t)=a_i(x,t,u^N,u^N_x)$ by virtue of estimates (6.47) and (6.49) have uniformly bounded norms in $L_{m'}(Q_T)$ and hence converge weakly in $L_{m'}(QT)$ to certain functions $A_i(x,t)\in L_{m'}(Q_T)$ (Lemma 2.2 of Chapter II). The functions $A^N(x,t)\equiv a(x,t,u^N,u^N_x)$, $(x,t)\in Q_T$, are uniformly bounded in $L_{q'}(Q_T)$ by virtue of (6.47) and (6.50). Therefore we can assume that they converge weakly in $L_{q'}(Q_T)$ to some function $A(x,t)$ from $L_{q'}(Q_T)$ (see Lemma 2.2 of Chapter II).

This together with the above-described convergence of the u^N to u gives as a limit relation for (6.51) the equality

$$\int_{Q_t}[-u\phi_t+A_i\phi_{x_i}+A\phi]dx\,dt+\int_\Omega u(x,T)\phi(x,T)dx-\int_\Omega\psi_0(x)\phi(x,0)\,dx=0. \tag{6.52}$$

It is valid for any ϕ from $\mathcal{P}_k$, $k=1,2,\cdots$, i.e. for any ϕ from $\mathcal{P}=\bigcup_{k=1}^\infty\mathcal{P}_k$. But the set $\mathcal{P}$ is dense in the space $\overset{\circ}{W}{}^{1,1}_{m,q'}(Q_T)$ consisting of the functions $\phi(x,t)$ equal to zero on S_T for which $\phi_t\in L_{q'}(Q_T)$ and $\phi_x\in L_m(Q_T)$. In view of this it is easily seen that identity (6.52) holds for any $\phi\in\overset{\circ}{W}{}^{1,1}_{m,q'}(Q_T)$. From this identity one can deduce that the function $u(x,t)$ is strongly continuous in t in $L_2(\Omega)$ and $h^{-1}\|\Delta_h u\|^2_{2,Q_{T-h}}\to 0$ where $\Delta_h u(x,t)\equiv u(x,t+h)-u(x,t)$, i.e. $u\in\overset{\circ}{V}{}^{1,1/2}_{m,2}(Q_T)$. Moreover, the norm $|||u|||^2_Q$ will be finite for u (see page 161), i.e. $u\in\overset{\circ}{W}{}^{1,1/2}_{m,2}(Q_T)$. This is proved in the same way as the corresponding assertions in §4 of Chapter III (see Lemmas 4.1 and 4.2 of Chapter III). Also from (6.52), in the same way as in §2 of Chapter III, one can prove the validity of the relation

$$\frac{1}{2}\|u\|_{2,\Omega}^2\Big|_{t=0}^{t=T}+\int_{Q_T}(A_iu_{x_i}+Au)\,dx\,dt=0. \tag{6.53}$$

Let us now prove that for any $\phi\in\mathfrak{P}$

$$\int_{Q_T}A_i\phi_{x_i}\,dx\,dt=\int_{Q_T}a_i(x,\,t,\,u,\,u_x)\phi_{x_i}\,dx\,dt. \tag{6.54}$$

To this end we make use of the monotonicity condition (6.44). Because of it

$$\int_{Q_T}[a_i(x,\,t,\,u^N,\,u_x^N)-a_i(x,\,t,\,u^N,\,\eta_x)](u_{x_i}^N-\eta_{x_i})\,dx\,dt\geq 0 \tag{6.55}$$

for any $\eta\in\mathfrak{P}$. We replace the integral $\int_{Q_T}a_i(x,\,t,\,u^N,\,u_x^N)\,(u_{x_i}^N-\eta_{x_i})\,dx\,dt$ in (6.55) by the expression equal to it from (6.51) with $\phi=u^N-\eta$, where $\eta\in\mathfrak{P}_N$ This gives the inequality

$$\int_{Q_t}[-u^N\eta_t-a(x,\,t,\,u^N,\,u_x^N)(u^N-\eta)$$
$$-a_i(x,\,t,\,u^N,\,\eta_x)(u_{x_i}^N-\eta_{x_i})]\,dx\,dt-\frac{1}{2}\|u^N\|_{2,\Omega}^2\Big|_{t=0}^{t=t}$$
$$+(u^N,\,\eta)\Big|_{t=0}^{t=t}\geqslant 0. \tag{6.56}$$

In it with $t=T$ one can pass to the limit as $N\longrightarrow\infty$ and as a result obtain

$$\int_{Q_T}[-u\eta_t-A(u-\eta)-a_i(x,\,t,\,u,\,\eta_x)(u_{x_i}-\eta_{x_i})]\,dx\,dt$$
$$-\frac{1}{2}\|u\|_{2,\Omega}^2\Big|_{t=0}^{t=T}+(u,\,\eta)\Big|_{t=0}^{t=T}\geq 0. \tag{6.57}$$

In fact, the $u^N(x,\,0)$ converge strongly in $L_2(\Omega)$ to $\psi_0=u(x,\,0)$, while the $u^N(x,\,t)$ converge weakly in $L_2(\Omega)$ to $u(x,\,t)$, so that $\lim_{N\to\infty}(u,\,\eta)=(u,\,\eta)$ and $\underline{\lim}_{N\to\infty}\|u^N(x,\,t)\|_{2,\Omega}\leq\|u(x,\,t)\|_{2,\Omega}$. Because of estimate (6.47) and the weak convergence of the u^N to u in $L_2(\Omega)$, which is uniform in $t\in[0,\,T]$, the functions u^N converge to u strongly in the norm of $L_{2,m}(Q_T)$, as is seen from inequality (2.27) in Lemma 2.4 of Chapter II applied to the difference $u^N-u^{N_1}$. And from the convergence of the u^N to u in $L_{2,m}(Q_T)$ follows the convergence of the u^N to u in $L_2(\Omega)$ for almost all t from $[0,\,T]$ and almost everywhere in Q_T. Furthermore, from it and the uniform boundedness of $\|u^N\|_{q,Q_T}$ (see (6.48)) follows the strong convergence of the u^N to u in $L_{\widetilde{q}}(Q_T)$ with any $\widetilde{q}<q$ and the weak convergence in $L_q(Q_T)$ (see Lemma 2.2 of Chapter II). From this, the

continuity of the $a_i(x, t, u, p)$ in (u, p) and estimate (6.49) it follows that the functions $a_i(x, t, u^N, \eta_x)$ converge strongly in $L_{m'}(Q_T)$ to $a_i(x, t, u, \eta_x)$ (Lemma 2.3 of Chapter II). The functions $f_k(x, t) \equiv A^k(x, t)$ and $u_k(x, t) \equiv u^k(x, t) - \eta(x, t)$, $k = 1, 2, \ldots$, satisfy all conditions of Lemma 2.2 of Chapter II (see (6.50)) and therefore $\int_{Q_T} f_k u_k \, dx\, dt \longrightarrow \int_{Q_T} A(u - \eta) dx\, dt$ when $k \longrightarrow \infty$. So (6.57) is proved.

We now combine inequality (6.57) with (6.53) and with (6.52), in which ϕ is taken equal to $-\eta \in \mathcal{P}_N$. As a result we obtain the inequality

$$\int_{Q_T} [A_i(x, t) - a_i(x, t, u, \eta_x)] (u_{x_i} - \eta_{x_i}) \, dx\, dt \geq 0. \tag{6.58}$$

It is deduced under the assumption that $\eta \in \mathcal{P}_N$, but since N is arbitrary, while $A_i - a_i(x, t, u, \eta_x) \in \overset{\circ}{L}_{m'}(Q_T)$ for $u, \eta \in \overset{\circ}{V}{}^{1,0}_{m,2}(Q_T)$, it follows that (6.58) is valid for any η from $V^{1,0}_{m,2}(Q_T)$. We put in (6.58) $\eta = u - \epsilon\zeta$, where $\zeta(x, t) \in \mathcal{P}$, while ϵ is an arbitrary positive number, and multiply both sides of the inequality by ϵ^{-1}. This gives

$$\int_{Q_T} [A_i(x, t) - a_i(x, t, u, u_x - \epsilon\zeta_x)] \, \zeta_{x_i} \, dx\, dt \geq 0,$$

from which, letting ϵ tend to zero, we obtain the inequality

$$\int_{Q_T} [A_i(x, t) - a_i(x, t, u, u_x)] \, \zeta_{x_i} \, dx\, dt \geq 0,$$

in which the equality sign actually holds, since together with ζ the function $-\zeta$ is also contained in $\mathcal{P}$. Thus (6.54) is proved and hence (6.52) can be written in the form

$$\int_{Q_T} [-u\phi_t + a_i(x, t, u, u_x)\phi_{x_i} + A\phi] dx\, dt + \int_\Omega u\phi \, dx \Big|_{t=0}^{t=T} = 0. \tag{6.59}$$

It remains to be proved that for almost all $(x, t) \in Q_T$

$$A(x, t) = a(x, t, u(x, t), u_x(x, t)). \tag{6.60}$$

We take a smooth function $\omega^\delta(t)$ on $(-\infty < t < \infty)$, which is equal to zero for $-\infty < t \leq \delta$ and for $t \geq T - \delta$, where δ is some small positive number, and equal to 1 for $t \in [2\delta, T - 2\delta]$. Let

$$v^N(x, t) = \begin{cases} u^N(x, t)\omega^\delta(t) & \text{for } t \in [0, T] \\ 0 & \text{for } t \leq 0 \text{ and } t \geq T \end{cases}, \quad \text{and } v(x, t) = \begin{cases} u(x, t)\omega^\delta(t) & \text{for } t \in [0, T] \\ 0 & \text{for } t \leq 0 \text{ and } t \geq T. \end{cases}$$

From (6.45) and (6.59) with $\phi = \omega^{\delta}\tilde{\phi}$ follow the relations

$$\int_Q \{v_t^N \tilde{\phi} + \omega^{\delta}(t) a_i(x, t, u^N, u_x^N) \tilde{\phi}_{x_i} + [\omega^{\delta}(t) a(x, t, u^N, u_x^N) - \omega_t^{\delta} u^N] \tilde{\phi}\} dx\, dt = 0 \tag{6.61}$$

and analogously

$$\int_Q \{- v\tilde{\phi}_t + \omega^{\delta} a_i(x, t, u, u_x)\tilde{\phi}_{x_i} + [\omega^{\delta} A - \omega_t^{\delta} u]\tilde{\phi}\}\, dx\, dt = 0 \tag{6.62}$$

where all of the factors under the integral signs are assumed to be equal to zero for $t \overline{\in} [0, T]$, and $\tilde{\phi}(x, t)$ in (6.61) is an arbitrary element of the form $\Sigma_{k=1}^N d_k(t)\psi_k(x)$ with continuous $d_k(t)$ having compact support, while in (6.62) it is any element of $\overset{\circ}{W}{}^{1,1}_{m,q'}(Q)$, $Q = \Omega \times (-\infty, +\infty)$ equal to zero for large $|t|$. From these identities and inequalities (6.47)–(6.50) the estimate

$$|||v^N|||_Q + |||v|||_Q \le c \tag{6.63}$$

is deduced in the way it was done in §4 of Chapter III (see pages 160–161).

We take a sequence $\{v_N\}$, $N = 1, 2, \cdots$, $v_N \in \mathfrak{P}_N$, such that the v_N converge to v strongly in $L_q(Q)$ and in the norm $|||\cdot|||_Q$, while the v_{Nx_i} converge to v_{x_i} strongly in $L_m(Q)$ (in other words the v_N converge to v in the norms of the spaces $L_q(Q)$ and $W'^{1,1/2}_{m,2}(Q)$). Because of (6.61) with $\tilde{\phi} = v^N - u^N$

$$\int_Q \{v_t^N (v^N - v) + \omega^{\delta}(t) a_i(x, t, u^N, u_x^N)(v^N_{x_i} - v_{x_i})$$
$$+ [\omega^{\delta}(t) a(x, t, u^N, u_x^N) - \omega_t^{\delta} u^N](v^N - v)\}\, dx\, dt$$
$$= - \int_Q \{v_t^N (v - v_N) + \omega^{\delta}(t) a_i(x, t, u^N, u_x^N)(v_{x_i} - v_{Nx_i})$$
$$+ [\omega^{\delta}(t) a(x, t, u^N, u_x^N) - \omega_t^{\delta} u^N](v - v_N)\}\, dx\, dt \equiv R_N \tag{6.64}$$

By the convergence of the v_N to v and estimates (6.47)–(6.50) and (6.63), the $R_N \to 0$ for $N \to \infty$ since

$$\left|\int_Q v_t^N (v - v_N)\, dx\, dt\right| = \left|\int_{-\infty}^{\infty}\int_{\Omega} \alpha \tilde{v}^N (\tilde{v} - \tilde{v}_N)\, dx\, d\alpha\right|$$
$$\le \left\{\int_{-\infty}^{\infty}\int_{\Omega} |\alpha|\, |\tilde{v}^N|^2\, dx\, d\alpha\right\}^{1/2} \left\{\int_{-\infty}^{\infty}\int_{\Omega} |\alpha|\, |\tilde{v} - \tilde{v}_N|^2\, dx\, d\alpha\right\}^{1/2}$$
$$= |||v^N|||_Q \cdot |||v - v_N|||_Q \to 0 \text{ for } N \to \infty.$$

On the other hand, from (6.62) follows the equality

$$\int_Q [\omega^\delta a_i(x, t, u, u_x) v_{x_i} + (\omega^\delta A - \omega_t^\delta u)v]\, dx\, dt = 0 \tag{6.65}$$

which is deduced in the same way as (6.53) from (6.52), or as (2.13) from (2.15) of Chapter III. We combine it with (6.44) and with (6.62) for $\tilde{\phi} = -v^N$. As a result, after elementary transformations of the left side of (6.64)

$$\begin{aligned}\int_Q \{&\omega^\delta[a_i(x, t, u^N, u_x^N) - a_i(x, t, u^N, u_x)](v_{x_i}^N - v_{x_i}) + \omega^\delta(t)[a_i(x, t, u^N, u_x) \\ &- a_i(x, t, u, u_x)](v_{x_i}^N - v_{x_i}) + \omega^\delta[a(x, t, u^N, u_x^N) - A](v^N - v) \\ &- \omega_t^\delta(u^N - u)(v - v^N)\}\, dx\, dt \equiv R_N.\end{aligned} \tag{6.66}$$

The second, third, and fourth members of the left side of (6.66) tend to zero as $N \to \infty$ (it is proved as above, when we derived (6.57) from (6.56)) and R_N also tends to zero. Therefore

$$j_N \equiv \int_Q \omega^\delta [a_i(x, t, u^N, u_x^N) - a_i(x, t, u^N, u_x)](v_{x_i}^N - v_{x_i})dx\, dt \to 0, N \to \infty$$

and because of (6.44)

$$\int_Q (\omega^\delta)^2 \nu(|u_x^N - u_x|, u^N, u_x, x, t)dx\, dt \to 0, N \to \infty$$

This gives $(\omega^\delta(t))^2 \nu(|u_x^N(x, t) - u_x(x, t)|, u^N(x, t), u_x(x, t)) \to 0$ almost everywhere in Q. According to the properties of $\nu(\cdot\cdot\cdot)$ and arbitrariness of $\delta > 0$, we can conclude that $|u_x^N(x, t) - u_x(x, t)| \to 0$ almost everywhere in Q_T. From this follows that $a(x, t, u^N(x, t), u_x^N(x, t))$ have as their limit $a(x, t, u(x, t), u_x(x, t))$. Equality (6.60) and Theorem 6.7 are thereby proved. (We had to choose in some places convergent subsequences from compact sequences, but we did not do this for the sake of conciseness.)

An assertion similar to Theorem 6.7 was proved by Lions in [134] with the use of an approximation of equation (6.40) by equations of elliptic type.

Theorem 6.7 admits different generalizations. For example, its assertions remain valid if the a_i and a have the form

$$a_i(x, t, u, p) = \bar{a}_i(x, t, u, p) + \bar{\bar{a}}_i(x, t, u, p),$$

$$a(x, t, u, p) = \bar{a}(x, t, u, p) + \bar{\bar{a}}(x, t, u, p) + \bar{\bar{\bar{a}}}(x, t, u, p),$$

where the functions $\bar{a}_i$ are subject to inequality (6.41), the function $\bar{a}$ is subject to inequality (6.42), the functions $\bar{\bar{a}}_i$ are subject to inequality (6.41), in which q^* is replaced by q, $\bar{\bar{a}}$ is subject to inequality (6.42), in which q^* is replaced by q, and $\bar{\bar{\bar{a}}}$ is subject to inequality (6.42), in which q^* is replaced by q and m^* by m, provided that in place of condition 3) it is known that for any u, $v \in \overset{\circ}{W}{}^1_m(\Omega)$

$$\int_\Omega [\bar{a}_i(x, t, v, v_x) - \bar{a}_i(x, t, v, u_x) + \bar{\bar{a}}_i(x, t, v, v_x) - \bar{\bar{a}}_i(x, t, u, u_x)]$$
$$\times (v_{x_i} - u_{x_i}) + [\bar{\bar{a}}(x, t, v, v_x) - \bar{\bar{a}}(x, t, v, u_x) + \bar{\bar{\bar{a}}}(x, t, v, v_x)$$
$$- \bar{\bar{\bar{a}}}(x, t, u, u_x)] (v - u)\, dx \geq \int_\Omega \nu(|v|, |u_x|)\, |v_x - u_x|\, dx, \tag{6.67}$$

where $\nu(r_1, r_2)$ is a continuous positive function for $r_1 \geq 0$, $r_2 \geq 0$.

Theorems analogous to Theorem 6.7 are also valid for quasi-linear parabolic equations of high order, as well as for quasi-linear strongly parabolic systems with principal part in divergence form, and they are proved in the same way as was just done for equations (6.40) of second order, it being possible in this connection to take as the boundary condition not only the first but also analogues of the second and third boundary conditions (see §5 of Chapter III).

The papers [11a–d; 134; 124d,e; 22] are devoted to a proof of the existence of generalized solutions of such problems. A construction is given in [124d,e] of generalized solutions having certain differential properties that are better than in [11a–d;134] (in the case of equation (6.40) they have the derivatives u_t and u_{xx}), which requires greater restrictions than a condition of type 1)–3) but in return dispenses with the weak passages to the limit through nonlinear functions. Such solutions are obtained as limits of approximate solutions constructed by a modification of Galerkin's method, which is similar to that given in [65k].

§7. OTHER BOUNDARY VALUE PROBLEMS

In this section we consider the second and third boundary value problems for equations of the form

$$\mathscr{L}u \equiv u_t - a_{ij}(x, t, u)\, u_{x_i x_j} + b(x, t, u, u_x) = 0, \tag{7.1}$$

which are a special case of equations (0.1). Namely, we will assign on the lateral surface of a cylinder Q_T the boundary condition

$$\mathscr{L}^{(S)}u \equiv [a_{ij}(x, t, u)\, u_{x_j} \cos(\mathbf{n}, x_i) + \psi(x, t, u)]\big|_{S_T} = 0, \tag{7.2}$$

and on the lower base of Q_T the condition

$$u\big|_{t=0} = \psi_0(x). \tag{7.3}$$

Here $n = n(x)$ is a vector of the outward normal to the surface S, which we assume to be sufficiently smooth.

We will investigate the solvability of problem (7.1)–(7.3) in the class $H^{2+\beta,1+\beta/2}(\overline{Q}_T)$. As a preliminary we establish all of the estimates that are necessary for this purpose. By the results of §§1–3, in obtaining the estimates we may restrict ourselves to small subdomains Q' that are adjacent to the lateral surface of the cylinder Q_T. We fix, for example, some part S' of the boundary S and denote by Ω' an arbitrary subdomain of Ω that is adjacent to S', and by Q'_T cylinders of the form $\Omega' \times (0, T)$. Without loss of generality one can assume that S' lies in the plane $x_n = 0$ and Ω is in the halfspace $x_n \le 0$ (this situation can always be achieved by means of a transformation of independent variables).

Keeping in mind the basic aim of this section, viz. a proof of the existence of a classical solution of problem (7.1)–(7.3), we will carry out all of the estimates for solutions $u(x, t)$ from $C^{2,1}(\overline{Q}_T)$, although many of them can also be obtained under weaker assumptions concerning $u(x, t)$, as was done in §§1–3.

We begin with a derivation of an estimate of the norm $|u|^{(\alpha)}_{Q'_T}$, assuming that $\max_{Q_T}|u(x, t)| \le M$ is already estimated. We will assume that for $(x, t) \in \overline{Q}_T$ and $|u| \le M$ the functions $a_{ij}(x, t, u)$ and $\psi(x, t, u)$ are differentiable with respect to x and u and satisfy the inequalities

$$\nu\xi^2 \leqslant a_{ij}(x, t, u)\xi_i\xi_j \leqslant \mu\xi^2, \quad \nu > 0, \tag{7.4}$$

$$\left|\frac{\partial a_{ij}(x, t, u)}{\partial u}, \frac{\partial a_{ij}}{\partial x}, \psi, \frac{\partial \psi}{\partial u}, \frac{\partial \psi}{\partial x}\right| \leqslant \mu, \tag{7.5}$$

while $b(x, t, u, p)$ for the same (x, t, u) and arbitrary p satisfies the inequality

$$|b(x, t, u, p)| \leqslant \mu(1 + p^2) \tag{7.6}$$

Suppose $u(x, t) \in C^{2,1}(\overline{Q}_T)$ is a solution of equation (7.1) satisfying condition (7.2). The following identity is valid for any function $\eta(x, t)$ from $V_2^1(Q_T)$:

$$\int_{t_0}^{t}\int_{\Omega}\Big[u_t\eta + a_{ij}(x, t, u)u_{x_j}\eta_{x_i} + \left(\frac{\partial a_{ij}}{\partial u}u_{x_i}u_{x_j} + \frac{\partial a_{ij}}{\partial x_i}u_{x_j} + b\right)\eta\Big]\,dx\,dt +$$

$$+\int_{t_0}^{t}\int_{S}\psi(x,\ t,\ u)\,\eta\,dx\,dt=0,\quad 0\leqslant t_0<t\leqslant T. \tag{7.7}$$

For functions $\eta(x,\ t)$ that are equal to zero on $S_T\setminus S'_T$, where $S'_T=S'\times[0,\ T]$, the latter integral can be transformed to the volume integral

$$\int_{t_0}^{t}\int_{S}\psi\eta\,dx\,dt=\int_{t_0}^{t}\int_{\Omega}\frac{d}{dx_n}(\psi\eta)\,dx\,dt$$

$$=\int_{t_0}^{t}\int_{\Omega}\left[\left(\frac{\partial\psi}{\partial u}u_{x_n}+\frac{\partial\psi}{\partial x_n}\right)\eta+\psi\frac{d\eta}{dx_n}\right]dx\,dt,$$

in which case (7.7) takes the form

$$\int_{t_0}^{t}\int_{\Omega}[u_t\eta+a_{ij}u_{x_j}\eta_{x_i}+a\eta+\psi\eta_{x_n}]\,dx\,dt=0, \tag{7.8}$$

where

$$a(x,\ t,\ u,\ u_x)=\frac{\partial a_{ij}(x,\ t,\ u)}{\partial u}u_{x_i}u_{x_j}$$

$$+\frac{\partial a_{ij}}{\partial x_i}u_{x_j}+b(x,\ t,\ u,\ u_x)+\frac{\partial\psi}{\partial u}u_{x_n}+\frac{\partial\psi}{\partial x_n}.$$

This equality is of the same form as (1.6), with the same conditions on the $a_i(x,\ t,\ u,\ p)=a_{ij}(x,\ t,\ u)p_j+\psi\delta_i^n$ and $a(x,\ t,\ u,\ p)$ as in (1.6), only in contrast to (1.6) the function $\eta(x,\ t)$ here must vanish not on the whole surface S_T but only on $S_T\setminus S'_T$.

We put $\eta=u^{(k)}(x,\ t)\,\xi^2(x,\ t)$ in (7.8), where the value of k is arbitrary, while ξ is an arbitrary smooth function with values between 0 and 1, that is equal to zero on the lateral surface of an arbitrarily chosen cylinder $Q(\rho,\ \tau)=K_\rho\times(t_0,\ t_0+\tau)$ with $0\leq t_0<t_0+\tau\leq T$, not intersecting $S_T\setminus S'_T$ even for $x\in K_\rho$. We carry out an estimate in (7.8) with the use of inequality (1.2) of Chapter II and conditions (7.4)–(7.6), as was done in §1 in deriving (1.11). As a result we obtain the inequality

$$\frac{1}{2}\int_{\Omega_\rho}[u^{(k)}(x,\ t)\,\zeta(x,\ t)]^2\,dx\,\Big|_{t_0}^{t}+\nu\int_{t_0}^{t}\int_{\Omega_\rho}(u_x^{(k)})^2\zeta^2\,dx\,dt\leqslant$$

$$\leqslant c(\mu)\int_{t_0}^{t}\int_{A_{k,\rho}(t)}[u_x^2u^{(k)}\zeta^2+\zeta^2+u^{(k)^2}\zeta_x^2+u^{(k)^2}|\zeta\zeta_t|]\,dx\,dt,$$

where $A_{k,\rho}(t)$ is the set of those x from $\Omega_\rho = K_\rho \cap \Omega$ for which $u(x, t) > k$. Hence for k satisfying the condition

$$\max_{Q(\rho,\tau)} u(x,t)-k\leqslant\delta=\frac{\nu}{2c(\mu)}, \tag{7.9}$$

we have the inequality

$$\|u^{(k)}(x,t_0+\tau)\zeta(x,t_0+\tau)\|_{2,\Omega_\rho}^2+\nu\|u_x^{(k)}\zeta\|_{2,Q(\rho,\tau)}^2$$

$$\leqslant\|u^{(k)}(x,t_0)\zeta(x,t_0)\|_{2,\Omega_\rho}^2+\gamma\left[\int_{t_0}^{t_0+\tau}\int_{\Omega_\rho}u^{(k)^2}[\zeta_x^2+\zeta|\zeta_t|]\,dx\,dt\right.$$

$$\left.+\int_{t_0}^{t_0+\tau}\int_{A_{k,\rho}(t)}\zeta\,dx\,dt\right],\quad 0\leqslant t_0\leqslant t_0+\tau\leqslant T,$$

which coincides with inequality (7.5) of Chapter II for $r = 2(n+2)/n$, $\kappa = 2/n$. The same inequality is also satisfied by the function $-u(x, t)$. As was noted in Remark 8.1 of Chapter II, this is sufficient to determine if the function $u(x, t)$ belongs to the class $\hat{\mathfrak{B}}_2(Q_T \cup \Gamma', M, \gamma, (2n+4)/n, \delta, 2/n)$, where Γ' is the union of S'_T and the lower base of the cylinder Q_T.

Hence, on the basis of Theorem 8.2 and Lemma 8.1 of Chapter II, one can draw the following conclusion.

Theorem 7.1. *For any solution $u(x, t)$ from $C^{2,1}(\overline{Q}_T)$ of problem* (7.1)–(7.3) *and an arbitrary cylinder $Q_\rho = K_\rho \times (t_0, t_0+\rho^2)$ not intersecting $S_T \setminus S'_T$ the estimates*

$$\operatorname{osc}\{u, Q_\rho\cap Q_T\}\leqslant c\rho^\alpha, \tag{7.10}$$

$$\int_{Q_\rho\cap Q_T}u_x^2\,dx\,dt\leqslant c_1\rho^{n+2\alpha} \tag{7.11}$$

are valid; here the exponent $\alpha > 0$ is determined only by the quantity $M = \max_{Q_T}|u|$ and the constants ν and μ from inequalities (7.4)–(7.6), *c is determined by M, ν, μ, $|u(x, 0)|_\Omega^{(\beta)}$ and the distance from the center of K_ρ to $S\setminus S'$, while c_1 depends on the same quantities as c and on* $\max_\Omega |u_x(x, 0)|$.

We now need a priori estimates of $\max_{Q'_T}|u_x|$ and $|u|^{(1+\delta)}_{Q_T}$. This will be done in the following way. We first establish an estimate of the integrals $\int_\Omega |u_t|^q\,dx$ with any $q \geq 1$. Then we make use of it for $q = n + 1$. Namely, we consider equation (7.1) for each t from $[0, T]$ as a quasi-linear elliptic equation depending on t as on a parameter:

$$a_{ij}(x, t, u)\,u_{x_ix_j} + \hat{a}(x, t, u, u_x) = 0,$$

for which the functions a_{ij} and $a(x, t, u, p) = -b(x, t, u, p) - u_t(x, t)$ satisfy conditions (7.4), (7.5) and

$$|\hat{a}(x, t, u, p)| \leqslant \mu p^2 + \varphi(x, t); \quad \max_{t\in[0, T]} \|\varphi\|_{n+1,\Omega} \leqslant \mu. \tag{7.12}$$

For solutions of such equations satisfying the boundary condition (7.2) the estimates

$$\max_\Omega |u_x| \leqslant c, \quad \operatorname{osc}\{u_x, \Omega_\rho\} \leqslant c\rho^\alpha, \tag{7.13}$$

are valid if the boundary S of the domain Ω belongs to the class O^2, while the function ψ from (7.2) is subject to condition (7.5). The constants c and $\alpha > 0$ in (7.13) are determined only by the quantities ν^{-1} and μ from (7.4), (7.5) and (7.12), $M = \max_\Omega |u|$ and the boundary S. Estimates (7.13) were derived in Chapter X of the book [65q] under somewhat more rigid restrictions, viz., the differentiability of a with respect to x, u, p and the boundedness of $\phi(x, t)$, but the proof given there can be carried over without essential changes to the case of conditions (7.12).

Thus, having estimates of the integrals $\int_\Omega |u_t|^{n+1}\,dx$, by virtue of (7.13) we will have estimates of $\max_{Q_T}|u_x|$ and the Hölder constants $\langle u_x(x, t)\rangle^{(\alpha)}_\Omega$, $t \in [0, T]$, with respect to the variables x. Using next the estimate for $\int_\Omega |u_t|^q\,dx$ with $q > 2n + 2$, as well as Theorem 2.1 and Lemma 3.1 of Chapter II, we arrive at the estimate

$$|u|^{(1+\delta)}_{Q_T} \leqslant c. \tag{7.14}$$

In deducing the subsequent estimates we require the following additional restrictions on the functions a_{ij}, b and ψ for $(x, t) \in \overline{Q}_T$ and $|u| \leq M$:

$$\left.\begin{aligned} &|\psi_{uu}(x, t, u), \psi_{ux}, \psi_{ut}, a_{ijt}, \psi_t| \leqslant \mu, \\ &|b_p|(1 + |p|) + |b_u| + |b_t| \leqslant \mu(1 + p^2), \end{aligned}\right\} \tag{7.15$_1$}$$

and

$$|a_{iju}, a_{ijut}, a_{ijux}, a_{ijxt}| \leqslant \mu. \tag{7.15$_2$}$$

Before estimating the integrals $\int |u_t|^q\, dx$, we will prove a lemma.

Lemma 7.1. *For any ball K_ρ with center in $\overline{\Omega}$ which, together with the concentric ball $K_{2\rho}$, does not intersect with $S\setminus S'$, the estimates*

$$\int_0^T \int_{\Omega_\rho} (u_t^2 + |u_x|^4 + |u_{xx}|^2)\, dx\, dt \leqslant c, \tag{7.16}$$

$$\max_{t\in[0,T]} \int_{\Omega_\rho} u_x^2(x, t)\, dx \leqslant c\rho^{n-2+2\alpha} \tag{7.17}$$

are valid; here the constants c and $\alpha > 0$ depend only on $M = \max_{Q_T} |u|$, ν and μ from (7.4)–(7.6) *and* (7.15), $\max_\Omega |u_x(x, 0)|$ *and the distance from the center of K_ρ to $S\setminus S'$.*

In (7.8) we take $t_0 = 0$ and $\eta = u_t(x, t)\zeta^2(x, t)$, where $\zeta(x, t)$ is a smooth nonnegative function that is equal to zero on the lateral surface of the cylinder $K_{2\rho} \times (0, T)$ and outside it.

The fact that $u_t\zeta^2 \in V_2^{1,0}(Q_T)$ (more precisely, the existence of the derivatives u_{tx_i} from $L_2(Q_T)$) will be proved below in Lemma 7.2. Noting that

$$a_{ij}u_{x_j}u_{tx_i}\zeta^2 = \frac{1}{2}\frac{d}{dt}(a_{ij}u_{x_i}u_{x_j}\zeta^2) - \frac{1}{2}\frac{d(a_{ij}\zeta^2)}{dt}u_{x_i}u_{x_j}$$

and

$$\psi u_{tx_n}\zeta^2 = \frac{d}{dt}(\psi u_{x_n}\zeta^2) - u_{x_n}\frac{d}{dt}(\psi\zeta^2),$$

we obtain from (7.8)

$$\int_0^t \int_{\Omega_{2\rho}} u_t^2\zeta^2\, dx\, dt + \frac{1}{2}\int_{\Omega_{2\rho}} (a_{ij}u_{x_i}u_{x_j}\zeta^2 + 2\psi u_{x_n}\zeta^2)\, dx \Big|_0^t$$

$$= \int_0^t \int_{\Omega_{2\rho}} \Big[(a_{iju}u_t\zeta^2 + a_{ij}2\zeta\zeta_t + a_{ijt}\zeta^2)\frac{1}{2}u_{x_i}u_{x_j} - a_{ij}u_{x_j}u_t 2\zeta\zeta_{x_i}$$

$$- au_t\zeta^2 + u_{x_n}(\psi_u u_t\zeta^2 + \psi_t\zeta^2 + \psi 2\zeta\zeta_t) - \psi u_t 2\zeta\zeta_{x_n}\Big]\, dx\, dt.$$

From this, by using assumptions (7.4)–(7.6) and (7.15) and inequality (1.2) of Chapter II, one easily deduces the inequality

$$\int_0^t \int_{\Omega_{2\rho}} u_t^2\zeta^2 + \int_{\Omega_{2\rho}} u_x^2(x, t)\zeta^2(x, t)\,dx$$
$$\leqslant c \int_{\Omega_{2\rho}} (u_x^2(x, 0)+1)\zeta^2(x, 0)\,dx + c_1 \int_0^t \int_{\Omega_{2\rho}} [|u_x|^4\zeta^2$$
$$+(1+u_x^2)(|\zeta\zeta_t| + \zeta_x^2 + \zeta^2)]\,dx\,dt. \tag{7.18}$$

Let us show that for the integral $\int_0^t \int_{\Omega_{2\rho}} |u_x|^4 \zeta^2\,dx\,dt$ we have the estimate

$$\int_0^t \int_{\Omega_{2\rho}} |u_x|^4 \zeta^2\,dx\,dt \leqslant c_2\rho^\alpha \int_0^t \int_{\Omega_{2\rho}} (u_{xx}^2\zeta^2 + u_x^2\zeta_x^2)\,dx\,dt. \tag{7.19}$$

Indeed, by an integration by parts we find

$$\int_{\Omega_{2\rho}} |u_x|^4\zeta^2\,dx = -\int_{\Omega_{2\rho}} (u-u_0)(\Delta u u_x^2\zeta^2 + 2u_{x_i}u_{x_k x_i}u_{x_k}\zeta^2$$
$$+ u_{x_i}u_x^2 2\zeta\zeta_{x_i})\,dx + \int_{K_{2\rho}\cap S} u_x^2(u-u_0)u_{x_n}\zeta^2\,ds. \tag{7.20}$$

Here $u_0 = u_0(t)$ is the value of $u(x, t)$ at the center of the ball $K_{2\rho}$. By virtue of estimate (7.10) we have $\max_{\Omega_{2\rho}} |u(x, t) - u_0(t)| \le c\rho^\alpha$, where c depends on the distance from the center of K_ρ to $S\backslash S'$. Therefore the boundary integral in (7.20) does not exceed

$$c_3\rho^\alpha \int_{K_{2\rho}\cap S'} |u_x|^3\zeta^2\,ds = c_3\rho^\alpha \int_{\Omega_{2\rho}} \frac{\partial}{\partial x_n}(|u_x|^3\zeta^2)\,dx$$
$$\leqslant c_4\rho^\alpha \int_{\Omega_{2\rho}} (|u_{xx}|u_x^2\zeta^2 + |u_x|^3\zeta|\zeta_x|)\,dx$$
$$\leqslant c_5\rho^\alpha \int_{\Omega_{2\rho}} \left(\varepsilon u_x^4\zeta^2 + \frac{1}{2\varepsilon}u_{xx}^2\zeta^2 + \frac{1}{2\varepsilon}u_x^2\zeta_x^2\right)dx.$$

The first integral in the right side of (7.20) is estimated analogously. Taking ϵ sufficiently small, we arrive at the estimate (7.19).

Let us now estimate the integral $\int_0^t \int_{\Omega_{2\rho}} u_{xx}^2\zeta^2\,dx\,dt$. To this end we put

$$\eta = \sum_{s=1}^{n-1} (u_{x_s}\zeta^2)_{x_s}$$

in the identity (7.8) and carry out an integration by parts. This gives

$$\sum_{s=1}^{n-1}\int_0^t\int_{\Omega_{2\rho}}\Big[(u_t+a)(u_{x_sx_s}\zeta^2+2u_{x_s}\zeta\zeta_{x_s}) - \Big(a_{ij}u_{x_sx_j}+a_{iju}u_{x_s}u_{x_j}+a_{ijx_s}u_{x_j}+\frac{d\psi}{dx_s}\delta_i^n\Big) \times(u_{x_sx_i}\zeta^2+u_{x_s}2\zeta\zeta_{x_i})\Big]\,dx\,dt=0.$$

From this, by using conditions (7.4)–(7.6) and inequality (1.2) of Chapter II, it is not difficult to deduce the inequality

$$\int_0^t\int_{\Omega_{2\rho}}\sum_{s=1}^{n-1}\sum_{i=1}^{n}u^2_{x_sx_i}\zeta^2\,dx\,dt \leqslant c_6\int_0^t\int_{\Omega_{2\rho}}[u_t^2\zeta^2+|u_x|^4\zeta^2+(u_x^2+1)(\zeta_x^2+\zeta^2)]\,dx\,dt. \tag{7.21}$$

On the other hand, by solving equation (7.1) with respect to the derivative $u_{x_nx_n}$:

$$u_{x_nx_n}=\frac{1}{a_{nn}}\Big(u_t-\sum_{s=1}^{n-1}\sum_{i=1}^{n}a_{si}u_{x_ix_s}+b\Big),$$

and taking into account assumptions (7.4) and (7.6), we see that

$$u^2_{x_nx_n}\leqslant c_7\Big(u_t^2+\sum_{s=1}^{n-1}\sum_{i=1}^{n}u^2_{x_sx_i}+|u_x|^4+1\Big). \tag{7.22}$$

From (7.22) and (7.21) it follows that

$$\int_0^t\int_{\Omega_{2\rho}}u^2_{xx}\zeta^2\,dx\,dt\leqslant c_8\int_0^t\int_{\Omega_{2\rho}}[u_t^2\zeta^2+|u_x|^4\zeta^2 +(u_x^2+1)(\zeta_x^2+\zeta^2)]\,dx\,dt. \tag{7.23}$$

Consider the inequalities (7.18), (7.19) and (7.23). For ρ not exceeding a sufficiently small number ρ_0 (the quantity ρ_0 is determined by the constants c_i and α from (7.18), (7.19) and (7.23) these inequalities imply the estimate

$$\int_{\Omega_{2\rho}} u_x^2(x, t)\zeta^2(x, t)\,dx + \int_0^t \int_{\Omega_{2\rho}} (|u_x|^4 + u_{xx}^2 + u_t^2)\zeta^2\,dx\,dt \leqslant c_9 \int_{\Omega_{2\rho}} [u_x^2(x, 0) + 1] \times \zeta^2(x, 0)\,dx + c_9 \int_0^t \int_{\Omega_{2\rho}} (u_x^2 + 1)(|\zeta_t|\zeta + \zeta_x^2 + \zeta^2)\,dx\,dt.$$

From it and inequalities (7.11) follow, as is easily seen, both of the assertions of Lemma 7.1.

We will prove the following proposition.

Lemma 7.2. *For any $q \geq 1$ it is possible to determine a constant $c(q)$ such that*

$$\max_{t \in [0, T]} \int_{\Omega'} |u_t|^q\,dx \leqslant c(q), \tag{7.24}$$

where Ω' is an arbitrary subdomain of Ω that is separated from $S \setminus S'$ by a positive distance d. The constant $c(q)$ depends only on q, M, the constants ν and μ from conditions (7.4)–(7.6) *and* (7.15), *then* $\max_\Omega |u_x(x, 0)|$, $\int_\Omega |u_t(x, 0)|^q\,dx$ *and* d.

We take the divided difference with respect to t of both sides of equalities (7.1) and (7.2):

$$\left(\frac{\Delta u}{\Delta t}\right)_t - \frac{\Delta (a_{ij} u_{x_i x_j})}{\Delta t} + \frac{\Delta b}{\Delta t} = 0, \tag{7.25}$$

$$\left[\frac{\Delta}{\Delta t}(a_{ij} u_{x_j}) \cos(n, x_i) + \frac{\Delta \psi}{\Delta t}\right]\Big|_{S_T} = 0, \tag{7.26}$$

then we multiply (7.25) by $|\Delta u/\Delta t|^{r-2}(\Delta u/\Delta t)\zeta^2(x)$, $r \geq 2$, integrate over Ω and over the interval $[0, t]$, and then carry out an integration by parts, taking into account condition (7.26). This gives

$$\frac{1}{r}\int_\Omega \left|\frac{\Delta u}{\Delta t}\right|^r \zeta^2\,dx\Big|_0^t + \int_0^t \int_{\Omega} \left[\frac{\Delta (a_{ij} u_{x_j})}{\Delta t} \frac{d}{dx_i}\left(\left|\frac{\Delta u}{\Delta t}\right|^{r-2} \frac{\Delta u}{\Delta t}\zeta^2\right) + \left(\frac{\Delta}{\Delta t}\left(\frac{da_{ij}}{dx_i} u_{x_j}\right) + \frac{\Delta b}{\Delta t}\right)\left|\frac{\Delta u}{\Delta t}\right|^{r-2} \frac{\Delta u}{\Delta t}\zeta^2\right]dx\,dt + \int_0^t \int_S \frac{\Delta \psi}{\Delta t}\left|\frac{\Delta u}{\Delta t}\right|^{r-2} \frac{\Delta u}{\Delta t}\zeta^2\,ds\,dt = 0. \tag{7.27}$$

For $r = 2$ and $\zeta \equiv 1$ it is not difficult to deduce from this equality, taking into account conditions (7.15) and (7.4)–(7.6) and the membership of $u(x, t)$ in

$C^{2,1}(\bar{Q}_T)$, the boundedness of the integrals

$$\int_0^t \int_\Omega \sum_{i=1}^n \left(\frac{\Delta u_{x_i}}{\Delta t}\right)^2 dx\, dt, \quad 0 \leqslant t \leqslant T-h,$$

which is uniform with respect to $|\Delta t| \leq h$, where h is any positive number. This, by virtue of Lemma 4.11 of Chapter II, implies the existence of generalized derivatives of the form $u_{x_i t}$ that are square-summable over the cylinder Q_T.

We will now assume that the function $\zeta(x)$ is a cutting function for some ball K_ρ not intersecting with $S \setminus S'$. In (7.27) we transform the latter integral into a volume integral, as was done in (7.7), and then pass to the limit for $\Delta t \to 0$. As a result we arrive at the equality

$$\frac{1}{r}\int_{\Omega_\rho} |u_t(x,t)|^r \zeta^2(x)\, dx \Big|_0^t + \int_0^t \int_{\Omega_\rho} \Big\{ (a_{ij} u_{x_j})_t \left(|u_t|^{r-2} u_t \zeta^2\right)_{x_i} + \frac{d}{dt}\left(\frac{d a_{ij}}{d x_i} u_{x_i} + b\right) |u_t|^{r-2} u_t \zeta^2 + \frac{d}{dx_n}\left[(\psi_u u_t + \psi_t) |u_t|^{r-2} u_t \zeta^2\right]\Big\}\, dx\, dt = 0.$$

We expand the expressions for the derivatives with respect to t and x_i:

$$\frac{1}{r}\int_{\Omega_\rho} |u_t(x,t)|^r \zeta^2(x)\, dx \Big|_0^t + \int_0^t \int_{\Omega_\rho} \Big[(a_{ij} u_{x_j t} + a_{iju} u_t u_{x_j} + a_{ijt} u_{x_j} + \psi_u \delta_n^i u_t + \psi_t \delta_n^i)\left((r-1)|u_t|^{r-2} u_{tx_i} \zeta^2 + |u_t|^{r-2} u_t 2\zeta \zeta_{x_i}\right) + (b_t + b_u u_t + b_{u_{x_i}} u_{tx_i} + \psi_{uu} u_t u_{x_n} + \psi_{ux_n} u_t + \psi_{ut} u_{x_n} + \psi_{tx_n} + 2 a_{iju} u_{x_i t} u_{x_j} + a_{ijuu} u_{x_i} u_{x_j} u_t + a_{ijut} u_{x_i} u_{x_j} + a_{ijx_i u} u_{x_j} u_t + a_{ijx_i t} u_{x_j}) |u_t|^{r-2} u_t \zeta^2 \Big]\, dx\, dt = 0.$$

From here, using conditions (7.4)–(7.5) and (7.15), it is not difficult to deduce the inequality

$$\int_{\Omega_\rho} |u_t|^r \zeta^2\, dx \Big|_0^t + \int_0^t \int_{\Omega_\rho} |u_t|^{r-2} u_{tx}^2 \zeta^2\, dx\, dt \leqslant c_1(r) \int_0^t \int_{\Omega_\rho} |u_t|^r (\zeta_x^2 + u_x^2 \zeta^2 + \zeta^2)\, dx\, dt, \tag{7.28}$$

in which $c_1(r)$, in general, tends to ∞ for $t \to \infty$. Let us show that the integral $\int_0^t \int_{\Omega_\rho} |u_t|^{r+1} \zeta^2 \,dx\,dt$ is also estimated in terms of the right side of this inequality and $\int_0^t \int_{\Omega_\rho} |u_t|^{r-2} u_{tx}^2 \zeta^2 \,dx\,dt$.

Indeed, putting $\eta = |u_t|^{r-1} u_t \zeta^2$ in (7.8), we obtain

$$\int_0^t \int_{\Omega_\rho} [|u_t|^{r+1}\zeta^2 + a_{ij}u_{x_j} r |u_t|^{r-1} u_{tx_i}\zeta^2 + a_{ij}u_{x_j}|u_t|^{r-1} u_t 2\zeta\zeta_{x_i} + a|u_t|^{r-1}u_t\zeta^2 + \psi r |u_t|^{r-1} u_{tx_n}\zeta^2 + \psi |u_t|^{r-1} u_t 2\zeta\zeta_{x_n}] \,dx\,dt = 0,$$

from which a zero estimate for $\int_0^t \int_{\Omega_\rho} |u_t|^{r+1} \zeta^2 \,dx\,dt$ follows by virtue of (7.4)–(7.6) and (7.28). Thus, in addition to (7.28), we also have the inequality

$$\int_{\Omega_\rho} |u_t|^r \zeta^2 \,dx \Big|_0^t + \int_0^t \int_{\Omega_\rho} (|u_t|^{r-2} u_{tx}^2 + |u_t|^{r+1}) \zeta^2 \,dx\,dt \leqslant c_1(r) \int_0^t \int_{\Omega_\rho} |u_t|^r (\zeta_x^2 + u_x^2\zeta^2 + \zeta^2) \,dx\,dt. \tag{7.29}$$

Consider the integral $\int_0^t \int_{\Omega_\rho} u_x^2 |u_t|^r \zeta^2 \,dx\,dt$ standing in the right side. For an estimate of it we take advantage of inequalities (7.17) (proved in Lemma 7.1) and Lemma 5.3 of Chapter II. According to this lemma

$$\int_{\Omega_\rho} u_x^2 |u_t|^r \zeta^2 \,dx \leqslant c\rho^{2\alpha} \int_{\Omega_\rho} \left(\left(|u_t|^{\frac{r}{2}}\zeta\right)_x\right)^2 dx,$$

and therefore

$$\int_0^t \int_{\Omega_\rho} u_x^2 |u_t|^r \zeta^2 \,dx\,dt \leqslant c_2(r)\rho^{2\alpha} \int_0^t \int_{\Omega_\rho} [|u_t|^{r-2} u_{tx}^2 \zeta^2 + |u_t|^r \zeta_x^2] \,dx\,dt. \tag{7.30}$$

The constant $c_2(r)$ here is determined by the quantities ν and μ from (7.4) and (7.6), M, r, the distance from the center of K_ρ to $S \setminus S'$ and $\max_\Omega |u_x(x, 0)|$.

From (7.30) and (7.29) it follows that for $\rho \le \rho_0(r)$, where $\rho_0(r)$ is chosen from the condition

$$c_2(r)\rho_0^{2\alpha}(r)c_1(r)=\frac{1}{2}, \tag{7.31}$$

the inequality

$$\int_{\Omega_\rho}|u_t|^r\zeta^2\,dx\Big|_0^t+\int_0^t\int_{\Omega_\rho}|u_t|^{r+1}\zeta^2\,dx\,dt \leqslant c_3(r)\int_0^t\int_{\Omega_\rho}|u_t|^r(\zeta^2+\zeta_x^2)\,dx\,dt \tag{7.32}$$

is fulfilled.

Suppose that we wish to estimate $\int|u_t|^q\,dx$ for some $q\ge 2$. We apply inequalities (7.32) to a sequence of contracting concentric balls K_{ρ_r}, $\rho_r=\rho_0(q)(1-r/2q)$, $r=2,\dots,q$, taking as $\zeta(x)$ for $\rho=\rho_r$ a cutting function for the ball $K_{\rho_{r-1}}$. Then, taking into account the estimate of $\int_0^T\int_{\Omega_\rho}u_t^2\,dx\,dt$ obtained in Lemma 7.1, we obtain the desired estimate:

$$\int_{\Omega_{\frac{\rho_0(q)}{2}}}|u_t(x,t)|^q\,dx\leqslant c(q).$$

Here t is an arbitrary value from $[0,T]$, while $\rho_0(q)$ is a sufficiently small but fixed (by condition (7.31)) number. Covering Ω' by a finite number of the balls $K_{\rho_0(q)/2}$, we arrive at estimate (7.24).

Lemma 7.2 is proved.

As noted above, estimates (7.13) and (7.14) follow from estimates (7.24) with $q>2n+2$. The constants c, α and δ in them are determined by $|u_x(x,0)|_\Omega^{(\alpha)}$, and also by the quantities determining $c(q)$ in Lemma 7.2, in particular, the quantity $\|u_t(x,0)\|_{q,\Omega}$. The latter can be estimated, using equation (7.1), in terms of $|u(x,0)|_\Omega^{(2)}$. We formulate this result in the form of a theorem.

Theorem 7.2. *Suppose the functions* $a_{ij}(x,t,u)$, $b(x,t,u,p)$ *and* $\psi(x,t,u)$ *for* $(x,t)\in\overline{Q}_T$, $|u|\le M$ *and arbitrary* p *are subject to conditions* (7.4)–(7.6) *and* (7.15), *and* $S\in O^2$. *Then for any solution* $u(x,t)$ *from* $C^{2,1}(\overline{Q}_T)$ *of equation* (7.1) *satisfying condition* (7.2) *and having* $\max_{Q_T}|u|\le M$ *one has the estimates*

$$\max_{Q_T} |u_x| \leqslant M_1, \quad |u|_{Q_T}^{(1+\delta)} \leqslant c, \tag{7.33}$$

where the constants M_1, c *and* $\delta > 0$ *depend only on* M, ν *and* μ *from* (7.4)–(7.6) *and* (7.15), *the norm* $|u(x, 0)|_\Omega^{(2)}$ *and the boundary* S.

Remark. It can be seen from the conclusion of the theorem that if one assumes in its conditions that not all of the surface S but only a part S' of it belongs to O^2 and that condition (7.2) is fulfilled only on $S'_T = S' \times [0, T]$, then estimates (7.33) will hold for any domain $Q' \subset Q_T$ that is separated from $S_T \setminus S'_T$ by a positive distance d, where the constants M_1 and c will in this case also depend on d.

We have assumed above in the derivation of all of the estimates that an estimate of $\max_{Q_T} |u(x, t)|$ is known. This latter fact is established by the same technique that was used in the linear case in §2 of Chapter I. Namely, the following proposition, generalizing Theorems 2.2 and 2.3 of Chapter I, holds.

Theorem 7.3. *Suppose the functions* $a_{ij}(x, t, u)$, $b(x, t, u, p)$ *and* $\psi(x, t, u)$ *are subject for arbitrary* u *to the conditions*

$$\left.\begin{aligned} &0 \leqslant a_{ij}(x, t, u)\xi_i\xi_j \leqslant \mu_1\xi^2 \quad \text{for} \quad (x, t) \in \bar{\Omega} \times (0, T], \\ &-ub(x, t, u, p) \leqslant c_0 p^2 + c_1 u^2 + c_2 \\ &\qquad\qquad\qquad\qquad\qquad \text{for} \quad (x, t) \in \bar{Q}_T \setminus \Gamma_T, \\ &\nu_1\xi^2 \leqslant a_{ij}(x, t, u)\xi_i\xi_j, \quad -u\psi(x, t, u) \leqslant c_3 u^2 + c_4 \\ &\qquad\qquad\qquad\qquad\qquad \text{for} \quad (x, t) \in S_T, \end{aligned}\right\} \tag{7.34}$$

where μ_1, $\nu_1 = \text{const} > 0$ *and* $c_i = \text{const} \geq 0$, $i = 0, \cdots, 4$. *Then the estimate*

$$\max_{Q_T} |u(x, t)| \leqslant \lambda_1 e^{\lambda T} \max\left\{\sqrt{c_2};\ \sqrt{c_4};\ \max_{\Omega} |u(x, 0)|\right\} \tag{7.35}$$

is valid for any solution $u(x, t)$ *from* $C^{2,1}(\bar{Q}_T \setminus \Gamma_T)$ *that is continuous in* $\bar{Q}_T$ *and has continuous (right up to* S_T*) derivatives* u_x, *where the constants* λ *and* λ_1 *are determined only by the quantities* ν_1, μ, c_0, c_1 *and* c_3 *from* (7.34) *and the boundary* S *of the domain* Ω, *which is assumed to belong to the class* O^2. *If* $S \in O^1$ *and in place of* (7.34) *the conditions*

$$\left.\begin{aligned} &a_{ij}(x, t, u)\xi_i\xi_j \geqslant 0 && \text{for} \quad (x, t) \in \bar{\Omega} \times (0, T], \\ &-ub(x, t, u, 0) \leqslant c_1 u^2 + c_2 && \text{for} \quad (x, t) \in \bar{Q}_T \setminus \Gamma_T, \\ &-u\psi(x, t, u) < 0 && \text{for} \quad |u| > 0, (x, t) \in S_T, \\ &\qquad c_1, c_2 = \text{const} \geqslant 0, \end{aligned}\right\} \tag{7.36}$$

are fulfilled, then

$$\max_{Q_T}|u(x,t)|\leqslant \min_{\lambda>c_1} e^{\lambda T}\max\left\{\sqrt{\frac{c_2}{\lambda-c_1}};\ \max_{\Omega}|u(x,0)|\right\}. \tag{7.37}$$

In order to prove the first part of the theorem we consider the function $w(x,t)=e^{-\lambda t}\phi(x)u(x,t)$, where $\phi(x)$ and λ are chosen from the conditions

$$\left.\begin{aligned}&\min_{\Omega}\varphi(x)\geqslant 1;\ A(x,t)\\&\equiv\min_u\left[-\frac{1}{\varphi(x)}a_{ij}(x,t,u)\cos(\mathbf{n},x_i)\varphi_{x_j}(x)\right]\Big|_S>c_3,\\&B(x,t)\equiv\min_u\left[\lambda+a_{ij}(x,t,u)\frac{\varphi_{x_ix_j}}{\varphi}\right.\\&\qquad\left.-2a_{ij}(x,t,u)\frac{\varphi_{x_i}\varphi_{x_j}}{\varphi^2}-\frac{c_0\varphi_x^2}{\varphi^2}\right]>c_1,\quad x\in\Omega\end{aligned}\right\} \tag{7.38}$$

(the existence of such ϕ and λ under our conditions on S and the a_{ij} is obvious).

Taking into consideration the relations

$$w_t=-\lambda w+e^{-\lambda t}\varphi u_t;\quad w_{x_i}=w\frac{\varphi_{x_i}}{\varphi}+e^{-\lambda t}\varphi u_{x_i};$$

$$w_{x_ix_j}=\frac{\varphi_{x_ix_j}}{\varphi}w+(\varphi_{x_i}u_{x_j}+\varphi_{x_j}u_{x_i})e^{-\lambda t}+e^{-\lambda t}\varphi u_{x_ix_j},$$

we conclude that $w(x,t)$ satisfies the equation

$$\begin{aligned}&w_t+\lambda w-a_{ij}(x,t,u)w_{x_ix_j}+a_{ij}\frac{\varphi_{x_ix_j}}{\varphi}w\\&+2a_{ij}\frac{\varphi_{x_i}}{\varphi}w_{x_j}-2a_{ij}\frac{\varphi_x\varphi_{x_j}}{\varphi^2}w+e^{-\lambda t}\varphi b(x,t,u,u_x)=0\end{aligned}$$

and the conditions

$$\begin{aligned}\Big[a_{ij}(x,t,u)w_{x_j}\cos(\mathbf{n},x_i)&-wa_{ij}\frac{\varphi_{x_j}}{\varphi}\cos(\mathbf{n},x_i)\\&+e^{-\lambda t}\varphi(x)\psi(x,t,u)\Big]\Big|_{S_T}=0,\end{aligned}$$

$$w|_{t=0}=\varphi(x)u_0(x).$$

Since the form $a_{ij}\xi_i\xi_j$ is positive definite on S_T the vector $\mathbf{l}$ with components $a_{ij}\cos(n, x_i)$, $j = 1, \cdots, n$, forms an acute angle with the outward normal n to the surface S. Suppose the maximum value of $w^2(x, t) > 0$ in $\overline{Q}_T$ is achieved at some point $(x_0, t_0) \in \overline{Q}_T$. If x_0 belongs to the surface S and $t_0 > 0$, then at this point

$$\frac{\partial w^2}{\partial l} = 2wa_{ij}w_{x_j}\cos(n, x_i) \geqslant 0,$$

and hence from the boundary condition for w and (7.34) we deduce

$$\left[w^2\left(-a_{ij}\frac{\varphi_{x_j}}{\varphi}\cos(n, x_i) - c_3\right) - c_4 e^{-2\lambda t}\varphi^2\right]\Bigg|_{(x_0, t_0)} \leqslant 0.$$

It follows by the choice of $\phi(x)$ that

$$|w(x_0, t_0)| \leqslant \sqrt{c_4}\,(A - c_3)^{-\frac{1}{2}}\varphi(x_0)e^{-\lambda t_0}.$$

If $(x_0, t_0) \in \Omega \times (0, T]$, then, by taking into account the relations

$$ww_t\big|_{(x_0, t_0)} \geqslant 0, \quad ww_{x_i}\big|_{(x_0, t_0)} = 0,$$

$$wa_{ij}w_{x_ix_j}\big|_{(x_0, t_0)} \leqslant 0, \quad \varphi u_{x_i}e^{-\lambda t}\big|_{(x_0, t_0)} = -w\frac{\varphi_{x_i}}{\varphi}\Big|_{(x_0, t_0)},$$

we obtain from the equation for w and conditions (7.34)

$$\begin{aligned} 0 \geqslant \Bigg[&\left(\lambda + \frac{a_{ij}\varphi_{x_ix_j}}{\varphi} - 2a_{ij}\frac{\varphi_{x_i}\varphi_{x_j}}{\varphi^2}\right)w^2 \\ &- e^{-2\lambda t}\varphi^2\left(c_0u_x^2 + c_1u^2 + c_2\right)\Bigg]\Bigg|_{(x_0, t_0)} \\ = \Bigg[&\left(\lambda + a_{ij}\frac{\varphi_{x_ix_j}}{\varphi} - 2a_{ij}\frac{\varphi_{x_i}\varphi_{x_j}}{\varphi^2} - c_0\frac{\varphi_x^2}{\varphi^2} - c_1\right)w^2 - c_2\varphi^2e^{-2\lambda t}\Bigg]\Bigg|_{(x_0, t_0)}, \end{aligned}$$

and this by virtue of (7.38) implies

$$|w(x_0, t_0)| \leqslant \sqrt{c_2}\,(B - c_1)^{-\frac{1}{2}}\varphi(x_0)e^{-\lambda t_0}.$$

Finally, if $t_0 = 0$, then $|w(x_0, t_0)| \leq \max_\Omega |u(x, 0)\phi(x)|$. From the three cases considered we obtain estimate (7.35) with constant

$$\lambda_1 = \max\left\{1;\ \max_{Q_T}[A(x, t) - c_3]^{-\frac{1}{2}}\varphi(x);\ \max_{Q_T}[B(x, t) - c_1]^{-\frac{1}{2}}\varphi(x)\right\}.$$

In order to prove the second part of the theorem it is sufficient to put $\phi(x)=1$ in the preceding arguments. It is easy to see that in the case of conditions (7.36) the function $w^2(x, t) = e^{-2\lambda t}u^2(x, t)$ cannot achieve a maximum on the surface S_T, while if $\max_{Q_T} w^2$ is achieved at a point $(x_0, t_0) \in \Omega \times (0, T]$, then

$$[(\lambda - c_1)w^2 - c_2 e^{-2\lambda t}]\big|_{(x_0, t_0)} \leqslant 0,$$

so that $|w(x_0, t_0)| \leq \sqrt{c_2/(\lambda - c_1)}$. By considering the last possibility for $\max_{Q_T} w^2 = w^2(x_0, 0)$, we arrive at estimate (7.37).

Let us make use of the a priori estimates established in this section to prove the unique solvability of problem (7.1)–(7.3) in the class $H^{2+\beta, 1+\beta/2}(\overline{Q}_T)$. We will assume that the initial condition (7.3) is homogeneous:

$$u\big|_{t=0} = 0 \tag{7.39}$$

(the general case of condition (7.3) reduces to this upon replacing $u(x, t)$ by $v(x, t) = u(x, t) - \psi(x)$) and $S \in H^{2+\beta}$.

In the same way as in §6, we include the problem (7.1), (7.2), (7.39) under investigation in a family of problems depending on a parameter $\tau \in [0, 1]$, and we will apply the Leray-Schauder Theorem. For simplicity we restrict ourselves to the specific case

$$\left.\begin{aligned} \mathscr{L}_\tau u &\equiv \tau \mathscr{L} u + (1-\tau)\mathscr{L}_0 u, \quad \tau \in [0, 1], \\ \mathscr{L}_\tau^{(S)} u &\equiv \tau \mathscr{L}^{(S)} u + (1-\tau)\mathscr{L}_0^{(S)} u = 0, \\ u\big|_{t=0} &= 0, \end{aligned}\right\} \tag{7.40}$$

where

$$\mathscr{L}_0 u \equiv u_t - \mu\Delta u, \quad \mathscr{L}_0^{(S)} u \equiv \mu\left(\frac{\partial u}{\partial n} + u\right)\Big|_{S_T},$$

with μ being a constant from inequalities (7.4)–(7.5).

Suppose that the compatibility condition for the initial and boundary data, i.e. $\psi(x, 0, 0)\big|_{x\in S} = 0$ is fulfilled for $\tau = 1$. Then, clearly, it will be fulfilled for all $\tau \in [0, 1]$. We assume in addition that the operators $\mathfrak{L}$ and $\mathfrak{L}^{(S)}$ satisfy the restrictions (7.34) or (7.36) of Theorem 7.3. Then by the same token the restrictions with the same constants c_i and with $\nu(\tau) = \min\{\nu_1, \mu\}$, $\mu(\tau) = \max\{\mu_1, \mu\}$ will obviously also be satisfied by the operators $\mathfrak{L}_\tau$ and $\mathfrak{L}_\tau^{(S)}$ for all τ from $[0, 1]$. Therefore it is possible to assert on the basis of Theorem 7.3 that the uniform (with respect to τ from $[0, 1]$) estimate

$$\max_{Q_T} |u(x, t, \tau)| \leq M \tag{7.41}$$

is valid for all possible solutions $u(x, t, \tau)$ from $C^{2,1}(\bar{Q}_T)$ of problems (7.40).

Further, we will assume that inequalities (7.4)–(7.6) and (7.15) are fulfilled for $(x, t) \in \bar{Q}_T$, $|u| \leq M$ and arbitrary p. It is easy to see that the functions $a^\tau_{ij} = \tau a_{ij} + (1-\tau)\delta^j_i\mu$, $b^\tau = \tau b$ and $\psi^\tau = \tau\psi + (1-\tau)\mu u$ satisfy inequalities of the form (7.4)–(7.6) and (7.15) for all $\tau \in [0, 1]$ with the same constants ν and μ as for $\tau = 1$. In view of this, Theorem 7.2 guarantees the estimates

$$\max_{Q_T} |u_x(x, t, \tau)| \leqslant M_1, \quad |u|^{(1+\delta)}_{Q_T} \leqslant c \tag{7.42}$$

for all solutions $u(x, t, \tau)$ of problem (7.40), where the constants M_1 and c are the same for all τ from $[0, 1]$. We now formulate a theorem.

Theorem 7.4. *Suppose the following conditions are fulfilled.*

a) *The functions $a_{ij}(x, t, u)$, $b(x, t, u, p)$ and $\psi(x, t, u)$ satisfy inequalities* (7.34) *or* (7.36).

b) *For $(x, t) \in \bar{Q}_T$, $|u| \leq M$, where M is a constant from estimate* (7.41), *and for arbitrary* p *the functions $a_{ij}(x, t, u)$, $b(x, t, u, p)$ and $\psi(x, t, u)$ are continuous in their arguments, possess the derivatives entering into conditions* (7.4)–(7.6) *and* (7.15), *and satisfy these conditions.*

c) *For $(x, t) \in \bar{Q}_T$, $|u| \leq M$ and $|p| \leq M_1$, where M_1 is a constant from inequality* (7.42), *the functions $a_{ijx}(x, t, u)$ are Hölder continuous in the variables x with exponent β, $\psi_x(x, t, u)$ is Hölder continuous in x and t with exponents β and $\beta/2$ respectively, and $b(x, t, u, p)$ is Hölder continuous in x with exponent β.*

d) $S \in H^{2+\beta}$; $\psi(x, 0, 0)|_{x\in S} = 0$.

Then for each τ from [0, 1] *problem* (7.40) *has a unique solution $u(x, t, \tau)$ in the class $H^{2+\beta, 1+\beta/2}(\bar{Q}_T)$.*

Proof. We consider the problems

$$\left.\begin{aligned} &v_t - [\tau a_{ij}(x, t, w) + (1-\tau)\delta^j_i\mu] v_{x_ix_j} \\ &\qquad\qquad + \tau b(x, t, w, w_x) = 0, \\ &\{[\tau a_{ij}(x, t, w) + (1-\tau)\delta^j_i\mu] v_{x_j} \cos(\mathrm{n}, x_i) \\ &\qquad + (1-\tau)\mu v + \tau\psi(x, t, w)\}|_{S_T} = 0, \\ &v|_{t=0} = 0. \end{aligned}\right\} \tag{7.43}$$

As the principal space B_α we take the subspace of $H^{1+\alpha,(1+\alpha)/2}(\overline{Q}_T)$ consisting of those functions that are equal to zero for $t = 0$. We put $\alpha = \min\{\delta; \beta\}$, where the index δ is taken from (7.42).

For any function $w(x, t)$ from B_α satisfying the conditions $\max_{Q_T}|w| \leq M$ and $\max_{Q_T}|w_x| \leq M_1$, problem (7.43) is a linear problem of the form (5.5) of Chapter IV, for which all of the conditions of Theorem 5.3 of Chapter IV are fulfilled. Namely, by virtue of the yet to be proved Theorem 7.4 the functions $a^\tau_{ij}(x, t) = \tau a_{ij}(x, t, w(x, t)) + (1-\tau)\delta^j_i\mu$ and $\psi^\tau(x, t) = \tau\psi(x, t, w(x, t))$ belong to $H^{1+\alpha,(1+\alpha)/2}(\overline{Q}_T)$, $a^\tau(x, t) = \tau b(x, t, w(x, t), w_x) \in H^{\alpha,\alpha/2}(\overline{Q}_T)$ and the compatibility condition for the initial and boundary data in problem (7.43) is fulfilled. It therefore follows from Theorem 5.3 of Chapter IV that to each function $w(x, t)$ possessing the properties just described there corresponds a unique solution $v(x, t, \tau)$ from $H^{2+\alpha,1+\alpha/2}(\overline{Q}_T)$ of problem (7.43), and this determines a completely continuous transformation $v = \Phi(w, \tau)$ in the space B_α, the fixed points of which are the solutions of problem (7.40). With the use of the estimates (7.33), (7.35), (7.37) established above it is possible to prove the existence of fixed points of this transformation and the membership of them in the space $H^{2+\beta,1+\beta/2}(\overline{Q}_T)$ in exactly the same way as in the case of the first boundary value problem in §6.

The uniqueness of the solution of problem (7.40) for each τ from $[0, 1]$ is easily established with the use of Theorems 2.2 and 2.3 of Chapter I.

The propositions proved in the present section admit certain generalizations and sharpenings within the limits of the methods set forth here. For example, one can assume the existence of singularities of the functions $a_{ij}(x, t, u)$ and $b(x, t, u, p)$ with respect to the variables x and t, as was done in §§1–4. Furthermore, all of the proofs given above survive almost without change for equations (0.1) of more general form with $a_i(x, t, u, p) = F_{p_i}(x, t, u, p)$ and the boundary conditions

$$a_i(x, t, u, u_x)\cos(\mathbf{n}, x_i) + \varphi(x, t, u)|_{S_T} = 0.$$

§8. THE CAUCHY PROBLEM

Let us consider the Cauchy problem for equation (0.1) in the layer $R_T = \{x \in E_n,\ t \in [0, T]\}$ with the initial condition

$$u|_{t=0} = \psi_0(x). \tag{8.1}$$

We will seek a solution in the class of bounded functions belonging to $H^{2+\beta,1+\beta/2}(\overline{Q}_T)$ for any finite subdomain Q_T of the layer R_T.

With respect to the functions $a_i(x, t, u, p)$ and $a(x, t, u, p)$, we assume that they satisfy in any finite domain $\overline{Q}_T \subset R_T$ conditions a)–c) of Theorem 6.1, where the constants in condition a) do not depend on the dimensions of Q_T. Furthermore, we suppose that $\psi_0(x) \in H^{2+\beta}(\Omega)$ in any bounded domain $\Omega \subset E_n$ and that $\max_{E_n} |\psi_0(x)| < \infty$.

The solution of the Cauchy problem can be obtained as the limit of a sequence of solutions of the first boundary value problem for equation (0.1) in cylinders $Q_T = \Omega \times (0, T)$ under an unlimited dilation of the domain Ω. Namely, we suppose that $\{\Omega^N\}$ is a sequence of expanding domains with smooth boundaries S^N that tend in the limit to all of E_n. In each of the cylinders $Q_T^N = \Omega^N \times (0, T)$ there exists by virtue of the theorems of §6 a unique solution $u^N(x, t)$ from

$$H^{2+\beta,\, 1+\frac{\beta}{2}}(\overline{Q}_T^N \setminus S_T^N) \cap H^{\alpha,\, \frac{\alpha}{2}}(\overline{Q}_T^N)$$

of equation (0.1) that satisfies the conditions

$$\left.\begin{aligned} u^N|_{t=0} &= \psi_0(x), \\ u^N|_{S_T^N} &= \psi_0(x)|_{S_T^N}. \end{aligned}\right\} \tag{8.2}$$

From the results of the preceding sections it follows that the solutions are uniformly bounded:

$$\max_{Q_T^N} |u^N(x, t)| \leqslant M, \tag{8.3}$$

and, furthermore, for any Q_T^m and u^N with $N > m$ we have the estimates

$$|u^N|_{Q_T^m}^{(2+\beta)} \leqslant c(m), \tag{8.4}$$

in which the constants $c(m)$ depend on m but not on N. By employing the usual diagonal process we can extract from $\{u^N\}$ a subsequence $\{u^{N_k}\}$ that converges together with the derivatives $u_x^{N_k}$, $u_{xx}^{N_k}$, $u_t^{N_k}$ at each point of R_T to some function u and its corresponding derivatives. It is clear that $u(x, t)$ does not exceed M, belongs to $H^{2+\beta,\, 1+\beta/2}(\overline{Q}_T)$ in each finite cylinder $\overline{Q}_T \subset R_T$ and satisfies equation (0.1) and the initial condition (8.1); in other words, it is a solution of the Cauchy problem for (0.1) in R_T.

If one assumes that the constants in conditions b)–c) of Theorem 6.1 do not depend on the dimensions of Q_T (i.e. conditions b) and c) are fulfilled in the whole strip R_T) and $\psi_0(x) \in H^{2+\beta}(E_n)$, then the constant $c(m)$ in (8.3) will not depend on m, and therefore the limit functions $u(x, t)$ for $\{u^N\}$ will belong to $H^{2+\beta, 1+\beta/2}(R_T)$.

In order to prove the uniqueness of such solutions we assume in addition that the functions

$$a_{ij}(x, t, u, p) \equiv \frac{\partial a_i(x, t, u, p)}{\partial p_j},$$
$$A(x, t, u, p) \equiv a(x, t, u, p) - \frac{\partial a_i}{\partial u} p_i - \frac{\partial a_i}{\partial x_i} \tag{8.5}$$

are differentiable with respect to u and p, where the derivatives $\partial a_{ij}/\partial p$, $\partial a_{ij}/\partial u$, $\partial A/\partial p$ are bounded in modulus, while $\partial A/\partial u$ is bounded from below in any finite domain of variation of (u, p) and $(x, t) \in R_T$.

Suppose, moreover, that

$$a_{ij}(x, t, u, p)\xi_i\xi_j \geq 0 \tag{8.6}$$

for $t \in (0, T]$ and arbitrary x, u, p. Under these conditions it is possible to apply estimate (2.23) in Theorem 2.5 of Chapter I to the difference $u(x, t) = u'(x, t) - u''(x, t)$ of two possible solutions u' and u'' of problem (0.1), (8.1) that are bounded, together with their derivatives of first and second orders.

Indeed, $u(x, t)$ is a bounded solution of the linear Cauchy problem

$$u_t - \hat{a}_{ij}(x, t) u_{x_i x_j} + \hat{a}_i(x, t) u_{x_i} + \hat{a}(x, t) u = 0,$$
$$u|_{t=0} = 0,$$

where

$$\hat{a}_{ij}(x, t) = a_{ij}(x, t, u', u'_x);$$

$$\hat{a}_i(x, t) = -u''_{x_i x_j} \int_0^1 \frac{\partial a_{ij}[x, t, u^\tau(x, t), u^\tau_x(x, t)]}{\partial u^\tau_{x_j}} d\tau + \int_0^1 \frac{\partial A[\dots]}{\partial u^\tau_{x_i}} d\tau;$$

$$\hat{a}(x, t) = -u''_{x_i x_j} \int_0^1 \frac{\partial a_{ij}[\dots]}{\partial u^\tau} d\tau + \int_0^1 \frac{\partial A[\dots]}{\partial u^{(\tau)}} d\tau;$$

$$u^\tau(x, t) = \tau u'(x, t) + (1 - \tau) u''(x, t),$$

with the functions $\hat{a}_{ij}$, $\hat{a}_i$, $\hat{a}$ being subject to the conditions of Theorem 2.5 of Chapter I, and therefore $u(x, t) \equiv 0$ in R_T by virtue of estimate (2.23) of Chapter I. We formulate the assertion just proved in the form of a theorem.

Theorem 8.1. *Suppose that the following conditions hold.*

a) $\psi_0(x) \in H^{2+\beta}(\Omega)$ *in any* $\Omega \subset E_n$ *and* $\max_{E_n} |\psi_0(x)| < \infty$.

b) *For* $t \in (0, T]$ *and arbitrary* x, u, p *inequality* (8.6) *and the inequality*

$$A(x, t, u, 0)\,u \geqslant -b_1 u^2 - b_2, \quad b_1, b_2 = \text{const} \geqslant 0,$$

are fulfilled for $A(x, t, u, p)$ *defined in* (8.5). *The latter inequality can be replaced by the more general inequality*

$$A(x, t, u, 0)\,u \geqslant -\Phi(|u|)\,|u| - b_2, \text{ where } \int_0^\infty \frac{d\tau}{\Phi(\tau)} = \infty;$$

c) *For any bounded cylinder* Q_T *of the layer* R_T *and* $|u| \le M$, *where* M *is a constant from* (8.3), *the functions* $a_i(x, t, u, p)$ *and* $a(x, t, u, p)$ *satisfy conditions* b) *and* c) *of Theorem* 6.1 *with the constants depending in general on* Q_T.

Then there exists at least one solution $u(x, t)$ *of the Cauchy problem* (0.1), (8.1) *in the strip* R_T *that does not exceed* M *in modulus and belongs to* $H^{2+\beta, 1+\beta/2}(\overline{Q}_T)$ *for any bounded cylinder* $Q_T \subset R_T$. *It will be an element of* $H^{2+\beta, 1+\beta/2}(R_T)$ *if in addition it is assumed that the constants in conditions* b)–c) *of Theorem* 6.1 *do not depend on* Q_T.

If the functions $a_{ij}(x, t, u, p)$ *and* $A(x, t, u, p)$ *defined by equalities* (8.5) *are differentiable with respect to* u *and* p *and obey conditions* (8.6) *and*

$$\max_{\substack{(x, t) \in R_T, \\ |u, p| \leqslant N}} \left| \frac{\partial a_{ij}(x, t, u, p)}{\partial u}, \ \frac{\partial a_{ij}(x, t, u, p)}{\partial p}, \ \frac{\partial A(x, t, u, p)}{\partial p} \right| \leqslant \mu_1(N),$$

$$\min_{\substack{(x, t) \in R_T, \\ |u, p| \leqslant N}} \frac{\partial A(x, t, u, p)}{\partial u} \geqslant -\mu_2(N),$$

for an arbitrary N *and some constants* μ_1, μ_2 *depending possibly on* N, *then problem* (0.1), (8.1) *in* R_T *has no more than one classical solution* $u(x, t)$ *that is bounded in* R_T *together with its derivatives of first and second orders.*

Remark 8.1. If condition a) is replaced by the weaker requirement of continuity and boundedness of $\psi_0(x)$ in E_n, then under the same assumptions b)–c) concerning the a_i and a there will exist a bounded solution $u(x, t)$ of problem (0.1), (8.1) that belongs to $H^{2+\beta, 1+\beta/2}(Q_T) \cap C(R_T)$ for any finite cylinder $Q_T \subset R_T$. In order to prove this assertion it is necessary to approximate $\psi_0(x)$ by smooth bounded functions $\psi_0^{(m)}(x)$ converging to $\psi_0(x)$ uniformly in E_n, and to take advantage of Theorem 8.1. The fact that the sequence of solutions $u^{(m)}(x, t)$ of the auxiliary problems will converge to a continuous solution in R_T of problem (0.1), (8.1) can be proved by means of the method of barriers.

Remark 8.2. It is not hard to see that the existence and uniqueness theorems for the Cauchy problem (0.1), (8.1) are preserved under the assumption that the known and unknown functions in problem (0.1), (8.1) have a certain growth when $|x| \longrightarrow \infty$ (see in connection with this the papers [123c, 63, 62d] etc.).

§9. ON THE STEFAN PROBLEM

Let us dwell a little on the Stefan problem (more precisely, on one of the Stefan problems). We will show that it can be reduced to the determination of generalized solutions of boundary value problems for certain quasi-linear equations of simplest form but with functions that are discontinuous in u. This is done in primarily the same way as in §13 of Chapter III for a diffraction problem. In either case we "hide" the first of the conditions on discontinuities in a requirement of continuity (in one sense or another) of the desired solution, and the second, in an integral identity in such a way that the surfaces of discontinuity do not appear in the identity. This circumstance in Stefan problems has significance since the surfaces of discontinuity in them are unknown.

Let us consider, for example, the problem of determining the temperature $u(x, t)$ in a domain Ω of variation of $x = (x_1, \cdots, x_n)$ for $t \in (0, T]$ when there are changes of phase at the temperatures $u_1, \cdots, u_m$. Let $u_1 < u_2 < \cdots < u_m$. In those parts of $Q_T = \Omega \times (0, T)$ where $u(x, t)$ does not have the values $u_1, \cdots, u_m$ it must satisfy the equation

$$\alpha(u)\, u_t - \left(k(u)\, u_{x_i}\right)_{x_i} = 0, \tag{9.1}$$

in which $\alpha(u)$ and $k(u)$ are known positive functions that are smooth on each of the intervals $[u_k, u_{k+1}]$ and have discontinuities of the first kind at the points $u = u_k$, $k = 1, \cdots, m$. On an interface $S^{(k)}$ of two phases we must satisfy the two conditions

$$[u]|_{S(k)} = 0 \tag{9.2}$$

and

$$b_k \cos(\mathrm{n},\ t) + [k(u)\, u_{x_i}] \cos(\mathrm{n},\ x_i)|_{S^{(k)}} = 0, \tag{9.3}$$

where the b_k are given positive numbers, n is the normal to the interface $S^{(k)} = \{(x,\ t) : u(x,\ t) = u_k\}$ in the direction of increasing u (along the gradient of u), and the saltus $[v]|_{S(k)}$ is the difference between the limiting value of v on $S^{(k)}$ when approached from the domain $\{(x,\ t) : u < u^k\}$ and the limiting value of v on $S^{(k)}$ when approached from the domain $\{(x,\ t) : u > u^k\}$.

Finally, $u(x,\ t)$ must satisfy the initial condition

$$u|_{t=0} = \psi_0(x) \tag{9.4}$$

and some boundary condition, for example,

$$u|_{S_T} = 0. \tag{9.5}$$

By the same (classical) statement of the problem it is required that the sets $S^{(k)}$ be piecewise smooth surfaces in the $(x,\ t)$-space. As a preliminary we simplify the problem by introducing in place of $u(x,\ t)$ a new unknown function

$$v(x,\ t) = \int_0^{u(x,\ t)} k(\xi)\, d\xi.$$

For this function we have, from (9.1),

$$\beta(v) v_t - \Delta v = 0 \tag{9.6}$$

with $\beta(v)$ possessing the same properties as $a(u)$ (i.e. $\beta(v)$ is a positive piecewise smooth function having discontinuities of the first kind at the points $v_1, \cdots, v_m$); the conditions on the surfaces $\{(x,\ t) : v(x,\ t) = v_k\}$ (these are the same surfaces $S^{(k)}$) take the form

$$[v]|_{S^{(k)}} = 0 \tag{9.7}$$

and

$$b_k \cos(\mathrm{n},\ t) + [v_{x_i}] \cos(\mathrm{n},\ x_i)|_{S^{(k)}} = 0, \tag{9.8}$$

while the initial and boundary conditions retain their original form

$$v|_{t=0} = \psi_1(x) \text{ and } v|_{S_T} = 0. \tag{9.9}$$

In the same way as in the problems of §13 of Chapter III we impart to equation

(9.6) a divergence form

$$\frac{\partial b(v)}{\partial t} - \Delta v = 0, \tag{9.10}$$

where $b(v)$ is a monotonically increasing piecewise smooth function, the derivative $b'(v)$ of which is equal to $\beta(v)$ on each of the intervals (v_k, v_{k+1}). In order to determine the behavior of $b(v)$ at a point of discontinuity $v = v_k$ we pass from equation (9.10) to an integral identity. To this end we multiply equation (9.10) by an $\eta(x, t)$ from $W_2^{1,1}(Q_T)$ that is equal to zero on the boundary of Q_T, and in each term we carry out a single integration by parts, transferring the derivatives $\partial/\partial t$ and $\partial/\partial x_i$ to η and bearing in mind that the function $b(v)$ can have discontinuities on the surfaces $S^{(k)} = \{(x, t): v(x, t) = v_k\}$, $k = 1, \cdots, m$ (which we assume to be piecewise smooth). This gives us

$$\int_{Q_T} \{-b(v)\eta_t + v_{x_i}\eta_{x_i}\}\,dx\,dt + \sum_{k=1}^{m} \int_{S^{(k)}} \{[b(v)]\cos(\mathbf{n}, t) - [v_{x_i}]\cos(\mathbf{n}, x_i)\}\,\eta\,ds = 0. \tag{9.11}$$

In this connection we have taken into account the choice of the direction of $\mathbf{n}$ and the rule of composition for the saltus [ ]. If the function $b(v)$ is defined so that $[b(v)]|_{S^{(k)}} = -b_k$, then all of the integrals with respect to the surfaces $S^{(k)}$ vanish by virtue of conditions (9.8), and the identity (9.11) takes the form

$$\int_{Q_T} \{-b(v)\eta_t + v_{x_i}\eta_{x_i}\}\,dx\,dt = 0. \tag{9.12}$$

Thus we have shown that if $v(x, t)$ is a classical solution of problem (9.6)–(9.9), and consequently has critical values only on the piecewise smooth surfaces $S^{(k)}$, then it satisfies the identity (9.12) with any $\eta(x, t)$ from $W_2^{1,1}(Q_T)$ that is equal to zero on the boundary of Q_T, and with any monotonically increasing function $b(v)$ having at the points v_k, $k = 1, \cdots, m$, the saltuses b_k, and on the intervals (v_k, v_{k+1}) a derivative $b'(v)$ equal to $\beta(v)$. The function $b(v)$ is determined by these requirements to within its values at the points v_k, $k = 1, \cdots, m$, and to within an arbitrary summand, which we fix by assigning a value to $b(v)$ at some point v_0 that is different from $v_1, \cdots, v_m$. We denote by $B(x, t, v)$ an arbitrary measurable function that is equal to $b(v)$ for $v \neq v_k$, $k = 1, \cdots, m$, and $(x, t) \in \overline{Q}_T$, and that for $v = v_k$ and $(x, t) \in \overline{Q}_T$ has values from the interval

$$\left[\lim_{v \to v_k - 0} b(v),\ \lim_{v \to v_k + 0} b(v)\right]$$

(they can be different at different points (x, t)). The classical solution $v(x, t)$ of problem (9.6)–(9.9) satisfies the identity

$$\int_{Q_T} \{-B(x, t, v)\eta_t + v_{x_i}\eta_{x_i}\}\, dx\, dt = 0 \tag{9.13}$$

with η from $W_2^{1,1}(Q_T)$ and equal to zero on the boundary of Q_T, and with any such function $B(x, t, v)$. Since the measure of the sets $S^{(k)}$ where $v(x, t)$ has the critical values $v_1, \cdots, v_m$ is equal to zero, the information which we obtain from (9.13) is the same for any of the functions $B(x, t, v)$.

By a *generalized solution of problem* (9.6)–(9.9) we will mean a bounded function v from $\overset{\circ}{W}{}_2^{1,1}(Q_T)$ that is equal to $\psi_1(x)$ for $t = 0$ and satisfies the identity (9.13) for any one of the functions of type $B(x, t, v)$ and the above-indicated η. We will show that problem (9.6)–(9.9) can have no more than one generalized solution. Moreover, we will do this for a wider class of generalized solutions, concerning which it is known a priori only that they are bounded, and not that they are elements of $\overset{\circ}{W}{}_2^{1,1}(Q_T)$. These solutions, in accordance with our conception of the definition of generalized solutions, are bounded functions satisfying the identity

$$\int_{Q_T} \{B(x, t, v)\eta_t + v\Delta\eta\}\, dx\, dt + \int_\Omega B(x, t, \psi_1(x))\eta(x, 0)\, dx = 0 \tag{9.14}$$

for any function η from $\overset{\circ}{W}{}_2^{2,1}(Q_T)$ that is equal to zero for $t = T$. Suppose there exist two such solutions v' and v'', and suppose the initial state $\psi_1(x)$ is such that the critical values v_k, $k = 1, \cdots, m$, are taken by $\psi_1(x)$ on a set of n-dimensional measure zero in the x-space. We subtract the identity (9.14) for v'' from the identity (9.14) for v', denoting the corresponding functions $B(x, t, v)$ by B'' and B', and we write the result in the form

$$\int_{Q_T} \{B'(x, t, v') - B''(x, t, v'')\} \times \left\{\eta_t + \frac{v' - v''}{B'(x, t, v') - B''(x, t, v'')}\Delta\eta\right\} dx\, dt = 0$$

or, more concisely,

$$\int_{Q_T} \tilde{B}(x, t)\{\eta_t + a(x, t)\Delta\eta\}\, dx\, dt = 0. \tag{9.15}$$

By virtue of the properties of the functions B' and B'' the function $a(x, t)$ is nonnegative and $\operatorname{vrai\,max}_{Q_T} a(x, t) \equiv a_0 < \infty$. We take as $\eta(x, t)$ the solution of the problem

$$\eta_t + (a(x, t) + \varepsilon)\Delta\eta = F(x, t), \quad \eta|_{t=T} = 0, \quad \eta|_{S_T} = 0, \tag{9.16}$$

in which ϵ is a small positive number and $F(x, t)$ is an arbitrary smooth function that is finite in Q_T. By virtue of what was said on pages 179 and 180 such a solution $\eta^\epsilon(x, t)$ exists and estimate (6.25) of Chapter III is valid for it, i.e.

$$\int_{Q_T} [(\eta_t^\varepsilon)^2 + (a + \varepsilon)(\Delta\eta^\varepsilon)^2]\, dx\, dt \leqslant c(T) \int_{Q_T} F_x^2\, dx\, dt \tag{9.17}$$

with $c(T)$ depending only on T and a_0. [1)]

Substituting η^ϵ in place of η in (9.15), we obtain

$$\int_{Q_T} \hat{B}(F - \varepsilon\Delta\eta^\varepsilon)\, dx\, dt = 0. \tag{9.18}$$

If we let ϵ tend to zero in this equality, we arrive at the equality

$$\int_{Q_T} \hat{B}F\, dx\, dt = 0, \tag{9.19}$$

since

$$\left| \int_{Q_T} \hat{B}\varepsilon\Delta\eta^\varepsilon\, dx\, dt \right| \leqslant \operatorname{vrai\,max}_{Q_T} |\hat{B}| \times \left[\int_{Q_T} \frac{\varepsilon^2}{a + \varepsilon}\, dx\, dt \int_{Q_T} (a + \varepsilon)(\Delta\eta^\varepsilon)^2\, dx\, dt \right]^{1/2} \to 0$$

for $\epsilon \to 0$. But from (9.19), since F is sufficiently arbitrary, it follows that $\hat{B}(x, t) \equiv 0$, and hence that v' and v'' coincide for almost all (x, t) from Q_T. The uniqueness theorem is proved.

The existence of a generalized solution for any bounded function ψ_1 from $\overset{\circ}{W}{}_2^1(\Omega)$ is obtained on the basis of some of the theorems in §6. To this end we approximate the function $b(v)$ by averagings $b_\rho(v)$ of it, obtained with the use of an infinitely differentiable nonnegative kernel $\omega_\rho(|v|)$ of radius ρ, and the

---

1) We note that if $S \in O^2$ then η^ϵ will be an element of $W_2^{2,1}(Q_T)$; otherwise it belongs to $W_2^{1,1}(Q_T)$ and the integral $\int_{Q_T} (\Delta\eta^\epsilon)^2 dx\, dt$ is finite for it. Such an $\eta^\epsilon(x, t)$, as is easily seen, can be taken as η in (9.15).

function $\psi_1(x)$ by some uniformly bounded and infinitely differentiable finite (in Ω) functions $\psi_1^\rho(x)$ that converge to it in the norm of $W_2^1(\Omega)$. If $S \in H^{2+\alpha}$ then we also approximate S by surfaces S^ρ of the class $H^{2+\alpha}$. The auxiliary problems

$$\frac{\partial b_\rho(v)}{\partial t} - \Delta v = 0, \quad v\big|_{S_T^\rho} = 0, \quad v\big|_{t=0} = \psi_1^\rho \tag{9.20}$$

have solutions v^ρ from $H^{2+\alpha,1+\alpha/2}(\overline{Q}_T^\rho)$ by virtue of Theorem 6.1. The uniform estimates

$$\left.\begin{aligned} &\max_{Q_T^\rho} |v^\rho| \leqslant \max_{x\in\Omega^\rho} |\psi_1^\rho| \leqslant c_1 \\ \text{and}\qquad & \\ &\|v_x^\rho\|_{2,Q_T^\rho} + \nu_1 \|v_t^\rho\|_{2,Q_T^\rho} \leqslant c \end{aligned}\right\} \tag{9.21}$$

are valid for them, where ν_1 is a lower bound of the functions $b'_\rho(v)$ on the interval $[-c, c]$. The first of them is a special case of inequality (2.31) of Chapter I, while the second is a special case of inequality (6.6) of Chapter III. The number ν_1 is positive by virtue of the strictly monotonic increase of the function $b(v)$. The inequalities (9.21) permit us to choose a subsequence $\{v^{\rho_k}\}$, $k = 1, 2, \cdots$, which converges almost everywhere in Q_T and weakly in $W_2^{1,1}(Q_T)$ to some bounded function v from $\overset{\circ}{W}{}_2^{1,1}(Q_T)$. Furthermore, the ρ_k are chosen so that the sequence of functions $b_{\rho_k}(v^{\rho_k}(x, t))$ converges weakly in $L_2(Q_T)$ to some function $\tilde{b}(x, t)$. This is possible since the functions $b_\rho(v^\rho(x, t))$ are uniformly bounded. The identity (9.12) is valid for each of the v^{ρ_k}. By passing to the limit in it as $k \longrightarrow \infty$, we arrive at the identity

$$\int_{Q_T} \{-\tilde{b}(x, t)\eta_t + v_{x_l}\eta_{x_l}\}\, dx\, dt = 0. \tag{9.22}$$

It is easily verified that $\tilde{b}(x, t)$ is a function of type $B(x, t, v(x, t))$ and consequently $v(x, t)$ is the desired generalized solution of problem (9.6)–(9.9).

We will show that outside of the sets $S^{(k)}$, where $v(x, t) = v_k$, the smoothness of the function $v(x, t)$ (and hence also of $u(x, t)$) is determined only by the smoothness of the function $b(v)$. Suppose the values of the function $v(x, t)$ on a set $\overline{Q}' = \overline{\Omega}' \times [t_1, t_2]$ belong to the interval $[v_k + \epsilon, v_{k+1} - \epsilon]$, $\epsilon > 0$, and $b(v)$ is a smooth function on it. On Q' the function $B(x, t, v(x, t))$ is equal to $b(v(x, t))$ and possesses a derivative with respect to t from $L_2(Q')$. Consequently, the identity (9.13) for $\eta(x, t)$ equal to zero outside of Q' can be transformed to the form

$$\int_{Q'} [b'(v) v_t \eta + v_{x_i} \eta_{x_i}]\, dx\, dt = 0.$$

This identity indicates that, for almost all t from $[t_1, t_2]$, the function v is a generalized solution from $W_2^1(\Omega')$ of the equation $\Delta v = b'(v) v_t$ with free term from $L_2(\Omega')$. By virtue of Theorem 10.1 of Chapter III in [65q], for such t the solution v has derivatives with respect to x of second order, and, for any interior subdomain Ω'' of the domain Ω',

$$\| v(x, t) \|_{2, \Omega''}^{(2)} \leqslant c(d'') \| b'(v) v_t \|_{2, \Omega'}, \tag{9.23}$$

where $c(d'')$ depends only on the distance from Ω'' to the boundary of Ω'. It follows that

$$\| v \|_{2, Q''}^{(2)} \leqslant c(d'') \| b'(v) v_t \|_{2, Q'} < \infty, \tag{9.24}$$

where $Q'' = \Omega'' \times [t_1, t_2]$. Thus v is an element of $W_2^{2,1}(Q'')$. For almost all points (x, t) from Q'' it satisfies the equation

$$b'(v) v_t - \Delta v = 0. \tag{9.25}$$

The function $w(x, t) = b(v(x, t))$, as is easily seen, satisfies by virtue of (9.25) the equation

$$w_t - \left(\frac{1}{b'(v)} w_{x_i} \right)_{x_i} = 0. \tag{9.26}$$

Since the function $b'(v(x, t))$ satisfies the inequalities $0 < \nu_1 \leq b'(v(x, t)) \leq \mu_1$ on Q'' and $w \in V_2^{1,0}(Q'')$, by virtue of Theorem 10.1 of Chapter III the function $w(x, t)$ is an element of $H^{\alpha, \alpha/2}(Q'')$ with some $\alpha > 0$. But then $v(x, t)$ also belongs to $H^{\alpha, \alpha/2}(Q'')$. If we assume that the function $b'(v)$ is Hölder continuous on the interval $[v_k + \epsilon, v_{k+1} - \epsilon]$, $\epsilon > 0$, then $b'(v(x, t))$ will be an element of $H^{\tilde{\alpha}, \tilde{\alpha}/2}(Q'')$ with some $\tilde{\alpha} > 0$, and hence $v(x, t)$, as a solution from $W_2^{2,1}(Q'')$ of the linear equation $v_t - (1/b'(v(x, t)))\Delta v = 0$ will be an element of $H^{2+\tilde{\alpha}, 1+\tilde{\alpha}/2}(Q'')$ (see Theorem 12.1 of Chapter III). The solution $v(x, t)$ will be a smoother function for smoother $b(v)$; this is guaranteed by Theorems 12.1 of Chapter III and 5.2 of Chapter IV on the solutions of linear equations.

Thus we see that the existence and uniqueness of generalized solutions of the Stefan problem (the case considered admits certain generalizations) is established in primarily the same way as the unique solvability of the diffraction problems in the generalized statement in [65q] (see §13 of Chapter III). This was done in [55] (see also [91d]). But the question whether or not these solutions are classical and, in particular, whether or not the sets $S^{(k)}$ of points at which $v(x, t)$ is equal to one of the singular values v_k form piecewise smooth n-dimensional surfaces in the (x, t)-space remains open and its solution is of unquestioned interest.

In conclusion let us mention the works [102, 108, 31, 58] etc., in which one-dimensional Stefan problems in their classical statement are studied. The methods of these works are different from the methods presented here. In them the problems are reduced to an investigation of nonlinear integral equations. In [31] and [102b] there are solutions of some of these problems in the large.

§10. ANOTHER METHOD OF ESTIMATING THE HÖLDER CONSTANT FOR SOLUTIONS

In the present section we will estimate Hölder constants for generalized solutions of equation (0.1) by a method that is different from the one presented in §1. This will be done without making use of the theorems of §§7–8 of Chapter II on functions of the classes $\mathfrak{B}_2$. The method to be used is an extension to the general case of the method described by the example of an elementary linear equation of parabolic type given at the end of §10 in Chapter III.

We will give here another proof of Theorem 1.1 (more precisely, of its first part) concerning interior estimates. Estimates near the surface S_T are established analogously, but with the enlistment of the lemmas of Chapter II relating to the behavior of functions in cylinders intersecting with the surface S_T.

Thus, suppose the conditions of Theorem 1.1, viz., the inequalities (1.1)–(1.5), are fulfilled, and suppose $u(x, t)$ is a bounded generalized solution of the class $V_2^{1,0}(Q_T)$ of equation (0.1), so that $u(x, t)$ satisfies the integral identity (1.6) for any bounded function $\eta(x, t)$ from $\overset{\circ}{W}{}_2^{1,1}(Q_T)$. From (1.6) and conditions (1.1)–(1.5) there follows the validity of inequality (1.7) for $u(x, t)$:

$$\int_\Omega u\eta\,dx\Big|_{t_1}^{t_2}+\int_{t_1}^{t_2}\int_\Omega [-u\eta_t+a_i(x,t,u,u_x)\eta_{x_i}]\,dx\,dt$$
$$\leqslant\int_{t_1}^{t_2}\int_\Omega [\mu_1 u_x^2+\varphi_2(x,t)]\,|\eta|\,dx\,dt,$$
$$0\leqslant t_1<t_2\leqslant T. \tag{10.1}$$

where $\eta(x, t)$ is an arbitrary bounded function from $\overset{\circ}{W}{}_2^{1,1}(Q_T)$, while the functions a_i and ϕ_2 satisfy restrictions (1.1)–(1.2), (1.4)–(1.5). As in §1, this inequality is the starting point for all of the subsequent arguments. In §1 we deduced from it that $w = \pm u(x, t)$ is subject to inequalities (7.5) of Chapter II:

$$\begin{aligned} \|w^{(k)}(x, t_0+\tau)\zeta(x, t_0+\tau)\|^2_{2, K_\rho} &+ \nu\|w_x^{(k)}\zeta\|^2_{2, Q(\rho,\tau)} \\ \leqslant \|w^{(k)}(x, t_0)\zeta(x, t_0)\|^2_{2, K_\rho} &+ \gamma_1\Bigg\{\int_{Q(\rho,\tau)}(w^{(k)})^2(\zeta_x^2+\zeta|\zeta_t|)\,dx\,dt \\ &+ \Bigg[\int_{t_0}^{t_0+\tau}\Bigg(\int_{A_{k,\rho}(t)}|\zeta|\,dx\Bigg)^{\frac{\hat{r}}{\hat{q}}}dt\Bigg]^{\frac{2}{\hat{r}}(1+\varkappa)}\Bigg\} \end{aligned} \tag{10.2}$$

for any values of $k \geq \operatorname{vrai\,max}_{Q(\rho,\tau)} w(x, t) - \delta$. Here

$$\varkappa = \frac{2}{n}\varkappa_1, \quad \hat{q} = \frac{2q}{q-1}(1+\varkappa), \quad \hat{r} = \frac{2r}{r-1}(1+\varkappa),$$

δ is a positive number determined by $M = \operatorname{vrai\,max}_{Q_T}|u|$, n, ν, μ, μ_1, μ_2, r and q, $Q(\rho, \tau) = K_\rho \times (t_0, t_0+\tau)$ is an arbitrary cylinder contained in Q_T, $A_{k,\rho}(t)$ is the set of those points x from K_ρ at which $w(x, t) > k$, and $\zeta(x, t)$ is an arbitrary smooth function that is equal to zero on the lateral surface of $Q(\rho, \tau)$. This guaranteed the membership of the solution $u(x, t)$ in the class $\mathfrak{B}_2(Q_T, M, \gamma, \hat{r}, \delta, \kappa)$, from which the desired conclusion was drawn concerning an estimate of the Hölder constant for $u(x, t)$.

As already noted above, we will not employ here the theorems on functions of the classes $\mathfrak{B}_2$. Of all of the propositions proved in §§7–8 of Chapter II we will make use of only the simplest of them, Lemma 7.1, which follows from inequalities (10.2).

The estimation of the Hölder constant will be carried out in the following way. We consider an arbitrary cylinder $Q_{2R} = K_{2R} \times (t_0 - \theta R^2, t_0)$, where the constant $\theta > 0$ will be fixed later, and the coaxial cylinder $Q_R = K_R \times (t_0 - \theta R^2/4, t_0)$ and we prove the validity of the inequality

$$\operatorname{osc}\{u, Q_R\} \leqslant (1-\delta_0)\operatorname{osc}\{u, Q_{2R}\} + R^{\varkappa_1}, \tag{10.3}$$

with the constant $\delta_0 > 0$ being determined by $M = \max_{Q_T}|u|$ and the known quantities from conditions (1.1)–(1.5). This by virtue of Lemma 5.8 of Chapter II gives the desired estimate for $\langle u\rangle_{Q'}^{(\alpha)}$, $Q' \subset Q_T$.

If $\operatorname{osc}\{u, Q_R\} \leq R^{\kappa 1}$, then (10.3) is valid. We assume that $\operatorname{osc}\{u, Q_R\} > R^{\kappa 1}$, and we introduce the notation $\omega = \operatorname{osc}\{u, Q_{2R}\}$. Obviously $\omega > R^{\kappa 1}$. Without loss of generality we can assume that in Q_{2R} the function $u(x, t)$ varies within the range $0 \leq u(x, t) \leq \omega$. At least one of the two following inequalities is always true:

$$\operatorname{mes}\left\{x \in K_R : u(x, t_0 - \theta R^2) \leqslant \frac{\omega}{2}\right\} \geqslant \frac{1}{2} \operatorname{mes} K_R \tag{10.4$_1$}$$

or

$$\operatorname{mes}\left\{x \in K_R : \omega - u(x, t_0 - \theta R^2) \leqslant \frac{\omega}{2}\right\} \geqslant \frac{1}{2} \operatorname{mes} K_R. \tag{10.4$_2$}$$

If the first of them holds, then we deduce from (10.4$_1$) and inequalities (10.1) with $\eta \geq 0$ that the difference in Q_R between $u(x, t)$ and ω is not less than a certain fraction of ω. If on the other hand (10.4$_2$) holds, then instead of $u(x, t)$ we take the function $\hat{u}(x, t) = \omega - u(x, t)$. Besides (10.4$_2$) inequalities of type (10.1) with $\eta \geq 0$ are valid for it, as follows from (10.1) for $u(x, t)$ and the fact that $\eta(x, t)$ can have any sign in (10.1). Therefore, as with $u(x, t)$ in the first case, the difference in Q_R between it and its maximum in Q_{2R} (i.e. ω) is not less than a certain fraction of ω. Thus in both cases we arrive at (10.3). Suppose, for example, (10.4$_1$) is true. Then a fortiori

$$\operatorname{mes}\left\{x \in K_R : u(x, t_0 - \theta R^2) \leqslant (1 - \mu_0)\omega\right\} \geqslant \frac{1}{2} \operatorname{mes} K_R,$$

where $\mu_0 = \min\{1/2, \delta/2M\} \leq \min\{1/2, \delta/\omega\}$. According to Lemma 7.1 of Chapter II there exist for any $\xi \in (1/\sqrt{2}, 1)$ positive numbers $\theta(\xi)$ and $b(\xi)$ such that for all $t \in [t_0 - \theta R^2, t_0]$

$$\operatorname{mes}\{x \in K_\rho : u(x, t) \leqslant (1 - \mu_0)\omega + \xi H\} \geqslant b(\xi) \operatorname{mes} K_R, \tag{10.5}$$

as long as

$$H = \max_{\substack{x \in K_\rho \\ t \in [t_0 - \theta R^2, t_0]}} u(x, t) - (1 - \mu_0)\omega$$

does not exceed δ and $H > R^{\kappa 1}$. The first of these requirements on H is fulfilled since $H \leq \omega - (1 - \mu_0)\omega = \mu_0\omega$, while $\mu_0 \leq \delta/\omega$. If the second requirement on H is not fulfilled, i.e. $H \leq R^{\kappa 1}$, then $\max_{Q_R} u(x, t) \leq (1 - \mu_0)\omega + R^{\kappa 1}$ and hence (10.3) is fulfilled with $\delta_0 = \mu_0$. If on the other hand $H > R^{\kappa 1}$, then we use the assertion of Lemma 7.1 of Chapter II. As ξ we take, for example, 7/8 and corresponding to it $\theta = \theta(\xi)$ and $b = b(\xi)$. Since $(1-\mu_0)\omega + 7H/8 \le \omega - \mu_0\omega/8$, it follows from (10.5) that

$$\operatorname{mes}\left\{x\in K_R : u(x,t) \leqslant \omega - \frac{\mu_0}{8}\omega\right\} \geqslant b \operatorname{mes} K_R,$$
$$t \in [t_0 - \theta R^2, t_0]. \tag{10.6}$$

Consider in Q_{2R} the function $v(x, t) = \psi(u(x, t))$, where

$$\psi(u) = -\ln 8\,\frac{\omega - u + R^{\varkappa_1}}{\mu_0\omega}.$$

It is bounded from below by the number $-\ln(16/\mu_0)$ and is obviously nonpositive at those points where $u(x, t) \le \omega - \mu_0\omega/8$, and this set of points constitutes according to (10.6) a definite fraction of K_ρ for any $t \in [t_0 - \theta R^2, t_0]$. If we prove that $\max_{Q_R} v(x, t) \le M_1$, then this will guarantee the estimate $\max_{Q_R} u(x, t) \le \omega(1 - \mu_0 e^{-M_1}/8) + R^{\kappa_1}$, and thereby also (10.3) with $\delta_0 = \mu_0 e^{-M_1}/8$.

Thus, in order to obtain estimate (10.3) it is sufficient to show that $v(x, t)$ is bounded in Q_R from above by a quantity determined by $M, \nu, \mu, \mu_1, \mu_2, q, r$ only and not depending on R. The estimate of $v(x, t)$ from above will be based on the use of Theorem 6.2 of Chapter II. Namely, we will establish the estimate

$$R^{-n-2}\int_{t_0-\frac{3}{4}\theta R^2}^{t_0}\int_{K_{\frac{3R}{2}}} v^2\,dx\,dt \leqslant c \tag{10.7}$$

and prove that the function $v(x, t)$ satisfies the inequalities

$$\begin{aligned}\max_{t\in[t_0-\tau,\,t_0]}\int_{K_\rho}(v^{(k)}\zeta)^2\,dx &+ \int_{t_0-\tau}^{t_0}\int_{K_\rho}|v_x^{(k)}|^2\zeta^2\,dx\,dt\\ \leqslant \gamma\Bigg\{&\int_{t_0-\tau}^{t_0}\int_{K_\rho}|v^{(k)}|^2(\zeta_x^2+\zeta|\zeta_t|)\,dx\,dt\\ &+R^{-n\varkappa}\left[\int_{t_0-\tau}(\operatorname{mes}A_{k,\rho}(t))^{\frac{\hat r}{\hat q}}\,dt\right]^{\frac{2}{\hat r}(1+\varkappa)}\Bigg\}\end{aligned} \tag{10.8}$$

for k exceeding a certain value k_0; here $A_{k,\rho}(t) = \{x \in K_\rho : v(x, t) > k\}$, $\zeta(x, t)$ is an arbitrary smooth nonnegative function with $\zeta(x, t) \le 1$ that is equal to zero on the lateral surface and lower base of the cylinder $Q(\rho, \tau) = K_\rho \times (t_0 - \tau, t_0)$, $\theta R^2/4 \le \tau \le 3\theta R^2/4$, $R \le \rho \le 3R/2$, and $\hat{q}$, $\hat{r}$ and κ are determined, as in (10.2), from the relations $\kappa = 2\kappa_1/n$, $\hat{q} = (2q/(q-1))(1+\kappa)$, $\hat{r} = (2r/(r-1))(1+\kappa)$. We can easily see, by using assumptions (1.5), that the parameters $\hat{r}$ and $\hat{q}$ are subject to the restrictions (3.3) of Chapter II.

The desired estimate for vrai $\max_{Q_R} v(x, t)$, and thereby inequality (10.3), will follow from (10.8) and (10.7) on the basis of Theorem 6.2 of Chapter II.

We begin with a derivation of inequality (10.7). To this end we first establish the following lemma.

Lemma 10.1. *If $u(x, t) \in V_2^{1,0}(K_{2R} \times (0, T)) \equiv V_2^{1,0}(\hat{Q}_T)$, vrai $\max_{\hat{Q}_T} |u| = M$ and for all $t_1 \in (0, T)$ inequalities (10.1) hold with an arbitrary bounded nonnegative $\eta \in \overset{\circ}{W}{}_2^{1,1}(\hat{Q}_T)$ while the a_i and ϕ_2 satisfy conditions (1.1)–(1.5), then for any $t_1 \le t_2$ from $[0, T]$ the estimate*

$$\int_{t_1}^{t_2} \int_{K_{2R}} u_x^2 \zeta^2 \, dx \, dt \leqslant cR^n + c(t_2 - t_1) R^n \left[\max \zeta_x^2 + (t_2 - t_1)^{-\frac{1}{r}} R^{-\frac{n}{q}} \right] \tag{10.9}$$

holds, where the quantity c depends only on n, M, ν, μ, μ_1, μ_2, q, and r, while $\zeta = \zeta(x)$ is a smooth function that is equal to zero on the boundary of the ball K_{2R}, with $0 \le \zeta \le 1$.

To prove this we put $\eta = (e^{\lambda u_h} \zeta^2)_{\bar{h}}$ in (10.1) and transform the term containing the derivative η_t in the following manner:

$$\begin{aligned} -\int_{t_1}^{t_2} \int_{K_{2R}} u (e^{\lambda u_h} \zeta^2)_{\bar{h}t} \, dx &= \int_{t_1}^{t_2} \int_{K_{2R}} u_{ht} e^{\lambda u_h} \zeta^2 \, dx - \int_{K_{2R}} [u(x, t+h) e^{\lambda u_h} \zeta^2]_{\bar{h}} \, dx \Big|_{t_1}^{t_2} \\ &= \frac{1}{\lambda} \int_{K_{2R}} e^{\lambda u_h} \zeta^2 \, dx \Big|_{t_1}^{t_2} - \int_{K_{2R}} [u(x, t+h) e^{\lambda u_h} \zeta^2]_{\bar{h}} \, dx \Big|_{t_1}^{t_2}. \end{aligned} \tag{10.10}$$

Now in (10.1) as transformed by means of (10.10) we can pass to the limit for $h \longrightarrow 0$. As a result of the passage to the limit and of estimates made with the use of conditions (1.1), (1.2), (1.4) and (1.5), we obtain

$$\frac{1}{\lambda}\int_{K_{2R}} e^{\lambda u}\zeta^2\,dx\Big|_{t_1}^{t_2} + \nu\lambda\int_{t_1}^{t_2}\int_{K_{2R}} e^{\lambda u}u_x^2\zeta^2\,dx\,dt$$

$$\leqslant \int_{t_1}^{t_2}\int_{K_{2R}} e^{\lambda u}[\lambda\varphi_0\zeta^2+2(\mu|u_x|+\varphi_1)\zeta|\zeta_x|+(\mu_1 u_x^2+\varphi_2)\zeta^2]\,dx\,dt$$

$$\leqslant \int_{t_1}^{t_2}\int_{K_{2R}} e^{\lambda u}\left[(\lambda\varphi_0+\varphi_1^2+\varphi_2)\zeta^2+\zeta_x^2+\frac{1}{\mu_1}\mu^2\zeta_x^2+2\mu_1 u_x^2\zeta^2\right]dx\,dt. \tag{10.11}$$

Choosing $\lambda = 4\nu^{-1}\mu_1$ in (10.11) and making use of the condition

$$\operatorname*{vrai\,max}_{Q_T}|u(x,t)| \le M$$

and (1.4), we obtain

$$\mu_1 e^{-\frac{4M\mu_1}{\nu}}\int_{t_1}^{t_2}\int_{K_{2R}} u_x^2\zeta^2\,dx\,dt \leqslant \frac{\nu}{4\mu_1}e^{\frac{4\mu_1 M}{\nu}}\varkappa_n 2^n R^n$$
$$+e^{\frac{4M\mu_1}{\nu}}\left[\mu_2\left(\frac{4\mu_1}{\nu}+2\right)(t_2-t_1)^{1-\frac{1}{r}}(\varkappa_n 2^n R^n)^{1-\frac{1}{q}}\right.$$
$$\left.+\left(1+\frac{\mu^2}{\mu_1}\right)\max_{K_{2R}}\zeta_x^2(t_2-t_1)\varkappa_n 2^n R^n\right],$$

from which follows (10.9). Lemma 10.1 is proved.

Let the ball K_{2R} have its center at x_0.

Consider the inequality (10.1) with $t_1 = t_0 - \theta R^2$ and $t_2 = t_0$. We put $\eta = (\psi'(u_h)\mathfrak{N}^2(x)\chi^2(t))_{\bar h}$ in it, where

$$\mathfrak{N}(x)=\mathfrak{N}(|x|)=\begin{cases}1, & 0\leqslant|x-x_0|\leqslant\frac{3}{2}R,\\ 2\left(2-\frac{|x-x_0|}{R}\right), & \frac{3}{2}R\leqslant|x-x_0|\leqslant 2R,\\ 0, & 2R\leqslant|x-x_0|;\end{cases}$$

$$\chi(t)=\begin{cases}0, & t\leqslant t_0-\theta R^2,\\ \frac{4}{\theta R^2}(t-t_0+\theta R^2), & t_0-\theta R^2\leqslant t\leqslant t_0-\frac{3}{4}\theta R^2,\\ 1, & t_0-\frac{3}{4}\theta R^2\leqslant t\leqslant t_0.\end{cases}$$

We carry out an integration by parts with respect to t in the second summand of (10.1) and we reduce its first term to the form

$$\int\limits_{Q_{2R}} u_{ht}\psi'(u_h)\,\mathfrak{N}^2(x)\,\chi^2(t)\,dx\,dt - \int\limits_{K_{2R}} (u^{+h}\psi(u_h)\,\mathfrak{N}^2\chi^2)_{\bar h}\,dx\Big|^{t_0}$$
$$= \int\limits_{K_{2R}} v^h\mathfrak{N}^2\chi^2\,dx\Big|^{t_0} - \int\limits_{K_{2R}} (u^{+h}\psi'(u_h)\,\mathfrak{N}^2\chi^2)_{\bar h}\,dx\Big|^{t_0}$$
$$- \int\limits_{Q_{2R}} 2v^h\mathfrak{N}^2\chi\chi'\,dx\,dt,$$

where $u^{+h} = u(x, t+h)$ and $v^h = \psi(u_h)$.

We then pass to the limit in (10.1) as $h \to 0$. The legitimacy of this passage is easily established by means of the lemmas in §4 of Chapter II. As a result we obtain the relation

$$\int\limits_{K_{2R}} v\mathfrak{N}^2\chi^2\,dx\Big|^{t_0} + \int\limits_{Q_{2R}} a_i\left(u_{x_i}\psi''\mathfrak{N}^2\chi^2 + 2\psi'\mathfrak{N}\mathfrak{N}_{x_i}\chi^2\right)dx\,dt$$
$$\leqslant \int\limits_{Q_{2R}} \left[\mu_1 u_x^2 + \varphi_2(x, t)\right]\psi'\mathfrak{N}^2\chi^2\,dx\,dt$$
$$+ \int\limits_{Q_{2R}} 2v\mathfrak{N}^2\chi\chi'\,dx\,dt. \tag{10.12}$$

We carry out estimates in (10.12) by means of inequalities (1.1)–(1.2) and (1.4)–(1.5), taking into account the definition of $v(x, t)$:

$$\int\limits_{K_{2R}} v\mathfrak{N}^2\chi^2\,dx\Big|^{t_0} + \nu\int\limits_{Q_{2R}} v_x^2\mathfrak{N}^2\chi^2\,dx\,dt$$
$$\leqslant \int\limits_{Q_{2R}} 2v\mathfrak{N}^2\chi\chi'\,dx\,dt$$
$$+ \int\limits_{Q_{2R}} \left[\varphi_0\psi''\mathfrak{N}^2\chi^2 + 2\psi'\mathfrak{N}\,|\mathfrak{N}_x|\,\chi^2(\mu\,|u_x| + \varphi_1)\right.$$
$$\left. + (\mu_1 u_x^2 + \varphi_2)\,\psi'\mathfrak{N}^2\chi^2\right]dx\,dt.$$

We estimate the first summand of the right side of this inequality by means of inequality (1.2) of Chapter II:

$$\int\limits_{Q_{2R}} 2v\mathfrak{N}^2\chi\chi'\,dx\,dt \leqslant \frac{\varepsilon}{R^2}\int\limits_{Q_{2R}} v^2\mathfrak{N}^2\chi^2\,dx\,dt + \frac{R^2}{\varepsilon}\int\limits_{Q_{2R}} \mathfrak{N}\chi'^2\,dx\,dt$$

and we take advantage of inequality (5.4) of Chapter II. By virtue of assumption

(10.6) and the fact that $v \geq -\ln(16/\mu_0)$, this gives us

$$\int_{Q_{2R}} 2v\mathfrak{N}^2\chi\chi'\,dx\,dt \leqslant \varepsilon\beta_3 \int_{Q_{2R}} v_x^2\mathfrak{N}^2\chi^2\,dx\,dt + \frac{c_1}{\varepsilon} R^n,$$

$$\beta_3 = \beta_2 (b\varkappa_n)^{-2} 2^{2n+2}.$$

The remaining terms are estimated as follows:

$$\int_{Q_{2R}} \varphi_0\psi''\mathfrak{N}^2\chi^2\,dx\,dt \leqslant R^{-2\varkappa_1} \int_{Q_{2R}} \varphi_0\mathfrak{N}^2\chi^2\,dx\,dt \leqslant c_2R^n \|\varphi_0\|_{q,r,Q_{2R}},$$

$$2\int_{Q_{2R}} \psi'\mathfrak{N}|\mathfrak{N}_x|\chi^2|u_x|\,dx\,dt \leqslant \varepsilon \int_{Q_{2R}} v_x^2\mathfrak{N}^2\chi^2\,dx\,dt$$

$$+\varepsilon^{-1} \int_{Q_{2R}} \mathfrak{N}_x^2\chi^2\,dx\,dt \leqslant \varepsilon \int_{Q_{2R}} v_x^2\mathfrak{N}^2\chi^2\,dx\,dt + \frac{c_3}{\varepsilon} R^n,$$

$$\int_{Q_{2R}} \psi'\mathfrak{N}|\mathfrak{N}_x|\chi^2\varphi_1\,dx\,dt \leqslant R^{-\varkappa_1} \max_{K_{2R}} |\mathfrak{N}_x| \int_{Q_{2R}} \varphi_1\,dx\,dt$$

$$\leqslant c_4R^n \|\varphi_1\|_{2q,2r,Q_{2R}},$$

$$\int_{Q_{2R}} u_x^2\psi'\mathfrak{N}^2\chi^2\,dx\,dt = \int_{Q_{2R}} |v_x|\,|u_x|\mathfrak{N}^2\chi^2\,dx\,dt$$

$$\leqslant \frac{\varepsilon}{2} \int_{Q_{2R}} v_x^2\mathfrak{N}^2\chi^2\,dx\,dt + \frac{1}{2\varepsilon} \int_{Q_{2R}} u_x^2\mathfrak{N}^2\chi^2\,dx\,dt.$$

By Lemma 10.1 it follows that

$$\int_{Q_{2R}} u_x^2\psi'\mathfrak{N}^2\chi^2\,dx\,dt \leqslant \frac{\varepsilon}{2} \int_{Q_{2R}} v_x^2\mathfrak{N}^2\chi^2\,dx\,dt + \frac{c_5}{\varepsilon} R^n;$$

$$\int_{Q_{2R}} \varphi_2\psi'\mathfrak{N}^2\chi^2\,dx\,dt \leqslant R^{-\varkappa_1} \int_{Q_{2R}} \varphi_2\mathfrak{N}^2\chi^2\,dx\,dt \leqslant c_6R^{n+\varkappa_1} \|\varphi_2\|_{q,r,Q_T}.$$

Upon reducing similar terms after these estimates, we obtain

$$\int_{K_{2R}} v\mathfrak{N}^2\chi^2\,dx\Big|^{t_0} + \nu \int_{Q_{2R}} v_x^2\mathfrak{N}^2\chi^2\,dx\,dt$$

$$\leqslant \varepsilon\left(\beta_3 + \mu + \frac{1}{2}\mu_1\right) \int_{Q_{2R}} v_x^2\mathfrak{N}^2\chi^2\,dx\,dt$$

$$+\left(\frac{c_1}{\varepsilon} + c_2\|\varphi_0\|_{q,r,Q_T} + \mu\frac{c_3}{\varepsilon} + 2c_4\|\varphi_1\|_{2q,2r,Q_T}\right.$$

$$\left. + \mu_1\frac{c_5}{\varepsilon} + c_6R^{\varkappa_1}\|\varphi_2\|_{r,q,Q_T}\right) R^n. \tag{10.13}$$

Let us choose $\epsilon = \nu/2(\beta_3 + \mu_1 + \mu/2)$ here. Then from (10.13) we obtain

$$\int_{Q_{2R}} v_x^2 \mathfrak{N}^2 \chi^2 \, dx \, dt \leqslant c_7 R^n,$$

which together with the inequality

$$\operatorname{mes}\{x \in K_R;\ v(x, t) \leqslant 0\} \geqslant b \operatorname{mes} K_R, \quad t \in [t_0 - \theta R^2, t_0],$$

implied by (10.6), the inequality (5.4) of Chapter II and the estimate $v \geq -\ln(16/\mu_0)$ gives us

$$\int_{t_0 - \frac{3\theta}{4}R^2}^{t_0} \int_{K_{2R}} v^2 \mathfrak{N}^2 \, dx \, dt \leqslant \left| \ln \frac{16}{\mu_0} \right| \frac{3\theta}{4} R^2 \operatorname{mes} K_{2R}$$

$$+ \beta_3 R^2 \int_{t_0 - \frac{3\theta}{4}R^2}^{t_0} \int_{K_{2R}} v_x^2 \mathfrak{N}^2 \, dx \, dt \leqslant c R^{n+2}$$

Thus estimate (10.7) is established. It remains to show the validity of inequalities (10.8). To this end we put in (10.1) $t_1 = t_0 - \tau$, $t_2 \in [t_0 - \tau, t_0]$ and

$$\eta(x, t) = [\psi'(u_h) \max\{\psi(u_h) - k, 0\} \zeta^2]_{\bar{h}} = [\psi'(u_h) v^{h(k)} \zeta^2]_{\bar{h}},$$

where ζ is a cutting function for the cylinder $Q(\rho, \tau) = K_\rho \times (t_0 - \tau, t_0)$, and we transform the term with η_t in (10.1) to the form

$$\int_{t_0-\tau}^{t_2} \int_{K_\rho} \frac{1}{2} \frac{\partial}{\partial t} [v^{h(k)}]^2 \zeta^2 \, dx \, dt - \int_{K_\rho} (\psi' v^{h(k)} \zeta^2)_{\bar{h}} \Big|^{t_2}$$

$$= \frac{1}{2} \int_{K_\rho} [v^{h(k)} \zeta]^2 \, dx \Big|^{t_2} - \int_{t_0-\tau}^{t_2} \int_{K_\rho} [v^{h(k)}]^2 \zeta \zeta_t \, dx \, dt$$

$$- \int_{K_\rho} (\psi' v^{h(k)} \zeta^2)_{\bar{h}} \Big|^{t_2};$$

we then pass to the limit as $h \to 0$ and carry out estimates in the resultant equality by means of conditions (1.1)–(1.2) and (1.4)–(1.5). As a result we will have an inequality of type (1.10), namely

$$\frac{1}{\tau}\max_{t_2\in[t_0-\tau,\,t_0]}\int_{K_\rho}[v^{(k)}\zeta]^2\,dx\,\Big|^{t_2}$$

$$+\frac{\nu}{2}\int_{Q(\rho,\,\tau)}\left[|v_x^{(k)}|^2 v^{(k)}\zeta^2+|v_x^{(k)}|^2\zeta^2\right]dx\,dt$$

$$\leqslant\int_{Q(\rho,\,\tau)}(v^{(k)})^2\zeta|\zeta_t|\,dx\,dt+\int_{t_0-\tau}^{t_0}\int_{A_{k,\,\rho}}\left[\varphi_0\zeta^2\left(v^{(k)}\psi''+\psi'^2\right)\right.$$

$$\left.+2\psi'v^{(k)}\zeta|\zeta_x|(\mu|u_x|+\varphi_1)+\psi'v^{(k)}\zeta^2(\mu_1u_x^2+\varphi_2)\right]dx\,dt, \tag{10.14}$$

where $A_{k,\,\rho}(t)=\{x\in K_\rho: v(x,\,t)>k\}$.

We estimate the summands of the right side of (10.14) in the following manner:

$$\int_{Q(\rho,\,\tau)}\varphi_0\zeta^2v^{(k)}\,\psi''(u)\,dx\,dt$$

$$\leqslant R^{-2\varkappa_1}\|\varphi_0\zeta\|_{q,\,r,\,Q(\rho,\,\tau)}\,\|v^{(k)}\zeta\|_{\hat q,\,\hat r,\,Q(\rho,\,\tau)}(\mu(k,\,\rho,\,\tau))^{\frac{1}{\hat r}(1+2\varkappa)}$$

$$\leqslant\frac{\varepsilon}{2}\|v^{(k)}\zeta\|^2_{\hat q,\,\hat r,\,Q(\rho,\,\tau)}+\frac{c_8}{2\varepsilon}R^{-2\varkappa_1}\|\varphi_0\zeta\|^2_{q,\,r,\,Q(\rho,\,\tau)}$$

$$\times(\mu(k,\,\rho,\,\tau))^{\frac{2}{\hat r}(1+\varkappa)}$$

(here and below

$$\mu(k,\,\rho,\,\tau)=\int_{t_0-\tau}^{t_0}\operatorname{mes}\{A_{k,\,\rho}(t)\}^{\frac{\hat r}{\hat q}}\,dt$$

with $\hat q$ and $\hat r$ defined above and $\kappa=2\kappa_1/n$),

$$\int_{t_0-\tau}^{t_0}\int_{A_{k,\,\rho}}\varphi_0\psi'^2\zeta^2\,dx\,dt\leqslant R^{-2\varkappa_1}\|\varphi_0\zeta\|_{q,\,r,\,Q(\rho,\,\tau)}\,\mu^{\frac{2}{\hat r}(1+\varkappa)}(k,\,\rho,\,\tau),$$

$$2\int_{Q(\rho,\,\tau)}\mu\psi'v^{(k)}\zeta|\zeta_x|\,|u_x|\,dx\,dt$$

$$\leqslant\varepsilon\int_{Q(\rho,\,\tau)}|v_x^{(k)}|^2\zeta^2\,dx\,dt+\frac{\mu^2}{\varepsilon}\int_{Q(\rho,\,\tau)}(v^{(k)})^2\zeta_x^2\,dx\,dt,$$

$$2\int_{Q(\rho,\tau)}\psi' v^{(k)}\zeta|\zeta_x|\varphi_1\,dx\,dt\leqslant\int_{Q(\rho,\tau)}(v^{(k)})^2\zeta_x^2\,dx\,dt$$

$$+R^{-2\varkappa_1}\int_{t_0-\tau}^{t_0}\int_{A_{k,\rho}(t)}\varphi_1^2\zeta^2\,dx\,dt\leqslant\int_{Q(\rho,\tau)}(v^{(k)})^2\zeta_x^2\,dx\,dt$$

$$+R^{-2\varkappa_1}\|\varphi_1\|^2_{2q,\,2r,\,Q(\rho,\tau)}\mu^{\frac{2}{r}(1+\varkappa)}(k,\rho,\tau),$$

$$\int_{Q(\rho,\tau)}\mu_1\psi' v^{(k)}\zeta^2u_x^2\,dx\,dt$$

$$=\mu_1\int_{Q(\rho,\tau)}(v_x^{(k)})^2\zeta^2v^{(k)}(\psi')^{-1}\,dx\,dt$$

$$\leqslant\mu_1\frac{\mu_0\omega}{8}\max_{Q(\rho,\tau)}[v^{(k)}e^{-v}]\int_{Q(\rho,\tau)}|v_x^{(k)}|^2\zeta^2\,dx\,dt.$$

Since $\mu_0\le\delta/\omega$ and $\max_{Q(\rho,\tau)}v^{(k)}e^{-v}\le e^{-1-k}$,

$$\int_{Q(\rho,\tau)}\mu_1\psi' v^{(k)}\zeta u_x^2\,dx\,dt\leqslant\frac{1}{8}\mu_1\delta e^{-1-k}\int_{Q(\rho,\tau)}(v_x^{(k)})^2\zeta^2\,dx\,dt;$$

$$\int_{Q(\rho,\tau)}\psi' v^{(k)}\zeta^2\varphi_2\,dx\,dt\leqslant R^{-\varkappa_1}\|v^{(k)}\zeta\|_{\hat q,\hat r,Q(\rho,\tau)}$$

$$\times\|\varphi_2\zeta\|_{q,r,Q(\rho,\tau)}\mu^{\frac{1}{r}(1+2\varkappa)}(k,\rho,\tau)\leqslant\frac{\varepsilon}{2}\|v^{(k)}\zeta\|^2_{\hat q,\hat r,Q(\rho,\tau)}$$

$$+\frac{c_0}{2\varepsilon}\|\varphi_2\|^2_{q,r,Q(\rho,\tau)}\mu^{\frac{2}{r}(1+\varkappa)}(k,\rho,\tau).$$

We substitute these estimates into (10.14), replace $\|v^{(k)}\zeta\|_{\hat q,\hat r,Q(\rho,\tau)}$ by the larger quantity $\beta|v^{(k)}\zeta|_{Q(\rho,\tau)}$ (see inequality (3.4) of Chapter II), and take ϵ equal to $(1/2(1+\beta^2))\min\{1/4;\ \nu/2\}$. After the reduction of similar terms, this gives us the inequality

$$\min\left\{\frac{1}{8};\ \frac{\nu}{4}\right\}\left[\max_{t\in[t_0-\tau,\,t_0]}\int_{K_\rho}[v^{(k)}\zeta]^2\,dx\,\Big|^t\right.$$

$$\left.+\int_{Q(\rho,\tau)}[v_x^{(k)}\zeta]^2\,dx\,dt\right]\leqslant\gamma_1\left[\int_{Q(\rho,\tau)}(v^{(k)})^2(\zeta|\zeta_t|+\zeta_x^{(2)})\,dx\,dt\right.$$

$$\left.+R^{-2\varkappa_1}\mu^{\frac{2}{r}(1+\varkappa)}(k,\rho,\tau)\right]+\frac{\delta}{8}\mu_1e^{-1-k}\int_{Q(\rho,\tau)}[v_x^{(k)}\zeta]^2\,dx\,dt\qquad(10.15)$$

which is valid for any k, with γ_1 determined only by M, ν, μ, μ_1, μ_2, q and r. We now impose on k the requirement, $k \geq k_0$, where $k_0 = \ln[(\delta/8)\mu_1 \max(8; 4/\nu)]$. Under this requirement inequalities (10.8) follow from (10.15). Thus the first part of Theorem 1.1 can be considered proved.

Remark 10.1. In the derivation of (10.3) presented above, instead of the assumption that $u(x, t)$ is a generalized solution of equation (0.1), it is only necessary that $u(x, t)$ satisfy inequality (10.1) for any bounded $\eta(x, t) \geq 0$ from $\overset{\circ}{W}{}_1^{1,1}(Q_T)$ and a_i and ϕ_2 subject to conditions (1.1)–(1.2), (1.4)–(1.5) (see in connection with this Remark 1.1 of the present chapter), and that condition (10.4_1) be fulfilled for the function $u(x, t)$. The latter assumption could be replaced by the condition

$$\operatorname{mes}\{x \in K_\rho : u(x, t_0 - \theta R^2) \leqslant \max_{Q_{2R}} u - \delta_2 \operatorname{osc}\{u, Q_{2R}\}\} \geqslant \delta_3 \operatorname{mes} K_R \tag{10.16}$$

with certain positive constants δ_2 and δ_3. The proof of estimate (10.3) from this goes through with only a few obvious changes.

By taking into account what has just been said we can assert the validity of the following lemma, which will be used in Chapter VI.

Lemma 10.2. *If a function* $u(x, t)$ *from* $V_2^{1,0}(Q_T)$ *has* vrai $\max_{Q_T}|u| = M < \infty$, *and satisfies inequality* (10.1) *for any bounded nonnegative* $\eta(x, t)$ *from* $\overset{\circ}{W}{}_2^{1,1}(Q_T)$ *with the* a_i *and* ϕ_2 *subject to restrictions* (1.1)–(1.2), (1.4)–(1.5), *and if condition* (10.16) *with* $\delta_2, \delta_3 > 0$ *is fulfilled for* $u(x, t)$, *then inequality* (10.3) *is valid for* $u(x, t)$ *with the constant* $\delta_0 > 0$ *determined only by the quantities* μ_1, M, ν, μ, μ_2, q, r *from conditions* (1.1)–(1.2), (1.4)–(1.5), δ_2 *and* δ_3.

CHAPTER VI

QUASI-LINEAR EQUATIONS OF GENERAL FORM

In the present chapter we investigate the quasi-linear equations of general form

$$u_t - a_{ij}(x, t, u, u_x)u_{x_ix_j} + a(x, t, u, u_x) = 0. \tag{0.1}$$

For their solutions $u(x, t)$ we establish a priori estimates of the norms $|u_x|_{Q_T}^{(\alpha)}$, which together with the results of Chapter IV on linear equations with smooth coefficients and the Leray-Schauder Theorem lead to the classical solvability in the large of the first boundary value problem for the equations (0.1). These estimates, as was shown by the examples in §3 of Chapter II, do not hold for the whole class of uniformly parabolic equations (0.1). To ensure their validity it is necessary that we impose certain restrictions on the behavior of the functions $a_{ij}(x, t, u, p)$ and $a(x, t, u, p)$ for large u and p. This is what we do in deriving estimates of $\max |u_x|$. (Estimates of $\max_{Q_T}|u|$ were given in §2 of Chapter I. For them it is also necessary that certain conditions be fulfilled.)

Estimates of the Hölder constant $\langle u_x\rangle_Q^{(\alpha)}$ in terms of $M = \max_{Q_T}|u|$ and $M_1 = \max_{Q_T}|u_x|$ and certain numerical characteristics of the known functions $a_{ij}(x, t, u, p)$ and $a(x, t, u, p)$ are given for the whole class of nonsingular parabolic equations. We begin the present chapter with a proof of this important fact (§§1 and 2). In §1 we present interior estimates of $\langle u_x\rangle_{Q'}^{(\alpha)}$ for all of the solutions $u(x, t)$ of equations (0.1) that are continuous in Q_T and have u_t from $L_2(Q_T)$ and bounded derivatives u_x from $V_2^{1,0}(Q_T)$. (The class of such generalized solutions u will be denoted by $\mathfrak{M}$.) We are able to obtain these estimates as consequences of the results of §9 of Chapter II on functions of the classes $\mathfrak{B}_2^N$ since, as is shown in §1, the vector function $u_x = (u_{x_1}, \cdots, u_{x_n})$ belongs to the class $\mathfrak{B}_2^{2n}$, the parameters of which can be calculated. In §2 we estimate

$\langle u_x \rangle_{Q_T}^{(\alpha)}$ in the whole cylinder Q_T for u satisfying the first boundary condition.

The following section (§3) is devoted to the derivation of local and total estimates of $\max |u_x|$. For solutions u of equations (0.1), as opposed to equations of the preceding chapter, we have not been able to give an estimate of the Hölder constant $\langle u \rangle_{Q_T}^{(\alpha)}$ (or at least of the modulus of continuity of u with respect to x) in terms of $\max_{Q_T} |u|$ and known quantities, and hence in proving the solvability of a boundary value problem it is necessary that we have an estimate of $\max |u_x|$ in terms of $\max |u|$ (and not in terms of $|u|_{Q_T}^{(\alpha)}$, as was done in Chapter V). In order to obtain such an estimate we have had to impose an additional condition on the functions $a_{ij}(x, t, u, p)$ and $a(x, t, u, p)$. It is removed in §5 for the case of a single space variable. Finally, in §4 we prove some existence theorems.

§1. A PROOF OF THE SMOOTHNESS OF GENERALIZED SOLUTIONS OF CLASS $\mathfrak{M}$ AND AN ESTIMATE OF $\langle u_x \rangle_{Q'}^{(\alpha)}$

Suppose $u(x, t)$ is an arbitrary generalized solution of equation (0.1) of class $\mathfrak{M}$, i.e. it is continuous, has generalized derivatives u_t, u_{xx} from $L_2(Q_T)$, its derivatives u_x are bounded in modulus and continuously depend on t in the norm of $L_2(\Omega)$, and it satisfies equation (0.1) almost everywhere. We will show that its smoothness with respect to (x, t) is actually determined only by the smoothness of the $a_{ij}(x, t, u, p)$ and $a(x, t, u, p)$ as functions of x, t, u and p, as long as equation (0.1) is parabolic with respect to the solution in question. The whole problem here really comes down to a proof of the Hölder continuity of the derivatives u_x, since if this is known, then the rest can be established on the basis of the results of Chapters III and IV on linear equations with smooth coefficients. In fact, it is possible to consider equation (0.1) as a linear equation with smooth coefficients $\hat{a}_{ij}(x, t) = a_{ij}(x, t, u(x, t), u_x(x, t))$ and with free term $\hat{a}(x) = a(x, t, u(x, t), u_x(x, t))$ and to reason in the following way.

If the $a_{ij}(x, t, u, p)$ and $a(x, t, u, p)$ are Hölder-continuous functions of their arguments, then the $\hat{a}_{ij}(x, t)$ and $\hat{a}(x, t)$ will belong to $H^{\alpha, \alpha/2}(Q_T)$ with some $\alpha > 0$, and hence (on the basis of Theorem 12.1 of Chapter III) each solution of the linear equation will actually be an element of $H^{2+\alpha, 1+\alpha/2}(Q_T)$. If the functions $a_{ij}(x, t, u, p)$ and $a(x, t, u, p)$ possess a greater smoothness, then the membership of $u(x, t)$ in $H^{2+\alpha, 1+\alpha/2}(Q_T)$ guarantees a greater smoothness of the functions $\hat{a}_{ij}(x, t)$ and $\hat{a}(x, t)$, and this, by the same Theorem 12.1, means a greater smoothness of any solution of the linear equation, i.e. of the function $u(x, t)$ in question.

In the present section we prove the smoothness of u_x inside Q_T and we obtain an estimate of $\langle u_x\rangle_{Q'}^{(\alpha)}$ for any cylinder $\bar{Q}'\subset\bar{Q}_T$ not intersecting S_T; namely, we prove

Theorem 1.1. *Suppose $u(x, t)$ is a generalized solution from the class $\mathfrak{M}$ of equation (0.1), for which the functions $a_{ij}(x, t, u, p)$ and $a(x, t, u, p)$ are defined, while the $a_{ij}(x, t, u, p)$ are also differentiable with respect to x, u and p in the domain*

$$\{(x, t)\in Q_T,\ |u|\leqslant \max_{Q_T}|u(x, t)|,\ |p|\leqslant \max_{Q_T}|u_x(x, t)|\equiv M_1\}$$

and satisfy the condition

$$\nu\xi^2\leqslant a_{ij}(x, t, u(x, t), u_x(x, t))\xi_i\xi_j\leqslant\mu\xi^2,\quad \nu, \mu=\text{const}>0. \tag{1.1}$$

If for $u=u(x, t)$ and $p=u_x(x, t)$

$$\begin{aligned}\max_{Q_T}\left|\frac{\partial a_{ij}(x, t, u, p)}{\partial p_k}\right|&\leqslant\mu_1,\qquad i, j, k=1, \dots, n,\\ \left|\frac{\partial a_{ij}}{\partial u}, \frac{\partial a_{ij}}{\partial x_k}, a\right|&\leqslant\varphi(x, t),\quad i, j, k=1, \dots, n,\end{aligned} \tag{1.2}$$

where

$$\|\varphi\|_{2q, 2r, Q_T}\leqslant\mu_1, \tag{1.3}$$

and q and r satisfy relations (7.2) of Chapter III, then the derivatives $u_{x_1}, \cdots, u_{x_n}$ are Hölder continuous in Q_T and for any $Q'\subset Q_T$ the quantity $\langle u_x\rangle_{Q'}^{(\alpha)}$ does not exceed a constant c depending only on M_1, n, ν, μ, μ_1, q and r from conditions (1.1)–(1.3) and on the distance from Q' to Γ_T. The quantity $\alpha>0$ is determined by M_1, n, ν, μ, μ_1, q and r.

If the norm $|u_x(x, 0)|_\Omega^{(\beta)}$, $\beta>0$, is bounded, then the estimates $\langle u_x\rangle_{Q'}^{(\alpha)}\leq c$, are also valid for cylinders $Q'=\Omega'\times[0, T]$, where Ω' is an arbitrary interior subdomain of Ω. In this case the constant c depends on M_1, n, ν, μ, μ_1, q and r from (1.1)–(1.3), on the distance from Ω' to S, and on $|u_x(x, 0)|_\Omega^{(\beta)}$ and β, where the exponent $\alpha\leq\beta$.

Without loss in generality we will take the functions u_{x_i}, $i=1, 2, \cdots, n$, to be normalized in the sense that their values do not fall outside of the interval $[0, 1]$: $0\leq u_{x_i}\leq 1$, $i=1, 2, \cdots, n$. Otherwise in equation (0.1) we could replace the functions $u(x, t)$ by

$$v(x, t) = \frac{u(x, t) + M_1 \sum_{i=1}^{n} x_i}{2M_1}$$

and conduct all of our arguments with the function v. It is obvious that v satisfies an equation of the same form as (0. 1) with the same properties of the a_{ij} and a.

Let us show that the vector function $\mathrm{U}(x, t)$ with components $u_{x_1}, \cdots, u_{x_n}$ belongs to the class $\mathfrak{B}_2^{2n}(Q_T, \hat{M}_1, \delta_1, \delta_2, \delta_3, \gamma, \hat{r}, \delta, \kappa)$, the parameters $\hat{M}_1$, γ, $\delta_1, \delta_2, \delta_3, \hat{r}, \delta$ and κ of which are determined only by the quantities M_1, ν, μ, μ_1, q and r. By virtue of Lemma 9.3 of Chapter II it is sufficient for this purpose to prove that the functions

$$w_+^l(x, t) = 10nu_{x_l}(x, t) + \sum_{i=1}^{n} u_{x_i}^2(x, t),$$

$$w_-^l(x, t) = 10n\left[1 - u_{x_l}(x, t)\right] + \sum_{i=1}^{n} u_{x_i}^2(x, t), \quad l = 1, \ldots, n,$$

satisfy for any bounded nonnegative function $\eta(x, t)$ from $\overset{\circ}{W}{}_2^{1,1}(Q_T)$ the inequalities

$$\int_\Omega w(x, t)\eta(x, t)\,dx\Big|_{t_0}^{t} + \int_{t_0}^{t}\int_\Omega \{-w\eta_t + [\hat{a}_{ij}(x, t)w_{x_j} + 2c_i^l(x, t)]\eta_{x_i}\}\,dx\,dt \leqslant \int_{t_0}^{t}\int_\Omega [\gamma w_x^2 + \varphi_2(x, t)]\eta\,dx\,dt, \tag{1.4}$$

$$0 \leqslant t_0 \leqslant t \leqslant T,$$

in which $\hat{a}_{ij}(x, t) = a_{ij}(x, t, u(x, t), u_x(x, t))$, while the $c_i^l(x, t)$ and $\phi_2(x, t)$ are subject to the conditions

$$\|(c_i^l)^2, \varphi_2\|_{q, r, Q_T} \leqslant \mu_2, \quad l, i = 1, \ldots, n. \tag{1.5}$$

Here the meaning of q and r is taken from (1.3). Inequalities (1.4) actually have the same form as inequalities (1.7) of Chapter V, with the same restrictions on the functions $a_i(x, t, u, p) = \hat{a}_{ij}(x, t)p_j + 2c_i^l(x, t)$ and $\phi_2(x, t)$ as in §1 of Chapter V. From them, as was shown in §1 of Chapter V (see Remark 1.3), follow inequalities (7.1) and (7.2) of Chapter II for $w = w_\pm^l$, $l = 1, \cdots, n$, which guarantee the membership of the function $\mathrm{U}(x, t)$ in the class $\mathfrak{B}_2^{2n}$.

Thus, we proceed to a derivation of (1.4). To this end we consider the integral identity

$$-\int_{t_0}^{t}\int_{\Omega}(\mathscr{L}u)\xi_{x_s}\,dx\,dt=0, \qquad s=1,\ldots,n, \tag{1.6}$$

where $\xi(x, t)$ is an arbitrary function from $\overset{\circ}{W}{}_2^{1,1}(Q_T)$. If we assume that the solution $u(x, t)$ is thrice continuously differentiable, we can transform (1.6) by an integration by parts into the form

$$\begin{aligned}0=\int_{t_0}^{t}\int_{\Omega}\Big[u_{x_st}\xi-a_{ij}u_{x_sx_ix_j}\xi-\Big(\frac{\partial a_{ij}}{\partial u_{x_m}}u_{x_mx_s}\\+\frac{\partial a_{ij}}{\partial u}u_{x_s}+\frac{\partial a_{ij}}{\partial x_s}\Big)u_{x_ix_j}\xi-a\xi_{x_s}\Big]\,dx\,dt\\=\int_{t_0}^{t}\int_{\Omega}\Big[u_{x_st}\xi+a_{ij}u_{x_sx_j}\xi_{x_i}\\+\Big(\frac{\partial a_{ij}}{\partial u_{x_m}}u_{x_mx_i}+\frac{\partial a_{ij}}{\partial u}u_{x_i}+\frac{\partial a_{ij}}{\partial x_i}\Big)u_{x_sx_j}\xi\\-\Big(\frac{\partial a_{ij}}{\partial u_{x_m}}u_{x_mx_s}+\frac{\partial a_{ij}}{\partial u}u_{x_s}+\frac{\partial a_{ij}}{\partial x_s}\Big)u_{x_ix_j}\xi-a\xi_{x_s}\Big]\,dx\,dt.\end{aligned} \tag{1.7}$$

We introduce the notation

$$\frac{\partial a_{ij}}{\partial u_{x_m}}-\frac{\partial a_{im}}{\partial u_{x_j}}\equiv a_{ij}^m,\quad \frac{\partial a_{ij}}{\partial u}u_{x_i}+\frac{\partial a_{ij}}{\partial x_i}=a^j,$$

$$-\frac{\partial a_{ij}}{\partial x_s}-\frac{\partial a_{ij}}{\partial u}u_{x_s}=b_{ij}^s.$$

Then the identity (1.7) will appear as follows:

$$\begin{aligned}\int_{t_0}^{t}\int_{\Omega}[u_{x_st}\xi+a_{ij}u_{x_sx_j}\xi_{x_i}+(a_{ij}^m u_{x_mx_i}u_{x_sx_j}+a^ju_{x_sx_j}\\+b_{ij}^s u_{x_ix_j})\xi-a\xi_{x_s}]\,dx\,dt=0,\qquad s=1,\ldots,n.\end{aligned} \tag{1.8}$$

Here we put

$$\xi(x, t)=u_{x_s}(x, t)\,\eta(x, t),$$

where $\eta(x, t)\in\overset{\circ}{W}{}_2^{1,1}(Q_T)$; and we sum the resultant equalities over s from 1 to n. Then, letting

$$v=\sum_{s=1}^{n} u_{x_s}^2,$$

we will have

$$\int_{t_0}^{t}\int_{\Omega}\Big[\frac{1}{2}v_t\eta+a_{ij}u_{x_sx_j}u_{x_sx_i}\eta+\frac{1}{2}a_{ij}v_{x_j}\eta_{x_i}$$
$$+\frac{1}{2}\left(a_{ij}^m v_{x_j}u_{x_mx_i}+a^j v_{x_j}+2b_{ij}^s u_{x_s}u_{x_ix_j}\right)\eta$$
$$-au_{x_sx_s}\eta-au_{x_s}\eta_{x_s}\Big]\,dx\,dt=0. \tag{1.9}$$

We combine this identity with the identity (1.8), in which s is taken equal to l, while $\xi=5n\eta(x,t)$, and we write the result as the following relation with respect to the function $w=w_+^l=10nu_{x_l}+v$:

$$\int_{t_0}^{t}\int_{\Omega}\Big[\frac{1}{2}w_t\eta+a_{ij}u_{x_sx_j}u_{x_sx_i}\eta+\frac{1}{2}a_{ij}w_{x_j}\eta_{x_i}$$
$$+\frac{1}{2}a_{ij}^m w_{x_j}u_{x_mx_i}\eta+\frac{1}{2}a^j w_{x_j}\eta+c_{ij}^l u_{x_ix_j}\eta+c_i^l\eta_{x_i}\Big]\,dx\,dt=0, \tag{1.10}$$

where

$$c_{ij}^l=b_{ij}^s u_{x_s}-a\delta_i^j+5nb_{ij}^l,$$
$$-c_i^l=au_{x_i}+5na\delta_i^l.$$

In the first term of equality (1.10) we carry out an integration by parts; this gives

$$\int_{\Omega} w(x,t)\,\eta(x,t)\,dx\Big|_{t_0}^{t}+\int_{t_0}^{t}\int_{\Omega}[-w\eta_t+2a_{ij}u_{x_sx_j}u_{x_sx_i}\eta$$
$$+(a_{ij}w_{x_j}+2c_i^l)\,\eta_{x_i}]\,dx\,dt$$
$$=-\int_{t_0}^{t}\int_{\Omega}[a_{ij}^m w_{x_j}u_{x_mx_i}+a^j w_{x_j}+2c_{ij}^l u_{x_ix_j}]\,\eta\,dx\,dt. \tag{1.11}$$

The terms in the right side are estimated by inequality (1.2) of Chapter II:

$$|a_{ij}^m w_{x_j}u_{x_mx_i}\eta|\leqslant|\eta|\left(\varepsilon u_{x_mx_i}^2+\frac{1}{4\varepsilon}|a_{ij}^m|^2 w_{x_j}^2\right),$$
$$|a^j w_{x_j}\eta|\leqslant\left(\frac{1}{2}w_{x_j}^2+\frac{1}{2}|a^j|^2\right)|\eta|,$$
$$|c_{ij}^l u_{x_ix_j}\eta|\leqslant\left(\varepsilon u_{x_ix_j}^2+\frac{1}{4\varepsilon}|c_{ij}^l|^2\right)|\eta|.$$

For $\epsilon = \nu/n$ and $\eta \geq 0$ inequalities (1.4) for the functions $w = w^l_+$, $l = 1, \cdots, n$, will follow from these estimates and (1.11) by virtue of assumptions (1.1)–(1.3). In this connection in (1.4)

$$\gamma = \frac{n}{4\nu} \max_{Q_T} \sum_{l,\,j,\,m} (a^m_{lj})^2 + \frac{1}{2},$$

$$\varphi_2(x, t) = \frac{1}{2} \sum_{j=1}^{n} (a^j)^2 + \frac{n}{4\nu} \sum_{l,\,j=1} (c^l_{lj})^2, \ a\ \eta(x, t),$$

and $\eta(x, t)$ is an arbitrary bounded nonnegative function from $\overset{\circ}{W}{}^{1,1}_2(Q_T)$.

In deriving the identity (1.11) and inequalities (1.4) following from it we have made use of the existence of the derivatives u_{xt}, $D^3_x u$ of $u(x, t)$. These derivatives do not appear in the identity (1.11) itself, nor in the original equality (1.6); the integrals there make sense for arbitrary functions ξ and bounded $\eta \geq 0$ from $\overset{\circ}{W}{}^{1,1}_2(Q_T)$ if $u(x, t)$ possesses only those properties which are prescribed by the conditions of Theorem 1.1.

We will show that (1.11), and hence (1.4), is also valid for any such solution. To this end we approximate $u(x, t)$ by thrice continuously differentiable functions $u^p(x, t)$, $p = 1, 2, \cdots$, in such a way that $0 \leq u^p_{x_i} \leq 1$, the functions u^p converge to u uniformly in Q_T, the derivatives $u^p_{x_i}$ converge to u_{x_i}, $i = 1, \cdots, n$, almost everywhere in Q_T, and the derivatives u^p_t and $u^p_{x_i x_j}$ converge to u_t and $u_{x_i x_j}$, $i, j = 1, \cdots, n$, respectively in $L_2(Q_T)$. From the u^p we compose the expressions

$$\mathcal{L}_p(u^p) = u^p_t \quad a_{ij}(x, t, u, u_x)\, u^p_{x_i x_j} + a(x, t, u, u_x).$$

It is clear that $\|\mathcal{L}_p(u^p) - \mathcal{L}u\|_{2, Q_T} \to 0$ for $p \to \infty$. In accordance with (1.6) we consider the identity

$$\int_{t_0}^{t} \int_{\Omega} [u^p_t - a_{ij}(x, t, u, u_x)\, u^p_{x_i x_j} + a(x, t, u, u_x)]\, \xi_{x_s}\, dx\, dt = \int_{t_0}^{t} \int_{\Omega} \mathcal{L}_p(u^p)\, \xi_{x_s}\, dx\, dt$$

and we transform its left side into a form similar to (1.8):

$$\int_{t_0}^{t} \int_{\Omega} \Big[u^p_{x_s t}\xi + a_{ij} u^p_{x_s x_j} \xi_{x_i} + \frac{\partial a_{ij}}{\partial u_{x_m}} (u_{x_m x_i} u^p_{x_s x_j} - u_{x_m x_s} u^p_{x_i x_j})\xi + (a^j u^p_{x_s x_j} + b^s_{ij} u^p_{x_i x_j})\xi - a\xi_{x_s} \Big] dx\, dt = \int_{t_0}^{t} \int_{\Omega} \mathcal{L}_p(u^p)\, \xi_{x_s}\, dx\, dt.$$

From here we deduce relations similar to (1.9), (1.10) and (1.11). The latter of these relations will appear as follows:

$$\int_\Omega w^p(x,t)\eta(x,t)\,dx\Big|_{t_0}^{t} + \int_{t_0}^{t}\int_\Omega [-w^p\eta_t + 2a_{ij}u^p_{x_s x_i}u^p_{x_s x_j}\eta$$
$$+ (a_{ij}w^p_{x_j} + 2c_i^{pl})\eta_{x_i}]\,dx\,dt$$
$$= -\int_{t_0}^{t}\int_\Omega \Big[\frac{\partial a_{ij}}{\partial u_{x_m}} u_{x_m x_i} w_{x_j} - 2u_{x_m x_s} u^p_{x_j x_i} u^p_{x_s}$$
$$- 10nu_{x_m x_l} u^p_{x_l x_j} + a^j w^p_{x_j} + 2c^{pl}_{ij} u^p_{x_i x_j}\Big]\eta\,dx\,dt$$
$$+ \int_{t_0}^{t}\int_\Omega \mathscr{L}_p(u^p)[(u^p_{x_s}\eta)_{x_s} + 5n\eta_{x_l}]\,dx\,dt. \qquad (1.12)$$

Here

$$w^p = w^{pl}_+ = 10nu^p_{x_l} + \sum_{i=1}^{n}(u^p_{x_i})^2,$$
$$-c_i^{pl} = au^p_{x_l} + 5na\delta_i^l,$$
$$c^{pl}_{ij} = -\left(\frac{\partial a_{ij}}{\partial x_s} + \frac{\partial a_{ij}}{\partial u}u_{x_s}\right)u^p_{x_s} - a\delta^l_j - 5n\left(\frac{\partial a_{ij}}{\partial x_l} + \frac{\partial a_{ij}}{\partial u}u_{x_l}\right).$$

If we pass to the limit in (1.12) as $p \to \infty$, we obtain equality (1.11). The possibility of such a passage can be easily proved if we take into account that vrai $\max_{Q_T}|u^p_x| \le 1$, the $u^p_{x_i}$ converge almost everywhere in Q_T to u_{x_i}, and $\|u^p_{x_ix_j} - u_{x_ix_j}\|_{2,Q_T} \to 0$ for $p \to \infty$. Let us consider, by way of illustration, one of the terms in (1.12):

$$j_p \equiv \int_{t_0}^{t}\int_\Omega \frac{\partial a_{ij}}{\partial u_{x_m}} u_{x_m x_s} u^p_{x_i x_j} u^p_{x_s}\eta\,dx\,dt$$

and show that

$$j_p \to j \equiv \int_{t_0}^{t}\int_\Omega \frac{\partial a_{ij}}{\partial u_{x_m}} u_{x_m x_s} \quad u_{x_i x_j} u_{x_s}\eta\,dx\,dt$$

for $p \to \infty$. Indeed, by virtue of (1.2)

$$|j - j_p| \leqslant c\,\|u_{xx}\|_{2,Q_T}\sum_{i,j,s}\|u^p_{x_i x_j}u^p_{x_s} - u_{x_i x_j}u_{x_s}\|_{2,Q_T}.$$

We take an arbitrary ϵ and, corresponding to it by Egorov's theorem, a set Q_T^ϵ with mes $Q^\epsilon_T < \epsilon$ such that on $Q_T \backslash Q^\epsilon_T$ the functions $u^p_{x_i}$ converge uniformly to u_{x_i}, $i = 1, \cdots, n$. It is clear that the norm

$$\| u^p_{x_i x_j} u^p_{x_s} - u_{x_i x_j} u_{x_s} \|_{2, Q_T}$$

does not exceed the sum of the norms

$$\| (u^p_{x_i x_j} - u_{x_i x_j}) u^p_{x_s} \|_{2, Q_T} + \| u_{x_i x_j} (u^p_{x_s} - u_{x_s}) \|_{2, Q_T \setminus Q_T^\varepsilon}$$

$$+ \| u_{x_i x_j} (u^p_{x_s} - u_{x_s}) \|_{2, Q_T^\varepsilon} \leqslant \| u^p_{xx} - u_{xx} \|_{2, Q_T}$$

$$+ \| u_{xx} \|_{2, Q_T} \max_{Q_T \setminus Q_T^\varepsilon} | u^p_{x_s} - u_{x_s} | + 2 \| u_{xx} \|_{2, Q_T^\varepsilon},$$

where in the right side the first two summands tend to zero for $p \to \infty$, while the last summand can be made arbitrarily small by the choice of ϵ. From what has been said it follows that $|j_p - j| \to 0$ for $p \to \infty$.

The fact that one can pass to the limit in the remaining terms of (1.12) as $p \to \infty$ is verified analogously.

Thus it is proved that equalities (1.11), and hence inequalities (1.4) with $\eta \geq 0$, hold for the solutions $u(x, t)$ of equation (0.1) considered in Theorem 1.1 and the functions

$$w = w^l_+ = 10 n u_{x_l} + \sum_{i=1}^n u^2_{x_i}, \quad l = 1, \ldots, n,$$

constructed with respect to them.

In completely the same way we can establish that inequalities (1.4) with $\eta \geq 0$ are satisfied by the functions

$$w = w^l_- = 10 n (1 - u_{x_l}) + \sum_{i=1}^n u^2_{x_i}, \quad l = 1, \ldots, n.$$

As was noted above, the existence of inequalities (1.4) with $\eta(x, t) \geq 0$ for $w = w^l_\pm$, $l = 1, \cdots, n$, permits us to conclude on the basis of Lemma 9.3 of Chapter II that the vector function $U(x, t) = (u_{x_1}(x, t), \cdots, u_{x_n}(x, t))$ belongs to the class $\mathfrak{B}_2^{2n}(Q_T, \hat{M}_1, \delta_1, \delta_2, \delta_3, \gamma, \hat{r}, \delta, \kappa)$, where $\hat{M}_1 = 11n$, $\delta_1 = 16n$, $\delta_2 = 1/4$, $\delta_3 = 1/2$ and the parameters γ, $\hat{r}$, δ and κ are determined by the quantities γ, q, r and μ_2 from inequalities (1.4) and (1.5). It follows by virtue of Theorems 9.1 and 9.2 of Chapter II that the assertions of Theorem 1.1 are valid.

Instead of reverting to results concerning functions of the classes $\mathfrak{B}_2^{N_1}$ we can take advantage of Lemma 10.2 of Chapter V in order to conclude from inequalities (1.4) and (1.5) that the assertions of Theorem 1.1 concerning an estimate of $\langle u_x\rangle^{(\alpha)}_{Q'}$ are valid. In fact, inequalities (1.4) are a special case of inequalities (10.1) of Chapter V (or, equivalently, inequalities (1.7) of Chapter V). For them $\mu_1 = \gamma$ and the functions $a_i(x, t, u, p) = a_{ij}(x, t, u, p)\, p_i + 2c_i^l$ and $\phi_2(x, t)$ satisfy by virtue of (1.5) and (1.1) the requirements of Lemma 10.2 of Chapter V. In Q_T we take an arbitrary cylinder $Q_{2R} = K_{2R} \times [t_0 - \theta R^2, t_0] \subset Q_T$ and the coaxial cylinder $Q_R = K_R \times [t_0 - \theta R^2/4, t_0]$, where θ is a number determined by n, M_1, ν, μ_1, r and q in accordance with Lemma 7.1 of Chapter II. Among the functions $u_{x_i}(x, t)$ we select that one which has the greatest oscillation in Q_{2R}; let it be u_{x_r}, so that $\omega^r = \max_{i=1,\dots,n} \omega^i = \max_i \operatorname{osc}\{u_{x_i}; Q_{2R}\}$. According to Lemma 9.3 of Chapter II,

$$\operatorname{osc}\{w^r_{\pm}, Q_{2R}\} \geqslant \delta_1 \omega^r, \; \delta_1 = \frac{8N_1}{M_1},$$

and the inequality

$$\operatorname{mes}\{x \in K_R : w^r(x, t_0 - \theta R^2) \leqslant \max_{Q_{2R}} w^r - \delta_2 \operatorname{osc}\{w^r, Q_{2R}\}\} \geqslant \delta_3 \operatorname{mes} K_R$$

is fulfilled for at least one of the functions $w^r = w^r_{\pm}$.

Therefore this function satisfies all of the requirements of Lemma 10.2 of Chapter V, and hence also the inequality

$$\operatorname{osc}\{w^r, Q_R\} \leqslant (1 - \delta_0) \operatorname{osc}\{w^r, Q_{2R}\} + R^{\varkappa_1} \tag{1.13}$$

with $\delta_0 > 0$ and $\kappa_1 > 0$ depending only on δ_2, δ_3 and the quantities in inequality (1.4). The desired estimate for $\langle u_x\rangle^{(\alpha)}_{Q'}$, $Q' \subset Q_T$, follows from (1.13) and Lemma 5.9 of Chapter II.

§2. AN ESTIMATE OF $\langle u_x\rangle^{(\alpha)}_{Q_T}$

In the preceding section estimates of $\langle u_x\rangle^{(\alpha)}_{Q'}$ were obtained for any interior cylinder $Q' = \Omega' \times (\epsilon, T), \epsilon \geq 0$, $\bar{\Omega}' \subset \Omega$. In order to prove a theorem on the solvability of a boundary value problem we must have estimates of $\langle u_x\rangle^{(\alpha)}_{Q_T}$ in the whole domain Q_T. This will be done according to the same plan that was used in the case of equations with principal part in divergence form in §5 of Chapter V. Namely, we will first establish the inequality $\max_{Q_T}|u_t| \leq c$, and then consider $u(x, t)$ as a solution of the elliptic equation

$$a_{ij}(x, t, u, u_x)\, u_{x_i x_j} + f(x, t) = 0 \tag{2.1}$$

with the free term $f(x, t) = -a(x, t, u(x, t), u_x(x, t)) - u_t(x, t)$ depending on t as on a parameter. The following assertion holds for the solutions of such equations (see Theorem 1.1 of Chapter VI in [65q]).

Theorem 2.1. *Suppose the function* $u(x)$ *belongs to* $W_2^2(\Omega)$, *has* vrai $\max_\Omega|u_x| \le M_1$ *and satisfies almost everywhere in* Ω *equation* (2.1) *(in which* t *is fixed:* $t=t_0$). *If the conditions*

$$\nu\xi^2 \leqslant a_{ij}(x, t_0, u(x), u_x(x))\xi_i\xi_j \leqslant \mu\xi^2, \quad \nu > 0,$$

$$\max_\Omega\left|\frac{\partial a_{ij}(x, t_0, u(x), u_x(x))}{\partial u_{x_k}}, \frac{\partial a_{ij}}{\partial u}, \frac{\partial a_{ij}}{\partial x_k}, f\right| \leqslant \mu_1$$

are fulfilled and if $u(x)|_S = \psi(x)|_S$, *where* $\psi(x)\in W_q^2(\Omega)$ *with* $q>n$, *and* $S\subset O^2$, *then* $u(x)\in H^{1+\beta}(\overline{\Omega})$, $\beta>0$, *and the quantity* $|u_x|_\Omega^{(\beta)}$ *is estimated from above by a constant determined only by* ν, μ, μ_1, M_1, q, $\|\psi\|_{q,\Omega}^{(2)}$ *and the boundary* S.

Theorem 2.1 gives an estimate of $\langle u_x\rangle_{x,\Omega}^{(\beta)}$ in terms of known quantities and $\max_{Q_T}|u_t|$, which together with the inequality $\max_{Q_T}|u_t|\le c$ permits one to estimate $\langle u_x\rangle_{Q_T}^{(\alpha)}$ (see Lemma 3.1 of Chapter II). We will first obtain an a priori estimate of $\max_{Q_T}|u_t|$. We assume that $u(x,t)$ is a classical solution of equation (0.1) and that $M_1=\max_{Q_T}|u_x|$ is already known to us. We further suppose that $\max_{\Gamma_T}|u_t|$ is known.

We will assume that the functions $a_{ij}(x,t,u,p)$ and $a(x,t,u,p)$ are continuous and continuously differentiable with respect to all of their arguments in a neighborhood of the manifold $\{(x,t)\in\overline{Q}_T,\ u=u(x,t),\ p=u_x(x,t)\}$. In order to obtain an estimate of $\max_{Q_T}|u_t|$ it is sufficient to estimate the maximum in Q_T of the expressions $v=u_t+\lambda\sum_{i=1}^n u_{x_i}^2$ and $w=-u_t+\lambda\sum_{i=1}^n u_{x_i}^2$, where $\lambda\ge 1$ is a certain constant. Let us consider, for example, $v(x,t)$.

If the function $u(x,t)$ has not only continuous derivatives of the form u_t and $u_{x_ix_j}$ but also the derivatives u_{tx_i}, D_x^3u, it is then possible to proceed as follows. We differentiate equation (0.1) with respect to t and x_l, $l=1,\cdots,n$, and form the sum

$$(\mathscr{L}u)_t+\lambda\sum_{l=1}^n(\mathscr{L}u)_{x_l}2u_{x_l}+\mathscr{L}u=0.$$

The result is written in the form of the following relation for v:

$$v_t - a_{ij}(x, t, u, u_x)v_{x_ix_j} + 2\lambda a_{ij}u_{x_lx_i}u_{x_lx_j}$$
$$+\left(-\frac{\partial a_{ij}}{\partial u_{x_m}}u_{x_ix_j}+\frac{\partial a}{\partial u_{x_m}}\right)v_{x_m}-\left(\frac{\partial a_{ij}}{\partial u}u_t+\frac{\partial a_{ij}}{\partial t}\right.$$
$$\left.+\frac{\partial a_{ij}}{\partial u}2\lambda\sum_{l=1}^n u_{x_l}^2+\frac{\partial a_{ij}}{\partial x_l}2\lambda u_{x_l}\right)u_{x_ix_j}+\frac{\partial a}{\partial u}u_t+\frac{\partial a}{\partial t}+$$

$$+\lambda\frac{\partial a}{\partial u}2\sum_{l=1}^{n}u_{x_l}^2+2\lambda\frac{\partial a}{\partial x_l}u_{x_l}+v-a_{ij}u_{x_ix_j}+a-\lambda\sum_{l=1}^{n}u_{x_l}^2=0.$$

We multiply this equality by $v-k$ and integrate the result first with respect to $x\in A_k(t)$, where $A_k(t)$ is the set of points $x\in\Omega$ at which $v(x,t)>k$, and then with respect to t from 0 to T. The number k is assumed to be positive and greater than $\max_{\Gamma_T} v$.

The term

$$J_1=\int_0^T\int_{A_k(t)}-a_{ij}v_{x_ix_j}(v-k)\,dt\,dt$$

is transfomed by means of an integration by parts into the form

$$J_1=\int_0^T\int_{A_k(t)}\left[a_{ij}v_{x_i}v_{x_j}+\left(\frac{\partial a_{ij}}{\partial u_{x_l}}u_{x_lx_i}+\frac{\partial a_{ij}}{\partial u}u_{x_i}+\frac{\partial a_{ij}}{\partial x_i}\right)v_{x_j}(v-k)\,dx\,dt.\right. \tag{2.2}$$

Then we obtain

$$\frac{1}{2}\int_{A_k(T)}[v(x,T)-k]^2\,dx+\int_0^T\int_{A_k(t)}[a_{ij}v_{x_i}v_{x_j}+2\lambda a_{ij}u_{x_lx_i}u_{x_lx_j}(v-k)$$

$$+v(v-k)]\,dx\,dt=-\int_0^T\int_{A_k(t)}\left\{\left(\frac{\partial a_{ij}}{\partial u_{x_l}}u_{x_lx_i}\right.\right.$$

$$\left.+\frac{\partial a_{ij}}{\partial u}u_{x_i}+\frac{\partial a_{ij}}{\partial x_i}\right)v_{x_j}+\left(-\frac{\partial a_{ij}}{\partial u_{x_l}}u_{x_ix_j}+\frac{\partial a}{\partial u_{x_l}}\right)v_{x_l}$$

$$-\left[\frac{\partial a_{ij}}{\partial u}(a_{lm}u_{x_lx_m}-a)+\frac{\partial a_{ij}}{\partial t}+\frac{\partial a_{ij}}{\partial u}2\lambda\sum_{l=1}^{n}u_{x_l}^2\right.$$

$$\left.+\frac{\partial a_{ij}}{\partial x_l}2\lambda u_{x_l}\right]u_{x_ix_j}+\frac{\partial a}{\partial u}(a_{ij}u_{x_ix_j}-a)+\frac{\partial a}{\partial t}$$

$$+2\lambda\frac{\partial a}{\partial u}\sum_{l=1}^{n}u_{x_l}^2+2\lambda\frac{\partial a}{\partial x_l}u_{x_l}-a_{ij}u_{x_ix_j}+a$$

$$\left.-\lambda\sum_{l=1}^{n}u_{x_l}^2\right\}(v-k)\,dx\,dt. \tag{2.3}$$

We estimate the right side in (2.3) from above, taking into account that $M_1 = \max_{Q_T} |u_x|$ is known and that we will take λ and k sufficiently large in the sequel. From (2.3) we have

$$\int_0^T \int_{A_k(t)} [a_{ij} v_{x_i} v_{x_j} + 2\lambda a_{ij} u_{x_l x_i} u_{x_l x_j} (v-k) + v(v-k)]\, dx\, dt$$
$$\leqslant c \int_0^T \int_{A_k(t)} [|u_{xx}||v_x| + |v_x| + |u_{xx}|^2 + \lambda |u_{xx}| + \lambda](v-k)\, dx\, dt$$
$$\leqslant c \int_0^T \int_{A_k(t)} \Big[\varepsilon v_x^2 + \frac{1}{4\varepsilon} |u_{xx}|^2 (v-k)^2 + \varepsilon v_x^2 + \frac{1}{4\varepsilon}(v-k)^2 + 2|u_{xx}|^2(v-k) + \frac{1}{4}\lambda^2(v-k) + \lambda(v-k)\Big]\, dx\, dt.$$

The constant c is determined by the quantities M_1 and

$$\max_{i,j,k=1,\ldots,n} \max_{Q_T} \left| a_{ij}(x,t,u(x,t),u_x(x,t)),\ a,\ \frac{\partial a_{ij}}{\partial u_{x_k}},\ \frac{\partial a}{\partial t},\ \frac{\partial a_{ij}}{\partial u},\ \frac{\partial a_{ij}}{\partial t},\ \frac{\partial a_{ij}}{\partial x_k},\ \frac{\partial a}{\partial u_{x_k}},\ \frac{\partial a}{\partial u},\ \frac{\partial a}{\partial x_k} \right| = M_2. \tag{2.4}$$

We take ϵ and λ so that

$$2c = \lambda \nu, \quad 2c\varepsilon = \frac{\nu}{2}.$$

Then, transferring the terms with v_x^2 and $u_{xx}^2(v-k)$ to the left, we obtain

$$\int_0^T \int_{A_k(t)} \Big[\frac{\nu}{2} v_x^2 + \lambda\nu |u_{xx}|^2 (v-k) + v(v-k)\Big]\, dx\, dt$$
$$\leqslant \frac{c}{4\varepsilon} \int_0^T \int_{A_k(t)} [|u_{xx}|^2 (v-k)^2 + (v-k)(v-k+\varepsilon\lambda^2 + 4\varepsilon\lambda)]\, dx\, dt. \tag{2.5}$$

The constants c, ϵ and λ are now fixed in their dependence on M_1 and M_2. The parameter k on the other hand is subject to only the one requirement

$$k > M_0 = \max_{\Gamma_T} v.$$

Let k_{max} be the maximum of v in Q_T and let it be greater than M_0. In (2.5) we take $k = k_{max} - \delta$, where $\delta = 4\epsilon\nu\lambda/c$. For such k inequality (2.5) implies

$$\int_0^T \int_{A_k(t)} v(v - k)\,dx\,dt \leqslant c \int_0^T \int_{A_k(t)} (v - k)\left(\frac{\delta}{4\varepsilon} + \frac{1}{4}\lambda^2 + \lambda\right) dx\,dt,$$

or, equivalently,

$$\int_0^T \int_{A_k(t)} (v - k)\left[v - c\left(\frac{\delta}{4\varepsilon} + \frac{\lambda^2}{4} + \lambda\right)\right] dx\,dt \leqslant 0,$$

$$k = k_{max} - \delta.$$

We only decrease the left side if we replace the v in brackets by $k_{max} - \delta$:

$$\left[k_{max} - \delta - c\left(\frac{\delta}{4\varepsilon} + \frac{\lambda^2}{4} + \lambda\right)\right] \int_0^T \int_{A_k(t)} (v - k)\,dx\,dt \leqslant 0,$$

i.e. we obtain

$$k_{max} \leqslant \delta + c\left(\frac{\delta}{4\varepsilon} + \frac{\lambda^2}{4} + \lambda\right).$$

Finally, we can write

$$\max_{Q_T} v \leqslant \max\{M_0;\ c_1\}, \tag{2.6}$$

where

$$c_1 = 2\frac{\nu}{c} + 2c + \frac{c^3}{\nu^2} + \frac{2c^2}{\nu}.$$

This is the desired estimate of v and, consequently, of u_t from above. The positive maximum of the function $w = -u_t + \lambda u_x^2$, which gives an estimate of u_t from below, is estimated analogously.

But all of our arguments are legitimate only for u possessing the derivatives u_{tx}, $D_x^3 u$ and the derivatives subordinate to them, even though the identity (2.3), from which we obtained estimate (2.6), contains only the derivatives $D_x^2 u$, u_{xt}, u_x.

Let us show that (2.3) is also valid for solutions of equation (0.1), when it is known only that they belong to $C^{2,1}(\overline{Q}_T)$. Such solutions have the generalized derivatives u_{tx}, and these derivatives belong to $L_2(Q_T)$. Indeed, we form a divided difference with respect to t from both sides of equation (0.1), then we multiply both sides of this equality by the divided difference $\Delta u/\Delta t$ from $u(x, t)$ and integrate the result over Q_{t_1}, $t_1 \leq T - \Delta t$. We obtain, by taking into consider-ration formulas (4.7)–(4.9) of Chapter II,

$$0=\int_0^{t_1}\int_\Omega \frac{\Delta(\mathscr{L}u)}{\Delta t}\frac{\Delta u}{\Delta t}\,dx\,dt=\int_0^{t_1}\int_\Omega\Big(\frac{\partial}{\partial t}\frac{\Delta u}{\Delta t}-a_{ij}\frac{\Delta u_{x_ix_j}}{\Delta t}$$

$$-u_{x_ix_j}(x,\,t+\Delta t)\frac{\Delta a_{ij}}{\Delta t}+\frac{\Delta a}{\Delta t}\Big)\frac{\Delta u}{\Delta t}\,dx\,dt$$

$$=\int_0^{t_1}\int_\Omega\Big\{\frac12\frac{\partial}{\partial t}\Big(\frac{\Delta u}{\Delta t}\Big)^2+a_{ij}\frac{\Delta u_{x_i}}{\Delta t}\frac{\Delta u_{x_j}}{\Delta t}+\frac{\Delta u_{x_j}}{\Delta t}\frac{\Delta u}{\Delta t}\frac{da_{ij}}{dx_i}$$

$$+\Big[-u_{x_ix_j}(x,\,t+\Delta t)\Big(\frac{\partial\hat a_{ij}}{\partial u_{x_l}}\frac{\Delta u_{x_l}}{\Delta t}+\frac{\partial \hat a_{ij}}{\partial u}\frac{\Delta u}{\Delta t}+\frac{\partial\hat a_{ij}}{\partial t}\Big)$$

$$+\Big(\frac{\partial\hat a}{\partial u_{x_l}}\frac{\Delta u_{x_l}}{\Delta t}+\frac{\partial\hat a}{\partial u}\frac{\Delta u}{\Delta t}+\frac{\partial\hat a}{\partial t}\Big)\Big]\frac{\Delta u}{\Delta t}\Big\}\,dx\,dt,$$

from which, by making use of the boundedness of $|u_x|$ and $|u_{xx}|$, we deduce

$$\frac12\int_\Omega\Big(\frac{\Delta u}{\Delta t}\Big)^2dx\,\Big|_0^{t_1}+\int_0^{t_1}\int_\Omega a_{ij}\Big(\frac{\Delta u}{\Delta t}\Big)_{x_i}\Big(\frac{\Delta u}{\Delta t}\Big)_{x_j}dx\,dt$$

$$\leqslant c\int_0^{t_1}\int_\Omega\Big[\Big|\frac{\Delta u}{\Delta t}\Big|\Big|\Big(\frac{\Delta u}{\Delta t}\Big)_x\Big|+\Big|\Big(\frac{\Delta u}{\Delta t}\Big)_x\Big|+\Big|\frac{\Delta u}{\Delta t}\Big|+\Big|\frac{\Delta u}{\Delta t}\Big|^2\Big]dx\,dt$$

$$\leqslant c\int_0^{t_1}\int_\Omega\Big[\varepsilon\Big|\Big(\frac{\Delta u}{\Delta t}\Big)_x\Big|^2+\Big(\frac1\varepsilon+1\Big)\Big|\frac{\Delta u}{\Delta t}\Big|^2+\Big(\frac1{2\varepsilon}+\varepsilon\Big)\Big]dx\,dt,$$

where ϵ is any number from the interval (0, 1). For $\epsilon=\nu/2c$ we obtain

$$2\int_\Omega\Big(\frac{\Delta u}{\Delta t}\Big)^2dx\,\Big|_0^{t_1}+\nu\int_0^{t_1}\int_\Omega\Big|\Big(\frac{\Delta u}{\Delta t}\Big)_x\Big|^2dx\,dt\leqslant c'\int_0^{t_1}\int_\Omega\Big(\frac{\Delta u}{\Delta t}\Big)^2dx\,dt+c'.$$

From this, on the basis of Lemma 5.5 of Chapter II, we deduce the uniform boundedness of the integral

$$\int_0^{t_1}\int_\Omega\Big|\Big(\frac{\Delta u}{\Delta t}\Big)_x\Big|^2dx\,dt\leqslant\text{const}$$

for all Δt and $t_1\le T-\Delta t$. Inequality (2.7) guarantees the existence of the generalized derivatives u_{x_it} of u and their square summability on Q_T.

In order to derive a relation of type (2.3) we take the function

$$v_\Delta = \frac{\Delta u}{\Delta t} + \lambda \sum_{l=1}^{n} \left(\frac{\Delta u}{\Delta x_l}\right)^2, \quad \Delta x_l = \Delta t = \Delta,$$

and proceed with it in the same way as above with the function v. Namely, we consider the relation

$$\frac{\Delta}{\Delta t}(\mathscr{L}u) + \lambda \sum_{l=1}^{n} \frac{\Delta(\mathscr{L}u)}{\Delta x_l} 2 \frac{\Delta u}{\Delta x_l} + \mathscr{L}u = 0 \tag{2.8}$$

in the domain $Q_{T,\Delta} = \Omega_\Delta \times [0, T-\Delta]$, where Ω_Δ still belongs to Ω after a translation by Δ along any of the axes. This relation can be transformed into a form similar to (2.3). Indeed, by virtue of (4.7)–(4.9) of Chapter II the equality (2.8) can be written as follows:

$$\begin{aligned} v_{\Delta t} &- a_{ij}(x, t, u(x, t), u_x(x, t)) v_{\Delta x_i x_j} \\ &+ 2\lambda a_{ij} \frac{\Delta u_{x_l}}{\Delta x_l} \frac{\Delta u_{x_j}}{\Delta x_l} - u_{x_i x_j}(x, t+\Delta t) \frac{\Delta a_{ij}}{\Delta t} + \frac{\Delta a}{\Delta t} \\ &- \lambda u_{x_i x_j}(x + \Delta x_l, t) \frac{\Delta a_{ij}}{\Delta x_l} 2 \frac{\Delta u}{\Delta x_l} + \frac{\Delta a}{\Delta x_l} 2\lambda \frac{\Delta u}{\Delta x_l} + \mathscr{L}u = 0. \end{aligned} \tag{2.9}$$

Let

$$M_0 = \max_{\Gamma_T} |v|.$$

By virtue of the continuity of $v(x, t)$ the set of points (x, t) from $\overline{Q}_T$ at which $v(x, t) > M_0 + 1$ is contained in a cylinder $Q_T^{(h)}$ whose distance from Γ_T is equal to $h > 0$. For sufficiently small $\Delta (\Delta < h)$ the functions v_Δ are defined in $Q_{T-\Delta}^{(h)}$ and the maximum of v_Δ on the lower base and lateral surface of $Q_{T-\Delta}^{(h)}$ does not exceed $M_0 + 2$.

We multiply (2.9) by $v_\Delta - k$, where $k \geq M_0 + 2$, and integrate the result first over the set $A_k^\Delta(t)$ of points x from Ω_Δ at which $v_\Delta(x, t) > k$, and then with respect to t from 0 to $t_1 \leq T - \Delta$. The second term is replaced in accordance with equality (2.2). This leads to a relation similar to (2.3):

$$\frac{1}{2} \int_{A_k^\Delta(t_1)} [v_\Delta(x, t_1) - k]^2 \, dx$$

$$+ \int_0^{t_1} \int_{A_k^\Delta(t)} \left[a_{ij} (v_\Delta)_{x_i} (v_\Delta)_{x_j} + 2\lambda a_{ij} \frac{\Delta u_{x_l}}{\Delta x_l} \frac{\Delta u_{x_j}}{\Delta x_l} (v_\Delta - k) + \mathscr{L}u (v_\Delta - k) \right] dx \, dt =$$

$$= -\int_0^{t_1}\int_{A_k^\Delta(t)} \Big[\frac{da_{lj}}{dx_l}(v_\Delta)_{x_j} - u_{x_l x_j}(x, t+\Delta t)\frac{\Delta a_{lj}}{\Delta t} + \frac{\Delta a}{\Delta t}$$

$$- \lambda u_{x_l x_j}(x+\Delta x_l, t)\frac{\Delta a_{lj}}{\Delta x_l} 2\frac{\Delta u}{\Delta x_l} + \frac{\Delta a}{\Delta x_l} 2\lambda \frac{\Delta u}{\Delta x_l}\Big](v_\Delta - k)\,dx\,dt. \quad (2.10)$$

We will pass to the limit in this equality by letting Δx_l and Δt tend to zero. The divided differences $\Delta u/\Delta x_l$, $\Delta u/\Delta t$, $\Delta u_{x_i}/\Delta x_l$ tend uniformly in $Q_{t_1}^{(h)}$ to $\partial u/\partial x_l$, $\partial u/\partial t$, $\partial^2 u/\partial x_i \partial x_l$ respectively, while the $\Delta u_{x_i}/\Delta t$ tend to $\partial^2 u/\partial x_i \partial t$ in $L_2(Q_t^{(h)})$ (see Lemma 4.10 of Chapter II). Consequently, the v_Δ converge uniformly in $Q_t^{(h)}$ to v, while the $(v_\Delta)_{x_i}$ converge to v_{x_i} in $L_2(Q_{t_1}^{(h)})$. By virtue of inequality (4.1) and Lemma 4.3 of Chapter II we conclude that the continuous functions

$$v_\Delta^{(k)}(x, t) = \max\{v_\Delta(x, t) - k;\ 0\}$$

tend uniformly in $Q_{t_1}^{(h)}$ to the continuous function

$$v^{(k)}(x, t) = \max\{v(x, t) - k;\ 0\}$$

and

$$\Big\|\big(v_\Delta^{(k)}\big)_{x_l} - v_{x_l}^{(k)}\Big\|_{2, Q_{t_1}^{(h)}} \to 0, \quad i = 1, \ldots, n, \quad \text{for} \quad \Delta \to 0.$$

All of the integrals in (2.10) are extended from $A_k^\Delta(t)$ onto the whole domain Ω by taking in place of $(v_\Delta - k)$ and $(v_\Delta)_{x_i}$ in them the functions $v_\Delta^{(k)}$ and $(v_\Delta^{(k)})_{x_i}$ respectively, so that the domain of integration will now be independent of Δ. From what has been said it follows that we can pass to the limit in all of the integrals in (2.10) as $\Delta t = \Delta x_l = \Delta \to 0$. After such a passage to the limit we obtain

$$\frac{1}{2}\int_{A_k(t_1)} [v(x, t_1) - k]^2\,dx$$

$$+ \int_0^{t_1}\int_{A_k(t)} [a_{ij} v_{x_i} v_{x_j} + 2\lambda a_{ij} u_{x_l x_i} u_{x_l x_j}(v-k) + \mathscr{L}u(v-k)]\,dx\,dt$$

$$= -\int_0^{t_1}\int_{A_k(t)} \Big[\frac{da_{lj}}{dx_l} v_{x_j} - u_{x_l x_j}\frac{da_{lj}}{dt} + \frac{da}{dt} - \lambda u_{x_l x_j}\frac{da_{lj}}{dx_l} 2u_{x_l} + \frac{da}{dx_l} 2\lambda u_{x_l}\Big](v-k)\,dx\,dt.$$

Here we take $t_1 = T$ and develop the total derivatives of the a_{ij} and a, while writing $\mathcal{L}u$ in the form

$$\mathcal{L}u = v - \lambda \sum_{l=1}^{n} u_{x_l}^2 - a_{lm}u_{x_l x_m} + a.$$

This gives

$$\begin{aligned}
&\frac{1}{2}\int_{A_k(T)} [v(x,\ T) - k]^2\,dx \\
&+ \int_0^T \int_{A_k(t)} \left[a_{ij}v_{x_i}v_{x_j} + 2\lambda a_{ij}u_{x_l x_i}u_{x_l x_j}(v-k) + v(v-k)\right] dx\,dt \\
&= -\int_0^T \int_{A_k(t)} \Bigg[-\lambda u_x^2 - a_{lm}u_{x_l x_m} + a \\
&+ \left(\frac{\partial a_{ij}}{\partial u_{x_l}} u_{x_l x_i} + \frac{\partial a_{ij}}{\partial u} u_{x_i} + \frac{\partial a_{ij}}{\partial x_i}\right) v_{x_j} \\
&- u_{x_i x_j}\left(\underline{\frac{\partial a_{ij}}{\partial u_{x_l}} u_{x_l t}} + \frac{\partial a_{ij}}{\partial u} u_t + \frac{\partial a_{ij}}{\partial t}\right) + \underline{\frac{\partial a}{\partial u_{x_l}} u_{x_l t}} \\
&+ \frac{\partial a}{\partial u} u_t + \frac{\partial a}{\partial t} - 2\lambda u_{x_i x_j} u_{x_l}\left(\underline{\frac{\partial a_{ij}}{\partial u_{x_m}} u_{x_m x_l}} + \frac{\partial a_{ij}}{\partial u} u_{x_l} + \frac{\partial a_{ij}}{\partial x_l}\right) \\
&\qquad - 2\lambda u_{x_l}\left(\underline{\frac{\partial a}{\partial u_{x_m}} u_{x_m x_l}} + \frac{\partial a}{\partial u} u_{x_l} + \frac{\partial a}{\partial x_l}\right)\Bigg](v-k)\,dx\,dt.
\end{aligned}$$

We combine the underlined terms, noting that

$$u_{x_l t} + 2\lambda u_{x_m x_l}u_{x_m} = v_{x_l},$$

and we replace the derivative u_t, using equation (0.1), by $a_{lm}u_{x_l x_m} - a$. As a result we obtain an equality which exactly coincides with (2.3). It is deduced by us for all values of k satisfying the condition

$$k \geqslant M_0 + 2.$$

In accordance with (2.6) we can assert the validity of the estimate

$$\max_{Q_T} v \leqslant \max\{M_0 + 2;\ c_1\}.$$

where the constant c_1 is the same as in (2.6).

Thus the estimate of $\max_{Q_T}|u_t|$ is established. The result obtained can be formulated as a theorem.

Theorem 2.2. *Let $u(x, t)$ be a solution of equation* (0.1) *belonging to* $C^{2,1}(\bar{Q}_T)$.

Suppose that equation (0.1) *is parabolic on* $u = u(x, t)$ (*i.e. inequality* (1.1) *is fulfilled*) *and the functions* $a_{ij}(x, t, u, p)$ *and* $a(x, t, u, p)$ *are continuous and continuously differentiable with respect to all of their arguments in the domain*

$$\left\{(x, t)\in \bar{Q}_T,\ |u|\leqslant \max_{Q_T}|u(x, t)|,\ |p|\leqslant M_1 = \max_{Q_T}|u_x(x, t)|\right\}$$

Then the quantity $\max_{Q_T}|u_t|$ *is estimated solely in terms of the constants* ν, μ *and* M_2 *from conditions* (1.1), (2.4), M_1 *and* $\max_{\Gamma_T}|u_t|$.

From this theorem, Theorem 2.1 and Lemma 3.1 of Chapter II follows

Theorem 2.3. *Suppose the conditions of Theorem* 2.2 *are fulfilled and* $S \in O^2$. *Then for some* $\alpha > 0$ *the quantity* $\langle u_x\rangle_{Q_T}^{(\alpha)}$ *is estimated by a constant depending only on* $M_1 = \max_{Q_T}|u_x(x, t)|$, *the quantities* ν, μ *and* M_2 *from conditions* (1.1) *and* (2.4), *the norm* $|u|_{\Gamma_T}^{(2)}$ *and the norms in* O^2 *of the functions defining the boundary* S. *These same quantities also determine the exponent* α.

The derivation of this theorem from Theorems 2.1 and 2.2 and from Lemma 3.1 of Chapter II is literally the same as for the equations with principal part in divergence form in §5 of Chapter V.

Remark 2.1. The assertions of the theorems remain in force for the solutions described in Remark 5.2 of Chapter V.

§3. THE ESTIMATION OF $\max|u_x|$

In §§3–4 of Chapter V we estimated $\max|u_x|$ in terms of $\max|u|$ for the solutions of equations with principal part in divergence form. Here we will do the same for the solutions of the general quasi-linear equations

$$\mathcal{L}u \equiv u_t - a_{ij}(x, t, u, u_x)\, u_{x_i x_j} + a(x, t, u, u_x) = 0 \tag{3.1}$$

under the assumption that they are uniformly parabolic:

$$\nu(|u|)(1+|p|)^{m-2}\xi^2 \leqslant a_{ij}(x, t, u, p)\xi_i\xi_j \leqslant \mu_0(|u|)(1+|p|)^{m-2}\xi^2. \tag{3.2}$$

As the examples in §3 of Chapter I show, for this purpose it is necessary that the order of growth of $a(x, t, u, p)$ with respect to p exceed the order of growth of the $a_{ij}(x, t, u, p)\xi_i\xi_j$ with respect to p by no more than two, i.e. that

$$|a(x, t, u, p)| \leqslant \mu(|u|)(1+|p|)^m. \tag{3.3}$$

In addition, it is necessary that we impose certain restrictions on the growth with respect to p of the partial derivatives of the functions $a_{ij}(x, t, u, p)$ and

$a(x, t, u, p)$ with respect to the variables x, u and p. These restrictions have usually been formulated as follows: under a differentiation of the a_{ij} and a with respect to x and u the orders of their growth with respect to x are not increased, while under a differentiation with respect to p they are decreased by at least one (it is clear that, in the case when the a_{ij} and a are polynomials in x and u for large $|p|$, these assumptions concerning their partial derivatives are a consequence of conditions (3.2) and (3.3)).

We will assume that these restrictions are fulfilled for the partial derivatives with respect to the p_k:

$$\left|\frac{\partial a_{ij}}{\partial p_k}\right|(1+|p|)^2+\left|\frac{\partial a}{\partial p_k}\right|\leqslant \mu_1(|u|)(1+|p|)^{m-1}. \tag{3.4}$$

We weaken them for the partial derivatives with respect to the x_k, assuming

$$\left|\frac{\partial a_{ij}}{\partial x_k}\right|(1+|p|)^2+\left|\frac{\partial a}{\partial x_k}\right| \leqslant [\varepsilon(|u|)+P(|p|,|u|)](1+|p|)^{m+1}, \tag{3.5}$$

where $\epsilon(r)$ is a continuous monotonically increasing nonnegative function of r, while $P(\rho, r)$ is a continuous function on the set $\{\rho\geq 0,\ r\geq 0\}$ that is monotonically increasing with respect to r and tends to zero for $\rho\to\infty$ uniformly with respect to r from $[0, r_1]$, where r_1 is any number. As for the partial derivatives with respect to u we will assume that

$$\begin{gathered}\left|\frac{\partial a_{ij}}{\partial u}\right|\leqslant [\varepsilon(|u|)+P(|p|,|u|)](1+|p|)^{m-2},\\ -\frac{\partial a(x,t,u,p)}{\partial u}\leqslant [\varepsilon(u)+P(|p|,|u|)](1+|p|)^{m},\end{gathered} \tag{3.6}$$

where $\epsilon(r)$ and $P(\rho, r)$ possess the same properties as in (3.5).

Under the fulfilment of conditions (3.2)–(3.6) there will be obtained an estimate of $\max|u_x|$ in terms of $\max_{Q_T}|u| = M$ and the known quantities

$$\nu=\nu(M),\quad \mu_0=\mu_0(M),\quad \mu_1=\mu_1(M),$$
$$\varepsilon=\varepsilon(M),\quad P(|p|)=P(|p|,M),$$

in (3.2)–(3.6), if $\epsilon(M)$ is sufficiently small. If the constant $\epsilon(M)$ is arbitrary (not small) in inequalities (3.5) and (3.6), then it is possible to estimate $\max|u_x|$ in terms of $|u|^{(\alpha)}_{Q_T}$ with any $\alpha>0$ (or in terms of $\max_{Q_T}|u|$ and the modulus of continuity of $u(x, t)$) and the same quantities ν, μ, μ_0, μ_1, ϵ and $P(|p|)$.

In the case of equations with principal part in divergence form (see Chapter V) the norm $|u|^{(\alpha)}_{Q_T}$ with some $\alpha>0$ was estimated in terms of $M=\max_{Q_T}|u|$ and known constants without any assumptions on the smallness of $\epsilon(M)$, and consequently $\max|u_x|$ of their solutions was estimated for any $\epsilon(M)$ from (3.5) and (3.6). It is not clear whether this can be done for all equations of form (3.1) under the fulfilment of just the conditions (3.2)–(3.6) with arbitrary $\epsilon(M)$.

Exactly the same situation exists in the case of equations of elliptic type (see [65q], Chapter VI, §2).

In this section we will follow the account of the analogous estimates in [65q]. Some difference will be caused, first, by the appearance of the term u_t in the equation and, secondly, by the fact that in comparison with [65q] we weaken the a priori assumptions concerning the smoothness of the solution $u(x, t)$. Namely, in [65q] it is assumed in obtaining estimates of $\max|u_x|$ for the solutions of the elliptic equations

$$a_{ij}(x, u, u_x)u_{x_ix_j}+a(x, u, u_x)=0$$

that $u(x) \in W_2^2(\Omega)$ and $\operatorname{vrai}\max_\Omega |u_x| < \infty$. From the results given below it follows that we can replace the requirement that $\operatorname{vrai}\max_\Omega |u_x|$ be finite by the condition

$$\int_\Omega |u_x|^{2s}(|u_{xx}|^2+1)\,dx<\infty, \tag{3.7}$$

where s is a sufficiently large number depending on the a_{ij} and a. In particular, all of the assertions from §2 of Chapter IV of [65q] concerning interior estimates of $\max|u_x|$ are valid for solutions $u(x)$ from $W_n^2(\Omega)$ with $n > 2$; for them, by virtue of the embedding theorems, the integrals (3.7) are finite for any s.

We will first estimate $|u_x|$ on the surface S_T. In this connection we will assume that the function $u(x, t) \in O^{1,0}(\overline{Q}_T)$ and at each point of $Q_T\backslash\Gamma_T$ has a second differential with respect to x, a derivative with respect to t and satisfies equation (3.1). We let

$$\max_{Q_T}|u|=M,\ \nu(M)=\nu,\ \mu(M)=\mu.$$

We will assume that

$$u|_{\Gamma_T}=\psi(x, t)|_{\Gamma_T}, \text{while } \psi(x, 0)\in O^1(\overline{\Omega}) \text{ and } \psi\in O^{2,1}(S_T). \tag{3.8}$$

Lemma 3.1. *Suppose the function u is a solution of equation (3.1), with u and ψ possessing the properties just described. Then $\max_{S_T}|u_x|$ is estimated from above by a constant depending only on M and $\mu(M)/\nu(M)$ from (3.2) and (3.3), on the boundary S of the domain Ω, which is assumed to belong to the* *class* O^2, *and on* $\max_{x\in\Omega}|\psi_x(x, 0)|$ *and* $|\psi|_{S_T}^{(2)}$.

Let us take a not large piece S^1 of the boundary of the domain Ω. Introducing new coordinates $y_1, \cdots, y_n$, we transform Ω so that S^1 lies in the plane $y_n = 0$ and the whole domain Ω is situated on the side of positive y_n. Equation (3.1) will have the same form and the same properties (3.2)–(3.3) in the variables $y_1, \cdots, y_n$, as long as the functions $y_i = y_i(x)$, $i = 1, \cdots, n$, have bounded derivatives up to the second order inclusively in $\overline{\Omega}$. Our assumptions concerning S imply that it is possible to divide S into a finite number of pieces and to introduce for each of them nonsingular coordinates $y_1, \cdots, y_n$ possessing the properties just described. In view of this we can assume without loss of generality that already in the coordinates $x_1, \cdots, x_n$ the piece S^1 of the boundary S lies in the plane $x_n = 0$ and the whole domain Ω is contained in the half space $x_n \geq 0$. In addition, we will consider the case of the homogeneous boundary condition

$$u|_{S_T} = 0.$$

The general case of condition (3.8) reduces to this case upon replacing $u(x, t)$ by $w(x, t) = u(x, t) - \hat{\psi}(x, t)$, where $\hat{\psi}(x, t)$ is a function from $O^{2,1}(\overline{Q}_T)$ that is equal to ψ on S_T.

Let us construct with respect to $u(x, t)$ a function $v(x, t)$ which will be equal to zero on S_T and for which the expression

$$(1 + |u_x|)^{2-m} \mathscr{L}_0 v \equiv (1 + |u_x|)^{2-m} [v_t - a_{ij}(x, t, u, u_x) v_{x_i x_j}]$$

will be less than some known constant. We introduce $v(x, t)$ by means of the equality

$$u = \varphi(v).$$

Then

$$u_t = \varphi' v_t, \quad u_{x_i} = \varphi' v_{x_i}, \quad u_{x_i x_j} = \varphi' v_{x_i x_j} + \varphi'' v_{x_i} v_{x_j},$$

and equation (3.1) gives

$$\mathscr{L}(u;\ v) \equiv v_t - a_{ij}(x, t, u, u_x) v_{x_i x_j} - \frac{\varphi''}{\varphi'} a_{ij} v_{x_i} v_{x_j} + \frac{1}{\varphi'} a(x, t, u, u_x) = 0. \tag{3.9}$$

If $\phi' > 0$ and $\phi'' < 0$, then

$$\begin{aligned} v_t - a_{ij}(x, t, u, u_x) v_{x_i x_j} &= \frac{\varphi''}{\varphi'} a_{ij} v_{x_i} v_{x_j} - \frac{a}{\varphi'} \\ &\leqslant (1 + |u_x|)^{m-2} \left[\frac{\varphi''}{\varphi'} \nu v_x^2 + \frac{1}{\varphi'} \mu (1 + |u_x|)^2\right] \\ &\leqslant \left[\left(\frac{\varphi''}{\varphi'} \nu + 2\mu\varphi'\right) v_x^2 + \frac{2\mu}{\varphi'}\right] (1 + |u_x|)^{m-2}. \end{aligned} \tag{3.10}$$

We select ϕ in such a way that

$$\frac{\varphi''}{\varphi'}\nu + 2\mu\varphi' \leqslant 0, \qquad \varphi(0) = 0.$$

For example, we take

$$\varphi(v) = \frac{\nu}{2\mu}\ln(1+v),$$

so that $v = -1 + e^{(2\mu/\nu)u}$ and $v|_{S_T} = 0$. Then from (3.10) it follows that

$$(1+|u_x|)^{2-m}\left[v_t - a_{ij}(x, t, u, u_x)v_{x_ix_j}\right] \leqslant \frac{2\mu}{\varphi'}$$
$$= \frac{4\mu^2}{\nu}(1+v) = \frac{4\mu^2}{\nu}e^{\frac{2\mu u}{\nu}} \leqslant \frac{4\mu^2}{\nu}e^{\frac{2\mu M}{\nu}} \equiv c.$$

For $t = 0$ we have

$$\max_\Omega |v_x(x, 0)| \leqslant \frac{2\mu}{\nu}\max_\Omega e^{\frac{2\mu u(x,0)}{\nu}} M_0 \leqslant \frac{2\mu M_0}{\nu}e^{\frac{2\mu M}{\nu}} \equiv c_1.$$

Now consider the function $v(x, t) + \lambda e^{-x_n}$. For $\lambda \geq c_1 e^d$, where d is the diameter of Ω, its maximal value on Γ_T is taken for $x \in S^1$, $t \in [0, T]$:

$$\max_{\Gamma_T}(v + \lambda e^{-x_n}) \leqslant (v + \lambda e^{-x_n})|_{x\in S^1} = \lambda. \tag{3.11}$$

Indeed,

$$\max_{S_T}(v + \lambda e^{-x_n}) = \max_{S_T}\lambda e^{-x_n} = \lambda e^{-x_n}|_{x\in S^1} = \lambda$$

and, furthermore,

$$\frac{\partial}{\partial x_n}[v(x, 0) + \lambda e^{-x_n}]|_{x\in\Omega} \leqslant c_1 - \lambda e^{-x_n}|_{x\in\Omega} \leqslant c_1 - \lambda e^{-d} \leqslant 0,$$

from which it follows that

$$\max_\Omega[v(x, 0) + \lambda e^{-x_n}] \leqslant \max_S[v(x, 0) + \lambda e^{-x_n}] \leqslant \lambda.$$

We put

$$\lambda = e^d \max\left\{c_1; \frac{2c}{\nu}\right\},$$

in which case

$$\frac{\partial(v + \lambda e^{-x_n})}{\partial t} - a_{ij}(x, t, u, u_x)(v + \lambda e^{-x_n})_{x_ix_j} < 0, \tag{3.12}$$

since

$$v_t - a_{ij}(x, t, u, u_x)(v + \lambda e^{-x_n})_{x_ix_j} \leqslant (1+|u_x|)^{m-2}c$$
$$- \lambda a_{nn}e^{-x_n} \leqslant (1+|u_x|)^{m-2}(c - \lambda\nu e^{-d}) < 0.$$

From (3.11) and (3.12) we conclude that the value of the function $v + \lambda e^{-x_n}$ for $x \in S^1$, $t \in [0, T]$ is its maximal value in $\overline{Q}_T$, and hence

$$\frac{\partial}{\partial x_n}(v + \lambda e^{-x_n})\bigg|_{x\in S^1} = \frac{\partial v}{\partial x_n}\bigg|_{x\in S^1} - \lambda \leqslant 0,$$

i.e.

$$\operatorname*{vrai\,max}_{\substack{x\in S^1\\ t\in[0,T]}} \frac{\partial v}{\partial x_n} \leqslant \lambda = e^{d+\frac{2\mu M}{\nu}} \max\left\{\frac{2\mu M_0}{\nu},\ \frac{8\mu^2}{\nu^2}\right\}. \tag{3.13}$$

This gives an estimate from above for $\partial u/\partial x_n$ on $S^1 \times [0, T]$, since

$$\left.\frac{\partial u}{\partial x_n}\right|_{x\in S^1} = \varphi'(v) \left.\frac{\partial v}{\partial x_n}\right|_{x\in S^1} = \frac{\nu}{2\mu} \left.\frac{\partial v}{\partial x_n}\right|_{x\in S^1} \leqslant \frac{\nu\lambda}{2\mu}$$

In order to estimate $\partial u(x, t)/\partial x_n$ from below for $x \in S^1$ it is sufficient to apply the preceding estimate to the solution $-u(x, t)$ of the equation

$$(-u)_t - a_{ij}(x, t, u, u_x)(-u)_{x_ix_j} - a(x, t, u, u_x) = 0.$$

From this we get

$$-\operatorname*{vrai\,min}_{\substack{x\in S^1\\ t\in[0,T]}} \frac{\partial u}{\partial n} = \operatorname*{vrai\,max}_{\substack{x\in S^1\\ t\in[0,T]}} \frac{\partial(-u)}{\partial n} \leqslant \frac{\nu\lambda}{2\mu}.$$

Thus we obtain a two-sided estimate of $\partial u/\partial x_n$ on $S^1 \times [0, T]$.

Estimates of this type of $\max|u_x|$ on S_T have their origin in the works of S. N. Bernšteĭn [8b]. In these works elliptic equations with two independent variables are considered, but the transition to any n and to equations of parabolic type does not present any difficulties for estimates of $\max|u_x|$ on a boundary. A different situation exists with interior estimates. In Bernstein's works such estimates were made with the use of the fact that at a maximum point of some derivative u_{x_i} it is possible to exclude from consideration all the derivatives of second order $u_{x_1x_1}$, $u_{x_1x_2}$ and $u_{x_2x_2}$ by expressing them in terms of the derivatives with respect to x of first order, using three relations: the two relations obtained by setting the derivatives $u_{x_ix_1}$ and $u_{x_ix_2}$ equal to zero and the equation itself. Such an exclusion is impossible for $n > 2$ in the elliptic case and for $n > 1$ in the parabolic case (it is clear that such an exclusion is not required for parabolic equations with one space variable $x = x_1$).

The first interior estimates of $\max|u_x|$ for quasi-linear parabolic and elliptic equations with arbitrary n were given in [65h]. The method proposed there for estimating $\max|u_x|$ is related to Bernšteĭn's method for making estimates by the idea of introducing in place of u a new function $v = \psi(u)$, for which it is impossible to estimate $\max|v_x|$ by using only the equation for u. In regard to the solutions investigated in [65h] it was assumed that they are continuous in Q_T together with their derivatives of the form u_{x_i}, u_t, u_{tx_i}, D_x^2u, D_x^3u. Let us modify the previous arguments in such a way that they become suitable for solutions $u(x, t)$

having only generalized derivatives entering into the equation (the case when these derivatives are continuous was handled in [65q]).

Thus, we will assume that the solution $u(x, t)$ of equation (3.1) is bounded in Q_T and has generalized derivatives of the form u_{x_i}, $u_{x_ix_j}$, u_t, in Q_T, with $|u_x|^{2s+2} \in V_2^{1,0}(Q_T)$ and

$$\int_{Q_T} \left[(1+|u_x|)^{4s+m}|u_{xx}|^2 + (1+|u_x|)^{4s+2}|u_t|^2 + |u_x|^{4s+m+4}\right] dx\,dt < \infty, \tag{3.14}$$

where s is a sufficiently large number, the magnitude of which will be further specified below as a function of the constants $\nu(M)$, $\mu_1(M)$ and $\epsilon(M)$ from conditions (3.2)–(3.6).

We will give two kinds of estimates of $|u_x|$: a global estimate for the whole cylinder at once and local estimates for subdomains $Q' \subset Q_T$. We begin with the first, assuming that $\max_{\Gamma_T}|u_x| = M_0$ is already known. We again introduce a function $v(x, t)$ by means of the equality

$$u = \varphi(v), \tag{3.15}$$

assuming $\phi'(v) > 0$. Relation (3.9) is valid for v and, consequently, so are the identities

$$\int_{Q_{t_1}} \mathscr{L}(u;\, v)\,\eta_{x_l}\,dx\,dt = 0, \quad l = 1, \dots n,\ t_1 \leqslant T, \tag{3.16}$$

where $\eta(x, t)$ is an arbitrary function that is continuously differentiable in $\overline{Q}_T$. If $\eta|_{\Gamma_T} = 0$, while the solution $u(x, t)$ has continuous derivatives u_x, u_{xx}, u_{xt}, $D_x^3 u$, then, by carrying out a double integration by parts in (3.16), we can transform it into the form

$$\int_{Q_{t_1}} \left\{ v_{x_l t}\eta + a_{ij}v_{x_lx_j}\eta_{x_i} + \left[\frac{da_{ij}}{dx_l}v_{x_lx_j} - \frac{da_{lj}}{dx_l}v_{x_ix_j} - \frac{d}{dx_l}\left(\frac{\varphi''}{\varphi'}a_{ij}v_{x_i}v_{x_j} - \frac{a}{\varphi'}\right)\right]\eta\right\} dx\,dt = 0, \quad l = 1, \dots, n. \tag{3.17}$$

Let $w = \sum_{l=1}^n v_{x_l}^2$ and $z = w^{s+1}$. We put

$$\eta(x, t) = 2v_{x_l}w^s z^{(k)}, \tag{3.18}$$

where

$$z^{(k)} = \max\{z(x, t) - k;\ 0\}$$

and

$$k \geqslant \operatorname{vrai\,max}_{\Gamma_T} z, \tag{3.19}$$

and in (3.17) we carry out a summation over l from 1 to n. Then, noting that

$$2v_{x_lx_j}v_{x_l}=w_{x_j};\quad 2v_{x_lt}v_{x_l}w^s=\frac{1}{s+1}z_t;\quad 2v_{x_lx_j}v_{x_l}w^s=\frac{1}{s+1}z_{x_j},$$

we deduce from (3.17) the equality

$$\frac{1}{s+1}\int_\Omega [z^{(k)}(x,t)]^2\,dx+2\int_{Q_{t_1}}\Big\{2a_{ij}v_{x_lx_i}v_{x_lx_j}w^sz^{(k)}+sa_{ij}w_{x_i}w_{x_j}w^{s-1}z^{(k)}+\frac{1}{s+1}a_{ij}z^{(k)}_{x_i}z^{(k)}_{x_j}$$
$$+\Big[\frac{da_{ij}}{dx_l}v_{x_ix_j}-\frac{da_{ij}}{dx_j}v_{x_lx_i}-\frac{d}{dx_l}\Big(\frac{\varphi''}{\varphi'}a_{ij}v_{x_i}v_{x_j}-\frac{a}{\varphi'}\Big)2v_{x_l}w^sz^{(k)}\Big]\Big\}\,dx\,dt=0,\quad t_1\in[0,T]. \tag{3.20}$$

In the intermediate transformations we have made use of the existence of the derivatives u_{xt}, D^3_xu of $u(x,t)$, although in equality (3.20) itself, as also the initial equality (3.16), the integrals have a meaning for any solution satisfying conditions (3.14) only. Let us show that (3.20) also holds for any such solution if the values of k are subject to inequality (3.19).

We introduce averagings of the function $v(x,t)$ of the form

$$v_\rho(x,t)=\int_{|x-y|\le\rho}\omega_\rho(|x-y|)\,v(y,t)\,dy,$$

$$v_h(x,t)=\frac{1}{h}\int_t^{t+h}v(x,\tau)\,d\tau$$

and in accordance with (3.16) we consider the identities

$$\int_{Q_{t_1}}\Big[v_{ht}-a_{ij}(x,t,u,u_x)v_{\rho x_ix_j}-\frac{\varphi''}{\varphi'}a_{ij}(x,t,u,u_x)v_{x_i}v_{x_j}+\frac{1}{\varphi'}a(x,t,u,u_x)\Big]\eta_{x_l}\,dx\,dt$$
$$=\int_{Q_{t_1}}\big[v_{ht}-v_t+a_{ij}(x,t,u,u_x)(v_{x_ix_j}-v_{\rho x_ix_j})\big]\eta_{x_l}\,dx\,dt,$$
$$l=1,\dots,n,\quad t_1\in[0,T-h], \tag{3.21}$$

with a smooth function $\eta(x,t)$ that is equal to zero in the vicinity of S_T. We carry out an integration by parts in the left side of (3.21) in the same way as was done in the derivation of (3.17) and then pass to the limit as $\rho\to0$. As a result we obtain the identities

$$\int_{Q_{t_1}} \left\{ v_{hx_l t}\eta + a_{ij} v_{x_l x_j}\eta_{x_i} + \left[\frac{da_{lj}}{dx_l} v_{x_l x_j} - \frac{d}{dx_l}\left(\frac{\varphi''}{\varphi'} a_{ij} v_{x_i} v_{x_j} - \frac{a}{\varphi'} \right) - \frac{da_{ij}}{dx_l} v_{x_i x_j} \right] \eta \right\} dx\,dt = \int_{Q_{t_1}} (v_{ht} - v_t)\,\eta_{x_l}\,dx\,dt,$$

$$l = 1, \ldots, n. \tag{3.22}$$

Here we can take as $\eta(x, t)$ any function from $V_2(Q_T)$ that is equal to zero on S_T and for which all of the integrals in (3.22) have a meaning. In particular, we can put

$$\eta(x, t) = 2v_{hx_l}(w^h)^s z^{h(k)}, \tag{3.23}$$

where

$$w^h = \sum_{l=1}^n v_{hx_l}^2, \quad z^h = (w^h)^{s+1},$$

and the number k is subject to condition (3.19). For such k the function $z^{h(k)}$ vanishes on S_{T-h}, since

$$\operatorname*{vrai\,max}_{S_{T-h}} z^h \leqslant \operatorname*{vrai\,max}_{S_{T-h}} \left[\frac{1}{h} \int_t^{t+h} |v_x(x, \tau)|^2 d\tau \right]^{s+1} \leqslant \operatorname*{vrai\,max}_{S_T} |v_x|^{2s+2} = \operatorname*{vrai\,max}_{S_T} z.$$

We substitute the indicated $\eta(x, t)$ into (3.22), carry out a summation over l and write the first term in the form

$$j_h = \int_{Q_{t_1}} v_{hx_l t} 2v_{hx_l}(w^h)^s z^{h(k)}\,dx\,dt = \frac{1}{2(s+1)} \int_\Omega [z^{h(k)}]^2 dx \Big|_0^{t_1}.$$

We let h tend to zero. In this connection we have

$$\| z^{h(k)} - z^{(k)} \|_{2,\Omega} \to 0,$$

by virtue of Lemma 4.7 of Chapter II, and since k is chosen from condition (3.19) it follows that

$$\lim_{h\to 0} \| z^{h(k)}(x, 0) \|_{2,\Omega} = 0.$$

Consequently,

$$j_h \to \frac{1}{2(s+1)} \int_\Omega [z^{(k)}(x, t_1)]^2 dx \quad \text{for } h \to 0.$$

The remaining terms in the left side of equality (3.22) with η from (3.23) are transformed in the limit into analogous terms with η from (3.18), while the right side of (3.22) tends to zero for $h \to 0$. In this way, on passing to the limit, we arrive at relation (3.20).

Let us put $t_1 = T$ in (3.20) and rewrite it as follows:

$$\frac{1}{s+1}\int_{A_k(T)}(z-k)^2\,dx+2\int_0^T\int_{A_k(t)}\Big\{2a_{ij}v_{x_lx_i}v_{x_lx_j}w^s(z-k)$$
$$+sa_{ij}w_{x_j}w_{x_i}w^{s-1}(z-k)+\frac{1}{s+1}a_{ij}z_{x_i}z_{x_j}$$
$$+\Big[\frac{da_{ij}}{dx_l}w_{x_j}-\frac{da_{ij}}{dx_l}2v_{x_i}v_{x_lx_j}-\Big(\frac{\varphi''}{\varphi'}\Big)'2wa_{ij}v_{x_i}v_{x_j}$$
$$-\frac{\varphi''}{\varphi'}\frac{da_{ij}}{dx_l}2v_{x_l}v_{x_i}v_{x_j}-\frac{\varphi''}{\varphi'}2a_{ij}w_{x_j}v_{x_i}-\frac{\varphi''}{\varphi'^2}2wa$$
$$+\frac{1}{\varphi'}\frac{da}{dx_l}2v_{x_l}\Big]w^s(z-k)\Big\}\,dx\,dt=0. \tag{3.24}$$

Here $A_k(t)$ is the set of points $x\in\Omega$ at which $z(x,t)>k$. We will choose a function $\phi(v)$ in such a way that $\phi'>0$ and $-(\phi''/\phi')'-(\phi''/\phi')^2>0$. In the left side of (3.24) we leave only the positive terms

$$\frac{1}{s+1}\int_{A_k(T)}(z-k)^2\,dx+2\int_0^T\int_{A_k(t)}\Big[2a_{ij}v_{x_lx_i}v_{x_lx_j}w^s(z-k)$$
$$+sa_{ij}w^{s-1}w_{x_i}w_{x_j}(z-k)+\frac{1}{s+1}a_{ij}z_{x_i}z_{x_j}$$
$$-2\Big(\frac{\varphi''}{\varphi'}\Big)'a_{ij}v_{x_i}v_{x_j}w^{s+1}(z-k)\Big]\,dx\,dt,$$

while the remaining terms are transferred to the right side and estimated from above with the use of inequalities (1.1)–(1.2) of Chapter II and conditions (3.3)–(3.6) as follows:

$$2\frac{\varphi''}{\varphi'}a_{ij}w_{x_j}v_x\leqslant\frac{a_{ij}w_{x_i}w_{x_j}}{w}+\Big(\frac{\varphi''}{\varphi'}\Big)^2a_{ij}v_{x_i}v_{x_j}w.$$

The total derivatives d/dx_l in (3.24) are represented in the form

$$\frac{da_{ij}}{dx_l}=\frac{\partial a_{ij}}{\partial u_{x_m}}u_{x_mx_l}+\frac{\partial a_{ij}}{\partial u}u_{x_l}+\frac{\partial a_{ij}}{\partial x_l}$$
$$=\frac{\partial a_{ij}}{\partial u_{x_m}}(\varphi'v_{x_mx_l}+\varphi''v_{x_m}v_{x_l})+\frac{\partial a_{ij}}{\partial u}\varphi'v_{x_l}+\frac{\partial a_{ij}}{\partial x_l}$$

and analogously for da/dx_l. Furthermore, we note that the derivatives da_{ij}/dx_l and da/dx_l in (3.24) are multiplied by $2v_{x_l}$ and summed over l from 1 to n. These sums are represented in the form

$$2\sum_{l=1}^{n}\frac{da_{ij}}{dx_l}v_{x_l}=\frac{\partial a_{ij}}{\partial u_{x_m}}(\varphi' w_{x_m}+2\varphi'' w v_{x_m})\cdot$$
$$+2\frac{\partial a_{ij}}{\partial u}\varphi' w+2\frac{\partial a_{ij}}{\partial x_l}v_{x_l},$$
$$2\sum_{l=1}^{n}\frac{da}{dx_l}v_{x_l}=\frac{\partial a}{\partial u_{x_m}}(\varphi' w_{x_m}+2\varphi'' w v_{x_m})$$
$$+2\frac{\partial a}{\partial u}\varphi' w+2\frac{\partial a}{\partial x_l}v_{x_l}.$$

As a result we deduce the following inequality from (3.24) with the use of conditions (3.2)–(3.6):

$$\nu\int_0^T\int_{A_k(t)}(1+|u_x|)^{m-2}\Big[2|v_{xx}|^2+(s-1)w^{-1}w_x^2-\Big(\frac{\varphi''}{\varphi'}\Big)^2w^2$$
$$-2\Big(\frac{\varphi''}{\varphi'}\Big)'w^2\Big]w^s(z-k)\,dx\,dt$$
$$\leqslant n^3\int_0^T\int_{A_k(t)}(1+|u_x|)^{m-2}\Big\{|w_x|\Big[\frac{\mu_1}{(1+\varphi'\sqrt{w})}(\varphi'|v_{xx}|$$
$$+|\varphi''|w)+(\varepsilon+P)\varphi'\sqrt{w}+(\varepsilon+P)(1+\varphi'\sqrt{w})\Big]$$
$$+\Big(|v_{xx}|+\frac{|\varphi''|}{\varphi'}w\Big)\Big[\frac{\mu_1}{(1+\varphi'\sqrt{w})}\Big(\varphi'|w_x|+|\varphi''|w^{\frac{3}{2}}\Big)$$
$$+2(\varepsilon+P)\varphi' w+2(\varepsilon+P)(1+\varphi'\sqrt{w})\sqrt{w}\Big]$$
$$+\frac{|\varphi''|}{\varphi'^2}2\mu(1+\varphi'\sqrt{w})^2w$$
$$+\frac{1}{\varphi'}\Big[\mu_1(1+\varphi'\sqrt{w})\Big(\varphi'|w_x|+2|\varphi''|w^{\frac{3}{2}}\Big)$$
$$+2(\varepsilon+P)(1+\varphi'\sqrt{w})^2\varphi' w$$
$$+2(\varepsilon+P)(1+\varphi'\sqrt{w})^3\sqrt{w}\Big]\Big\}w^s(z-k)\,dx\,dt. \qquad (3.25)$$

Consider the terms in the braces. We estimate those of them which contain $|v_{xx}|$ by the inequality $2ab\le\delta a^2+\delta^{-1}b^2$, $\delta>0$, assigning a small δ to the factor v_{xx}:

$$|w_x|\frac{\mu_1}{(1+\varphi'\sqrt{w})}\varphi'|v_{xx}|\leqslant\frac{\delta}{2}|v_{xx}|^2+\frac{\mu_1^2}{2\delta}\frac{w_x^2\varphi'^2}{(1+\varphi'\sqrt{w})^2}\leqslant\frac{\delta}{2}|v_{xx}|^2+\frac{\mu_1^2}{2\delta}\frac{w_x^2}{w};$$

$$|v_{xx}|\left[\frac{\mu_1}{(1+\varphi'\sqrt{w})}\left(\varphi'|w_x|+\varphi''w^{\frac{3}{2}}\right)+2(\varepsilon+P)\varphi'w+2(\varepsilon+P)(1+\varphi'\sqrt{w})\sqrt{w}\right]\leqslant 3\delta|v_{xx}|^2+\frac{\mu_1^2}{2\delta}\frac{w_x^2}{w}$$
$$+\frac{\mu_1^2}{2\delta}\left(\frac{\varphi''}{\varphi'}\right)^2w^2+\frac{1}{\delta}(\varepsilon+P)^2\varphi'^2w^2+\frac{2}{\delta}(\varepsilon+P)^2(1+\varphi'^2w)w.$$

The terms with $|w_x|$ are estimated as follows:

$$|w_x|(\varepsilon+P)(1+2\varphi'\sqrt{w})\leqslant\frac{w_x^2}{2w}+(\varepsilon+P)^2(1+4\varphi'^2w)w;$$

$$|\varphi''|w\frac{\mu_1}{(1+\varphi'\sqrt{w})}|w_x|\leqslant\frac{w_x^2}{2w}+\frac{\mu_1^2}{2}\left(\frac{\varphi''}{\varphi'}\right)^2w^2;$$

$$\mu_1(1+\varphi'\sqrt{w})|w_x|\leqslant\frac{\mu_1^2}{2\varepsilon}\frac{w_x^2}{w}+\varepsilon(1+\varphi'^2w)w.$$

The remaining terms do not exceed the expressions

$$\mu_1\left(\frac{\varphi''}{\varphi'}\right)^2w^2+4(\varepsilon+P)|\varphi''|w^2+2(\varepsilon+P)\frac{|\varphi''|}{\varphi'}w^{\frac{3}{2}}$$
$$+4\mu|\varphi''|w^2+4\mu\frac{\varphi''}{\varphi'^2}w+2\mu_1|\varphi''|w^2+2\mu_1\frac{|\varphi''|}{\varphi'}w^{\frac{3}{2}}$$
$$+4(\varepsilon+P)(\varphi'^2w^2+w)+16(\varepsilon+P)\left(\frac{\sqrt{w}}{\varphi'}+\varphi'^2w^2\right).$$

We choose δ equal to $\nu/2n^3$, while s is fixed so that

$$n^3\left(\frac{2\mu_1^2n^3}{\nu}+1+\frac{\mu_1^2}{2\varepsilon}\right)\leqslant\frac{s\nu}{2}. \tag{3.26}$$

We substitute the above estimates into the right side of inequality (3.25) and then transfer the terms with $|v_{xx}|^2$ and w_x^2 into the left side of the inequality and combine similar terms. In the right side the principal terms will be the terms containing w^{s+2}, with the remaining terms containing smaller powers of w, namely $w^{s+3/2}$ and lower. We will take $k\geq 1$; then the function z, and hence also w, will be not less than 1 on the set $A_k(t)$, i.e. all of the smaller powers of w can be estimated in terms of $w^{s+3/2}$. By virtue of all that has been said we will have from (3.25)

$$\int_0^T \int_{A_k(t)} (1+|u_x|)^{m-2}\left[\frac{s-1}{2}\frac{w_x^2}{w}+\frac{1}{4}|v_{xx}|^2 \right.$$
$$\left. -2\left(\frac{\varphi''}{\varphi'}\right)' w^2\right] w^s(z-k)\,dx\,dt$$
$$\leqslant \int_0^T \int_{A_k(t)} (1+|u_x|)^{m-2}\left\{c_1\left[\left(\frac{\varphi''}{\varphi'}\right)^2+|\varphi''|(1+\varepsilon+\hat{P})\right.\right.$$
$$\left.\left.+(\varepsilon^2+P^2+2\varepsilon+P)\varphi'^2\right]w^2+c_2 w^{\frac{3}{2}}\right\} w^s(z-k)\,dx\,dt. \tag{3.27}$$

Here

$$\hat{P}=\max_{|p|\geqslant 0} P(|p|).$$

The constant c_1 is determined only by n and the values of $\nu(M)=\nu$, $\mu(M)=\mu$, $\mu_1(M)=\mu_1$ from (3.3)–(3.4), while the constant c_2 also depends on $\epsilon(M)=\epsilon$, $\hat{P}$ and the function $\phi(v)$, which we will choose in a special way. Namely, we will choose $\phi(v)$ so that

$$-\left(\frac{\varphi''}{\varphi'}\right)' \geqslant c_1\left[\left(\frac{\varphi''}{\varphi'}\right)^2+|\varphi''|(1+\varepsilon+\hat{P})+(2\varepsilon^2+3\varepsilon)\varphi'^2\right], \tag{3.28}$$

$$\varphi'(v)>0. \tag{3.29}$$

In addition, it is necessary that the inverse function for $\phi(v)$ be defined on the interval $[-M, M]$. The construction of such a function $\phi(v)$ will be taken up below, but now, assuming that such a function has already been found, we continue with the estimation. Consider inequality (3.27) for $k \geq k_1$, where $k_1 \geq 1$ and satisfies the condition

$$P\left(k^{\frac{1}{2s+2}} \min_{-M\leqslant u\leqslant M}\varphi'(v)\right)+P^2\left(k^{\frac{1}{2s+2}} \min_{-M\leqslant u\leqslant M}\varphi'(v)\right)\leqslant \varepsilon \text{ for } k \geqslant k_1. \tag{3.30}$$

Such a k_1 exists, since $P(\rho)\to 0$ for $\rho\to\infty$ and

$$\min_{-M\leqslant u\leqslant M}\varphi'(v)\geqslant \text{const}>0.$$

It follows from (3.27) by virtue of (3.30) and (3.28) that

$$\nu_1\int_0^T \int_{A_k(t)} (1+|u_x|)^{m-2} w^{s+2}(z-k)\,dx\,dt \leqslant$$

$$\leqslant c_2 \int_0^T \int_{A_k(t)} (1+|u_x|)^{m-2} w^{s+\frac{3}{2}} (z-k)\, dx\, dt, \tag{3.31}$$

where

$$\nu_1 = \min_{-M \leqslant u \leqslant M} \left(-\frac{\varphi''}{\varphi'}\right)' > 0.$$

From (3.31) we obviously have

$$\left(\nu_1 k^{\frac{1}{2(s+1)}} - c_2\right) \int_0^T \int_{A_k(t)} (1+|u_x|)^{m-2} w^{s+\frac{3}{2}} (z-k)\, dx\, dt \leqslant 0.$$

If the value of k exceeds

$$k_2 = \max\left\{k_1;\ \left(\frac{c_2}{\nu_1}\right)^{2(s+1)}\right\},$$

then this inequality is possible only in the case when the integral is equal to zero. In other words, the measure of the set of points (x, t) from Q_T at which $z > k_2$ is equal to zero, i.e.

$$\operatorname{vrai}\max_{Q_T} z(x, t) \leqslant \max\left\{k_1;\ \left(\frac{c_2}{\nu}\right)^{2(s+1)};\ \max_{\Gamma_T} z\right\}. \tag{3.32}$$

This yields an estimate of

$$\operatorname{vrai}\max_{Q_T} w^{s+1}(x, t) = \operatorname{vrai}\max_{Q_T} |v_x|^{2s+2},$$

and hence also of $\operatorname{vrai}\max_{Q_T} |u_x|$.

For the completion of the proof it remains to find the function ϕ. As $\phi(v)$ we take the function

$$\varphi(v) = -2M + 6Me \int_0^v e^{-\xi^q}\, d\xi.$$

We will verify that it satisfies all of the requirements for sufficiently large q. Let us determine the range of variation $[v_1, v_2]$ of v when $u = \phi(v)$ varies from $-M$ to M. The integral

$$\int_0^v e^{-\xi^q}\, d\xi,$$

will obviously vary in the interval $[1/6e, 1/2e]$, so that

$$\int_0^{v_1} e^{-\xi^q}\, d\xi = \frac{1}{6e}, \quad \int_0^{v_2} e^{-\xi^q}\, d\xi = \frac{1}{2e}.$$

For any $q \geq 1$ the values of the function $e^{-\xi^q}$ are included between e^{-1} and 1 on the interval $0 \leq \xi \leq 1$ and between 0 and 1 for all $\xi \geq 0$; thus for all $q \geq 1$

$$\frac{1}{6e} < \int_0^{v_1} d\xi = v_1 \text{ and } \frac{1}{2e} = \int_0^{v_2} e^{-\xi^q}\, d\xi = \int_0^{\frac{1}{2}} e^{-\xi^q}\, d\xi$$
$$+ \int_{\frac{1}{2}}^{v_2} e^{-\xi^q} d\xi > \frac{1}{2e} + \int_{\frac{1}{2}}^{v_2} e^{-\xi^q}\, d\xi,$$

i.e. $v_2 < 1/2$.

Thus, for all $q \geq 1$

$$\frac{1}{6e} < v_1 < v_2 < \frac{1}{2}.$$

Let us now verify relations (3.28) and (3.29):

$$\varphi' = 6Mee^{-v^q}, \qquad \varphi'' = -6Meqv^{q-1}e^{-v^q},$$
$$\frac{\varphi''}{\varphi'} = -qv^{q-1}, \quad \left(\frac{\varphi}{\varphi'}\right)' = -q(q-1)v^{q-2}.$$

Inequality (3.29) is obviously fulfilled. To verify (3.28) we calculate its right side:

$$c_1\big[q^2v^{2q-2} + (1+\varepsilon+\hat{P})\,6Meqv^{q-1}e^{-v^q}$$
$$+ (3\varepsilon + 2\varepsilon^2)\,36M^2e^{2-2v^q}\big] \leqslant c_1\big[q^2v^{q-2}2^{-q}$$
$$+ (1+\varepsilon+\hat{P})\,6Meqv^{q-1}e^{-(6e)^{-q}} + (3\varepsilon + 2\varepsilon^2)\,36M^2e^{2-2(6e)^{-q}}\big].$$

From a comparison of this expression with $-(\phi''/\phi')'$ it is clear that inequality (3.28) will be fulfilled if ϵ is sufficiently small and q is taken sufficiently large. It is here (and only here!) that we encounter the need to assume that ϵ is a sufficiently small number. Its maximal possible value can be calculated exactly, but we will not do this here. Thus an estimate of $\operatorname{vrai\,max}_{Q_T} |u_x(x,t)|$ has been obtained. We formulate it in the fom of a theorem.

Theorem 3.1. *Suppose a function $u(x,t) \in W_2^{2,1}(Q_T)$ is subject to conditions* (3.14) *for all s up to the order indicated in* (3.26), *and satisfies equation* (3.1) *almost everywhere in Q_T, and suppose $M = \operatorname{vrai\,max}_{Q_T}|u| < \infty$ and $M_0 = \operatorname{vrai\,max}_{\Gamma_T}|u_x| < \infty$. Then, if conditions* (3.2)–(3.6) *are fulfilled and $\epsilon(M)$ in* (3.5) *and* (3.6) *does not exceed a certain number ϵ_0 determined only by M, $\nu = \nu(M)$, $\mu = \mu(M)$, $\mu_1 = \mu_1(M)$ and*

$$\hat{P} = \max_{|p|\geqslant 0} P(|p|) = \max_{|p|\geqslant 0} P(|p|,\, M)$$

from (3.2)–(3.6), *it follows that* vrai $\max_{Q_T}|u_x|$ *does not exceed a certain constant depending only on* M, ν, μ, μ_1, ϵ, $P(|p|)$ *and* M_0.

From this theorem and Lemma 3.1 we get

Theorem 3.2. *Suppose a function* $u(x, t)$ *belongs to*

$$O^{1,0}(\bar{Q}_T) \cap W_2^{2,1}(Q_T)$$

and at each point of the cylinder Q_T *it has a second differential with respect to* x, *a derivative with respect to* t *and satisfies equation* (3.1). *Suppose further that conditions* (3.2)–(3.6) *are fulfilled with a sufficiently small* $\epsilon(M)$ *determined by* $M = \max_{Q_T}|u|$, $\nu(M)$, $\mu(M)$, $\mu_1(M)$ *and*

$$M = \max_{Q_T}|u|, \ \nu(M), \ \mu(M), \ \mu_1(M) \text{ and } \hat{P} = \max_{|p|>0} P(|p|, M).$$

Then, if $u|_{\Gamma_T} = \psi(x, t)|_{\Gamma_T}$, *where* $\psi(x, 0) \in O^1(\bar{\Omega})$ *and* $\psi(x, t) \in O^{2,1}(S_T)$, *with* S *belonging to the class* O^2, *it follows that* vrai $\max_{Q_T}|u_x|$ *does not exceed a constant depending only on* M, $\nu(M)$, $\mu(M)$, $\mu_1(M)$, $\epsilon(M)$, $\hat{P}(|p|, M)$, *the boundary* S *of the domain* Ω, $\max_{x\in\Omega}|\psi_x(x, 0)|$ *and the norm* $|\psi|_{S_T}^{(2)}$.

The assertions of Theorems 3.1 and 3.2 remain in force if instead of the smallness of $\epsilon(M)$ one requires a sufficient smallness in the oscillation of u in Q_T.

Indeed, suppose $\epsilon(M)$ is arbitrary. We denote by u_1 and u_2 the least and greatest values of u in $\bar{Q}_T$, and by $\omega = u_2 - u_1$ the oscillation of u in $\bar{Q}_T$. We will verify that conditions (3.28) and (3.29) and the requirement that the inverse function $v = \phi^{-1}(u)$ be defined on the interval $[u_1, u_2]$ are satisfied by the function

$$\varphi(v) = \frac{3u_1 - u_2}{2} + 3\omega r \int_0^v e^{-\xi^2} d\xi \tag{3.33}$$

for sufficiently large $r > 0$, as long as ω is sufficiently small (the maximal possible value of ω is determined by the constants c_1, $\hat{P}$ and ϵ from (3.28)).

Let us determine the range v_1, v_2 of variation of v when $u = \phi(v)$ varies on the interval $[u_1, u_2]$. From (3.33) we will have

$$\int_0^{v_1} e^{-\xi^2} d\xi = \frac{1}{6r}, \quad \int_0^{v_2} e^{-\xi^2} d\xi = \frac{1}{2r}.$$

Since

$$\int_0^1 e^{-\xi^2} d\xi > \frac{1}{e}$$

it follows that the numbers v_1 and v_2 will be small for sufficiently large r:

$1/6r < v_1 < v_2 < e/2r$. The derivative $\phi'(v) = 3\omega re^{-v^2}$ is everywhere positive. The difference between the left and right sides of (3.28) for the function (3.33) (for arbitrarily fixed constants c_1, $\hat{P}$ and ϵ)

$$2 - c_1[4v^2 + (1+\varepsilon+\hat{P})\,6\omega re^{-v^2}v + (3\varepsilon + 2\varepsilon^2)\,9\omega^2r^2e^{-2v^2}]$$

will be greater than zero if r is taken sufficiently large, for example so that v^2 is less than $1/2\sqrt{c_1}$, and ω is assumed to be sufficiently small.

Thus we have proved the following theorem.

Theorem 3.3. *The assertions of Theorems 3.1 and 3.2 remain valid if instead of the smallness of $\epsilon(M)$ in (3.5) and (3.6) one requires a sufficient smallness in the oscillation ω of the function $u(x, t)$ in Q_T.*

An admissible value of ω is determined by the constants n, $M = \max_{Q_T}|u|$, $\nu(M)$, $\mu(M)$, $\mu_1(M)$, $\epsilon(M)$ and

$$\hat{P} = \max_{|p|\geq 0} P(|p|, M)$$

from (3.2)–(3.6) and it can easily be calculated explicitly, starting from the method just given for estimating $|u_x|$.

We now proceed to obtain local estimates of $|u_x|$. We will show, for any interior cylinder $Q' = \Omega' \times (\epsilon, T)$, $\epsilon > 0$, that $\operatorname{vrai}\max_{Q'}|u_x|$ can be estimated in terms of n, the distance from Q' to Γ_T, $M = \max_{Q_T}|u|$ and the constants m, $\nu = \nu(M)$, $\mu_0 = \mu_0(M)$, $\mu = \mu(M)$, $\mu_1 = \mu_1(M)$, $\epsilon = \epsilon(M)$ and $P(|p|, M)$ from conditions (3.2)–(3.6), if $m \geq 1$. These estimates do not depend on the properties of S and the values of $u(x, t)$ on Γ_T. If on the other hand the boundedness of $\operatorname{vrai}\max_{\Omega}|u_x(x, 0)| = M_0$ is presupposed, then $\operatorname{vrai}\max_{Q'}|u_x|$ with $Q' = \Omega' \times (0, T)$ can be estimated for any m in terms of M_0, M, ν, μ_0, μ, μ_1, ϵ and $P(|p|)$.

Without loss of generality we take as Q' an arbitrary cylinder $Q_\rho = K_\rho \times (\tau - \rho^2, \tau)$, where $K_\rho \subset \Omega$ and $0 \leq \tau - \rho^2 < \tau \leq T$. Let $\zeta(x, t)$ be a cutting function for Q_ρ that is equal to one in the coaxial cylinder $Q_{\rho/2} = K_{\rho/2} \times (\tau - \rho^2/4, \tau)$ and satisfies the condition

$$|\zeta_x| \leqslant c\rho^{-1}, \quad |\zeta_{xx}|, \quad |\zeta_t| \leqslant c\rho^{-2}.$$

We will estimate $\operatorname{vrai}\max_{Q_T} z$, where

$$z(x, t) = (\zeta^2 w)^{s+1} = \left(\zeta^2 \sum_{i=1}^{n} v_{x_i}^2\right)^{s+1}$$

Let us deduce an equality for $z(x, t)$ that is analogous to (3.24). To this end, in the identity (3.22) with $t_1 = \tau$ we take the function $\eta(x, t)$ equal to

$$2v_{hx_l}\zeta^{2s+2}(w^h)^s z^{h(k)}$$

where k is an arbitrary positive number, and carry out a summation over l from 1 to n. Then the first term in (3.22) will have the form

$$\int_{Q_\tau} v_{hx_lt}2v_{hx_l}\zeta^{2s+2}(w^h)^s z^{h(k)}\,dx\,dt$$
$$=\frac{1}{2(s+1)}\int_\Omega [z^{h(k)}(x,\tau)]^2\,dx-\int_{Q_\tau}(w^h)^{s+1}\zeta^{2s+1}\zeta_t z^{h(k)}\,dx\,dt,$$

and therefore, passing to the limit as $h\to 0$, we obtain the following equality, similar to (3.24):

$$\frac{1}{s+1}\int_{A_{k,\rho}(\tau)}(z-k)^2\,dx+2\int_{\tau-\rho^2}^{\tau}\int_{A_{k,\rho}(t)}\{-w^{s+1}\zeta^{2s+1}\zeta_t(z-k)$$
$$+2a_{ij}v_{x_tx_j}v_{x_tx_l}\zeta^{2s+2}w^s(z-k)+sa_{ij}w_{x_l}w_{x_j}w^{s-1}\zeta^{2s+2}(z-k)$$
$$+(2s+2)a_{ij}w_{x_j}\zeta^{2s+1}\zeta_{x_l}w^s(z-k)+a_{ij}w_{x_j}w^s\zeta^{2s+2}z_{x_l}$$
$$+[\ldots]w^s\zeta^{2s+2}(z-k)\}\,dx\,dt=0. \tag{3.34}$$

Here the contents of the brackets $[\cdots]$ are the same as in (3.24), while $A_{k,\rho}(t)$ is the set of points x from K_ρ at which $z(x,t)>k$.

We carry out estimates in (3.34) with the use of conditions (3.2)–(3.6), imposing the restrictions (3.26) on s, as was done above in deriving (3.27). This leads us to the inequality

$$\int_{\tau-\rho^2}^{\tau}\int_{A_{k,\rho}(t)}(1+|u_x|)^{m-2}\left[\frac{s-1}{2}\frac{w_x^2}{w}+\frac{1}{4}|v_{xx}|^2-2\left(\frac{\varphi''}{\varphi'}\right)'w^2\right]$$
$$\times\zeta^{2s+2}w^s(z-k)\,dx\,dt\leqslant\frac{1}{\nu}\int_{\tau-\rho^2}^{\tau}\int_{A_{k,\rho}(t)}[w^{s+1}\zeta^{2s+1}|\zeta_t|(z-k)$$
$$+(2s+2)\mu_0(1+|u_x|)^{m-2}|w_x|\zeta^{2s+1}|\zeta_x|w^s(z-k)$$
$$-a_{ij}w_{x_j}w^s\zeta^{2s+2}z_{x_l}]\,dx\,dt+\int_{\tau-\rho^2}^{\tau}\int_{A_{k,\rho}(t)}(1+|u_x|)^{m-2}$$
$$\times\left\{c_1\left[\left(\frac{\varphi''}{\varphi'}\right)^2+|\varphi''||1+\varepsilon+\hat{P}|+(\varepsilon^2+P^2+2\varepsilon+P)\varphi'^2\right]w^2\right.$$
$$\left.+c_2w^{\frac{3}{2}}\right\}\zeta^{2s+2}w^s(z-k)\,dx\,dt; \tag{3.35}$$

here it is assumed that $k \geq 1$. The function $\phi(v)$ is determined from the conditions formulated above (see (3.28), (3.29) and following). For it

$$-2\left(\frac{\varphi''}{\varphi'}\right)' - c_1\Big[\left(\frac{\varphi''}{\varphi'}\right)^2 + |\varphi''|(1+\varepsilon+\hat{P})$$
$$+(2\varepsilon+2\varepsilon^2+P(|u_x|)+2P^2(|u_x|))\varphi'^2\Big] \geqslant c_3 > 0,$$

as long as the level k in (3.35) is taken sufficiently large, namely $k \geq k_1$, where k_1 is chosen from the condition

$$P(|u_x|)+2P^2(|u_x|) \leqslant \varepsilon. \tag{3.36}$$

Since $P(\rho)$ by definition tends to zero for $\rho \longrightarrow \infty$, while

$$|u_x| = \varphi'(v)|v_x| = \varphi'(v)w^{\frac{1}{2}} = \varphi'(v)\frac{z^{\frac{1}{2(s+1)}}}{\zeta} \geqslant \text{const}\, z^{\frac{1}{2(s+1)}},$$

i.e. for an unbounded increase in z we have an unbounded increase in $|u_x|$, it follows that inequality (3.36) can be satisfied by taking the levels k larger than a certain number $k_1 \geq 1$. For such levels k we will have from (3.35)

$$\int_{\tau-\rho^2}^{\tau}\int_{A_{k,\rho}(t)} (1+|u_x|)^{m-2}\left[\frac{s-1}{2}\frac{w_x^2}{w}+\frac{1}{4}|v_{xx}|^2+c_3w^2\right]$$
$$\times \zeta^{2s+2}w^s(z-k)\,dx\,dt \leqslant \frac{1}{\nu}\int_{\tau-\rho^2}^{\tau}\int_{A_{k,\rho}(t)}\Big[w^{s+1}\zeta^{2s+1}|\zeta_t|(z-k)$$
$$+(2s+2)\mu_1(1+|u_x|)^{m-2}|w_x|\zeta^{2s+1}\zeta_x w^s(z-k)$$
$$-a_{ij}w_{x_j}w^s\zeta^{2s+2}z_{x_i}\Big]\,dx\,dt$$
$$+c_2\int_{\tau-\rho^2}^{\tau}\int_{A_{k,\rho}(t)} (1+|u_x|)^{m-2}w^{s+\frac{3}{2}}\zeta^{2s+2}(z-k)\,dx\,dt.$$

We concern ourselves with a transformation of the integral

$$I_1 \equiv -\int_{\tau-\rho^2}^{\tau}\int_{A_{k,\rho}(t)} a_{ij}w_{x_j}w^s\zeta^{2s+2}z_{x_i}\,dx\,dt,$$

bearing in mind that

$$z=(w\zeta^2)^{s+1},\quad z_{x_i}=(s+1)\left[w^s w_{x_i}\zeta^{2s+2}+w^{s+1}2\zeta^{2s+1}\zeta_{x_i}\right]:$$

$$I_1=\int_{\tau-\rho^2}^{\tau}\int_{A_{k,\rho}(t)}\left(-\frac{1}{s+1}a_{ij}z_{x_j}z_{x_i}+a_{ij}w^{s+1}2\zeta^{2s+1}\zeta_{x_j}z_{x_i}\right)dx\,dt=$$

$$= \int_{\tau-\rho^2}^{\tau} \int_{A_{k,\rho}(t)} \Big[\frac{1}{s+1} a_{ij} z_{x_i} z_{x_j} + (z-k)$$

$$\times \frac{d}{dx_i}\Big(a_{ij} w^{s+1} 2\zeta^{2s+1}\zeta_{x_j}\Big)\Big] dx dt = -\int_{\tau-\rho^2}^{\tau} \int_{A_{k,\rho}(t)} \Big\{\frac{1}{s+1} a_{ij} z_{x_i} z_{x_j}$$

$$+ 2\Big[\frac{da_{ij}}{dx_i}\zeta^{2s+1}\zeta_{x_j} + a_{ij}(s+1) w^{-1} w_{x_i}\zeta^{2s+1}\zeta_{x_j}$$

$$+ (2s+1) a_{ij}\zeta^{2s}\zeta_{x_i}\zeta_{x_j} + a_{ij}\zeta^{2s+1}\zeta_{x_i x_j}\Big] w^{s+1}(z-k)\Big\} dx\, dt.$$

We estimate this expression from above, using assumptions (3.2)–(3.6):

$$I_1 \leqslant 2 \int_{\tau-\rho^2}^{\tau} \int_{A_{k,\rho}(t)} (1+|u_x|)^{m-2} \Big\{\Big[\frac{\mu_1}{1+\varphi'\sqrt{w}}(\varphi'|v_{xx}| + \varphi'' w)$$

$$+ (\varepsilon + P)\varphi'\sqrt{w} + (\varepsilon+P)(1+\varphi'\sqrt{w})\Big]\zeta^{2s+1}|\zeta_x|$$

$$+ (s+1)\mu_0 \frac{|w_x|}{w}\zeta^{2s+1}|\zeta_x| + \mu_0\zeta^{2s+1}|\zeta_{xx}|\Big\} w^{s+1}(z-k)\, dx\, dt.$$

From this and from (3.37) we get

$$\int_{\tau-\rho^2}^{\tau} \int_{A_{k,\rho}(t)} (1+|u_x|)^{m-2}\Big[\frac{s-1}{2}\frac{w_x^2}{w} + \frac{1}{4}v_{xx}^2 + c_3 w^2\Big]$$

$$\times w^s \zeta^{2s+2}(z-k)\, dx\, dt \leqslant \frac{1}{\nu}\int_{\tau-\rho^2}^{\tau} \int_{A_{k,\rho}(t)} w^{s+1}\zeta^{2s+1}|\zeta_t|(z-k)$$

$$+ \int_{\tau-\rho^2}^{\tau} \int_{A_{k,\rho}(t)} (1+|u_x|)^{m-2}\Big[c_4\Big(|w_x|\zeta^{2s+1}|\zeta_x| w^s$$

$$+ |v_{xx}| w^{s+\frac{1}{2}}\zeta^{2s+1}|\zeta_x| + w^{s+\frac{3}{2}}\zeta^{2s+1}|\zeta_x| + \zeta^{2s+1}|\zeta_{xx}| w^{s+\frac{3}{2}}\Big)$$

$$+ c_2 w^{s+\frac{3}{2}}\zeta^{2s+2}\Big](z-k)\, dx\, dt.$$

We replace the terms

$$c_4|w_x|\zeta^{2s+1}|\zeta_x| w^s \text{ and } c_4|v_{xx}| \times w^{s+\frac{1}{2}}\zeta^{2s+1}|\zeta_x|$$

in the right side by the larger terms

$$\frac{s-1}{2}w_x^2\zeta^{2s+2}w^{s-1}+\frac{c_4^2}{2(s-1)}w^{s+1}\zeta_x^2\zeta^{2s} \quad \text{and} \quad \frac{1}{4}|v_{xx}|^2w^s\zeta^{2s+2}+c_4^2w^{s+1}\zeta_x^2\zeta^{2s}$$

respectively. After this we transfer the terms containing $|v_{xx}|^2$ and w_x^2 to the left side and reduce similar terms.

As a result of all that has been said we obtain the inequality

$$\begin{aligned}\int_{\tau-\rho^2}^{\tau}\int_{A_{k,\rho}(t)}&(1+|u_x|)^{m-2}w^{s+2}\zeta^{2s+2}(z-k)\,dx\,dt\\ \leqslant\int_{\tau-\rho^2}^{\tau}\int_{A_{k,\rho}(t)}&(1+|u_x|)^{m-2}\Big[cw^{s+\frac{3}{2}}\zeta^{2s+2}+c_\rho\Big(w^{s+1}\zeta^{2s}\\ &+w^{s+\frac{3}{2}}\zeta^{2s+1}\Big)+c_\rho'w^{s+1}\zeta^{2s+1}w^{\frac{2-m}{2}}\Big](z-k)\,dx\,dt. \qquad (3.38)\end{aligned}$$

The constants c_ρ and c'_ρ depend on ζ_x^2, $|\zeta_{xx}|$ and $|\zeta_t|$ and tend to infinity like ρ^{-2} as $\rho\to 0$, but this is not important to us since, although ρ can be taken to be small, it is fixed. Noting that $w^{s+1}\zeta^{2s+2}=z$, we can rewrite (3.38) as follows:

$$\begin{aligned}\int_{\tau-\rho^2}^{\tau}\int_{A_{k,\rho}(t)}&(1+|u_x|)^{m-2}wz(z-k)\,dx\,dt\\ \leqslant\int_{\tau-\rho^2}^{\tau}\int_{A_{k,\rho}(t)}&(1+|u_x|)^{m-2}w\Bigg[cz^{\frac{s+\frac{1}{2}}{s+1}}+c_\rho\Big(z^{\frac{s}{s+1}}+z^{\frac{s+\frac{1}{2}}{s+1}}\Big)\\ &+c_\rho'z^{\frac{s+\frac{1}{2}}{s+1}}w^{\frac{1}{2}-\frac{m}{2}}\Bigg](z-k)\,dx\,dt.\end{aligned}$$

Since $k\geq 1$ it follows that $z\geq 1$ and $w\geq 1$ on the sets $A_{k,\rho}(t)$, and therefore $w^{(1-m)/2}\leq 1$ for $m\geq 1$. Thus inequality (3.39) is consistent with inequality (3.31) for $m\geq 1$. It yields an estimate of vrai $\max_{Q_\rho}z$ in the same way as (3.31) yields an estimate of vrai $\max_{Q_T}w$. The quantity vrai $\max_{Q_\rho}z$ is determined by the constants c, c_ρ and c'_ρ in (3.39) and by the height of the level $k_1\geq 1$, beginning with that at which (3.39) holds. Knowing vrai $\max_{Q_\rho}z$, we will also know estimates of vrai max w and vrai max $|u_x|$ in the cylinder $Q_{\rho/2}$.

Thus, by introducing the cutting function $\zeta(x,t)$, we have estimated $|u_x|$ in any cylinder $Q_\rho=K_\rho\times(\tau-\rho^2,\tau)$, using the value of $\max|u|$ in a somewhat larger cylinder $K_{\rho_1}\times(\tau-\rho_1^2,\tau)$, $\rho_1>\rho$. In this connection we must assume that ϵ

is sufficiently small in inequalities (3.5)–(3.6). This condition of smallness can be dropped if it is assumed that we know the modulus of continuity of the function $u(x, t)$ in Q_T, i.e. the function $\omega(\rho) = \sup \operatorname{osc} \{u; Q_\rho \cap Q_T\}$, where the supremum is taken over all of Q_ρ, and $\omega(\rho) \to 0$ for $\rho \to 0$. As was shown in the proof of Theorem 3.3, it is possible to pick out the necessary function $\phi(v)$ (see (3.28), (3.29) and following) for any ϵ and c_1 if it is known that the oscillation of $u(x,t)$ in the domain under consideration (in this case the cylinder $Q_{\rho/2}$) does not exceed a known quantity ω_0. Therefore, assuming ρ so small that $\omega(\rho) \le \omega_0$, we can estimate $\max |u_x|$ in the cylinder $Q_{\rho/2}$ without any assumptions on the smallness of ϵ. If it is assumed that we know $\max |u|$ and the modulus of continuity of $u(x, t)$ for the whole cylinder Q_T, then it follows from what has just been said that we can estimate $\max|u_x|$ for any cylinder of the form $Q' = \Omega' \times (\delta, T)$, where $\Omega' \subset \Omega$ and $\delta > 0$. To do so it is necessary to cover Q' by a finite number of cylinders $Q_{\rho/2}$ with sufficiently small ρ (so that $\omega(\rho) \le \omega_0$) and to make use of the estimate of $\max |u_x|$ for each of these cylinders $Q_{\rho/2}$.

If $\max_{K_\rho}|u_x(x, 0)| = M_0$ is already known, then we can also give estimates of $\max|u_x|$ for the cylinders $Q' = K_\rho \times (0, \tau)$, $\tau \le T$, and in this case we can assume that m is arbitrary. Indeed, in this case it is necessary in equality (3.34) and below to subject k to the restriction $k \ge M_0$ and to take a function ζ that is independent of t. The term containing c_ρ' in (3.39) will then be equal to zero, and this was the only term whose estimate above required the restriction $m \ge 1$.

We can also give local estimates of $\max |u_x|$ for cylinders intersecting S_T. If it is known that $\max_{\Gamma_T} |u_x| \le M_0$, then it is necessary to assume in (3.34) and below that $k \ge \max_{Q_\rho \cap \Gamma_T} z$. We note that it is also possible to make an estimate of $|u_x|$ on S_T not for the whole surface S_T at once (as was done above in Lemma 3.1) but locally, on a separately taken piece $S_\rho = S_T \cap Q_\rho$ of it, without the use of any information on the behavior of $u(x, t)$ in the vicinity of the remaining part of S_T.

We formulate what has been proved on local estimates in the form of two theorems.

Theorem 3.4. *Suppose a function $u(x, t)$ is bounded in the cylinder Q_T, has generalized derivatives of the form u_x, u_{xx}, u_t in Q_T and satisfies equation (3.1) almost everywhere in $\overline{Q}_T \setminus \Gamma_T$. Suppose further that for $(x, t) \in \overline{Q}_T$ the coefficients of the equation are subject to inequalities (3.2)–(3.6) with $m \ge 1$ and that for some s satisfying inequality (3.26) conditions (3.14) are fulfilled with* *respect to $u(x, t)$. Then for any cylinder $Q' = \Omega' \times (\delta, T)$, $\Omega' \subset \Omega$, $\delta > 0$, the quantity* $\operatorname{vrai\,max}_{Q'}|u_x(x, t)|$ *does not exceed a certain constant depending only on the distance from Q' to Γ_T, n, $M = \max_{Q_T}|u|$, $\nu = \nu(M)$, $\mu_0 = \mu_0(M)$, $\mu = \mu(M)$, $\mu_1 = \mu_1(M)$, $\epsilon = \epsilon(M)$, $P(|p|) = P(|p|, M)$ and m from inequalities (3.2)–(3.6), where $\epsilon(M)$ must be sufficiently small. If $u(x, t)$ is continuous in $\overline{Q}_T$, then* $\operatorname{vrai\,max}_{Q'}|u_x(x, t)|$ *can be estimated in terms of these same quantities and the modulus of continuity $\omega(\rho)$ of $u(x, t)$ in Q_T for an arbitrary (not necessarily small) number $\epsilon(M)$.*

If $\operatorname{vrai\,max}_{\Omega}|u_x(x, 0)| \le M_0$, *then, regardless of the value of m in conditions (3.2)–(3.6),* $\operatorname{vrai\,max}|u_x|$ *in the cylinder $Q' = \Omega' \times (0, T)$, where Ω' is an arbitrary interior subdomain of Ω, is estimated in terms of the distance from Ω' to S, n, M_0, M, ν, μ_0, μ, μ_1, ϵ and $P(|p|)$ if $\epsilon = \epsilon(M)$ is small, and also in terms of the modulus of continuity $\omega(\rho)$ of $u(x, t)$ in Q_T if $\epsilon(M)$ is arbitrary but u is continuous in $\overline{Q}_T$.*

Theorem 3.5. *Suppose all of the assumptions of Theorem 3.2 are fulfilled except the conditions on the boundary S and on the function $\psi(x, t)$, which are replaced by the following conditions: a section S' of the boundary S belongs to the class O^2 and on a certain part $S_T' \subset S' \times [0, T]$ the function $u(x, t)|_{S_T'} = \psi(s, t) \in O^{2,1}(S_T')$. Suppose further that the number m in conditions (3.2)–(3.6) is not less than one. Then, for any part Q' of the cylinder $\overline{Q}_T$ that is separated from $\Gamma_T \backslash S_T'$ by a positive distance d, the quantity* $\operatorname{vrai\,max}_{Q'}|u_x|$ *is estimated by a constant depending only on S', $|\psi|_{S_T'}^{(2)}$, d, n, M, m, $\nu(M)$, $\mu_0(M)$, $\mu(M)$, $\mu_1(M)$, $\epsilon(M)$ and $P(|p|, M)$ from (3.2)–(3.6) if $\epsilon(M)$ is sufficiently small, and also on the modulus of continuity $\omega(\rho)$ of $u(x, t)$ in $\overline{Q}_T$ if $\epsilon(M)$ is arbitrary but u is continuous in $\overline{Q}_T$.*

Let $S_\tau' = S' \times [0, \tau]$, $\tau \le T$ and $\operatorname{vrai\,max}_{\Omega}|u_x(x, 0)| \le M_0$, *and let Γ_τ' denote the union of the lower base of the cylinder $\overline{Q}_T$ and the section S_τ' of the surface S_T. Then, for any subdomain Q' of the cylinder Q_T that is separated from $S_\tau \backslash S_\tau'$ by a positive distance d,* $\operatorname{vrai\,max}_{Q'}|u_x|$ *does not exceed a constant depending only on S', $|\psi|_{S_\tau'}^{(2)}$, n, d, M_0, M, m, $\nu(M)$, $\mu_0(M)$, $\mu(M)$, $\mu_1(M)$, $\epsilon(M)$, and $P(|p|, M)$ if $\epsilon(M)$ is small, and also on the modulus of continuity $\omega(\rho)$ of $u(x, t)$ in Q_T if $\epsilon(M)$ is arbitrary but u is continuous in $\overline{Q}_T$.*

Remark 3.1. For certain classes of equations (0.1) it is possible to obtain an estimate of $\max_{Q_T}|u_x|$ in terms of $\max_{\Gamma_T}|u_x|$ or the known quantities in the problem by imposing no restrictions on the orders of growth of the functions a_{ij}

and a with respect to p or by imposing weaker restrictions than the conditions of Theorems 3.1–3.5. As an example we take an equation of the form

$$\mathscr{L}u = u_t - a_{ij}(t, u_x)u_{x_ix_j} + a(t, u_x) = 0 \tag{3.40}$$

and we will assume for the sake of simplicity that all functions and solutions are sufficiently smooth.

From the equality

$$0 = (\mathscr{L}u)_{x_l} u_{x_l} = \frac{1}{2}\frac{\partial u_x^2}{\partial t} - \frac{1}{2}a_{ij}\frac{\partial^2 u_x^2}{\partial x_i \partial x_j} + a_{ij}u_{x_ix_l}u_{x_lx_j} - \frac{1}{2}\frac{\partial a_{ij}}{\partial x_k}u_{x_ix_j}\frac{\partial u_x^2}{\partial x_k} + \frac{1}{2}\frac{\partial a}{\partial u_{x_k}}\frac{\partial u_x^2}{\partial x_k}$$

we get the inequality

$$v_t - a_{ij}v_{x_ix_j} + A_k v_{x_k} \leqslant 0,$$

in which $v = u_x^2$, while

$$A_k = -\frac{\partial a_{ij}}{\partial u_{x_k}}u_{x_ix_j} + \frac{\partial a}{\partial u_{x_k}},$$

and it implies, as is known (see §2 of Chapter I), that v takes its greatest value on Γ_T, i.e.

$$v(x, t) \leqslant \max_{\Gamma_T} v. \tag{3.41}$$

This inequality reduces the problem of estimating $\max_{Q_T} u_x^2$ to that of estimating the maximum of u_x^2 on Γ_T. But in order to estimate $\max_{\Gamma_T} u_x^2$ we must take into account the boundary and initial conditions. (See in this connection §7 of Chapter V and Lemma 3.1 of this section.)

§4. EXISTENCE THEOREMS

In §6 of the preceding chapter we showed how the solvability of nonlinear problems is established by means of (i) the Leray-Schauder principle, (ii) a priori estimates of $|u_x|_{Q_T}^{(\alpha)}$ for all possible solutions of these nonlinear problems, and (iii) solvability theorems in Hölder classes for linear problems. Here the question as to whether or not the equation in question has its principal part in divergence form is not important. The arguments presented in §6 of Chapter V for equations with principal part in divergence form remain valid for the general quasi-linear equations

$$\mathscr{L}u \equiv u_t - a_{ij}(x, t, u, u_x)u_{x_ix_j} + a(x, t, u, u_x) = 0, \tag{4.1}$$

$$u|_{\Gamma_T} = \psi|_{\Gamma_T}. \tag{4.2}$$

The only difference lies in the conditions under which the a priori estimates of the solutions of the various equations are established, and in the equation with which the given equation is connected by a parameter. In view of this we will not repeat here all of the arguments in §6 of Chapter V but formulate only the final results (and only those that are fundamental, without their corollaries), using the a priori estimates obtained in the present chapter for the solutions of problem (4.1), (4.2), and give whatever explanations are necessary.

In the same way as in §6 of Chapter V, we consider problem (4.1), (4.2) not by itself but along with a one-parameter family of problems of the same type:

$$\mathscr{L}_\tau u \equiv u_t - \left[\tau a_{ij}(x, t, u, u_x) + (1-\tau)(1+u_x^2)^{\frac{m}{2}-1}\delta_i^j\right] u_{x_i x_j}$$
$$+\tau a(x, t, u, u_x) - (1-\tau)\left[\psi_t - (1+\psi_x^2)^{\frac{m}{2}-1}\Delta\psi\right] = 0,$$
$$u|_{\Gamma_T} = \psi|_{\Gamma_T}, \quad 0 \leqslant \tau \leqslant 1, \text{ assuming } \psi \in H^{2+\beta, 1+\frac{\beta}{2}}(\bar{Q}_T). \tag{4.3}$$

The linear problem

$$v_t - \left[\tau a_{ij}(x, t, w, w_x) + (1-\tau)(1+w_x^2)^{\frac{m}{2}-1}\delta_i^j\right] v_{x_i x_j}$$
$$+\tau a(x, t, w, w_x) - (1-\tau)\left[\psi_t - (1+\psi_x^2)^{\frac{m}{2}-1}\Delta\psi\right] = 0,$$
$$v|_{\Gamma_T} = \psi|_{\Gamma_T}, \tag{4.4}$$

determines a transformation $\Phi(w, \tau) = v$ of w into v, to which the Leray-Schauder principle is applied.

Concerning the functions $a_{ij}(x, t, u, p)$ and $a(x, t, u, p)$ we assume fulfilled either conditions (2.29), (2.30) of Chapter I or, more generally, conditions (2.29), (2.32), (2.33) of Chapter I. Either of these conditions, as was proved in Theorem 2.9 of Chapter I, yield the estimate

$$\max_{Q_T} |u^\tau| \leqslant M \tag{4.5}$$

for all possible classical solutions of problems (4.3).

We suppose further that for $(x, t) \in \bar{Q}_T$, $|u| \leq M$ and arbitrary p the functions $a_{ij}(x, t, u, p)$, $a(x, t, u, p)$ are continuous in x, t, u, p, continuously differentiable with respect to x, u and p, and satisfy the inequalities

$$\nu(1+|p|)^{m-2}\xi^2 \leqslant a_{ij}(x, t, u, p)\xi_i\xi_j \leqslant \mu(1+|p|)^{m-2}\xi^2,$$
$$\nu, \mu = \text{const} > 0, \tag{4.6}$$
$$\left|\frac{\partial a_{ij}}{\partial p_k}\right|(1+|p|)^3 + |a| + \left|\frac{\partial a}{\partial p_k}\right|(1+|p|) \leqslant \mu_1(1+|p|)^m. \tag{4.7}$$

$$\left|\frac{\partial a_{ij}}{\partial x_k}\right|(1+|p|)^2+\left|\frac{\partial a}{\partial x_k}\right|$$
$$\leqslant[\varepsilon+P(|p|)](1+|p|)^{m+1}, \quad \varepsilon>0, \tag{4.8}$$
$$\left|\frac{\partial a_{ij}}{\partial u}\right|\leqslant[\varepsilon+P(|p|)](1+|p|)^{m-2}, \tag{4.9}$$
$$-\frac{\partial a}{\partial u}\leqslant[\varepsilon+P(|p|)](1+|p|)^m, \tag{4.10}$$

where $P(\rho)$ is a nonnegative continuous function that tends to zero for $\rho\to\infty$ and m is an arbitrary number. One can easily verify that the functions

$$a^{\tau}_{ij}(x,t,u,p)=\tau a_{ij}(x,t,u,p)+(1-\tau)(1+p^2)^{\frac{m-2}{2}}\delta^j_i$$

and

$$-(1-\tau)\left[\psi_t-(1+\psi_x^2)^{\frac{m}{2}-1}\Delta\psi\right]+\tau a\ (x,\ t,\ u,\ p),$$

are subject to inequalities of the same type, where the constants ν, μ, μ_1, ϵ and the functions $P(|p|)$ in them can be taken to be the same for all $\tau\in[0,1]$. In view of this, as was proved in Theorem 3.2, when $S\in O^2$ the $|u^\tau_x|$ are uniformly bounded for all solutions u^τ possessing the smoothness indicated in this theorem:

$$\max_{Q_T}|u^\tau_x|\leqslant M_1, \tag{4.11}$$

as long as the number ϵ in (4.8)–(4.10) is less than a certain number determined by the numbers M, ν, μ, μ_1 and

$$\hat{P}=\max_{\rho\geqslant 0}P(\rho),$$

or if all of the solutions u^τ are known to be equicontinuous in $\bar{Q}_T$.

We now formulate the existence theorems.

Theorem 4.1. *Suppose that the following conditions are fulfilled.*

a) *For* $(x,t)\in\bar{Q}_T$ *and arbitrary* u *either conditions* (2.29), (2.30) *or, more generally,* (2.29), (2.32), (2.33) *of Chapter* I *are fulfilled.*

b) *For* $(x,t)\in\bar{Q}_T$, $|u|\leq M$ (*where* M *is taken from estimate* (4.5)) *and arbitrary* p *the functions* $a_{ij}(x,t,u,p)$ *and* $a(x,t,u,p)$ *are continuous and differentiable with respect to* x, u *and* p *and satisfy inequalities* (4.6)–(4.10) *with a sufficiently small* ϵ *determined by the numbers* M, ν, μ, μ_1 *and*

$$\hat{P}=\max_{\rho\geqslant 0}P(\rho).$$

c) *For* $(x,t)\in\bar{Q}_T$, $|u|\leq M$ *and* $|p|\leq M_1$ (*where* M_1 *is taken from* (4.11)) *the functions* $a_{ij}(x, t, u, p)$ *and* $a(x, t, u, p)$ *are continuously differentiable with respect to all of their arguments.*

d) *The boundary condition* (4.2) *is given by a function* $\psi(x, t)$ *belonging to* $H^{2+\beta, 1+\beta/2}(\bar{Q}_T)$ *and satisfying on* $S_0 = \{(x, t): x \in S, t = 0\}$ *equation* (4.1), *i.e.*

$$\psi_t - a_{ij}(x, 0, \psi(x, 0), \psi_x(x, 0))\psi_{x_ix_j}(x, 0) + a(x, 0, \psi(x, 0)\psi_x(x, 0))|_{x\in S} = 0 \tag{4.12}$$

(*in other words, the compatibility conditions of zero and first orders are assumed to be fulfilled*).

e) $S \in H^{2+\beta}$.

Then there exists a unique solution of problem (4.1), (4.2) *in the space* $H^{2+\beta,1+\beta/2}(\bar{Q}_T)$. *This solution has derivatives* u_{tx_i} *from* $L_2(Q_T)$.

Using this theorem, the theorem on estimating $\max_{Q_T}|u_x|$ and theorems on interior estimates of the solutions in stronger norms, we see that the following theorem is valid.

Theorem 4.2. *Suppose conditions* a) *and* b) *of Theorem* 4.1 *are fulfilled, and that*

c) *For* $(x, t) \in \bar{Q}_T$, $|u| \leq M$ *and* $|p| \leq M_1$ *the functions* $a_{ij}(x, t, u, p)$ *and* $a(x, t, u, p)$ *are continuously differentiable with respect to* x, u *and* p *and Hölder continuous in* t *with exponent* $\beta/2$.

d) $$\psi|_{S_T} \in O^{2,1}(S_T); \quad \max_{\Omega} |\psi_x(x, 0)| < \infty, \text{ and}$$

$$\psi \in C(\bar{Q}_T)$$

e) $S \in O^2$.

Then there exists a solution $u(x, t)$ *of problem* (4.1), (4.2) *which belongs to* $C(\bar{Q}_T) \cap H^{2+\beta,1+\beta/2}(Q_T)$ *and has a bounded* $\sup_{Q_T'}|u_x|$

Theorem 4.1 is proved analogously to Theorem 6.1 of Chapter V. Under the conditions of Theorems 4.1 and 4.2 the requirement that ϵ be small in inequalities (4.8)–(4.10) can be replaced, following Theorem 5.5, by the condition that all of the solutions u^τ be equicontinuous in $\bar{Q}_T$. This change in Theorems 4.1 and 4.2 leads to the following existence theorem.

Theorem 4.3. *The assertions of Theorems* 4.1 *and* 4.2 *remain valid if the condition in them that* ϵ *be small in inequalities* (4.8)–(4.10) *is replaced by the* *requirement that all of the possible solutions of problems* (4.3) *be equicontinuous in* $\overline{Q}_T$.

Also we can weaken our assumptions on S and ψ at a cost in smoothness of the solution in $\overline{Q}_T$. For example, the following theorem holds:

Theorem 4.4. *Suppose conditions* a)–c) *of Theorem* 4.1 *are fulfilled with* $m \geqslant 1$

d) $\psi \in C(\overline{Q}_T) \cap H^{2+\beta,\, 1+\frac{\beta}{2}}(Q_T)$.

e) *Each point of* S *can be touched from without by a ball* (*or cone*) *of fixed size in such a way that the ball* (*cone*) *does not have any points in common with* Ω.

Then problem (4.1), (4.2) *has a solution from* $C(\overline{Q}_T) \cap H^{2+\beta,1+\beta/2}(Q_T)$.

Theorems 4.2 and 4.4 are proved in the same way as Theorems 6.2–6.4 of Chapter V, by means of an approximation of the equation, the boundary and initial conditions and the domain itself by a smoother problem satisfying the conditions of Theorem 4.1, and with the use of the theorems of the preceding sections on a priori estimates in $\overline{Q}_T$ and in $Q' \in Q_T$. The specific character of the conditions on S and ψ in them is such that we will now have for closed domains $\overline{Q}^m$ a single uniform estimate

$$\max_{Q^m} |u_m(x, t)| \leqslant M \tag{4.13}$$

instead of $|u_m|^{(\alpha)}_{Q^m} \leq M$ as in previous cases. Estimate (4.13) does not yield the conclusion that the limit function for the u_m is continuous in $\overline{Q}_T$. This must be proved separately by using the fact that the u_m are solutions of parabolic equations. In this regard one employs the classical method of barriers, which was developed in the first decades of our century. The method of barriers for parabolic equations differs little from that for equations of elliptic type. In either case it is based on the "maximum principle," which is inherent in equations of second order of both types. We will not present the proof of Theorem 4.4, but refer to §3 of Chapter VI in our book [65q], where this is done in detail for quasi-linear elliptic equations of general form.

§5. EQUATIONS WITH ONE SPACE VARIABLE

For the case of equations

$$\mathscr{L}u \equiv u_t - a_{11}(x, t, u, u_x)\, u_{xx} + a(x, t, u, u_x) = 0 \tag{5.1}$$

with one space variable the arguments and results mentioned above admit certain simplifications and sharpenings. As above, we confine ourselves to the domains $Q_T = (-l, l) \times (0, T)$, although the methods presented here are applicable without any essential change to domains of the form $\{(x, t): \phi_1(t) \le x \le \phi_2(t)\}$ with $\phi_i'(t) \neq \infty$. Moreover, in this section, to save space we confine ourselves to classical solutions and continuous functions $a_{11}(x, t, u, p)$, and $a(x, t, u, p)$. From what has been stated above it is clear as to what generalizations of the propositions given below are possible in regard to the admission of various singularities of the functions $a_{11}(x, t, u, p)$ and $a(x, t, u, p)$ and of the solutions $u(x, t)$.

For a smooth function $a_{11}(x, t, u, p)$ equations (5.1) can be reduced to the form of equations with principal part in divergence form, namely

$$u_t - \frac{d}{dx} a_1(x, t, u, u_x) + b(x, t, u, u_x) = 0, \tag{5.2}$$

in which

$$a_1(x, t, u, p) = \int_0^p a_{11}(x, t, u, \tau)\, d\tau,$$

$$b(x, t, u, p) = a(x, t, u, p) + \frac{\partial a_1(x, t, u, p)}{\partial u} p + \frac{\partial a_1(x, t, u, p)}{\partial x}.$$

In view of this the theorems proved in Chapters V and VI can be applied to equations (5.1). Moreover, for their solutions u it is possible to estimate $|u_x|^{(\alpha)}_{Q_T}$ with respect to the whole domain Q_T without a preliminary estimate of $\max_{Q_T}|u_t|$ and hence without any assumptions concerning the differentiability with respect to t of the functions making up the equation. Let us demonstrate this fact.

Theorem 5.1. *Suppose that $u(x, t)$ is a solution of equation (5.1) that belongs to $C^{2,1}(\bar{Q}_T)$ and suppose that for $(x, t) \in \bar{Q}_T$*

$$\begin{gathered} \nu \leqslant a_{11}(x, t, u(x, t), u_x(x, t)) \leqslant \mu, \\ |a(x, t, u(x, t), u_x(x, t)| \leqslant \mu, \quad \nu, \mu = \text{const} > 0 \end{gathered} \tag{5.3}$$

Then it is possible to estimate the quantity $\langle u_x\rangle^{(\alpha)}_{Q'}$, $\alpha > 0$, for any rectangle $Q' = (-l + \epsilon, l - \epsilon) \times (\epsilon_1, T)$ with $\epsilon, \epsilon_1 > 0$, from above by a constant depending only on $\nu, \mu, M_1 = \max_{Q_T}|u_x|$, ϵ and ϵ_1. For $\epsilon_1 = 0$ the quantity $\langle u_x\rangle^{(\alpha)}_{Q'}$ is estimated from above by a constant depending only on ν, μ, M_1, ϵ, $\langle u_x(x, 0)\rangle^{(\beta)}_{(-l,l)}$ and $\beta > 0$. For the whole rectangle $\langle u_x\rangle^{(\alpha)}_{Q_T}$ is estimated from above by a constant depending only on $\nu, \mu, M_1, \langle u_x(x, 0)\rangle^{(\beta)}_{(-l,l)}$, β and $\max_{t\in[0,T]}(|u_t(-l,t)|, |u_t(l, t)|)$. The constant α is determined by ν, μ, M_1 and β.

For a proof we consider the relation

$$-\int_{t_0}^{t}\int_{-l}^{l} \mathscr{L}u\,(u_x^{(k)}\zeta^2)_x\, dx\, dt = 0, \tag{5.4}$$

where $u_x^{(k)}(x, t) = \max\{u_x(x, t) - k; 0\}$, k is an arbitrary number, $\zeta(x, t)$ is a smooth function with values from $[0, 1]$ that is equal to zero outside of $K_\rho = \{x : |x - x_0| < \rho\}$, while $(x_0, t_0) \in \overline{Q}_T$ and $\rho \leq l$. If the segment K_ρ is contained in $(-l, l)$, then, by carrying out an integration by parts in (5.4), we obtain

$$\frac{1}{2}\int_{-l}^{l}[u_x^{(k)}(x, t)\zeta(x, t)]^2 dx\Big|_{t_0}^{t} + \int_{t_0}^{t}\int_{-l}^{l}[-(u_x^{(k)})^2\zeta\zeta_t + (a_{11}u_{xx} - a)(u_{xx}^{(k)}\zeta^2 + 2u_x^{(k)}\zeta\zeta_x)]\,dx\,dt = 0. \tag{5.5}$$

But if one of the ends of the interval $[-l, l]$, for example $x = l$, is contained in K_ρ then we transform the first term of (5.4) as follows:

$$-\int_{t_0}^{t}\int_{-l}^{l} u_t(u_x^{(k)}\zeta^2)_x\,dx\,dt$$

$$= \int_{t_0}^{t}\int_{-l}^{l}\{[-u_t(x, t) + u_t(l, t)](u_x^{(k)}\zeta^2)_x - u_t(l, t)(u_x^{(k)}\zeta^2)_x\}\,dx\,dt$$

$$= \int_{t_0}^{t}\int_{-l}^{l}[u_{tx}u_x^{(k)}\zeta^2 - u_t(l, t)(u_x^{(k)}\zeta^2)_x]\,dx\,dt$$

$$= \frac{1}{2}\int_{-l}^{l}(u_x^{(k)}\zeta)^2 dx\Big|_{t_0}^{t} - \int_{t_0}^{t}\int_{-l}^{l}[(u_x^{(k)})^2\zeta\zeta_t + u_t(l, t)(u_x^{(k)}\zeta^2)_x]\,dx\,dt.$$

As a result we arrive at equality (5.5), in which $a(x, t, u, p)$ is replaced by $\tilde{a}(x, t, u, p) = a(x, t, u, p) + u_t(l, t)$. The fact that in deriving (5.5) it was assumed that the solution has derivatives u_{xt} from $L_2(Q_T)$ is nonessential. As was explained many times above, this can be avoided in various ways; for example, one takes instead of (5.4) the relation

$$\int_{t_0}^{t}\int_{-l}^{l}[-u_{ht} + (a_{11}u_{xx} - a)](u_{hx}^{(k)}\zeta^2)_x\,dx\,dt = \int_{t_0}^{t}\int_{-l}^{l}(u_t - u_{ht})(u_{hx}^{(k)}\zeta^2)_x\,dx\,dt,$$

where

$$u_h(x, t) = \frac{1}{h}\int_{t}^{t+h} u(x, \tau)\,d\tau,$$

transforms its first term in the same way as above, and then passes to the limit as $h \to 0$.

From (5.5) and the assumptions of Theorem 5.1, as can easily be seen, we get the inequality

$$\int_{\Omega_\rho} (u_x^{(k)}\zeta)^2\,dx\Big|_{t_0}^{t} + \nu\int_{t_0}^{t}\int_{\Omega_\rho} (u_{xx}^{(k)}\zeta)^2\,dx\,dt$$

$$\leqslant \gamma_1\left[\int_{t_0}^{t}\int_{\Omega_\rho} (u_x^{(k)})^2(\zeta_x^2+\zeta|\zeta_t|)\,dx\,dt + \int_{t_0}^{t} \operatorname{mes} A_{k,\rho}(t)\,dt\right],$$

in which $\Omega_\rho = K_\rho\times(-l, l)$, $A_{k,\rho}(t)$ is the set of those points x from Ω_ρ at which $u_x(x, t) > k$, while the constant γ_1 depends only on ν and μ for $K_\rho \subset (-l, l)$, but also on $\max_{[t_0, t]} |u_t(\pm l, t)|$ if $x = l$ or $x = -l$ lies in K_ρ. Analogous inequalities are valid for the function $-u_x(x, t)$. In view of this (see §§7 and 8 of Chapter II) the function $u_x(x, t)$ belongs to the class $\hat{\mathfrak{B}}_2(\overline{Q}_T, M_1, \gamma, 6, \infty, 2)$ with $\gamma = 2\gamma_1/\min\{1; \nu\}$, and therefore the assertions of Theorem 5.1 hold for u_x by virtue of Theorems 7.1, 8.1 and 8.2 of Chapter II.

Let us revert to the estimates of $\max|u_x|$ for solutions of equation (5.1). In §3 we gave estimates of $\max|u_x|$ directly for any $n \geq 1$. The assumptions under which such estimates are possible can be weakened somewhat for $n = 1$. If $n = 1$, $m = 2$ in conditions (3.2), then instead of conditions (3.3)–(3.6) we can take the condition

$$\left|\frac{\partial a_{11}(x, t, u, p)}{\partial p}\right|(1+|p|)^3 + \left|\frac{\partial a_{11}(x, t, u, p)}{\partial u}\right|(1+|p|)^2$$
$$+\left|\frac{\partial a_{11}(x, t, u, p)}{\partial x}\right|(1+|p|) + |a(x, t, u, p)|$$
$$\leqslant \mu(|u|)(1+|p|)^2. \tag{5.6}$$

for under these assumptions equation (5.1), when written in the form (5.2), satisfies all of the conditions in §§3 and 4 of Chapter V, and therefore the theorems concerning the estimation of $\max_{Q'}|u_x|$ and $\max_{Q_T}|u_x|$ established in these sections are valid for it.

For $n = 1$ and $m \neq 2$ we can eliminate from assumptions (3.2)–(3.6) of this chapter the requirement that $\epsilon(|u|)$ be small in condition (3.6) with respect to $\partial a_{11}/\partial u$. This can easily be seen by tracing the estimates of the corresponding terms in inequality (3.25) and noting that for $n = 1$ it is possible to represent the function $|v_{xx}|$ in the form $(1/2)w^{-1/2}|w_x|$ and to cancel the terms corresponding to it at the expense of a term existing in the left side of the inequality.

We will now formulate a theorem on the solvability of the first boundary value problem for equation (5.1), which follows from the just-described estimates of $\max_{Q_T}|u_x|$ and $\langle u_x\rangle^{(\alpha)}_{Q_T}$. As above, we take a family of boundary value problems in Q_T:

$$\mathscr{L}_\tau u \equiv u_t - \left[\tau a_{11}(x, t, u, u_x) + (1-\tau)(1+u_x^2)^{\frac{m}{2}-1}\right]u_{xx}$$
$$+\tau a(x, t, u, u_x) - (1-\tau)\left[\psi_t - (1+\psi_x^2)^{\frac{m}{2}-1}\psi_{xx}\right] = 0,$$
$$u|_{t=0} = \psi(x, 0), \quad u(-l, t) = \psi(-l, t), \quad u(l, t) = \psi(l, t),$$
$$0 \leqslant \tau \leqslant 1, \tag{5.7}$$

where $\psi(x, t)$ is a known function defined in $\overline{Q}_T$. The following proposition is valid for them:

Theorem 5.2. *Suppose the following conditions are fulfilled.*

a) *For $(x, t) \in \overline{Q}_T$ and arbitrary u*

$$a_{11}(x, t, u, 0) \geqslant 0 \tag{5.8}$$

and

$$-a(x, t, u, 0)u \leqslant |u|\Phi(|u|), \quad \int_1^\infty \frac{d\tau}{\Phi(\tau)} = \infty \tag{5.9}$$

(this yields the a priori estimate $\max_{Q_T}|u^\tau| \leq M$ for all possible (classical) solutions of problems (5.7)).

b) *For $(x, t) \in \overline{Q}_T$, $|u| \leq M$ and arbitrary p, the functions $a_{11}(x, t, u, p)$ and $a(x, t, u, p)$ are continuous, differentiable with respect to x, u and p and satisfy the inequalities*

$$\nu(1+|p|)^{m-2} \leqslant a_{11}(x, t, u, p) \leqslant \mu(1+|p|)^{m-2}, \quad \nu > 0, \tag{5.10}$$

$$\left|\frac{\partial a_{11}(x, t, u, p)}{\partial p}\right|(1+|p|)^3 + \left|\frac{\partial a_{11}}{\partial u}\right|(1+|p|)^2$$
$$+\left|\frac{\partial a}{\partial p}\right|(1+|p|) + |a| \leqslant \mu_1(1+|p|)^m, \tag{5.11}$$

$$\left|\frac{\partial a_{11}}{\partial x}\right|(1+|p|)^2 + \left|\frac{\partial a}{\partial x}\right| \leqslant [\varepsilon + P(|p|)](1+|p|)^{m+1} \tag{5.12}$$

$$-\frac{\partial a(x, t, u, p)}{\partial u} \leqslant [\varepsilon + P(|p|)](1+|p|)^m, \tag{5.13}$$

where m is an arbitrary number, $P(\rho)$ is a nonnegative continuous function that tends to zero for $\rho \to \infty$, and ϵ is a positive number that is less than a certain *number determined by the quantities* M, ν, μ, μ_1 *and*

$$\hat{P} = \max_{\rho > 0} P(\rho)$$

(inequalities (5.10)–(5.13) *guarantee the estimate* $\max_{Q_T} |u_x^\tau| \le M_1$).

c) *For* $(x, t) \in \bar{Q}_T$, $|u| \le M$ *and* $|p| \le M_1$ *the functions* $a_{11}(x, t, u, p)$ *and* $a(x, t, u, p)$ *are Hölder continuous in the variable* t *with exponent* $\beta/2$, *and in the variables* x, u *and* p *with exponent* β.

d) *The function* $\psi(x, t) \in H^{2+\beta, 1+\beta/2}(\bar{Q}_T)$ *and*

$$[\psi_t(x, 0) - a_{11}(x, 0, \psi(x, 0), \psi_x(x, 0))\psi_{xx}(x, 0) + a(x, 0, \psi(x, 0), \psi_x(x, 0))]|_{x=\pm l} = 0. \tag{5.14}$$

Then problems (5.7) *are uniquely solvable in* $H^{2+\beta, 1+\beta/2}(\bar{Q}_T)$ *for all* τ *from* $[0, 1]$.

Remark 5.1. Condition b) in Theorem 5.2 can be replaced by the following condition.

b′) *For* $(x, t) \in \bar{Q}_T$, $|u| \le M$ *and arbitrary* p, *the functions* $a_{11}(x, t, u, p)$ *and* $a(x, t, u, p)$ *are continuous*, $a_{11}(x, t, u, p)$ *is differentiable with respect to* x, u *and* p, *and the inequalities* (5.10) *with* $m = 2$ *and* (5.6) *are satisfied.*

Remark 5.2. Condition b) in Theorem 5.2 can also be replaced by the following condition.

b″) *For* $(x, t) \in \bar{Q}_T$, $|u| \le M$ *and arbitrary* p, *the functions* $a_{11}(x, t, u, p)$ *and* $a(x, t, u, p)$ *are thrice continuously differentiable with respect to* x, t, u *and* p, $a_{11}(x, t, u, p) \ge \nu$, $\nu = \text{const} > 0$, *and the function* $q(x, t, u, p) = a(x, t, u, p)/a_{11}(x, t, u, p)$ *is subject to the inequalities*

$$|q(x, t, u, p)| + \left|\frac{\partial q(x, t, u, p)}{\partial t}\right| + \left|\frac{\partial q(x, t, u, p)}{\partial u}\right| + (1 + |p|)\left|\frac{\partial q(x, t, u, p)}{\partial p}\right| \leqslant \mu(1 + |p|)^2, \tag{5.15}$$

$$\left|\frac{\partial q(x, t, u, p)}{\partial x}\right| \leqslant [\varepsilon + P(|p|)](1 + |p|)^3, \tag{5.16}$$

where ϵ *and* $P(|p|)$ *are the same as in* (5.12).

Upon comparing assumption b) with b″) it is seen that in the latter more stringent conditions are imposed on the smoothness of the functions $a_{11}(x, t, u, p)$ and $a(x, t, u, p)$, while the restrictions on the growth of a_{11} and a with respect to p in b″) are somewhat weaker than in b) (namely, it is not required that ϵ be small in (5.13)). On the other hand assumption b′) is the weakest of all the conditions b), b′) and b″), but it involves only the case $m = 2$.

In order to show that Remark 5.2 is valid we must prove that condition b″) ensures an estimate of $\max_{Q_T}|u_x^\tau|$. Estimates for $\max_{t\in[0,T]}|u_x(\pm l, t)|$ follow from Lemma 3.1 of the present chapter. And an estimate of $|u_x(x, t)|$ at interior points of Q_T under conditions b″) has been established by A. F. Filippov in the note [29]. We will give a proof of it.

Lemma 5.1. *If assumptions* b″) *and* (5.10) *are fulfilled, then* $\max_{Q_T}|u_x|$ *is estimated from above by a constant depending only on* ν, μ, ϵ, $P(\rho)$ *and* $\max_{\Gamma_T}|u_x|$.

We introduce an auxiliary function $w(x, t)$ by means of the equality

$$w(x, t) = \varphi(x, t, u(x, t), u_x(x, t)), \tag{5.17}$$

where $\phi(x, t, u, p)$ is a certain twice continuously differentiable function of its arguments in the domain $\{(x, t, u, p);\ (x, t) \in \overline{Q}_T,\ |u| \leq M,\ -\infty < p < \infty\}$ that is strictly monotonically increasing with respect to p and tends to $+\infty$ for $p \to +\infty$ and to $-\infty$ for $p \to -\infty$. In the following we will impose several other restrictions on $\phi(x, t, u, p)$ and then construct it.

In order to simplify the calculations we assume that $q(x, t, u, p)$ does not depend on x. Then we can also take ϕ to be independent of x: $\phi = \phi(t, u, p)$. From (5.17) we have

$$w_x = \varphi_u u_x + \varphi_p u_{xx}, \tag{5.18}$$

$$w_t = \varphi_t + \varphi_u u_t + \varphi_p u_{xt}. \tag{5.19}$$

From (5.18) and equation (5.1) it follows that

$$u_t = \frac{a_{11}}{\varphi_p}(w_x - \varphi_u u_x - \varphi_p q),$$

and from this and (5.19) we get

$$\begin{aligned} w_t = \varphi_t + \frac{\varphi_u a_{11}}{\varphi_p}[w_x - (\varphi_u u_x + \varphi_p q)] \\ + \varphi_p \frac{\partial}{\partial x}\left[\frac{a_{11}}{\varphi_p} w_x - \frac{a_{11}}{\varphi_p}(\varphi_u u_x + \varphi_p q)\right]. \end{aligned} \tag{5.20}$$

We require that for large $|p|$ the function $\phi(t, u, p)$ satisfy the equation

$$\varphi_u(t, u, p)\, p + \varphi_p(t, u, p)\, q(t, u, p) = 0 \tag{5.21}$$

in partial derivatives of first order. It is well known that such an equation is satisfied by the first integrals of the ordinary differential equation

$$\frac{dp}{du} = \frac{q(t, u, p)}{p}, \tag{5.22}$$

in which t enters as a parameter. The right side of equation (5.22) has a singularity for $p = 0$. We therefore take instead of (5.22) a somewhat different equation:

$$\frac{dp}{du} = F(t, u, p), \tag{5.23}$$

where $F(t, u, p)$ coincides with $q(t, u, p)/p$ for $|p| \geq 1$ and is a thrice continuously differentiable function of its arguments in the whole domain

$$\{(t, u, p): t \in [0, T], |u| \leqslant M, -\infty < p < \infty\}.$$

We will take as $\phi(t, u, p)$ one of the first integrals of equation (5.23) that satisfies for $t \in [0, T]$, $|u| \leq M$ and arbitrary w the inequality

$$|\varphi_t(t, u, p(t, u, w))| \leqslant \Phi(|w|), \quad \int_1^\infty \frac{d\tau}{\Phi(\tau)} = \infty, \tag{5.24}$$

where $p = p(t, u, w)$ is a solution of the equation

$$w = \varphi(t, u, p). \tag{5.25}$$

We will construct such a function $\phi(t, u, p)$ below, but let us now show how it is used to estimate $|u_x(x, t)|$. We consider relation (5.20) as an equation for w:

$$w_t - a_{11} w_{xx} - b(x, t\ w, w_x) = 0, \tag{5.26}$$

in which

$$b(x, t, w, w_x) = \varphi_t + \left(\frac{\varphi_u a_{11}}{\varphi_p} + \varphi_p \frac{\partial}{\partial x} \frac{a}{\varphi_p}\right) w_x \tag{5.27}$$

for large $|p|$ and

$$\begin{aligned} b(x, t, w, w_x) = \varphi_t + \left(\frac{\varphi_u a_{11}}{\varphi_p} + \varphi_p \frac{\partial}{\partial x} \frac{a}{\varphi_p}\right) w_x \\ - \frac{\varphi_u a_{11}}{\varphi_p} (\varphi_u u_x + \varphi_p q) - \varphi_p \frac{\partial}{\partial x} \left[\frac{a_{11}}{\varphi_p} (\varphi_u u_x + \varphi_p q)\right] \end{aligned} \tag{5.28}$$

for all other p. Here in the right sides of (5.27) and (5.28) we must assume that u_{xx} has been replaced by $(w_x - \phi_u u_x)/\phi_p$, that p and u_x have been replaced by the function $p(t, u, w)$ which is a solution of equation (5.25), and that u has been replaced by the solution $u(x, t)$ being investigated. For $w_x = 0$ and large w (i.e. large p) the function $b(x, t, w, 0)$ satisfies by virtue of (5.24) the condition

$$|b(x, t, w, 0)| \leqslant \Phi(|w|), \quad \int_1^\infty \frac{d\tau}{\Phi(\tau)} = \infty, \tag{5.29}$$

and therefore $\max_{Q_T}|w|$ for a solution w of equation (5.26) can be estimated from above by a constant depending only on $\Phi(r)$ and $\max_{\Gamma_T}|w|$, i.e. on known quantities (see Theorem 2.9 of Chapter I). This in turn yields the desired estimate for $\max_{Q_T}|u_x|$.

Thus it only remains to construct a function $\phi(t, u, p)$ having the properties enumerated above. Let $p = p(t, u, p_0)$ be a solution of equation (5.23) satisfying the initial condition

$$p\,|_{u=-M} \equiv p(t, -M, p_0) = p_0. \tag{5.30}$$

Such a solution exists on the whole interval $-M \leq u \leq M$ for any p_0, and the curves $p = p(t, u, p_0)$ fill without intersections the whole strip $-M \leq u \leq M$ of the (u, p)-plane. These assertions are valid on the basis of (i) a well-known theorem on the unique solvability of the Cauchy problem for ordinary differential equations, and (ii) the properties of the function $F(t, u, p)$, namely its smoothness and its not greater than linear growth with respect to p (see [117, 55]). The latter follows from assumption (5.15), more precisely from the fact that

$$|F| = \left|\frac{q(t, u, p)}{p}\right| \leqslant \mu \frac{(1 + |p|)^2}{|p|} \leqslant 4\mu |p|,$$

for $|p| \geq 1$, while for $|p| \leq 1$ we have

$$|F(t, u, p)| \leqslant c \equiv \max_{t \in [0, T],\ |u| \leqslant M,\ |p| \leqslant 1} \left|F, \frac{\partial F}{\partial p}, \frac{\partial F}{\partial t}\right|.$$

Owing to these inequalities we have, as is easily calculated, the following estimate for the solutions $p(t, u, p_0)$:

$$|p(t, u, p_0)| \leqslant e^{8\mu M}\left(p_0 + \frac{c}{4\mu}\right) \quad \text{for } |u| \leqslant M. \tag{5.31}$$

The function $p(t, u, p_0)$ is a strictly monotonically increasing function of p_0 (for fixed t and u); this follows directly from the way in which the integral curves $p = p(t, u, p_0)$ are distributed in the strip $-M \leq u \leq M$. Therefore the equation $p = p(t, u, p_0)$ can be uniquely solved with respect to p_0:

$$p_0 = p_0(t, u, p), \tag{5.32}$$

where $p_0 \to \pm\infty$ when $p \to \pm\infty$ respectively. We will show that the function $p_0(t, u, p)$ can be taken as the desired function $\phi(t, u, p)$.

As is well known, $\partial p(t, u, p_0)/\partial p_0$ satisfies on the integral curve $p = p(t, u, p_0)$ the linear equation

$$\frac{d}{du}\frac{\partial p}{\partial p_0} = \frac{\partial F}{\partial p}\frac{\partial p}{\partial p_0}, \tag{5.33}$$

and becomes equal to unity for $u = -M$, i.e.

$$\left.\frac{\partial p}{\partial p_0}\right|_{u=-M} = 1. \tag{5.34}$$

Therefore $\partial p/\partial p_0$ and hence also $\partial p_0/\partial p = (\partial p/\partial p_0)^{-1}$ are strictly positive.

Moreover, we can obtain an estimate from below for $\partial p/\partial p_0$, starting from (5.33) and (5.34) and the assumptions (5.15). Namely, from (5.15) it follows that

$$\left|\frac{\partial F}{\partial p}\right| \leqslant \left|\frac{q_p}{p} - \frac{q}{p^2}\right| \leqslant 6\mu \quad \text{for } |p| \geqslant 1$$

and

$$\left|\frac{\partial F}{\partial p}\right| \leqslant c \quad \text{for } |p| \leqslant 1,$$

and hence

$$\frac{\partial p(t, u, p_0)}{\partial p_0} = \exp\left\{\int_{-M}^{u} \frac{\partial F}{\partial p}\, d\tau\right\} \geqslant e^{-6\mu(u+M)-c(u+M)} \geqslant e^{-2M(6\mu+c)}. \tag{5.35}$$

Further, on the basis of theorems from the theory of ordinary differential equations on the nature of the dependence of the solutions of the Cauchy problem (5.23), (5.30) on the argument u, the initial value p_0 and the parameter t (see [117, 55]) we can conclude that $p(t, u, p_0)$ is a twice continuously differentiable function of all its arguments, and from the positiveness of $\partial p/\partial p_0$ it follows that these same properties are possessed by the function $p_0(t, u, p)$ inverse to it.

We will show, finally, that property (5.24) is valid for $\phi(t, u, p) \equiv p_0(t, u, p)$, or, more precisely, that

$$\left|\frac{\partial p_0(t, u, p)}{\partial t}\right|\Bigg|_{p=p(t, u, p_0)} \leqslant c_1(1 + |p_0|). \tag{5.36}$$

Indeed, upon differentiating with respect to t the equation

$$p = p(t, u, p_0(t, u, p))$$

(considered as an identity in the independent arguments t, u and p), we obtain

$$0 = \frac{\partial p}{\partial t} + \frac{\partial p}{\partial p_0}\frac{\partial p_0}{\partial t},$$

i.e.

$$\frac{\partial p_0}{\partial t} = -\frac{\partial p}{\partial t}\left(\frac{\partial p}{\partial p_0}\right)^{-1}. \tag{5.37}$$

Owing to (5.23) the derivative $\partial p/\partial t$ satisfies on the trajectory $p = p(t, u, p_0)$ the equation

$$\frac{d}{du}\frac{\partial p}{\partial t} = \frac{\partial F}{\partial p}\frac{\partial p}{\partial t} + \frac{\partial F}{\partial t} \tag{5.38}$$

and the initial condition

$$\left.\frac{\partial p}{\partial t}\right|_{u=-M} = 0, \tag{5.39}$$

and since by virtue of (5.15) and (5.31)

$$\left|\frac{\partial F}{\partial p}\right| \leqslant 6\mu + c, \quad \left|\frac{\partial F}{\partial t}\right| \leqslant 4\mu\,|\,p\,| + c \leqslant c_2\,|\,p_0\,| + c_3,$$

where $c_2 = 4\mu e^{8\mu M}$, and $c_3 = c + ce^{8\mu M}$, it follows that

$$\left|\frac{\partial p}{\partial t}\right| \leqslant \frac{c_2\,|\,p_0\,| + c_3}{6\mu + c}\, e^{2(6\mu + c)M}.$$

This together with (5.37) and (5.35) yields the desired inequality (5.36). Thus we have shown that $p_0(t, u, p)$ possesses all of the properties which are required of the function $\phi(t, u, p)$, and this, as was shown above, implies that it is possible to estimate $\max_{Q_T}|w|$, and hence also $\max_{Q_T}|u_x|$, in terms of known quantities. Lemma 5.1 is proved.

CHAPTER VII

SYSTEMS OF LINEAR AND QUASI-LINEAR EQUATIONS

We will consider in this chapter a particular class of systems of differential equations of second order and of parabolic type, linear and quasi-linear, and we will investigate the same questions for it that were investigated for one equation in the preceding chapters. It is selected on the basis of the following criterion: the systems are those for which a weakened form of the maximum principle is valid. Namely, if one rejects the minor terms from all of the equations of the system and freezes the coefficients in the major terms, then, for the modulus $|\mathbf{u}(x, t)|$ of the classical solutions $\mathbf{u}(x, t)$ of such a simplified system the maximum principle holds in the same form as for the heat equation. All of the equations of such systems have the same principle parts. It turns out that all of the basic assertions established in the preceding chapters for one equation are also valid for these systems (see [650]). We will prove them, but only in detail at those points that essentially distinguish systems from the case of one equation.

§§8–10 contain a survey of the basic results that have been obtained on general boundary value problems for linear systems of equations of parabolic type.

§1. GENERALIZED SOLUTIONS OF LINEAR SYSTEMS

We will consider in Q_T systems of the form

$$\mathbf{u}_t - \frac{\partial}{\partial x_i}\left(a_{ij}(x, t)\,\mathbf{u}_{x_j} + A_i(x, t)\,\mathbf{u}\right) + B_i(x, t)\,\mathbf{u}_{x_i} + A(x, t)\,\mathbf{u} = \frac{\partial \mathbf{f}_i}{\partial x_i} - \mathbf{f}, \qquad (1.1)$$

where $\mathbf{u}(x, t)$, $\mathbf{f}_i(x, t)$, $\mathbf{f}(x, t)$ are the N-dimensional vectors $(u^1(x, t), \ldots, u^N(x, t))$, $(f_i^1, \cdots, f_i^N)$, $(f^1, \cdots, f^N)$ respectively, $A_i(x, t)$, $B_i(x, t)$, $A(x, t)$ are $N \times N$ matrices with elements $a_i^{kl}(x, t)$, $b_i^{kl}(x, t)$, $a^{kl}(x, t)$ respectively, and $a_{ij}(x, t)$ are scalars.

We will use the following notation in this chapter:

$$\mathbf{v}\mathbf{w}=\sum_{k=1}^{N} v^k w^k, \quad |\mathbf{v}|=\sqrt{\mathbf{v}^2}=\sqrt{\mathbf{v}\mathbf{v}},$$

$$|\mathbf{v}_x|^2=\mathbf{v}_x^2=\sum_{k=1}^{N}\sum_{i=1}^{n}(v_{x_i}^k)^2,$$

$$|A|=\left(\sum_{k=1}^{N}\sum_{l=1}^{N}|a^{kl}|^2\right)^{1/2}, \quad \|A\|_{q,r,Q_T}=\||A|\|_{q,r,Q_T}.$$

The membership of a vector function in some space will mean that all of its components belong to this space.

We will assume that the system (1.1) is such that the inequalities

$$\nu\xi^2\leqslant a_{ij}\xi_i\xi_j\leqslant\mu\xi^2, \quad \nu>0, \tag{1.2}$$

are valid for any real $\xi=(\xi_1,\cdots,\xi_n)$. Condition (1.2) means that the system (1.1) is parabolic.

For (1.1) we will consider the first boundary value problem, i.e. the problem of finding a function $\mathbf{u}(x,t)=(u^1,\cdots,u^N)$ satisfying the system (1.1) and the conditions

$$\mathbf{u}|_{S_T}=\boldsymbol{\psi}(s,t), \tag{1.3$_1$}$$

$$\mathbf{u}|_{t=0}=\boldsymbol{\psi}_0(x). \tag{1.3$_2$}$$

By a generalized solution from $V_2(Q_T)$ (from $V_2^{1,0}(Q_T)$ or $V_2^{1,1/2}(Q_T)$) of problem (1.1), (1.3) we will mean a vector function $\mathbf{u}(x,t)\in V_2(Q_T)$ ($V_2^{1,0}(Q_T)$ or $V_2^{1,1/2}(Q_T)$), satisfying condition (1.3$_1$) and such that for any vector function $\eta(x,t)\in \overset{\circ}{W}{}_2^{1,1}(Q_T)$ the integral identity

$$\int_\Omega \mathbf{u}(x,t)\boldsymbol{\eta}(x,t)\,dx+\int_0^t\int_\Omega[-\mathbf{u}\boldsymbol{\eta}_t+(a_{ij}\mathbf{u}_{x_j}+A_i\mathbf{u}+\mathbf{f}_i)\boldsymbol{\eta}_{x_i}$$
$$+(B_i\mathbf{u}_{x_i}+A\mathbf{u}+\mathbf{f})\boldsymbol{\eta}]\,dx\,dt-\int_\Omega\boldsymbol{\psi}_0(x)\boldsymbol{\eta}(x,0)\,dx=0$$

is fulfilled for almost all (for all) t from $[0,T]$.

In this section we will discuss the unique solvability of problem (1.1), (1.3) in the space $V_2^{1,1/2}(Q_T)$.

As was shown in §2 of Chapter I, in order to retain the basic properties inherent in the problem in its classical statement it is necessary to impose certain additional restrictions on the coefficients of the system (1.1), which we will formulate as usual in terms of their membership in the spaces $L_{q,r}(Q_T)$. These restricitions are the following:

$$\left\| \sum_{i=1}^{n} |A_i|^2; \quad \sum_{i=1}^{n} |B_i|^2; \quad |A| \right\|_{q,r,Q_T} \leqslant \mu_1,$$

where $1/r + n/2q = 1$, and

$$q \in \left(\frac{n}{2}, \infty\right]; \quad r \in [1, \infty) \quad \text{for} \quad n \geqslant 2,$$
$$q \in [1, \infty]; \quad r \in [1, 2] \quad \text{for} \quad n = 1, \qquad (1.5)$$

$$\mathbf{f}_i \in L_2(Q_T), \quad \mathbf{f} \in L_{q_1, r_1}(Q_T),$$

where $1/r_1 + n/2q_1 = (n+4)/4$, and

$$q_1 \in \left[\frac{2n}{n+2}, 2\right], \quad r_1 \in [1, 2] \quad \text{for} \quad n \geqslant 3,$$
$$q_1 \in (1, 2], \quad r_1 \in [1, 2) \quad \text{for} \quad n = 2, \qquad (1.6)$$
$$q_1 \in [1, 2], \quad r_1 \in \left[1, \frac{4}{3}\right] \quad \text{for} \quad n = 1.$$

In conditions (1.5) and in the second of conditions (1.6) the indices q and r can be assumed to be different for each of the elements of the matrices and each of the components of the vectors on which these conditions are imposed. None of the assertions is essentially changed in this connection; their formulations and proofs are merely lengthened.

In addition, we will assume for the sake of simplicity that $\psi(s, t) = 0$.

Just as for one equation, the following theorem is valid for systems of the form (1.1).

Theorem 1.1. *Under the fulfilment of conditions* (1.2), (1.5), (1.6) *and the initial condition* $\mathbf{u}|_{t=0} = \psi_0(x) \in L_2(\Omega)$ *the system* (1.1) *has a unique generalized solution* $\mathbf{u}(x, t)$ *from* $\overset{\circ}{V}{}_2^{1,1/2}(Q_T)$. *Any solution from* $\overset{\circ}{V}_2(Q_T)$ *belongs to* $\overset{\circ}{V}{}_2^{1,1/2}(Q_T)$. *If furthermore the* $|B_i|^2$ *and* $|A|$ *belong to* $L^*_{r,q}(Q_T)$, $r \geqslant q$ *and* $\mathbf{f} \in L^*_{r_1,q_1}(Q_T)$, $r_1 \geq q_1$, *then* $\mathbf{u} \in \overset{\circ}{W}{}_2^{1,1/2}(Q_T)$.

The proof is carried out in exactly the same way as in the case of one equation, and is based on the a priori energy estimate

$$|\mathbf{u}|_{Q_T}^2 \leqslant c \left[\sum_{i=1}^{n} \|\mathbf{f}_i\|_{2,Q_T}^2 + \|\mathbf{f}\|_{q_1,r_1,Q_T}^2 + \|\psi_0\|_{2,\Omega}^2 \right]. \qquad (1.7)$$

which holds for an arbitrary generalized solution from $V_2(Q_T)$ of problem (1.1), (1.3) with $\psi(s, t) \equiv 0$.

The constant c in this inequality is determined only by n, ν, μ, the norms μ_1 of the coefficients and the indices q and r in conditions (1.5).

The proof of inequality (1.7) does not differ essentially from the proof of inequality (2.2) in Chapter III; it is only necessary to replace the usual multiplicatioı of functions by the scalar multiplication of vector functions.

Uniqueness in $V_2(Q_T)$ follows at once from inequality (1.7) applied to the difference of two possible solutions, while existence can be proved, for example, by Galerkin's method.

§ 2. ON THE BOUNDEDNESS OF $\max_{Q_T} |\mathbf{u}|$

In this section we will estimate the maximum of the modulus of a solution $\mathbf{u}(x, t)$ of the system (1.1). According to the examples of Chapter I, for this to be possible we must impose additional restrictions on the coefficients of (1.1) and on the $\mathbf{f}_i$ and $\mathbf{f}$. The conditions under which the boundedness of generalized solutions from $V_2^{1,0}(Q_T)$ will be proved are formulated as follows:

$$\left\| \sum_{i=1}^{n} |A_i|^2,\ \sum_{i=1}^{n} |B_i|^2,\ \sum_{i=1}^{n} |\mathbf{f}_i|^2,\ |A|,\ |\mathbf{f}| \right\|_{q,r,Q_T} \leqslant \mu_1, \tag{2.1}$$

where $1/r + n/2q = 1 - \kappa_1$, with

$$\left.\begin{aligned} &q \in \left[\frac{n}{2(1-\varkappa_1)}, \infty\right], \quad r \in \left[\frac{1}{1-\varkappa_1}, \infty\right], \\ &\qquad\qquad 0 < \varkappa_1 < 1, \quad \text{for} \quad n \geqslant 2, \\ &q \in [1, \infty], \quad r \in \left[\frac{1}{1-\varkappa_1}, \frac{2}{1-2\varkappa_1}\right], \\ &\qquad\qquad 0 < \varkappa_1 < \frac{1}{2}, \quad \text{for} \quad n = 1. \end{aligned}\right\} \tag{2.2}$$

The numbers q and r can assumed to be different for each of the elements of the vectors $\mathbf{f}_i$ and $\mathbf{f}$ and for each of the elements of the matrices A_i, B_i and A in conditions (2.1). It is sufficient that they be subject to condition (2.2). For the sake of simplicity we will assume that $\mathbf{u}|_{S_T} = 0$. The basic idea for obtaining an estimate of $\max_{Q_T} |\mathbf{u}|$ is the same as for one equation in § 7 of Chapter III. But its realization for a system of equations is complicated by the fact that our a priori assumptions on the solution (that $\mathbf{u} \in V_2^{1,0}(Q_T)$) do not permit a development of it in the same way as for one equation. Let us explain. For one equation the inequalities underlying the definition of the class $\mathfrak{A}(Q_T, \cdots)$ and thereby permitting a determination of the boundedness of u were obtained from a basic integral identity satisfied by the function u, in which the arbitrary function $\eta(x, t)$ is taken equal to $u^{(k)}(x, t) = \max\{u(x, t) - k; 0\}$. For a system of equations quite similar inequalities are obtained from the identity (1.4) if the vector function $\eta(x, t)$ in it is taken equal to $\mathbf{u}(x, t) \max\{v(x, t) - k; 0\}$, where $v(x, t) = \mathbf{u}^2(x, t)$. But we are not justified in doing this if we do not know in advance, for example, that $|\mathbf{u}|$ is bounded or at least that $\mathbf{u}^2 \in V_2^{1,0}(Q_T)$. In view of this the function η must be taken in a more complicated form, by introducing a "slicing off" of $v(x, t)$ at large values of $|\mathbf{u}(x, t)|$. In addition, as an intermediate step, it is necessary here as elsewhere to carry out a smoothing with respect to t. Let us show how this is done.

First we note that, in the same way as for one equation, the identity (1.4) implies the identity

$$\int_{Q_{t_1}} [\mathbf{u}_{ht}\eta + (a_{ij}\mathbf{u}_{x_j} + A_i\mathbf{u} + \mathbf{f}_i)_h \eta_{x_i} + (B_i\mathbf{u}_{x_i} + A\mathbf{u} + \mathbf{f})_h \eta]\, dx\, dt = 0 \qquad (2.3)$$

with any vector function η from $\overset{\circ}{V}{}_2^{1,0}(Q_T)$ and $t_1 \le T - h$ (see §2 of Chapter III). The notation $(\ )_h$ means an averaging over t (see (2.10) of Chapter III). In (2.3) we put

$$\eta(x, t) = 2\mathbf{u}_h(x, t) \max\{v_M^h(x, t) - k; 0\} \equiv 2\mathbf{u}_h v_M^{h(k)},$$

where $\mathbf{u}_h(x, t)$ is an averaging of $\mathbf{u}$ over t, $v(x, t) = \mathbf{u}^2(x, t)$, $v^h = (\mathbf{u}_h)^2$, the function $v_M^h(x, t)$ is equal to $v^h(x, t)$ for $v^h(x, t) < M$, and to M for $v^h(x, t) \ge M$, while k is an arbitrary nonnegative number. We transform the first term obtained in this connection in accordance with the equality

$$2\mathbf{u}_{ht}\mathbf{u}_h v_M^{h(k)} = v_t^h v_M^{h(k)} = \frac{\partial}{\partial t}(v^{h(k)} v_M^{h(k)}) - \frac{1}{2}\frac{\partial}{\partial t}(v_M^{h(k)})^2, \qquad (2.4)$$

which can be verified directly from the definition of the functions v^h, $v^{h(k)}$ and $v_M^{h(k)}$. After substituting the right side of (2.4) in place of $\mathbf{u}_{ht}\eta$ in (2.3) and carrying out an integration with respect to t in these terms, we then pass to the limit in all of the integrals of (2.3) as $h \to 0$. This, as can be easily verified (see Chapter III), is admissible and leads to the relation

$$\int_\Omega v^{(k)} v_M^{(k)}\, dx \Big|_0^{t_1} - \frac{1}{2}\int_\Omega (v_M^{(k)})^2\, dx \Big|_0^{t_1}$$
$$+2\int_{Q_{t_1}(k)} \Big[(a_{ij}\mathbf{u}_{x_j} + A_i\mathbf{u} + \mathbf{f}_i)(\mathbf{u}v_M^{(k)})_{x_i}$$
$$+(B\mathbf{u}_{x_i} + A\mathbf{u} + \mathbf{f})\,\mathbf{u}v_M^{(k)}\Big]\, dx\, dt = 0, \tag{2.5}$$

where $Q_{t_1}(k)$ denotes the set of points (x, t) of the cylinder Q_{t_1} at which $v_M(x, t) > k$. In order to estimate the different terms of (2.5) we make use of the following inequalities:

$$a_{ij}\mathbf{u}_{x_j}(\mathbf{u}v_M^{(k)})_{x_i} = a_{ij}\mathbf{u}_{x_j}\mathbf{u}_{x_i}v_M^{(k)} + \frac{1}{2}a_{ij}v_{x_j}v_{Mx_i}^{(k)}$$
$$\geqslant \nu\mathbf{u}_x^2 v_M^{(k)} + \frac{\nu}{2}(v_{Mx}^{(k)})^2,$$
$$2\left|A_i\mathbf{u}(\mathbf{u}v_M^{(k)})_{x_i}\right| \leqslant 2|A_i|\left(|\mathbf{u}||\mathbf{u}_{x_i}|v_M^{(k)} + \sqrt{vv_M}\,|v_{Mx_i}^{(k)}|\right)$$
$$\leqslant \varepsilon\mathbf{u}_x^2 v_M^{(k)} + \frac{1}{\varepsilon}\sum_{i=1}^n |A_i|^2\, v_M^{(k)} + \varepsilon(v_{Mx}^{(k)})^2 + \frac{1}{\varepsilon}\sum_{i=1}^n |A_i|^2 vv_M,$$
$$2\left|\mathbf{f}_i(\mathbf{u}v_M^{(k)})_{x_i}\right| \leqslant 2|\mathbf{f}_i|\left(|\mathbf{u}_{x_i}|v_M^{(k)} + \sqrt{v_M}\,|v_{Mx_i}^{(k)}|\right)$$
$$\leqslant \varepsilon\mathbf{u}_x^2 v_M^{(k)} + \frac{1}{\varepsilon}\sum_{i=1}^n |\mathbf{f}_i|^2 v_M^{(k)} + \varepsilon(v_{Mx}^{(k)})^2 + \frac{1}{\varepsilon}\sum_{i=1}^n |\mathbf{f}_i|^2 v_M$$
$$\leqslant \varepsilon\mathbf{u}_x^2 v_M^{(k)} + \varepsilon(v_{Mx}^{(k)})^2 + \frac{1}{\varepsilon}\sum_{i=1}^n |\mathbf{f}_i|^2(vv_M + 1),$$
$$2\left|B_i\mathbf{u}_{x_i}\mathbf{u}v_M^{(k)}\right| \leqslant 2|B_i|\left|\mathbf{u}_{x_i}\sqrt{v_M^{(k)}}\right|\left|\mathbf{u}\sqrt{v_M^{(k)}}\right|$$
$$\leqslant \varepsilon\mathbf{u}_x^2 v_M^{(k)} + \frac{1}{\varepsilon}\sum_{i=1}^n |B_i|^2 vv_M^{(k)},$$
$$2|A\mathbf{u}\mathbf{u}v_M^{(k)}| \leqslant 2|A||vv_M^{(k)}|,$$
$$2|\mathbf{f}\mathbf{u}v_M^{(k)}| \leqslant 2|\mathbf{f}||\mathbf{u}|v_M^{(k)} \leqslant 2|\mathbf{f}|(vv_M)^{\frac{3}{4}} \leqslant 2|\mathbf{f}|(vv_M + 1).$$

Here we have taken into account that $v^m v_{Mx_i}^{(k)} = v_M^m v_{Mx_i}^{(k)}$ for any m and $v_M \leq v$.

By virtue of these estimates (2.5) implies the inequality

$$\frac{1}{2}\int_\Omega v^{(k)}v_M^{(k)}\, dx\Big|^{t_1} + \nu\int_{Q_{t_1}(k)} \left[2\mathbf{u}_x^2 v_M^{(k)} + (v_{Mx}^{(k)})^2\right] dx\, dt \leqslant$$

$$\leqslant \frac{1}{2}\int_\Omega (v^{(k)})^2\,dx\Big|^{t=0} + \int_{Q_{t_1}(k)}\Big[3\varepsilon u_x^2 v_M^{(k)} + 2\varepsilon (v_{Mx}^{(k)})^2$$
$$+\frac{1}{\varepsilon}\Big(2\sum_{i=1}^n |A_i|^2 + \sum_{i=1}^n |B_i|^2 + 2\varepsilon|A|$$
$$+\sum_{i=1}^n |f_i|^2 + 2\varepsilon|f|\Big)(vv_M + 1)\Big]\,dx\,dt, \tag{2.6}$$

from which for $\epsilon = \nu/4$ we in turn obtain

$$\frac{1}{2}\int_\Omega v^{(k)} v_M^{(k)}\,dx\Big|^{t_1} + \frac{\nu}{2}\int_{Q_{t_1}(k)} [2u_x^2 v_M^{(k)} + (v_{Mx}^{(k)})^2]\,dx\,dt$$
$$\leqslant \frac{1}{2}\int_\Omega (v^{(k)})^2\,dx\Big|^{t=0} + \int_{Q_{t_1}(k)} \mathfrak{D}(vv_M + 1)\,dx\,dt, \tag{2.7}$$

where

$$\mathfrak{D} = \frac{4}{\nu}\Big(2\sum_{i=1}^n |A_i|^2 + \sum_{i=1}^n |B_i|^2 + \frac{\nu}{2}|A| + \sum_{i=1}^n |f_i|^2 + \frac{\nu}{2}|f|\Big),$$

and from (2.7) the inequality

$$\nu_1\Big\{\max_{0\leqslant t\leqslant t_1}\int_\Omega v^{(k)} v_M^{(k)}\,dx + \int_{Q_{t_1}(k)} [2u_x^2 v_M^{(k)} + (v_{Mx}^{(k)})^2]\,dx\,dt\Big\}$$
$$\leqslant \|v^{(k)}(x,0)\|_{2,\Omega}^2 + 2\int_{Q_{t_1}(k)} \mathfrak{D}(vv_M+1)\,dx\,dt \leqslant \|v^{(k)}(x,0)\|_{2,\Omega}^2$$
$$+2\|\mathfrak{D}\|_{q,r,Q_{t_1}(k)}\Big(\|\sqrt{vv_M}\|_{\bar q,\bar r,Q_{t_1}(k)}^2 + \|1\|_{\bar q,\bar r,Q_{t_1}(k)}^2\Big), \tag{2.8}$$

in which $\nu_1 = \min\{½, \nu/2\}$, while $\bar q = 2q/(q-1)$, $\bar r = 2r/(r-1)$. From this, by virtue of the inequality

$$\sum_{i=1}^n \big[(\sqrt{vv_M})_{x_i}\big]^2 = \sum_{i=1}^n \frac{1}{4vv_M}(v_{x_i} v_M + vv_{Mx_i})^2$$
$$\leqslant \frac{1}{2vv_M}(v_x^2 v_M^2 + v^2 v_{Mx}^2) \leqslant u_x^2 v_M + \frac{1}{2} v_{Mx}^2 \tag{2.9}$$

and the fact that $v^{(0)} = v$, for $k = 0$ and $t_1 = T$ we get

$$\nu_1 |\sqrt{vv_M}|_{Q_{t_1}}^2 \leqslant \|v(x,0)\|_{2,\Omega}^2$$
$$+2\|\mathfrak{D}\|_{q,r,Q_{t_1}}\Big(\|\sqrt{vv_M}\|_{\bar q,\bar r,Q_{t_1}}^2 + \|1\|_{\bar q,\bar r,Q_{t_1}}^2\Big). \tag{2.10}$$

This inequality has the same character as inequality (7.8) of Chapter III for a fixed k. From it, by estimating the right side in the same way as in §7 of Chapter III (see formulas (7.9)–(7.11)), we establish the uniform (with respect to M) boundedness of the integral $|\sqrt{vv_M}|_{Q_{t_1}}$, and hence also of the integral $|v|_{Q_{t_1}}$, as long as the height t_1 does not exceed a certain number τ (see the type of condition (7.13)). By repeating the argument for the cylinders $Q_{\tau,2\tau}=\Omega\times(\tau,2\tau)$, $Q_{2\tau,3\tau}=\Omega\times(2\tau,3\tau)$ and so on, we exhaust the whole cylinder Q_T in a finite number of steps and find that the integral $|v|_{Q_T}$ does not exceed a certain constant, which can easily be calculated from the quantities n, ν, μ_1, r, q, $\operatorname{mes}\Omega$ and T.

After this we return to inequality (2.8) with an arbitrary $k\geq 0$. Its right side for any $M\geq 0$ does not exceed

$$\|v^{(k)}(x,0)\|^2_{2,\Omega}+2\|\mathfrak{D}\|_{q,r,Q_{t_1}(k)}\left(\|v\|^2_{\bar q,\bar r,Q_{t_1}(k)}+\|1\|^2_{\bar q,\bar r,Q_{t_1}(k)}\right),$$

while the supremum of the left side with respect to $M\in[0,\infty)$ is

$$\nu_1\left\{\max_{0\leqslant t\leqslant t_1}\|v^{(k)}(x,t)\|^2_{2,\Omega}+\int_{Q_{t_1}(k)}\left[2\mathrm{u}_x^2v^{(k)}+(v_x^{(k)})^2\right]dx\,dt\right\}$$
$$\geqslant\nu_1|v^{(k)}|^2_{Q_{t_1}}.$$

Therefore

$$\nu_1|v^{(k)}|^2_{Q_{t_1}}\leqslant\|v^{(k)}(x,0)\|^2_{2,\Omega}+2\|\mathfrak{D}\|_{q,r,Q_{t_1}(k)}\left(\|v\|^2_{\bar q,\bar r,Q_{t_1}(k)}+\|1\|^2_{\bar q,\bar r,Q_{t_1}(k)}\right).\tag{2.11}$$

For $k\geq\hat k=\max_{x\in\Omega}v(x,0)+1$ we obtain from (2.11) the inequality

$$\nu_1|v^{(k)}|^2_{Q_{t_1}}\leqslant 6\|\mathfrak{D}\|_{q,r,Q_{t_1}(k)}\left(\|v-k\|^2_{\bar q,\bar r,Q_{t_1}(k)}+\|k\|^2_{\bar q,\bar r,Q_{t_1}(k)}\right)\tag{2.12}$$

which essentially coincides with inequality (7.8) of Chapter III, from which in §7 we deduced the boundedness from above of the function u entering into it. In our case this implies the boundedness of v in Q_{t_1} and an estimate for $\operatorname{vrai}\max_{Q_{t_1}}v$ in terms of $\hat k$ and the known parameters from conditions (1.2), (2.1) and (2.2), as long as t_1 does not exceed τ. Repeating the argument for the cylinders $Q_{\tau,2\tau}$, $Q_{2\tau,3\tau}$ and so on, we obtain the desired estimate of $\operatorname{vrai}\max_{Q_T}v$ and hence also of $\operatorname{vrai}\max_{Q_T}|\mathrm{u}|$. Thus the following theorem has been proved.

Theorem 2.1. *Suppose* $\mathrm{u}(x,t)$ *is a generalized solution from* $V_2^{1,0}(Q_T)$ *of the system* (1.1) *that is equal to zero on* S_T *and is bounded for* $t=0$,[1] *and sup-*

1) The assumption that $\mathrm{u}(x,t)$ is equal to zero on S_T is not essential; it can be replaced by the assumption that $\mathrm{u}|_{S_T}$ is bounded.

pose conditions (1.2), (2.1) *and* (2.2) *are fulfilled for the system* (1.1). *Then* $u(x, t)$ *is a bounded function in* Q_T, *and* $\operatorname{vrai\,max}_{Q_T}|u|$ *is estimated from above by a constant depending only on* n, $\operatorname{vrai\,max}_{x\in\Omega}|u(x, 0)|$, T, $\operatorname{mes}\Omega$ *and the parameters* ν, μ_1, q *and* r *from conditions* (1.2) *and* (2.1).

Under conditions (1.2), (2.1) and (2.2) an arbitrary generalized solution u of system (1.1) belonging to $V_2^{1,0}(Q_T)$ (by virtue of Theorem 1.1, any solution from $V_2(Q_T)$) will be a bounded function in any subdomain $Q' \subset Q_T$ that is separated by a positive distance from that part $\Gamma' \subset \Gamma_T$ on which u is not bounded. We formulate this in the form of a theorem.

Theorem 2.2. *Suppose conditions* (1.2), (2.1) *and* (2.2) *are fulfilled for the system* (1.1). *Then for any generalized solution* u *of it from* $V_2^{1,0}(Q_T)$ *the quantity* $\operatorname{vrai\,max}_{Q'}|u|$, *where* Q' *is a subdomain of* Q_T *that is separated from* Γ_T *by a positive distance* d, *does not exceed a certain constant determined only by* n, ν, μ, μ_1, q *and* r *from conditions* (1.2) *and* (2.1), *the norm* $\|u\|_{2,Q_T}$ *and* d. *If also* $\operatorname{vrai\,max}_{x\in\Omega}|u(x, 0)| = M_0 < \infty$, *then* $\operatorname{vrai\,max}_{\Omega'\times[0,T]}|u|$ *is finite and is estimated from above by a constant determined by* M_0 *and the same quantities as above, only this time the role of* d *will be played by the distance from* Ω' *to* S.

Theorem 2.2 is proved in basically the same way as Theorem 2.1. The modifications of the arguments connected with the necessity of introducing a cutting function are analogous to those given in the case of one equation (see §8 of Chapter III). We will not repeat them here, assuming that the reader can do this independently.

§3. AN ESTIMATE OF $|u|_{Q_T}^{(\alpha)}$

Let us show that under the fulfilment of inequalities (1.2), (2.1) and (2.2) it is possible to estimate not only $\max_{Q_T}|u|$ but also the Hölder norms for each component of u with some positive exponent $\alpha > 0$. To this end it is sufficient to prove that bounded generalized solutions of the system (1.1) from $V_2^{1,0}(Q_T)$ belong to the class $\mathfrak{B}_2^{2N}(\overline{Q}_T, M_1, \delta_1, \cdots, \delta_6, \gamma, \hat{r}, \delta, \kappa)$. Suppose $u(x, t) \in V_2^{1,0}(Q_T)$ and $\operatorname{vrai\,max}_{Q_T}|u(x, t)| = M < \infty$. For simplicity we will assume that $0 \le u^i(x, t) \le 1$, $i = 1, \cdots, N$. This can easily be achieved by introducing in place of $u^i(x, t)$ in (1.1) the function $\hat{u}^i(x, t) = (u^i + M)/2M$. We assume that this has already been done.

As the functions $\phi^l(u)$ for $u(x, t)$ we take

$$\varphi^l_+(\mathbf{u}) = 10Nu^l + \sum_{l=1}^N (u^l)^2.$$

$$\varphi^l_-(\mathbf{u}) = 10N(1-u^l) + \sum_{l=1}^N (u^l)^2.$$

By virtue of Lemma 9.3 of Chapter II it is sufficient to verify that the functions $w^l_\pm(x,t) = \phi^l_\pm(\mathbf{u}(x,t))$ satisfy conditions 3) and 4) in the definition of the class $\mathfrak{B}_2^{2N}(\overline{Q}_T, M_1, \delta_1, \cdots, \delta_6, \gamma, \delta, \hat{r}, \kappa)$.

For this purpose, in the identity (2.3) we put $\eta = 5N\Phi(x,t)\mathbf{e}^{(l)}$ and $\eta = \mathbf{u}_h(x,t)\Phi(x,t)$, and add the resulting identities. Here $\Phi(x,t)$ is an arbitrary bounded function from $\overset{\circ}{V}{}_2^{1,0}(Q_T)$ and $\mathbf{e}^{(l)}$ is the vector whose lth component is equal to 1 and whose remaining components are equal to zero. This gives us

$$\int_0^{t_1}\int_\Omega \mathbf{u}_{ht}(5N\mathbf{e}^{(l)}\Phi + \mathbf{u}_h\Phi)\,dx\,dt$$
$$+ \int_0^{t_1}\int_\Omega \big[(a_{ij}\mathbf{u}_{x_j} + A_i\mathbf{u} + \mathbf{f}_i)_h(5N\mathbf{e}^{(l)}\Phi + \mathbf{u}_h\Phi)_{x_i}$$
$$+ (B_i\mathbf{u}_{x_i} + A\mathbf{u} + f)_h(5N\mathbf{e}^{(l)}\Phi + \mathbf{u}_h\Phi)\big]\,dx\,dt = 0. \tag{3.1}$$

We then put $\Phi(x,t) = 2\zeta^2\max\{w^{hl}_+(x,t) - k; 0\} \equiv 2w^{hl(k)}_+(x,t)\zeta^2$, where k is some number, $\zeta^2(x,t)$ is a smooth function that is nonnegative in some strictly interior cylinder $Q(\rho,\tau) = K_\rho \times (t_0 - \tau, t_0)$ and equal to zero for $x \overline{\in} K_\rho$, and

$$w^{hl}_+ = 10Nu^l_h + |\mathbf{u}_h|^2.$$

We subtract the equality obtained with $t_1 = t_0 - \tau$ from the same equality with $t_1 = t_0$. As a result we will have for the first summand

$$\frac{1}{2}\int_\Omega [w^{hl(k)}_+\zeta]^2\,dx\,\Big|_{t_0-\tau}^{t_0} - \int_{Q(\rho,\tau)} \zeta\zeta_t[w^{hl(k)}_+]^2\,dx\,dt.$$

Taking this transformation into account, we pass to the limit for $h \to 0$. In what follows we will write simply $w(x,t)$ in place of $w^l_+(x,t)$. As a result of the limit passage we obtain

$$\frac{1}{2}\int_{A_{k,\rho}(t)} [w^{(k)}(x,t)\zeta(x,t)]^2\,dx\,\Big|_{t_0-\tau}^{t_0}$$
$$+ \int_{Q(\rho,\tau)} (2a_{ij}\mathbf{u}_{x_i}\mathbf{u}_{x_j}w^{(k)}\zeta^2 + a_{ij}w^{(k)}_{x_i}w^{(k)}_{x_j}\zeta^2)\,dx\,dt =$$

$$= \int_{Q(\rho,\tau)} \{(w^{(k)})^2\zeta\zeta_t - 2a_{ij}w^{(k)}_{x_j}w^{(k)}\zeta\zeta_{x_i}$$
$$-(A_i u + f_i)[2u_{x_i}w^{(k)}\zeta^2 + (10Ne^{(l)} + 2u)\, w^{(k)}_{x_i}\zeta^2$$
$$+ 2\,(10Ne^{(l)} + 2u)\, w^{(k)}\zeta\zeta_{x_i}] -$$
$$-(B_i u_{x_i} + Au + f)(10Ne^{(l)} + 2u)\, w^{(k)}\zeta^2\}\, dx\, dt. \tag{3.2}$$

The left side of (3.2) is estimated from below by the expression

$$\frac{1}{2}\int_{A_{k,\rho}(t)} [w^{(k)}(x,t)\zeta(x,t)]^2\, dt\,\Big|_{t_0-\tau}^{t_0}$$
$$+ 2\nu \int_{Q(\rho,\tau)} \left[u_x^2 w^{(k)}\zeta^2 + \frac{1}{2}(w_x^{(k)})S^2\right] dx\, dt.$$

Here and above $A_{k,\rho}(t) = \{x \in K_\rho : w(x,t) > k\}$. All of the summands of the right side are estimated in terms of the second and third summands of the left side of (3.2) and in terms of

$$\int_{Q(\rho,\tau)} (w^k)^2\zeta_x^2\, dx\, dt, \quad \int_{Q(\rho,\tau)} (w^{(k)})^2|\zeta||\zeta_t|\, dx\, dt, \quad \mu^{\frac{2}{\hat r}(1+\varkappa)}(k,\rho,\tau),$$

where

$$\mu(k,\rho,\tau) = \int_{t_0-\tau}^{t_0} [\operatorname{mes} A_{k,\rho}(t)]^{\frac{\hat r}{\hat q}}\, dt, \qquad \hat q = \frac{2q}{q-1}(1+\varkappa),$$

$$\hat r = \frac{2r}{r-1}(1+\varkappa), \text{ and } \varkappa = \frac{2\varkappa_1}{n}.$$

By virtue of (2.2) the parameters $\hat q$ and $\hat r$ are subject to conditions (3.3) of Chapter II. The estimates are carried out in almost the same way as in the proof of the boundedness of $v(x,t)$ in §2.[1] In a certain sense they are even simpler, since we already know that $u(x,t)$ is bounded. As a result we obtain from (3.2) the inequalities

$$\int_{A_{k,\rho}(t_0)} [w^{(k)}(x,t_0)\zeta(x,t_0)]^2\, dx + \nu \int_{Q(\rho,\tau)} [w^{(k)}\zeta]^2\, dx\, dt$$
$$\leqslant \int_{A_{k,\rho}(t_0-\tau)} (w^{(k)}\zeta)^2\, dx$$
$$+\gamma\left[\max_{Q(\rho,\tau)}(|\zeta_x|^2 + |\zeta_t|)\int_{Q(\rho,\tau)}(w^{(k)})^2\, dx\, dt + \mu^{\frac{2}{\hat r}(1+\varkappa)}(k,\rho,\tau)\right] \tag{3.3}$$

---

1) They are also very similar to estimates in §10 of Chapter III. For this reason our presentation of them is quite brief.

for $\rho \leq \rho_0$ or $r \leq r_0$ and any k with constant γ being determined only by n, M, ν^{-1}, μ, the indices q and r and the norms of the known functions from conditions (2.1). The same inequalities also hold for $w = w_-^l$. Inequalities (3.3) indicate the fulfilment of condition 3) from the definition of the class $\mathfrak{B}_2^{2N}(Q_T, \cdots)$. By the same token the membership of $\mathbf{u}(x, t)$ in $\mathfrak{B}_2^{2N}(Q_T, \cdots)$ is proved. By virtue of Theorem 9.1 of Chapter II $\mathbf{u}$ will belong to $H^{\alpha,\alpha/2}(Q_T)$ with some $\alpha > 0$ and the norms $|\mathbf{u}|_{Q'}^{(\alpha)}$ for any strictly interior subdomain $Q' \subset Q_T$ are estimated in terms of known quantities and the distance from Q' to Γ_T. In order to obtain an estimate of the Hölder norm of $\mathbf{u}$ in the whole cylinder Q_T it is necessary to obtain inequalities (3.3) when $k \geq \max_{Q(\rho,r) \cap \Gamma_T} w(x, t)$ for $w = w_\pm^l$ and in the cylinders $Q(\rho, r)$ intersecting the surface Γ_T. Such inequalities actually hold and their derivation is exactly the same as the above derivation of inequalities (3.3), since

$$\Phi(x, t) = w^{(k)}(x, t)\zeta^2(x, t) \in \overset{\circ}{V}{}_2^{1,0}(Q_T)$$

for any smooth function $\zeta(x, t)$ that is equal to zero in a neighborhood of some circular cylinder $Q(\rho, r)$ intersecting the boundary Γ_T, and this is all that one needs to know about $\Phi(x, t)$ in order to derive inequalities (3.3).

Thus $\mathbf{u}(x, t)$ belongs to the class $\mathfrak{B}_2^{2N}(Q_T \cup \Gamma_T, \cdots)$ and, consequently, we have

Theorem 3.1. *Suppose conditions* (1.2), (2.1), *and* (2.2) *are fulfilled for the system* (1.1). *Then any bounded generalized solution* $\mathbf{u}(x, t)$ *of the system* (1.1) *from* $V_2^{1,0}(Q_T)$ *belongs to* $H^{\alpha,\alpha/2}(Q_T)$ *with some* $\alpha > 0$ *and the norm* $|\mathbf{u}|_{Q'}^{(\alpha)}$, *where* Q' *is an arbitrary strictly interior subdomain of* Q'_T, *is estimated from above by a constant depending only on* n, $M = \operatorname{vrai\,max}_{Q_T} |\mathbf{u}|$, N, *the quantities* ν, μ, μ_1, q *and* r *from conditions* (1.2), (2.1) *and* (2.2) *and the distance from* Q' *to* Γ_T. *If also* S *satisfies condition* (A) *and* $\mathbf{u}|_{S_T} \in H^{\beta,\beta/2}(\Gamma_T)$, *then* $\mathbf{u} \in H^{\alpha,\alpha/2}(\overline{Q}_T)$ *and* $|\mathbf{u}|_{Q_T}^{(\alpha)}$ *is estimated from above by a constant depending only on* n, N, M, ν, μ, μ_1, q, r *from* (1.2), (2.1), (2.2), β, $|\mathbf{u}|_{\Gamma_T}^{(\beta)}$ *and the constants* a_0 *and* θ_0 *in the formulation of condition* (A). *The number* $\alpha > 0$ *is determined in both cases by the values* n, N, ν, μ, μ_1, β, *the norms of the known functions and the indices* q *and* r *from conditions* (2.1).

§4. ON ESTIMATES OF $|\mathbf{u}_x|_Q^{(\alpha)}$ AND OF OTHER HIGHER NORMS OF THE SOLUTIONS

In §§ 2 and 3 we showed that, if the coefficients and free term of a system of equations possess somewhat better properties than the properties enumerated in § 1, then any generalized solution of the system from the class $V_2^{1,0}(Q_T)$ will be a Hölder continuous function of (x, t). A further improvement in the properties of the coefficients and free term of the system creates a further increase in the smoothness of all of its generalized solutions, with the form of this dependence being just the same as for one equation in Chapter III. We will dwell only on an a priori estimate for $|\mathbf{u}_x|_{Q'}^{(\alpha)}$. All of the remaining propositions are proved either in the same way as for one equation or actually follow from the corresponding propositions established by us in Chapters III and IV for one equation. An estimate of $|\mathbf{u}_x|_{Q_T}^{(\alpha)}$ for the whole cylinder Q_T will be proved in the next section directly for quasi-linear equations. The proof given there naturally simplifies for the linear case.

Thus, suppose the coefficients and free terms of the system satisfy, in addition to (1.2), the conditions

$$\left\| \frac{\partial a_{ij}}{\partial x_k},\ |A_i|,\ \left|\frac{\partial A_i}{\partial x_k}\right|,\ |B_i|,\ |A|,\ \mathbf{f}_i,\ \frac{\partial \mathbf{f}_i}{\partial x_i},\ \mathbf{f} \right\|_{2q,\,2r,\,Q_T} \leqslant \mu_1, \tag{4.1}$$

in which the numbers q and r are the same as in (2.2). As to the solution $\mathbf{u}$, we assume that it is bounded and its derivatives $\mathbf{u}_x$ are elements of $V_2^{1,0}(Q_T)$. We will show that an estimate of $|\mathbf{u}_x|_{Q'}^{(\alpha)}$ can be obtained on the basis of Theorems 2.2 and 3.1 established in this chapter. To this end let us demonstrate that the vector function $\mathbf{U} = (u_{x_1}^1, u_{x_2}^1, \cdots, u_{x_n}^N)$ can be considered as a generalized solution from $V_2^{1,0}(Q_T)$ of a system of the form (1.1), the coefficients and free terms of which satisfy the conditions of Theorems 2.2 and 3.1.

In the integral identity (1.4) we take $\eta(x, t) = \xi_{kx_k}(x, t)$, where $\xi_k(x, t)$ is an arbitrary function from $W_2^{2,1}(Q_T)$ that together with its first derivatives with respect to x is equal to zero on S_T, and then we carry out an integration by parts in the first, second, third and last terms, transferring the derivative $\partial/\partial x_k$ from the second factor to the first. After this the equalities obtained for $k = 1, \cdots, n$ are added and the result is written as an identity for $\mathbf{U}$ and $\Xi = (\xi_1^1, \xi_2^1, \cdots, \xi_n^N)$:

$$\int_\Omega \mathbf{U}(x, t)\,\Xi(x, t)\,dx + \int_0^t \int_\Omega [-\mathbf{U}\Xi_t + (a_{ij}\mathbf{U}_{x_j} + \hat{A}_i\mathbf{U} + \mathbf{F}_i)\,\Xi_{x_i}]\,dx\,dt - \int_\Omega \psi_{0x_k}^l \xi_k^l(x, 0)\,dx = 0, \tag{4.2}$$

in which the $\hat{A}_i$ are the matrices with elements

$$a_l^{dk,\,lj} = \delta_l^k b_j^{dl} + a_l^{dl}\delta_j^k + \frac{\partial a_{lj}}{\partial x_k}\delta_d^l,$$

$$l,\ j,\ k = 1, \ldots, n; \quad d,\ l = 1, \ldots, N, \tag{4.3}$$

and the $\mathbf{F}_i$ are the vectors with components

$$\mathbf{F}_l^{d,\,k} = \delta_l^k\left(f^d + a^{dl}u^l\right) + \frac{\partial f_l^d}{\partial x_k} + \frac{\partial a_l^{dl}}{\partial x_k}u^l,$$

$$l,\ k = 1, \ldots, n; \quad d = 1, \ldots, N.$$

The identity (4.2) is valid, as is easily seen, for any function Ξ from $\overset{\circ}{W}{}_2^{1,1}(Q_T)$. It implies that U is a generalized solution from $V_2^{1,0}(Q_T)$ of the system

$$\mathbf{U}_t - \frac{\partial}{\partial x_l}\left(a_{lj}\mathbf{U}_{x_j} + \hat{A}_l\mathbf{U} + \mathbf{F}_l\right) = 0, \tag{4.4}$$

which is a special case of the systems of form (1.1). By virtue of assumptions (1.2) and (4.1) and the already known boundedness of $|\mathbf{u}(x, t)|$ the conditions of Theorems 2.1 and 3.1 are fulfilled for the system (4.4), and thus their inferences concerning the boundedness of $|\mathbf{U}|_{Q'}^{(\alpha)}$ are valid. Thus we have proved

Theorem 4.1. *Suppose conditions* (1.2) *and* (4.1) *are fulfilled and suppose* $\mathbf{u}$ *is a generalized solution of the system* (1.1) *from* $V_2^{1,0}(Q_T)$ *such that* $\mathbf{u}_{x_k} \in V_2^{1,0}(Q_T)$ *for all* $k = 1, \cdots, n$. *Then* $\mathbf{u}_x \in H^{\alpha,\alpha/2}(Q_T)$, $\alpha > 0$, *with the norms* $|\mathbf{u}_x|_{Q'}^{(\alpha)}$, $Q' = \Omega' \times (\epsilon, T)$, $\epsilon > 0$, *being estimated from above by a constant determined only by the norm* $\|\mathbf{u}\|_{2,Q_T}$, n, N, ν *and* μ *from condition* (1.2), *the norms of the coefficients and the indices* q *and* r *from condition* (4.1), ϵ *and the distance from* Ω' *to* S. *If it is also known that* $\mathbf{u}(x, 0) \in H^{1+\beta}(\Omega)$, *then* $\mathbf{u}_x \in H^{\alpha,\alpha/2}(\Omega \times [0, T])$, $0 < \alpha \le \beta$, *with the norm* $|\mathbf{u}_x|_{Q'_T}^{(\alpha)}$, $Q'_T = \Omega' \times [0, T]$ *being estimated from above by a constant determined by* β, *the norm* $|\mathbf{u}(x, 0)|_{\Omega}^{(1+\beta)}$ *and the above-mentioned quantities, except* ϵ.

The exponent α *is determined in both cases by the same quantities as* $|\mathbf{u}_x|^{(\alpha)}$, *except* $\|\mathbf{u}\|_{2,\Omega}$, d, ϵ *and* $|\mathbf{u}(x, 0)|_{\Omega}^{(1+\beta)}$.

After an estimate for $|\mathbf{u}_x|^{(\alpha)}$ has been obtained it is possible to consider the system (1.1) as a collection of separate equations of the form $u_t^k - \partial a_{ij}u_{x_j}^k/\partial x_i = \Phi^k$ and to utilize further results of Chapters III and IV for their solutions u^k. We will not formulate all that this leads to in connection with the system (1.1), in particular, when problem (1.1), (1.3) is uniquely solvable in $W_q^{2m,m}(Q_T)$, $m \ge 1$, or in $H^{l,l/2}(\overline{Q}_T)$, $l > 2$, but only mention that the final results are the same as for one equation.

§ 5. QUASI-LINEAR PARABOLIC SYSTEMS. ESTIMATES OF THE NORMS $|u|_{Q_T}^{(l+\alpha)}$, $l \geq 1$, IN TERMS OF $\max_{Q_T} |u, u_x|$

We will consider in §§ 5–7 quasi-linear parabolic systems of the form

$$u_t - a_{ij}(x, t, u)u_{x_i x_j} + a(x, t, u, u_x) = 0, \tag{5.1}$$

in which $u(x, t) = (u^1(x, t), u^2(x, t), \dots, u^N(x, t))$ is an unknown vector function defined in $\overline{Q}_T$. For systems of the form (5.1) a theorem will be proved, under certain conditions on the smoothness of the known functions and restrictions on their behavior for large values of the arguments, on the existence of a unique classical solution of the first boundary value problem. As a preliminary we will obtain the a priori estimates needed for a proof of this theorem.

Suppose $u(x, t)$ is a classical solution of the system (5.1), i.e. $u(x, t)$ belongs to $C^{2,1}(\overline{Q}_T)$ and satisfies the system of equations (5.1). If the estimates

$$\max_{Q_T} |u| = M, \quad \max_{Q_T} |u_x| = M_1$$

are already known, then all further a priori estimates follow from the results obtained in Chapters III and IV for one equation. Indeed, each of the equations of the system (5.1) can be considered as a linear equation for determining $u^l(x, t)$:

$$u_t^l - \frac{d}{dx_i}\left[\hat{a}_{ij}(x, t)u_{x_j}^l\right] + \hat{a}^{(l)}(x, t) = 0, \tag{5.1'}$$

where

$$\hat{a}_{ij}(x, t) = a_{ij}(x, t, u(x, t)), \quad \hat{a}^{(l)}(x, t)$$
$$= a^l(x, t, u(x, t), u_x(x, t)) + \frac{\partial a_{ij}(x, t, u(x, t))}{\partial u^m} u_{x_i}^m u_{x_j}^l + \frac{\partial a_{ij}}{\partial x_i} u_{x_j}^l.$$

(We assume that the functions in equations (5.1) are sufficiently smooth and, often without saying so, we will assume that all of the functions and their derivatives involved in the calculations are continuous in their arguments.)

Equality (5.1′) represents an equation for u^l with bounded coefficients and bounded derivatives

$$\frac{d\hat{a}_{ij}}{dx_k} = \frac{\partial a_{ij}}{\partial u^m} u_{x_k}^m + \frac{\partial a_{ij}}{\partial x_k}.$$

From this, as was shown in § 3 of Chapter V, there follows an estimate of $|u_x^l|_{Q'}^{(\alpha)}$. We then obtain an estimate of $u^l(x, t)$ in the norm of $H^{2+\alpha, 1+\alpha/2}(Q')$ on the basis of a theorem of A. Friedman (Theorem 5.2 of Chapter IV) on linear equations, since each of the equations (5.1) can be considered as a linear equation with respect to $u^l(x, t)$ with Hölder continuous coefficients.

Thus, by taking into consideration the conditions of the theorems referred to above, we obtain the following theorem.

Theorem 5.1. *Let* $\mathbf{u}(x, t)$ *be a solution of the system* (5.1) *belonging to* $C^{2,1}(\overline{Q}_T)$, *and suppose the system* (5.1) *is parabolic on it, i.e. the inequality*

$$a_{ij}(x, t, \mathbf{u}(x, t))\xi_i\xi_j \geqslant \nu\xi^2, \quad \nu > 0, \tag{5.2}$$

is fulfilled in Q_T. *Suppose that the functions* $a_{ij}(x, t, \mathbf{u})$ *are differentiable with respect to the* x_k *and* u^l *in the domain* $\mathfrak{M} = \{(x, t) \in \overline{Q}_T, |\mathbf{u}| \leq M = \max_{Q_T}|\mathbf{u}(x, t)|, |\mathbf{p}| \leq M_1 = \max_{Q_T}|\mathbf{u}_x(x, t)|\}$, *and that these same functions and their derivatives with respect to the* x_k *and* u^l *and the functions* $a^l(x, t, \mathbf{u}, \mathbf{p})$ *are continuous on* $\mathfrak{M}$. *An upper bound of their moduli is denoted by* M_2. *Then for* $Q' \subset Q_T$ *the norms* $|\mathbf{u}_x|_{Q'}^{(\alpha)}$ *with some* $\alpha > 0$ *are estimated only in terms of the quantities* M_1, M_2, ν *from condition* (5.2) *and the distance from* Q' *to* Γ_T. *These same norms are estimated in terms of the quantities* M_1, M_2, ν, *the distance from* Q' *to* S_T *and* $|\mathbf{u}_x(x, 0)|_{\Omega}^{(\alpha)}$.

If, in addition, the functions $a_{ij}(x, t, \mathbf{u})$ *and* $a^l(x, t, \mathbf{u}, \mathbf{p})$ *satisfy a Hölder condition in* $\mathfrak{M}$ *in the arguments* $x, t, \mathbf{u}, \mathbf{p}$ *with exponents* $\beta, \beta/2, \beta, \beta$ *respectively, then the norms* $|\mathbf{u}|_{Q'}^{(2+\beta)}$ *are estimated in terms of* M_1, M_2, ν, *the Hölder norms of the functions* a_{ij} *and* a^l *in* $\mathfrak{M}$, *and either the distance from* Q' *to* Γ_T *or* $|\mathbf{u}(x, 0)|_{\Omega}^{(2+\beta)}$ *and the distance from* Q' *to* S_T.

Estimates of $|\mathbf{u}_x|_{Q_T}^{(\alpha)}$ (and along with them estimates of $|\mathbf{u}|_{Q_T}^{(2+\beta)}$) for the whole cylinder $\overline{Q}_T$ are obtained as corollaries of Theorem 6.1 of Chapter IV in [65q] and Lemma 3.1 of Chapter II if $\max_{Q_T}|\mathbf{u}_t|$ has been estimated beforehand.

In order to obtain this estimate we form a divided difference with respect to t for the system (5.1) and write the result in the form

$$\mathbf{v}_t - \frac{d}{dx_i}[a_{ij}\mathbf{v}_{x_j} + A_i\mathbf{v} + \mathbf{f}_i] + B_i\mathbf{v}_{x_i} + A\mathbf{v} + \mathbf{f} = 0, \tag{5.3}$$

where

$$\mathbf{v} = \frac{1}{\Delta t}[\mathbf{u}(x, t + \Delta t) - \mathbf{u}(x, t)],$$

$$a_i^{lm} = u^l_{x_j}\int_0^1 \frac{\partial a_{ij}[x, t + \tau\Delta t, \tau\mathbf{u}(x, t + \Delta t) + (1 - \tau)\mathbf{u}(x, t)]}{\partial u^m}\,d\tau,$$

$$f_i^l = u^l_{x_j}\int_0^1 \frac{\partial a_{ij}[\ldots]}{\partial t}\,d\tau,$$

$$b_l^{lm} = \int_0^1 \frac{\partial \hat{a}^l [x,\ t+\tau \Delta t,\ \tau u(x,\ t+\Delta t)+(1-\tau) u(x,\ t)]}{\partial u^m_{x_i}}\, d\tau,$$

$$a^{lm} = \int_0^1 \frac{\partial \hat{a}^l [\ldots]}{\partial u^m}\, d\tau, \qquad f^l = \int_0^1 \frac{\partial \hat{a}^l [\ldots]}{\partial t}\, d\tau,$$

$$\hat{a}^l (x,\ t,\ u,\ p) = a^l (x,\ t,\ u,\ p) + \frac{\partial a_{ij}(x,\ t,\ u)}{\partial u^m} p_i^m p_j^l + \frac{\partial a_{ij}}{\partial x_i} p_j^l.$$

(5.3) can be considered as a linear system with respect to v with bounded coefficients. Bounds for these coefficients will be known to us if we know the quantity ν from (5.2), $M = \max_{Q_T} |u|$, $M_1 = \max_{Q_T} |u_x|$ and a bound M_3 for

$$\left| a_{ij}(x,\ t,\ u),\ a^l (x,\ t,\ u,\ p),\ \frac{\partial a_{ij}}{\partial u^m},\ \frac{\partial a_{ij}}{\partial t},\ \frac{\partial \hat{a}_i^l}{\partial u^m},\ \frac{\partial \hat{a}^l}{\partial p_i^m},\ \frac{\partial \hat{a}^l}{\partial t} \right|$$

in the domain $\mathfrak{M}$. These bounds can be taken independently of Δt. By Theorem 2.1 we can estimate $\max_{Q_T} |v|$ solely in terms of M_1, M_3 and $\max_{\Gamma_T} |v|$. Passing to the limit as $\Delta t \to 0$, we obtain an estimate of $\max_{Q_T} |u_t|$ in terms of M_1, M_3 and $\max_{\Gamma_T} |u_t|$. Thus for the system (5.1) we have the following result.

Theorem 5.2. *Suppose all of the conditions of Theorem 5.1 are fulfilled and suppose the functions*

$$\frac{\partial a_{ij}}{\partial t},\ \frac{\partial^2 a_{ij}}{\partial u^m \partial u^l},\ \frac{\partial^2 a_{ij}}{\partial u^m \partial t},\ \frac{\partial^2 a_{ij}}{\partial u^m \partial x_k},\ \frac{\partial^2 a_{ij}}{\partial x_k \partial t},\ \frac{\partial a^l}{\partial u^m},\ \frac{\partial a^l}{\partial p_i^m},\ \frac{\partial a^l}{\partial t}$$

are continuous on $\mathfrak{M}$. *A bound for them is denoted by* M_3. *Then* $\max_{Q_T} |u_t|$ *and* $|u_x|_{Q_T}^{(\alpha)}$ *with some* $\alpha > 0$ *are estimated in terms of* ν^{-1}, M_1, M_2, M_3, $\max_{\Gamma_T} |u_t|$ *and the norms in* $H^{1+\beta,(1+\beta)/2}(\Gamma_T)$ *of the values of* u *on* Γ_T.

If, in addition, it is known that the $a^l(x,\ t,\ u,\ p)$ *satisfy a Hölder condition on* $\mathfrak{M}$ *in* x *with exponent* β *and the corresponding Hölder constant does not exceed* M_3, *then* $u \in H^{2+\beta,1+\beta/2}(\overline{Q}_T)$ *and the norm* $|u|_{Q_T}^{(2+\beta)}$ *is estimated in terms of the same quantities and the norms in* $H^{2+\beta,1+\beta/2}(\Gamma_T)$ *of the values of* u *on* Γ_T. *Furthermore, all of the indicated norms of* u *depend on the norms of the functions defining the boundary* S: *in the first case they depend on the norms of these functions in* O^2; *in the second, on their norms in* $H^{2+\beta}$.

Thus Theorems 5.1 and 5.2 reduce the problem concerning a priori estimates of the solutions of systems of form (5.1) to the problem of obtaining estimates of $\max_{Q_T} |u|$ and $\max_{Q_T} |u_x|$.

§6. AN ESTIMATE OF $\max_{Q_T} |\mathbf{u}_x|$

The examples in §3 of Chapter I show that in order to obtain estimates of $\max_{Q_T} |\mathbf{u}_x|$ we must have agreement in the orders of growth of the coefficients of the equation with respect to $|\mathbf{p}| = \sqrt{\Sigma_{k,i} (p_i^k)^2}$ for $|\mathbf{p}| \to \infty$. In our case the condition of parabolicity has the form

$$\nu(|\mathbf{u}|)\xi^2 \leqslant a_{ij}(x, t, \mathbf{u})\xi_i\xi_j \leqslant \mu(|\mathbf{u}|)\xi^2, \quad \nu > 0, \tag{6.1}$$

and therefore a necessary requirement on the order of growth of a with respect to $|\mathbf{p}|$ is the following:

$$|\mathbf{a}(x, t, \mathbf{u}, \mathbf{p})| \leqslant \mu(|\mathbf{u}|)(1 + |\mathbf{p}|)^2.$$

However, as the example of the system

$$\begin{cases} u_t^1 - u_{xx}^1 = u^1 [(u_x^1)^2 + (u_x^2)^2], \\ u_t^2 - u_{xx}^2 = u^2 [(u_x^1)^2 + (u_x^2)^2], \end{cases}$$

$$u^1 = \cos mx \text{ and } u^2 = \sin mx$$

given by E. Heinz shows, this restriction on $\mathbf{a}(x, t, \mathbf{u}, \mathbf{p})$ is not in itself sufficient for deriving such an estimate. But if the terms having quadratic growth with respect to $\mathbf{p}$ have a special form, then an estimate becomes possible. Namely, we will assume that the system has the form

$$\mathbf{u}_t - a_{ij}(x, t, \mathbf{u})\,\mathbf{u}_{x_i x_j} + b_i(x, t, \mathbf{u}, \mathbf{u}_x)\,\mathbf{u}_{x_i} + \mathbf{b}(x, t, \mathbf{u}, \mathbf{u}_x) = 0, \tag{6.2}$$

where the functions b_i and $\mathbf{b}$ are subject to the conditions

$$\left.\begin{aligned} &|b_i(x, t, \mathbf{u}, \mathbf{p})| \leqslant \mu(|\mathbf{u}|)(1 + |\mathbf{p}|), \\ &|\mathbf{b}(x, t, \mathbf{u}, \mathbf{p})| \leqslant [\varepsilon(|\mathbf{u}|) + P(|\mathbf{p}|, |\mathbf{u}|)](1 + |\mathbf{p}|)^2, \end{aligned}\right\} \tag{6.3}$$

where $P(|\mathbf{p}|, |\mathbf{u}|) \to 0$ for $|\mathbf{p}| \to \infty$, while $\epsilon(M)$ is a sufficiently small number determined only by the quantities M, $\nu(M)$ and $\mu(M)$ from (6.1) and (6.3). Of the $a_{ij}(x, t, \mathbf{u})$ we will require that they satisfy inequality (1.2), be differentiable with respect to the x_k and u^l and satisfy the inequality

$$\left| \frac{\partial a_{ij}(x, t, \mathbf{u})}{\partial x_k}, \frac{\partial a_{ij}(x, t, \mathbf{u})}{\partial u^l} \right| \leqslant \mu(|\mathbf{u}|). \tag{6.4}$$

We will assume that $\mathbf{u}(x, t)|_{S_T} = 0$. (This assumption is not essential; i.e. if the boundary S and boundary values of $\mathbf{u}|_S$ are sufficiently smooth, then we can extend $\mathbf{u}|_S$ with preservation of smoothness into Q_T and by a subtraction achieve homogeneous boundary conditions.) Let us first estimate $\max_{S_T}|u_x|$ or, what is the same, $|\partial u/\partial n|$ on the surface S_T.

Lemma 6.1. *Let* $u(x, t)$ *be a solution of the system* (6.2) *that belongs to* $C^{2,1}(Q_T)$, *is continuous together with its derivatives* u_x *in* $\overline{Q}_T$ *and is equal to zero on* S_T. *Suppose inequalities* (6.1) *and* (6.3) *are fulfilled for* $(x, t) \in \overline{Q}_T$, $u = u(x, t)$ *and arbitrary* p. *Then* $\max_{S_T}|u_x(x, t)|$ *is estimated by a constant depending only on* $M = \max|u(x, t)|$, *the quantities* $\nu(M), \mu(M), \epsilon(M)$ *and* $P(|p|, M)$ *from* (6.1) *and* (6.3), $\max_\Omega|u_x(x, 0)|$ *and the boundary* S, *which is assumed to belong to the class* O^2.

Proof. It is clear that under the conditions of the lemma

$$|u_x^l|_{S_T} = \left|\frac{\partial u^l}{\partial n}\right|_{S_T}.$$

Let us take a point (x_0, t_0) on S_T and a number r such that

$$\left|\frac{\partial u^r(x_0, t_0)}{\partial n}\right| = \max_{l=1,\dots,N}\ \max_{S_T}\left|\frac{\partial u^l}{\partial n}\right| = M_0.$$

If $t_0 = 0$, then $M_0 \le \max_\Omega|u_x(x, 0)|$.

Suppose $t_0 > 0$ and suppose for definiteness that $\partial u^r(x_0, t_0)/\partial n < 0$. We take the function

$$w^r = u^r + \sum_{l=1}^N (u^l)^2$$

(it is clear that $\partial w^r/\partial n|_{S_T} = \partial u^r/\partial n|_{S_T}$) and consider the equality

$$\mathcal{L}^r(u) + 2\sum_{l=1}^N u^l \mathcal{L}^l(u) = 0,$$

where $\mathfrak{L}^l$ is the left side of the lth equation of the system (6.2) – in case $\partial u^r(x_0, t_0)/\partial n > 0$ it is necessary to consider the function

$$-u^r + \sum_{l=1}^N (u^l)^2 .$$

It can be written in the form

$$w_t^r - a_{ij}(x, t, u)w_{x_ix_j}^r + 2a_{ij}(x, t, u)u_{x_i}^l u_{x_j}^l + b_i w_{x_i}^r + c_r = 0, \tag{6.5}$$

where $c_r = 2b^l(x, t, u, u_x)u^l + b^r(x, t, u, u_x)$, while the $b^l(x, t, u, p)$ are the components of $b(x, t, u, p)$. Let us introduce in place of w^r a function v^r by means of the equality $w^r = \phi(v^r)$. The function ϕ will be chosen below, but in such a way that $\phi'(y) > 0$. We substitute in (6.5)

$$w^r_t = \varphi' v^r_t; \quad w^r_{x_i} = \varphi' v^r_{x_i}; \quad w^r_{x_i x_j} = \varphi' v^r_{x_i x_j} + \varphi'' v^r_{x_i} v^r_{x_j}.$$

This gives us

$$v^r_t - a_{ij}(x, t, \mathbf{u}) v^r_{x_i x_j} - \frac{\varphi''}{\varphi'} a_{ij} v^r_{x_i} v^r_{x_j} + \frac{2}{\varphi'} a_{ij} u^l_{x_i} u^l_{x_j} + b_i v^r_{x_i} + \frac{1}{\varphi'} c_r = 0.$$

Hence by virtue of (6.3) we will have

$$v^r_t - a_{ij} v^r_{x_i x_j} - \frac{\varphi''}{\varphi'} a_{ij} v^r_{x_i} v^r_{x_j} + \frac{2}{\varphi'} a_{ij} u^l_{x_i} u^l_{x_j} \leqslant \left[\mu (1 + |\mathbf{u}_x|) |v^r_x| + \frac{2M+1}{\varphi'} (1 + \mathbf{u}^2_x)(\varepsilon + P(|\mathbf{u}_x|)) \right], \tag{6.6}$$

where $\mu = \mu(M)$, $\epsilon = \epsilon(M)$ and $P(|\mathbf{p}|) = P(|\mathbf{p}|, M)$ are quantities from conditions (6.1) and (6.3).

Applying Cauchy's inequality (2.1) of Chapter II, we estimate the right side of (6.6) from above by the expression

$$\frac{2(M+1)}{\varphi'} [\varepsilon + P(|\mathbf{u}_x|)] \mathbf{u}^2_x + \frac{\mu^2 \varphi'}{2\varepsilon} |v^r_x|^2 + \frac{2(M+1)}{\varphi'} (2\varepsilon + \hat{P}).$$

Here $\epsilon = \epsilon(M)$ from conditions (6.3), while $\hat{P} = \max_{|\mathbf{p}| \geq 0} P(|\mathbf{p}|)$.

By assumption $\epsilon(M)$ is small, while $P(|\mathbf{p}|) \longrightarrow 0$ for $|\mathbf{p}| \longrightarrow \infty$. We assume that

$$(M+1)\varepsilon(M) \leqslant \frac{\nu}{2} \tag{6.7}$$

and $|\mathbf{u}_x(x_0, t_0)| \geq k_0$, where the value of k_0 is determined by the inequality

$$(M+1) P(k, M) \leqslant \frac{\nu}{2} \quad \text{for} \quad k \geqslant k_0. \tag{6.8}$$

It then follows from (6.6) that at the point (x_0, t_0)

$$v^r_t - a_{ij} v^r_{x_i x_j} - \frac{\varphi''}{\varphi'} a_{ij} v^r_{x_i} v^r_{x_j} \leqslant c \left[\varphi' |v^r_x|^2 + (\varphi')^{-1} \right], \tag{6.9}$$

where c is a known constant.

Let us select $\phi(y)$ in such a way as to fulfill for $y \geq 0$ the conditions

$$-\frac{\varphi''}{\varphi'} \nu(M) - c\varphi' \geqslant 0 \quad \text{and} \quad \varphi(0) = 0,$$

where $\nu(M)$ is taken from (6.1). These conditions are satisfied by the function

$$\varphi(y) = \frac{\nu(M)}{c} \ln(1 + y).$$

For this $\phi(y)$ inequality (6.9) implies

$$v_t^r - a_{ij}v^r_{x_ix_j} \leqslant c_1,$$

$$c_1 = \frac{c^2}{\nu}\max_{Q_T}(1+v^r) \leqslant \frac{c^2}{\nu}e^{\frac{c}{\nu}(M+M^2)}. \tag{6.10}$$

The function v^r, as well as w^r, is equal to zero on S^T, and

$$\left.\frac{\partial v^r}{\partial n}\right|_{S_T} = \frac{1}{\varphi'(v^r)}\left.\frac{\partial w^r}{\partial n}\right|_{S_T} = \frac{c}{\nu(M)}\left.\frac{\partial u^r}{\partial n}\right|_{S_T}$$

achieves its maximum on S_T at the same point (x_0, t_0) as $\partial u^r/\partial n$.

Let us now construct a barrier function $\psi(x)$, i.e. a function satisfying the relations

$$-a_{ij}\psi_{x_ix_j} < -c_1, \quad x \in \Omega,$$

$$\max_{\Omega}[v^r(x, 0) + \psi(x)] = v^r(x_0, 0) + \psi(x_0),$$

$$\max_{S}\psi(x) = \psi(x_0).$$

As such a function we can take, for example, the function $\psi(x) = me^{-k\Phi(x)}$ with sufficiently large k and m and a twice continuously differentiable function $\Phi(x)$ having the following properties:

1) it is greater than zero in Ω;

2) $|\Phi_x| \geq \text{const} > 0$ in Ω;

3) the surface $\Phi(x) = 0$ contains the point x_0.

If the domain Ω is situated entirely on one side of the tangent plane to S at the point x_0 (let this be the plane $x_n = x_n^{(0)}$ and suppose $x_n > x_n^{(0)}$ in Ω), then as Φ we can take $\Phi(x) = x_n - x_n^{(0)}$. Otherwise it is necessary to transform Ω so that it is situated with respect to x_0 in the indicated manner.

For the function $v^r(x, t) + \psi(x)$ we have

$$(v^r + \psi)_t - a_{ij}(v^r + \psi)_{x_ix_j} < 0.$$

Consequently, its maximum occurs on Γ_T. Moreover, this maximum is achieved at all points of the interval $\{x = x_0, t \in [0, T]\}$, since

$$\max_{\Omega}[v^r(x, 0) + \psi(x)] = v^r(x_0, 0) + \psi(x_0) = \psi(x_0)$$

and

$$\max_{S_T}(v + \psi) = \max_{S_T}\psi = \psi(x_0).$$

Therefore

$$\frac{\partial\,[v^r(x,t)+\psi(x)]}{\partial \mathrm{n}}\Big|_{x=x_0}\geqslant 0$$

and in particular

$$-\frac{\partial v^r}{\partial \mathrm{n}}\Big|_{(x_0,\,t_0)}\leqslant\frac{\partial\psi}{\partial \mathrm{n}}\Big|_{x_0}.$$

This implies the desired estimate for the quantity

$$M_0=\max_{l=1,\,\dots,\,N}\ \max_{S_T}\left|\frac{\partial u^l}{\partial \mathrm{n}}\right|.$$

The lemma is proved.

Let us now estimate the modulus of continuity of the function $\mathbf{u}(x, t)$ in Q_T.

Lemma 6.2. *Let* $\mathbf{u}(x, t)$ *be a solution of the system* (6.2) *belonging to* $C^{2,1}(\overline{Q}_T)$ *and let* $\max_{Q_T}|\mathbf{u}(x, t)| = M$. *Suppose the functions*

$$a_{ij}(x,\ t,\ \mathbf{u}),\ b_i(x,\ t,\ \mathbf{u},\ \mathbf{p}),\ b^l(x,\ t,\ \mathbf{u},\ \mathbf{p}),\ \frac{\partial a_{ij}}{\partial x_k},\ \frac{\partial a_{ij}}{\partial u^l}$$

are continuous and satisfy inequalities (6.1), (6.3) *and* (6.4) *for* $(x, t) \in Q_T$, $|\mathbf{u}| \leq M$ *and arbitrary* $\mathbf{p}$. *Then for any interior cylinder* $Q' \subset Q_T$ *the norms* $|\mathbf{u}|^{(\alpha)}_{Q'}$ *with some* $\alpha > 0$ *are estimated only in terms of* M, $\nu(M)$, $\mu(M)$, $\epsilon(M)$ *and* $P(|\mathbf{p}|, M)$ *from* (6.1), (6.3) *and* (6.4) *and the distance from* Q' *to* Γ_T.

If, in addition, S *satisfies condition* (A), *then the norm* $|\mathbf{u}|^{(\alpha)}_{Q_T}$ *is estimated in terms of not only the indicated quantities* M, $\nu(M)$, $\mu(M)$, $\epsilon(M)$ *and* $P(|\mathbf{p}|, M)$ *but also* $|\mathbf{u}|^{(\beta)}_{\Gamma_T}$, β *and the constants* a_0 *and* θ_0 *from condition* (A).

The proof of this lemma is analogous to the proof of Theorem 3.1 on solutions of linear systems and reduces to verifying that the solution belongs to the function class $\mathfrak{B}_2^{2N}(\overline{Q}_T, M_1, \delta_1, \dots, \delta_6, \gamma, r, \delta, \kappa)$ with δ now different from infinity, $r = 2(n+2)/n$ and $\kappa = 2/n$. In view of this we will not carry out the proof of the lemma here.

Let us now proceed to the formulation and proof of the basic fact of the present section: an a priori estimate of $\max_{Q_T}|\mathbf{u}_x(x, t)|$.

Theorem 6.1. *Suppose the conditions of Lemma* 6.2 *are fulfilled. Then the quantity* $\max_{Q'}|\mathbf{u}_x(x, t)|$ *in* $Q' = \Omega' \times (0, T)$ *is estimated in terms of* $M = \max_{Q_T}|\mathbf{u}|$, *the constants* $\nu(M)$, $\mu(M)$ *and* $\epsilon(M)$ *and the quantity* $P(|\mathbf{p}|, M)$ *from inequalities* (6.1), (6.3) *and* (6.4), $\max_\Omega|\mathbf{u}_x(x, 0)|$ *and the distance from* Q' *to* S_T.

If, in addition, $\mathbf{u}|_S = 0$ and $S \in O^2$, *then* $\max_{Q_T}|\mathbf{u}_x|$ *is estimated in terms of the same quantities* M, $\nu(M)$, $\mu(M)$, $\epsilon(M)$, $P(|\mathbf{p}|, M)$ *and* $\max_\Omega|\mathbf{u}_x(x, 0)|$ *and the* *norm in C^2 of the functions defining the boundary S.*

Let us first consider the interior estimates of $|u_x|$. As a preliminary we will estimate $\int_{Q_T}|u_x|^2\,dx\,dt$. To this end, in the identity

$$\int_0^t\int_\Omega\Big[u_t^l\eta+a_{ij}(x,t,u)u_{x_i}^l\eta_{x_j}+\Big(\frac{da_{ij}}{dx_j}u_{x_i}^l+b_iu_{x_i}^l\Big)\eta+b^l\eta\Big]dx\,dt=0,\qquad t\in[0,T],\tag{6.11}$$

which is valid for any $\eta(x,t)$ from $\overset{\circ}{V}{}_2^{1,0}(Q_T)$, we put $t=T$ and $\eta=u^le^{\lambda v}\zeta^2(x)$, where $v=|u|^2=\sum_{r=1}^N(u^r)^2$ and ζ is a cutting function for Ω. If $u|_{S_T}=0$, then we take $\zeta\equiv 1$. In (6.11) we carry out a summation over l from 1 to N. This gives us

$$\frac{1}{2\lambda}\int_\Omega e^{\lambda v(x,t)}\zeta^2dx\Big|_0^T+\int_{Q_T}e^{\lambda v}\Big\{\frac{\lambda}{2}a_{ij}v_{x_i}v_{x_j}\zeta^2+a_{ij}v_{x_i}\zeta\zeta_{x_j}+\zeta^2\Big[a_{ij}u_{x_i}^lu_{x_j}^l+\Big(\frac{\partial a_{ij}}{\partial u^m}u_{x_j}^m+\frac{\partial a_{ij}}{\partial x_j}+b_i\Big)\frac{1}{2}v_{x_i}+b^lu^l\Big]\Big\}dx\,dt=0.$$

From here, by making use of (6.1), (6.3) and (6.4) (with ϵ sufficiently small in (6.3)) and taking λ sufficiently large, it is not difficult to obtain the estimate

$$\int_{Q_T}\sum_{l=1}^N\sum_{i=1}^n(u_{x_i}^l)^2\zeta^2dx\,dt\leqslant c.\tag{6.12}$$

Let us put $\eta=(u_{x_k}^l\xi)_{x_k}$ in (6.11), where $\xi(x,t)$ together with its first derivatives is assumed to be equal to zero on S_T. Then after the obvious transformations by means of an integration by parts and a summation over k and l we arrive at the identity

$$\int_0^t\int_\Omega\Big[\frac{1}{2}V_t\xi+a_{ij}u_{x_kx_i}^lu_{x_kx_j}^l\xi+\frac{1}{2}a_{ij}V_{x_i}\xi_{x_j}+\frac{da_{ij}}{dx_k}u_{x_i}^lu_{x_kx_j}^l\xi+\frac{da_{ij}}{dx_k}u_{x_i}^lu_{x_k}^l\xi_{x_j}-\Big(\frac{da_{ij}}{dx_j}u_{x_i}^l+b_iu_{x_i}^l+b^l\Big)(\Delta u^l\xi+u_{x_k}^l\xi_{x_k})\Big]dx\,dt=0,\tag{6.13}$$

where $V=\sum_{l=1}^N\sum_{k=1}^n(u_{x_k}^l)^2$. In (6.13) we put $\xi=2V^s\zeta^2$, where $\zeta(x)$ is a cutting function for a ball $K_\rho\subset\Omega$ and $s\geq 0$. The first three terms of the integrand yield the basic positive terms, while the remaining terms can be estimated in modulus from above with the use of inequalities (6.1), (6.3) and (6.4). This leads to the inequality

$$\frac{1}{s+1}\int_\Omega V^{s+1}\zeta^2\,dx\Big|_0^t + \nu\int_0^t\int_\Omega\left[2u_{xx}^2V^s\zeta^2 + sV^{s-1}V_x^2\zeta^2\right]dx\,dt \leqslant c(s)\int_0^t\int_\Omega\left(V^{s+1}\zeta_x^2 + V^{s+2}\zeta^2 + \zeta_x^2\right)dx\,dt \tag{6.14}$$

with a certain constant $c(s)$ depending on known constants and the number s. On the other hand,

$$\int_{K_\rho} V^{s+2}\zeta^2\,dx = \int_{K_\rho} V^{s+1}\zeta^2\sum_{k,l} u_{x_k}^l u_{x_k}^l\,dx$$
$$= -\int_{K_\rho}(u^l - u_0^l)\left[V^{s+1}\zeta^2\Delta u^l + (1+s)V^s\zeta^2 u_{x_k}^l 2u_{x_p}^m u_{x_p x_k}^m\right.$$
$$\left. + V^{s+1}u_{x_k}^l 2\zeta\zeta_{x_k}\right]dx \leqslant \max_{l,K_\rho}|u^l - u_0^l|\,c_1(s)\int_{K_\rho}\left[V^{s+2}\zeta^2\right.$$
$$\left. + V^s u_{xx}^2\zeta^2 + V^{s+1}\zeta_x^2\right]dx.$$

If we take as u_0^l the value of $u^l(x, t)$ at the center of the ball K_ρ, then by virtue of Lemma 6.2

$$\max_{l,K_\rho}|u^l - u_0^l| \leqslant c\rho^\alpha,\quad \alpha > 0.$$

Taking ρ sufficiently small, we obtain

$$\int_{K_\rho} V^{s+2}\zeta^2\,dx \leqslant c\rho^\alpha\int_{K_\rho}\left(V^s|u_{xx}|^2\zeta^2 + V^{s+1}\zeta_x^2\right)dx. \tag{6.15}$$

This inequality together with (6.14) implies for sufficiently small $\rho \le \rho(s)$ that

$$\frac{1}{s+2}\int V^{s+1}\zeta^2\,dx\Big|_0^t + \frac{\nu}{2}\int_0^t\int_{K_\rho}\left(|u_{xx}|^2V^s\zeta^2 + V^{s+2}\zeta^2\right)dx\,dt \leqslant c(s)\int_0^t\int_{K_\rho}\left(1 + V^{s+1}\right)\zeta_x^2\,dx\,dt. \tag{6.16}$$

Inequalities (6.16) are valid for $s = 0, 1, 2, \cdots$. Together with the original inequality (6.12) they permit us to give successively with respect to s the estimates

$$\max_{t\in[0,T]}\int_{\Omega'} V^{s+1}\,dx\,dt\leqslant c(s,\Omega'). \tag{6.17}$$

Let us return to the identity (6.11). We take $\eta=\xi_{x_k}$ in it, where $\xi(x,t)$ is an arbitrary smooth function that is equal to zero in the vicinity of S_T, and we carry out an integration by parts. As a result (6.11) takes the form

$$\int_0^t\int_\Omega\left[u^l_{x_k t}\xi+\left(a_{lj}u^l_{x_k x_l}+f_j^{k,l}\right)\xi_{x_j}\right]dx\,dt=0, \tag{6.18}$$

where

$$f_j^{k,l}=\frac{\partial a_{lj}}{\partial x_k}u^l_{x_l}+\frac{\partial a_{lj}}{\partial u^m}u^m_{x_k}u^l_{x_l}+\left(\frac{da_{lr}}{dx_r}+b_l\right)u^l_{x_l}\delta^j_k+b^l\delta^j_k.$$

From here it is seen that each of the functions $v=u^l_{x_k}$, $k=1,\cdots,n$, $l=1,\cdots,N$, can be considered as a solution from $V_2^{1,0}(Q_T)$ of the equation

$$v_t-\frac{\partial}{\partial x_l}\left(a_{lj}v_{x_j}+f_j^{k,l}\right)=0.$$

By virtue of assumptions (6.1), (6.3) and (6.4) and inequality (6.17) with $s=n$ it is possible to assert that the conditions

$$\|f_j^{k,l}\|_{q,r,Q'}\leqslant\mu_1,\quad q=n+1,\quad r=\infty,\quad Q'=\Omega'\times(0,T),$$

are fulfilled for the a_{ij} and $f_j^{k,l}$. It therefore follows on the basis of Theorem 7.1 of Chapter III that $\max_{Q''}|u^l_{x_k}|$ in $Q''\subset Q'$ is estimated in terms of ν, μ, μ_1 and the distance from Q'' to the lateral surface of the cylinder Q'. Thus the first part of Theorem 6.1 can be considered proved.

It remains for us to verify the analogous estimate in cylinders intersecting the boundary. The quantity $\max_{\Gamma_T}|u_x|=M_0$ is already estimated. An estimate of $\|u_x\|_{2,Q_T}$ is given for all of Q_T. We must prove estimate (6.17) for domains $\overline{\Omega}_\rho$ intersecting the boundary S. For $s\geq 1$ we obtain inequalities (6.14)–(6.17) in the same way as above as long as we replace $\xi=V^s\zeta^2$ in (6.13) by the function

$$\xi=\begin{cases}2\left(V-M_0^2\right)^s\zeta^2 & \text{for}\quad V>M_0^2,\\ 0 & \text{for}\quad V\leqslant M_0^2.\end{cases}$$

This is admissible, since such a ξ is equal to zero on all of the boundary of the domain $K_\rho\cap\Omega$. In the case $s=0$ we take in (6.9)

$$\xi=\begin{cases}0 & \text{for}\quad V\leqslant M_0^2,\\ 2\left(V-M_0^2\right)\zeta^2 & \text{for}\quad M_0^2\leqslant V\leqslant M_0^2+1,\\ 2\zeta^2 & \text{for}\quad V\geqslant M_0^2+1.\end{cases}$$

Then after a series of transformations and estimates analogous to those made in the proof of Theorem 4.1 of Chapter V for one equation we arrive at an estimate for $\max_{t\in[0,T]}\int_{\Omega_\rho} V^{s+1}\,dx\,dt$.

In this way one proves the second part of Theorem 6.1

§7. AN EXISTENCE THEOREM FOR QUASI-LINEAR SYSTEMS

The a priori estimates obtained in the preceding sections for a solution $\mathbf{u}(x, t)$ of the system (5.1) permit us to prove the solvability of the first boundary value problem

$$\left.\begin{aligned} \mathbf{u}_t - a_{ij}(x, t, \mathbf{u})\,\mathbf{u}_{x_ix_j} + b_i(x, t, \mathbf{u}, \mathbf{u}_x)\mathbf{u}_{x_i} & \\ + \mathbf{b}(x, t, \mathbf{u}, \mathbf{u}_x) = 0, & \\ \mathbf{u}|_{S_T} = 0, \quad \mathbf{u}|_{t=0} = \boldsymbol{\psi}_0(x) & \end{aligned}\right\} \tag{7.1}$$

in the cylinder $Q_T = \Omega\times(0, T)$.

As a preliminary we note that $\max_{Q_T}|\mathbf{u}(x, t)|$ for solutions of problem (7.1) is estimated in the same way as for one equation if

$$a_{ij}(x, t, \mathbf{u})\xi_i\xi_j \geqslant 0 \tag{7.2}$$

and

$$b^l(x, t, \mathbf{u}, \mathbf{p})u^l \geqslant -c_1|\mathbf{u}|^2 - c_2, \quad c_i = \text{const} \geqslant 0, \quad i = 1, 2. \tag{7.3}$$

Namely,

$$\max_{Q_T}|\mathbf{u}(x, t)| \leqslant \min_{c>c_1} e^{cT}\left[\max_{\Omega}|\mathbf{u}(x, 0)| + \sqrt{\frac{c_2}{c-c_1}}\right] \equiv M. \tag{7.4}$$

A priori estimates of $\max_{Q_T}|\mathbf{u}(x, t)|$ are also possible under certain other conditions on the coefficients.

The following theorem is valid.

Theorem 7.1. *Suppose the following conditions are fulfilled for problem* (7.1).

a) *Inequalities* (7.2) *and* (7.3) *are valid for* $(x, t)\in \overline{Q}_T\setminus\Gamma_T$ *and arbitrary* $\mathbf{u}$.

b) *For* $(x, t)\in Q_T$, $|\mathbf{u}|\le M$, *where* M *is defined by* (7.4), *and arbitrary* $\mathbf{p}$ *the functions*

$$a_{ij}(x, t, \mathbf{u}),\ b_i(x, t, \mathbf{u}, \mathbf{p}),\ b^l(x, t, \mathbf{u}, \mathbf{p}),\ \frac{\partial a_{ij}}{\partial x_k},\ \frac{\partial a_{ij}}{\partial u^l}$$

are continuous and satisfy inequalities (6.1), (6.3) *and* (6.4). (*This gives an a priori estimate for the solution:* $\max_{Q_T}|\mathbf{u}_x|\le M_1$, *where* M_1 *is determined by the quantities in conditions* a) *and* b), S *and* $\max_\Omega |u_x(x, 0)|$.)

c) *For* $(x, t) \in \overline{Q}_T$, $|u| \le M$ *and* $|p| \le M_1$ *the first derivatives of the functions* $a_{ij}(x, t, u)$, $b_i(x, t, u, p)$, $b^l(x, t, u, p)$ *with respect to* x, t, u, p *and the derivatives* $\partial^2 a_{ij}/\partial u^l \partial u^m$, $\partial^2 a_{ij}/\partial u^l \partial x_k$, $\partial^2 a_{ij}/\partial u^l \partial t$, $\partial^2 a_{ij}/\partial x_k \partial t$ *are continuous.*

d) $\psi_0(x) \in H^{2+\beta}(\overline{\Omega})$ *and satisfies the compatibility conditions*

$$\psi_0|_{x\in S} = 0,$$

$$\left[-a_{ij}(x, 0, 0)\psi^l_{0x_ix_j} - b_i(x, 0, 0, \psi_{0x})\psi^l_{0x_i} + b^l(x, 0, 0, \psi_{0x})\right]\Big|_{x\in S} = 0, \qquad l = 1, \dots, N.$$

e) $S \in H^{2+\beta}$.

Then in the class $H^{2+\beta, 1+\beta/2}(\overline{Q}_T)$ *there exists a unique solution* $u(x, t)$ *of problem* (7.1).

The proof of this theorem for systems is carried out in the same way as that of the corresponding existence theorem in the case of one equation (Theorem 6.1 of Chapter V), by means of the method of continuity in a parameter and an application of the Leray-Schauder fixed point theorem.

§8. LINEAR PARABOLIC SYSTEMS OF GENERAL FORM

In 1938 I. G. Petrovskiĭ in the paper [93c] introduced a very extensive class of parabolic systems of linear differential equations, which were subjected to a detailed analysis. It turned out that many of the characteristic properties of solutions of parabolic equations of second order carry over to these systems. Afterwards Petrovskiĭ's definition was generalized in various directions and, in a number of cases involving more general parabolic systems than those of Petrovskiĭ, certain complete results were obtained.

We will begin with Petrovskiĭ's definition of a parabolic equation of high order. Let $L(x, t, \partial/\partial x, \partial/\partial t)$ be a linear differential operator[1] of arbitrary order with complex coefficients depending on x and t in some domain $Q \subset E_{n+1}$. It is clear that at any point $(x, t) \in Q$ the function $L(x, t, i\xi, p)$, where ξ is an n-dimensional vector with components $\xi_1, \dots, \xi_n$ and p is a scalar complex parameter, is a polynomial in the ξ_j and p. Let b be some positive integer and let the degree of the polynomial $L(x, t, i\xi\lambda, p\lambda^{2b})$ in λ be equal to $2br$, where r is

1) In §§8–10 we will frequently use the term "differential operator" or even simply "operator" in the sense of "differential expression."

a positive integer. We denote by L_0 the principal part of the polynomial L, i.e. the sum of those terms of L for which

$$L_0(x, t, i\xi\lambda, p\lambda^{2b}) = \lambda^{2br} L_0(x, t, i\xi, p).$$

Definition 1. *An operator L is said to be parabolic ($2b$-parabolic) at a point (x, t) if for any real ξ the roots p_s of the polynomial $L_0(x, t, i\xi, p)$ in the variable p satisfy the condition*

$$\operatorname{Re} p_s \leqslant -\delta |\xi|^{2b} \quad (\delta > 0). \tag{8.1}$$

An operator L is said to be uniformly parabolic in a domain Q_T if it is parabolic at each point of this domain and inequality (8.1) *is fulfilled at each point $(x, t) \in Q_T$ with one and the same number $\delta > 0$.*

For equations of second order with real coefficients $b = 1$, we have $L_0(x, t, i\xi, p) = p + \sum_{i,j=1}^n a_{ij}(x, t)\xi_i\xi_j$ and, by virtue of condition (2.5) of Chapter I,

$$\operatorname{Re} p_s = p_s = -\sum_{i,j=1}^{n} a_{ij}(x, t)\xi_i\xi_j \leqslant -\nu |\xi|^2,$$

so that $\delta = \nu$.

From the definition of parabolicity we see that a single differentiation with respect to t of a parabolic operator is equivalent in force to a $2b$-order differentiation with respect to the space variables x (one says that a differentiation with respect to t has weight $2b$). The weight must always be an even integer; otherwise the parabolicity condition (8.1) will not be fulfilled for any real ξ.

Let us explain this in more detail. We assume that $R(i\xi, p)$ is a polynomial that is homogeneous in the following sense:

$$R(i\xi\lambda^q, p\lambda^s) = \lambda^N R(i\xi, p). \tag{8.2}$$

Here q, s and N are positive integers, while λ is any complex number. Let ξ_0 and p_0 be roots of the polynomial R, where ξ_0 is a real vector and $\operatorname{Re} p_0 < 0$. It follows from (8.2) that, along with ξ_0 and p_0, $\xi_0\lambda^q$ and $p_0\lambda^s$ will also be roots of the polynomial R. If $s/q \neq 2b$, then we can select λ so that λ^q is real, while $\operatorname{Re} p_0\lambda^s > 0$, and in this case condition (8.1) will not be fulfilled for all of the roots p_s for all real ξ. Indeed, if $s/q \neq 2b$, then for any $\vartheta \in (1/2, 3/2)$ we can find integers k and m such that

$$2m + \frac{3}{2} - \vartheta < \frac{ks}{q} < 2m + \frac{5}{2} - \vartheta. \tag{8.3}$$

Let $\arg p_0 = \vartheta\pi \in (\pi/2, 3\pi/2)$ and $\lambda = e^{i\pi k/q}$. Then it follows from inequality (8.3) that

$$\left(2m+\frac{3}{2}\right)\pi < \arg(p_0\lambda^s) < \left(2m+\frac{5}{2}\right)\pi,$$

i.e. $\operatorname{Re} p_0\lambda^s > 0$.

But if $s/q = 2b$, then such a λ cannot be selected, since $\lambda^s = (\lambda^q)^{2b} > 0$ if λ^q is real.

Thus, if the polynomial R is homogeneous in the sense of (8.2), then a parabolicity condition of type (8.1) can only be fulfilled for it in the event $s = 2bq$.

Let us assume that condition (8.1) is fulfilled for a homogeneous polynomial R. Then by virtue of what has just been proved

$$R(i\xi\lambda,\ p\lambda^{2b}) = \lambda^{N_1} R(i\xi,\ p). \tag{8.4}$$

It is not difficult to show that

$$N_1 = 2br,$$

where r is an integer and the coefficients of p^r and ξ_j^{2br} are different from zero. Indeed, putting $\xi = 0$ in (8.4), we obtain

$$R(0,\ p\lambda^{2b}) = \lambda^{N_1} R(0,\ p).$$

If $N_1 \neq 2br$, then this equality can only be fulfilled if both sides of it are equal to zero for all p.

But then

$$R(0,\ p) = 0$$

for all p, in particular for $\operatorname{Re} p > 0$, which contradicts condition (8.1). This proves that $N_1 = 2br$ and $R(0, p) = \gamma p^r$, $\gamma \neq 0$. It can be shown in an analogous manner that the coefficients of the ξ_j^{2br} $(j = 1, \cdots, m)$ in the polynomial R are different from zero. From what has been proved it follows that the numbers b and r for a parabolic operator L are uniquely determined: r is the degree of the polynomial $L(x, t, i\xi, p)$ in p, while $2br$ is its degree in the variables ξ_j.

Let us proceed to Petrovskiĭ's definition of parabolic systems.

Definition 2. *A matrix differential operator* $\mathfrak{L}(x, t, \partial/\partial x, \partial/\partial t)$ *with elements* $L_{kj}(x, t, \partial/\partial x, \partial/\partial t)$ $(k, j = 1, \cdots, m)$ *is said to be parabolic in the sense of I. G. Petrovskiĭ if:*

1) *the operator*

$$L\left(x, t, \frac{\partial}{\partial x}, \frac{\partial}{\partial t}\right) = \det \mathscr{L}\left(x, t, \frac{\partial}{\partial x}, \frac{\partial}{\partial t}\right)$$

is 2b-parabolic in the sense of Definition 1;

2) *the degree of the polynomials* $L_{kj}(x, t, i\xi\lambda, p\lambda^{2b})$ *in* λ *does not exceed* $2br_j$ *and*

$$L_{kj}(x, t, i\xi, p) = \delta_k^j p^{r_j} + L'_{kj}(x, t, i\xi, p),$$

where L'_{kj} *is a polynomial not containing* p^{r_j}.

Thus the systems $\mathfrak{L}u = f$ of I. G. Petrovskiĭ are solved with respect to the leading derivatives with respect to t, i.e. the derivatives $\partial^{r_j}u^j/\partial t^{r_j}$.

An important (and the most-studied) special case of systems that are parabolic in the sense of Petrovskiĭ is furnished by the systems in which all the $r_j = 1$; they have the form

$$\mathscr{L}u = \frac{\partial u}{\partial t} + \mathscr{A}\left(x, t, \frac{\partial}{\partial x}\right)u = f, \tag{8.5}$$

where $\mathscr{A}$ is an operator of order $2b$.

By the principal part of the polynomial $L_{kj}(x, t, i\xi, p)$ we will mean the sum L^0_{kj} of all those terms of L_{kj} for which the condition of homogeneity

$$L^0_{kj}(x, t, i\xi\lambda, p\lambda^{2b}) = \lambda^{2br_j} L^0_{kj}(x, t, i\xi, p)$$

is fulfilled, while the matrix $\mathfrak{L}_0$ composed of the L^0_{kj} will be called the principal part of the matrix $\mathfrak{L}$. It is obvious that the polynomial

$$L_0(x, t, i\xi, p) = \det \mathscr{L}_0(x, t, i\xi, p)$$

is the principal part of the polynomial L.

A characteristic of Petrovskiĭ systems is the fact that the degree of homogeneity of the polynomials L^0_{kj}, which is equal to $2br_j$, does not depend on k. Other definitions of parabolicity have been proposed in which condition 1) of Petrovskiĭ's definition is retained but the degree of homogeneity of the polynomials L^0_{kj} is permitted to depend on k as well as j. For hyperbolic systems such a generalization was made even earlier by Leray, and for elliptic systems, by Douglis and Nirenberg. In these definitions the order of the operators L_{kj} is determined by a sum $s_k + t_j$, where the s_k and t_k $(k = 1, \cdots, n)$ are certain collections of integers. For parabolic systems one of these generalizations of Petrovskiĭ's definition was proposed by T. Shirota [109b].

Definition 3. *A system of form* (8.5) *is said to be parabolic in the sense of Shirota if:*

1) *the operator* $L = \det \mathfrak{L}$ *is* $2b$*-parabolic;*

2) *there exist integers* s_k $(k = 1, \cdots, m)$ *such that the order of the operator* A_{kj} *does not exceed the quantity* $s_k - s_j + 2b$.

Although these systems contain only derivatives of first order with respect to t, while Petrovskiĭ systems contain derivatives of arbitrarily high order, the latter are a subclass of the former in the following sense. If in a system that is parabolic in the sense of Petrovskiĭ one introduces new unknown functions in place of the $\partial^k u_j/\partial t^k$ $(k = 1, \cdots, r_j - 1)$ in the case $r_j > 1$, then a system of form (8.5) is obtained as a result, and it can be shown that it is parabolic in the sense of Shirota.

An even wider class of parabolic systems has been introduced in the papers [116c, e].

Definition 4. *An operator* $\mathfrak{L}$ *will be called parabolic if:*

1) *the operator* $L = \det \mathfrak{L}$ *is* $2b$*-parabolic in the sense of Definition* 1;

2) *there exist integers* s_k *and* t_k $(k = 1, \cdots, m)$ *such that the degree of the polynomial* $L_{kj}(x, t, i\xi\lambda, p\lambda^{2b})$ *does not exceed* $s_k + t_j$ (*if* $s_k + t_j < 0$, *then* $L_{kj} = 0$) *and, in addition,*

$$\sum_{k=1}^{m} (s_k + t_k) = 2br,$$

where r *is the degree of the polynomial* $L(x, t, i\xi, p)$ *in the variable* p.

By the principal part of $\mathfrak{L}$ we will mean the matrix $\mathfrak{L}_0$ whose elements are the polynomials $L^0_{kj}(x, t, i\xi, p)$, i.e. the principal parts of the polynomials L_{kj}. These are the sums of all those terms of the L_{kj} for which the condition of homogeneity

$$L^0_{kj}(x, t, i\xi\lambda, p\lambda^{2b}) = \lambda^{s_k + t_j} L^0_{kj}(x, t, i\xi, p)$$

is fulfilled. If the degree of the polynomial $L_{kj}(x, t, i\xi\lambda, p^{2b})$ is less than $s_k + t_j$, then $L^0_{kj} = 0$. It is obvious that the polynomial $L_0 = \det \mathfrak{L}_0$ is the principal part of the polynomial L.

Petrovskiĭ and Shirota systems are subclasses of the systems just defined. Petrovskiĭ systems are those parabolic systems for which $s_1 = \cdots = s_m = 0$ and $t_j = 2br_j$. Further, if a system that is parabolic in the sense of Definition 4 has the form (8.5), then for the diagonal operators L^0_{jj}

$$L^0_{jj}(x, t, i\xi\lambda, p\lambda^{2b}) = p\lambda^{2b} + A^0_{jj}(x, t, i\xi\lambda)$$
$$= \lambda^{s_j+t_j} L^0_{jj}(x, t, i\xi, p),$$

from which

$$s_j + t_j = 2b, \quad t_j = 2b - s_j.$$

Consequently, this system is parabolic in the sense of Shirota.

In connection with Definition 4 there arises the following question. Suppose a matrix differential operator $\mathfrak{L}$ is given such that the operator $L = \det \mathfrak{L}$ is parabolic in the sense of Definition 1. Is it always possible to select the numbers s_j and t_j so that the operator $\mathfrak{L}$ is parabolic in the sense of Definition 4, and will they be unique? Let a_{kj} be the degree of the polynomial $L_{kj}(x, t, i\xi\lambda, p\lambda^{2b})$ in λ. The numbers s_j and t_j must satisfy the following requirements:

$$s_k + t_j \geqslant a_{kj},$$
$$\sum_{k=1}^{m} (s_k + t_k) = 2br, \tag{8.6}$$

where b and r are numbers uniquely determined by the polynomial L.

L. R. Volevič [125a, c] showed that *such numbers always exist but are not uniquely defined.* First of all, if certain numbers s_k and t_k satisfy conditions (8.6), then these conditions will also be satisfied by the numbers $s_k + a$ and $t_j - a$. We fix the constant a by the requirement that $\max_i s_i = 0$. But even under this condition the numbers s_k and t_k are sometimes not determined uniquely from (8.6).

A system of two heat equations is a simple example of a system for which it is possible to give infinitely many collections of numbers s_k and t_k. For it

$$\mathscr{L}(i\xi, p) = \mathscr{L}_0(i\xi, p) = \begin{pmatrix} p + \xi^2, & 0 \\ 0 & p + \xi^2 \end{pmatrix}.$$

The operator $\mathfrak{L}$ is obviously parabolic according to Petrovskiĭ and we can put $s_1 = s_2 = 0$, $t_1 = t_2 = 2$. But along with this we can also take

$$s_1 = 0, \quad s_2 = -k, \quad t_1 = 2, \quad t_2 = 2 + k,$$
$$s_1 = -k, \quad s_2 = 0, \quad t_1 = 2 + k, \quad t_2 = 2,$$

where k is a positive integer.

We note, finally, that Definition 4 is somewhat different in form from the definition given in [116c], although it is completely equivalent to it. The definition given here is more constructive. The possibility of such an improvement in the definition (when applied to elliptic systems) was pointed out by L. R. Volevič [125a].

In Definitions 2–4 there is one common feature, namely the condition imposed on the determinant. This condition has also been subjected to generalizations.

Definition 5. *The system*

$$\frac{\partial u}{\partial t}+\mathcal{A}\left(x,\ t,\ \frac{\partial}{\partial x}\right)u=0$$

is said to be parabolic in the sense of G. E. Šilov if for all real ξ *the roots* p_s *of the polynomial*

$$L(x,\ t,\ i,\ \xi,\ p)=\det\left[pE+\mathcal{A}(x,\ t,\ i,\ \xi)\right] \tag{8.7}$$

have the property

$$\operatorname{Re}p_s\leqslant -c|\xi|^h+c_1,$$

where c *and* h *are positive constants.*

This class of systems is more extensive than the class of systems that are parabolic in the sense of T. Shirota. These systems have been investigated in the papers [10, 110] for the case when the coefficients do not depend on t. The Cauchy problem has been solved for them. The basic results are cited in the monograph [25a] (Appendix 1).

Let us cite in addition the definition of 2b-parabolic systems introduced by S. D. Eĭdel'man [25a]. In these systems differentiation with respect to different x_k has in general a different weight $1/2b_k$ with respect to differentiation with respect to t. By b is meant the vector $(b_1, \cdots, b_n)$.

Definition 6. *An operator* $\mathfrak{L}(x,\ t,\ \partial/\partial x,\ \partial/\partial t)$ *is said to be* 2b*-parabolic if there exist positive numbers* b_k $(k=1,\cdots,n)$ *such that:*

1) *the functions* $L_{kj}(x,\ t,\ i\xi_1\lambda^{1/2b_1},\cdots,i\xi_n\lambda^{1/2b_n},\ p\lambda)$ *are linear combinations of powers of* λ, *with the maximal power not exceeding* r_j *and*

$$L_{kj}(x,\ t,\ i\xi,\ p)=\delta_k^j p^{r_j}+L'_{kj}(x,\ t,\ i\xi,\ p),$$

where L'_{kj} *is a polynomial not containing* p^{r_j};

2) *the polynomial*

$$L_0(x,\ t;\ i\xi,\ p)=\det\mathcal{L}_0(x,\ t,\ i\xi,\ p),$$

where $\mathfrak{L}_0$ *is the matrix composed of the principal parts* L^0_{kj} *of the polynomials* L_{kj}, *which are the sums of all those terms of the* L_{kj} *for which*

$$L^0_{kj}\left(x,\ t,\ i\xi_1\lambda^{\frac{1}{2b_1}},\ \dots,\ i\xi_n\lambda^{\frac{1}{2b_n}},\ p\lambda\right)=\lambda^{r_j}L^0_{kj}(x,\ t,\ i\xi,\ p),$$

possesses the following property: for all real ξ its roots p_s satisfy the inequality

$$\operatorname{Re} p_s\leqslant-\delta\left(\xi_1^{2b_1}+\dots+\xi_n^{2b_n}\right),\quad \delta>0.$$

As was pointed out in [25a], it is possible to construct a fundamental matrix (fundamental solution) for these systems and with its help to investigate the Cauchy problem in roughly the same way as for systems that are parabolic in the sense of I. G. Petrovskiĭ.

In conclusion we cite the definition of strongly parabolic systems.

Definition 7. *A system of form* (8.5) *is said to be strongly parabolic if the operator A is strongly elliptic* [124a], *i.e. its order is equal to $2b$ and its principal part A_0 satisfies the condition*

$$\operatorname{Re}[\mathcal{A}_0(x,\ t,\ i\xi)\eta,\ \eta]\geqslant\delta|\xi|^{2b}|\eta|^2, \tag{8.7}$$

where

$$\eta=(\eta_1,\ \dots,\ \eta_m),\quad [\zeta,\ \eta]=\sum_{l=1}^m\zeta_l\bar{\eta}_l.$$

It is obvious that strongly parabolic systems are parabolic in the sense of Petrovskiĭ. Property (8.7) permits us to use many of the methods set forth in Chapter III in investigating the first boundary value problem for strongly parabolic systems.

§9. STATEMENT OF THE BOUNDARY VALUE PROBLEMS AND THE CAUCHY PROBLEM FOR PARABOLIC SYSTEMS

We will consider general boundary value problems for systems that are parabolic in the sense of Definition 4. The boundary conditions in them are given by the equality

$$\mathscr{B}\left(x,\ t,\ \frac{\partial}{\partial x},\ \frac{\partial}{\partial t}\right)\mathbf{u}\Big|_{S_T}=\mathbf{\Phi}, \tag{9.1}$$

where $\mathscr{B}$ is a matrix differential operator and $\mathbf{\Phi}$ is a vector arbitrarily prescribed on the surface S_T.

Let us determine the number of rows that the matrix $\mathscr{B}$ must have (i.e. the number of conditions that must be given on the boundary) and the algebraic conditions it must satisfy in order for condition (9.1) to provide us with a well-stated boundary value problem.

We will first define the principal part $\mathfrak{B}_0$ of the matrix $\mathfrak{B}$.

We denote by B_{qj} $(q = 1, \cdots, N;\ j = 1, \cdots, m)$ the elements of the matrix $\mathfrak{B}$, and by β_{qj} the degree of the polynomial $B_{qj}(x, t, i\xi\lambda, p\lambda^{2b})$ in λ. If $B_{qj} = 0$, we take any integer as β_{qj}. Let

$$\sigma_q = \max_j (\beta_{qj} - t_j),$$

so that

$$\sigma_q + t_j \geqslant \beta_{qj}.$$

By the principal part of the polynomial B_{qj} we will mean the sum B^0_{qj} of all terms of it satisfying the homogeneity condition

$$B^0_{qj}(x, t, i\xi\lambda, p\lambda^{2b}) = \lambda^{\sigma_q + t_j} B^0_{qj}(x, t, i\xi, p).$$

The matrix with elements B^0_{qj} is $\mathfrak{B}_0$. Thus the choice of the principal part $\mathfrak{B}_0$ of the matrix $\mathfrak{B}$ depends on the numbers t_j and can be ambiguous.

Let us fix arbitrary admissible numbers s_k, t_j and σ_q. We take any point $(x_0, t_0) \in S_T$, introduce local coordinates $\{y\}$ in the space $E_n(x_1, \cdots, x_n)$ at the point $x_0 \in S$, with the y_n-axis being directed along the inward normal $\nu(x_0)$, and consider in the half space $y_n > 0$ the problem

$$\begin{aligned} &\mathcal{L}_0\left(x_0, t_0, \frac{\partial}{\partial x}, \frac{\partial}{\partial t}\right) \mathbf{u} = 0 \ (-\infty < y_1, \ldots, y_{n-1}, t < +\infty,\ y_n > 0), \\ &\mathcal{B}_0\left(x_0, t_0, \frac{\partial}{\partial x}, \frac{\partial}{\partial t}\right) \mathbf{u}\Big|_{y_n = 0} = \mathbf{\Phi}. \end{aligned} \tag{9.2}$$

The coefficients of the operators $\mathfrak{L}_0$ and $\mathfrak{B}_0$ in this problem are "frozen" at the point (x_0, t_0), while all minor terms are discarded. Following Ja. B. Lopatinskiĭ [77], we assume that the following local condition is fulfilled: no matter what point $(x_0, t_0) \in S_T$ we take, problem (9.2) will always be uniquely solvable for any smooth vector function Φ with a compact support in the class of functions admitting the transformation (1.19) of Chapter IV (x' denotes the coordinates in the tangent plane to S at the point x_0).

Let us express this condition in algebraic form. Going over in (9.2) to the coordinates $\{y\}$, we obtain the problem

$$\begin{aligned} &\mathfrak{L}\left(\frac{\partial}{\partial y}, \frac{\partial}{\partial t}\right) \mathbf{u} = 0, \\ &\mathfrak{B}\left(\frac{\partial}{\partial y}, \frac{\partial}{\partial t}\right) \mathbf{u}\Big|_{y_n = 0} = \mathbf{\Phi}(y', t). \end{aligned} \tag{9.3}$$

It is easy to see that the operator $\mathfrak{L}$ is parabolic with the same constant δ in condition (8.1).

Problem (9.3) can be reduced to a boundary value problem for a system of ordinary differential equations by taking the Fourier transform with respect to the variables $y' = (y_1, \cdots, y_{n-1})$ and then the Laplace transform with respect to t ((1.19) of Chapter IV).

As a result of this transformation we obtain

$$\left.\begin{aligned} &\mathfrak{L}\left(i\zeta, \frac{d}{dy_n}, p\right)\tilde{\mathrm{u}}(y_n) = 0, \\ &\mathfrak{B}\left(i\zeta, \frac{d}{dy_n}, p\right)\tilde{\mathrm{u}}(y_n)\Big|_{y_n=0} = \tilde{\Phi}, \\ &|\tilde{\mathrm{u}}(y_n)| \xrightarrow[y_n\to\infty]{} 0. \end{aligned}\right\} \tag{9.4}$$

Here $\zeta = (\zeta_1, \cdots, \zeta_{n-1})$ is an $(n-1)$-dimensional vector. The necessity of the third condition is proved in the same way as in the case of the corresponding condition in (1.23) of Chapter IV.

Disregarding for the moment the dependence of the operators $\mathfrak{L}$ and $\mathfrak{B}$ on ζ and p, we will consider on the half line $z > 0$ a problem of form (9.4) for a certain system of ordinary differential equations (the number of equations is denoted as usual by m):

$$\mathscr{P}\left(\frac{d}{dz}\right)\mathrm{w}(z) = 0, \tag{9.5}$$

$$\mathscr{A}\left(\frac{d}{dz}\right)\mathrm{w}(z)\Big|_{z=0} = \mathrm{h}, \tag{9.6}$$

$$|\mathrm{w}(z)| \xrightarrow[z\to+\infty]{} 0. \tag{9.7}$$

Let us determine the conditions under which this problem is uniquely solvable for any numerical vector h. We denote by $\mathfrak{P}_0$ the space of the solutions of system (9.5) satisfying conditions (9.7). This space is obviously finite dimensional and the number of its dimensions is equal to the number of roots of the polynomial

$$P(\lambda) = \det \mathscr{P}(\lambda),$$

having a negative real part or, equivalently, the number of roots of the polynomial $P(i\tau)$ having a positive imaginary part (with the multiplicity of the roots being taken into account). We denote these roots by τ_s^+, and their number by r^+. Further, the components of any vector $\mathrm{v}(z) \in \mathfrak{P}_0$ satisfy the equation

$$P^+\left(-i\frac{d}{dz}\right)v^k(z) = 0, \tag{9.8}$$

where

$$P^{+}(\tau)=\prod_{s=1}^{r^{+}}(\tau-\tau_s^{+}),$$

since they are linear combinations of the exponential functions $e^{ir_s^+ z}$, which are possibly multiplied by polynomials (in the case of multiple roots r_s^+). All this can be easily shown by passing from the system (9.5) to an equivalent triangular system (see [50]).

Let $\mathbf{w}_s(z)$ denote some vectors forming a basis in the space $\mathfrak{P}_0$. Then the solution of problem (9.5)–(9.7) has the form

$$\mathbf{w}(z)=\sum_{s=1}^{r^{+}} c_s \mathbf{w}_s(z),$$

where the c_s are certain constants. We will find them from conditions (9.6):

$$\sum_{j=1}^{m}\sum_{s=1}^{r^{+}} A_{qj}\left(\frac{d}{dz}\right) w_s^j(z)\Big|_{z=0} c_s=h_q \quad (q=1,\ \ldots,\ N).$$

We have obtained a linear system of equations for the c_s. In order that they be uniquely solvable for any h_q it is necessary and sufficient that

$$N=r^{+} \tag{9.9}$$

and

$$\det \mathfrak{A}\neq 0, \tag{9.10}$$

where $\mathfrak{A}$ is the matrix with elements

$$\mathfrak{A}_{qs}=\sum_{j=1}^{m} A_{qj}\left(\frac{d}{dz}\right) w_s^j(z)\Big|_{z=0}. \tag{9.11}$$

By virtue of (9.9) the matrix $\mathfrak{A}$ must have r^+ rows.

Let us now give another equivalent formulation to condition (9.10). To this end we will prove that every vector $\mathbf{w}(z)\in\mathfrak{P}_0$ is representable in the form

$$\mathbf{w}(z)=\hat{\mathscr{P}}\left(\frac{d}{dz}\right)\mathbf{v}(z), \tag{9.12}$$

where $\hat{\mathscr{P}}$ is the adjoint of $\mathscr{P}$ ($\hat{\mathscr{P}}=\mathscr{P}^{-1}\det\mathscr{P}$) and $\mathbf{v}(z)$ is a vector whose components satisfy equation (9.8) (it does not in general belong to $\mathfrak{P}_0$).

Indeed, as $\mathbf{v}(z)$ we can take the vector

$$\mathbf{v}(z)=\mathscr{P}\left(\frac{d}{dz}\right)\mathbf{u}(z), \tag{9.13}$$

where $\mathbf{u}(z)$ is a vector whose components are solutions of the equations

$$P\left(\frac{d}{dz}\right) u^k(z) = w^k(z)$$

and have the same structure as the $w^k(z)$, i.e. are linear combinations of the exponential functions $e^{i\overset{+}{r}_s z}$ multiplied by polynomials. From the elementary theory of ordinary differential equations we know that functions u^k with the required properties can be found.

Let us show that the vector $\mathbf{v}(z)$ of (9.13) satisfies all of the above-mentioned requirements. Letting the operator $\mathcal{P}(d/dz)$ act on both sides of (9.13), we obtain equality (9.12). In order to prove that the vector $\mathbf{v}(z)$ of (9.13) satisfies equation (9.8) we let the operator $\mathcal{P}(d/dz)$ act on both sides of (9.12). This gives us

$$P\left(\frac{d}{dz}\right) v^i(z) = 0.$$

From this equality we obtain the stronger equality (9.8) since $v^i(z)$ is expressed only in terms of the exponential functions $e^{i\overset{+}{r}_s z}$.

Thus the desired representation (9.12) for any vector $\mathbf{w}(z) \in \mathfrak{P}_0$ has been proved.

The degree of the polynomial P^+ is equal to r^+, and therefore equation (9.8) has r^+ linearly independent solutions. It is not difficult to see that we can take as these solutions the functions

$$V_s(z) = \int_{\gamma^+} \frac{\tau^{s-1} e^{i\tau z}}{P^+(\tau)} d\tau \quad (s = 1, \ldots, r^+),$$

where γ^+ is a simple contour in the complex r-plane enclosing all of the $\overset{+}{r}_s$. Hence every solution of equation (9.8) is representable in the form

$$v(z) = \int_{\gamma^+} \frac{Q(\tau)}{P^+(\tau)} e^{i\tau z} d\tau,$$

while a solution of the system (9.5) that is decreasing for $z \to +\infty$ is by virtue of (9.12) representable in the form

$$w^k(z) = \int_{\gamma^+} \sum_{j=1}^{m} \hat{P}_{kj}(i\tau) Q_j(\tau) \frac{e^{i\tau z}}{P^+(\tau)} d\tau, \tag{9.14}$$

where Q and Q_j are polynomials of degree $r^+ - 1$. In particular, the basis vectors $\mathbf{w}_s(z)$ have the form

$$w_s^k(z) = \int_{\gamma^+} \sum_{j=1}^{m} \hat{P}_{kj}(i\tau) Q_j^{(s)}(\tau) \frac{e^{i\tau z}}{P^+(\tau)} d\tau. \tag{9.15}$$

No matter what polynomials $Q_j(\tau)$ $(j = 1, \cdots, m)$ of degree $r^+ - 1$ we take there exist numbers a_s $(s = 1, \cdots, r^+)$ such that for any $k = 1, \cdots, m$

$$\int_{\gamma^+} \sum_{j=1}^{m} \hat{P}_{kj}(i\tau)\, Q_j(\tau) \frac{e^{i\tau z}}{P^+(\tau)}\, d\tau = \sum_{s=1}^{r^+} a_s w_s^k$$

$$= \int_{\gamma^+} \sum_{j=1}^{m} \hat{P}_{kj}(i\tau) \sum_{s=1}^{r^+} a_s Q_j^{(s)}(\tau) \frac{e^{i\tau z}}{P^+(\tau)}\, d\tau. \tag{9.16}$$

Let us now consider the functions (9.11). As a consequence of (9.15)

$$\mathfrak{A}_{qs} = \int_{\gamma^+} \sum_{j=1}^{m} D_{qj}(i\tau)\, Q_j^{(s)}(\tau) \frac{d\tau}{P^+(\tau)},$$

where

$$D_{qj} = \sum_{k=1}^{m} A_{qk} \hat{P}_{kj}.$$

By virtue of condition (9.10), from the equation

$$\sum_{q=1}^{r^+} d_q \mathfrak{A}_{qs} = 0 \qquad (s = 1, \ldots, r^+), \tag{9.17}$$

it follows that $d_q = 0$ $(q = 1, \cdots, r^+)$. Let

$$D_j(i\tau) = \sum_{q=1}^{r^+} d_q D_{qj}(i\tau).$$

Equality (9.17) can be written in the form

$$\int_{\gamma^+} \sum_{j=1}^{m} D_j(i\tau)\, Q_j^{(s)}(\tau) \frac{d\tau}{P^+(\tau)} = 0 \qquad (s = 1, \ldots, r^+). \tag{9.18}$$

This is equivalent to the fact that

$$\int_{\gamma^+} \sum_{j=1}^{m} D_j(i\tau)\, Q_j(\tau) \frac{d\tau}{P^+(\tau)} = 0 \tag{9.19}$$

for any polynomials $Q_j(\tau)$ of degree $r^+ - 1$. A proof of (9.19) is obtained by letting the operator

$$\sum_{q=1}^{r^+} d_q A_{qk}\left(\frac{d}{dz}\right).$$

act on both sides of (9.16), then summing the resultant equalities over k from 1 to m, putting $z = 0$ and taking into consideration (9.18).

Since the polynomials Q_j are arbitrary, (9.19) is equivalent to the fact that for any j and any polynomial $Q(\tau)$ of degree $r^+ - 1$ we have

$$\int_{\gamma^+} D_j(i\tau)Q(\tau)\frac{d\tau}{P^+(\tau)}=0,$$

which is obviously equivalent to the equality

$$\sum_{q=1}^{r^+} d_q D_{qj}(i\tau)=D_j(i\tau)=p_j(\tau)P^+(\tau), \tag{9.20}$$

where p_j are certain polynomials.

Thus we have shown that conditions (9.17) and (9.20) are equivalent, so that from (9.20) it follows that all $d_q=0$. This property of the matrix $\mathfrak{D}$ with elements D_{qj} can be formulated in the following manner: *the rows of the matrix*

$$\mathscr{D}(i\tau)=\mathscr{A}(i\tau)\hat{\mathscr{P}}(i\tau)$$

are linearly independent modulo the polynomial $P^+(\tau)$.

In order to formulate this condition for problem (9.4) we must know what the number r^+ is equal to for the polynomial

$$L_0(i\zeta,\ i\tau,\ p)=\det\mathscr{L}(i\zeta,\ i\tau,\ p).$$

The answer to this question is given by the following theorem.

Theorem 9.1. *If the polynomial* L_0 *satisfies condition* (8.1), *then for any real* $\zeta_1,\cdots,\zeta_{n-1}$ *and any complex* p *satisfying the conditions*

$$\operatorname{Re}p\geqslant-\delta_1|\zeta|^{2b}\qquad(0<\delta_1<\delta),\qquad |p|+|\zeta|>0, \tag{9.21}$$

the polynomial $L_0(i\zeta,\ i\tau,\ p)$ *as a function of* τ *has* br *roots with positive, and* br *roots with negative, imaginary part.*

Thus under condition (9.21) $r^+=br$.

Theorem 9.1 was first proved by T. Ja. Zagorskiĭ [129] under the assumption that the polynomial L has real coefficients; it has been proved without this assumption in [25a] and [116c].

We can now formulate the desired algebraic condition for problem (9.3) and, consequently, for the original problem (9.2). We will formulate it straight off for problem (9.2). Let $\xi=(\xi_1,\cdots,\xi_n)$ be an arbitrary real vector. It can be uniquely represented in the form

$$\xi=\zeta+\tau\nu,$$

where ζ is a vector lying in the tangent plane to the surface S at the point x_0, ν is the unit inward normal at x_0 and τ is a real parameter. Consider the polynomial

$$L_0(x_0, t_0, i(\zeta+\tau\nu), p)$$

as a function of τ on the whole complex plane. By virtue of Theorem 9.1, under condition (9.21) this polynomial has br roots with a positive imaginary part (we denote them by $\tau_s^+(x_0, t_0, \zeta, p)$) and br roots with a negative imaginary part. Let

$$L^+(x_0, t_0, \zeta, p, \tau)=\prod_{s=1}^{br}\left[\tau-\tau_s^+(x_0, t_0, \zeta, p)\right]. \tag{9.22}$$

We will assume that the matrices $\mathfrak{L}_0$ and $\mathfrak{B}_0$ satisfy the following condition, which is called the " complementing condition:" *for any point $(x_0, t_0) \in S_T$ the rows of the matrix*

$$\mathscr{B}_0(x_0, t_0, i(\zeta+\tau\nu), p)\hat{\mathscr{L}}_0(x_0, t_0, i(\zeta+\tau\nu), p)$$

are linearly independent modulo the polynomial L^+ as a polynomial in τ if the vector ζ and the number p satisfy condition (9.21) *for some $\delta_1 \in (0, \delta)$.*

For systems of form (8.5) that are parabolic in the sense of Petrovskiĭ (for them $r = m$) the complementing condition is formulated in most papers in exactly the same way as in the paper [77] of Ja. B. Lopatinskiĭ: *the rank of the matrix*

$$\int_{\gamma^+}\mathscr{B}_0(x_0, t_0, i(\zeta+\tau\nu, p)\mathscr{L}_0^{-1}(x_0, t_0, i(\zeta+\tau\nu, p)\mathscr{M}(\tau)\,d\tau, \tag{9.23}$$

where

$$\mathscr{M}=(\mathscr{E}, \tau\mathscr{E}, \ldots, \tau^{2b-1}\mathscr{E}),$$

with $\mathfrak{E}$ being the unit matrix of order m, is equal to bm at each point $(x_0, t_0) \in S_T$ and for any tangential ζ and any p satisfying (9.21).

These two formulations are equivalent [115c].

As was mentioned above, the choice of the principal part $\mathfrak{B}_0$ of the matrix $\mathfrak{B}$ is dictated by the numbers t_j and can be ambiguous. In this connection it can turn out that for one choice of the numbers s_k and t_j, and consequently of the principal parts of the matrices $\mathfrak{L}$ and $\mathfrak{B}$ the complementarity condition holds, while for another choice it does not.

As an example we consider in the half space $x_2 > 0$ of the space $E_3(x_1, x_2, t)$ a parabolic system decomposing into two separate heat equations:

$$\mathscr{L}(i\zeta, i\tau, p)=\mathscr{L}_0(i\zeta, i\tau, p)=\begin{pmatrix} p+\zeta^2+\tau^2 & 0 \\ 0 & p+\zeta^2+\tau^2\end{pmatrix}.$$

For this system $b = 1$, $r = 2$ and, as we saw in §8, the following collections of numbers s_k and t_j are possible:

$$s_1 = -k, \quad s_2 = 0, \quad t_1 = 2 + k, \quad t_2 = 2 \qquad (k > 0).$$
$$s_1 = 0, \quad s_2 = -k, \quad t_1 = 2, \quad t_2 = 2 + k \qquad (k \geqslant 0).$$

Let

$$\mathcal{B}(i\zeta, i\tau, p) = \begin{pmatrix} 1 & 0 \\ i\zeta & 1 \end{pmatrix}.$$

For this operator the numbers σ_1 and σ_2 take the following values: if $t_1 = 2 + k$, $t_2 = 2$ $(k > 0)$, then $\sigma_1 = -2 - k$, $\sigma_2 = -2$; while if $t_1 = 2$, $t_2 = 2 + k$ $(k \geq 0)$, then $\sigma_1 = -2$, $\sigma_2 = -1$. It is not difficult to verify that in the first case the complementarity condition is fulfilled. In the second case

$$\mathcal{B}_0 = \begin{pmatrix} 1 & 0 \\ i\zeta & 0 \end{pmatrix}.$$

and since the rows of this matrix, and hence also of the matrix $\mathfrak{B}_0\hat{\mathfrak{L}}_0$, are linearly dependent, the complementarity condition does not hold. In particular, it is not fulfilled if we regard the operator $\mathfrak{L}$ as an operator that is parabolic in the sense of Petrovskiĭ $(s_1 = s_2 = 0,\ t_1 = t_2 = 2)$.

We now turn to the problem of assigning initial conditions for parabolic systems. For systems parabolic in the sense of Petrovskiĭ the initial conditions are assigned in the following way: if the highest order of the differentiation with respect to t of a function u_j is equal to r_j, then all derivatives of this function with respect to t of lower order are given for $t = 0$. But in the general case, as will be seen from what follows, the assignment of initial conditions in this way is impossible.

We will assign them in the form

$$\mathfrak{C}\left(x, \frac{\partial}{\partial x}, \frac{\partial}{\partial t}\right) \mathbf{u}\Big|_{t=0} = \varphi(x), \tag{9.24}$$

where $\mathfrak{C}$ is a matrix differential operator with elements $C_{\alpha k}(\alpha = 1, \cdots, N_1;\ k = 1, \cdots, m)$ and $\phi(x)$ is an arbitrary vector function.

We define the principal part of the matrix $\mathfrak{C}$ in exactly the same way as for the matrix $\mathfrak{B}$. Let $\gamma_{\alpha k}$ denote the degree of the polynomial $C_{\alpha k}(x, i\xi\lambda, p\lambda^{2b})$ in λ (and any integer for $C_{\alpha k} = 0$), and let $\rho_\alpha = \max_k (\gamma_{\alpha k} - t_k)$. The sum $C^0_{\alpha k}$ of all terms of the polynomial $C_{\alpha k}$ satisfying the homogeneity condition

$$C^0_{\alpha k}(x, i\xi\lambda, p\lambda^{2b}) = \lambda^{\rho_\alpha + t_j} C^0_{\alpha k}(x, i\xi, p), \tag{9.25}$$

will be called the principal part of $C_{\alpha k}$, while the matrix $\mathfrak{C}_0$ composed of the $C^0_{\alpha k}$ will be called the principal part of the matrix $\mathfrak{C}$.

Let us determine what the number N_1 must be and what conditions the matrix $\mathfrak{C}$ must satisfy if we are to be able to determine uniquely the value of an derivative of any function u_j for $t = 0$ from the system

$$\mathscr{L}\mathbf{u} = \mathbf{f} \tag{9.26}$$

and the initial condition (9.24) by merely differentiation and solving linear algebraic systems (assuming, of course, that these functions and the coefficients of the operators $\mathfrak{L}$ and $\mathfrak{C}$ are sufficiently smooth). We shall also require that the condition have the same formulation for all parabolic systems and that it involve only the principal parts of the matrices $\mathfrak{L}$ and $\mathfrak{C}$, i.e. $\mathfrak{L}_0$ and $\mathfrak{C}_0$, as in the definition of parabolicity and in the complementarity condition. Consequently, the same properties must be possessed by the system

$$\mathscr{L}_0\left(x,\ 0,\ \frac{\partial}{\partial x},\ \frac{\partial}{\partial t}\right)\mathbf{u} = \mathbf{f} \tag{9.27}$$

and the initial conditions

$$\mathscr{C}_0\left(x,\ \frac{\partial}{\partial x},\ \frac{\partial}{\partial t}\right)\mathbf{u}\Big|_{t=0} = \varphi. \tag{9.28}$$

We will substitute into the left sides of (9.27) and (9.28) all possible vector functions $\mathbf{u}(t)$ not depending on x. For them (9.27) and (9.28) go over into

$$\mathscr{L}_0\left(x,\ 0,\ 0,\ \frac{d}{dt}\right)\mathbf{u}(t) = \mathbf{f}, \tag{9.29}$$

$$\mathscr{C}_0\left(x,\ 0,\ \frac{d}{dt}\right)\mathbf{u}(t)\Big|_{t=0} = \varphi. \tag{9.30}$$

In order for it to be possible to uniquely determine $d^k u_j(t)/dt^k|_{t=0}$ $(k \geq 0)$ from these equalities it is necessary that (9.29) and (9.30) constitute a Cauchy problem for the system of ordinary differential equations (9.29) that is uniquely solvable for any $\mathbf{f}$ and ϕ and any fixed $x \in \Omega$.

We have

$$\det \mathscr{L}_0(x,\ 0,\ 0,\ p) = L(x,\ 0,\ 0,\ p) = \gamma(x) p^r,$$

in which by virtue of the condition of parabolicity (1.1) $\gamma(x) \neq 0$.

From the theory of linear systems of ordinary differential equations (see, for example, [50]) we know that the number of arbitrary constants in the general solution of the system (9.29) is equal to r, i.e. the degree of the determinant L. Consequently, among the conditions (9.30) there must be r independent ones, and since we wish to assign the vector ϕ arbitrarily, the matrix $\mathfrak{C}$ must have precisely r rows. Thus

$$N_1 = r.$$

From this it follows in particular that one cannot generally assign the initial data for systems that are parabolic in the sense of Definition 4 in the same way as for Petrovskiĭ systems, since the system may contain very high derivatives of the unknown functions with respect to t. This occurs, for example, in the case of a parabolic system with

$$\mathcal{L}_0(i\xi, p) = \begin{pmatrix} p+\xi^2 & p^{10} \\ 0 & p+\xi^2 \end{pmatrix}$$

$$(b=1,\ r=2,\ s_1=0,\ s_2=-8,\ t_1=2,\ t_2=10).$$

For the purpose of deriving an algebraic condition that will serve as a criterion for the unique solvability of problem (9.29), (9.30) for any f and ϕ it is sufficient to consider this problem for $f=0$. Reasoning in exactly the same way as in the derivation of the complementarity condition we see that in order for the initial data (9.24) to possess all of the desired properties it is necessary that the following condition be fulfilled: *The rows of the matrix*

$$\mathcal{D}(x, p) = \mathcal{C}_0(x, 0, p)\hat{\mathcal{L}}_0(x, 0, 0, p)$$

are linearly independent modulo the polynomial

$$p^r = \frac{1}{\gamma(x)} L(x, 0, 0, p)$$

for any $x \in \Omega$.

As was shown in the paper [116c], this condition is also sufficient for the matrix $\mathcal{C}$ to have the desired properties. In addition, the following fact was proved in [116c]. The formulated condition (we will call it the complementing condition for the matrix $\mathcal{C}$) permits one to uniquely determine, for a given parabolic system and collections of numbers s_k and t_k for it, the possible numbers ρ_a, which turn out to be negative ($\rho_a \leq -2b$). Furthermore, this condition for a certain wide class of domains Ω (in particular, for all domains homeomorphic to a ball) permits one to determine the matrix $\mathcal{C}_0(x, 0, p)$ to within unimportant algebraic transformations. For other Ω the construction of the matrix $\mathcal{C}_0$ is impossible and it is then necessary to give the initial conditions not in the form (9.24) but in a certain different form ([116c], Appendix 4).

The complementing condition does not involve

$$\mathcal{C}_0'(x, i\xi, p) = \mathcal{C}_0(x, i\xi, p) - \mathcal{C}_0(x, 0, p),$$

which also contains principal terms. This matrix, like the minor terms, can be arbitrary, and the only thing that we require of it is the homogeneity condition (9.25) for its elements.

Thus the general boundary value problem for parabolic systems consists in finding a vector $\mathbf{u} = (u^1, \cdots, u^m)$ for which

$$\left.\begin{aligned} &\mathscr{L}\left(x, t, \frac{\partial}{\partial x}, \frac{\partial}{\partial t}\right)\mathbf{u} = \mathbf{f} \quad ((x, t) \in Q_T), \\ &\mathscr{B}\left(x, t, \frac{\partial}{\partial x}, \frac{\partial}{\partial t}\right)\mathbf{u}\Big|_{S_T} = \Phi, \\ &\mathscr{C}\left(x, \frac{\partial}{\partial x}, \frac{\partial}{\partial t}\right)\mathbf{u}\Big|_{t=0} = \varphi, \end{aligned}\right\} \tag{9.31}$$

where the matrices $\mathfrak{B}$ and $\mathfrak{C}$ satisfy the complementing conditions formulated above. The operator $\mathfrak{C}$ is uniquely determined by this condition to within $\mathfrak{C}_0'$, minor terms and unimportant algebraic transformations. The operator $\mathfrak{B}$ can be very arbitrary, but, as will be shown below, a dropping of the complementing condition for the matrix $\mathfrak{B}_0$ can lead to a lack of correctness in the corresponding boundary value problem.

§ 10. BASIC RESULTS ON THE SOLVABILITY OF THE CAUCHY PROBLEM AND OF THE GENERAL BOUNDARY VALUE PROBLEMS FOR PARABOLIC SYSTEMS

The algebraic conditions formulated in § 9 for the matrices $\mathfrak{B}$ and $\mathfrak{C}$ are sufficient for the solvability of problem (9.31). This was established in the book [116c] with the use of the method presented in §§5–9 of Chapter IV.

For a formulation of the results we require the function spaces $H^{l, l/2b}(\overline{Q}_T)$ and $W_q^{2bm, m}(Q_T)$, which in many respects are analogous to the spaces $H^{l, l/2}(\overline{Q}_T)$ and $W_q^{2m, m}(Q_T)$ repeatedly used above. The norms in them are defined in the following way. For any integral $j > 0$ let

$$\langle u \rangle_{Q_T}^{(j)} = \sum_{2br+s=j} \langle D_t^r D_x^s u \rangle_{Q_T}^{(0)},$$

and for nonintegral $l > 0$ let

$$\langle u \rangle_{x, Q_T}^{(l)} = \sum_{2br+s=[l]} \langle D_t^r D_x^s u \rangle_{x, Q_T}^{(l-[l])},$$

$$\langle u \rangle_{t, Q_T}^{\left(\frac{l}{2b}\right)} = \sum_{0 < l-2br-s < 2b} \langle D_t^r D_x^s u \rangle_{t, Q_T}^{\frac{l-2br-s}{2b}},$$

$$\langle u \rangle_{Q_T}^{(l)} = \langle u \rangle_{x, Q_T}^{(l)} + \langle u \rangle_{t, Q_T}^{\left(\frac{l}{2b}\right)}.$$

where $\langle v\rangle_{Q_T}^{(0)}$, $\langle v\rangle_{x,Q'}^{(\alpha)}$, $\langle v\rangle_{t,Q_T}^{(\alpha)}$ for $\alpha<1$ are the same as in §1 of Chapter I. In the space $H^{l,l/2b}(\overline{Q}_T)$ we introduce the norm

$$|u|_{Q_T}^{(l)}=\langle u\rangle_{Q_T}^{(l)}+\sum_{j=0}^{[l]}\langle u\rangle_{Q_T}^{(j)}.$$

The norm

$$\|u\|_{q,Q_T}^{(2bm)}=\sum_{j=0}^{2bm}\sum_{2br+s=j}\|D_t^rD_x^su\|_{q,Q_T}$$

is the norm in the space $W_q^{2bm,m}(Q_T)$. Its principal part is denoted in the following way:

$$\langle\langle u\rangle\rangle_{q,Q_T}^{(2bm)}=\sum_{2br+s=2bm}\|D_t^rD_x^su\|_{q,Q_T}.$$

The restrictions of the function $u\in W_q^{2bm,m}(Q_T)$ and its derivatives $D_t^rD_x^su$ $(2br+s<2bm)$ to the surface S_T belong to the space $W_q^{k-1/q,(k-1/q)/2b}(S_T)$, where $k=2bm-2br-s$.

In order to define the norm in $W_q^{l,l/2b}(S_T)$ we must vary the definition given in §3 of Chapter II of the norm in $W_q^{l,l/2}(S_T)$ in exactly the same way as the norm of $H^{l,l/2}(\overline{Q}_T)$ was varied in the definition of $H^{l,l/2b}(\overline{Q}_T)$ (i.e. the number 2 is replaced every where by $2b$).

The spaces $\overset{\circ}{H}{}^{l,l/2b}(\overline{Q}_T)$, $\overset{\circ}{W}{}_q^{2bm,m}(Q_T)$, $\overset{\circ}{W}{}_q^{s-1/q,1/2b(s-1/q}(S_T)$, where l is a nonintegral positive number and m and s are integral positive numbers, are defined in exactly the same way as in §4 of Chapter III (with obvious variations). For example, $\overset{\circ}{H}{}^{l,l/2b}(\overline{Q}_T)$ is the space of functions from $H^{l,l/2b}(\overline{Q}_T)$ satisfying the null initial conditions

$$\left.\frac{\partial^ku}{\partial t^k}\right|_{t=0}=0\quad\left(k=0,\ \dots,\ \left[\frac{l}{2b}\right]\right).$$

We now proceed to a formulation of the basic results of [116c].

Theorem 10.1. *Suppose the coefficients of the operators L_{kj}, B_{qj} and C_{aj} belong to the classes $H^{l-s_k,(l-s_k)/2b}(\overline{Q}_T)$, $H^{l-\sigma_q,(l-\sigma_q)/2b}(\overline{S}_T)$ and $H^{l-\rho_a}(\overline{\Omega})$ respectively and $S\in H^{l+t_{\max}}$, where l is a nonintegral number satisfying the condition*

$$l>\max(0,\ \sigma_1,\ \dots,\ \sigma_{br}).$$

Then problem (9.31) has a unique solution, in the class of vector functions with $u^j\in H^{l+t_j,(l+t_j)/2b}(\overline{Q}_T)$, for any $f^j\in H^{l-s_j,(l-s_j)/2b}(\overline{Q}_T)$, $\Phi^q\in$

$H^{l-\sigma_q,(l-\sigma_q)/2b}(\overline{S}_T)$ *and* $\phi^\alpha \in H^{l-\rho_\alpha}(\overline{\Omega})$[1]) *satisfying the compatibility conditions that are necessary for the existence of a solution from the indicated class. The solution is subject to the inequality*

$$\sum_{j=1}^{m} |u^j|_{Q_T}^{(l+t_j)} \leqslant c\Big(\sum_{j=1}^{m} |f^j|_{Q_T}^{(l-s_j)} + \sum_{q=1}^{br} |\Phi^q|_{S_T}^{(l-\sigma_q)} + \sum_{\alpha=1}^{r} |\varphi^\alpha|_{\Omega}^{(l-\rho_\alpha)}\Big). \tag{10.1}$$

The relation indicated in the theorem between the smoothness of the known functions and the smoothness of the solution is exact, while the restrictions imposed on the coefficients are necessary in order for each term of any equation of the system to belong to the same class as that containing the free term (the same remark holds for the boundary and initial conditions).

The compatibility conditions for problem (9.31) consist in the fact that the functions $\partial^k u^j/\partial t^k|_{t=0}$ $(k \geq 0)$, which can be calculated from the system and the initial data, must satisfy for $x \in S$ the boundary conditions and relations obtained as a result of differentiating the boundary conditions with respect to t. In the theorem it is required that the vectors $\mathbf{f}$, $\mathbf{\Phi}$, ϕ satisfy the compatibility conditions of minimal order, i.e. only those conditions which $\mathbf{f} = \mathfrak{L}\mathbf{u}$, $\mathbf{\Phi} = \mathfrak{B}\mathbf{u}|_{S_T}$ and $\phi = \mathfrak{C}\mathbf{u}|_{t=0}$ satisfy for $u^j \in H^{l+t_j,(l+t_j)/2b}(Q_T)$.

We now formulate an analogous theorem for the Cauchy problem in the domain $D_{n+1}^{(T)}$:

$$\left.\begin{aligned} \mathscr{L}\mathbf{u} &= \mathbf{f}, \\ \mathfrak{C}\mathbf{u}|_{t=0} &= \varphi. \end{aligned}\right\} \tag{10.2}$$

Theorem 10.2. *Let $l > 0$ be a nonintegral number. If the coefficients of the operators L_{kj} and $C_{\alpha j}$ belong to the classes $H^{l-s_k,(l-s_k)/2b}(\overline{D_{n+1}^{(T)}})$ and $H^{l-\rho_\alpha}(\overline{E}_n)$ respectively, then for any $f^j \in H^{l-s_j,(l-s_j)/2b}(\overline{D_{n+1}^{(T)}})$ and $\phi^\alpha \in H^{l-\rho_\alpha}(\overline{E}_n)$ problem (10.2) has a unique solution in the class of vector functions $\mathbf{u} = (u^1, \dots, u^m)$ with $u^j \in H^{l+t_j,(l+t_j)/2b}(\overline{D_{n+1}^{(T)}})$, and this solution is subject to the inequality*

$$\sum_{j=1}^{m} |u^j|_{D_{n+1}^{(T)}}^{(l+t_j)} \leqslant c\Big(\sum_{j=1}^{m} |f^j|_{D_{n+1}^{(T)}}^{(l-s_j)} + \sum_{\alpha=1}^{r} |\varphi^\alpha|_{E_n}^{(l-\rho_\alpha)}\Big). \tag{10.3}$$

The relation between the smoothness of the given functions and that of the solutions of problems (9.31) and (10.2) is determined by the numbers s_j, t_j, ρ_α and σ_q. In §8 we saw that for certain systems the numbers s_j and t_j are uniquely

1) Since $s_j \leq 0$, $\rho_\alpha < 0$ it follows that $l - s_j > 0$, $l - \rho_\alpha > 0$.

determined. For such systems Theorems 10.1 and 10.2 hold for any admissible s_j and t_j.

We consider as an example the Cauchy problem for a system of two heat equations, which decomposes into two separate problems:

$$\frac{\partial u^1}{\partial t} - \Delta u^1 = f^1, \quad \frac{\partial u^2}{\partial t} - \Delta u^2 = f^2,$$
$$u^1|_{t=0} = \varphi^1(x), \quad u^2|_{t=0} = \varphi^2(x).$$

As was determined in §8, there exist for this system infinitely many collections of numbers s_1, s_2, t_1, t_2, in particular,

$$s_1 = 0, \ s_2 = -k, \ t_1 = 2, \ t_2 = 2 + k,$$

where k is an arbitrary nonnegative integer. Here $\rho_1 = -2$, $\rho_2 = -2-k$. According to Theorem 10.2, for any $f^1 \in H^{l,l/2}(\overline{D_{n+1}^{(T)}})$, $f^2 \in H^{l+k,(l+k)/2}(\overline{D_{n+1}^{(T)}})$, $\phi^1 \in H^{l+2}(\overline{E}_n)$, $\phi^2 \in H^{l+2+k}(\overline{E}_n)$ there exists a solution of the problem with the following properties: $u^1 \in H^{l+2,(l+2)/2}(\overline{D_{n+1}^{(T)}})$, $u^2 \in H^{l+2+k,(l+2+k)/2}(\overline{D_{n+1}^{(T)}})$. It is seen that the ambiguity in the definition of the numbers s_j and t_j is connected with the fact that we can solve the Cauchy problem separately for the functions u_1 and u_2 in different classes: $H^{l+2,(l+2)/2}(\overline{D_{n+1}^{(T)}})$ and $H^{l+2+k,(l+2+k)/2}(\overline{D_{n+1}^{(T)}})$.

The basic argument in the proof of Theorems 10.1 and 10.2 is exactly the same as in the proof of Theorems 5.1–5.3 of Chapter IV. But it differs in certain essential details, for example, in the function spaces in which problems (9.31) and (10.2) are investigated. We will therefore present the basic steps in the proof of Theorem 10.1. It is cen ered on a study of the following problem with null initial conditions: find the vector function $\mathbf{u} = (u^1, \ldots, u^m)$ with $u^j \in \overset{\circ}{H}{}^{l+t_j,(l+t_j)/2b}(\overline{Q}_\tau)$, where $\tau \le T$, such that

$$\begin{aligned} &\mathscr{L}\left(x, t, \frac{\partial}{\partial x}, \frac{\partial}{\partial t}\right) \mathbf{u} = \mathbf{f}, \\ &\mathscr{B}\left(x, t, \frac{\partial}{\partial x}, \frac{\partial}{\partial t}\right) \mathbf{u}|_{S_T} = \mathbf{\Phi} \end{aligned} \tag{10.4}$$

for $f^j \in \overset{\circ}{H}{}^{l-s_j,(l-s_j)/2b}(\overline{Q}_\tau)$, $\Phi^q \in \overset{\circ}{H}{}^{l-\sigma_q,(l-\sigma_q)/2b}(\overline{S}_\tau)$. This problem can also be written in the form

$$A\mathbf{u} = \mathbf{h},$$

where A is an operator acting from the space

$$\mathfrak{H}_1^{(l)} = \prod_{j=1}^{m} \overset{\circ}{H}{}^{l+t_j, \frac{1}{2b}(l+t_j)}(\overline{Q}_\tau)$$

into the space

$$\mathfrak{H}_2^{(l)}=\prod_{j=1}^{m} \overset{\circ}{H}{}^{l-s_j,\ \frac{1}{2b}(l-s_j)}(\bar{Q}_\tau)\times\prod_{q=1}^{br} \overset{\circ}{H}{}^{l-\sigma_q,\ \frac{1}{2b}(l-\sigma_q)}(\bar{S}_\tau)$$

according to the formula

$$A\mathbf{u}=(\mathscr{L}\mathbf{u},\ \mathscr{B}\mathbf{u}|_{S_\tau}),$$

while $\mathbf{h}=(\mathbf{f},\ \Phi)\in\mathfrak{H}_2^{(l)}$.

Theorem 10.3. *Suppose the coefficients of the operators* $\mathfrak{L}$ *and* $\mathfrak{B}$ *and the boundary* S *satisfy the conditions of Theorem* 10.1. *If* $\tau\le\tau_0\le T$, *where* τ_0 *is a certain number depending on* $\mathfrak{L}$, $\mathfrak{B}$ *and the various characteristics of* Ω *and* S, *then problem* (10.4) *is uniquely solvable in the class* $\mathfrak{H}_1^{(l)}$ *and the solution is subject to the inequality*

$$\sum_{j=1}^{m}|u^j|_{Q_\tau}^{(l+t_j)}\le c\left(\sum_{j=1}^{m}|f^j|_{Q_\tau}^{(l-s_j)}+\sum_{q=1}^{br}|\Phi^q|_{S_\tau}^{(l-\sigma_q)}\right). \qquad (10.5)$$

To prove this theorem we construct an operator R which acts from $\mathfrak{H}_2^{(l)}$ into $\mathfrak{H}_1^{(l)}$, is bounded and is such that for any $\mathbf{h}\in\mathfrak{H}_2^{(l)}$, $\mathbf{v}\in\mathfrak{H}_1^{(l)}$

$$ARh=\mathbf{h}+T\mathbf{h},$$
$$RAh=\mathbf{v}+W\mathbf{v},$$

where T and W are contraction operators if $\tau\le\tau_0$. The operator R is given by formula (7.8) of Chapter IV, in which the functions $v^{(k)}$ are replaced by vectors $\mathbf{v}^{(k)}(x,\ t)$ defined in terms of the solutions of the following model problems: the Cauchy problem with null initial conditions

$$\left.\begin{aligned}&\mathscr{L}_0\left(\frac{\partial}{\partial x},\ \frac{\partial}{\partial t}\right)\mathbf{u}=\mathbf{f},\ f^j\in\overset{\circ}{H}{}^{l-s_j,\ \frac{1}{2b}(l-s_j)}(\overline{D_{n+1}^{(\tau)}}),\\ &u^j\in\overset{\circ}{H}{}^{l+t_j,\ \frac{1}{2b}(l+t_j)}(\overline{D_{n+1}^{(\tau)}})\end{aligned}\right\} \qquad (10.6)$$

and the boundary value problem in the domain $R^{(\tau)}$

$$\left.\begin{aligned}&\mathscr{L}_0\left(\frac{\partial}{\partial x},\ \frac{\partial}{\partial t}\right)\mathbf{u}=\mathbf{f},\ f_j\in\overset{\circ}{H}{}^{l-s_j,\ \frac{1}{2b}(l-s_j)}(\overline{R^{(\tau)}}),\\ &\mathscr{B}_0\left(\frac{\partial}{\partial x},\ \frac{\partial}{\partial t}\right)\mathbf{u}|_{x_n=0}=\Phi,\ \Phi_q\in\overset{\circ}{H}{}^{l-\sigma_q,\ \frac{1}{2b}(l-\sigma_q)}(\overline{\widetilde{D}_n^{(\tau)}}),\\ &u_j\in\overset{\circ}{H}{}^{l+t_j,\ \frac{1}{2b}(l+t_j)}(\overline{R^{(\tau)}}).\end{aligned}\right\} \qquad (10.7)$$

The operators $\mathfrak{L}_0$ and $\mathfrak{B}_0$ in these problems are obtained from the principal parts of $\mathfrak{L}$ and $\mathfrak{B}$ by freezing the coefficients and, if necessary, carrying out an orthogonal transformation of the space variables x.

The solutions of problems (10.6) and (10.7) are expressed in terms of volume and surface potentials, the kernels of which are elements of two fundamental matrices of the parabolic system $\mathfrak{L}_0(\partial/\partial x, \partial/\partial t)\, u = 0$: a matrix $\Gamma(x, t)$ with m rows and m columns satisfying the system

$$\mathscr{L}_0\left(\frac{\partial}{\partial x}, \frac{\partial}{\partial t}\right)\Gamma(x, t) = \delta(x, t)\,\mathscr{E}^{(m)}$$

($\mathfrak{E}^{(m)}$ is the unit matrix of order m) and a matrix $\mathfrak{G}(x, t)$ with br rows and m columns, which is a solution of a boundary value problem in the half space $x_n > 0$ with a δ-function in the boundary condition:

$$\mathscr{L}_0\left(\frac{\partial}{\partial x}, \frac{\partial}{\partial t}\right)\mathscr{G}(x, t) = 0,$$

$$\mathscr{B}_0\left(\frac{\partial}{\partial x}, \frac{\partial}{\partial t}\right)\mathscr{G}(x, t)|_{x_n=0} = \delta(x', t)\,\mathscr{E}^{(br)}. \tag{10.8}$$

The following estimates are valid for the elements Γ_{ij}, G_{iq} of the matrices Γ and $\mathfrak{G}$:

$$|D_t^k D_x^s \Gamma_{ij}| \leqslant c_{ks}\,(t-\tau)^{\frac{1}{2b}(s_i+t_j-n-2b-2bk-s)} \exp\left(-C_{ks}\frac{|x|^{\frac{2b}{2b-1}}}{t^{\frac{1}{2b-1}}}\right)$$

$$(i, j = 1, \ldots, m), \tag{10.9}$$

$$|D_t^k D_x^s G_{iq}| \leqslant c_{ks}\,(t-\tau)^{-\frac{1}{2b}(n-1+2b-t_i-\sigma_q+s+2bk)} \exp\left(-C_{ks}\frac{|x|^{\frac{2b}{2b-1}}}{t^{\frac{1}{2b-1}}}\right)$$

$$(i = 1, \ldots, m;\ q = 1, \ldots, br;\ x_n > 0). \tag{10.10}$$

A considerable part of the difficulties arising in the investigation of parabolic equations of high order and especially of parabolic systems is connected with the proof of these estimates. Estimates (10.9) and (10.10) permit us to estimate the volume and surface potentials $(\Gamma * \mathbf{f})$ and $(\mathfrak{G} *_2 \Phi)$ in roughly the same way as in Chapter IV, and to prove in this way estimate (10.5) for the solution of problem (10.7) and an analogous estimate for the solution of problem (10.6). The constants in these estimates depend on r but they remain bounded for finite r.

A proof of the boundedness of the operator R, estimates of the norms of the operators T and W, and a derivation of Theorem 10.1 from Theorem 10.3 are made in the same way as in Chapter IV.

Theorem 10.1 is also valid for noncylindrical Q_T, with the method of proof given here being retained in the case of noncylindrical Q_T without any essential changes [116c].

We will now formulate a theorem on the solvability of problems (9.31) in the class of vector functions $u(x, t)$ with $u^j \in W_q^{l+t_j, (l+t_j)/2b}(Q_T)$ for integral $l + t_j$ and $(l + t_j)/2b$. From the latter condition it is seen that we must confine ourselves to only those systems of which $t_j = 2bt_j'$ for integral t_j' (it is easily seen that in this case $s_j = 2bs_j'$ for integral s_j').

Theorem 10.4. *Let $l = 2bk$, $k \geq 0$, be an integer, with $l > \sigma_{\max}$.* Let $s_k = 2bs_k'$, $t_k = 2bt_k'$. *Suppose that $S \in C^{l+t_{\max}}$, the coefficients of the operators L_{kj} (L_{kj}^0) have bounded (continuous) derivatives $D_t^\mu D_x^\nu$ for $2b\mu + \nu \leq l - s_k$, and the coefficients of the operators B_{sj} and C_{aj} belong to the classes $H^{l-\sigma_s-1/q-\epsilon,\ (l-\sigma_s-1/q+\epsilon)/2b}(\overline{S}_T)$ and $H^{l-\rho_\alpha-2b/q+\epsilon}(\overline{\Omega})$ (ϵ is an arbitrarily small positive number). Suppose further that the numbers*

$$\frac{l-\sigma_s}{2b}-\frac{1}{q}\left(\frac{1}{2b}+1\right) \tag{10.11}$$

are nonintegral. Then problem (9.31) has a unique solution in the class of vector functions for which $u^j \in W_q^{l+t_j, (l+t_j)/2b}(Q_T)$ for any

$$f^j \in W_q^{l-s_j,\ \frac{1}{2b}(l-s_j)}(Q_T),\quad \Phi^s \in W_q^{l-\sigma_s-\frac{1}{q},\ \frac{1}{2b}\left(l-\sigma_s-\frac{1}{q}\right)}(S_T),$$

and $\phi^\alpha \in B_q^{l-\rho_\alpha-2b/q}(\Omega)$ satisfying the compatibility conditions that are necessary for the existence of a solution from the class indicated above.

The following inequality holds:

$$\sum_{j=1}^{m}\|u^j\|_{q,Q_T}^{(l+t_j)} \leqslant c\left(\sum_{j=1}^{m}\|f^j\|_{q,Q_T}^{(l-s_j)}+\sum_{s=1}^{br}\|\Phi^s\|_{q,S_T}^{\left(l-\sigma_s-\frac{1}{q}\right)} + \sum_{\alpha=1}^{r}\|\varphi^\alpha\|_{B_q^{l-\rho_\alpha-\frac{2b}{q}}(\Omega)}\right). \tag{10.12}$$

The spaces $B_q^{l-\rho_\alpha-2b/q}(\Omega)$ differ from $W_q^{l-\rho_\alpha-2b/q}(\Omega)$ only for integral $l-\rho_\alpha-2b/q$ and $q \neq 2$ (see in this connection §2 of Chapter II). In the case $b > 1$ the number $l-\rho_\alpha-2b/q$ can be integral for $q \neq 2$ (in contrast to the case considered in Chapter IV).

The restriction on the numbers (10.11) is an analog of the condition $q \neq 3/2$ in Theorem 9.1 of Chapter IV.

The requirements in Theorem 10.4 on the coefficients in the minor terms of the operators L_{kj}, B_{sj} and C_{aj} are excessive. As a matter of fact it is sufficient to require of the coefficients in the minor terms of the operator L_{kj}, for example, that $L_{kj}u^j \in W_q^{l-s_k,\,(l-s_k)/2b}(Q_T)$ for $u^j \in W_q^{l+t_j,\,(l+t_j)/2b}(\overline{Q}_T)$. As in §9 of Chapter IV, it is possible to show that for this to be the case it is sufficient that each coefficient in the minor terms of the operator L_{kj}, and possibly certain of its derivatives, have norms (9.1) of Chapter IV, which are finite and tend to zero for $\tau \to 0$, with a certain $r > 1$ depending on which derivative of which function the coefficient is with. The same remarks apply to the operators B_{sj} and C_{aj}. It is also possible to require less of the boundary S.

We have always assumed that the matrix $\mathfrak{B}$ satisfies the complementarity condition. But if this condition is violated at even one point $(\zeta, p) \neq 0$, with $\operatorname{Im}\zeta_j = 0$ and $\operatorname{Re} p \geq 0$, then estimates (10.1) and (10.12) cannot hold with the constants c being independent of $\mathbf{u}$. Consider in the domain $R^{(T)}$ the problem

$$
\begin{aligned}
\mathscr{L}_0\left(\frac{\partial}{\partial x}, \frac{\partial}{\partial t}\right)\mathbf{u}(x, t) &= \mathbf{f}(x, t),\\
\mathscr{B}_0\left(\frac{\partial}{\partial x}, \frac{\partial}{\partial t}\right)\mathbf{u}\Big|_{x_n=0} &= \boldsymbol{\Phi}(x, t),\\
\mathscr{C}_0\left(\frac{\partial}{\partial x}, \frac{\partial}{\partial t}\right)\mathbf{u}\Big|_{t=0} &= \boldsymbol{\varphi}(x),
\end{aligned}
\tag{10.13}
$$

in which the operators $\mathfrak{L}_0$, $\mathfrak{B}_0$, $\mathfrak{C}_0$ with constant coefficients contain only principal terms. We assume that the complementarity condition is violated for some real ζ_0 and some p_0 with $\operatorname{Re} p_0 \geq 0$. In this case the problem

$$
\begin{aligned}
\mathscr{L}_0\left(i\zeta_0, \frac{d}{dx_n}, p_0\right)\tilde{\mathbf{u}}(x_n) &= 0,\\
\mathscr{B}_0\left(i\zeta_0, \frac{d}{dx_n}, p_0\right)\tilde{\mathbf{u}}(x_n)\Big|_{x_n=0} &= 0,\\
\tilde{\mathbf{u}}(x_n) &\xrightarrow[x_n\to+\infty]{} 0
\end{aligned}
\tag{10.14}
$$

has a solution different from zero. It is not difficult to see that for any $\lambda > 0$ the vector $\mathbf{u}_\lambda(x, t) = (u_\lambda^1, \dots, u_\lambda^m)$, where

$$
u_\lambda^j(x, t) = \lambda^{-l-t_j} e^{p_0\lambda^{2b}(t-T)} e^{i\lambda\zeta_0 x'} \tilde{u}^j(\lambda x_n)
$$

is a solution of problem (10.14) for $\mathbf{f} = 0$, $\boldsymbol{\Phi} = 0$ and

$$
\boldsymbol{\varphi}(x) = (\varphi_\lambda^1, \dots, \varphi_\lambda^r),
$$

$$
\varphi_\lambda^a = \lambda^{-l+\rho_a} e^{-p_0\lambda^{2b}T} e^{i\lambda\zeta_0 x'} \sum_{j=1}^m C_{aj}^0\left(i\zeta_0, \frac{d}{dz}, p_0\right)\tilde{u}^j(z)\Big|_{z=\lambda x_n}.
$$

We first assume that $\operatorname{Re} p_0 > 0$. In this case for any λ

$$\sum_{\alpha=1}^{r}|\varphi_\lambda^\alpha|_{D_n}^{(l-p_\alpha)}\leqslant ce^{-k\lambda^{2b}},\quad k>0,$$

while

$$\sum_{j=1}^{m}|u_\lambda^j(x,t)|_{R^{(T)}}^{(l+t_j)}\geqslant\sum_{j=1}^{m}|u_\lambda^j(x,T)|_{D_n}^{(l+t_j)}$$

$$\geqslant\sum_{j=1}^{m}\langle u_\lambda^j(x,T)\rangle_{D_n}^{(l+t_j)}=\sum_{j=1}^{m}\langle e^{i\zeta_0 x'}\tilde{u}^j(x_n)\rangle_{D_n}^{(l+t_j)}>0.$$

Therefore inequality (10.1) with a constant independent of u cannot hold for a solution of problem (10.13). For sufficiently large λ not even the estimate

$$\sum_{j=1}^{m}\langle u_\lambda^j(x,T)\rangle_{D_n}^{(0)}\leqslant c\sum_{\alpha=1}^{r}|\varphi_\lambda^\alpha|_{D_n}^{(l-p_\alpha)},\tag{10.15}$$

for the functions u_λ^j holds, no matter what constant c we take and how small the number $T>0$ is. This means that problem (10.13) is incorrect.

It is not difficult to verify that when $\operatorname{Re}p_0=0$ inequality (10.1) for problem (10.13) is violated in the case of large λ by the vector

$$\mathrm{U}_\lambda(x,t)=\mathrm{u}_\lambda(x,t)\zeta(t),$$

where $\zeta(t)$ is a smooth function satisfying the conditions $0\le\zeta(t)\le 1$ and

$$\zeta(t)=\begin{cases}0 & \text{for}\quad t\leqslant\dfrac{T}{3}\\[2mm] 1 & \text{for}\quad t\geqslant\dfrac{2T}{3}.\end{cases}$$

Vectors violating inequality (10.12) for problem (10.13) are constructed in an analogous manner.

We have thus proved that the complementarity condition is necessary for the validity of estimates (10.1) and (10.12) for the solutions of problem (10.13). The passage to the more general problem (9.31) is made in the same way as for elliptic equations [65q].

Thus the complementarity condition distinguishes a class of well-stated problems for equations and systems of parabolic type, for which Theorems 10.1 and 10.4 are valid. This condition is fulfilled, in particular, for all of the basic boundary value problems for equations of second order that were considered in the preceding chapters, and also for problem (5.15). It is also fulfilled in the case of the first boundary value problem for strongly parabolic systems; as was shown in [65f], inequality (10.12) with $q=2$ is valid for the solutions of such problems.

The situation is different in the case of the first boundary value problem for systems that are parabolic in the sense of Petrovskiĭ, even when they are of first order with respect to t. Consider in the domain $R^{(T)}$ the first boundary value problem for a system with constant coefficients

$$\frac{\partial \mathbf{u}}{\partial t} + \mathcal{A}_0\left(\frac{\partial}{\partial x}\right)\mathbf{u} = \mathbf{f}(x, t),$$

$$\mathbf{u}\big|_{x_n=0} = \Phi^{(1)}(x', t), \ \ldots, \ \frac{\partial^{b-1}\mathbf{u}}{\partial x_n^{b-1}}\bigg|_{x_n=0} = \Phi^{(b)}(x', t), \tag{10.16}$$

$$\mathbf{u}\big|_{t=0} = \varphi(x).$$

The operator $\mathfrak{A}_0$ contains only terms of order $2b$.

The problem corresponding to (10.14) has the form

$$p\tilde{\mathbf{u}} + \mathcal{A}_0\left(i\zeta, \frac{d}{dx_n}\right)\tilde{\mathbf{u}} = 0,$$

$$\tilde{\mathbf{u}}\big|_{x_n=0} = \ldots = \frac{d^{b-1}\tilde{\mathbf{u}}}{dx_n^{b-1}}\bigg|_{x_n=0} = 0, \tag{10.17}$$

$$\tilde{\mathbf{u}} \xrightarrow[x_n\to+\infty]{} 0.$$

If $m = 1$, then it is completely obvious that this problem has only the zero solution. This is also true in the case $m > 1$, $n = 1$. Indeed, for $n = 1$

$$\mathcal{A}_0\left(i\zeta, \frac{d}{dx_n}\right)\tilde{\mathbf{u}} = A_0\frac{d^{2b}\tilde{\mathbf{u}}}{dx^{2b}},$$

where A_0 is a numerical matrix. Let T be a transformation taking A_0 into a Jordan form:

$$TA_0T^{-1} = A_0' = \begin{pmatrix} \alpha_1 & \lambda_1 & & & \\ & \ddots & \ddots & & 0 \\ & & \ddots & \ddots & \\ & & & \ddots & \lambda_{m-1} \\ 0 & & & & \alpha_m \end{pmatrix}.$$

By virtue of the parabolicity condition $\alpha_j \neq 0$.

We introduce the new vector

$$\tilde{\mathbf{v}}(x) = T\tilde{\mathbf{u}}(x).$$

Multiplying (10.17) from the left by T, we obtain the following problem for $\tilde{\mathbf{v}}$:

$$p\tilde{\mathbf{v}} + A_0'\frac{d^{2b}\tilde{\mathbf{v}}}{dx^{2b}} = 0, \tag{10.18}$$

$$\tilde{\mathbf{v}}\big|_{x=0} = \ldots = \frac{d^{b-1}\tilde{\mathbf{v}}}{dx^{b-1}}\bigg|_{x=0} = 0, \quad \tilde{\mathbf{v}} \xrightarrow[x\to+\infty]{} 0 \tag{10.19}$$

Since

$$\det\left[pE + A_0'(i\tau)^{2b}\right] = \det\left[pE + A_0(i\tau)^{2b}\right] = \prod_{s=1}^{m}\left[p + (i\tau)^{2b}\alpha_s\right],$$

each operator

$$L^{(s)}\left(\frac{\partial}{\partial x}, \frac{\partial}{\partial t}\right) = \frac{\partial}{\partial t} + \alpha_s \frac{\partial^{2b}}{\partial x^{2b}}$$

is parabolic. From (10.18) and (10.19) it follows that the function $\tilde{v}^m$ is a solution of the problem

$$L^{(m)}\left(\frac{d}{dx}, p\right)\tilde{v}^m = 0,$$

$$\tilde{v}^m\big|_{x=0} = \dots = \frac{d^{b-1}\tilde{v}^m}{dx^{b-1}}\bigg|_{x=0} = 0, \quad \tilde{v}^m \xrightarrow[x\to+\infty]{} 0.$$

Therefore $\tilde{v}^m = 0$ and the next to last equation in (10.18) has the form

$$L^{(m-1)}\left(\frac{d}{dx}, p\right)\tilde{v}^{m-1} + \lambda_{m-1}\tilde{v}^m = L^{(m-1)}\left(\frac{d}{dx}, p\right)\tilde{v}^{m-1} = 0.$$

Consequently $\tilde{v}^{m-1} = 0$, and so on. Thus problem (10.18), (10.19) and problem (10.17) have only the zero solution.

However, this assertion does not hold for $n > 1$. Consider problem (10.17) for a system of two equations, the matrix $\mathcal{A}_0(i\zeta, i\tau)$ of which has the form

$$\mathcal{A}_0(i\zeta, i\tau) = \begin{pmatrix} (\tau - i\sqrt{1+\lambda}\,\zeta_{n-1})^2 + \zeta'^2 - \lambda\zeta_{n-1}^2 & -4(1+\lambda)\zeta_{n-1}^2 \\ -(\tau - i\sqrt{1+\lambda}\,\zeta_{n-1})^2 & (\tau + i\sqrt{1+\lambda}\,\zeta_{n-1})^2 + \zeta'^2 - \lambda\zeta_{n-1}^2 + 4\zeta_{n-1}^2(1+\lambda) \end{pmatrix} \tag{10.20}$$

(here $\zeta'^2 = \sum_{i=1}^{n-2}\zeta_i^2$, $\lambda \geq 0$). This system is parabolic with $b = 1$ since

$$\det\left(pE + \mathcal{A}_0(i\zeta, i\tau)\right) = (p + \zeta'^2 + \zeta_{n-1}^2 + \tau^2)^2 = (p + \zeta^2 + \tau^2)^2.$$

The complementarity condition is violated in this problem for $\zeta' = 0$, $p = \lambda\zeta_{n-1}^2$ and any $\zeta_{n-1} > 0$, since for the indicated values of p and ζ problem (10.17) with the matrix $\mathcal{A}_0$ of (10.20) has the nonzero solution

$$\tilde{\mathbf{u}}(x_n) = \left(x_n e^{-\sqrt{1+\lambda}\,\zeta_{n-1}x_n}, 0\right).$$

Thus for $n > 1$ there exist systems parabolic in the sense of Petrovskiĭ for which the first boundary value problem is poorly stated.

Let us return to the question of the solvability of the boundary value problem (9.31) when the complementarity condition is fulfilled.

For systems that are parabolic according to Petrovskiĭ some of the above-mentioned results of [116c] were obtained in earlier papers by other authors. For example, Theorem 10.4 is proved by a different method in the papers [2] of M. S. Agranovič and M. I. Višik, where the general boundary value problems for systems that are parabolic according to Petrovskiĭ are investigated in the spaces $W_2^{2bl,l}$ with integral and nonintegral $l > 0$. Briefly, this method is as follows. The case when the coefficients of the operators $\mathfrak{L}$ and $\mathfrak{B}$ depend only on x is considered first. These problems are solved by application of the Laplace transform with respect to t. The passage to the general case is effected by successive steps in small intervals $t_k \leq t \leq t_{k+1}$.

Exact estimates of solutions of boundary value problems in the norms of $W_2^{2bl,l}$ for (Petrovskiĭ) parabolic systems of first order with respect to t (without an existence theorem) were announced by L. N. Slobodeckiĭ [111b].

In the case $q = 2$ exact estimates are obtained with the use of Parseval's equality, which removes the need for estimates (10.9) and (10.10).

Exact estimates in Hölder norms have been established by A. Friedman [31a] for solutions of (Petrovskiĭ) parabolic systems of form (8.5), interior to the domain in which these solutions are defined. Friedman proved these estimates by generalizing a method proposed by Douglis and Nirenberg for elliptic systems.

We proceed now to a brief account of the results, obtained by the methods of potential theory, on the solvability of the Cauchy problem and of the boundary value problems for systems parabolic in the sense of Petrovskiĭ. Systems of form (8.5) are the ones that have been most studied by these methods.

A fundamental matrix $Z(x, \xi, t, \tau)$ for such systems was first constructed by S. D. Èĭdel'man [25b(I)] on the assumption that the principal coefficients of the operator $\mathfrak{A}$ are sufficiently smooth functions. These results have been carried over in [25b(II)], [135] and [132c] to the case when the coefficients of $\mathfrak{A}$ satisfy only a Hölder condition. In [25b(II)] a fundamental matrix has also been constructed for (Petrovskiĭ) parabolic systems of general form. It is required in [25b(II)] that the coefficients of the operator $\mathfrak{A}$ satisfy a Hölder condition only with respect to the variables x and be continuous in the variable t. Further progress in this direction has been achieved in the note [79], where it is required that the coefficients of $\mathfrak{A}$ satisfy the more general (than a Hölder condition) Dini condition.

If the coefficients of the operator $\mathfrak{A}$ satisfy a Holder condition in $D_{n+1}^{(T)}$ with respect to x with exponent α, and are continuous in t, then the elements Z_{ij} of the matrix Z are subject to the following inequalities:

$$|D_x^k Z_{ij}(x, \xi, t, \tau)| \leqslant c(t-\tau)^{-\frac{n+k}{2b}} \exp\left(-C\frac{|x-\xi|^{\frac{2b}{2b-1}}}{(t-\tau)^{\frac{1}{2b-1}}}\right),$$

$$k \leqslant 2b;$$

$$|D_x^{2b} Z_{ij}(x, \xi, t, \tau) - D_{x'}^{2b} Z_{ij}(x', \xi, t, \tau)|$$

$$\leqslant c|x-x'|^{\beta}(t-\tau)^{-\frac{n+2b+\beta}{2b}}\left[\exp\left(-C\frac{|x-\xi|^{\frac{2b}{2b-1}}}{(t-\tau)^{\frac{1}{2b-1}}}\right) + \exp\left(-C\frac{|x'-\xi|^{\frac{2b}{2b-1}}}{(t-\tau)^{\frac{1}{2b-1}}}\right)\right],$$

$$0 \leqslant \beta \leqslant \alpha.$$

The differential properties of the functions Z_{ij} are improved together with those of the coefficients of the operator $\mathfrak{A}$: if the coefficients have continuous derivatives $D_t^{k_0} D_x^k$ satisfying a Hölder condition with respect to x, then there exist derivatives $D_t^{m_0} D_x^m Z_{ij}(x, \xi, t, \tau)$ $(2bm_0 + m = 2b + r,\ r = 2bk_0 + k)$ and also $D_\xi^r Z_{ij}$ [25a].

The fundamental matrix Z has been used to investigate the Cauchy problem

$$\begin{aligned} &\frac{\partial \mathbf{u}}{\partial t} + \mathcal{A}\mathbf{u} = \mathbf{f}, \\ &\mathbf{u}\,|_{t=0} = \varphi(x) \end{aligned} \tag{10.21}$$

in various function classes, including those in which the functions are increasing exponentially for $|x| \to \infty$ (see [25a; 111a, c]); in particular, exact estimates of type (10.3) have been obtained for the solution of this problem. A detailed study of the properties of the matrix Z has also made it possible to investigate the stability and behavior for $t \to +\infty$ of the solution of problem (10.21), to study the Cauchy problem for operators with increasing coefficients and to prove theorems of Liouville type for solutions of parabolic systems. All of these and many other results, some of which have also been proved for (Petrovskiĭ) parabolic systems of general form, are presented in the monograph [25a], which also contains an extensive bibliography.

Substantially less complete results have been obtained by the methods of potential theory for boundary value problems. Problems of the form

$$\begin{aligned} &\frac{\partial \mathbf{u}}{\partial t} + \mathcal{A}\left(x, t, \frac{\partial}{\partial x}\right)\mathbf{u} = \mathbf{f}, \\ &\mathcal{B}\left(x, t, \frac{\partial}{\partial x}\right)\mathbf{u}\Big|_{S_T} = \mathbf{\Phi}, \\ &\mathbf{u}\,|_{t=0} = \varphi(x) \end{aligned} \tag{10.22}$$

have been considered by these methods with the assumptions that the order of the operators B_{qj}, which can depend on q, is less than $2b$ (the order of the operator $\mathfrak{A}$) and that the following condition of Ja.B. Lopatinskiĭ is fulfilled: the rank of the matrix (9.23) is equal to bm.

The general plan for solving problem (10.22) in the indicated papers is as follows. First the matrix Z is used to reduce this problem to the same problem but with $\mathbf{f}=0$ and $\phi=0$. The solution of this latter problem is then sought in the form of a sum of potentials

$$u^j(x, t)=\sum_{s=1}^{bm}\int_0^t d\tau\int_S G_{js}(x, \xi, t, \tau)\,\mu^s(\xi, \tau)\,dS_\xi, \tag{10.23}$$

where the G_{js} are kernels having the following properties.

1) For each $s=1, \cdots, bm$ the vector $\mathbf{G}^{(s)}=(G_{1s}, \cdots, G_{ms})$ is a solution of the homogeneous system

$$\frac{\partial \mathbf{G}^{(s)}}{\partial t}+\mathcal{A}\left(x, t, \frac{\partial}{\partial x}\right)\mathbf{G}^{(s)}=0$$

in a domain Q_T.

2) The principal parts of the G_{js} with respect to order of singularity are functions $G_{js}^{(0)}(x-\xi, t-\tau)$ forming the solution of a boundary value problem of type (10.8) in the half space $x\nu(\xi)\geq 0$ bounded by the tangent plane at the point $(\xi, \tau)\in S_T$ to the surface S_T. The functions $G_{js}^{(0)}$ are subject to inequalities (10.10) in this half space:

$$\begin{aligned}&|D_t^r D_x^s G_{jq}^{(0)}(x, t)|\\ &\leqslant c\,(t-\tau)^{-\frac{1}{2b}(n-1+2b-\beta_q+s+2br)}\exp\left(-C\frac{|x|^{\frac{2b}{2b-1}}}{t^{\frac{1}{2b-1}}}\right),\end{aligned} \tag{10.24}$$

where β_q is the order of the operators B_{qj}.

The functions $\mu^s(\xi, \tau)$ in (10.23) are determined from a system of Volterra integral equations obtained from the boundary condition. An exact and explicit derivation of these integral equations has at present been given only under one of the following assumptions: 1) the orders β_q of all operators B_{qj} are the same and 2) $n=1$, although the functions G_{js} with the above-mentioned properties have been constucted without these additional assumptions.

The method of potential theory was the first method to be used in investigating boundary value problems of form (10.22). Beginning in 1956 T. Ja. Zagorskiĭ published a series of papers and the monograph [129], in which problem (10.22) is considered in bounded convex domains. But these papers and the monograph contain important errors (see in this connection [116c]). S. D. Èĭdel'man [25a] obtained the exact estimates (10.24) for the kernels $G_{js}^{(0)}$, constructed the functions G_{js} in nonconvex domains and proved the solvability of problem (10.22) under the above-mentioned additional assumptions. Also, in the note [74] Èĭdel'man and G. Ja. Lipko announced certain results on the solvability of boundary value problems of form (10.22) in noncylindrical domains and in domains with infinite boundaries.

Estimate (10.24) and all of the results following from it were obtained by Èĭdel'man for those systems for which the complementarity condition is fulfilled in the case of the first boundary value problem. The example of (10.20) shows that this class of systems does not coincide with the class of all (Petrovskiĭ) parabolic systems of the form (8.5). It should be noted that this restriction on the class of systems is in no way connected with the representation of a solution in the form (10.23) or with the determination of the densities μ^s. It is connected with the method of proof of estimate (10.24). In [116c] the analogous estimate (10.10) for more general systems has been proved by another method. Therefore all of the subsequent arguments and final results of Èĭdel'man are valid for any (Petrovskiĭ) parabolic system of form (8.5).

BIBLIOGRAPHY

[1] S. Agmon, A. Douglis and L. Nirenberg, *Estimates near the boundary for solutions of elliptic partial differential equations satisfying general boundary conditions.* I, Comm. Pure Appl. Math. 12(1959), 623–727; Russian transl., IL, Moscow, 1962. MR 23 #A2610.

[2] M. S. Agranovič and M. I. Višik, a) *Parabolic boundary problems*, Uspehi Mat. Nauk 18(1963), no. 1(109), 206–208. (Russian)
b) *Elliptic problems with a parameter and parabolic problems of general type*, Uspehi Mat. Nauk 19(1964), no. 3(117), 53–161 = Russian Math. Surveys 19(1964), no. 3, 53–157. MR 33 #415.

[3] A. D. Aleksandrov, a) *Certain estimates concerning the Dirichlet problem*, Dokl. Akad. Nauk SSSR 134(1960), 1001–1004 = Soviet Math. Dokl. 1(1960), 1151–1154. MR 26 #5290.
b) *Investigation on the maximum principle.* IV, V, VI, Izv. Vysš. Učebn. Zaved. Matematika 1960, no. 3(16), 3–15; ibid., 1960, no. 5(18), 16–26; ibid. 1961, no. 1(20), 3–20. MR 24 #A3400d, e, f.
c) *Uniqueness conditions and bounds for the solution of the Dirichlet problem*, Vestnik Leningrad. Univ. Ser. Mat. Meh. Astronom. 18(1963), no. 3, 5–29; English transl., Amer. Math. Soc. Transl. (2) 68(1968), 89–119. MR 29 #1434.

D. G. Aronson—see [132].

[4] N. Aronszajn, *Boundary values of functions with finite Dirichlet integral*, Conference on Partial Differential Equations, Univ. of Kansas Press, 1955, pp. 77–93.

[5] V. M. Babič, a) *On the extension of functions*, Uspehi Mat. Nauk 8(1953), no. 2 (54), 111–113. (Russian) MR 15, 110.
V. M. Babič and L. N. Slobodeckiĭ, b) *On the boundedness of the Dirichlet integral*, Dokl. Akad. Nauk SSSR 106(1956), 604–606. (Russian) MR 17,959.

[6] G. I. Barenblatt and M. I. Višik, *On finite velocity of propagation in problems of non-stationary filtration of a liquid or gas*, Prikl. Mat. Meh. 20(1956), 411–417. (Russian) MR 18, 256.

[7] R. B. Barrar, *Some estimates for solutions of parabolic equations*, J. Math. Anal. Appl. 3(1961), 373–397. MR 27 #448.

[8] S. N. Bernšteĭn, a) *A limitation on the moduli of a sequence of derivatives of solutions of equations of parabolic type*, Dokl. Akad. Nauk SSSR 18(1938), 385–388. (Russian)
b) *Collected works.* III: *Differential equations, calculus of variations and geometry* (1903–1947), Izdat. Akad. Nauk SSSR, Moscow, 1960. (Russian) MR 23 #A32.

[9] O. V. Besov, *An investigation of a family of function spaces in connection with embedding and extension theorems*, Trudy Mat. Inst. Steklov. 60 (1961), 42–81; English transl., Amer. Math. Soc. Transl. (2) 40 (1964), 85–126. MR 24 #A3501.

[10] V. M. Borok, a) *On a characteristic property of parabolic systems*, Dokl. Akad. Nauk SSSR 110 (1956), 903–905. (Russian) MR 19, 150.
b) *On the numerical characteristics of systems that are correct according to I. G. Petrovskiĭ*. Izv. Vysš. Učebn. Zaved. Matematika 1959, no. 1(8), 16–22. (Russian) MR 24 #A737.

[11] F. E. Browder, a) *Estimates and existence theorems for elliptic boundary value problems*, Proc. Nat. Acad. Sci. U. S. A. 45 (1959), 365–372. MR 24 #A2749.
b) *Nonlinear parabolic boundary value problems of arbitrary order*, Bull. Amer. Math. Soc. 69 (1963), 858–861. MR 27 #6049.
c) *Strongly non-linear parabolic boundary value problems*, Amer. J. Math. 86 (1964), 339–357. MR 29 #3764.
d) *Non-linear initial value problems*, Ann. of Math. (2) 82 (1965), 51–87. MR 34 #7923.
e) *Existence and uniqueness theorems for solutions of nonlinear boundary value problems*, Proc. Sympos. Appl. Math., vol 17, Amer. Math. Soc., Providence, R. I., 1965, pp. 24–49. MR 33 #6092.

[12] A. Calderón and A. Zygmund, *On the existence of certain singular integrals*, Acta Math. 88 (1952), 85–139. MR 14, 637.

[13] C. Ciliberto, a) *Sulle equazioni non lineari di tipo parabolico in due variabili*, Ricerche Mat. 3 (1954), 129–165. MR 16, 1028.
b) *Formule de maggiorazione e teoremi di esistenza per le soluzioni delle equazioni paraboliche in due variabili*, Ricerche Mat. 3 (1954), 40–75. MR 16, 139.
c) *Sulle equazioni quasi-lineari di tipo parabolico in due variabili*, Ricerche Mat. 5 (1956), 97–125. MR 18, 46.
d) *Nuovi contributi alla teoria dei problemi al contorno per le equazioni paraboliche non lineari in due variabili*, Ricerche Mat. 5 (1956), 206–225. MR 18, 742.

[14] E. M. Čistjakova, *An application of the method of finite differences to the solution of mixed boundary value problems for certain hyperbolic and parabolic systems*, Candidate's dissertation, Leningrad. Gos. Ped. Inst., Leningrad, 1956. (Russian)

[15] R. Courant, K. Friedrichs and H. Lewy, a) *Über die partiellen Differentialgleichungen der mathematischen Physik*, Math. Ann. 100 (1928), 32–74; English transl., New York University Courant Inst. Math Sciences Research Dept. N. Y. O.–7689.

R. Courant and D. Hilbert, b) *Methoden der mathematischen Physik*. Vol. I, Springer–Verlag, Berlin, 1931; reprint, Interscience , New York, 1943; Russian transl., 3rd ed., GITTL, Moscow, 1951; English transl., Interscience, New York, 1953. MR 5, 97; MR 13, 800; MR 16, 426.

[16] Čžou Juĭ-lin' (Zhou Yu-lin, Chou Yu-lin) (see also Kalašnikov), *Boundary value problems for non-linear parabolic equations*, Dokl. Akad. Nauk SSSR 117(1957), 195–198; Mat. Sb. 47(89) (1959), 431–484; English transl., Amer. Math. Soc. Transl. (2) 41(1964), 155–213. MR 20 #3374; MR 22 #1751.

[17] E. De Giorgi, *Sulla differenziabilità e l'analiticità delle estremali degli integrali multipli regolari*, Mem. Accad. Sci. Torino. Cl. Sci. Fis. Mat. Nat. (3) 3(1957), 25–43. MR 20 #172.

[18] G. Doetsch, *Theorie und Anwendung der Laplace-Transformation*, Dover, New York, 1943. MR 5, 119.

[19] J. Douglas, Jr., *A uniqueness theorem for the solution of a Stefan problem*, Proc. Amer. Math. Soc. 8(1957), 402–408. MR 19, 1060.

[20] A. Douglis. See Agmon.

[21] F. G. Dressel, *The fundamental solution of the parabolic equation*. I, Duke Math. J. 7(1940), 186–203. MR 2, 204.

[22] Ju. A. Dubinskiĭ, a) *Weak convergence in nonlinear elliptic and parabolic equations*, Mat. Sb. 67(109) (1965), 609–642; English transl., Amer. Math. Soc. Transl. (2) 67(1968), 226–258. MR 32 #7958.
b) *On nonlinear parabolic equations having nondivergence form*, Dokl. Akad. Nauk SSSR 163(1965), 805–808 = Soviet Math. Dokl. 6(1965), 1009–1013. MR 35 #3270.

[23] A. S. Dynin, *n-dimensional elliptic boundary value problems with a single unknown function*, Dokl. Akad. Nauk SSSR 141(1961), 285–287 = Soviet Math. Dokl. 2(1961), 1431–1432. MR 26 #6564.

[24] E. B. Dynkin, *Markov processes*, Fizmatgiz, Moscow, 1963; English transl., Vols. I, II, Academic Press, New York and Springer-Verlag, Berlin, 1963. MR 33 #1886; MR 33 #1887.

[25] S. D. Èĭdel'man (see also Lipko and Matiĭčuk), a) *Parabolic systems*, "Nauka", Moscow, 1964; English transl., Noordhoff, Groningen and North-Holland, Amsterdam (to appear). MR 29 #4998.
b) *On fundamental solutions of parabolic systems*. I, Mat. Sb. 38(80) (1956), 51–92; English transl., Amer. Math. Soc. Transl. (2) 41(1964), 1–48; II, Mat. Sb. 53(95), (1961), 73–136; English transl., Amer. Math. Soc. Transl. (2) 41(1964), 49–120. MR 17, 857; MR 24 #A925.

[26] S. Faedo, *Un nuovo metodo per l'analisi esistenziale e quantativa dei problemi di propagazione*, Ann. Scuola Norm. Sup. Pisa (3) 1(1947), 1–41. MR 11, 363.

[27] W. Feller, *Zur Theorie der stochastischen Prozesse*, Math. Ann. 113(1936), 113–160.

[28] G. M. Fihtengol'c, *A course in differential and integral calculus*, Vol. I, 6th ed., "Nauka", Moscow, 1966; German transl., VEB Deutscher Verlag, Berlin, 1964; 2nd ed., Hochschulbücher für Math., Band 61, 1966.

[29] A. F. Filippov (see also Rjaben'kiĭ), *On conditions for the existence of a solution of a quasi-linear parabolic equation*, Dokl. Akad. Nauk SSSR 141 (1961), 568–570 = Soviet Math. Dokl. 2(1961), 1517–1519. MR 24 #A3428.

[30] G. E. Forsythe and P. C. Rosenbloom, *Numerical analysis and partial differential equations*, Wiley, New York, 1958.

[31] A. Friedman, a) *Interior estimates for parabolic systems of partial differential equations*, J. Math. Mech. 7(1958), 393–417. MR 21 #7362.
b) *Partial differential equations of parabolic type*, Prentice-Hall, Englewood Cliffs, N. J., 1964. MR 31 #6062.

[32] K. Friedrichs. See Courant.

[33] E. Gagliardo, a) *Problema al contorno per equazioni differentiali lineari di tipo parabolico in n variabili*, Ricerche Mat. 5(1956), 169–205. MR 18, 742.
b) *Teoremi di esistenza e di unicità per problemi al contorno relativi ad equazioni paraboliche lineari e quasi lineari in n variabili*, Ricerche Mat. 5(1956), 239–257. MR 18, 742.
c) *Caratterizzioni delle tracce sulla frontiera relative ad alcune classi di funzioni in n variabili*, Rend. Sem. Mat. Univ. Padova 27(1957), 284–305. MR 21 #1525.
d) *Proprietà di alcune classi di funzioni in più variabili*, Ricerche Mat. 7 (1958), 102–137. MR 21 #1526.

[34] I. M. Gel'fand and G. E. Šilov, a) *Generalized functions*. Vol. III: *Some questions in the theory of differential equations*, Fizmatgiz, Moscow, 1958; English transl., Academic Press, New York, 1967. MR 21 #5142b.
b) *The Fourier transforms of rapidly increasing functions and questions of uniqueness of the solution of the Cauchy problem*, Uspehi Mat. Nauk 8(1953), no. 6(58), 3–54; English transl., Amer. Math. Soc. Transl. (2) 5(1957), 221–274. MR 15, 867; MR 18, 736.

[35] M. Gevrey, *Sur les équations aux dérivées partielles du type parabolique*, J. Math. Pures Appl. (6) 9 (1913), 305–471; (6) 10(1914), 105–148.

[36] N. M. Gjunter (N. M. Gunther), *La théorie du potentiel et ses applications aux problèmes fondamentaux de la physique mathématique*, Gauthier-Villars, Paris, 1934; Russian transl., GITTL, Moscow, 1953. MR 16, 357.

[37] S. K. Godunov, *On non-uniqueness for parabolic systems*, Dokl. Akad. Nauk SSSR 145(1962), 498–500 = Soviet Math. Dokl. 3(1962), 1020–1023. MR 28 #2361.

[38] K. K. Golovkin, a) *On embedding theorems*, Dokl. Akad. Nauk SSSR 134(1960), 19–22 = Soviet Math. Dokl. 1(1960), 998–1001. MR 22 #12374.
b) *Two classes of inequalities for sufficiently smooth functions of n variables*, Dokl. Akad. Nauk SSSR 138(1961), 22–25 = Soviet Math. Dokl. 2(1961), 510–513. MR 22 #12189.
K. K. Golovkin and V. A. Solonnikov, c) *Estimates for integral operators in translation invariant norms*. I, Trudy Mat. Inst. Steklov. 70(1964), 47–58; English transl., Amer. Math. Soc. Transl. (2) 61(1967), 97–112. MR 29 #470.

[39] J. W. Green, *An expansion method for parabolic partial differential equations*, J. Research Nat. Bur. Standards 51(1953), 127–132. MR 15, 322.

[40] A. V. Grekov, *The Dirichlet problem for certain quasi-linear parabolic equations*, Dokl. Akad. Nauk SSSR 134(1960), 255–258 = Soviet Math. Dokl. 1 (1960), 1043–1046. MR 22 #1748.

[41] Fr. Guglielmino, a) *Sulla regolarizzazione delle soluzioni deboli dei problemi al contorno per operatori parabolici*, Ricerche Mat. 12(1963), 44–66. MR 28 #357a.
b) *Ulteriori contributi alla regolarizzazione delle soluzioni deboli dei problemi parabolici*, Ricerche Mat. 12(1963), 140–150. MR 28 #357b.
c) *Sulle equazione paraboliche del secondo ordine di tipo non-variazionale*, Ann. Mat. Pure Appl. (4) 65(1964), 127–151. MR 32 #4395.

N. M. Gunther – see [36].

[42] G. H. Hardy, J. E. Littlewood and G. Pólya, *Inequalities*, Cambridge Univ. Press, New York, 1934; 2nd ed., 1952; Russian transl., IL, Moscow, 1948. MR 13, 727; MR 18, 722.

[43] D. Hilbert. See Courant.

[44] E. Hille, a) *Functional analysis and semigroups*, Amer. Math. Soc. Colloq. Publ., vol. 31, Amer. Math. Soc., Providence, R. I., 1948. MR 9, 594.
E. Hille and R. S. Phillips, b) *Functional analysis and semi-groups*, Amer. Math. Soc. Colloq. Publ., vol. 31, rev. ed., Amer. Math. Soc., Providence, R. I., 1957. MR 19, 664.

[45] E. Holmgren, *Über Systeme von linearen partiellen Differentialgleichungen*, Öfvers. Kongl. Vetens. -Akad. Förk. 58(1901), 91–103.

[46] E. Hopf, a) *Elementare Bemerkungen über die Lösungen partieller Differentialgleichungen zweiter Ordnung vom elliptischen Typus*, S. -B. Preuss. Akad. Wiss. Berlin 19(1927), 147–152.
b) *Über die Anfangswertaufgabe für die hydrodynamischen Grundgleichungen*, Math. Nachr. 4(1951), 213–231. MR 14, 327.

[47] A. M. Il'in, a) *On the fundamental solution of a parabolic equation*, Dokl. Akad. Nauk SSSR 147(1962), 768–771 = Soviet Math. Dokl. 3(1962), 1697–1700. MR 29 #1453.

A. M. Il'in, A. S. Kalašnikov and O. A. Oleĭnik, b) *Linear equations of second order of parabolic type*, Uspehi Mat. Nauk 17(1962), no. 3(105), 3–146 = Russian Math. Surveys 17(1962), no. 3, 1–143. MR 25 #2328.

[48] V. A. Il'in, *On the solvability of mixed boundary value problems for hyperbolic and parabolic equations*, Uspehi Mat. Nauk 15 (1960), no. 2(92), 97–154 = Russian Math. Surveys 15(1960), no. 2, 85–142. MR 22 #9721.

[49] V. P. Il'in, a) *Some inequalities in function spaces and their application to the investigation of the convergence of variational methods*, Trudy Mat. Inst. Steklov. 53(1959), 64–127. (Russian) MR 22 #9738.

b) *Properties of certain classes of differentiable functions of many variables given in an n-dimensional domain*, Trudy Mat. Inst. Steklov. 66(1962), 227–363. (Russian) MR 27 #3750.

V. P. Il'in and V. A. Solonnikov, c) *Some properties of differentiable functions of many variables*, Trudy Mat. Inst. Steklov. 66(1962), 205–226. (Russian) MR 27 #2768.

[50] E. L. Ince, *Ordinary differential equations*, Longmans, Green & Co., London, 1926; reprint, Dover, New York, 1944. MR 6, 65.

[51] A. V. Ivanov, a) *On some properties of generalized solutions of linear equations of elliptic and parabolic type*, Vestnik Leningrad. Univ. Ser. Mat. Meh. Astronom. 19 (1964), no. 2, 5–15. (Russian) MR 30 #1292.

b) *On a class of functions*, Trudy Mat. Inst. Steklov. 73(1964), 159–171. (Russian) MR 33 #6154.

c) *A priori estimates for solutions of linear second-order equations of elliptic and parabolic types*, Dokl. Akad. Nauk SSSR 161(1965), 1270–1273 = Soviet Math. Dokl. 6(1965), 583–587. MR 32 #1443.

d) *On certain properties of generalized solutions of parabolic equations of second order*. I, Izv. Akad. Nauk SSSR Ser. Mat. (to appear). (Russian)

e) *An investigation of the properties of generalized solutions of quasilinear parabolic equations of second order*, Candidate's dissertation, Leningrad. Gos. Univ., 1966 (see also: Dokl. Akad. Nauk SSSR 173(1967), 752–754 = Soviet Math. Dokl. 8(1967), 463–466. MR 35#4598).

A. V. Ivanov, O. A. Ladyženskaja, N. N. Ural'ceva and A. L. Treskunov, f) *Certain properties of generalized solutions of parabolic equations of second order*, Dokl. Akad. Nauk SSSR 168(1966), 17–20 = Soviet Math. Dokl. 7(1966), 579–583; Trudy Mat. Inst. Steklov. 92(1966), 57–92 = Proc. Steklov Inst. Math. 92(1966), 63–104. MR 35 #577.

[52] F. John, *On integration of parabolic equations by difference methods*. I, *Linear and quasi-linear equations for the infinite interval*, Comm. Pure Appl. Math. 5(1952), 155–211. MR 13, 947.

[53] J. Kadlec and R. Výborný, *Strong maximum principle for weakly nonlinear parabolic equations*, Comment. Math. Univ. Carolinae 6(1965), 19–20. MR 31 #1467.

[54] A. S. Kalašnikov (see also A. M. Il'in), O. A. Oleĭnik and Čžou Juĭ-lin'; *The Cauchy problem and boundary value problems for equations of non-stationary filtration type*, Izv. Akad. Nauk SSSR Ser. Mat. 22(1958), 667–704. (Russian) MR 20 #6271.

[55] S. L. Kamenomostskaja, a) *On the Stefan problem*, Naučn. Dokl. Vysš. Školy. 1(1958), 60–62. (Russian)
b) *On the Stefan problem*, Mat. Sb. 53(95) (1961), 489–514. (Russian) MR 25 #5292.

[56] L. I. Kamynin, a) *On the existence of a solution of boundary-value problems for a parabolic equation with discontinuous coefficients*, Izv. Akad. Nauk SSSR Ser. Mat. 28(1964), 721–744. (Russian) MR 29 #2534 (see also: Dokl. Akad. Nauk SSSR 139(1961), 1048–1051 = Soviet Math. Dokl. 2(1961), 1043–1046. MR 24 #A1519).
L. I. Kamynin and V. N. Maslennikova, b) *Boundary estimates of the solution of the third boundary value problem for a parabolic equation*, Dokl. Akad. Nauk SSSR 153(1963), 526–529 = Soviet Math. Dokl. 4(1953), 1711–1714. MR 29 #1447.
———, c) *Boundary estimates of the solution of a problem with directional derivative for a parabolic equation in a noncylindrical domain*, Dokl. Akad. Nauk SSSR 160 (1965), 527–529 = Soviet Math. Dokl. 6(1965), 128–131. MR 31 #2516.

[57] T. Kato, a) *Integration of the equation of evolution in a Banach space*, J. Math. Soc. Japan 5(1953), 208–304. MR 15, 437.
b) *On the linear differential equations in Banach spaces*, Comm. Pure Appl. Math. 9(1956), 479–486. MR 19, 279.
c) *Nonlinear evolution equations in Banach spaces*, Proc. Sympos. Appl. Math., vol. 17, Amer. Math. Soc., Providence, R. I., 1965, pp. 50–67. MR 32 #1573.

[58] I. I. Kolodner, *Free boundary problem for the heat equation with applications to problems of change of phase. I, General method of solution*, Comm. Pure Appl. Math. 9(1956), 1–31. MR 19, 285.

[59] A. S. Kompaneec. See Zel'dovič.

[60] M.A. Krasnosel'skiĭ, S. G. Kreĭn and P. E. Sobolevskiĭ, a) *On differential equations with unbounded operators in a Banach space*, Dokl. Akad. Nauk SSSR 111(1956), 19–22. (Russian) MR 19, 550.
b) *On differential equations with unbounded operator coefficients in a Hilbert space*, Dokl. Akad. Nauk SSSR 112(1957), 990–993. (Russian) MR 19, 747.

[61] S. G. Kreĭn and P. E. Sobolevskiĭ (see also Krasnosel'skiĭ), *Differential equations with abstract elliptic operators in a Hilbert space*, Dokl. Akad. Nauk SSSR 118(1958), 233–236. (Russian) MR 20 #6043.

[62] S. N. Kružkov, a) *On the a priori estimation of solutions of linear parabolic equations and of solutions of boundary value problems for a certain class of quasi-linear parabolic equations*, Dokl. Akad. Nauk SSSR 138(1961), 1005–1008 = Soviet Math. Dokl. 2(1961), 764–767. MR 24 #A2147.
b) *A priori estimates and certain properties of the solutions of elliptic and parabolic equations of second order*, Dokl. Akad. Nauk SSSR 150(1963), 748–751 = Soviet Math. Dokl. 4(1963), 757–760. MR 27 #1687.
c) *A priori estimates and certain properties of the solution of elliptic and parabolic equations*, Mat. Sb. 65(107) (1964), 522–570; English transl., Amer. Math. Soc. Transl. (2) 68(1968), 169–220. MR 30 #1317.
d) *The Cauchy problem in the large for nonlinear equations and certain quasi-linear systems of first order with several variables*, Dokl. Akad. Nauk SSSR 155(1964), 743–746 = Soviet Math. Dokl. 5(1964), 493–496. MR 29 #1436.
S. N. Kružkov and O. A. Oleĭnik, e) *Quasi-linear parabolic equations of second order with many independent variables*, Uspehi Mat. Nauk 16(1961), no. 5 (101), 115–155 = Russian Math. Surveys 16(1961), no. 5, 105–146. MR 25 #5289.

[63] M. Krzyzanski, *Sur l'unicité des solutions des second et troisième problèmes de Fourier relatifs á l'équation linéaire normale du type parabolique*, Ann. Polon. Math. 7(1960), 201–208. MR 22 #1749.

M. Kurihara — see [133].

[64] A. Kživickiĭ. See Ladyženskaja.

[65] O. A. Ladyženskaja (see also Ivanov and Višik), a) *On the uniqueness of the solution of Cauchy's problem for a linear parabolic equation*, Mat. Sb. 27(69) (1950), 175–184. (Russian) MR 12, 709.
b) *On the closure of an elliptic operator*, Dokl. Akad. Nauk SSSR 79(1951), 723–725. (Russian) MR 14, 280.
c) *The mixed problem for a hyperbolic equation*, GITTL, Moscow, 1953. (Russian) MR 17, 160.
d) *On the solvability of the basic boundary value problems for equations of parabolic and hyperbolic type*, Dokl. Akad. Nauk SSSR 97(1954), 395–398. (Russian) MR 17, 495.
e) *On the solution of nonstationary operator equations of various types*, Dokl. Akad. Nauk SSSR 102(1955), 207–210. (Russian) MR 17, 161.
f) *On non-stationary operator equations and their applications to linear problems of mathematical physics*, Mat. Sb. 45(87) (1958), 123–158. (Russian) MR 22 #12290.
g) *On the solution of the general diffraction problem*, Dokl. Akad. Nauk SSSR 96(1954), 433–436. (Russian) MR 16, 1117.
h) *The solution in the large of the first boundary-value problem for quasi-linear parabolic equations*, Dokl. Akad. Nauk SSSR 107(1956), 636–639; Trudy Moskov. Mat. Obšč. 7(1958), 149–177. (Russian) MR 22 #4880.

i) *A simple proof of the solvability of the fundamental boundary problems and of a problem on eigenvalues for linear elliptic equations*, Vestnik Leningrad. Univ. Ser. Mat. Meh. Astron. 10(1955), no. 11, 23–29. (Russian) MR 17, 855.

j) *The method of finite differences in the theory of partial differential equations*, Uspehi Mat. Nauk 12(1957), no. 5 (77), 123–148; English transl., Amer. Math. Soc. Transl. (2) 20(1962), 77–104. MR 20 #3395; MR 25 #342.

k) *On integral estimates, convergence, approximate methods and solution in functionals for elliptic operators*, Vestnik Leningrad. Univ. 13(1958), no. 7, 60–69. (Russian) MR 20 #5353.

l) *On the solvability "in the large" of the boundary-value problems for linear and quasi-linear parabolic equations and the Navier-Stokes equations*, Proc. Fourth All-Union Math. Congress 1961, vol. 1, Izdat. Akad. Nauk SSSR, Leningrad, 1963, pp. 134–157. (Russian) MR 29 #386.

O. A. Ladyženskaja and A. Kživickiĭ, m) *The method of nets for non-stationary Navier-Stokes equations*, Trudy Mat. Inst. Steklov. 92(1966), 93–99 = Proc. Steklov Inst. Math. 92(1966), 105–112. MR 34 #4651.

O. A. Ladyženskaja and N. N. Ural′ceva n) *A boundary value problem for linear and quasi-linear parabolic equations*, I, Dokl. Akad. Nauk SSSR 139 (1961), 544–547 = Soviet Math. Dokl. 2(1961), 969–972. MR 25 #5288; I, II, Izv. Akad. Nauk SSSR Ser. Mat. 26(1962), 5–52, 753–780; English transl., Amer. Math. Soc. Transl. (2) 47(1965), 217–267, 268–299. MR 31 #6063.

———, o) *A boundary value problem for linear and quasi-linear equations and systems of parabolic type*. III, Izv. Akad. Nauk SSSR Ser. Mat. 27(1963), 161–240; English transl., Amer. Math. Soc. Transl. (2) 56(1966), 103–192. MR 31 #6063.

———, p) *On the Hölder continuity of solutions and their derivatives for linear and quasi-linear equations of elliptic and parabolic types*, Dokl. Akad. Nauk SSSR 155(1964), 1258–1261 = Soviet Math. Dokl. 5(1964), 565–568. MR 28 #5256; Trudy Mat. Inst. Steklov. 73(1964), 172–220; English transl., Amer. Math. Soc. Transl. (2) 61(1967), 207–269. MR 30 #4063.

———, q) *Linear and quasilinear equations of elliptic type*, "Nauka", Moscow, 1964; English transl., Academic Press, New York, 1968. MR 35 #1955.

O. A. Ladyženskaja, V. Ja. Rivkind and N. N. Ural′ceva, r) *On the classical solvability of diffraction problems for equations of elliptic and parabolic types*, Dokl. Akad. Nauk SSSR 158(1964), 513–515 = Soviet Math. Dokl. 5(1964), 1249–1252. MR 29 #4973; Trudy Mat. Inst. Steklov. 92 (1966), 116–146 = Proc. Steklov Inst. Math. 92(1966), 132–166.

[66] E. M. Landis, *Some questions in the qualitative theory of elliptic and parabolic equations*, Uspehi Mat. Nauk 14(1959), no. 1(85), 21–85; English transl., Amer. Math. Soc. Transl. (2) 20(1962), 173–238. MR 22 #3870.

[67] P. D. Lax and A. N. Milgram, *Parabolic equations. Contributions to the theory of partial differential equations*, Ann. of Math. Studies, no. 33, Princeton Univ. Press, Princeton, N. J., 1954. MR 16, 709.

[68] J. Leray and J. L. Lions, a) *Quelques résultats de Višik sur les problèmes elliptiques non linéaires par les méthodes de Minty-Browder*, Bull. Soc. Math. France 93(1965), 97–107. MR 33 #2939.

J. Leray and J. Schauder, b) *Topologie et équations fonctionnelles*, Ann. Sci. École Norm. Sup. (3) 51(1934), 45–78; Russian transl., Uspehi Mat. Nauk 1(1946), no. 3–4 (13–14), 71–95. MR 9, 606.

[69] E. E. Levi, *Sulle equazioni lineari totalmente ellittiche alle derivate parziali*, Rend. Circ. Mat. Palermo 24(1907), 275–317.

[70] H. Lewy. See Courant.

[71] L. Lichtenstein, *Die neue Entwicklung in der Theorie der partiellen Differentialgleichungen zweiter Ordnung*, Enzyklopädie der math. Wissenschaften, Band II, Teubner, Leipzig, 1924, Chapter 8, pp. 1277–1334.

[72] V. B. Lidskiĭ, a) *The Fourier series expansion in terms of the principal functions of a non-selfadjoint elliptic operator*, Mat. Sb. 57(99) (1962), 137–150. (Russian) MR 26 #460.

b) *On the summability of series in the principal vectors of nonselfadjoint operators*, Trudy Moskov. Mat. Obšč. 11(1962), 3–35; English transl., Amer. Math. Soc. Transl. (2) 40(1964), 193–228. MR 26 #1760.

[73] J. L. Lions (see also [134]). (see also Leray), *Équations différentielles operationelles et problèmes aux limites*, Die Grundlehren der math. Wissenshaften, Bd. 111, Springer-Verlag, Berlin, 1961. MR 27 #3935.

[74] B. Ja. Lipko and S. D. Èĭdel'man, *On boundary value problems for parabolic systems in regions of general type*, Dokl. Akad. Nauk SSSR 150(1963), 58–61 = Soviet Math. Dokl. 4(1963), 614–618. MR 29 #1455.

[75] J. E. Littlewood. See Hardy.

[76] V. È. Ljance, *On a boundary value problem for parabolic systems of differential equations with a strongly elliptic right-hand side*, Mat. Sb. 35 (77) (1954), 357–368. Mat. Sb. 39(81) (1956), 525. (Russian) MR 16, 709.

[77] Ja. B. Lopatinskiĭ, *On a method for reducing boundary problems for systems of differential equations of elliptic type to regular integral equations*, Ukrain. Mat. Ž. 5(1953), 123–151. (Russian) MR 17, 494.

[78] V. N. Maslennikova. See Kamynin.

[79] M. I. Matiĭčuk and S. D. Èĭdel'man, *On parabolic systems with coefficients satisfying a Dini condition*, Dokl. Akad. Nauk SSSR 165(1965), 482–485 = Soviet Math. Dokl. 6(1965), 1461–1464. MR 32 #7962.

[80] N. N. Meĭman, a) *On the theory of partial differential equations*, Dokl. Akad. Nauk SSSR 97 (1954), 593–596. (Russian) MR 16, 254.
b) *On the equation of heat conduction*, Dokl. Akad. Nauk SSSR 99 (1954), 209–212. (Russian) MR 16, 485.

[81] V. P. Mihaĭlov, a) *On the Dirichlet problem and the first mixed problem for a parabolic equation*, Dokl. Akad. Nauk SSSR 140 (1961), 303–306 = Soviet Math. Dokl. 2 (1961), 1204–1207. MR 26 #487.
b) *On the Dirichlet problem for a parabolic equation*. I, Mat. Sb. 61 (103) (1963), 40–64; II, Mat. Sb. 62 (104) (1963), 140–159. (Russian) MR 27 #1710; MR 28 #1410.

[82] S. G. Mihlin, *Fourier integrals and multiple singular integrals*, Vestnik Leningrad. Univ. Ser. Mat. Meh. Astronom. 12 (1957), no. 7, 143–155. (Russian) MR 19, 546.

[83] A. N. Milgram. See Lax.

[84] G. Minty, a) *Monotone (nonlinear) operators in Hilbert space*, Duke Math. J. 29 (1962), 341–346. MR 29 #6319.
b) *Two theorems on nonlinear functional equations in Hilbert space*, Bull. Amer. Math. Soc. 69 (1963), 691–692. MR 32 #8188.
c) *On a "monotonicity" method for the solution of nonlinear equations in Banach spaces*, Proc. Nat. Acad. Sci. U.S.A. 50 (1963), 1038–1041. MR 28 #5358.

[85] S. Mizohata, *Hypoellipticité des équations paraboliques*, Bull. Soc. Math. France 85 (1957), 15–50. MR 20 #3381.

[86] G. M. Mjuntc, a) *Zum dynamischen Wärmeleitungsproblem*, Math. Z. 38 (1934), 323–338.
b) *Integral equations*, Moscow, 1934. (Russian)

[87] J. Moser, a) *A new proof of de Giorgi's theorem concerning the regularity problem for elliptic differential equations*, Comm. Pure Appl. Math. 13 (1960), 457–468. MR 30 #332.
b) *A Harnack inequality for parabolic differential equations*, Comm. Pure Appl. Math. 17 (1964), 101–134. MR 28 #2357.

[88] J. Nash, *Continuity of solutions of parabolic and elliptic equations*, Amer. J. Math. 80 (1958), 931–954. MR 20 #6592.

[89] S. M. Nikol'skiĭ, *The properties of certain classes of functions of many variables on differentiable manifolds*, Mat. Sb. 33 (75) (1953), 261–326. (Russian) MR 16, 453.

[90] L. Nirenberg (see also Agmon), a) *A strong maximum principle for parabolic equations*, Comm. Pure Appl. Math. 6 (1953). 167–177. MR 14, 1089; MR 16, 1336.
b) *On elliptic partial differential equations*, Ann. Scuola Norm. Sup. Pisa

(3) 13 (1959), 115–162. MR 22 #823.

[91] O. A. Oleĭnik (see also A. Il'in, Kalašnikov, Kružkov and Ventcel'), a) *On a method for solving the general diffraction problem.* (Russian)
b) *Boundary-value problems for linear elliptic and parabolic equations with discontinuous coefficients*, Izv. Akad. Nauk SSSR Ser. Mat. 25 (1961), 3–20; English transl., Amer. Math. Soc. Transl. (2) 42(1964), 175–194. MR 23 #A1136.
c) *On quasi-linear parabolic equations with many independent variables*, Dokl. Akad. Nauk SSSR 138 (1961), 43–46 = Soviet Math. Dokl. 2 (1961), 529–532. MR 24 #A924.
d) *On a method for solving the general Stefan problem*, Dokl. Akad. Nauk SSSR 135 (1960), 1054–1057 = Soviet Math. Dokl. 1(1960), 1350–1354. MR 23 #A2644.

[92] M. Pagni, *Su un problema al contorno tipico per l'equazione del calore*, Ann. Scuola Norm. Sup. Pisa (3) 11(1957), 73–115. MR 25 #332.

[93] I. G. Petrovskiĭ, a) *Lectures on partial differential equations*, GITTL, Moscow, 1950; 3rd augmented ed., Fizmatgiz, Moscow, 1961; English transl., Interscience, New York, 1964. MR 16, 478; MR 25 #2308.
b) *Zur ersten Randwertaufgabe der Wärmeleitungsgleichung*, Compositio Math. 1(1935), 383–419.
c) *On the Cauchy problem for systems of linear partial differential equations in a domain of nonanalytic functions*, Bjull. Moskov. Gos. Univ. Ser. A., 1 (1938), no. 7, 1–72. (Russian)
d) *A new proof of the existence of a solution of the Dirichlet problem by the method of finite differences*, Uspehi Mat. Nauk 8 (1941), 161–170. (Russian) MR 3, 123.

R. S. Phillips. See [44].

[94] W. Pogorzelski (see also [135]), a) *Étude de la solution fondamentale de l'équation parabolique*, Ricerche Mat. 5 (1956), 25–57. MR 18, 47.
b) *Problèmes aux limites pour l'équation parabolique normale*, Ann. Polon. Math. 4(1957), 61–92. MR 21 #3656a.
c) *An investigation of the integrals of a parabolic equation and of boundary value problems in an unbounded domain*, Mat. Sb. 47 (89) (1959), 397–430. (Russian) MR 22 #2791.
d) *Étude d'une fonction de Green et du problème aux limites pour l'équation parabolique normale*, Ann. Polon. Math. 4 (1958), 288–307. MR 21 #5078.

[95] G. Pólya. See Hardy.

[96] G. Prodi, *Teoremi di esistenze per equazioni allé derivate parziali non lineari di tipo parabolico.* I, II, 1st Lombardo Sci. Lett. Rend. Cl. Sci. Mat. Nat. (3) 17(86) (1953), 3–26, 27–47. MR 16, 259.

[97] R. D. Richtmyer, *Difference methods for initial value problems*, Interscience Tracts in Pure and Appl. Math., no. 4, Interscience, New York, 1957; Russian transl., IL, Moscow, 1960. MR 20 #438; MR 22 #3864.

[98] V. Ja. Rivkind. See Ladyženskaja.

[99] V. S. Rjaben'kiĭ and A. F. Filippov, *On the stability of difference equations*, GITTL, Moscow, 1956; German transl., Mathematik für Natur. und Technik, Band 3, VEB Deutscher Verlag, Berlin, 1960. MR 19, 865; MR 23 #A437.

[100] P. C. Rosenbloom. See Forsythe.

[101] E. Rothe, *Wärmeleitungsgleichung mit nichtkonstanten Koeffizienten*, Math. Ann. 104 (1931), 340–362.

[102] L. I. Rubinšteĭn, a) *On a certain nonlinear problems provided by an equation of Fourier*, Doctoral dissertation, Moskov. Gos. Univ., Moscow, 1957. (Russian)
b) *A two-phase Stefan problem on an interval with one-phase initial state of a heat-conducting medium*, Latvijas Valsts Univ. Zinātn. Raksti 58(1964), vyp. 2, 111–148. (Russian) MR 31 #6066.

[103] E. S. Sabinina, a) *On the Cauchy problem for the equation of nonstationary gas filtration in several space variables*, Dokl. Akad. Nauk SSSR 136(1961), 1034–1037 = Soviet Math. Dokl. 2(1961), 166–169. MR 28 #1416.
b) *On a class of nonlinear degenerating parabolic equations*, Dokl. Akad. Nauk SSSR 143(1962), 794–797 = Soviet Math. Dokl. 3(1962), 495–498. MR 24 #A2762.

[104] A. A. Samarskiĭ, a) *An efficient difference method for solving a multi-dimensional parabolic equation in an arbitrary domain*, Ž. Vyčisl. Mat. i Mat. Fiz. 2(1962), 787–811 = U.S.S.R. Comput. Math. and Math. Phys. 1963 (1964), no. 5, 894–926; 3(1963), 416. MR 32 #609.
A. A. Samarskiĭ and A. N. Tihonov, b) *On finite difference schemes for equations with discontinuous coefficients*, Dokl. Akad. Nauk SSSR 108 (1956), 393–396. (Russian) MR 18, 938.

[105] V. K. Saul'ev, *The integration of equations of parabolic type by the method of nets*, Fizmatgiz, Moscow, 1960; English transl., Internat. Series in Pure and Appl. Math., vol. 54, Pergamon Press, New York, 1964. MR 23 #A428; MR 33 #6153.

[106] J. Schauder. See Leray.

[107] L. Schwartz, *Théorie des distributions*. Tomes I, II, Actualités Sci. Indust., nos. 1091, 1122, Hermann, Paris, 1950, 1951. MR 12, 31; MR 12, 833.

[108] G. Sestini, a) *Esistenza di una soluzione in problemi analoghi a quello di Stefan*, Rivista Mat. Univ. Parma 3(1952), 3–23. MR 14, 381.
b) *Esistenza ed unicità nel problema di Stefan relativo a campi dotati di simmetria*, Rivista Mat. Univ. Parma 3(1952), 103–113. MR 14, 476.

[109] T. Shirota, a) *The initial value problem for linear differential equations with variable coefficients.* I, II, III, Proc. Japan Acad. 33(1957), 31–36, 103–104, 457–461. MR 21 #3665; MR 21 #3666; MR 21 #3667.
b) *On Cauchy problem for linear partial differential equations with variable coefficients*, Osaka Math. J. 9(1957), 43–59. MR 21 #3668.

[110] G. E. Šilov (see also Gel'fand), *On conditions of correctness of the Cauchy problem for systems of partial differential equations with constant coefficients*, Uspehi Mat. Nauk 10(1955), no. 4(66) 89–100. (Russian) MR 17, 495.

[111] L. N. Slobodeckiĭ (see also Babič), a) *Generalized solutions of parabolic and elliptic systems*, Izv. Akad. Nauk SSSR Ser. Mat. 21(1957), 809–834. (Russian) MR 21 #5084.
b) *The estimation of solutions of elliptic and parabolic systems*, Dokl. Akad. Nauk SSSR 120(1958), 468–471. (Russian) MR 21 #5060.
c) *On the fundamental solution and the Cauchy problem for parabolic systems*, Mat. Sb. 46(88) (1958), 229–258; English transl., Amer. Math. Soc. Transl. (2) 41(1964), 121–154. MR 22 #830.
d) *Estimates in L_p of solutions of elliptic systems*, Dokl. Akad. Nauk SSSR 123(1958), 616–619. (Russian) MR 21 #5061.

[112] V. I. Smirnov, *A course in higher mathematics*. Vol. IV, OGIZ, Moscow, 1941; 2nd ed., GITTL, Moscow, 1951; English transl., Pergamon Press, Oxford and Addison-Wesley, Reading, Mass., 1964. MR 6, 42; MR 14, 145; MR 31 #1333.
A course in higher mathematics. Vol. V, OGIZ, Moscow, 1947; rev. ed., Fizmatgiz, Moscow, 1959; English transl., Pergamon Press, Oxford and Addison-Wesley, Reading, Mass., 1964. MR 9, 574; MR 25 #5141; MR 29 #5964.

[113] S. L. Sobolev, a) *Applications of functional analysis in mathematical physics*, Izdat. Leningrad. Gos. Univ., Leningrad, 1950; English transl., Transl. Math. Monographs, vol. 7, Amer. Math. Soc., Providence, R. I., 1963. MR 14, 565; MR 29 #2624.
b) *The equations of mathematical physics*, OGIZ, Moscow, 1947; 2nd ed., GITTL, Moscow, 1950; 4th ed., "Nauka", Moscow, 1966; English transl., *Partial differential equations of mathematical physics*, Pergamon Press, New York and Addison-Wesley, Reading, Mass., 1964. MR 13, 42; MR 31 #2478.

[114] P. E. Sobolevskiĭ (see also Krasnosel'skiĭ and Kreĭn), a) *On differential equations of parabolic type in a Hilbert space and on the approximate solution of them by the Bubnov-Galerkin method*, Candidate's dissertation, Leningrad. Gos. Univ., Leningrad, 1958. (Russian)
b) *Equations of parabolic type in a Banach space*, Trudy Moskov. Mat. Obšč. 10(1961), 297–350; English transl., Amer. Math. Soc. Transl. (2)

49(1965), 1–62. MR 25 #5297.
c) *Local and nonlocal existence theorems for nonlinear second-order parabolic equations*, Dokl. Akad. Nauk SSSR 136(1961), 292–295 = Soviet Math. Dokl. 2(1961), 63–66. MR 25 #5291.

[115] M. Z. Solomjak, a) *An application of the theory of semigroups to differential equations in a Banach space*, Dokl. Akad. Nauk SSSR 122(1958), 766–769. (Russian) MR 21 #3775.
b) *The analytic character of semigroups generated by elliptic operators in* L_p, Dokl. Akad. Nauk SSSR 127(1959), 37–39. (Russian) MR 22 #12398.
c) *On a condition of Ja. B. Lopatinskiĭ for the solvability of boundary value problems*, Vestnik Leningrad. Univ. 20(1965), no. 1, 143–144. (Russian)

[116] V. A. Solonnikov (see also Golovkin and V. P. Il'in), a) *On a priori estimates for certain boundary value problems*, Dokl. Akad. Nauk SSSR 138 (1961), 781–784 = Soviet Math. Dokl. 2(1961), 723–727. MR 28 #2352.
b) *A priori estimate for equations of second order of parabolic type*, Trudy Mat. Inst. Steklov. 70(1964), 133–212; English transl., Amer. Math. Soc. Transl. (2) 65(1967), 51–137. MR 28 #5267.
c) *On boundary value problems for linear parabolic systems of differential equations of general form*, Trudy Mat. Inst. Steklov. 83(1965) = Proc. Steklov. Inst. Math. 83(1965), MR 35 #1965.
d) *On general boundary value problems for systems of differential equations of elliptic and parabolic types*, Outlines Joint Sympos. Partial Differential Equations (Novosibirsk 1963), Acad. Sci. USSR Siberian Branch, Moscow, 1963, pp. 246–252. MR 34 #4675.
e) *On boundary value problems for linear parabolic systems of differential equations of general form*, Doctoral dissertation, Leningrad. Gos. Univ., Leningrad, 1964. (Russian)

[117] V. V. Stepanov, *A course in differential equations*, Fizmatgiz, Moscow, 1958; German transl., 2nd rev. ed., Hochschulbücher für Mathematik, Band 20, VEB Deutscher Verlag, Berlin, 1963. MR 31 #1418.

[118] S. Täcklind, *Sur les classes quasianalytiques des solutions des équations aux dérivées partielles du type parabolique*, Nord Acta Regial Soc. Sci. Uppsaliensis 4(1936), no. 10.

[119] A. N. Tihonov (see also Samarskiĭ), a) *A uniqueness theorem for the heat equation*, Mat. Sb. 2(1935), 199–216. (Russian)
b) *On the heat equation in several variables*, Vestnik Moskov. Gos. Univ. Ser. A. 1(1937), no. 9, 1–49. (Russian)
c) *On the third boundary value problem for an equation of parabolic type*, Izv. Akad. Nauk SSSR Ser. Geograf. Geofiz. 14(1950), 193–198. (Russian) MR 12, 29.

[120] E. C. Titchmarsh, *An introduction to the theory of Fourier integrals*, Clarendon Press, Oxford, 1937; 2nd ed., 1948; Russian transl., GITTL, Moscow, 1948.

[121] A. L. Treskunov. See Ivanov.

N. S. Trudinger – see [136].

[122] N. N. Ural'ceva (see also Ivanov and Ladyženskaja), a) *On the regularity of solutions of multidimensional elliptic equations and variational problems*, Dokl. Akad. Nauk SSSR 130 (1960), 1206–1209 = Soviet Math. Dokl. 1 (1960), 161–164. MR 23 #A4036.
b) *Boundary value problems for quasi-linear elliptic equations and systems with principal part in divergence form*, Dokl. Akad. Nauk SSSR 147 (1962), 313–316 = Soviet Math. Dokl. 3 (1962), 1615–1618. MR 26 #453.

[123] T. D. Ventcel', a) *The first boundary problem and the problem of Cauchy for quasi-linear parabolic equations with several space variables*, Mat. Sb. 41 (83) (1957), 499–520. (Russian) MR 19, 657.
b) *On certain quasi-linear parabolic systems*, Dokl. Akad. Nauk SSSR 117 (1957), 21–24. (Russian) MR 21 #205.
c) *On certain quasi-linear parabolic systems with increasing coefficients*, Dokl. Akad. Nauk SSSR 140 (1961), 284–286 = Soviet Math. Dokl. 2 (1961), 1185–1187. (Russian)
d) *Quasi-linear parabolic systems with increasing coefficients*, Vestnik Moskov. Univ. Ser. I Mat. Meh. 1963, no. 6, 34–44. (Russian) MR 28 #4252.
T. D. Ventcel' and O. A. Oleĭnik, e) *Cauchy's problem and the first boundary problem for a quasilinear equation of parabolic type*, Dokl. Akad. Nauk SSSR 97 (1954), 605–608. (Russian) MR 16, 259.
———, f) *The first boundary problem and the Cauchy problem for quasi-linear equations of parabolic type*, Mat. Sb. 41 (83) (1957), 105–128. (Russian) MR 19, 149.

[124] M. I. Višik (see also Agranovič and Barenblatt), a) *On strongly elliptic systems of differential equations*, Dokl. Akad. Nauk SSSR 74 (1950), 881–884; Mat. Sb. 29 (71) (1951), 615–676. (Russian) MR 13, 134; MR 14, 174.
b) *Mixed boundary-value problems for equations containing a first derivative with respect to time and an approximate method for solving them*, Dokl. Akad. Nauk SSSR 99 (1954), 189–192. (Russian) MR 17, 859.
c) *The Cauchy problem for equations with operator coefficients; mixed boundary value problem for systems of differential equations and approximation methods for solving them*, Mat. Sb. 39 (81) (1956), 51–148; English transl., Amer. Math. Soc. Transl. (2) 24 (1963), 173–278. MR 18, 215.
d) *On boundary-value problems for quasi-linear parabolic systems of equations and on the Cauchy problem for hyperbolic equations*, Dokl. Akad. Nauk SSSR 140 (1961), 998–1001 = Soviet Math. Dokl. 2 (1961), 1292–1295. MR 28 #360.

e) *On the solvability of boundary problems for quasilinear parabolic equations of higher order*, Mat. Sb. 59(101) (1962), 289–325; English transl., Amer. Math. Soc. Transl. (2) 65(1967), 1–40. MR 28 #361.

M. I. Višik and O. A. Ladyženskaja, f) *Boundary value problems for partial differential equations and certain classes of operator equations*, Uspehi Mat. Nauk 11(1956), no. 6(72), 41–97; English transl., Amer. Math. Soc. Transl. (2) 10(1958), 223–281. MR 20 #1091; MR 20 #1092.

[125] L. R. Volevič, a) *On general systems of differential equations*, Dokl. Akad. Nauk SSSR 132(1960), 20–23 = Soviet Math. Dokl. 1(1960), 458–461. MR 24 #A899.

b) *Local properties of solutions of quasi-elliptic systems*, Mat. Sb. 59(101) (1962), supplement, 3–52. (Russian) MR 27 #446.

c) *On a problem of linear programming occurring in differential equations*, Uspehi Mat. Nauk 18(1963), no. 3(111), 155–162. (Russian) MR 28 #5248.

[126] N. D. Vvedenskaja, a) *An example of nonuniqueness of generalized solutions of quasi-linear systems of equations*, Dokl. Akad. Nauk SSSR 136(1961), 532–533 = Soviet Math. Dokl. 2(1961), 89–90. MR 23 #A1154.

b) *Some examples of parabolic systems of equations*, Uspehi Mat. Nauk 16 (1961), no. 5, 218–219. (Russian)

[127] R. Výborný (see also Kadlec), *On the properties of the solutions of certain boundary value problems for equations of parabolic type*, Dokl. Akad. Nauk SSSR 117(1957), 563–565. (Russian) MR 23 #A3918.

[128] K. Yosida, a) *On the differentiability and the representation of one-parameter semi-group of linear operators*, J. Math. Soc. Japan 1(1948), 15–21. MR 10, 462.

b) *On the integration of diffusion equations in Riemannian spaces*, Proc. Amer. Math. Soc. 3(1952), 864–879. MR 14, 560.

[129] T. Ja. Zagorskiĭ, *The mixed problem for a system of partial differential equations of parabolic type*, L'vov, 1961. (Russian)

[130] Ju. V. Zel'dovič and A. S. Kompaneec, *On the theory of propagation of heat with the heat conductivity depending upon the temperature*. Collection in honor of the seventieth birthday of Academician A. F. Ioffe, Izdat. Akad. Nauk SSSR, Moscow, 1950, pp. 61–71. MR 16, 1029.

[131] A. Zygmund (see also Calderón), *On a theorem of Marcinkiewicz concerning interpolation of operations*, J. Math. Pures Appl. (9) 35(1956), 223–248. MR 18, 321.

References added by the authors for the English translation

[132] D. G. Aronson, a) *On the Green's function for second order parabolic differential equations with discontinuous coefficients*, Bull. Amer. Math. Soc. 69(1963), 841–847. MR 27 #5049.

b) *Bounds for the fundamental solution of a parabolic equation* (preprint).

c) *On the initial value problem for parabolic systems of differential equations*, Bull. Amer. Math. Soc. 65(1959), 310–318. MR 22 #3888.
D. G. Aronson and I. Serrin, d) *Local behavior of solutions of quasi-linear parabolic equations*, Arch. Rational Mech. Anal. 25(1967), 81–122.

[133] M. Kurihara, *On Harnack's inequality for parabolic partial differential equations*. I, II, Funkcialaj Ekvacioj (Serio Japana) 17(1965), 3–57; 18 (1966), 3–32. (Japanese)

[134] J. L. Lions, *Sur certaines équations paraboliques non linéaires*, Bull. Soc. Math. France 93(1965), 155–175. MR 33 #2966.

[135] W. Porgorzelski, *Propriétés des solutions du système parabolique d'équations aux dérivées partielles*, Bull. Acad. Polon. Sci. Sér. Sci. Math. Astr. Phys. 6(1958), 369–373. MR 21 #4301.

[136] N. S. Trudinger, *Quasilinear elliptic partial differential equations in n variables*, Doctoral Dissertation, Stanford Univ., 1966.

Other Titles in This Series

(*Continued from the front of this publication*)

106 **G.-C. Wen,** Conformal mappings and boundary value problems, 1992
105 **D. R. Yafaev,** Mathematical scattering theory: General theory, 1992
104 **R. L. Dobrushin, R. Kotecký, and S. Shlosman,** Wulff construction: A global shape from local interaction, 1992
103 **A. K. Tsikh,** Multidimensional residues and their applications, 1992
102 **A. M. Il'in,** Matching of asymptotic expansions of solutions of boundary value problems, 1992
101 **Zhang Zhi-fen, Ding Tong-ren, Huang Wen-zao, and Dong Zhen-xi,** Qualitative theory of differential equations, 1992
100 **V. L. Popov,** Groups, generators, syzygies, and orbits in invariant theory, 1992
99 **Norio Shimakura,** Partial differential operators of elliptic type, 1992
98 **V. A. Vassiliev,** Complements of discriminants of smooth maps: Topology and applications, 1992 (revised edition, 1994)
97 **Itiro Tamura,** Topology of foliations: An introduction, 1992
96 **A. I. Markushevich,** Introduction to the classical theory of Abelian functions, 1992
95 **Guangchang Dong,** Nonlinear partial differential equations of second order, 1991
94 **Yu. S. Il'yashenko,** Finiteness theorems for limit cycles, 1991
93 **A. T. Fomenko and A. A. Tuzhilin,** Elements of the geometry and topology of minimal surfaces in three-dimensional space, 1991
92 **E. M. Nikishin and V. N. Sorokin,** Rational approximations and orthogonality, 1991
91 **Mamoru Mimura and Hirosi Toda,** Topology of Lie groups, I and II, 1991
90 **S. L. Sobolev,** Some applications of functional analysis in mathematical physics, third edition, 1991
89 **Valeriĭ V. Kozlov and Dmitriĭ V. Treshchëv,** Billiards: A genetic introduction to the dynamics of systems with impacts, 1991
88 **A. G. Khovanskiĭ,** Fewnomials, 1991
87 **Aleksandr Robertovich Kemer,** Ideals of identities of associative algebras, 1991
86 **V. M. Kadets and M. I. Kadets,** Rearrangements of series in Banach spaces, 1991
85 **Mikio Ise and Masaru Takeuchi,** Lie groups I, II, 1991
84 **Đào Trọng Thi and A. T. Fomenko,** Minimal surfaces, stratified multivarifolds, and the Plateau problem, 1991
83 **N. I. Portenko,** Generalized diffusion processes, 1990
82 **Yasutaka Sibuya,** Linear differential equations in the complex domain: Problems of analytic continuation, 1990
81 **I. M. Gelfand and S. G. Gindikin, Editors,** Mathematical problems of tomography, 1990
80 **Junjiro Noguchi and Takushiro Ochiai,** Geometric function theory in several complex variables, 1990
79 **N. I. Akhiezer,** Elements of the theory of elliptic functions, 1990
78 **A. V. Skorokhod,** Asymptotic methods of the theory of stochastic differential equations, 1989
77 **V. M. Filippov,** Variational principles for nonpotential operators, 1989
76 **Phillip A. Griffiths,** Introduction to algebraic curves, 1989
75 **B. S. Kashin and A. A. Saakyan,** Orthogonal series, 1989
74 **V. I. Yudovich,** The linearization method in hydrodynamical stability theory, 1989
73 **Yu. G. Reshetnyak,** Space mappings with bounded distortion, 1989
72 **A. V. Pogorelev,** Bendings of surfaces and stability of shells, 1988
71 **A. S. Markus,** Introduction to the spectral theory of polynomial operator pencils, 1988
70 **N. I. Akhiezer,** Lectures on integral transforms, 1988
69 **V. N. Saliĭ,** Lattices with unique complements, 1988
68 **A. G. Postnikov,** Introduction to analytic number theory, 1988
67 **A. G. Dragalin,** Mathematical intuitionism: Introduction to proof theory, 1988

(See the AMS catalog for earlier titles)